AF335413

Mouse Genetics after the Mouse Genome

Editor

Silvia Garagna, Pavia

108 figures, 61 in color, and 36 tables, 2004

Basel · Freiburg · Paris · London · New York ·
Bangalore · Bangkok · Singapore · Tokyo · Sydney

Reprint of **Cytogenetic and Genome Research** (ISSN 1424–8581)
Vol. 105, No. 2–4, 2004

Cover illustration

Alfred Gropp's model of chromosomal variability, the tobacco mouse (Fatio 1869), is just an example of the many developed during the last century. The sequencing of the entire mouse genome has favored a new age for this animal model, the results of which have already flooded over all biological disciplines.

S. Karger
Medical and Scientific Publishers
Basel · Freiburg · Paris · London
New York · Bangalore · Bangkok
Singapore · Tokyo · Sydney

Drug Dosage
The authors and the publisher have exerted every effort to ensure that drug selection and dosage set forth in this text are in accord with current recommendations and practice at the time of publication. However, in view of ongoing research, changes in government regulations, and the constant flow of information relating to drug therapy and drug reactions, the reader is urged to check the package insert for each drug for any change in indications and dosage and for added warnings and precautions. This is particularly important when the recommended agent is a new and/or infrequently employed drug.

KARGER

Fax + 41 61 306 12 34
E-Mail karger@karger.ch
www.karger.com

Contents

Preface

The existence of mouse chromosomal variants in the tobacco mouse, first discovered in the Poschiavo valley by Alfred Gropp, was a breakthrough for the study of speciation and even more so, for reinforcing the establishment of new mouse models for research on developmental processes.

The completion of a draft sequence of the mouse genome in December 2002 was a major landmark in biology, opening new research opportunities that will provide insights into the human genetic code and will make the mouse an even better tool for biomedical research.

When I was asked by Michael Schmid to edit a special issue of *Cytogenetic and Genome Research* devoted to our favourite experimental model, I immediately thought of two earlier books, the *Biology of the House Mouse*, edited by R.J. Berry (1981) and the *Genetic variants and strains of the laboratory mouse*, edited by Antony G. Searle and Mary F. Lyon (1981). These books, that encompassed nearly all the possible topics on mouse research, have been excellent companions during all my studies anticipating areas of interest which are still very important. Following these editors' example, I have tried to collect a wide-ranging overview of all current aspects of mouse genetics, with the aim of making the book a suitable tool of investigation for the wide population of researchers that use this model animal.

The house mouse holds a unique place among model organisms, representing an indispensable tool for most investigators in many areas, particularly biomedical research. The amount of information on the physiology, reproduction and development of the mouse is greater than that for any other mammal, with the possible exception of man. High fertility, short gestation, genetic variations, susceptibility to diseases and genetic tractability, are just a few of the reasons that have led researchers to exploit the mouse. Comparative genomics analyses showed that less than 1% of genes have no detectable homologs in humans even though mice diverged from the human lineage about 75 million years ago. Thus, it is not a surprise that mice are involved in about 90% of all experiments carried out in USA and represent a USD 200 million business per year.

Mouse genetics encompasses nearly 100 years of research, from William Castle's studies of Granby mice (1901–1908) and the establishment of the first inbred strain (DBA, 1909) to the publication of the mouse genome sequence (2002). In between there have been milestone achievements which have revolutionized biomedical science leading the mouse to be *the* experimental model for human biology: 1915, first linkage study; 1937, Peter Gorer's description of the H-2 locus; late 1940s, George Snell bred the first congenic strains; 1961, Mary Lyon's X-inactivation hypothesis; during the 1960s the lack of thymus was discovered and nude mouse strains were established. The golden achievements of the 1980s are: 1980, first transgenic mouse created by pronuclear injection of fertilized eggs; 1981, first murine embryonic stem cell line; 1983, a SCID mouse strain established; 1987, site-directed mutagenesis by gene targeting in embryonic stem cells; 1987, first knockout mouse; 1992, a p53 knockout mouse; 1998, Ryuzo Yanagimachi's mouse clones.

Genomics is a new and fascinating area of biology that takes a broad approach to molecular biology and evolution by studying the complete genome functioning. Now that the mouse genome sequence has been published, the "mouse community" has a powerful tool for dissecting the complexity of biological processes which require to be tackled at both molecular and organism level. For many researchers the availability of the full mouse genome sequence is really exciting; for all scientists it is a watershed that forces us to re-consider our conceptual tools and the way we do research. From now on, we all need to frame our research topics within genomics and be familiar with genome sequences, transcriptome profiling data and high-throughput technologies.

This issue features 37 peer-reviewed articles and brings together contributions that cover all major aspects of mouse genetics. The guest editor's preface usually offers a very short presentation of the articles, a truly impossible task when dealing with such a large number of articles on topics ranging from nuclear transfer and embryonic development to cancer, from reproduction to speciation and to functional genomics technologies. I shall let the articles speak for themselves!

I hope the readers will share my excitement for the huge impact that the knowledge of the mouse genome already has on all fields of biological research and human health, since the mouse genome encodes an experimentally tractable organism, a feature that has been uniquely exploited in this small mammal.

My warmest thanks to all the authors, the reviewers and the editors who have devoted their time and enthusiasm to explain current knowledge on the mouse genetics and offer glimpses of the exciting future.

Silvia Garagna
Pavia, April 2004

Cytogenet Genome Res 105:166–171 (2004)
DOI: 10.1159/000078186

The mouse "tool box" for meiotic studies

T. Ashley

Department of Genetics, Yale University School of Medicine, New Haven, CT (USA)

Abstract. Besides availability, there are numerous advantages to using mice for meiotic studies: (1) techniques exist for enriching the population of spermatocytes at particular meiotic stages; (2) spermatogenesis in mice is highly synchronized and testis sections afford a highly accurate method of staging meiotic events; (3) knock-out mice provide a rich source of meiotic mutants. Coupled with antibody localization techniques these tools, singularly and in combination, provide multiple means of temporally and physically dissecting meiosis.

Copyright © 2004 S. Karger AG, Basel

Introduction

Meiosis, the cell division that reduces the chromosome number from diploid to haploid, is a defining characteristic of all eukaryotes. In fact, meiosis is two cell divisions without an intervening replication. In animals, the meiotic products then differentiate into eggs or sperm. Fertilization restores the chromosome number to diploid, a state that is maintained in somatic cells by successive mitotic divisions. Meiotic errors in early prophase generally lead to apoptosis and even sterility, while errors in either of the two divisions lead to aneuploidy. Therefore there are more than academic reasons for studying meiosis.

It might seem that the best biological system to use for meiotic studies would be the organism of most interest – human. However for obvious reasons, human spermatocytes and oocytes are not readily available. This is particularly the case for oocytes in early meiotic prophase, a stage that occurs prenatally in females. In contrast, a breeding colony of mice provides a continuous and reliable source of meiotic cells from both sexes. However, ethics and convenience are not the only advantages of using mice as a model system.

Received 25 September 2003; manuscript accepted 14 October 2003.

Request reprints from: Dr. Terry Ashley
Department of Genetics, Yale University School of Medicine
333 Cedar Street, New Haven, CT 06510 (USA)
telephone: +1-203-785-4473; fax: +1-203-785-4958
e-mail: Terry.Ashley@yale.edu

Duration of meiosis

Mammalian meiosis is an extremely protracted process, requiring over two weeks in male mice (Oakberg, 1956) compared to approximately 80 min in the budding yeast, *Saccharomyces cerevisiae*. Most of this time is spent in meiotic prophase, which is usually divided into 5 stages: leptonema, zygonema, pachynema, diplonema, and diakinesis. For reasons discussed below under "Mouse testis sections and synchronization of meiotic prophase events", estimates of the duration of specific stages and substages are especially accurate for mouse spermatocytes. All times given here are from Van der Meer et al. (1993). Preceding prophase the spermatocytes undergo bulk DNA synthesis, a stage referred to as "preleptonema" or premeiotic S phase. This replication takes 41.4 h. Prophase begins as the chromosomes begin to condense and homologues move into close proximity during leptonema, a stage that lasts 23.6 h. During zygonema, estimated to take another 39.5 h, homologous autosomes synapse along their entire length. Although usually referred to as separate stages, leptonema/zygonema overlap in the mouse. Throughout pachynema, which lasts an extraordinary 186.6 h, homologues remain synapsed. Although recombination (crossover) occurs during pachynema, transcription and considerable chromatin remodeling also occurs during this time. In diplonema, which lasts 20.4 h, considerable condensation occurs and the homologues begin to separate, remaining connected only at chiasmata (sites where crossovers have occurred). In diakinesis, the final stage of prophase, further condensation occurs and the bivalents (pairs of homologues) begin to move onto the metaphase I plate. Homologues move to opposite poles during anaphase I and sister chromatids segregate to opposite poles during anaphase II. Both divisions are completed during a further 15.5-hour period.

Meiotic prophase is initiated in the fetal mouse ovary over a period of several days and progression is synchronous, although probably not as tightly controlled as in males. Oocytes in leptonema-zygonema can be found in female embryos 15–16 days post conception (pc) and oocytes in pachynema can be found in embryos 17 days pc until a day or two after birth. The duration of meiosis is shorter in females and most of the difference can be accounted for by a shorter pachynema. When the oocytes reach diplonema, they arrest and remain in this state until stimulated to undergo growth and maturation. Because of the greater ease of obtaining material, the rest of the review will focus on males. However, as we will see in the discussion of knockout mice, there are often male/female differences in meiosis that should not be ignored.

As cytogeneticists most of us are accustomed to thinking about spatial resolution, but we seldom give the same thought to temporal resolution. The remarkable duration of meiotic prophase in mouse provides an exceptional opportunity to temporally dissect meiotic events. This is particularly the case during pachynema, which lasts over a week.

Meiotic-specific structures

During the first meiotic division homologous chromosomes must find one another, synapse, recombine and disjoin to opposite poles. The second division, accomplished without an intervening S phase, is similar to a mitotic division and segregates sister chromatids. The unique events of the first division are facilitated by several meiotic-specific structures: the synaptonemal complex (SC) and early and late meiotic nodules. The most prominent of these is the SC, a proteinaceous structure located between the homologously synapsed chromosomes. The SC consists of two axial elements (AEs) that form between sister chromatids during leptonema. As homologues come into close (approximately 100 nm) proximity, transverse filaments form between the homologous AEs. Despite the morphological description of a central element between the AEs (Fawcett, 1956; Moses, 1956), no protein component of the central element has yet been identified. In many organisms, AE formation is completed before homologous chromosomes begin to synapse. However in mouse spermatocytes, SC formation begins before AE formation along the length of the chromosomes is finished. Consequently, as mentioned above, there is no clear distinction between late leptonema and early zygonema in the mouse.

Two types of meiotic nodules were first observed by electron microscopy in meiotic prophase: early nodules on asynapsed AEs and newly synapsed SCs, present during zygonema and early pachynema (Albini and Jones, 1987; Anderson and Stack, 1988) and late nodules present later in pachynema (Carpenter, 1975). Early nodules (ENs) were postulated to be involved in the check for homology that accompanies synapsis (Stack et al., 1989), while late nodules (LNs or RNs) were postulated to be involved in reciprocal recombination or crossover (Carpenter, 1975). Consistent with the posited role in recombination, the number of LNs in wild type *Drosophila* was shown to correspond to the expected number of crossovers (Carpenter, 1979b) and a *Drosophila* mutant defective in recombination was shown to have a decreased number of LNs of abnormal size and morphology (Carpenter, 1979a). Although these nuclear organelles were originally identified in electron micrographs, as discussed below, protein components of each of these meiotic-specific structures have been identified.

Spreads and antibody localization

Early meiotic studies were carried out on sectioned material examined by electron microscopy (for review see Moses, 1968; Wettstein, 1984). In 1973 Counce and Meyer (1973) introduced a technique that spreads the nuclear contents of a spermatocyte (or oocyte) on the surface of a hypotonic solution. This procedure allowed all the bivalents in a nucleus to be visualized in the same low-magnification electron micrograph. Although the dimensions of the SC are theoretically below the limit of resolution of light microscopy, they also discovered that SCs could be visualized with air-interface (non-oil immersion) phase optics. Shortly thereafter silver staining techniques were employed that facilitated examination of microspreads by both light and electron microscopy (Dresser and Moses, 1979; Fletcher, 1979).

A modification of this spreading technique (Peters et al., 1997) has proven to be an excellent method for studying localization of various proteins in meiotic nuclei. Since this review focuses on the mouse, I will restrict my description of SC components to mammalian studies. SCP3 (Dobson et al., 1994; Lammers et al., 1994) and SCP2 (Offenberg et al., 1998; Schalk et al., 1998) have been shown to be components of the AE. Antibodies to these proteins allow visualization of the axial elements and lateral elements (as the AEs are called after homologues synapse) from the beginning of their assembly in early leptonema until their disassembly in diplonema. In addition both proteins remain at the kinetochore regions until the anaphase II (Dobson et al., 1994; Offenberg et al., 1998). SCP1 is a component of the transverse filaments (Dobson et al., 1994; Liu et al., 1996) and can be used to follow synapsis. Prior to the appearance of SCP2 and SCP3, several meiotic cohesions begin to form between the sister chromatids (Eijpe et al., 2003).

A number of proteins have been localized to sites along the asynapsed AEs including RAD51, a Rec A homologue (Plug et al., 1996; Barlow et al., 1997; Moens et al., 1997). BRCA1, the protein product of a gene mutated in many cases of familial breast and ovarian cancers (Scully et al., 1997), and ATR, a protein involved in the damage surveillance network of proteins (Keegan et al., 1996; Moens et al., 1999; Baart et al., 2000). Several proteins have also been localized to the newly synapsed SCs in late zygonema and early pachynema, of which only RAD51 appears to localize on both the asynapsed and synapsed axes (Moens et al., 1997). (Given the fact that synapsis occurs over a 36-hour period, turnover is unlikely to be instantaneous. Thus, it is logical to expect some delay in loss of these pre-synaptic proteins). Once homologues synapse, different proteins are seen to localize on the newly synapsed SCs including RPA (Plug et al., 1997b, 1998), the primary eukaryotic DNA single-strand binding protein; BLM (Walpita et al., 1999;

Moens et al., 2000), the helicase mutated in individuals with Bloom syndrome (Ellis et al., 1995); POLβ (Plug et al., 1997a), a gap-filling DNA polymerase; and CDK4 (Ashley et al., 2001), a cell cycle protein.

It can be argued that absolute proof of inclusion of a protein in ENs requires either demonstration by immunogold localization of the protein to one of these particles by electron microscopy, or isolation of the nodule itself and biochemical analysis of a purified fraction. No one has yet isolated either ENs or RNs. However, immunogold localization of RAD51 and DMC1, a meiotic-specific RecA homologue, has been reported (Anderson et al., 1997; Moens et al., 1997). Based on biochemical evidence of direct binding and interactions between RAD51 and BRCA1 (Scully et al., 1997) and BRCA1 and ATR (Tibbetts et al., 2000) it can be inferred that these proteins are also components of presynaptic ENs. However, ATR (Keegan et al., 1996; Moens et al., 1999; Baart et al., 2000) and BRCA1 (Scully et al., 1997) both accumulate on asynapsed axes when there are chromosome aberrations or mutations that delay or prevent synapsis.

MLH1, a mismatch repair protein, was inferred to be a component of RNs based on the same set of criteria that were used to establish that RNs were involved in reciprocal recombination in the first place. Namely, the number and distribution of MLH1 foci on the synaptonemal complexes in male and female mice corresponds to the number and distribution of crossovers inferred genetically and cytologically (see Baker et al., 1996). This inference has since been confirmed by immunogold localization of MLH1 antibody to RNs (Moens et al., 2002). MLH3, another mismatch repair protein that forms a dimer with MLH1 co-localizes with it during pachynema (Lipkin et al., 2002).

Juvenile male mice and synchronization of development

Anyone who has made microspread preparations from testis of adult males of any species is all too familiar with a corollary of Murphy's Law, "If there is a beautiful well-spread spermatocyte, a sperm tail will be lying across the middle of it". There is a simple way of circumventing this particular cytological plague in mice: use testis of juvenile males that are undergoing the first meiotic wave. Besides avoiding sperm, selecting an animal of a particular age postpartum has additional advantages. Goetz and colleagues (1984) made spread preparations from juvenile males on successive days postpartum and constructed a table that provides data on the most advanced stage or substage of meiotic prophase and the frequency of the earlier stages. For example, they found that the first few spermatocytes enter leptonema/zygonema on day 9 postpartum and by the next day 52% of the spermatocyte population has reached this stage (Goetz et al., 1984). By day 17 postpartum the first spermatocytes (13% of the total population) have reached diplonema, while 7% are classified as preleptonema, 22% as leptonema/zygonema, 28.3% early pachynema, 21.3% mid-pachynema, and 7.7% late-pachynema. If you know the stage, or substage of meiotic prophase in which you are most interested, you can consult their chart and select an appropriately aged animal that will provide the maximum percentage of spermatocytes in that stage.

Staging using vitamin A-deficient male mice

Using juvenile mice for staging suffers one major disadvantage – all earlier stages are also present in the testis. More synchronized staging can be obtained by placing males on a vitamin A-deficient (VAD) diet which results in arrest of spermatogenesis at G1 of A1 spermatogonia (van Pelt and de Rooij, 1990). Following injection of retinoic acid and return to a normal diet, spermatogenesis resumes with all cells within the testis proceeding through meiosis in a highly synchronous cohort (van Pelt and de Rooij, 1990) and at the same pace as in nontreated animals (van Pelt and de Rooij, 1991). There is less than 36 h difference between the ages of all spermatocytes within these synchronized testes (van Pelt and de Rooij, 1990, 1991).

Mouse testis sections and synchronization of meiotic prophase events

The developmental biology of spermatogenesis in mice provides an additional powerful tool for meiotic studies. Development of cells within the mouse testis is highly synchronized (Oakberg, 1956; Van der Meer et al., 1993). Within a cross section of the testis, spermatogonial cells lie around the periphery; spermatocytes slightly more interior; and developing spermatozoa (spermatids and sperm) are found progressively more toward the lumen of the tubule as they differentiate. All spermatocytes within a cross section develop synchronously with the other spermatocytes within the tubule, as do the spermatogonia and spermatids. Moreover, the maturation of cells between cell types is synchronous. Therefore if the developmental stage of one cell type (spermatid or spermatogonia) can be established, the stage of all the spermatocytes within the same tubule can be accurately deduced (Russel et al., 1990). Cross sections with the progression of cell associations can be arranged in the natural order of spermatogenesis to show the developmental sequence. This sequence is called the cycle of seminiferous epithelium and in mouse this cycle is divided into twelve stages (Russel et al., 1990). As will be discussed below, this is a powerful tool for two types of studies. While all cell types develop synchronously within the mouse testis, this is not the case in human. In man the cycle can only be divided into six stages. There is a mixed and apparently random population of cells types and even different stages of the same cell type within the tubules (Johnson et al., 1996). This randomness and asynchrony give an additional value to the mouse as a "model organism"!

One important use of testis sections is for the assessment of epithelial stage. Staging is most accurate when the sections are stained with periodic acid Schiff (PAS) and hematoxylin. When antibody detection is carried out in conjunction with this classic staining procedure for light microscopy, detection is dependent on a peroxidase reaction rather than fluorescent antibody staining. Accuracy is most critical when evaluating meiotic mutants (see below). However, for staging of normal meiotic events fluorescent labeling has also been sucessfully employed (Eijpe et al., 2003). Evaluation of sections is the best method of determining the time of appearance of proteins in early meiotic prophase before the axial elements and SC are fully developed (Eijpe et al., 2003).

Combining testis section analysis, VAD staging, and antibody localization on microspread preparations provide an even more powerful tool for determining when during prophase certain events are initiated or completed. For this type of study, mice were made vitamin A deficient, injected with retinoic acid and returned to a normal diet. On successive days post injection, an animal was sacrificed and one testis was fixed for histological evaluation of the testis sections and one was used for microspreads and immunostaining (Ashley, Gaeth, Creemer, Hack and de Rooij, submitted). This analysis showed that MLH1 did not appear until 13 days post-injection when the spermatocytes were in epithelial stage II (early-mid pachynema). Most spermatocytes (77.6%) on day 13 had no MLH1 foci, and of those that did have foci, the numbers ranged from 1 to 5, suggesting MLH1 was only beginning to bind (directly, or indirectly) to a recombination intermediate. Interestingly, one day later 94.4% of the spermatocytes had at least 19 MLH1 foci per nucleus (a theoretical one per bivalent). This suggests that acquisition of MLH1 occurs relatively rapidly in the transition between early and mid-pachynema.

Cell separation procedures

Spermatocytes in various stages of meiotic prophase can be separated on the basis of differences in their size and mass using methods such as elutriation and density centrifugation in Percoll (Heyting and Dietrich, 1991). Although the yield of spermatocytes in mid-late pachynema can be as high as 95%, the similar size and density of spermatogonia and spermatocytes in premeiotic S phase reduces the yield of these early spermatocytes to less that 50% purity (Eijpe et al., 2003). However using spermatocytes from VAD mice in which the differentiated spermatogonia have not fully repopulated the testis can increase the yield of early spermatocytes. These techniques provide an excellent starting point for Western blot or immunoprecipitation experiments on stage-specific differences in meiotic proteins.

Knockout mice

For many years mammalian studies were severely handicapped by a paucity of meiotic mutants (Handel, 1988). However the advent of techniques for targeted disruption of specific genes has provided a growing list of mutations with meiotic effects. A recent review summarizes the data on knock-out mice with detrimental effects in spermatogenesis (de Rooij, 2003), so only a few examples that halt meiotic progression will be mentioned here. In some of these mutants the cytological phenotype is immediately obvious as was the case with *Dmc1$^{-/-}$*, in which the autosomes fail to synapse (Pittman et al., 1998; Yoshida et al., 1998), or *Mlh1$^{-/-}$*, in which the spermatocytes arrest at metaphase I with univalent (achiasmatic) chromosomes (Baker et al., 1996). In other cases, such as *Mei1*, in which the gene has no known homologue and the protein domains offer no clues to function, the phenotype (in this case asynapsis) is more puzzling (Libby et al., 2002).

Knockout mice and testis sections

The evaluation of testis sections is crucial for determining the time of arrest of a knockout mouse. The importance of this technique is best illustrated by the *Msh5$^{-/-}$* mouse. Cytological examination of spread preparations reveals that AEs form along the entire length of each chromosome, but that little, or no synapsis occurs and the spermatocytes arrest and become apoptotic (de Vries et al., 1999; Edelmann et al., 1999). By cytological definition, this synaptic failure places the time of arrest at zygonema. However the testis sections tell a different story.

Although these mutant spermatocytes lack the late prophase stages and post meiotic cells (spermatids) helpful for staging, this is not an insurmountable handicap, so long as the spermatogonial stage associated with the arrested cells can be determined (de Boer, 1986; de Vries et al., 1999; Van der Meer et al., 1993). In the case of the *Msh5$^{-/-}$* males, there are surviving spermatocytes in tubules with intermediate spermatogonia (epithelial stage III) and the mutant spermatocytes do not become apoptotic until epithelial stage IV (de Vries et al., 1999). This is at least 60 h after the usual zygotene to pachytene transition and a time by which normal spermatocytes have reached mid-pachynema. This is not a trivial difference since spermatocytes arrive at stage IV 74 h (3 days) after the usual zygotene to pachytene transition. The asynaptic phenotype has been referred to as an arrest. However, the time lapse between failure to synapse and apoptosis suggests that this term may be inappropriate. Rather than "arrested", it is more likely that the cells are engaged in attempts to circumvent the blockage. True arrest only occurs when time runs out and the checkpoint pulls the apoptotic plug.

Antibody localization on knock-out mice

As we have seen above in the case of VAD mice, testis sections and spreads, combining techniques greatly increases the amount of information that can be obtained from meiotic analysis. *Atm* is the gene disrupted in individuals with the autosomal recessive disorder ataxia telangiectasia. ATM is a component of the damage surveillance network of checkpoint proteins responsible for maintaining the integrity of the genome. ATM detects double-strand breaks and signals downstream targets that in turn halt cell cycle progression until the damage can be repaired or, if repair is impossible, shunts the cell into an apoptotic pathway (Hoekstra, 1997; Shiloh, 2001). A mutation in a checkpoint gene is expected to abrogate the checkpoint and allow the cell to proceed through the cell cycle with unrepaired damage. Therefore it came as some surprise that the *Atm$^{-/-}$* knockout mice were male sterile (Barlow et al., 1996; Xu et al., 1996). Homologous chromosomes began to synapse but fragmented, and the spermatocytes underwent apoptosis (Xu et al., 1996). Yet a likely explanation was soon apparent. ATR is a related protein kinase (Cimprich et al., 1996; Keegan et al., 1996) and ATR accumulated on asynapsed axial elements (Keegan et al., 1996). Every *Atm$^{-/-}$* spermatocyte nucleus has asynapsed AEs, and it is apparently the continued presence of ATR that triggers the meiotic checkpoint (Keegan et al., 1996; Plug et al., 1997b).

Both the special and temporal resolution afforded in microspread meiotic nuclei from mammals far exceeds that of similar preparations from yeast. This resolving power has to date been under-utilized. Failure of an antibody to localize to a protein in a spermatocyte from a knockout mouse when it does localize in a wild type mouse suggests that the protein product of the mutated gene is required for localization of the non-mutated protein.

As discussed above, a meiotic phenotype should be considered a roadblock. The mutant may seek a detour to override the obstruction. Therefore, if localization of a protein in the knockout is different from its localization in wild type spermatocytes, the data may be indicating that the cell is pursuing an alternate, perhaps more somatic, pathway around the blockage.

Summary

In this review I have provided a brief summary of some of the tools available for meiotic studies, emphasizing those that are unique for mouse, or similar small mammals, such as the rat. I have tried to emphasize that while individual tools are useful, two or more tools in combination are much more powerful. Meiosis is complex. The wider the variety of approaches to dissecting the process, the greater the likelihood of meaningful insights.

References

Albini SM, Jones GH: Synaptonemal complex spreading in *Allium cepa* and *A. fistulosum* I. The initiation and sequence of pairing. Chromosoma 95:324–338 (1987).

Anderson LK, Stack SM: Nodules associated with axial cores and synaptonemal complexes during zygotene in *Psilotum nudum.* Chromosoma 97:96–100 (1988).

Anderson LK, Offenberg HH, Verkuijlen WHMC, Heyting C: RecA-like proteins are components of early meiotic nodules in lily. Proc Natl Acad Sci USA 94:6868–6873 (1997).

Ashley T, Walpita D, de Rooij DG: Localization of two mammalian cyclin dependent kinases during mammalian meiosis. J Cell Sci 114:685–693 (2001).

Baart EB, de Rooij DG, Keegan KS, de Boer P: Distribution of ATR protein in primary spermatocytes of a mouse chromosomal mutant: a comparison of preparation techniques. Chromosoma 109:139–147 (2000).

Baker SM, Plug AW, Prolla TA, Bronner CE, Harris AC, Yao X, Christie D-M, Monell C, Arnheim N, Bradley A, Ashley T, Liskay RM: Involvement of mouse Mlh1 in DNA mismatch repair and meiotic crossing over. Nat Genet 13:336–342 (1996).

Barlow AL, Benson FE, West SC, Hultén MA: Distribution of Rad51 recombinase in human and mouse spermatocytes. EMBO J 16:5207–5215 (1997).

Barlow C, Hirotsune S, Paylor R, Liyanage M, Eckhaus M, Collins F, Shiloh Y, Crawley JN, Ried T, Tangle D, Wynshaw-Boris A: Atm-deficient mice: a paradigm of ataxia telangiectasia. Cell 86:159–171 (1996).

Carpenter ATC: Electron microscopy of meiosis in *Drosophila melanogaster* females. II. The recombination nodule- a recombination-associated structure in pachytene? Proc Natl Acad Sci USA 72:3186–3189 (1975).

Carpenter ATC: Recombination nodules and synaptonemal complex in recombination-defective females of *Drosophila melanogaster*. Chromosoma 75:259–292 (1979a).

Carpenter ATC: Synaptonemal complex and recombination nodules in the wild-type *Drosophila melanogaster* females. Genetics 92:511–541 (1979b).

Cimprich KA, Shin TB, Keith CT, Schreiber SL: cDNA cloning and gene mapping of a candidate human cell cycle checkpoint protein. Proc Natl Acad Sci USA 93:2850–2855 (1996).

Counce SJ, Meyer GF: Differentiation of the synaptonemal complex and the kinetochores in *Locusta* spermatocytes studies by whole mount electron microscopy. Chromosoma 44:231–253 (1973).

de Boer P: Chromosomal causes for fertility reduction in mammals, in de Serres FJ (ed): Chemical Muta gens, pp 427–467 (Plenum, New York, London 1986).

de Rooij DG, de Boer P: Specific arrest in spermatogenesis in genetically modified and mutant mice. Cytogenet Genome Res 103:267–276 (2003).

de Vries SS, Baart EB, Dekker M, Siezen A, de Rooij DG, de Boer P, te Riele H: Mouse MutS-like protein MSH5 is required for proper chromosome synapsis in male and female meiosis. Genes Dev 13:523–531 (1999).

Dobson MJ, Pearlman RE, Karaiskakis A, Spyropoulos B, Moens PB: Synaptonemal complex proteins, epitope mapping and chromosome disjunction. J Cell Sci 107:2749–2760 (1994).

Dresser ME, Moses MJ: Silver staining of synaptonemal complexes in surface spreads for light and electron microscopy. Exp Cell Res 121:416–419 (1979).

Edelmann W, Cohen PE, Kneitz B, Winand N, Lia M, Heyer J, Kolodner R, Pollard JW, Kucherlapati R: Mammalian MutS homologue 5 is required for chromosome pairing in meiosis. Nat Genet 21:123–127 (1999).

Eijpe M, Offenberg HH, Jessberger R, Revenkova E, Heyting C: Meiotic cohesin REC8 marks the axial elements of rat synaptonemal complexes before cohesins SMC1beta and SMC3. J Cell Biol 160:657–670 (2003).

Ellis NA, Groden J, Ye TZ, Straughen J, Lennon DJ, Ciocci S, Proytcheva M, German J: The Bloom's syndrome gene product is homologous to RecQ helicases. Cell 83:655–666 (1995).

Fawcett DW: The fine structure of chromosomes in the meiotic prophase of vertebrate spermatocytes. J Biophys Biochem Cytol 2:403–406 (1956).

Fletcher JM: Light microscope analysis of meiotic prophase chromosomes by silver staining. Chromosoma 72:241–248 (1979).

Goetz P, Chandley AC, Speed RM: Morphological and temporal sequence of meiotic prophase development at puberty in male mouse. J Cell Sci 65:249–263 (1984).

Handel MA: Genetic control of spermatogenesis in mice, in Hennig W (ed): Results and Problems in Cell Differentiation, Spermatogenesis: Genetic Aspects, pp 1–62 (Springer-Verlag, Berlin 1988).

Heyting C, Dietrich AJ: Meiotic chromosome preparation and protein labeling. Methods Cell Biol 35:177–202 (1991).

Hoekstra MF: Responses to DNA damage and regulation of cell cycle checkpoints by the ATM protein kinase family. Curr Opin Genet Dev 7:170–175 (1997).

Johnson L, McKenzie KS, Snell JR: Partial wave in human seminiferous tubules appears to be a random occurence. Tissue Cell 28:127–136 (1996).

Keegan KS, Holtzman DA, Plug AW, Christenson ER, Brainerd EE, Flaggs G, Bentley NJ, Taylor EM, Meyn MS, Moss SB, Carr AM, Ashley T, Hoekstra MF: The ATR and ATM protein kinases associate with different sites along meiotically pairing chromosomes. Genes Dev 10:2423–2437 (1996).

Lammers JHM, Offenberg HH, van Aalderen M, Vink ACG, Dietrich AJJ, Heyting C: The gene encoding a major component of the lateral element of the synaptonemal complex of the rat is related to X-linked lymphocyte-regulated genes. Mol Cell Biol 14:1137–1146 (1994).

Libby BJ, de la Fuente R, O'Brien MJ, Wigglesworth K, Cobb J, Inselman A, Eaker S, Handel MA, Eppig JJ, Schimenti JC: The mouse meiotic mutant mei1 disrupts chromosome synapsis with sexually dimorphic consequences for meiotic progression. Dev Biol 242:423–444 (2002).

Lipkin SM, Moens PB, Wang V, Lenzi M, Shanmugarajah D, Gilgeous A, Thomas J, Cheng J, Touchman JW, Green ED, Schwartzberg P, Collins FS, Cohen PE: Meiotic arrest and aneuploidy in MLH3-deficient mice. Nat Genet 31:385–390 (2002).

Liu J-G, Yuan L, Brundell E, Bjorkroth B, Daneholt B, Hoog C: Localization of the N-terminus of SCP1 to the central element of the synaptonemal complex and evidence for direct interactions between the N-termini of SCP1 molecules organized head-to-head. Exp Cell Res 226:11–19 (1996).

Moens PB, Chen DJ, Shen Z, Kolas N, Tarsounas M, Heng HHQ, Spyropoulos B: RAD51 immunocytology in rat and mouse spermatocytes and oocytes. Chromosoma 106:207–215 (1997).

Moens PB, Tarsounas M, Morita T, Habu T, Rottinghaus ST, Freire R, Jackson SP, Barlow C, Wynshaw-Boris A: The association of ATR protein with mouse meiotic chromosome cores. Chromosoma 108:95–102 (1999).

Moens PB, Freire R, Tarsounas M, Spyropoulos B, Jackson SP: Expression and nuclear localization of BLM, a chromosome stability protein mutated in Bloom's syndrome, suggests a role in recombination during meiotic prophase. J Cell Sci 113:663–672 (2000).

Moens PB, Kolas NK, Tarsounas M, Marcon E, Cohen PE, Spyropoulos B: The time course and chromosomal localization of recombination-related proteins at meiosis in the mouse are compatible with models that can resolve the early DNA-DNA interactions without reciprocal recombination. J Cell Sci 115:1611–1622 (2002).

Moses MJ: Chromosomal structures in crayfish spermatocytes. J Biophys Biochem Cytol 2:215–218 (1956).

Moses MJ: Synaptonemal complex. Annu Rev Genet 2:363–412 (1968).

Oakberg EF: Duration of spermatogenesis in the mouse and timing of stages of the cycle in the seminiferous epithelium. Am J Anat 99:507–516 (1956).

Offenberg HH, Schalk JAC, Meuwissen RLJ, van Aalderen M, Kester HA, Dietrich AJJ, Heyting C: SCP2: a major protein component of the axial elements of synaptonemal complexes of the rat. Nucl Acids Res 26:2572–2579 (1998).

Peters AHFM, Plug AW, van Vugt MJ, de Boer P: A drying-down technique for spreading of mammalian meiocytes from the male and female germ line. Chromosome Res 5:66–71 (1997).

Pittman DL, Cobb J, Schimenti KJ, Wilson LA, Cooper DM, Brignull E, Handel MA, Schimenti JC: Meiotic prophase arrest with failure of chromosome synapsis in mice deficient for Dmc1, a germline-specific RecA homolog. Mol Cell 1:697–705 (1998).

Plug AW, Xu J, Reedy G, Golub EI, Ashley T: Presynaptic association of RAD51 protein with selected sites in meiotic chromatin. Proc Nat Acad Sci USA 93:5920–5924 (1996).

Plug AW, Clairmont CA, Sapi E, Ashley T, Sweasy JB: Evidence for a role for DNA polymerase B in mammalian meiosis. Proc Natl Acad Sci USA 94:1327–1331 (1997a).

Plug AW, Peters AHFM, Xu Y, Keegan KS, Hoekstra MF, Baltimore D, de Boer P, Ashley T: ATM and RPA in meiotic chromosome synapsis and recombination. Nat Genet 17:457–461 (1997b).

Plug AW, Peters AHFM, van Breuklen B, Keegan KS, Hoekstra M, de Boer P, Ashley T: Changes in protein composition of meiotic nodules during mammalian meiosis. J Cell Sci 111:413–423 (1998).

Russel LD, Ettlin RA, Hikim APS, Clegg ED: Histological and Histopathological Evaluation of the Testes (Cache River Press, Clearwater 1990)

Schalk JA, Dietrich AJ, Vink AC, Offenberg HH, van Aalderen M, Heyting C: Localization of SCP2 and SCP3 protein molecules within synaptonemal complexes of the rat. Chromosoma 107:540–548 (1998).

Scully R, Chen J, Plug A, Xiao Y, Weaver D, Feunteun J, Ashley T, Livingston DM: Association of BRCA1 with RAD51 in mitotic and meiotic cells. Cell 88:265–275 (1997).

Shiloh Y: ATM and ATR: network in cellular responses to DNA damage. Curr Opin Genet Dev 11:71–77 (2001).

Stack SM, Anderson LK, Sherman JD: Chiasmata and recombination nodules in *Lilium longifolium*. Genome 32:486–498 (1989).

Tibbetts RS, Cortez D, Brumbaugh KM, Scully R, Livingston DM, Elledge SJ, Abraham RT: Functional interactions between BRCA1 and the checkpoint kinase ATR during genotoxic stress. Genes Dev 14:2989–3002 (2000).

Van der Meer Y, Cattanach BC, de Rooij DG: The radiosensitivity of spermatogonial stem cells in C3H/101 F1 hybrid mice. Mut Res 290:201–210 (1993).

van Pelt AM, de Rooij DG: Synchronization of the seminiferous epithelium after vitamin A replacement in vitamin A-deficient mice. Biol Reprod 43: 363–367 (1990).

van Pelt AM, de Rooij DG: Retinoic acid is able to reinitiate spermatogenesis in vitamin A-deficient rats and high replicate doses support the full development of spermatogenic cells. Endocrinology 128: 697–704 (1991).

Walpita D, Plug AW, Neff N, German J, Ashley T: Bloom's syndrome protein (BLM) co-localizes with RPA in meiotic prophase nuclei of mammalian spermatocytes. Proc Natl Acad Sci USA 96:5622–5627 (1999).

Wettstein Dv: The synaptonemal complex and genetic segregation, in Evans CW, Dickinson HG (eds): Controlling Events in Meiosis, pp 195–231 (Company of Biologists LTD, Cambridge, UK 1984).

Xu Y, Ashley T, Brainerd EE, Bronson RT, Meyn MS, Baltimore D: Targeted disruption of ATM leads to growth retardation, chromosomal fragmentation during meiosis, immune defects, and thymic lymphoma. Genes Dev 10:2411–2422 (1996).

Yoshida K, Kondoh G, Matsuda Y, Habu T, Nishimune Y, Marita T: The mouse RecA-like gene *Dmc1* is required for homologous chromosome synapsis during meiosis. Mol Cell 1:707–718 (1998).

Cytogenet Genome Res 105:172–181 (2004)
DOI: 10.1159/000078187

The initiation of homologous chromosome synapsis in mouse fetal oocytes is not directly driven by centromere and telomere clustering in the bouquet

M. Tankimanova, M.A. Hultén, and C. Tease

Department of Biological Sciences, University of Warwick, Coventry (UK)

Abstract. We investigated the behaviour of centromeres and distal telomeres during the initial phases of female meiosis in mice. In particular, we wished to determine whether clustering of centromeres and telomeres (bouquet formation) played the same crucial role in homologous chromosome pairing in female meiosis as it does in the male. We found that synapsis (intimate homologous chromosome pairing) is most frequently initiated in the interstitial regions of homologous chromosomes, apparently ahead of the distal regions. The proximal ends of the chromosomes appear to be disfavoured for synaptic initiation. Moreover, initiation of synapsis occurred in oocytes that showed little or no evidence of bouquet formation. A bouquet was present in a substantial proportion of cells at mid to late zygotene, and was still present in some pachytene oocytes. This pattern of bouquet formation and pairing initiation is in stark contrast to that previously described in the male mouse. We propose that although dynamic movements of centromeres and telomeres to form clusters may facilitate alignment of homologues or homologous chromosome segments during zygotene, in the female mouse positional control of synaptic initiation is dependent on some other mechanism.

Copyright © 2004 S. Karger AG, Basel

Meiosis is the cell division process used by sexually reproducing organisms to produce haploid gametes from diploid progenitor cells and to increase the genetic variability in gametes by means of recombination and segregation. Meiosis has three unique features: homologous chromosome pairing; reciprocal recombination (chiasma formation), and chromosome, as opposed to chromatid, segregation. Many aspects of homologous chromosome pairing are still not entirely understood. Meiotic chromosome pairing can be partitioned into three steps: homologue searching and recognition; juxtaposition (alignment) of homologues, and synapsis, the intimate association of homologous chromosomes along their axes. Synapsis is mediated by a meiosis-specific proteinaceous structure, called the synaptonemal complex (SC), consisting of two lateral elements (LEs), derived from the axial elements (AEs) of unpaired chromosomes, and a central element (CE) consisting largely of the protein SYCP1. Although the general features of pairing show considerable evolutionary conservation, the detail of the process varies between different organisms (reviewed by Roeder, 1997; Zickler and Kleckner, 1998, 1999).

The mechanism of homologous chromosome pairing at meiosis in germ cells of female mammals is of interest for both theoretical and practical reasons. Failure of pairing, for example, is correlated with increased oocyte atresia (cell death) in both humans and mice (Burgoyne and Baker, 1984; Speed, 1988). Additionally, as pairing and recombination are intimately linked, errors in pairing can impact on the numbers and distributions of crossovers (chiasmata) between homologues and consequently affect the risk of nondisjunction in mature oocytes (see Hassold and Hunt, 2001).

This work was supported by a grant to M.A.H. from the Wellcome Trust.

Received 31 October 2003; manuscript accepted 26 November 2003.

Request reprints from: Dr. M. Tankimanova
  Department of Biological Sciences, University of Warwick
  Gibbet Hill Road, Coventry CV4 7AL (UK)
  telephone: +44(0)-2476-522444; fax: +44(0)-2476-572748
  e-mail: M.Tankimanova@warwick.ac.uk

In mammals, most observations on meiotic chromosome behaviour are obtained from male germ cells, principally because of the greater ease of their availability. However, as the pattern of germ cell development differs greatly between male and female mammals it is questionable whether chromosome behaviour is identical in both sexes. Scherthan and colleagues (Scherthan et al., 1996), for example, showed that synaptic initiation occurs predominantly from chromosome ends in both human and mouse spermatogenesis. In contrast, it has been reported that chromosome ends can be relatively late pairing in female meiosis (e.g. Speed, 1982, 1985). This inter-sex difference may indicate that particular chromosomal segments, such as centromeres and telomeres, play a distinctive role in the early phases of homologous chromosome pairing in male and female meiosis. Support for this supposition comes from some recent studies in cattle (Pfeifer et al., 2001, 2003). In contrast to predominantly terminal initiation of synapsis in spermatocytes, synapsis in oocytes can initiate both terminally and interstitially. Pfeifer and colleagues (Pfeifer et al., 2001, 2003) also noted that in both sexes the initiation of synapsis coincided with the clustering of telomeres in a small region of the nuclear membrane (bouquet formation; see Scherthan et al., 1996) and that the duration of this clustering was much longer in oocytes than in spermatocytes.

Telomere clustering has been reported in a range of species from yeasts (Chikashige et al., 1997; Trelles-Sticken et al., 1999, 2000), through plants (Bass et al., 1997, 2000; Armstrong et al., 2001; Carlton and Cande, 2002) to invertebrate and vertebrate animals (Moens, 1969; Kezer et al., 1989; Scherthan et al., 1996, 1998; Moens et al., 1998; Scherthan and Schönborn, 2001; Pfeifer et al., 2001, 2003). The almost universal occurrence of this process has led to the presumption that it plays a similar role in all of these organisms either directly to initiate pairing or indirectly by promoting homologue alignment.

The present study was undertaken to further examine chromosome behaviour during the early phases of meiosis in female germ cells of a model species, the mouse. In particular, we investigated the behaviour of centromeres and distal telomeres during the crucial phases of homology searching and synaptic initiation to determine whether these chromosomal regions were as important for homologous chromosome pairing in female meiosis as they are in male meiosis.

Materials and methods

Mice

Female meiosis initiates during fetal development in the mouse. Germ cells first enter meiosis on day 12 or 13 of gestation (day of copulation plug = day 0). Oocytes pass through meiosis in a partially synchronised wave. The earliest stages of prophase I of meiosis are seen in oocytes of day-13 fetuses. With increasing gestational age, the ovaries contain cells at increasingly later stages of prophase I. By the time of parturition, the majority of oocytes have reached diplotene, at which stage they enter a prolonged arrest phase. It is very difficult to accurately stage fetal oocytes unless they have been prepared as two-dimensional spreads on microscope slides. This constraint places considerable difficulties in the investigation of chromosome behaviour in cells prepared in a way to retain their three-dimensional structure. We can overcome this problem to a considerable extent by the simple expedient of sampling ovaries from fetuses at different gestational ages; in this way, it is possible to sample cell populations in which the predominant number of cells are at a known stage of prophase I.

In the present study, we used mice from the random bred TO strain and F_1 hybrid fetuses from a cross between C3H/HeH females and 101/H males. Pregnant females were killed either by CO_2 anaesthesia or cervical dislocation. Ovaries were removed from female fetuses and placed in culture medium before preparation for microscopy.

Two-dimensional spreads

Fetal oocytes from 9 fetuses (day 13–14) were prepared for immunocytochemistry, essentially as described by Barlow and Hultén (1996) and Moens et al. (1997). In brief, the ovaries were placed in a small drop of F10 Ham tissue culture medium (Sigma), the oocytes were liberated mechanically using fine needles, and then 2 drops of hypotonic solution (3 % sucrose) were added. The cells were allowed to settle for approximately 30 min before the addition of 10 drops of 2 % formaldehyde/PBS solution. After 20 min, the slides were briefly dipped in distilled water and washed in 0.1 % Triton X-100/PBS 3 times for 10 min.

Conventional chromosome spreads

Conventional chromosome preparations were used in combination with FISH (whole chromosome paints). These spreads were prepared by placing ovaries in a 0.56 % KCl hypotonic solution for 4 h. The ovaries were then fixed in 3:1 methanol-acetic acid. Cell preparations were made by mechanically disrupting the ovaries in a 6:3:1 acetic acid:water:methanol mixture, drying the suspension on a warm hotplate and removing excess acetic acid by a 5-min wash in absolute ethanol (Tease and Fisher, 1998). The slides were then air dried. Meiotic stages were identified after DAPI counterstaining, using the criteria described in earlier publications (Speed, 1982; O'Keeffe et al., 1997).

Cryosectioning

For preparing cryosections, the ovaries from eleven 14-day fetuses were embedded in OCT (RaLamb), rapidly frozen in dry ice and stored at –70 °C until use. Sections, 10 μm thick, were cut from frozen blocks, mounted on polysilane-coated slides (BDH), air-dried and fixed with 4 % formaldehyde/ PBS. Slides were then given three 10-min washes in 0.1 % Triton X-100/PBS at room temperature (RT).

Immunofluorescence staining

The polyclonal rabbit antibody COR1, that principally detects the synaptonemal complex protein SYCP3 in the axial elements of unpaired chromosomes and lateral elements of the SC, and the antibody SYN1, that detects the protein SYCP1 present in the central element of the SC, were used at a 1:1000 dilution. The human CREST antibody, which recognises kinetochore-associated proteins, was used at a 1:100 dilution.

Mouse chromosomes, with the exception of the Y, are highly acrocentric with no visible short arm. At the proximal end, the telomere is abutted by centromeric heterochromatin. Therefore, staining for the kinetochores (centromeres) effectively also identifies the proximal telomeres.

Slides with spread oocytes and frozen sections were incubated overnight at RT with a combination of antibodies, either COR1 and CREST or SYN1 and CREST. After three 10-min washes in 0.1 % Triton X-100/PBS appropriate secondary antibodies, conjugated with different fluorochromes, were applied to the slides. COR1 and SYN1 were detected using Texas Red (TR)-conjugated goat anti-rabbit antibody at 1:200 dilution (Vector). CREST was detected using FITC-conjugated goat anti-human antibody at 1:200 dilution (Vector). After incubation for 30 min at 37 °C and final washes, slides were mounted in Vectashield/DAPI (Vector). Following analysis (see below), the slides were washed in 0.1 % Triton X-100/PBS and subjected to FISH.

Fluorescence in situ hybridisation

We used a FITC-conjugated, telomere-specific PNA probe (DAKO) on two-dimensional cell spreads and frozen sections subsequent to immunostaining. Two locus-specific probes (ID Labs, Canada) were used on cryosections: D6Mit15 that identifies a distal locus, described to be positioned at 67 cM on chromosome 6; and D17Mit129 that identifies an interstitial locus at 49 cM along chromosome 17. FISH procedures were performed according to the manufacturer's recommendations. Slides were mounted in Vectashield/DAPI (Vector).

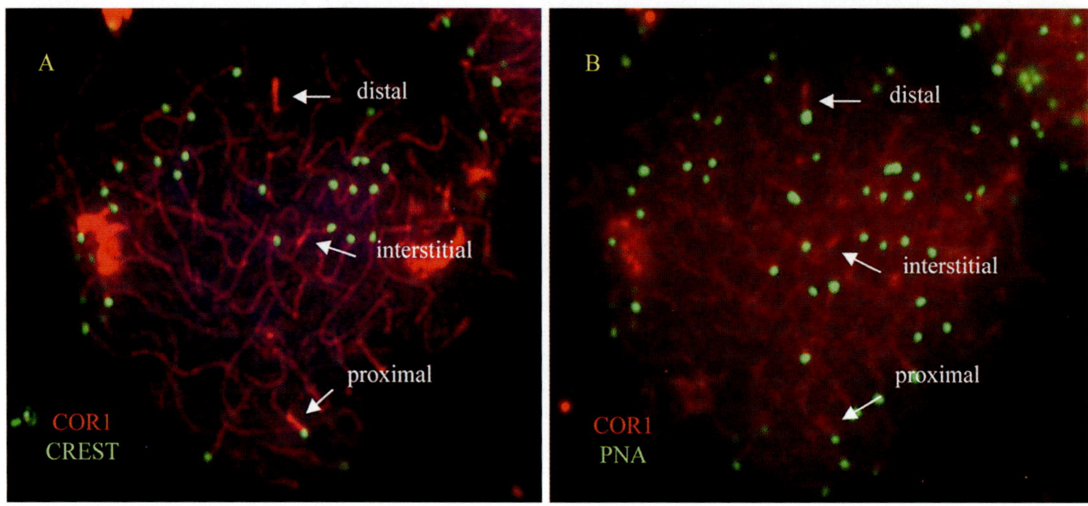

Fig. 1. Mouse oocyte at early zygotene stage labelled with anti-SC antibody COR1 (red), CREST antibody (green) and DAPI counterstain (blue) (**A**) and after telomere PNA-FISH (green) (**B**). (**A**) Note the more intensely labelled synapsed regions (arrows) than unsynapsed by COR1 antibody. The synapsed region at the proximal end is adjacent to the centromere signal (CREST). (**B**) The synapsed region at the distal end is abutted to the telomere signal.

Whole chromosome-specific mouse paints for chromosomes 2 and 4 (Cambio), in combination with minor satellite and telomere probes, were used on conventional air-dried chromosome spreads. Minor satellite and telomere probes were produced by PCR (Ijdo et al., 1991; Weier et al., 1991), labelled with digoxigenin-11-dUTP (Boehringer Mannheim) or biotin-14-dATP (Gibco BRL) and used according to the standard FISH protocol. The whole chromosome paints were used according to the manufacturer's recommendations. Nuclei were counterstained with DAPI in Vectashield mountant solution (Vector).

Microscopy and image analysis

Immunostained, two-dimensionally spread preparations were analysed using a Zeiss Axioskop fluorescence microscope equipped with a cooled CCD camera (Photometrics) and Quips SmartCapture image acquisition software (Vysis). Measurements were performed on selected images using IPLab software. Cryosections were analysed using a Zeiss fluorescence microscope equipped with a XYZ motorised stage (Prior) for spatial series acquisition, a digital CCD camera (Hamamatsu) and the IMSTAR CytogenTM Imaging system.

Results

1. Initiation of synapsis

Immunostained two-dimensional spreads

Oocytes prepared as two-dimensional spreads and immunostained for SCs proteins provide an excellent approach to observe and analyse the fine structure of SCs. Using this approach, we investigated the initiation and progression of synapsis in fetal oocytes from 13- and 14-day fetuses (the gestational ages when synaptic initiation predominates).

A total of 92 zygotene oocytes were screened; they were subjectively assigned to early, mid and late sub-stages on the basis of extent of SC formation. Cells with many unpaired AEs and limited SC formation were classified as early zygotene. The stretches of SCs stained more intensely than the AEs. This enabled us to detect sites of synaptic initiation along unpaired axes (Fig. 1A). The number of synaptic initiation sites was determined in 54 early zygotene oocytes.

As mentioned above, mouse chromosomes, with the exception of the Y, are highly acrocentric with no visible short arm. At the proximal end, the telomere is abutted by centromeric heterochromatin. Sites of proximal synaptic initiation were distinguished from those at distal telomeres by the presence of SC stretches adjacent to the centromeres (Fig. 1A). SC stretches, not adjacent to either the centromeres or distal telomeres, were classified as interstitial. The number of synaptic initiation sites per nucleus and their location on AEs varied between early zygotene oocytes (Fig. 2). Some cells had only one or two synaptic initiation sites, whereas several synapsis initiation sites with different location on AEs were observed in others. Overall, early zygotene cells showed a limited number of synaptic sites per nucleus ranging from 1 to 9. This inter-cell variation presumably reflects differences in the state of progression of the synaptic process in these cells. However, it is notable that none of these cells showed synaptic initiation in the majority of chromosome pairs; this indicates considerable asynchrony in synaptic initiation within a nucleus. The positions of synaptic sites were screened in these early zygotene cells. Overall, 17 of 54 cells had one or more sites of proximal initiation, 49 had interstitial sites, and 39 had distal sites (Fig. 2). The sample of 54 early zygotene nuclei contained a total of 324 short SC stretches. Of these, 25 (8%) were proximal, 187 (58%) were interstitial, and 112 (34%) distal (Fig. 2B). Our observations clearly show that synaptic initiation is not limited to chromosome ends in the female mouse, but regularly occurs at interstitial chromosomal regions.

In mid and late zygotene the progression of synapsis continued to be asynchronous. The extent of CE formation varied considerably between chromosome pairs within cells. Some chromosomes appeared as fully synapsed bivalents while others were still present as univalents.

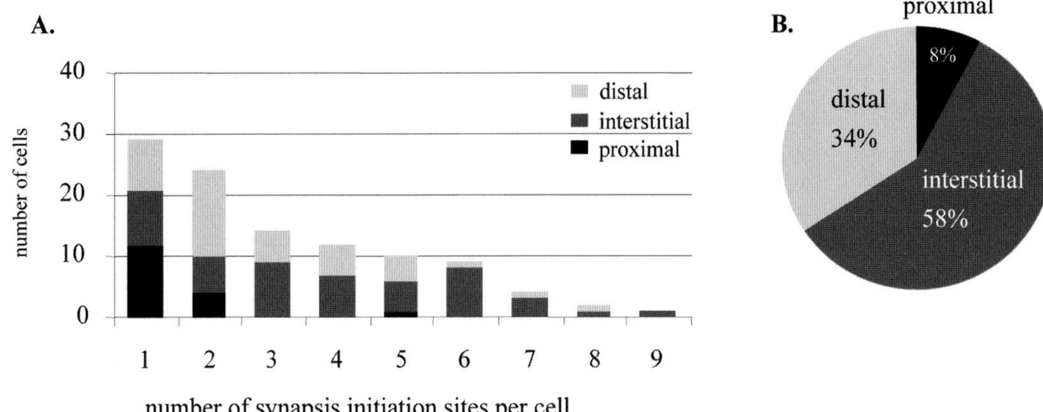

Fig. 2. Frequency of synapsis initiation in oocytes at early zygotene. (**A**) Synapsis initiation sites at proximal, interstitial and distal regions of AEs were scored in 54 cells. The number of synapsis initiation sites per cell ranged from 1 to 9. (**B**) The total number of synaptic sites observed at proximal, interstitial and distal regions of AEs.

Table 1. Rates of association of chromosomes 2 and 4 at different stages of early meiosis

Chromosome	Stage	Total number of cells	Associated, %
2	leptotene	70	32.9
	early zygotene	94	46.8
	mid zygotene	55	83.6
	late zygotene	56	100
4	leptotene	202	22.8
	early zygotene	144	48.6
	mid zygotene	52	92.3
	late zygotene	39	97.4

FISH on conventional chromosome spreads

As detailed in Materials and methods, we analysed the patterns of pairing of two target chromosomes, namely chromosomes 2 and 4. We found that in the majority of leptotene oocytes from day-14 and -15 foetuses, the homologues of the target chromosomes were separate (Table 1). The occurrence of homologue association in a small proportion of leptotene cells could be due to (1) random positioning of homologous domains within the nucleus; (2) a consequence of the flattening of the cells during preparation for microscopy, or (3) an indication of a slight tendency for homologues to be in adjacent domains within cells. As anticipated, the proportion of cells with associated homologues increased through zygotene. However, chromosomes 4 were found to remain separate even at late zygotene in a few cells (Table 1).

Chromosomes in leptotene oocytes were generally too diffuse to analyse in detail, therefore, homologous associations were analysed in early to late zygotene cells. At early zygotene, the extent of contact varied from end-to-end association only to over half the chromosome length paired (Fig. 3). As anticipated, the extent of pairing increased through zygotene (Table 2). It was also clear that the time and rate of pairing varied considerably between cells. Thus, while the majority of late zygotene cells contained fully paired homologues, some still had restricted lengths of homologous association. There appeared to be a slight difference in the rate of association of the two pairs of chromosomes studied. However, as the cell sample sizes are not large, it is possible that this apparent difference is a sampling artefact.

Zygotene cells with homologue associations were also assessed as to whether the centromeres and distal telomeres were paired. This analysis is summarised in Table 3. Centromeric and/or distal telomeric associations were relatively frequent at early zygotene, but initial contacts often involved interstitial chromosomal segments with "zipping up" of pairing from an interstitial position.

Locus-specific FISH on cryosections

The two probes used on tissue sections, D6Mit15 and D17Mit129, identify a distal locus and an interstitial locus on chromosomes 6 and 17, respectively. The relative positions of the loci were assessed in cells from preleptotene to pachytene and scored as unpaired, aligned or paired. Cells were categorized as unpaired when the distance between two signals was more than two diameters of a signal and "aligned" when the distance between the loci was equal to or less than two diameters of a signal. Nuclei with signals touching each other or fused were scored as paired. In total, 341 oocytes for D6Mit15 and 251 oocytes for D17Mit129 were analysed (Table 4). Prophase I sub-stages were identified by immunostaining AEs/SCs.

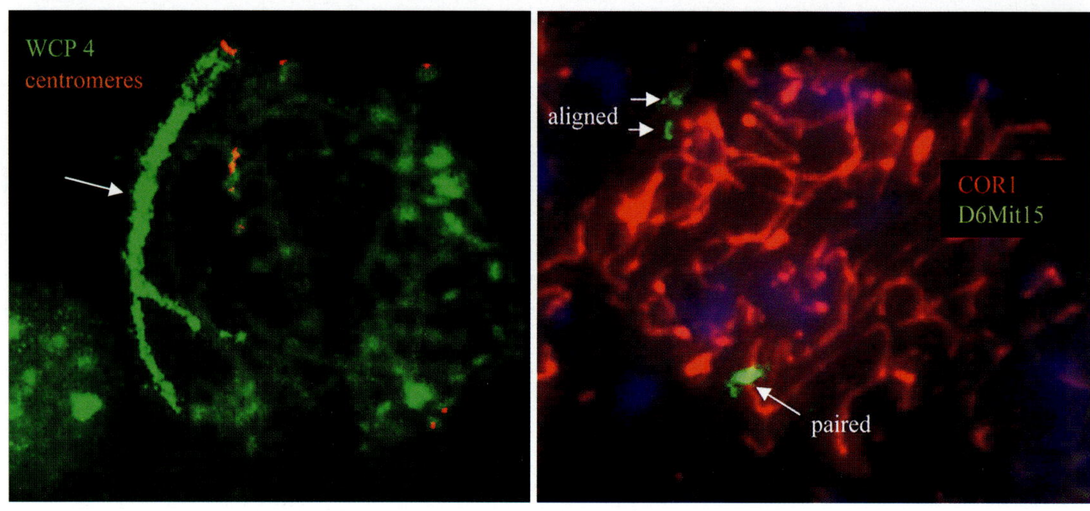

Fig. 3. Early zygotene oocyte after FISH with chromosome 4 paint (green). The initiation of pairing often involves interstitial chromosomal segments with homologous ends being separate.

Fig. 4. Immunostaining of central elements of SCs (red) together with FISH using a locus-specific chromosome 6 probe (green). Two nuclei at late zygotene show aligned and paired signals.

Table 2. The proportions of oocytes with different extents of homologous association

Chromosome	Stage	<1/2 paired, %	>1/2 paired, %	Fully paired, %	Total number of cells
2	early zygotene	75.0	25.0	0	44
	mid zygotene	17.4	65.2	17.4	46
	late zygotene	0	19.6	80.4	56
4	early zygotene	90.0	10.0	0	70
	mid zygotene	33.3	52.1	14.6	48
	late zygotene	10.5	31.6	57.9	38

Table 3. The frequencies of different patterns of centromeric (cen) and distal telomeric (telo) associations in early, mid- and late-zygotene oocytes where chromosomes 2 or 4 were paired

Chromosome	Stage	Cen separate/ telo separate, %	Cen associated/ telo separate,%	Cen separate/ telo associated,%	Cen associated/ telo associated, %	Total number of cells
2	early zygotene	38.6	20.5	27.3	13.6	44
	mid zygotene	13.3	13.0	30.4	43.5	46
	late zygotene	0	1.8	10.7	87.5	56
4	early zygotene	52.9	18.6	22.9	5.7	70
	mid zygotene	27.1	14.6	31.3	27.1	48
	late zygotene	5.3	13.2	15.8	65.8	38

Table 4. Distribution of D6Mit15 and D17Mit129 signals in mouse female germ cells

Stage	Probes[a]	Unpaired, %	Aligned, %	Paired, %	Total number of cells	χ^2 (df=2)
preleptotene/leptotene	D6Mit15	83.7	11.6	4.7	86	4.5
	D17Mit129	70.3	18.7	11.0	91	
early zygotene	D6Mit15	73.0	16.2	10.8	148	6.7[b]
	D17Mit129	57.4	23.4	19.2	94	
mid zygotene	D6Mit15	45.7	35.7	18.6	70	6.8[b]
	D17Mit129	25.5	38.3	36.2	47	
late zygotene	D6Mit15	21.6	29.7	48.7	74	7.1[b]
	D17Mit129	5.2	31.6	63.2	19	

[a] Mouse locus-specific probes with distal location on chromosome 6 (D6Mit15) and interstitial location on chromosome 17 (D17Mit129).

[b] $P<0.05$.

At preleptotene and leptotene, a large majority of cells had unpaired signals for both chromosomes 17 (70%) and 6 (84%). Nevertheless, some cells did contain aligned (11 and 19%) or apparently paired signals (5 and 11%) for both chromosomes 6 and 17, respectively. The proportions of paired, aligned and unpaired signals at early zygotene were not significantly different from preleptotene and leptotene cells for both chromosomes 6 and 17 (Table 4). At early zygotene stage onwards, the FISH signals displayed an elongated morphology, which was different from the compact signals observed in preleptotene and leptotene nuclei (Fig. 4). Significantly high levels of pairing were detected at mid and late zygotene ($P < 0.001$). Paired signals for chromosome 6 were present in 18% of mid zygotene cells and 49% of cells at late zygotene. The frequency of paired signals for chromosome 17, however, was higher and observed in 36 and 63% of mid and late zygotene oocytes, respectively. The proportions of cells with aligned signals were almost the same at mid and late zygotene for chromosome 6 (30%) and for chromosome 17 (36%).

Interestingly, the two target chromosomes show different patterns in the timing of their associations. A comparison of the proportions of cells with unpaired, aligned and paired signals for the two probes revealed significant differences from early zygotene onwards (Table 4). At all sub-stages of zygotene, this difference was due to a higher frequency of aligned and paired signals for chromosome 17 than for chromosome 6 ($P < 0.05$). At mid zygotene, for example, the interstitial region of chromosome 17 was paired in 36% of cells, whereas only 18% showed pairing in the distal region of chromosome 6. The distal end of chromosome 6 remained unpaired in a large proportion (20%) of late zygotene cells.

Thus, our observations show a rapid increase in the number of paired signals from the mid to the late zygotene for both locus-specific probes. Significant differences between two loci of chromosome 6 and 17 in progression of pairing were observed at mid and late zygotene. The interstitial segment of chromosome 17 was more advanced in pairing than the distal end of chromosome 6.

2. Centromere and distal telomere distribution

Two-dimensional spreads

Analysis of centromere and telomere distributions was carried out on 300 cells. Preleptotene stage oocytes were identified by the presence of COR1-positive aggregates with no visible AE stretches. In all 92 analysed preleptotene cells, centromere and telomere signals were scattered throughout the nucleus with no tendency to form associations. A total of 113 leptotene nuclei were screened and subjectively assigned to early, mid and late sub-stages on the basis of extent of AE formation. Sixty-two cells had no indication of centromere associations; the remaining 51 showed clusters with variable numbers of centromeres involved in these aggregates. In 20 of these 51 cells, centromeres were tightly associated in these clusters (Fig. 5A). The distal telomeres displayed a very different arrangement to the centromeres. The vast majority of cells (91 of 113) contained single, discrete signals with no evidence of clustering (Fig. 5B).

In general terms, the behaviour of the centromeres and distal telomeres at early zygotene did not differ radically from that described above for leptotene cells. Thus, the centromeres were associated in variable numbers of clusters involving an inconstant number of chromosomes. The distal telomeres were generally dispersed throughout the nucleus. In 24 of 54 cells screened, the centromeres, distal telomeres or both appeared either to be peripherally located or to show some degree of polarisation. At mid zygotene, evidence of a tendency toward peripheral distribution or polarization of the signals was present in 16 of 25 cells, but only 4 cells displayed tight clustering. Similar indications of a restricted distribution of chromosome ends were also present in 9 of 13 late zygotene oocytes, whereas only 3 cells had tight centromere associations. Our observations on two-dimensional spreads, therefore, demonstrate that centromeres and distal telomeres cluster during meiotic prophase I. However, there were no statistically significant differences between leptotene and zygotene cells in the progression of telomere clustering. The most probable explanation for this is that the spreading of the cells and hypotonic pre-treatment partially disrupt the nuclear architecture of the cells and, consequently, alter the centromere and distal telomere distributions. Therefore, in order to circumvent this problem, we decided to study their distributions in cryosections as well.

Cryosections

Ovarian cryosections were stained with COR1 and CREST antibodies to aid identification of prophase I sub-stages. A total of 791 oocytes from eleven 14-day fetuses were analysed. Cells displayed a wide range of centromeric distribution patterns (Table 5). For the purposes of analysis, cells were subjectively assigned to one of three categories depending on the presence and degree of centromere clustering: dispersed, grouped or tightly grouped. Cells with centromeres scattered throughout the nucleus and with little to no indication of centromeric associations were classified as "dispersed". The "grouped" category included centromere associations varying from cells with closely apposed centromeres in variably sized clusters to those with centromeres involved in loose groupings and located in one half of the nucleus. Cells with tightly clustered centromeres in one to three places were classified as "tightly grouped". The same classification system for patterns of distal telomere distribution was used as for the centromeres.

A dispersed pattern of centromere distribution was observed in the majority (79%) of preleptotene cells. The remaining cells displayed a loose grouping (Table 5). The centromere distribution was markedly different at leptotene nuclei. A total of 329 cells was analysed: 24% of the cells showed a dispersed pattern of centromere distribution; 58% had centromeres arranged in small clusters and located around the nuclear periphery, and 18% displayed tightly clustered centromeres (Table 5). At zygotene, centromere clustering had further progressed: 10 cells (3%) had dispersed centromeres; 139 (46%) had grouped centromeres, and 157 (51%) displayed tightly clustered centromeres (Fig. 5C).

After analysis of centromere distributions, the sections were subjected to PNA-FISH to identify the telomeres. The same population of cells, previously analysed for centromere posi-

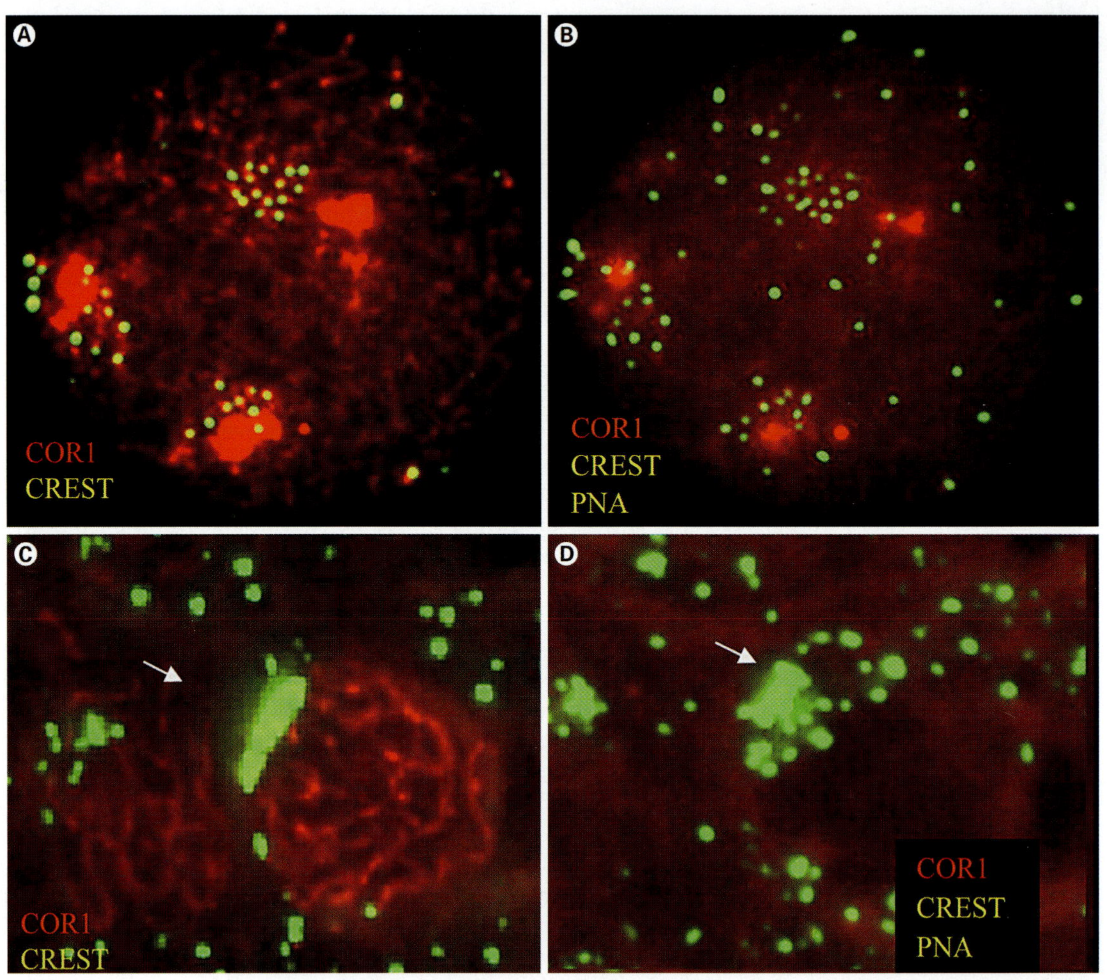

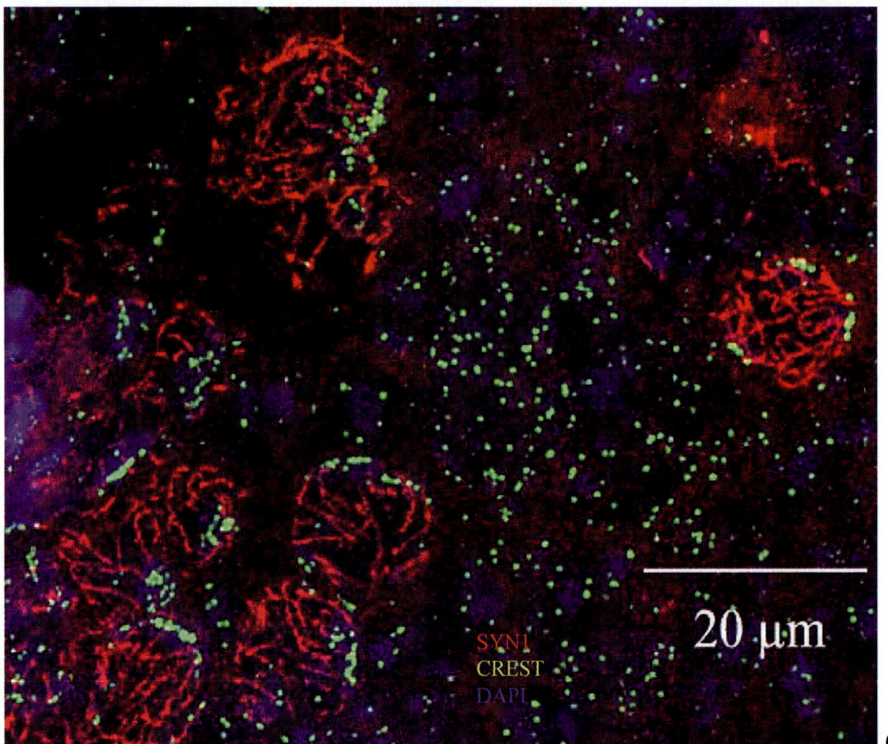

Fig. 5. Proximal (centromere) and distal telomere distribution in oocytes stained with COR1 (red) and CREST (green) antibodies (**A, C**) and after telomere PNA-FISH (**B, D**). (**A**) Leptotene oocyte. Centromeres are tightly clustered and co-localized with axial element protein aggregations. (**B**) Proximal telomeres are tightly clustered at the same positions as centromeres, whereas the distal telomeres are dispersed throughout the nucleus. (**C**) Zygotene oocyte with tightly clustered centromeres. (**D**) Same zygotene oocyte after telomere PNA-FISH showing clustered telomeres ("bouquet"). The distal telomeres display less clustering than the centromeres.

Fig. 6. Centromere distribution and assembly of central elements of SCs identified by immunostaining with CREST and SYN1 antibodies in cryosections of fetal mouse ovaries. Zygotene and pachytene oocytes display tight centromere clustering, while SYN1 negative cells show dispersed centromere signals.

Table 5. Distribution of centromeres and distal telomeres in female mouse germ cells stained with COR1, CREST antibodies and telomere PNA probe

Stage	Centromeres and distal telomeres	Dispersed (%)	Grouped (%)	Tightly grouped (%)	Total number of cells
preleptotene	centromeres	123 (79)	33 (21)	0	156
	distal telomeres	151 (97)		0	
leptotene	centromeres	80 (24)	190 (58)	59 (18)	329
	distal telomeres	245 (75)	84 (25)	0	
zygotene	centromeres	10 (3)	139 (46)	157 (51)	306
	distal telomeres	71 (23)	124 (40)	113 (37)	

Table 6. Percentage distribution of centromeres and distal telomeres in female mouse germ cells stained with SYN1 and CREST antibodies and the telomere PNA probe

Stage	Centromeres and distal telomeres	Dispersed (%)	Grouped (%)	Tightly grouped (%)	Total cells	χ^2 (df=2)[a]
early zygotene	centromeres	31 (17)	98 (52)	58 (31)		
	distal telomeres	63 (34)	82 (44)	42 (22)	187	
mid zygotene	centromeres	2 (2)	52 (48)	54 (50)		20.01***
	distal telomeres	43 (40)	38 (35)	27 (25)	108	2.17
late zygotene	centromeres	3 (5)	29 (46)	31 (49)		9.59*
	distal telomeres	14 (22)	18 (29)	31 (49)	63	16.30***
pachytene	centromeres	11 (10)	44 (41)	53 (49)		9.83*
	distal telomeres	27 (25)	28 (26)	53 (49)	108	22.65***

[a] *$P<0.05$; *** $P<0.001$

tions, were analysed again for telomere distribution. By using the recorded images of these cells, we were able to distinguish proximal and distal telomeres. The timing and extent of distal telomeres clustering were considerably different from centromere clustering (Table 5). Distal telomeres were scattered throughout the nucleus with no visible grouping in almost all preleptotene (97%) and in the majority of leptotene cells (75%). Significant alteration in their distribution was observed at zygotene. The frequency of cells with a "tightly grouped" pattern rapidly increased up to 37%, and 40% of cells showed a varied grouping of distal telomeres. However, 23% of cells still displayed a dispersed pattern of distal telomere distribution at zygotene.

We also carried out a second, general survey of centromere and distal telomere distributions in ovarian cryosections stained with the SYN1 and CREST antibodies (Fig. 6) followed by PNA-FISH. As the SYN1 antibody binds to the central element of the SC, then by definition it can only identify prophase I stages in which there is some degree of homologous chromosome synapsis, i.e. zygotene to diplotene. Zygotene oocytes were subjectively assigned to early, mid or late sub-stages on the basis of the lengths of the central elements. Cells with complete SC formation were classified as pachytene cells. A total of 358 zygotene and 108 pachytene cells were analysed (Table 6).

The proportions of centromere distribution patterns at early zygotene were significantly different from those at preleptotene and leptotene (Tables 5, 6). At zygotene, the proportion of cells with "tightly grouped" centromeres increased to 31% and only 17% of cells displayed centromeres scattered throughout the nucleus. The majority of early zygotene cells (52%) showed a variety of centromere distribution patterns. In some cells, centromere signals appeared to be restricted to roughly one-half of

the nucleus, whereas in other cells, centromeres were arranged in several clusters and the clusters were randomly localised on the nuclear surface. Further changes in centromere distribution were apparent at mid and late zygotene, and pachytene. Overall, these stages displayed similar patterns of centromere distribution; this pattern differed significantly from that present at early zygotene ($P<0.001$; $P<0.05$). The number of cells with dispersed centromeres fell after early zygotene (2% for mid and 5% for late zygotene cells). Also, there was a concomitant rise (50%) in the proportion of cells with a "tightly grouped" centromere pattern (Table 6).

Overall, throughout zygotene and into pachytene, distal telomeres displayed less of a tendency to cluster than centromeres (Tables 5 and 6). The proportions of dispersed, grouped and tightly grouped patterns were almost the same at early and mid zygotene. A significant alteration of these proportions was evident at late zygotene and persisted at pachytene ($P<0.001$). The proportion of cells with "dispersed" and "grouped" distal telomeres decreased at late zygotene and pachytene to 22–29%, whereas the percentage of cells with a "tightly grouped" pattern rapidly increased up to 49% (Table 6). Thus, 31 late zygotene and 53 pachytene cells showed tight clustering of centromeres and distal telomeres with location in one ("bouquet") to a few places on the nuclear surface. Seven nuclei of these late zygotene cells and 10 of the pachytene cells displayed a "bouquet" (Fig. 5C, D).

In summary, our observations showed the occurrence of dynamic rearrangements of centromere and distal telomere positions between leptotene and pachytene stages in mouse female meiosis. The timing and extent of clustering were considerably different for centromeres and distal telomeres, with the latter showing less of a tendency to cluster. A high proportion of cells had tightly clustered centromeres at mid zygotene, while clustering of the distal telomeres did not become abundant until late zygotene.

Discussion

The observations reported here, in combination with another study in female mice (O'Keeffe et al., 1997), provide evidence of a variety of chromosome behaviours at early meiosis in female germ cells. The timing of pairing and synapsis shows asynchrony for a given chromosome pair and considerable inter-cell variation; the number of chromosomes involved in the centromeres and distal telomeres aggregations vary between cells at the same meiotic stage and also between cells at different meiotic stages There is evidence that these clusters are not dependent on homology; the timing of formation of the clusters differs for the centromeres and distal telomeres with the former generally occurring earlier; and, the ends of chromosomes, in particular the centromeres, are regularly late pairing.

Analysis of homologous chromosome pairing in this study further showed the absence of pre-meiotic alignment in female mice. This fact is in agreement with previous studies in mammals (Scherthan et al., 1996, 1998; Scherthan and Schönborn, 2001). Moreover, the observation on chromosomes 2 and 4 showed that the homologues, prior to the initiation of synapsis and even during the progression of synapsis, frequently were not aligned along their entire lengths. The occurrence of alignment and then synapsis on a segment-by-segment basis has been reported in other organisms (reviewed by Zickler and Kleckner, 1999).

Our investigation also demonstrated that, in female mice, synapsis most frequently initiates in interstitial regions of chromosomes and, to a slightly lesser degree, in distal regions. The proximal ends of the chromosomes appear to be disfavoured for synaptic initiation. A number of other investigations have similarly shown that interstitial initiation of synapsis can occur and further that its prevalence is strongly organism-dependent (see Roeder, 1997; Zickler and Kleckner, 1998, 1999; Moens et al., 1998). Variability between sexes in patterns of synaptic initiation has also been reported. For example, in human and bovine oocytes, synapsis initiates more often in the interstitial region of chromosomes than in spermatocytes (Bojko, 1983; Barlow and Hultén, 1998; Pfeifer et al., 2003).

Scherthan and colleagues (Scherthan et al., 1994, 1996) suggested that the clustering behaviour of both centromeres and telomeres played an important role in homologous chromosome pairing. The observations made here and previously (O'Keeffe et al., 1997) indicate that clustering of chromosome ends does not promote homologous associations in female meiosis, thereby facilitating pairing interactions. At leptotene, the proximal ends of the X chromosomes were often present in centromere clusters; however, they were most frequently seen to be in separate rather than the same cluster (O'Keeffe et al., 1997). In a similar fashion, our observation of homologous chromosomes 2 and 4 pairing also showed that the ends of chromosomes, particularly the centromeres, regularly pair later than the interstitial segments. In addition, the ends of the chro-

mosomes are often associated with centric aggregations but with each of the homologues involved with a different cluster. Pairing of the distal end of chromosome 6 was also considerably delayed compared with the specific interstitial region of chromosome 17 at mid and late zygotene.

The fact that initiation of homologous chromosome pairing varies between the sexes is puzzling. It seems reasonable to suppose that the same batteries of "meiotic" genes are switched on in both sexes as the germ cells enter meiosis. Why these should lead to the initiation of chromosome pairing from telomeres in male germ cells but result in a more variable pattern in oocytes requires further investigation.

Bouquet formation precedes the initiation of chromosome synapsis in mouse spermatogenesis (Scherthan et al., 1996). This is in complete contrast to the observations reported here for female meiosis in the mouse. Our findings clearly indicate that synaptic initiation occurs in cells that show little to no evidence of bouquet formation. Our results also indicate a further difference between male and female meiosis. In the male, bouquet formation is a transitory phenomenon that occurs at leptotene/early zygotene. However, in the female, bouquet formation peaked at mid to late zygotene, and was still present in a substantial proportion of pachytene oocytes. Interestingly, despite the apparent prolongation of the bouquet state in female meiosis, we did not observe any prophase I stage in which this pattern was present in all, or even most, cells. Thus, it seems unlikely that bouquet formation promotes the first specific contacts between homologous chromosomes in female mice meiosis. Rather, our observations on mouse oocytes favour the proposition that, at best, clustering facilitates alignment of homologues or homologous chromosome segments during zygotene, and that the positional control of synaptic initiation is dependent on some other mechanism.

A similar sex-specific difference has been reported recently in cattle (Pfeifer et al., 2003). In contrast to the male, telomere clustering in females is established during leptotene and persists through zygotene to early pachytene. The authors have suggested that tighter telomere clustering and a prolonged bouquet stage contributes to pre-alignment and interstitial synapsis initiation in female. Although our results in mice similarly show a clear inter-sex difference, the timing of bouquet formation differs and suggests that the behaviour of chromosome ends during synapsis can show inter-species differences. We found that clustering of centromeres was generally much tighter than that of distal telomeres (e.g. Fig. 5) and that the timing of cluster formation differed between centromeres and distal telomeres. The fact that the centric and distal ends of the chromosomes behave differently indicates that centromere clusters may be the result of some effect other than that causing association of distal telomeres. Overall, the role of centromeric and telomeric clusters in homologous chromosome pairing initiation in female mouse meiosis remains obscure.

Acknowledgements

We thank Drs Peter Moens and Barbara Spyropoulos, and Professors Christa Heyting and Bill Earnshaw for providing the antibodies used in this study.

References

Armstrong SJ, Franklin FC, Jones GH: Nucleolus-associated telomere clustering and pairing precede meiotic chromosome synapsis in *Arabidopsis thaliana*. J Cell Sci 114:4207–4217 (2001).

Barlow AL, Hultén MA: Combined immunocytogenetic and molecular cytogenetic analysis of meiosis I human spermatocytes. Chromosome Res 4:562–573 (1996).

Barlow AL, Hultén MA: Combined immunocytogenetic and molecular cytogenetic analysis of meiosis I oocytes from normal human females. Zygote 6:27–38 (1998).

Bass HW, Wallace FM, Sedat JW, Agard DA, Cande WZ: Telomeres cluster de novo before the initiation of synapsis: a three-dimensional spatial analysis of telomere positions before and during meiotic prophase. J Cell Biol 137:5–18 (1997).

Bass HW, Riera-Lizarazu O, Ananiev EV, Bordoli SJ, Rines HW, Phillips RL, Sedat JW, Agard DA, Cande WZ: Evidence for the coincident initiation of homologue pairing and synapsis during the telomere-clustering (bouquet) stage of meiotic prophase. J Cell Sci 113:1033–1042 (2000).

Bojko M: Human meiosis. VIII. Chromosome pairing and formation of the synaptonemal complex in oocytes. Carlsberg Res 48:457–483 (1983).

Burgoyne PS, Baker TG: Meiotic pairing and gametogenic failure. Symp Soc Exp Biol 38:349–362 (1984).

Carlton PM, Cande WZ: Telomeres act autonomously in maize to organize the meiotic bouquet from a semipolarized chromosome orientation. J Cell Biol 157:231–242 (2002).

Chikashige Y, Ding DQ, Imai Y, Yamamoto M, Haraguchi T, Hiraoka Y: Meiotic nuclear reorganization: switching the position of centromeres and telomeres in the fission yeast *Schizosaccharomyces pombe*. EMBO J 16:193–202 (1997).

Hassold T, Hunt P: To err (meiotically) is human: the genesis of human aneuploidy. Nat Rev Genet 2:280–291 (2001).

Ijdo JW, Wells RA, Baldini A, Reeders ST: Improved telomere detection using a telomere repeats probe $(TTAGGG)_n$ generated by PCR. Nucleic Acids Res 19:4780 (1991).

Kezer J, Sessions SK, Leon P: The meiotic structure and behaviour of the strongly heteromorphic X/Y sex chromosomes of neotropical plethodontid salamanders of the genus *Oedipina*. Chromosoma 98:433–442 (1989).

Moens PB: The fine structure of meiotic chromosome polarization and pairing in *Locusta migratoria* spermatocytes. Chromosoma 28:1–25 (1969).

Moens PB, Chen DJ, Shen Z, Kolas N, Tarsounas M, Heng HH, Spyropoulos B: Rad51 immunocytology in rat and mouse spermatocytes and oocytes. Chromosoma 106:207–215 (1997).

Moens PB, Pearlman RE, Heng HH, Traut W: Chromosome cores and chromatin at meiotic prophase. Curr Top Dev Biol 37:241–262 (1998).

O'Keeffe C, Hultén MA, Tease C: Analysis of proximal X chromosome pairing in early female mouse meiosis. Chromosoma 106:276–283 (1997).

Pfeifer C, Thomsen PD, Scherthan H: Centromere and telomere redistribution precedes homologue pairing and terminal synapsis initiation during prophase I of cattle spermatogenesis. Cytogenet Cell Genet 93:304–314 (2001).

Pfeifer C, Scherthan H, Thomsen PD: Sex-specific telomere redistribution and synapsis initiation in cattle oogenesis. Dev Biol 255:206–215 (2003).

Roeder GS: Meiotic chromosomes: it takes two to tango. Genes Dev 11:2600–2621 (1997).

Scherthan H, Schönborn I: Asynchronous chromosome pairing in male meiosis of the rat *(Rattus norvegicus)*. Chromosome Res 9:273–282 (2001).

Scherthan H, Bahler J, Kohli J: Dynamics of chromosome organization and pairing during meiotic prophase in fission yeast. J Cell Biol 127:273–285 (1994).

Scherthan H, Weich S, Schwegler H, Heyting C, Harle M, Cremer T: Centromere and telomere movements during early meiotic prophase of mouse and man are associated with the onset of chromosome pairing. J Cell Biol 134:1109–1125 (1996).

Scherthan H, Eils R, Trelles-Sticken E, Dietzel S, Cremer T, Walt H, Jauch A: Aspects of three-dimensional chromosome reorganization during the onset of human male meiotic prophase. J Cell Sci 111:2337–2351 (1998).

Speed RM: Meiosis in the foetal mouse ovary. I. An analysis at the light microscope level using surface spreading. Chromosoma 85:427–437 (1982).

Speed RM: The prophase stages in human foetal oocytes studied by light and electron microscopy. Hum Genet 69:69–75 (1985).

Speed RM: The possible role of meiotic pairing anomalies in the atresia of human fetal oocytes. Hum Genet 78:260–266 (1988).

Tease C, Fisher G: Analysis of meiotic chromosome pairing in the female mouse using a novel minichromosome. Chromosome Res 6:269–276 (1998).

Trelles-Sticken E, Loidl J, Scherthan H: Bouquet formation in budding yeast: initiation of recombination is not required for meiotic telomere clustering. J Cell Sci 112:651–658 (1999).

Trelles-Sticken E, Dresser ME, Scherthan H: Meiotic telomere protein Ndj1p is required for meiosis-specific telomere distribution, bouquet formation and efficient homologue pairing. J Cell Biol 151:95–106 (2000).

Weier HU, Kleine HD, Gray JW: Labelling of the centromeric region on human chromosome 8 by in situ hybridization. Hum Genet 87:489–494 (1991).

Zickler D, Kleckner N: The leptotene-zygotene transition of meiosis. Annu Rev Genet 32:619–697 (1998).

Zickler D, Kleckner N: Meiotic chromosomes: integrating structure and function. Annu Rev Genet 33:603–754 (1999).

Cytogenet Genome Res 105:182–188 (2004)
DOI: 10.1159/000078188

Male mouse meiotic chromosome cores deficient in structural proteins SYCP3 and SYCP2 align by homology but fail to synapse and have possible impaired specificity of chromatin loop attachment

N.K. Kolas,[a] L. Yuan,[b] C. Hoog,[b] H.H.Q. Heng,[c] E. Marcon,[d] and P.B. Moens[d]

[a] Department of Molecular Genetics, Albert Einstein College of Medicine, Bronx, NY (USA);
[b] Department of Cell and Molecular Biology, The Medical Nobel Institute, Karolinska Institute, Stockholm (Sweden);
[c] Center for Molecular Medicine and Genetics, Department of Pathology, Karmanos Cancer Institute,
Wayne State University School of Medicine, Detroit, MI (USA);
[d] Department of Biology, York University, Toronto, Ontario (Canada)

Abstract. The targeted deletion of the meiotic chromosome core component MmSYCP3 results in chromosome synaptic failure at male meiotic prophase, extended meiotic chromosomes, male sterility, oocyte aneuploidy and absence of the MmSYCP2 chromosome core component. To test the functions of SYCP2 and SYCP3 proteins in the cores, we determined the effect of their deletion on homology recognition by whole chromosome painting and the effect on chromatin loop attachment to the cores with endogenous and exogenous sequences. Because we observed that the alignment of cores is between homologs, it suggested that alignment is not a function of the chromosome core components but might be mediated by chromatin-chromatin interactions. The alignment function therefore appears to be separate from intimate synapsis function of homologous cores that is observed to be defective in the SYCP3$^{-/-}$ males. To examine the functions of the SYCP2 and 3 core proteins in chromatin loop attachment, we measured the loop sizes of the centromeric major satellite chromatin and the organization of an exogenous transgene in SYCP3$^{+/+}$ and SYCP3$^{-/-}$ males. We observed that these satellite chromatin loops have a normal appearance in SYCP3$^{-/-}$ males, but the loop regulation of a 2-Mb exogenous λ phage insert appears to be altered. Normally the insert fails to attach to the core except by flanking endogenous sequences, but in the absence of SYCP2 and SYCP3, there appears to be multiple attachments to the core. This suggests that the selective preference for the attachment of mouse sequences to the chromosome core in the wild-type male is impaired in the SYCP3$^{-/-}$ male. Apparently the SYCP2 and SYCP3 proteins function in the specificity of chromatin attachment to the chromosome core.

Copyright © 2004 S. Karger AG, Basel

A number of proteins have been identified as components of mammalian meiotic chromosome cores. These include the 30- and 33-kDa SYCP3 proteins, the 190-kDa SYCP2 protein, the 111-kDa SYCP1 protein (Offenberg et al., 1998), and the cohesins, SMC1α and β, SMC3, STAG3 and REC8 (Eijpe et al., 2000; Pezzi et al., 2000; Prieto et al.,2001; Revenkova et al., 2001) (Fig. 6A). It has been shown that the deletion of the *Mus musculus MmSYCP3* gene leads to spermatocyte defects in synapsis of homologous chromosomes, spermatocyte loss, male infertility, aneuploid oocytes and loss of the SYCP2 component from the core (Yuan et al., 2000, 2002; Pelttari et al., 2001). Remarkably, the loss of the SYCP3/SYCP2 components from the core does not severely affect the structure of the core in terms of cohesin components, associated recombination proteins or female fertility. A more subtle effect appears to be a lengthening of the SYCP3/SYCP2$^{-/-}$ cores. Conversely, deletion of the meiosis-specific cohesin SMC1β results in a shorten-

NSERC of Canada generously supports P.B.M.'s research financially.

Received 8 October 2003; accepted 22 October 2003.

Request reprints from: Dr. P.B. Moens, Department of Biology, York University
4700 Keele Street, Toronto, Ontario, M3J 1P3 (Canada)
telephone: +1-416-736-5358; fax: +1-416-736-5731; e-mail: moens@yorku.ca

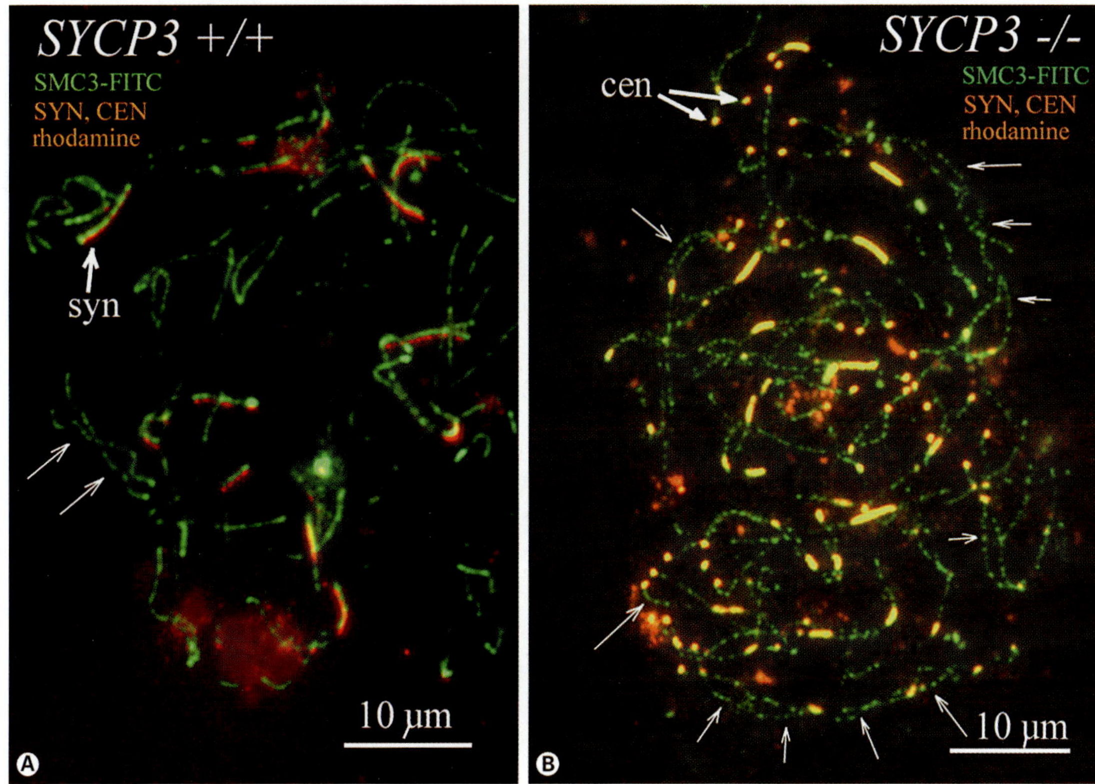

Fig. 1. Anti-SMC3 immunostaining of meiotic chromosome cores. (**A**) *SYCP3*[+/+] spermatocyte at the zygotene stage of meiotic prophase with partially synapsed chromosome cores. The antibody effectively recognizes chromosome cores (FITC, green, narrow arrows) and the alignment of cores is evident (narrow arrows). The SYCP1 protein at synapsed segments is recognized by the anti-SYN1 antibody (rhodamine, red). It follows that the antibodies are proper for the use in *SYCP3*[-/-] spermatocytes. (**B**) An *SYCP3*[-/-] spermatocyte immunostained with anti-SMC3 to visualize the chromosome cores (FITC, green), anti-SYN1 to mark the synapsed core segments (rhodamine, yellow in combination with green cores) and centromeres (cen) (rhodamine, yellow dots). Most spermatocytes arrest at this stage without completing synapsis. The narrow arrows mark co-aligned cores. The cores are notably longer than the aligned cores in spermatocytes from wild-type males as in Fig. 1A.

ing of the chromosome cores (Jessberger, Eijpe, Heyting, personal communications). To further investigate the effects of the *SYCP3* deletion on the functions of the chromosome cores, we examined homology recognition by the use of whole chromosome paint and chromatin loop association with the chromosome cores in pachytene spermatocytes of wild-type and mutant males.

Using fluorescent in situ hybridization (FISH) in wild-type mice, we have defined the chromatin loop organization of a number of endogenous sequences and exogenous transgenes (Heng et al., 1996; Moens et al., 1997). A mouse endogenous sequence probed with a 720-kb YAC probe paints several chromatin loops on a meiotic chromosome core, indicating several attachment sites, possibly one for every 180 kb. An 11.4-Mb transgene consisting of bacterial sequences interspersed with mouse β-globin sequences has multiple core attachments, but loop sizes are somewhat longer than wild-type loops, presumably extended by intervening bacterial sequences that are not attached to the core (Heng et al., 1994). The failure of foreign sequences to associate with the chromosome core is particularly evident for a 2 Mb λ phage insert which forms a single loop that is attached to the core by the flanking mouse sequences. Appar-

ently the attachment of chromatin to the cores during meiosis involves some kind of sequence-based discrimination by the core. The unique terminal attachments of the λ transgene provide a sensitive assay for the detection of possible attachment modifications in SYCP3[-/-] males. This report indicates that the discrimination is less effective in SYCP3[-/-] spermatocytes.

The technical difficulty in fluorescence immunocytology in *SYCP3*[-/-] males is that the convenient immuno-fluorescent labeling of the cores with anti-SYCP3 antibodies is not feasible. In addition, the SYCP2 component of the core is not evident in the cores of *SYCP3*[-/-] mice and therefore the anti-SYCP2 antibodies are not effective either. There are, however, two solutions. First, there is still a moderate amount of synapsis taking place in the *SYCP3*[-/-] spermatocytes that can be visualized with antibodies to the 111-kDa synaptic protein, SYN1/SYCP1. Second, antibodies against the cohesins SMC3 and STAG3 can be used as an alternative to visualize the chromosome cores in the absence of SYCP3 and SYCP2 protein (Fig. 1A and B) (Pelttari et al., 2001). With these methods, we used antibodies and FISH to examine the alignment and homology of chromosome cores, the organization of major satellite chromatin and the loop organization of the exogenous λ transgene.

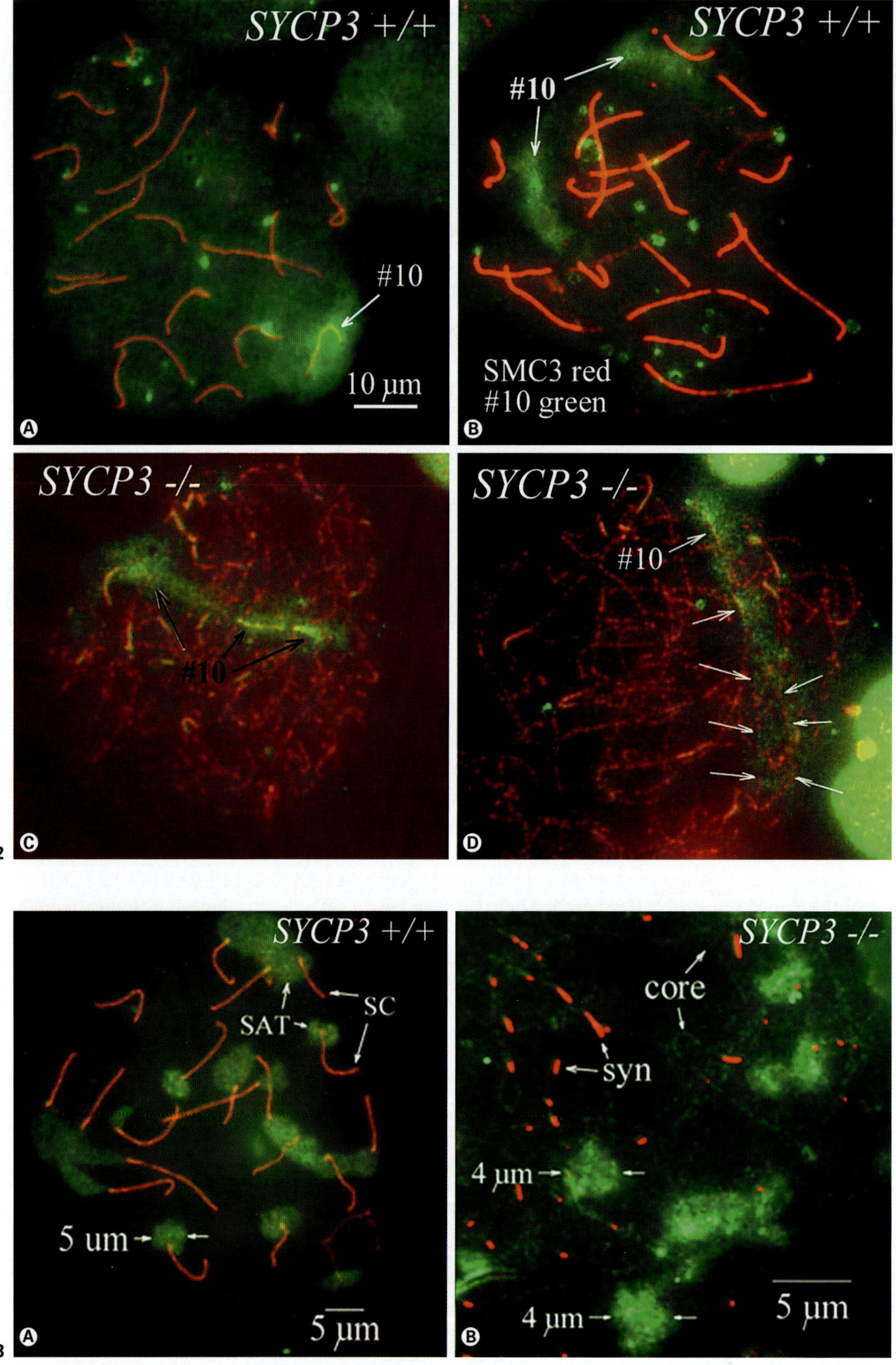

SYCP3 +/+
#10
10 µm
A
SYCP3 +/+
#10
SMC3 red
#10 green
B
SYCP3 -/-
#10
C
SYCP3 -/-
#10
D
2
SYCP3 +/+
SAT
SC
5 um
A
SYCP3 -/-
core
syn
4 µm
4 µm
5 µm
B
3

Materials and methods

The *SYCP3⁻ᐟ⁻* mice were generated by Li Yuan with the assistance of Jian-Guo Liu and provided by Christer Hoog (Yuan et al., 2000). *SYCP3⁻ᐟ⁻* females were crossed with *SYCP3⁻ᐟ⁻* males. Both were hemizygous or homozygous for the λ transgene from the Stratagene Muta™ Mouse which has a concatamer of a 47-kb λgt10 shuttle vector in the distal portion of mouse chromosome #3. Since the FISH signal of Muta™ Mouse is notably larger than the signal of the 1.7 Mb λ transgene of the Stratagene Big Blue™ mouse, we previously estimated the size of the Muta™ Mouse λ transgene to be in the order of at least 2 Mb (Moens et al., 1997). Testicular cells were hypotonically disrupted and the nuclei settled and adhered to 1% albumin-coated multiwell glass slides.

2% paraformaldehyde fixation was followed by washes and blocking with serum. Primary antibodies were applied overnight and rhodamine or FITC-conjugated secondary antibodies were then applied for 1 h following washes (Dobson et al., 1994). The rabbit and mouse polyclonal antibodies to the hamster 30-kDa core protein COR1/SYCP3 and to the hamster 111-kDa synaptic protein SYN1/SYCP1 have been characterized (Dobson et al., 1994). The anti-SMC3 antibodies were gifts of K. Yokomori (James et al., 2002) and of R. Jessberger (Eijpe et al., 2000).

Primers against the 200-base-pair *Mus musculus* major γ satellite were designed using sequence from the NCBI data base, clone pMG2-33. Forward primer: 5′-CGT GAT ACC TGG ACA TGG AA-3′. Reverse primer: 5′-TTT CAA GTC GTC AAG TGG ATG-3′. These primers were used to amplify the satellite DNA from mouse total RNA, isolated using Trizol reagent (Invitrogen) according to manufacturer's protocol. The PCR product was purified (Roche) and biotinylated using a Biotin Nick-Translation Kit (Roche) according to manufacturer's protocol. 5 μl of resulting biotinylated DNA was mixed with salmon sperm DNA and *E. coli* tRNA, precipitated, dissolved in 20 μl of 100% deionized formamide and 20 μl of hybridization buffer, denatured and applied to slides for overnight hybridization. For fluorescent in situ hybridization (FISH), the major satellite and λ probes were biotinylated by nick translation and applied to the antibody-stained cells. Biotinylated mouse chromosome #10 whole-chromosome probe was obtained from ID laboratories (London, ON) and used according to the manufacturer's protocol at a half to one quarter of the recommended concentration. The probes were visualized with FITC-conjugated avidin (Heng et al., 1996).

Results and discussion

Homologous alignment

In the absence of SYCP3 and SYCP2 proteins, the organization and behaviour of chromosome cores at meiotic prophase can be visualized with antibodies to cohesins and antibodies to the synaptic protein SYN1/SYCP1 in the synapsed portions of the chromosome cores. In the wild-type *SYCP3⁺ᐟ⁺* spermatocyte nuclei, the early meiotic prophase zygotene stage is characterized by the alignment of cores and the initiation of synapsis (Fig. 1A). The figure demonstrates with antibodies to the cohesin SMC3, the effective visualization of the unpaired cores (thin arrows), the alignment of unpaired cores (thin arrows), and the synapsed segments (syn). At the later stage, fully synapsed pachytene chromosomes can be seen in Figs. 2A, B, and 3A. Given the adequacy of the SMC3 immunocytology for vizualizing cores, the behaviour of chromosome cores can be analyzed in *SYCP3*-deleted spermatocytes (Fig. 1B). Evidently, the cohesin SMC3 component of the chromosome cores is intact and the cores show a degree of alignment (thin arrows) and some synapsed segments. This is not a zygotene stage but a typical end point of mostly defective synapsis in *SYCP3⁻ᐟ⁻* spermatocytes (Pelttari et al., 2001).

Given the defect in synapsis, we questioned if the lack of SYCP2 and SYCP3 proteins in the cores might interfere with the homology of the alignment process. To test the extent of homologous alignment, we made use of whole-chromosome paint for mouse chromosome #10. In the wild-type *SYCP3⁺ᐟ⁺* male, the chromatin of a single bivalent, #10, is selectively recognized by the probe (Fig. 2A, #10). The probe also recognizes the chromatin of chromosomes #10 when they are not synapsed (Fig. 2B, #10) which is a necessary attribute for the analysis of aligned chromosomes that are not synapsed. Incidentally, the separation of the two chromosomes #10 in the nucleus of Fig. 2B is a fortuitous, low-frequency event. Observations on *SYCP3⁻ᐟ⁻* spermatocytes invariably show that there is only a single strip of chromosome #10-positive chromatin (Fig. 2C and D, #10 thin arrows) giving clear evidence for the homologous alignment of chromosomes #10. The arrows in Fig. 2D indicate that the alignment varies from close proximity to well separated cores.

Apparently, although homologous chromosome synapsis is severely curtailed in *SYCP3⁻ᐟ⁻* males, homologous alignment is intact within the constraints of observed whole chromosome painting. The separation of the two aspects of chromosome behavior suggests that alignment and synapsis are separate functions. Possibly alignment depends on chromatin interactions while synapsis involves the structure of the chromosome cores. In females, synapsis is less defective but high levels of aneuploid oocytes, near 50%, suggest that there is limited chiasma formation possibly due to extensive misaligned synapsis (Yuan et al., 2002).

SYCP3 and SYCP2 are required for chromosome compaction

Comparison of the chromosome cores of wild-type and *SYCP3⁻ᐟ⁻* spermatocytes in Figs. 1A and B indicates that the absence of SYCP2 and SYCP3 proteins correlates with a great increase in the length of the chromosome cores. The difference is not likely the effect of different prophase stages. Figure 1A shows a wild-type spermatocyte at early zygotene with few synapsed segments at which time the cores are still relatively long. The length of the pair of aligned chromosome cores marked by thin arrows is about 12 μm. The spermatocyte nucleus shown in Fig. 1B is probably at an arrested later stage, judging by the

Fig. 2. Mouse chromosome #10 painting. (**A**) In the wild-type pachytene spermatocyte, the paint recognizes the chromatin of chromosome #10 as a single entity. The green halo gives an estimate of about 5–6 μm for the size of the chromatin loops. (**B**) In an occasional nucleus, homologs fail to synapse, in this case the #10 chromosomes where the two univalents are well visualized by the paint. This characteristic is valuable in the recognition of unpaired chromosomes in the mutant mice. (**C, D**) The single green strip across the nucleus gives evidence that the #10 homologs are co-aligned in the *SYCP3⁻ᐟ⁻* spermatocytes. The aligned chromosomes are some 40–50 μm in length, greatly exceeding the 10–15 μm length of wild-type meiotic prophase chromosomes (Fig. 1A).

Fig. 3. Major satellite chromatin loop sizes. (**A**) The diameter of the FISH major satellite halos (SAT) at the centromeric end of the SCs (SC) is in the order of 5 μm in the wild-type pachytene spermatocytes. (**B**) In the *SYCP3⁻ᐟ⁻* spermatocytes with asynapsed cores (core, FITC, green) and the short synapsed segments (syn, rhodamine, red), the major satellite halos are very similar in size compared to those observed in the wild-type males, about 4 μm to 5 μm.

numerous synapsed segments, and it is evident that the aligned cores marked by thin arrows are 20 µm or longer. Similarly, the length of painted chromosomes #10 shown in Figs. 2C and D is about 40–50 µm. In female meiosis where synapsis is complete, the normal average SC length is about 10 µm whereas the average SC length in *SYCP3*⁻/⁻ is in the order of 22 µm (Yuan et al., 2002). Apparently the cohesins of the cores are incapable by themselves of organizing the normal core structure and the SYCP2 and SYCP3 components are required to control the level of chromosome compaction. As is shown below, the reduced compaction does not seem to affect the sizes of the endogenous chromatin loops (Fig. 3) which suggests that both wild-type and *SYCP3*⁻/⁻ chromosomes have similar numbers of loops of approximately the same size but the loops of the *SYCP3*⁻/⁻ chromosomes may be spaced further apart on the deficient cores.

The chromatin loops of major satellite DNA

To what extent the absence of SYCP2 and SYCP3 proteins from the meiotic chromosome cores affects the organization of the chromatin loops relative to the chromosome cores can be evaluated by the appearance of the satellite DNA loops at the centromeric ends of the mouse cores/SCs. With a biotinylated major satellite DNA probe and FITC-conjugated avidin, the organization of the loops at the proximal ends of the SCs can be visualized in the wild-type mouse (Fig. 3A). The size of the green halos is approximately 4–5 µm, but frequently the satellites of several chromosomes are joined together giving somewhat larger dimensions. In the *SYCP3*⁻/⁻ male, the chromosomes have limited synapsis (red segments in Fig. 3B) and the cores are not as well defined as in the wild-type mouse, but the sizes of the major satellite chromatin halos are very similar in size to the wild-type mouse, about 4–5 µm (Fig. 3B). This suggests that the chromatin loop organization of endogenous sequences is not much affected by the lack of SYCP2 and SYCP3 protein in the cores.

Attachment of the λ transgene to the cores/SCs in the SYCP3⁺/⁺ male

The Stratagene Muta™ Mouse has a transgene consisting of some 30–40 repeats of a 47-kb λgt shuttle vector that is located in the distal portion of chromosome #3 (Fig. 4) (Moens et al., 1997). Whereas most of the endogenous chromatin loops are in the order of about 3–5 µm, the λ loops are many times that length (Fig. 4). Apparently the transgene has no sequences capable of attachment to the mouse chromosome core. Consequently, the flanking mouse sequences are the first available segments that can attach to the core (Fig. 4A and D). The mice used for these observations are hemizygous for the transgene so that only one of a pair of sister chromatids has the 2-Mb insert. The question that we address in this investigation is to what extent is this organization maintained in the absence of chromosome core proteins SYCP3 and SYCP2.

Attachment of the λ transgene to the core/SC in the SYCP3⁻/⁻ males

In the *SYCP3*⁻/⁻ spermatocytes, neither the SYCP3 protein nor the SYCP2 protein can be visualized with immunofluores-

cence. In addition, because the chromosome synapsis is severely limited, only short stretches of SCs can be visualized with antibodies to the pairing protein SYN1/SYCP1 (Figs. 1B, 2C and D). An alternative to the visualization of the meiotic chromosome cores is the use of antibodies to the cohesins that are an integral component of the cores/SCs in wild-type (Figs. 1A, 2A and B) and *SYCP3*⁻/⁻ mice (Figs. 1B, 2C, D, and 5). The fully paired pachytene chromosomes have a solid and continuous immunofluorescence of the SCs (Fig. 2A and B). Because of the low level of synapsis in *SYCP3*⁻/⁻ mice, most of the cores are unpaired (Fig. 1B, 2C and D).

Occasionally the λ insert lies within a synapsed segment and the organization of the insert relative to the SC/core can be observed (Fig. 5A and B). If the absence of SYCP2 and SYCP3 proteins from the core results in failure of the chromatin loops to attach properly, then the transgene could be expected to have a disorderly distribution throughout the nuclear volume. Clearly such is not the case. From Fig. 5 it is evident that the transgene has a well-defined domain and that it is inserted on the SC.

A notable difference was observed in the organization of the λ transgene in the wild-type versus the *SYCP3*⁻/⁻ spermatocytes. Whereas the lengthy insert meanders through much of the nuclear volume in the wild-type spermatocyte (Fig. 4), in the *SYCP3*⁻/⁻ mutant, the inserts are compact and there appears to be a number of loops emanating from the core/SCs (Fig. 5). Because this observation was consistent over several dozen recorded pachytene nuclei, we suspect that there is a real difference between the wild type and the mutant. The numerous loops in the *SYCP3*⁻/⁻ spermatocyte nuclei suggest that there are several attachment sites resulting in several loops, unlike the terminal attachments in the wild type that leaves the long insert free in the nuclear volume. Our interpretation is illustrated in the diagram in Fig. 6. If this is the correct observation and interpretation, then it follows that in the absence of SYCP2 and SYCP3 protein, the sequence preference that prevents the λ sequences from associating with the mouse chromosome core in *SYCP3*⁺/⁺ males is lost in *SYCP3*⁻/⁻ males.

Summary

The painting of chromosome #10 chromatin (Figs. 2C and D) provides experimental support for the postulate that in *SYCP3*⁻/⁻ males the chromosomes are homologously aligned (Yuan et al., 2000; Pelttari et al., 2001). The failure of the aligned homologs to complete intimate synapsis in the male indicates that alignment and synapsis are separable functions. Presumably the recognition and interactions of homologous chromatin domains are intact. The causes of subsequent synaptic failure are speculative. The four-fold increase of chromosome length when comparing aligned cores in the wild-type male (Fig. 1A) with the *SYCP3*⁻/⁻ chromosome length (Figs. 2C and D) could contribute to faulty juxtaposition of homologous sequences and result in failure to achieve extensive intimate synapsis. Alternatively, or contributing, the organization of the chromatin loops may interfere with synapsis but our observations on endogenous loop organization seem not to support that hypothesis. That is, the painted major satellite chromatin and

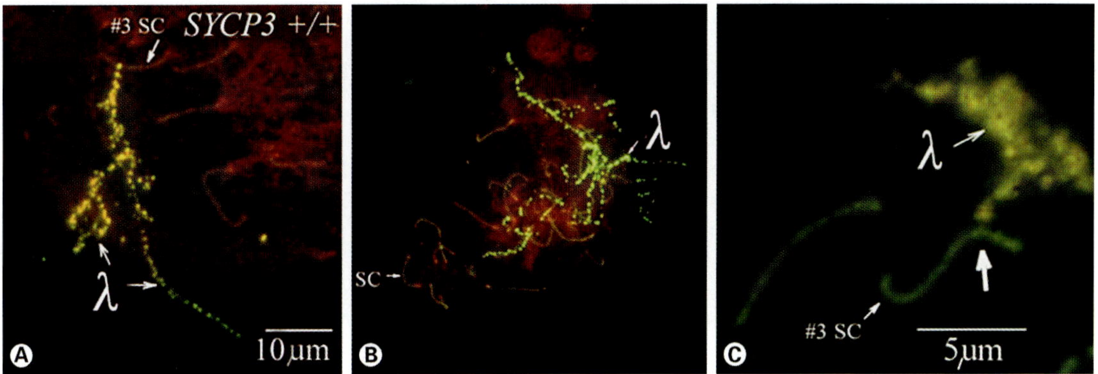

Fig. 4. Organization of the roughly 40-Mb λ transgene in the wild-type spermatocyte. (**A**) Notably all of the transgene, except for a single attachment to the SC, is dispersed in the nuclear volume. It appears that the λ sequence and the core cannot interact and attachment to the core is probably via flanking mouse sequences. (**B**) Another example of the usual appearance of the λ transgene in a spermatocyte nucleus. An excess of 50 images of this type have been collected and many more have been observed. (**C**) At late pachytene and diplotene, the transgene becomes quite compact but maintains only a single attachment to the SC.

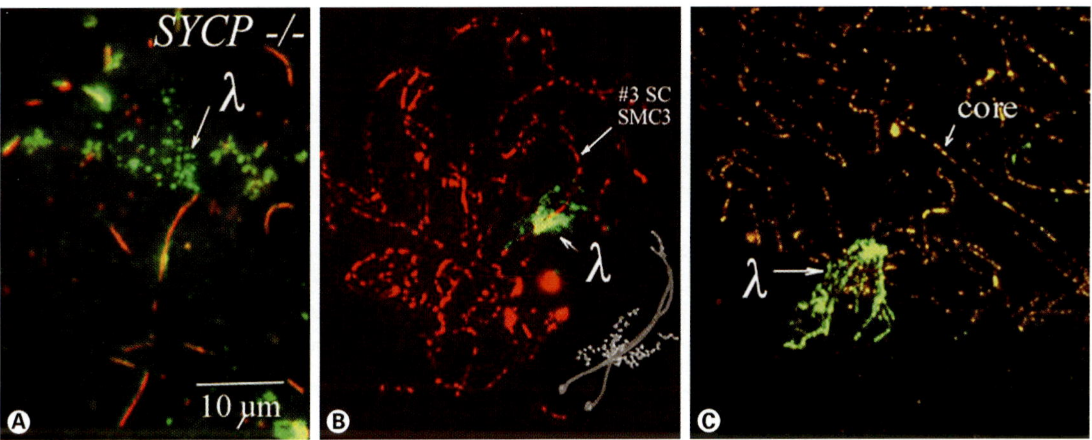

Fig. 5. Organization of the λ insert in the *SYCP3⁻/⁻* spermatocyte. (**A**) There appears to be a single attachment site to the core, probably dictated by the flanking mouse sequences. However, unlike the wild-type spermatocytes, there appear to be several much shorter loops, suggesting that in addition to the flanking sequence attachments, there are also λ sequences attached to the core. The transgene is attached to the end of a segment of synapsed cores (red). (**B**) The compact arrangement of the transgene in the *SYCP3⁻/⁻* spermatocyte in association with the #3 chromosome (insert) is vastly different from the meandering insert in the wild-type spermatocyte (Fig. 4A and B). (**C**) Further evidence that the compaction of the λ transgene is probably the result of multiple attachment sites resulting in several short loops.

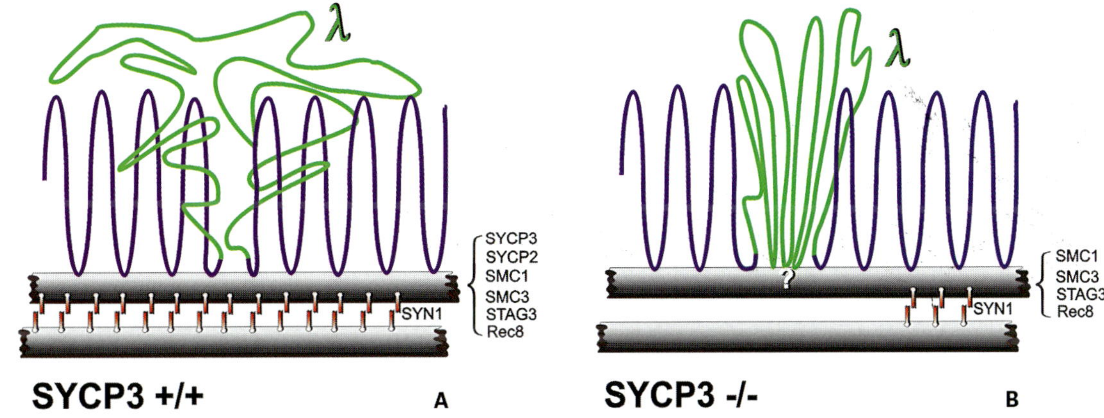

Fig. 6. A diagrammatic representation of the proposed arrangement of the λ transgene in the wild-type spermatocyte (**A**), and the SYCP3⁻/⁻ spermatocyte (**B**). The components of the chromosome cores are listed in A and components in the absence of SYCP2 and SYCP3 are listed in B.

the chromatin loops of #10 chromosome in wild-type and mutant male spermatocytes appear quite similar. This occurs in spite of the much elongated chromosome in the mutant suggesting that the numbers of attachment sites are not greatly altered and therefore the loop sizes are minimally affected. On the other hand, the λ transgene appears to acquire additional attachment sites in the mutant. This may indicate a loss of specificity of sequence attachment in the absence of SYCP2/SYCP3 protein from the cores. If such is the case, the repetitive sequences that normally are associated with the cores (Pearlman et al., 1992) and that have recently been correlated with preferences for recombination (J. Greally, personal communication), may be misaligned, thereby interfering with synapsis. If the mechanisms of alignment versus synapsis are to be resolved, additional information on sequence specificity is required.

Acknowledgements

We thank Jian-Guo Liu for help in generating the SYCP3[+/−] heterozygotes and Barbara Spyropoulos (York University) for her assistance with the project and manuscript preparation. Dr. John Heddle (York University) kindly provided the Stratagene Muta™ Mouse with the λ transgene. Antibody gifts from Rolf Jessberger (Mount Sinai, New York) and Kyoto Yokomori (University of California, Irvine, CA) made the work with cohesin antibodies possible.

References

Eijpe M, Heyting C, Gross B, Jessberger R: Association of mammalian SMC1 and SMC3 proteins with meiotic chromosomes and synaptonemal complexes. J Cell Sci 113:673–682 (2000).

Dobson MJ, Pearlman RE, Karaiskakis A, Spyropoulos B, Moens PB: Synaptonemal complex proteins: occurrence, epitope mapping and chromosome disjunction. J Cell Sci 107:2749–2760 (1994).

Heng HHQ, Tsui L-C, Moens PB: Organization of heterologous DNA inserts on the mouse meiotic chromosome core. Chromosoma 103:401–407 (1994).

Heng HHQ, Chamberlain JW, Shi XM, Spyropoulos B, Tsui LC, Moens PB: Regulation of meiotic chromatin loop size by chromosomal position. Proc Natl Acad Sci USA 93:2795–2800 (1996).

James RD, Schmiesing JA, Peters AH, Yokomori K, Disteche CM: Differential association of SMC1alpha and SMC3 proteins with meiotic chromosomes in wild-type and SPO11-deficient male mice. Chromosome Res 10:549–560 (2002).

Moens PB, Heddle JA, Spyropoulos B, Heng HHQ: Identical megabase transgenes on mouse chromosomes 3 and 4 do not promote ectopic pairing or synapsis at meiosis. Genome 40:770–773 (1997).

Offenberg HH, Schalk JA, Meuwissen RL, van Aalderen M, Kester HA, Dietrich AJ, Heyting C: SCP2: a major protein component of the axial elements of synaptonemal complexes of the rat. Nucleic Acids Res 26:2572–2579 (1998).

Pelttari J, Hoja MR, Yuan L, Liu JG, Brundell E, Moens P, Santucci-Darmanin S, Jessberger R, Barbero JL, Heyting C, Hoog C: A meiotic chromosomal core consisting of cohesin complex proteins recruits DNA recombination proteins and promotes synapsis in the absence of an axial element in mammalian meiotic cells. Mol Cell Biol 21:5667–5677 (2001).

Pearlman RE, Tsao N, Moens PB: Synaptonemal complexes from DNase-treated rat pachytene chromosomes contain (GT)n and LINE/SINE sequences. Genetics 130:865–872 (1992).

Pezzi N, Prieto I, Kremer L, Perez Jurado LA, Valero C, Del Mazo J, Martinez-A C, Barbero JL: STAG3, a novel gene encoding a protein involved in meiotic chromosome pairing and location of STAG3-related genes flanking the Williams-Beuren syndrome deletion. FASEB J 14:581–592 (2000).

Prieto I, Suja JA, Pezzi N, Kremer L, Martinez AC, Rufas JS, Barbero JL: Mammalian STAG3 is a cohesin specific to sister chromatid arms in meiosis I. Nat Cell Biol 3:761–766 (2001).

Revenkova E., Eijpe M, Heyting C, Gross B, Jessberger R: Novel meiosis-specific isoform of mammalian SMC1. Mol Cell Biol 21:6984–6998 (2001).

Yuan L, Liu JG, Zhao J, Brundell E, Daneholt B, Hoog C: The murine SCP3 gene is required for synaptonemal complex assembly, chromosome synapsis, and male fertility. Mol Cell 5:73–83 (2000).

Yuan L, Liu JG, Hoja MR, Wilbertz J, Nordqvist K, Hoog C: Female germ cell aneuploidy and embryo death in mice lacking the meiosis-specific protein SCP3. Science 296(5570):1115–1118 (2002).

Cytogenet Genome Res 105:189–202 (2004)
DOI: 10.1159/000078189

Role of retinoid signaling in the regulation of spermatogenesis

S.S.W. Chung[a] and D.J. Wolgemuth[a,b,c,d,e]

Departments of [a] Genetics and Development and [b] Obstetrics and Gynecology; [c] The Center for Reproductive Sciences;
[d] The Institute of Human Nutrition, and [e] The Herbert Irving Comprehensive Cancer Center, Columbia University College of
Physicians and Surgeons, New York, NY (USA)

Abstract. While the need for vitamin A for the normal progression of male germ cell differentiation has been known for many years, the molecular mechanisms underlying this requirement are poorly understood. This review will explore the aspects of the effects on spermatogenesis of dietary deprivation of vitamin A, in particular as to how they compare to the male sterility that results from the genetic ablation of function of the retinoid receptor RARα. The effects of other genes involved with retinoid synthesis, transport, and degradation are also considered. The possible cellular mechanisms that may be affected by the lack of retinoid signaling are discussed, in particular, cell cycle regulation and cell-cell interaction, both of which are critical for normal spermatogenesis.

Copyright © 2004 S. Karger AG, Basel

I. Overview of vitamin A and spermatogenesis

Historical perspective and goals of this review

The need for dietary retinol or vitamin A for normal spermatogenesis has been recognized for decades (Wolbach and Howe, 1925; Howell et al., 1963; reviewed in Eskild and Hansson, 1994; Packer and Wolgemuth, 1999). While considerable insight has been obtained as to the molecular basis for this requirement, in particular from targeted mutagenesis of genes important at various levels of retinoid metabolism, transport, and receptor activity, the mechanisms underlying specific requirement for retinol and its active metabolite *all trans*-retinoic acid (ATRA) in regulating spermatogenesis remain to be elucidated. Mutagenesis of the mouse RARα receptor gene resulted in disruptions in spermatogenesis that were noted to be similar to those observed in the vitamin A-deficient (VAD) rat testis (Lufkin et al., 1993), although detailed comparisons have not yet been reported in the mouse models. Further, it is still not clear whether the abnormalities observed in testes deprived of ATRA signaling from conception (RARα-deficient mice) would be phenocopied by the induction of VAD in the animal after spermatogenesis has been established. Conversely, it is equally unclear as to whether all of the effects of VAD would be manifested in animals deficient for a single retinoic acid receptor, as in the RARα-deficient mice. While many potential targets of RARα have been identified from studies screening for ATRA-induced up-regulation or down-regulation of genes or for genes containing retinoic acid response elements (RAREs), those which might be important for spermatogenesis, a key physiological target of vitamin A function, remain unknown.

This review will not attempt to provide a comprehensive overview of vitamin A function during development and differentiation or of the mechanisms of retinoic acid signaling. Rather, it will explore what has been learned about these functions that are important specifically during spermatogenesis and that depend upon signaling via RARα in particular. It will also consider the basic cellular processes within testicular cells that may be regulated by retinol and ATRA, with a focus on cells within the seminiferous tubules. For discussion of retinoid signaling in other compartments, the reader is referred to Livera et al. (2000, 2002) and Lopez-Fernandez and del Mazo (1997).

This work was supported by P01 DK54057, Project 5, to D.J.W. and fellowships from the Croucher Foundation, Hong Kong to S.S.W.C.

Received 15 November 2003; manuscript accepted 22 December 2003.

Request reprints from: Dr. Debra J. Wolgemuth
Department of Genetics & Development
Columbia University College of Physicians & Surgeons
630 W. 168th street, New York NY 10032 (USA)
telephone: +1-212-305-7900; fax: +1-212-305-6084
e-mail: djw3@columbia.edu

Brief overview of mammalian spermatogenesis

Spermatogenesis is a highly regulated process of differentiation and complex morphologic alterations that leads to the formation of sperm in the seminiferous epithelium. In adult male mammals, it can be subdivided into three main phases: spermatogonial proliferation, meiosis of spermatocytes, and spermiogenesis of haploid spermatids. In rodent testes, differentiating spermatogenic cells form defined associations called stages of the cycle of the seminiferous epithelium (stages I to XII in mouse and stages I to XIV in rat) that can be identified by morphologic criteria (Oakberg, 1956; Russell et al., 1990). The length of the cell cycle and pattern of cell associations, as well as the time necessary to produce spermatozoa, vary greatly among species (Russell et al., 1990). However, genetic control of the timing appears to be fixed, intrinsic to the germ cell, and cell-autonomous. This was clearly evidenced by the rat-specific timing of spermatogenesis observed in rat germ cells transplanted into mouse testes (Franca et al., 1998).

During spermatogonial proliferation, undifferentiated type A spermatogonia, subdivided into $A_{isolated}$ (A_{iso}), A_{paired} (A_{pr}), and $A_{aligned}$ (A_{al}) spermatogonia according to their topographical arrangement on the basement membrane and associated cells, divide mitotically. In response to unknown signals, they then form A_1 spermatogonia, which are the first generation of differentiating spermatogonia (de Rooij, 1998, 2001; Russell et al., 1990). In mouse, these differentiated diploid A_1 spermatogonia synchronously go through a series of six divisions, forming sequentially A_2, A_3, A_4, intermediate, and type B spermatogonia. After the last mitosis of type B spermatogonia, preleptotene spermatocytes are formed. They then initiate meiosis and give rise to leptotene and zygotene spermatocytes. These cells differentiate into pachytene and diplotene spermatocytes, followed by two meiotic divisions and formation of haploid step 1 spermatids. Thereafter, spermatids undergo spermiogenesis, during which the nucleus of the germ cell is remodeled and compacted into the form that is found in mature spermatozoa. Haploid spermatids are morphologically classified into 16 steps in the mouse and 19 steps in the rat (Russell et al., 1990). Spermatogenesis culminates in spermiation, when spermatozoa are released from Sertoli cells into the lumen of the tubule.

What happens as a result of vitamin A deficiency in testis?

Early studies of the specific functions of vitamin A in reproduction were complicated by the overall poor health of VAD animals. The discovery that ATRA could alleviate most of the symptoms of VAD except defects in vision and male fertility has allowed the exploration of the effects on spermatogenesis upon deprivation of vitamin A from the diet (Dowling and Wald, 1960; Thompson et al., 1964). The changes that occur in the VAD rat testis have been studied extensively (reviewed in detail by de Rooij et al., 1989; Griswold et al., 1989; Kim and Wang, 1993; Eskild and Hansson, 1994). For example, examination of the sequence and kinetics of spermatogenic cell disappearance following the onset of VAD revealed that all stages of spermatids in VAD rat testes decreased abruptly from day 2 following the initial loss of body weight (designated as the growth retardation phase of VAD). Spermatids disappeared from the tubules by day 10, while primary spermatocytes

decreased markedly during days 5–12. Degeneration of spermatids has also been observed (Mitranond et al., 1979; Sobhon et al., 1979), along with a disruption of Sertoli cell-spermatid association (Huang et al., 1988) and a delay in spermiation (Huang and Marshall, 1983; Morales and Griswold, 1991). The rapid disappearance of spermatids and spermatocytes from the tubules at the onset of growth retardation phase of VAD suggests that the mechanisms responsible for spermiogenesis, completion of spermiation, and differentiation of spermatocytes are extremely sensitive to change in the status in vitamin A. During the same time period following the onset of growth retardation, the decrease in the number of spermatogonia was reported to be gradual and comparatively low (Mitranond et al., 1979; Sobhon et al., 1979). However, a reduction of spermatogonial population (A and B type) was obviously noted by day 16–20, where only approximately 25% of the spermatogonia remained. This suggested the requirement of vitamin A in the maintenance of the spermatogonial population as well.

It is important to consider whether the effects of VAD occur directly on the germ cells or are a result of disrupted Sertoli cell-germ cell interactions. It is clear that any failure in proper functioning of the specialized junctions found between Sertoli cells and germ cells or between Sertoli cells would alter the microenvironment in the adluminal compartment of the tubules, wherein lie the spermatocytes and spermatids. Alteration of the microenvironment in the adluminal compartment due to improper Sertoli cell function could therefore be responsible for the degeneration of these spermatogenic cells. Ismail and Morales (1992) studied the effect of VAD on tight junction formation and found that it remained intact even during the severe regression three to four weeks after the onset of growth retardation. Although Huang and colleagues (1988) reported that deprivation of vitamin A has resulted in a disruption of inter-Sertoli cell tight junctions in rats on a VAD diet as early as at around 10 days after growth retardation phase, abnormalities in germ cells were observed before this period. This suggests that the degeneration of germ cells occurring in VAD rats was not due to a breakdown of the inter-Sertoli cell tight junctions, but instead, was an immediate consequence of the absence of vitamin A. Recent studies support the effect of VAD on Sertoli cell tight junctions, noting that lanthanum nitrate penetrated deeply into the seminiferous cords in VAD rat testis (3–9 weeks after the onset of growth retardation) (Morales and Cavicchia, 2002). The Sertoli cell barrier could be restored by vitamin A replenishment following prolonged VAD (7–9 weeks after the onset of growth retardation). Interestingly, in testes of animals in which vitamin A was restored and spermatogenesis was resumed, intercellular tracer continued to freely penetrate the inter-Sertoli spaces surrounding not only preleptotene spermatocytes, but also the zygotene and pachytene spermatocytes (Morales and Cavicchia, 2002). Zygotene and pachytene spermatocytes were thus forming without an intact Sertoli cell tight junctional barrier, although some of these cells exhibited apoptosis.

Subsequent, more detailed analyses revealed defects at specific stages of spermatogonial proliferation and differentiation and at the entry into meiotic prophase, i.e. the transition of spermatocytes from preleptotene to leptotene (de Rooij and

Cytogenet Genome Res 105:189–202 (2004)

Table 1. Retinoid receptors and testicular function

Receptors	Phenotype of knockout males	Germ cells[a]	Sertoli cells
RARα	Sterile Morphology similar to VAD testis (Lufkin et al., 1993)	B-SG, PL, L, Z, RS, ES (Akmal et al., 1997) Nu (early and late SP and ES) (Dufour and Kim, 1999)	SC throughout postnatal development Cy (SC) in young animals and partially translocated into Nu in adults (Dufour and Kim, 1999)
		PS (Kim and Griswold, 1990) RS (Eskild et al., 1991)	SC (Kim and Griswold, 1990)
		A-SG (de Rooij et al., 1994) RS (Kim and Akmal, 1996) A-SG, In, early SP, RS (Akmal et al., 1997)	SC (Akmal et al., 1997; Kim and Akmal, 1996)
RARβ	Normal spermatogenesis (Luo et al., 1995)	Cy (SG, early meiotic and PS) throughout development (days 5–52) (Dufour and Kim, 1999)	SC throughout postnatal development Cy (SC) in young animals & partially translocated into Nu in adults (Dufour and Kim, 1999)
		(-) (Kim and Griswold, 1990)	SC (Kim and Griswold, 1990)
		---	Cy (SC) (de Rooij et al., 1994)
RARγ	Normal spermatogenesis Sterile due to abnormal seminal vesicles and prostate glands (Lohnes et al., 1993)	(-)(Dufour and Kim, 1999)	First in SC in rats at 30–35 days of age Mainly Nu (SC) (Dufour and Kim, 1999)
		PS, RS, ES (Huang et al., 1994)	SC (Huang et al., 1994)
		A-SG (de Rooij et al., 1994)	(-) (de Rooij et al., 1994)
RXRα	Died in utero between embryonic days 13.5 & 16.4 (Kastner et al., 1994)	Cy (SG, early and late SP, RS) throughout development (days 5-52) (Dufour and Kim, 1999) A-SG, SP, ST (Gaemers et al., 1998)	SC throughout postnatal development Mainly in Cy (SC) in the young animals & partially translocate into Nu in adults (Dufour and Kim, 1999)
		RS (Kastner et al., 1996)	---
RXRβ	Sterile because of abnormal spermiogenesis (Kastner et al., 1996)	(-) (Dufour and Kim, 1999)	First in SC in rats at 30–35 days of age Mainly Nu (SC) (Dufour and Kim, 1999; Kastner et al., 1996)
		(-) (Kastner et al., 1996)	SC (Kastner et al., 1996)
RXRγ	Normal spermatogenesis (Krezel et al., 1996)	Nu (SG, early and late SP and RS, ES) (Dufour and Kim, 1999) A-SG, PS (Gaemers et al., 1998)	SC throughout postnatal development Mainly Nu (SC) (Dufour and Kim, 1999)

[a] Cy, cytoplasm; Nu, nucleus; A-SG, A-type spermatogonia; B-SG, B-type spermatogonia; SP, spermatocytes; PL, preleptotene spermatocytes; L, leptotene spermatocytes; Z, zygotene spermatocytes; PS, pachytene spermatocytes; RS, round spermatids; ES, elongated spermatids; (-) no expression; --- unknown expression; white background, protein expression by immunohistochemical analysis; light gray background, mRNA expression by Northern blot analysis; dark gray background, mRNA expression by in situ hybridization

Ismail et al., 1990; van Pelt and de Rooij, 1990b; de Rooij et al., 1994; van Dissel-Emiliani, 1997; Packer and Wolgemuth, 1999). At the onset of the growth retardation phase of VAD in rats, the production of A_2 spermatogonia was arrested, and there was also a temporary arrest of preleptotene spermatocytes (Griswold et al., 1989; de Rooij et al., 1989; Ismail et al., 1990; Van Pelt and de Rooij, 1990a; de Rooij et al., 1994). Administration of either vitamin A (Griswold et al., 1989; de Rooij et al., 1989; Ismail et al., 1990; Van Pelt and de Rooij, 1990a) or intraperitoneal injection of ATRA (van Pelt and de Rooij, 1991) caused a massive and synchronous production of A_1 spermatogonia and subsequent spermatogenic stages. The formation of A_1 spermatogonia from A_{al} can be seen in response to vitamin A or ATRA even after long periods of VAD (de Rooij, 1998, 2001). Further, *RARα (Rara)* mRNA was readily detected by in situ hybridization in Sertoli cells and A-spermatogonia 6 h after injection of ATRA in VAD mice (de Rooij et al., 1994). The differentiation of A_{al} into A_1 spermatogonia appears to be a rather vulnerable stage because it can be blocked in a number of different situations, including VAD, elevated testosterone levels, and high testicular temperature (reviewed in de Rooij and Grootegoed, 1998; de Rooij, 2001). This stage also appears to be affected in specific genetic lesions, including the *Sl17H/Sl17H, W/W^v, jsd/jsd,* and *Dazl^{-/-}* strains of mice (Brannan et al., 1992; Matsumiya et al., 1999; Schrans-Stassen et al., 2001; Tohda et al., 2001; Ohta et al., 2003). It is not yet known whether the action of ATRA in inducing differentiation of A_{al} into A_1 is direct, or indirect via Sertoli cells, since both spermatogonia and Sertoli cells possess nuclear receptors for retinoids (Akmal et al., 1997; Gaemers et al., 1998; Cupp et al., 1999) (Table 1).

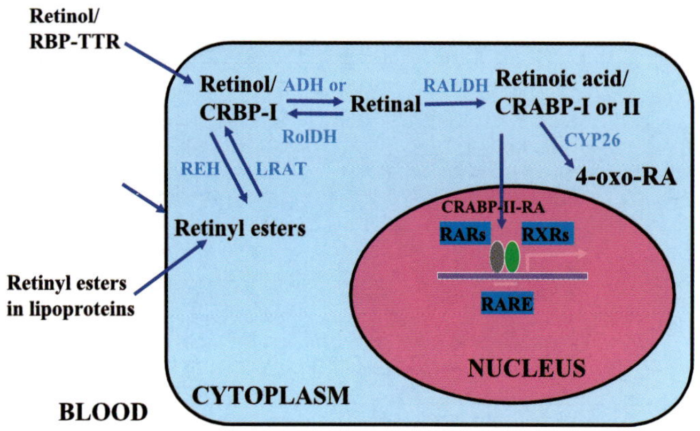

Fig. 1. Vitamin A delivery and its metabolic activation to retinoic acid. RBP, retinol-binding protein; TTR, transthyretin; CRBP-I, cellular retinol-binding protein, type I; CRABP-I, cellular retinoic acid-binding protein, type I; CRABP-II, cellular retinoic acid-binding protein, type II; ADH, medium-chain alcohol dehydrogenases; RolDH, short-chain alcohol dehydrogenase or short-chain dehydrogenase/reductase; RALDH, retinal dehydrogenase; LRAT, lecithin:retinol acyltransferase; REH, retinyl ester hydrolase. Adapted from Gottesman et al. (2001).

Collectively, at least four major defects in spermatogenesis have been identified in the adult VAD rat testis. They include failure of the production of A_2 spermatogonia from A_1 spermatogonia at the onset of VAD, a delay in the onset of and an abnormality in the progression of meiotic prophase, spermatid degeneration and a breakdown of inter-Sertoli cell tight junctions.

II. How does vitamin A exert its function in the testis?

What happens to vitamin A after ingestion?

In omnivores (including humans), all vitamin A in the body must be acquired from the diet as either preformed vitamin A (primarily retinyl esters and retinol from animal sources) or provitamin A carotenoids (such as β-carotene from plants), which are subsequently converted in the body to retinal and ATRA (reviewed in Vogel et al., 1999; Gottesman et al., 2001; Li and Tso, 2003) (also Fig. 1). In brief, within the small intestine, dietary retinyl esters (REs) are hydrolyzed to retinol by

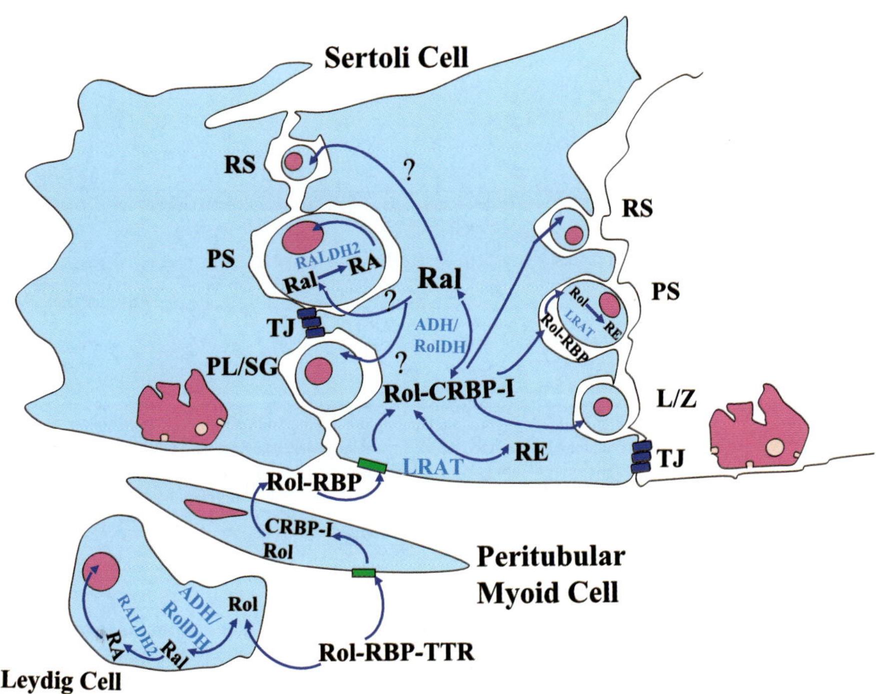

Fig. 2. Proposed retinoid metabolism, transport, and action in the seminiferous tubules of testis. Circulating retinol (Rol) is bound to RBP and complexed with TTR. It is then internalized in the peritubular myoid cells and then passes to Sertoli cells and spermatogenic cells outside the blood-testis barrier. CRBP-I is present in Sertoli cells and may facilitate retinol update from circulating retinol-RBP complexes. Rol-CRBP in Sertoli cells then oxidizes to retinal by ADH in Sertoli cells or certain spermatogenic cells (see section V). By unknown processes, Ral is transferred to spermatogenic cells. Then, RALDH1 or RALDH2 in various spermatogenic cells (SG, early meiotic spermatocytes, PS, RS) oxidizes retinal to retinoic acid; however, neither RALDH1 nor RALDH2 has been found in Sertoli or peritubular myoid cells. Alternatively, the retinol will convert to retinyl esters for storage. CRABP-I is found in the cytoplasm of spermatogenic cells where it may sequester retinoic acid in the cytoplasm and prevent ligand-dependent activation of the nuclear retinoid receptor. Accordingly, inside the blood-testis barrier, no CRABP-I was found in any cells. CRABP-II, is however, found in Sertoli cells. The expression of degrading enzymes is not well described, but is basically characterized as "being expressed in the testis". Similar pathways of uptake and metabolism are believed to function in Leydig cells as in Sertoli cells. RBP, retinol-binding protein; TTR, transthyretin; CRBP-I, cellular retinol-binding protein, type I; CRABP-I, cellular retinoic acid-binding protein, type I; CRABP-II, cellular retinoic acid-binding protein, type I; ADH, medium-chain alcohol dehydrogenases; RolDH, short-chain alcohol dehydrogenase or short-chain dehydrogenase/reductase; RALDH, retinal dehydrogenase; LRAT, lecithin:retinol acyltransferase; REH, retinyl ester hydrolase. PL/SG, preleptotene spermatocytes or spermatogonia; L/Z, leptotene/zygotene spermatocytes; PS, pachytene spermatocytes; RS, round spermatids; TJ, tight junction, Sertoli cell barrier. Adapted from Livera et al. (2002) and Kim and Akmal (1996).

one or more of the retinyl ester hydrolases (REHs). Provitamin A carotenoids are cleaved to retinal which can be then reduced to retinol. Retinol is bound to cellular retinol-binding protein, type II (CRBP-II) and subsequently esterified to REs through the action of lecithin:retinol acyltransferase (LRAT). REs are then packaged along with other dietary lipids into chylomicrons and secreted into the lymphatic system. The chylomicron REs are internalized in the liver by hepatocytes. Within hepatocyte and hepatic stellate cells, retinol is bound to cellular retinol-binding protein, type I (CRBP-I). CRBP-I has been proposed to carry retinol to newly synthesized serum retinol-binding protein (RBP). The RBP-retinol complex is then secreted into the circulation to meet tissue vitamin A needs. Alternatively, LRAT in the hepatocytes can esterify retinol to RE, which is the primary form for storage. Hepatic RE levels in mice are particularly high as compared to other animals, a fact that has complicated VAD studies in mice (McCarthy and Cerecedo, 1952; Smith et al., 1987).

Within cells, retinol is either esterified for storage or converted to active metabolites, such as retinals and retinoic acids (Fig. 1, reviewed in Blaner et al., 1999; Gottesman et al., 2001). Conversion of retinol to retinal is reversible, but conversion of retinal to ATRA is irreversible. Animals maintained on a VAD diet that are supplemented with ATRA are relieved of virtually all the symptoms of VAD, except that they are blind and the males are sterile (Howell et al., 1963). The requirement for dietary retinol, not ATRA, for the maintenance of spermatogenesis is likely due to the existence of the blood-testis barrier or Sertoli cell barrier. In the rat testis, it has in fact been shown that less than 1% of ATRA is derived from the plasma pool indicating that testicular ATRA must be derived from retinol that is taken up from the serum (Kurlandsky et al., 1995) (Fig. 2).

Role of retinoid-specific binding proteins

To maintain solubility, retinoids must be bound to proteins in an aqueous environment (Noy, 1999). Many retinoid-specific binding proteins have been identified, some of which are intracellular while others are extracellular. CRBP-I and CRBP-II and cellular retinoic acid-binding protein, types I and II (CRABP-I and CRABP-II) are exclusively intracellular, whereas RBP, transthyretin (TTR) and interphotoreceptor retinoid-binding protein (IRBP) are extracellular. It has been postulated that CRABPs could act as "buffers" to control spatiotemporally the actual level of "free" intracellular retinoic acid available for binding to the nuclear receptors (Mangelsdorf et al., 1994). RBP is the sole specific transport protein for retinol in the circulation and was proposed to have as its physiological function the delivery of retinol to tissues. TTR is an abundant serum protein composed of four 14-kDa monomers found in 1:1 association with RBP and has been shown to prevent glomerular filtration of RBP (Monaco, 2000).

RBP bound to TTR has been shown to participate in the transport of serum retinol to Sertoli cells via the peritubular myoid cells that surround the seminiferous tubules. In vitro studies suggest that retinol is delivered to the peritubular cells in a complex with both RBP and TTR, where it is then re-secreted bound to RBP synthesized by the peritubular cells themselves (Davis and Ong, 1995). RBP protein was also shown to be synthesized and secreted by rat Sertoli cells in culture (Davis and Ong, 1992). The observation of the uptake of tritiated retinol from retinol-RBP or retinol-RBP-TTR complexes by cultured Sertoli cells led to the hypothesis that serum retinol is chaperoned by RBP and TTR first to the peritubular cells, and finally to the Sertoli cells (reviewed in Eskild and Hansson, 1994; Kim and Akmal, 1996). Although it is assumed that retinol is taken up by Sertoli cells bound by RBP-TTR, the ablation of either *Rbp* or *Ttr* by gene targeting has not revealed any effects on spermatogenesis (Episkopou et al., 1993; Quadro et al., 1999) (discussed in section V).

Using autoradiography, Rajguru and colleagues (1982) demonstrated that labeled retinol is primarily localized to three cellular sites within the adult rat testis: macrophages of the interstitial tissue, lipid droplets of the Sertoli cells, and in spermatids in association with Golgi saccules. However, the blood-testis barrier mediated by junctions in the Sertoli cells prevents the diffusion of retinol or its metabolites into the adluminal compartment which contains the spermatocytes and spermatids (Fig. 2). That is, retinol-RBP cannot be delivered directly to these cells nor can ATRA diffuse to them. Retinol complexed with RBP can only reach Sertoli cells and spermatogenic cells outside the barrier, i.e. the basal compartment (Fig. 2). Thus, the mechanism by which the labeled retinol reached spermatids remains unknown.

CRBP facilitates retinol uptake by cells from circulating retinol-RBP complexes (Fig. 2). Among this family, only CRBP-I has been demonstated in testis, with expression restricted to Sertoli and peritubular myoid cells (reviewed in Eskild and Hansson, 1994). Thus, only Sertoli cells in the testis take up retinol from retinol-RBP complexes in the blood stream (Fig. 2). The cytoplasmic expession of CRBP-I in rat Sertoli cells has been shown to be stage specific, with highest expression at stage XII-XIII, and the lowest at stage VI-VIII. Recently, two additional CRBPs have been isolated and characterized. CRBP-III is distributed mainly in kidney and liver (Folli et al., 2001), while CRBP-IV has been shown to be expressed primarily in kidney, heart and transverse colon, suggesting that different intracellular mediators of retinol metabolism function in different tissues (Folli et al., 2002).

CRABP-I was found to be particularly abundant in the male and female reproductive tracts, including the testis (Ong et al., 1994). Within cells, ATRA is known to be rapidly metabolized to a number of retinoid metabolites, some of which are inactive, in a tissue-dependent manner (Napoli and McCormick, 1981). Therefore, it is likely that retinoic acid must be generated just before its action and near the site of action, probably within the target cells. The detection of both synthesizing and degradation enzymes of retinoic acid in the testis supports this hypothesis (discussed in section V). CRABPs have been shown to act to deliver ATRA to the nucleus (Takase et al., 1986). Recent studies on the molecular mechanism of CRABP function have shown that in COS-7 and MCF-7 cells, movement of ATRA to the RARα receptor involves the direct channeling of ATRA between CRABP-II (and not CRABP-I) and RARα (Dong et al., 1999; Budhu and Noy, 2002).

In the testis, CRABP-I was exclusively localized to the cytoplasm of embryonic gonocytes and in spermatogonia of the post-natal and adult testis, including intermediate and B-type spermatogonia, and was never observed in Sertoli cells (Rajan et al., 1991; Zheng et al., 1996). The presence of CRABP-I exclusively in the cytoplasm of gonocytes and spermatogonia may indicate its possible role in sequestering ATRA in the cytoplasm of these mitotically dividing germ cells that are outside the blood-testis barrier mediated by Sertoli cells, thereby preventing ligand-dependent activation of the nuclear retinoic acid receptors (Zheng et al., 1996). The lack of expression of CRABP-I in the more advanced spermatogenic cells may suggest that it is not needed in cells beyond the barrier, which successfully sequesters their exposure to ATRA.

In contrast to CRABP-I, CRABP-II expression in the embryonic testis was definitively localized to Leydig and Sertoli cells using a well-characterized CRABP-II antibody (reviewed in Ong et al., 1994; Zheng et al., 1996). CRABP-II protein was detected as early as the day 16 rat fetus by immunostaining, and relatively high levels of CRABP-II mRNA were found at postnatal day (pnd) 4, with levels steadily decreasing to undetectable levels at pnd 20 (Zheng et al., 1996). The developmental pattern of expression of CRABP-II in fetal and prepubertal Sertoli cells overlaps almost exactly with the developmental timing of Sertoli cell proliferation in the rat testis (Orth, 1982; Walker, 2003; Buzzard et al., 2003), which is maximal at day 20 of gestation and decreases until pnd 21. These correlations in expression pattern led Zheng and colleagues (Zheng et al. 1996) to propose that CRABP-II is involved in the retinoic acid-dependent autocrine or paracrine regulation of Sertoli cell proliferation.

Discovery of the receptors gave insight into the mechanisms by which the metabolites of retinol could exert their functions

The two major active isoforms of retinoic acid – ATRA and *9-cis* retinoic acid (*9-cis* RA) – both exert the pleiotropic biological effects of vitamin A by transcriptional regulation of target genes via a class of nuclear receptors comprised of two subfamilies: the retinoic acid receptors (RARs) and the retinoid X receptors (RXRs) (reviewed in Chambon, 1995; Chambon, 1996; Leid et al., 1992). Each class of receptors contains three major subtypes (α, β, and γ) whose distribution and function is partly overlapping and partly unique, as illustrated by gene knockout strategies (reviewed in Kastner et al., 1995; Packer and Wolgemuth, 1999). RARs are activated by either ATRA and *9-cis* RA while RXRs are exclusively activated by *9-cis* RA. The RARs and RXRs bind to RAREs to achieve activation or repression of target genes as RAR/RXR heterodimers or as RXR homodimers (reviewed in Chambon, 1996; Mangelsdorf and Evans, 1995; Piedrafita and Pfahl, 1999). The RAREs consist of two direct repeats [PuG(G/T)TCA] usually separated by a spacer of 1, 2, or 5 nucleotides. Over the last quarter century, more than 532 genes have been put forward as regulatory targets of ATRA (reviewed in Balmer and Blomhoff, 2002; Gudas et al., 1994; Nagpal and Chandraratna, 1998). These genes include those encoding growth factors and their receptors, hormones, protein kinases, components of the extracellular matrix, enzymes involved in intermediary metabolism, proto-oncoproteins, and HOX proteins. In some cases, this control is direct, driven by a ligand-complexed heterodimer of retinoid receptors to a RARE; in others, it is indirect, reflecting the actions of intermediate transcription factors, non-classical associations of receptors with other proteins, or even more distant mechanisms.

In normal mouse and rat testes, the cellular localization of the six retinoid receptors has been extensively studied by Northern blot analysis, in situ hybridization and immunohistochemical analysis (Table 1; reviewed in de Rooij et al., 1989; Griswold et al., 1989; Kim and Wang, 1993; Eskild and Hansson, 1994). The expression patterns of these retinoid receptors in Sertoli cells and spermatogenic cells led to several hypotheses of their possible functions in spermatogenesis; however, only RARα or RXRβ, but not other RARs or RXRs, have been shown by genetic ablation approaches to have effects on spermatogenesis (Lufkin et al., 1993; Kastner et al., 1996).

In *Rara⁻/⁻* testes at 4–5 months of age, severe degeneration of the germinal epithelium was observed, with some tubules containing few or no germ cells while adjacent tubules in the same testis sample contained almost all the characteristic cell types, from spermatogonia to fully elongated spermatids (Lufkin et al., 1993). The epididymis of mutant animals contained only a few abnormal spermatozoa and the animals were sterile (Lufkin et al., 1993). No information was provided as to when the abnormalities in spermatogenesis were first observed, complicating our understanding the primary site of action of RARα in the testis. That is, it is possible that RARα is required at only a few steps of differentiation, perhaps even in either only the germ cell or somatic cell lineages, but the severe degeneration that was observed resulted from the disruption of the overall spermatogenic process. In such a case, the loss of other germ cell types would not be a primary result of the loss of the protein, but rather would occur as the key cell-cell interactions between various stages of germ cells and the supporting Sertoli cells are disrupted.

RXRβ-deficient males are sterile, owing to oligo-asthenoteratozoospermia (Kastner et al., 1996). Failure of spermatid release in the seminiferous epithelium and abnormal acrosomes and tails of spermatozoa in the epididymis were also noted. There was a progressive accumulation of lipids within the mutant Sertoli cells, characterized as unsaturated triglycerides. RXRβ has been shown to be expressed exclusively in Sertoli cells; hence, these cells are likely the primary site of RXRβ signaling during spermatogenesis.

Disruption of RARγ also resulted in male sterility but not due to abnormal spermatogenesis; rather, there were abnormalities in the seminal vesicles and prostate glands (Lohnes et al., 1993). A squamous metaplasia of the glandular epithelia of the prostate and seminal vesicles was observed, which was also noted to be a characteristic symptom of VAD (Thompson et al., 1964). *Rara⁻/⁻Rarg⁻/⁻* double mutant males also showed severely abnormal genital ducts (Lohnes et al., 1994). RARγ-mediated ATRA signaling is likely to perform an essential role at some stages in the ontogeny of reproductive organs, but the specific defects have not been well characterized.

34

What testicular cell types require retinoid signaling via RARα?

RARα transcripts and protein have been reported to be expressed in most cell types in the testis, suggesting that it could play a role in processes as diverse as Sertoli cell function, meiotic prophase, and spermiogenesis (Kim and Griswold, 1990; Eskild et al., 1991; van Pelt et al., 1992; Kim and Wang, 1993; Lufkin et al., 1993; Akmal et al., 1997; Dufour and Kim, 1999). In situ hybridization analysis of developing rat testes has shown that RARα transcripts increased between pnd 5 and 10 (Akmal et al., 1997). The mRNA level peaked from pnd 10 to 15 and was found primarily towards the center of seminiferous epithelium (Akmal et al., 1997). The levels then declined in pnd 20 testis. The rat testis at pnd 9 contains only type A spermatogonia and Sertoli cells, B type spermatogonia do not develop until 11 days of age, primary spermatocytes are first apparent on pnd 15, and round spermatids appear on pnd 25 (Wing and Christensen, 1982; Yang et al., 1990; Dym et al., 1995; Malkov et al., 1998). By immunohistochemical analysis, RARα proteins were also found in a similar developmental profile (Dufour and Kim, 1999). The highest expression in Sertoli and spermatogenic cells was found in the seminiferous tubules of rats at pnd 5–15 (Akmal et al., 1997). RARα protein was detected in the nuclei of type A and B spermatogonia, and was high in the nuclei of preleptotene, leptotene and zygotene spermatocytes (Akmal et al., 1997; Dufour and Kim, 1999). In Sertoli cells, RARα protein was reported to be present in cytoplasm throughout development and to be partially translocated into nuclei in the adult testis (Dufour and Kim, 1999).

To what extent does dietary deprivation of vitamin A phenocopy loss of RARα function?

In a recent study to determine the testicular cells first exhibiting abnormalities during spermatogenic differentiation in the absence of RARα-mediated signaling, we observed striking changes in spermatogenic cell associations that appeared to be primary as opposed to secondary events (Chung et al., 2004). The absence of the RARα receptor resulted in an increased number of degenerating pachytene spermatocytes and a temporary arrest in spermiogenic progression, at step 8–9 spermatids, in the first wave of spermatogenesis. There was also a delay in the onset of the second wave, as well as a temporary arrest and delay in the appearance of preleptotene and leptotene spermatocytes in the first, second and third waves. In vivo BrdU labeling revealed a notable decrease in germ cell proliferation in both juvenile and adult *Rara*[−/−] testes and confirmed the arrest at step 8–9 spermatids observed morphologically in the first wave of spermatogenesis. The net result of these abnormalities was a profound asynchrony of spermatogenic progression in *Rara*[−/−] semininferous tubules. These findings implicated RARα as being essential for synchronous development of spermatogenic cells in the testis. We speculated that RARα receptor-regulated factors were required for A-spermatogonia to initiate a precise series of divisions, for preleptotene spermatocytes to traverse a normal meiotic prophase, and for step 8–9 spermatids to continue to undergo spermiogenesis, respective-

ly. Interestingly, in VAD rat testes, examples of temporary arrest in development of spermatogenesis have also been demonstrated, further suggesting the tightly controlled spermatogenic cell cycle can be altered by retinoids (Huang and Hembree, 1979; de Rooij, 1983; Morales and Griswold, 1987; Ismail et al., 1990; de Rooij et al., 1994).

Another striking feature of the *Rara*[−/−] testes were spermatids undergoing apoptosis, apparently engulfed by Sertoli cells situated at the basal lamina (Chung, Wang, and Wolgemuth, manuscript in preparation). Huang and Marshall (Huang and Marshall, 1983) also observed the presence of late spermatids and residual bodies at the luminal edge of epithelia at stages later than stage VIII of the spermatogenic cycle in VAD rat testis, consistent with the defective spermiation of these spermatids. In normal testis, spermatogonia and meiotic spermatocytes are the primary cells that undergo apoptosis and apoptotic haploid germ cells are rarely detected (Billig et al., 1995; Callard et al., 1995; Hikim et al., 1995; Shetty et al., 1996). The death of spermatozoa, which have little cytoplasm, may not involve the classical apoptotic machinery, since caspases do not appear to be used (Weil et al., 1998). Prolonged VAD is known to cause germ cell degeneration (Mason, 1933; Thompson et al., 1964) and the phagocytosis of degenerating germ cells by Sertoli cells is normal (Fawcett, 1975). The similar findings of defects in spermiation in both the VAD and RARα-deficient models suggested that mechanisms leading to completion of spermiation are extremely sensitive to change in the status in vitamin A, most likely through an RARα receptor-mediated pathway. Thus, the failure of the production of A_2 spermatogonia from A_1 spermatogonia at the onset of VAD, a delay in the onset of and an abnormality in the progression of meiotic prophase, and spermatid degeneration are defects in spermatogenesis shared in the VAD and RARα-deficient models.

IV. Potential cellular processes regulated by vitamin A signaling during spermatogenesis revealed by mutation of *RARα*

Control of cell cycle progression by retinoids

Cell cycle progression is regulated by the sequential activity of various cyclin-dependent kinases (Cdks) (Roberts, 1999; Sherr and Roberts, 1999; reviewed in Miller and Cross, 2001; Nurse, 2002; Wolgemuth et al., 2002). Cyclins are key components of the core cell cycle machinery. These proteins constitute regulatory subunits that bind, activate, and provide substrate specificity for their associated Cdks. Regulation of cell cycle progression in mammalian cells by external factors such as retinoids generally occurs in the G1 phase of the cycle (reviewed in Rogers, 1997; Harvat and Jetten, 1999; Sherr and Roberts, 1999). Three classes of cyclins operate during the G1 to S phase of mitosis in mammalian cells: D-type cyclins (D1, D2, and D3), A-type (A2), and E-type cyclins (E1 and E2) (Sherr and Roberts, 1999). Mammalian cells rely on assorted strategies of negative regulation to enforce the order of events, to provide optimal timing of phase transitions, and to maintain genomic integrity. The simplest way to negatively regulate the cell cycle is through inhibition of an essential Cdk. The activities of G1

Cdks can be blocked by Cdk inhibitors (CKIs). The INK4 class (inhibitors of CDK4) consists of four members (p16[Ink4a], p15[Ink4b], p18[Ink4c], and p19[Ink4d]) that exclusively bind to and inhibit the cyclin D-dependent catalytic subunits Cdk4 and Cdk6. The Cip/Kip family includes three members (p21[Cip1], p27[Kip1], and p57[Kip2]) that bind to both cyclins and Cdks to preferentially inhibit cyclin E- and A-dependent Cdk2.

The possibility that retinoid signaling could be important for initiation of and progression through the cell cycle in spermatogonia, spermatocytes, and spermatids (a specialized telophase) is intriguing. Retinoids have been shown to directly affect expression of cell cycle-regulatory genes at the G1 phase of mitosis (reviewed in Rogers, 1997; Harvat and Jetten, 1999; Pestell et al., 1999; Altucci and Gronemeyer, 2001; Boyle, 2001; de Rooij, 2001). For example, in a negative regulatory mode of cell cycle progression, vitamin D and ATRA can induce expression of p21[Cip1], which blocks Cdk activity, resulting in G1 arrest of treated U937 cells (Liu et al., 1996). The p21[Cip1] gene is a retinoic acid-responsive target gene, with an RARE in the promoter that is required to confer ATRA induction through RAR/RXR heterodimers (Liu et al., 1996). During G1-S phase, another direct inhibitor of Cdks, p27[Kip1], is phosphorylated and subjected to ubiquitin-mediated degradation (Reed, 2002). In human lung squamous carcinoma CH27 cells, ATRA-mediated G1 arrest is associated with induction of p27[Kip1] (Hsu et al., 2000). Interestingly, in p27[Kip1] knockout mice, aberrations in the spermatogenic process were observed (Fero et al., 1996; Kiyokawa et al., 1996; Nakayama et al., 1996; Beumer et al., 1999). First, a 50% increase in the number of A spermatogonia at stage VIII was found. Second, abnormal (pre)leptotene spermatocytes were observed, some of which seemingly tried to enter a mitotic division instead of entering meiotic prophase. By the use of immunohistochemistry, p27[Kip1] expression was seen in gonocytes from E16.5 to pnd 2, while in adult mouse testis staining was found only in Sertoli cells. This suggests a role of p27[Kip1] and possibly vitamin A signaling in the regulation of spermatogonial proliferation, and/or apoptosis, and possibly the onset of meiotic prophase in preleptotene spermatocytes. In the adult testis, the effects would be indirect, via Sertoli cells, however.

Changes in the expression of cell cycle-regulatory genes by retinol replenishment after VAD have been reported. For example, Cdc25A, whose activity is required for entry into S phase, is expressed in spermatogonia, pre- and post-meiotic rat germ cells and is upregulated in response to retinol (Mizoguchi and Kim, 1997). There is increased expression of cyclin D2 with retinol replenishment of VAD testis (Beumer et al., 2000). Free E2F-1 has been shown to inhibit expression of retinoic acid-responsive promoters and to inhibit RARα transactivation function in P19 embryonal carcinoma cells (Costa et al., 1996); however, its effects during spermatogenesis on retinoid signaling are not known. Studies on the possible defects on these cell cycle-regulatory gene expression in the absence of RARα receptor-mediated pathway may provide insight into the regulation of the cell cycle in germ cells and hence, the regulation of synchronization of the spermatogenic cycle, which was perturbed in Rara[-/-] mice (Chung et al., 2004).

The down-regulation of cyclin D1 protein in bronchial epithelial cells in response to ATRA is regulated at the posttranslational level, likely through increased ubiquitin-dependent proteasome degradation (Langenfeld et al., 1997). In these studies, the proteasome inhibitors calpain inhibitor I and lactacystin prevented the decrease in cyclin D1 protein expression, suggesting that ATRA somehow induced proteolysis of cyclin D1 (Langenfeld et al., 1997). Similar ATRA-mediated post-translational mechanisms have been reported to contribute to the changes in activity of Cdk2 and Cdk4, which, in the case of Cdk4, involve the ubiquitin-proteasome pathway (Sueoka et al., 1999). ATRA and 9-*cis* RA were also shown to inhibit the expression of cyclin D1 and D3 and the kinase activities of Cdk2 and Cdk4 in human breast carcinomal cells in vitro (Zhou et al., 1997). Interestingly, targeted disruption of Cdk4, the primary catalytic partner for the D-type cyclins, which is expressed in spermatogonia and early stage spermatocytes, resulted in smaller testis with reduced numbers of spermatogonia/spermatocytes with perturbed layer formation (Kang et al., 1997; Tsutsui et al., 1999; Zhang et al., 1999).

In normal testis, the ratio between self-renewal and differentiation of spermatogonial stem cells should be about 1.0. The regulatory mechanism controlling the ratio between stem cell renewal and differentiation in the testis may involve glial cell line-derived neurotrophic factor (GDNF), as overexpression of GDNF resulted in clusters of mostly single A spermatogonia in 3-week-old testes (Meng et al., 2000). GDNF, which is produced by Sertoli cells, promotes stem cell renewal (Meng et al., 2000; de Rooij, 2001). ATRA has shown to act through RARα to induce GDNF responsiveness in rat superior cervical ganglia neurons, by upregulating the expression of ligand-specific receptor for GDNF (GFRα-1) at both the mRNA and protein levels (Thang et al., 2000). Whether GDNF may be a possible target in RARα-mediated pathway in testis is unknown, but examining its expression in mutant and VAD testis is of considerable interest.

Cell-cell interactions affected by loss of RARα mediated signaling

As mentioned before, in the VAD rat testis model, spermatid degeneration is associated with a delay in spermiation (Mitranond et al., 1979; Sobhon et al., 1979; Huang and Marshall, 1983; Morales and Griswold, 1991) and a disruption of Sertoli cell tight junctions has been reported (Huang et al., 1988; Morales and Cavicchia, 2002). Our studies on the *Rara* mutant model revealed a delay and temporary arrest at step 8–9 spermatids and a failure of spermatids to align next to the lumen for spermiation (Chung et al., 2004). The similar findings of defects in spermiation in both VAD rat testis and *Rara[-/-]* mouse testis suggested that mechanisms involved in spermiogenesis and completion of spermiation are extremely sensitive to change in the status in vitamin A, most likely through an RARα receptor-mediated pathway.

Several studies have suggested the involvement of retinoid receptor-mediated signaling for induction of both tight junction (TJ)-associated molecules and barrier function. *Stra6* and *Stra8*, whose testicular expression is affected by retinoids, are ATRA-inducible genes identified in P19 EC cells (Bouillet et al., 1995, 1997). *Stra6* appears to encode a novel integral membrane protein whose expression localized to blood-organ barriers in several tissues (Bouillet et al., 1997). In the testis, *Stra6* is expressed in a spermatogenic cycle-dependent pattern in the basal lamina membranes of Sertoli cells, particularly in stage VI-VII tubules. In *Rara*^–/– testes, *Stra6* is expressed in almost all the tubules, suggesting that RARα is required for the selective expression of *Stra6* at stage VI-VII (Bouillet et al., 1997).

Treatment of rat lung endothelial cells with 1 µM ATRA for 4 days was shown to induce 7H6 antigen, a TJ peripheral protein. It was preferentially localized at the cell border and simultaneously enhanced the barrier function two-fold, as assessed by transepithelial electrical resistance and fluxes of albumin and dextran (Satoh et al., 1996). 7H6 antigen rapidly disappeared in parallel with a decrease in the paracellular barrier function during the course of cellular ATP depletion (Zhong et al., 1994), suggesting a close association between the phosphorylation and localization of 7H6 antigen at TJ. Recently, it was reported that ATRA induced the TJ structure and expression of several TJ-associated molecules, such as ZO-1, occludin, claudin-6, and claudin-7, as well as a barrier function in F9 murine EC cells. This induction was further shown to be mediated specifically by the RXRα-RARγ receptor pair (Kubota et al., 2001). A delay in the assembly of ZO-1, a peripheral component protein of TJ, into the Sertoli cell barrier has been detected in *Rara*^–/– testis (SSWC and DJW, unpublished observations). Similarly, gap junctional permeability has been shown to be blocked by retinol or ATRA in other cells (Walder and Lutzelschwab, 1984; Mehta et al., 1986; Pitts et al., 1986; Brummer et al., 1991; Weiler et al., 1999). The nature of the gap junctions has not been studied in either VAD testis or in RARα-deficient testis, but given the observed abnormalities in Sertoli-Sertoli or Sertoli-germ cell interactions, they are likely to be affected.

V. Additional genetic manipulations to fine-tune retinoid action – any effects on spermatogenesis?

Synthesizing enzymes

Vitamin A signaling depends on enzymes capable of catalyzing the two-step conversion of retinol to ATRA, with ATRA then functioning as a ligand for retinoid receptors (reviewed in Blaner et al., 1999; Duester, 2000, 2001; Duester et al., 2003). Enzymes that catalyze the oxidation of retinol include the family of cytosolic medium-chain alcohol dehydrogenases (ADHs) and short-chain dehydrogenase/reductases (SDR). ADH oxidizes free retinol, but not retinol bound to CRBP-I. Four classes of ADHs have been reported including ADH1, ADH2, ADH3, and ADH4. In situ hybridization and immunohistochemical analysis revealed that in the testis, ADH1 is localized in Sertoli cells and Leydig cells (Fig. 2) and ADH4 in late spermatids (Deltour et al., 1997). ADH3 is ubiquitously expressed as shown by Northern blot analysis (Zgombic-Knight et al., 1995). ADH2 has a very limited tissue distribution, being found only in adult liver and possibly human skin (Duester, 2000). Reviews with a particular focus on the mouse SDR family including several microsomal enzymes able to oxidize retinal to reti-

naldehyde in vitro may be found in the papers by Duester (2000) and Duester et al. (2003).

Targeted disruption of the *Adh1, Adh3* and *Adh4* genes showed normal growth and fertility (Deltour et al., 1999a, b). However, the mice exhibited marked differences in clearance of ethanol and formaldehyde: ADH1 and AHD4 both demonstrated functions in ethanol and retinol metabolism in vivo, whereas ADH3 functions as a formaldehyde dehydrogenase (Deltour et al., 1999a, b). ADH1 and ADH4 double null mutant mice were viable and fertile, demonstrating no additive effects of loss of both ADH1 and ADH4 (Molotkov et al., 2002). However, studies on single or double knockout mice indicated that ADH1 provided considerable protection against vitamin toxicity, whereas ADH4 promoted survival during VAD, thus revealing largely non-overlapping functions for enzymes in retinoid metabolism (Molotkov et al., 2002). Whether the testes of the mutant strains would exhibit enhanced sensitivity to excess vitamin A or more pronounced response to VAD remains to be determined.

Retinaldehyde dehydrogenases (RALDHs) catalyze the final step of ATRA synthesis and include RALDH1, RALDH2, RALDH3 and RALDH4 (Duester et al., 2003). In testis, pachytene spermatocytes expressed RALDH1 most intensely, while spermatogonia, early meiotic spermatocytes and round spermatids expressed much less (Zhai et al., 2001). RALDH2 was highly expressed in various spermatogenic cell types including spermatogonia, early meiotic spermatocytes and pachytene spermatocytes, as well as Leydig cells (Fig. 2). Neither RALDH1 nor RALDH2 expression was detected in Sertoli or peritubular myoid cells (Zhai et al., 2001). A null mutation of Raldh2 *(Aldhla2)* is embryonic lethal, eliminating most mesodermal RA synthesis (Niederreither et al., 1999), whereas loss of Raldh1 *(Aldhla1)* eliminates ATRA synthesis only in the embryonic dorsal retina with no obvious effect on development (Fan et al., 2003). Further, RALDH1-deficient mice show no apparent gross abnormalities or defects in survival or growth, and are relatively healthy and fertile (Fan et al., 2003). The genes encoding RALDH3 (Suzuki et al., 2000) and RALDH4 (Lin et al., 2003) have been recently cloned, but the effects of targeted mutation of both genes has not yet been reported. Given their pattern of expression – Raldh3 *(Aldhla3)* is expressed in the ventral region of the retina (Suzuki et al., 2000) and Raldh4 *(Aldhla4)* in adult mouse liver and kidney (Lin et al., 2003) and not in testis – neither gene would be predicted to be important for spermatogenesis.

Serum binding proteins

As mentioned earlier, mutant mice that lack functional RBP are viable and fertile (Quadro et al., 1999). However, they showed reduced levels of plasma retinol (12.5% of wild type), markedly impaired retinal function with abnormal vision throughout the first several months of life, and abnormal retinol mobilization with an inability to mobilize retinol from hepatic stores. Although the hepatic levels of retinol reached significantly higher levels in *Rbp*^–/– mice at 5 months of age, short-term exposure to a VAD diet did not result in the expected drop in hepatic levels in *Rbp*^–/– mice (Quadro et al., 1999). This indicated that *Rbp*^–/– mice cannot mobilize hepatic retinoid stores and may become VAD more quickly (W.S. Blan-

37

er, personal communication). As such, these RBP-deficient mice represent a tool for studying the onset of abnormalities of VAD during the first wave of spermatogenesis.

Although disruption of the *Ttr* gene resulted in mice with depressed levels of plasma retinol and thyroid hormone, the mice were healthy and fertile, and showed no apparent developmental defects (Episkopou et al., 1993). The mutation affected neither the uptake nor storage of dietary retinol. Further, the concentration of retinol and RE in testis, kidney, spleen and eye cups of mutant mice were normal, suggesting no requirement of TTR in the delivery of retinol to these tissues. TTR could be important when animals are stressed via malnutrition such as on a VAD diet, an idea that has not yet been studied.

Subcellular binding proteins

The genes for CRBP I and CRBP II have both been disrupted and the physiological actions of these proteins determined. Mice lacking CRBP I appear physiological normal, with no embryonic lethality, impaired fertility, or shortened life span (Ghyselinck et al., 1999). However, these mice have impaired storage of retinyl esters in hepatic stellate cells, indicating that CRBP I plays an important role in the transfer of retinol to and/or esterification of retinol in these cells (Ghyselinck et al., 1999). Under VAD, CRBP I-null mice exhaust their RE stores six times faster than wild-type mice, and develop abnormalities characteristic of hypovitaminosis A. Mice lacking CRBP I and maintained on a VAD diet for 14 weeks exhibited testicular degeneration with sloughing of germ cells and the appearance of irregular vacuoles in the tubules (Ghyselinck et al., 1999). The coexistence of morphological normal tubules with tubules with reduced size or devoid of germ cells, which is similar to the phenotype of the RARα-deficient testes, was frequently observed. At 23 weeks, it was reported that the seminiferous tubules displayed a Sertoli cell-only morphology (Ghyselinck et al., 1999). Mice lacking CRBP II also have reduced hepatic stores of vitamin A, but grow and reproduce normally (E et al., 2002). However, under a VAD diet, there is increased neonatal mortality, indicating a role of CRBP II in ensuring adequate delivery of vitamin A to the developing fetus when dietary vitamin A is limiting. The VAD phenotype on a CRBP II-null background and effects on spermatogenesis have not been reported to date.

The genes for CRABP I and CRABP II have also been disrupted and the physiological actions of these proteins have been determined in studies using single and double knockouts. CRABP I-null mice are normal in development, fertility, and life span, demonstrating that under normal homeostasis, this protein is dispensable (Gorry et al., 1994). With the exception of a minor, partially penetrant, limb malformation, CRABP II-deficient mice were also normal with regard to development, fertility, life span and behavior (Lampron et al., 1995). In order to explore the possibility of redundant function, CRABP I/II-double mutant mice were generated. These mice were essentially normal, except for an enhanced penetrance of abnormalities in limb formation (Lampron et al., 1995). CRABPs could still be important for maintaining physiological levels of intracellular retinoic acid under conditions of limited supply of vitamin A and it is further not known whether the testis will be affected in knockout mice under VAD.

Degrading enzymes

ATRA is a crucial signaling molecule involved in tissue morphogenesis during embryonic development. The distribution and concentration of ATRA is precisely regulated during embryogenesis by balanced complementary activities of synthesizing (RALDH) and metabolizing (CYP26) enzymes. Three classes of CYP26 have been reported including A1, B1 and B2 (also designated C1). Cytochrome P450 (now designated as CYP26A1) was first identified as an ATRA-inducible gene (White et al., 1997). ATRA can cause direct and rapid induction of *Cyp26a1* due to the presence of a highly conserved 32-bp element containing a canonical, direct repeat-5 type of RARE in the 5′ upstream promoter region (reviewed in Ross, 2003). *Cyp26b1* transcripts were detected in the developing lung, kidney, spleen, thymus and testis, whereas *Cyp26a1* transcripts were found in the diaphragm and outer stomach mesenchyme (Abu-Abed et al., 2002). In the fetal testis, only *Cyp26b1* transcripts were detected, specifically in cells located outside of the developing testicular cords (Abu-Abed et al., 2002).

By targeted mutagenesis, CYP26A1-deficient mice die during mid-late gestation with several major morphogenetic defects, leading in extreme cases to a sirenomelia ("mermaid tail") phenotype (Abu-Abed et al., 2001; Sakai et al., 2001). *Cyp26a1*[−/−] mice are phenotypically rescued by lowering the embryonic levels of ATRA by heterozygous disruption of the RALDH2 gene *(Aldhla2)* (Niederreither et al., 2002). Some of *Cyp26A1*[−/−]*Aldhla2*[+/−] mutants can reach adulthood without any phenotypic defect apart from their abnormal tails and both males and females were fertile (Niederreither et al., 2002). Compound heterozygotes (*Cyp26A1*[+/−]*Aldhla2*[+/−]) were healthy and fertile. This implied that retinoid homeostasis was crucial for embryonic development and presumably, for subsequent tissue function. Further, genetic ablation of *Rarg* also rescued mice lacking CYP26A1 from caudal regression and embryonic lethality (Abu-Abed et al., 2003). However, *Cyp26a1*[−/−]*Rarg*[−/−] double mutant males and females both failed to reproduce normally (Abu-Abed et al., 2003). *Cyp26c1* has been recently cloned and characterized (Tahayato et al., 2003; Taimi et al., 2003). It was expressed in the hindbrain, inner ear, first branchial arch and tooth buds during murine development (Tahayato et al., 2003). Disruption of the *Cyp26b1* or *Cyp26c1* genes has not yet reported.

Unanswered questions

There is still much to be learned regarding the targets of retinoid signaling in spermatogenic cells. It is not known whether there would be differences in the spermatogenic phenotypes observed in animals in which RARα is only in the germ line or only in somatic cells of the testis, as compared to the present model wherein it is mutated in all the cells in the testis since the onset of gonadal development. To address this question, it would be necessary to ablate RARα function specifically in one lineage or the other. One possible approach would be to achieve cell-specific ablation using the "Cre recombinase/loxP recognition sequence" system for generating conditional mutations.

This approach is made feasible by the availability of spermatogenic stage-specific promoters that are well characterized with regard to their specificity and activity in transgenic mice (Wolgemuth and Watrin, 1991; Wolgemuth et al., 1995). Alternatively, one could consider expressing a dominant negative form of RARα in specific spermatogenic stages in transgenic mice and ask if this inhibits the progression of spermatogenesis. Conversely, one could ask if expression of RARα in specific spermatogenic stages would result in cell-specific rescue of the RARα-deficient phenotype. We anticipate that this approach will be very powerful in highlighting specific functions and pathways in which the receptor acts, pathways that are obscured when function is blocked in a ubiquitous manner, as in total null animals.

Retinol metabolism in the seminiferous tubules plays an important role in spermatogenesis, suggesting that synthesizing and degrading enzymes and the various intracellular and extracellular binding proteins should similarly play critical roles during homeostasis in the testis. However, except for those that are embryonic lethal, male mice lacking either the metabolic enzymes or the binding proteins have no fertility problem. Whether this is due to functional redundancy among the enzymes and binding proteins remains to be determined. This effort will benefit from better characterization of their patterns of expression in both control and mutant testes and studies in which dietary retinol is limiting or in excess. Further, the determination of exact pathway of retinol delivery from the circulation to specific spermatogenic cells remains to be elucidated.

References

Abu-Abed S, Dolle P, Metzger D, Beckett B, Chambon P, Petkovich M: The retinoic acid-metabolizing enzyme, CYP26A1, is essential for normal hindbrain patterning, vertebral identity, and development of posterior structures. Genes Dev 15:226–240 (2001).

Abu-Abed S, MacLean G, Fraulob V, Chambon P, Petkovich M, Dolle P: Differential expression of the retinoic acid-metabolizing enzymes CYP26A1 and CYP26B1 during murine organogenesis. Mech Dev 110:173–177 (2002).

Abu-Abed S, Dolle P, Metzger D, Wood C, MacLean G, Chambon P, Petkovich M: Developing with lethal RA levels: genetic ablation of Rarg can restore the viability of mice lacking *Cyp26a1*. Development 130:1449–1459 (2003).

Akmal KM, Dufour JM, Kim KH: Retinoic acid receptor alpha gene expression in the rat testis: potential role during the prophase of meiosis and in the transition from round to elongating spermatids. Biol Reprod 56: 549–556 (1997).

Altucci L, Gronemeyer H: Nuclear receptors in cell life and death. Trends Endocrinol Metab 12:460–468 (2001).

Balmer JE, Blomhoff R: Gene expression regulation by retinoic acid. J Lipid Res 43:1773–1808 (2002).

Beumer TL, Kiyokawa H, Roepers-Gajadien HL, van den Bos LA, Lock TM, Gademan IS, Rutgers DH, Koff A, de Rooij DG: Regulatory role of p27kip1 in the mouse and human testis. Endocrinology 140:1834–1840 (1999).

Beumer TL, Roepers-Gajadien HL, Gademan IS, Kal HB, de Rooij DG: Involvement of the D-type cyclins in germ cell proliferation and differentiation in the mouse. Biol Reprod 63:1893–1898 (2000).

Billig H, Furuta I, Rivier C, Tapanainen J, Parvinen M, Hsueh AJ: Apoptosis in testis germ cells: developmental changes in gonadotropin dependence and localization to selective tubule stages. Endocrinology 136:5–12 (1995).

Blaner WS, Piantedosi R, Sykes A, Vogel S: Retinoic acid synthesis and metabolism, in Nau H, Blaner WS (eds): Retinoids: The Biochemical and Molecular Basis of Vitamin A and Retinoid Action, pp 117–149 (Springer-Verlag, Berlin 1999).

Bouillet P, Oulad-Abdelghani M, Vicaire S, Garnier JM, Schuhbaur B, Dolle P, Chambon P: Efficient cloning of cDNAs of retinoic acid-responsive genes in P19 embryonal carcinoma cells and characterization of a novel mouse gene, *Stra1* (mouse LERK-2/Eplg2). Dev Biol 170:420–433 (1995).

Bouillet P, Sapin V, Chazaud C, Messaddeq N, Decimo D, Dolle P, Chambon P: Developmental expression pattern of *Stra6*, a retinoic acid-responsive gene encoding a new type of membrane protein. Mech Dev 63:173–186 (1997).

Boyle JO: Retinoid mechanisms and cyclins. Curr Oncol Rep 3:301–305 (2001).

Brannan CI, Bedell MA, Resnick JL, Eppig JJ, Handel MA, Williams DE, Lyman SD, Donovan PJ, Jenkins NA, Copeland NG: Developmental abnormalities in Steel17H mice result from a splicing defect in the steel factor cytoplasmic tail. Genes Dev 6:1832–1842 (1992).

Brummer F, Zempel G, Buhle P, Stein JC, Hulser DF: Retinoic acid modulates gap junctional permeability: a comparative study of dye spreading and ionic coupling in cultured cells. Exp Cell Res 196:158–163 (1991).

Budhu AS, Noy N: Direct channeling of retinoic acid between cellular retinoic acid-binding protein II and retinoic acid receptor sensitizes mammary carcinoma cells to retinoic acid-induced growth arrest. Mol Cell Biol 22:2632–2641 (2002).

Buzzard JJ, Wreford NG, Morrison JR: Thyroid hormone, retinoic acid, and testosterone suppress proliferation and induce markers of differentiation in cultured rat sertoli cells. Endocrinology 144:3722–3731 (2003).

Callard GV, Jorgensen JC, Redding JM: Biochemical analysis of programmed cell death during premeiotic stages of spermatogenesis in vivo and in vitro. Dev Genet 16:140–147 (1995).

Chambon P: The molecular and genetic dissection of the retinoid signaling pathway. Recent Prog Horm Res 50:317–332 (1995).

Chambon P: A decade of molecular biology of retinoic acid receptors. FASEB J 10:940–954 (1996).

Chung SSW, Sung YK, Wang XY, Wolgemuth DJ: RARalpha is required for synchronization of spermatogenic cycles and its absence results in progressive breakdown of the spermatogenic process. Dev Dyn, in press (2004).

Costa SL, Pratt MA, McBurney MW: E2F inhibits transcriptional activation by the retinoic acid receptor. Cell Growth Differ 7:1479–1485 (1996).

Cupp AS, Dufour JM, Kim G, Skinner MK, Kim KH: Action of retinoids on embryonic and early postnatal testis development. Endocrinology 140:2343–2352 (1999).

Davis JT, Ong DE: Synthesis and secretion of retinol-binding protein by cultured rat Sertoli cells. Biol Reprod 47:528–533 (1992).

Davis JT, Ong DE: Retinol processing by the peritubular cell from rat testis. Biol Reprod 52:356–364 (1995).

de Rooij DG: Proliferation and differentiation of undifferentiated spermatogonia in the mammalian testis, in Potten CS (ed): Stem Cells: Their Identification and Characterization, pp 89–117 (Churchill Livingstone, Edinburgh 1983).

de Rooij DG: Stem cells in the testis. Int J Exp Pathol 79:67–80 (1998).

de Rooij DG: Proliferation and differentiation of spermatogonial stem cells. Reproduction 121:347–354 (2001).

de Rooij DG, Grootegoed JA: Spermatogonial stem cells. Curr Opin Cell Biol 10:694–701 (1998).

de Rooij DG, van Dissel-Emiliani FMF: Regulation of proliferation and differentiation of stem cells in the male germ line, in Potten CS (ed): Stem Cells, pp 283–313 (Academic Press, London 1997).

de Rooij DG, Van Dissel-Emiliani FM, Van Pelt AM: Regulation of spermatogonial proliferation. Ann NY Acad Sci 564:140–153 (1989).

de Rooij DG, van Pelt AMM, Van de Kant HJG, van der Saag PT, Peters AHFM, Heyting C, de Boer P: Role of retinoids in spermatogonial proliferation and differentiation and the meiotic prophase, in Bartke A (ed): Function of Somatic Cells in the Testis, p 345 (Springer-Verlag, Berlin 1994).

Deltour L, Haselbeck RJ, Ang HL, Duester G: Localization of class I and class IV alcohol dehydrogenases in mouse testis and epididymis: potential retinol dehydrogenases for endogenous retinoic acid synthesis. Biol Reprod 56:102–109 (1997).

Deltour L, Foglio MH, Duester G: Impaired retinol utilization in Adh4 alcohol dehydrogenase mutant mice. Dev Genet 25:1–10 (1999a).

Deltour L, Foglio MH, Duester G: Metabolic deficiencies in alcohol dehydrogenase Adh1, Adh3, and Adh4 null mutant mice. Overlapping roles of Adh1 and Adh4 in ethanol clearance and metabolism of retinol to retinoic acid. J Biol Chem 274:16796–16801 (1999b).

Dong D, Ruuska SE, Levinthal DJ, Noy N: Distinct roles for cellular retinoic acid-binding proteins I and II in regulating signaling by retinoic acid. J Biol Chem 274:23695–23698 (1999).

Dowling JE, Wald G: The biological function of vitamin A acid. Proc Natl Acad Sci USA 46:487 (1960).

Duester G: Families of retinoid dehydrogenases regulating vitamin A function: production of visual pigment and retinoic acid. Eur J Biochem 267:4315–4324 (2000).

Duester G: Genetic dissection of retinoid dehydrogenases. Chem Biol Interact 130–132:469–480
(2001).

Duester G, Mic FA, Molotkov A: Cytosolic retinoid
dehydrogenases govern ubiquitous metabolism of
retinol to retinaldehyde followed by tissue-specific
metabolism to retinoic acid. Chem Biol Interact
143–144:201–210 (2003).

Dufour JM, Kim KH: Cellular and subcellular localization of six retinoid receptors in rat testis during
postnatal development: identification of potential
heterodimeric receptors. Biol Reprod 61:1300–
1308 (1999).

Dym M, Jia MC, Dirami G, Price JM, Rabin SJ, Moc-
chetti I, Ravindranath N: Expression of c-kit receptor and its autophosphorylation in immature rat
type A spermatogonia. Biol Reprod 52:8–19 (1995).

E X, Zhang L, Lu J, Tso P, Blaner WS, Levin MS, Li E:
Increased neonatal mortality in mice lacking cellular retinol-binding protein II. J Biol Chem
277:36617–36623 (2002).

Episkopou V, Maeda S, Nishiguchi S, Shimada K, Gai-
tanaris GA, Gottesman ME, Robertson EJ: Disruption of the transthyretin gene results in mice
with depressed levels of plasma retinol and thyroid
hormone. Proc Natl Acad Sci USA 90:2375–2379
(1993).

Eskild W, Hansson V: Vitamin A functions in the
reproductive organs, in Blomhoff R (ed): Vitamin
A in Health and Disease, pp 531–559 (Dekker,
New York 1994).

Eskild W, Ree AH, Levy FO, Jahnsen T, Hansson V:
Cellular localization of mRNAs for retinoic acid
receptor-alpha, cellular retinol-binding protein,
and cellular retinoic acid-binding protein in rat testis: evidence for germ cell-specific mRNAs. Biol
Reprod 44:53–61 (1991).

Fan X, Molotkov A, Manabe S, Donmoyer CM, Deltour L, Foglio MH, Cuenca AE, Blaner WS, Lipton
SA, Duester G: Targeted disruption of Aldh1a1
(Raldh1) provides evidence for a complex mechanism of retinoic acid synthesis in the developing
retina. Mol Cell Biol 23:4637–4648 (2003).

Fawcett DW: Ultrastructure and function of the Sertoli
cell, in Hamilton DW, Greep RO (eds): Handbook
of Physiology Sec 7, Endocriology, Male Reproductive System, pp 21–55 (American Physiological
Society, Bethesda MD 1975).

Fero ML, Rivkin M, Tasch M, Porter P, Carow CE,
Firpo E, Polyak K, Tsai LH, Broudy V, Perlmutter
RM, Kaushansky K, Roberts JM: A syndrome of
multiorgan hyperplasia with features of gigantism,
tumorigenesis, and female sterility in p27(Kip1)-
deficient mice. Cell 85:733–744 (1996).

Folli C, Calderone V, Ottonello S, Bolchi A, Zanotti G,
Stoppini M, Berni R: Identification, retinoid binding, and x-ray analysis of a human retinol-binding
protein. Proc Natl Acad Sci USA 98:3710–3715
(2001).

Folli C, Calderone V, Ramazzina I, Zanotti G, Berni R:
Ligand binding and structural analysis of a human
putative cellular retinol-binding protein. J Biol
Chem 277:41970–41977 (2002).

Franca LR, Ogawa T, Avarbock MR, Brinster RL, Russell LD: Germ cell genotype controls cell cycle during spermatogenesis in the rat. Biol Reprod 59:
1371–1377 (1998).

Gaemers IC, van Pelt AM, van der Saag PT, Hoogerbrugge JW, Themmen AP, de Rooij DG: Differential expression pattern of retinoid X receptors in
adult murine testicular cells implies varying roles
for these receptors in spermatogenesis. Biol Reprod 58:1351–1356 (1998).

Ghyselinck NB, Bavik C, Sapin V, Mark M, Bonnier D,
Hindelang C, Dierich A, Nilsson CB, Hakansson
H, Sauvant P, Azais-Braesco V, Frasson M, Picaud
S, Chambon P: Cellular retinol-binding protein I is
essential for vitamin A homeostasis. EMBO J 18:
4903–4914 (1999).

Gorry P, Lufkin T, Dierich A, Rochette-Egly C, Decimo D, Dolle P, Mark M, Durand B, Chambon P:
The cellular retinoic acid binding protein I is dispensable. Proc Natl Acad Sci USA 91:9032–9036
(1994).

Gottesman ME, Quadro L, Blaner WS: Studies of vitamin A metabolism in mouse model systems. Bioessays 23:409–419 (2001).

Griswold MD, Bishop PD, Kim KH, Ping R, Siiteri JE,
Morales C: Function of vitamin A in normal and
synchronized seminiferous tubules. Ann NY Acad
Sci 564:154–172 (1989).

Gudas LJ, Sporn MB, Roberts AB: Cellular biology and
biochemistry of the retinoids, in Sporn MB, Roberts AB, Goodman DS (eds): The Retinoid: Biology, Chemistry, and Medicine, 2nd ed, pp 443–520
(Raven Press, New York 1994).

Harvat B, Jetten AM: Growth control by retinoids: regulation of cell cycle progression and apoptosis. In
Nau H, Blaner WS (eds): Retinoids: The Biochemical and Molecular Basis of Vitamin A and Retinoid
Action, pp 239–276 (Springer-Verlag, Berlin
1999).

Hikim AP, Wang C, Leung A, Swerdloff RS: Involvement of apoptosis in the induction of germ cell
degeneration in adult rats after gonadotropin-
releasing hormone antagonist treatment. Endocrinology 136:2770–2775 (1995).

Howell JM, Thompson JN, Pitt GAJ: Histology of the
lesions produced in the reproductive tract of animals fed a diet deficient in vitamin A alcohol but
containing vitamin A acid, I. The male rat. J
Reprod Fertil 5:159–167 (1963).

Hsu SL, Hsu JW, Liu MC, Chen LY, Chang CD: Retinoic acid-mediated G1 arrest is associated with
induction of p27(Kip1) and inhibition of cyclin-
dependent kinase 3 in human lung squamous carcinoma CH27 cells. Exp Cell Res 258:322–331
(2000).

Huang HF, Hembree WC: Spermatogenic response to
vitamin A in vitamin A deficient rats. Biol Reprod
21:891–904 (1979).

Huang HF, Marshall GR: Failure of spermatid release
under various vitamin A states – an indication of
delayed spermiation. Biol Reprod 28:1163–1172
(1983).

Huang HF, Yang CS, Meyenhofer M, Gould S, Bocca-
bella AV: Disruption of sustentacular (Sertoli) cell
tight junctions and regression of spermatogenesis
in vitamin-A-deficient rats. Acta Anat 133:10–15
(1988).

Huang HF, Li MT, Pogach LM, Qian L: Messenger
ribonucleic acid of rat testicular retinoic acid receptors: developmental pattern, cellular distribution, and testosterone effect. Biol Reprod 51:541–
550 (1994).

Ismail N, Morales CR: Effects of vitamin A deficiency
on the inter-Sertoli cell tight junctions and on the
germ cell population. Microsc Res Tech 20:43–49
(1992).

Ismail N, Morales C, Clermont Y: Role of spermatogonia in the stage-synchronization of the seminiferous epithelium in vitamin-A-deficient rats. Am J
Anat 188:57–63 (1990).

Kang MJ, Kim MK, Terhune A, Park JK, Kim YH,
Koh GY: Cytoplasmic localization of cyclin D3 in
seminiferous tubules during testicular development. Exp Cell Res 234:27–36 (1997).

Kastner P, Grondona JM, Mark M, Gansmuller A,
LeMeur M, Decimo D, Vonesch JL, Dolle P,
Chambon P: Genetic analysis of RXR alpha developmental function: convergence of RXR and RAR
signaling pathways in heart and eye morphogenesis. Cell 78:987–1003 (1994).

Kastner P, Mark M, Chambon P: Nonsteroid nuclear
receptors: what are genetic studies telling us about
their role in real life? Cell 83:859–869 (1995).

Kastner P, Mark M, Leid M, Gansmuller A, Chin W,
Grondona JM, Decimo D, Krezel W, Dierich A,
Chambon P: Abnormal spermatogenesis in RXR
beta mutant mice. Genes Dev 10:80–92 (1996).

Kim KH, Akmal KM: Role of vitamin A in male germ-
cell development, in Desjardins C (ed): Cellular
and Molecular Regulation of Testicular Cells, pp
83–98 (Springer-Verlag, Berlin 1996).

Kim KH, Griswold MD: The regulation of retinoic acid
receptor mRNA levels during spermatogenesis.
Mol Endocrinol 4:1679–1688 (1990).

Kim KH, Wang ZQ: Action of vitamin A on testis: role
of the Sertoli cell, in Griswold MD, Rusell LD
(eds): The Sertoli Cell, pp 514–535 (Cache River
Press, Clearwater 1993).

Kiyokawa H, Kineman RD, Manova-Todorova KO,
Soares VC, Hoffman ES, Ono M, Khanam D, Hay-
day AC, Frohman LA, Koff A: Enhanced growth of
mice lacking the cyclin-dependent kinase inhibitor
function of p27(Kip1). Cell 85:721–732 (1996).

Krezel W, Dupe V, Mark M, Dierich A, Kastner P,
Chambon P: RXR gamma null mice are apparently
normal and compound RXR alpha +/–/RXR beta
–/–/RXR gamma –/– mutant mice are viable. Proc
Natl Acad Sci USA 93:9010–9014 (1996).

Kubota H, Chiba H, Takakuwa Y, Osanai M, Tobioka
H, Kohama G, Mori M, Sawada N: Retinoid X
receptor alpha and retinoic acid receptor gamma
mediate expression of genes encoding tight-junc-
tion proteins and barrier function in F9 cells during visceral endodermal differentiation. Exp Cell
Res 263:163–172 (2001).

Kurlandsky SB, Gamble MV, Ramakrishnan R, Blaner
WS: Plasma delivery of retinoic acid to tissues in
the rat. J Biol Chem 270:17850–17857 (1995).

Lampron C, Rochette-Egly C, Gorry P, Dolle P, Mark
M, Lufkin T, LeMeur M, Chambon P: Mice deficient in cellular retinoic acid binding protein II
(CRABPII) or in both CRABPI and CRABPII are
essentially normal. Development 121:539–548
(1995).

Langenfeld J, Kiyokawa H, Sekula D, Boyle J, Dmi-
trovsky E: Posttranslational regulation of cyclin D1
by retinoic acid: a chemoprevention mechanism.
Proc Natl Acad Sci USA 94:12070–12074 (1997).

Leid M, Kastner P, Chambon P: Multiplicity generates
diversity in the retinoic acid signalling pathways.
Trends Biochem Sci 17:427–433 (1992).

Li E, Tso P: Vitamin A uptake from foods. Curr Opin
Lipidol 14:241–247 (2003).

Lin M, Zhang M, Abraham M, Smith SM, Napoli JL:
Mouse retinal dehydrogenase 4 (RALDH4), molecular cloning, cellular expression, and activity in 9-
cis-retinoic acid biosynthesis in intact cells. J Biol
Chem 278:9856–9861 (2003).

Liu M, Iavarone A, Freedman LP: Transcriptional activation of the human p21(WAF1/CIP1) gene by
retinoic acid receptor. Correlation with retinoid
induction of U937 cell differentiation. J Biol Chem
271:31723–31728 (1996).

Livera G, Rouiller-Fabre V, Durand P, Habert R: Multiple effects of retinoids on the development of Sertoli, germ, and Leydig cells of fetal and neonatal
rat testis in culture. Biol Reprod 62:1303–1314
(2000).

Livera G, Rouiller-Fabre V, Pairault C, Levacher C,
Habert R: Regulation and perturbation of testicular functions by vitamin A. Reproduction 124:
173–180 (2002).

Lohnes D, Kastner P, Dierich A, Mark M, LeMeur M,
Chambon P: Function of retinoic acid receptor
gamma in the mouse. Cell 73:643–658 (1993).

Lohnes D, Mark M, Mendelsohn C, Dolle P, Dierich A,
Gorry P, Gansmuller A, Chambon P: Function of
the retinoic acid receptors (RARs) during development (I). Craniofacial and skeletal abnormalities in
RAR double mutants. Development 120:2723–
2748 (1994).

Lopez-Fernandez LA, del Mazo J: The cytosolic aldehyde dehydrogenase gene *(Aldh1)* is developmentally expressed in Leydig cells. FEBS Lett 407:225–229 (1997).

Lufkin T, Lohnes D, Mark M, Dierich A, Gorry P, Gaub MP, LeMeur M, Chambon P: High postnatal lethality and testis degeneration in retinoic acid receptor alpha mutant mice. Proc Natl Acad Sci USA 90:7225–7229 (1993).

Luo J, Pasceri P, Conlon RA, Rossant J, Giguere V: Mice lacking all isoforms of retinoic acid receptor beta develop normally and are susceptible to the teratogenic effects of retinoic acid. Mech Dev 53:61–71 (1995).

Malkov M, Fisher Y, Don J: Developmental schedule of the postnatal rat testis determined by flow cytometry. Biol Reprod 59:84–92 (1998).

Mangelsdorf DJ, Evans RM: The RXR heterodimers and orphan receptors. Cell 83:841–850 (1995).

Mangelsdorf DJ, Umesono K, Evans RM: The Retinoid Receptors, in Sporn MB, Roberts AB, Goodman DS (eds): The Retinoid, Biology, Chemistry, and Medicine, 2nd ed, pp 319–349 (Raven Press, New York 1994).

Mason KE: Differences in testes injury and repair after vitamin A deficiency, Vitamin E deficiency and inanition. Am J Anat 52:153–239 (1933).

Matsumiya K, Meistrich ML, Shetty G, Dohmae K, Tohda A, Okuyama A, Nishimune Y: Stimulation of spermatogonial differentiation in juvenile spermatogonial depletion (jsd) mutant mice by gonadotropin-releasing hormone antagonist treatment. Endocrinology 140:4912–4915 (1999).

McCarthy PT, Cerecedo LR: Vitamin A deficiency in the mouse. J Nutr 46:361–376 (1952).

Mehta PP, Bertram JS, Loewenstein WR: Growth inhibition of transformed cells correlates with their junctional communication with normal cells. Cell 44:187–196 (1986).

Meng X, Lindahl M, Hyvonen ME, Parvinen M, de Rooij DG, Hess MW, Raatikainen-Ahokas A, Sainio K, Rauvala H, Lakso M, Pichel JG, Westphal H, Saarma M, Sariola H: Regulation of cell fate decision of undifferentiated spermatogonia by GDNF. Science 287:1489–1493 (2000).

Miller ME, Cross FR: Cyclin specificity: how many wheels do you need on a unicycle? J Cell Sci 114:1811–1820 (2001).

Mitranond V, Sobhon P, Tosukhowong P, Chindaduangrat W: Cytological changes in the testes of vitamin-A-deficient rats. I. Quantitation of germinal cells in the seminiferous tubules. Acta Anat 103:159–168 (1979).

Mizoguchi S, Kim KH: Expression of cdc25 phosphatases in the germ cells of the rat testis. Biol Reprod 56:1474–1481 (1997).

Molotkov A, Deltour L, Foglio MH, Cuenca AE, Duester G: Distinct retinoid metabolic functions for alcohol dehydrogenase genes *Adh1* and *Adh4* in protection against vitamin A toxicity or deficiency revealed in double null mutant mice. J Biol Chem 277:13804–13811 (2002).

Monaco HL: The transthyretin-retinol-binding protein complex. Biochim Biophys Acta 1482:65–72 (2000).

Morales A, Cavicchia JC: Spermatogenesis and blood-testis barrier in rats after long-term Vitamin A deprivation. Tissue Cell 34:349–355 (2002).

Morales C, Griswold MD: Retinol-induced stage synchronization in seminiferous tubules of the rat. Endocrinology 121:432–434 (1987).

Morales CR, Griswold MD: Variations in the level of transferrin and SGP-2 mRNAs in Sertoli cells of vitamin A-deficient rats. Cell Tissue Res 263:125–130 (1991).

Nagpal S, Chandraratna RA: Vitamin A and regulation of gene expression. Curr Opin Clin Nutr Metab Care 1:341–346 (1998).

Nakayama K, Ishida N, Shirane M, Inomata A, Inoue T, Shishido N, Horii I, Loh DY: Mice lacking p27(Kip1) display increased body size, multiple organ hyperplasia, retinal dysplasia, and pituitary tumors. Cell 85:707–720 (1996).

Napoli JL, McCormick AM: Tissue dependence of retinoic acid metabolism in vivo. Biochim Biophys Acta 666:165–175 (1981).

Niederreither K, Subbarayan V, Dolle P, Chambon P: Embryonic retinoic acid synthesis is essential for early mouse post-implantation development. Nat Genet 21:444–448 (1999).

Niederreither K, Abu-Abed S, Schuhbaur B, Petkovich M, Chambon P, Dolle P: Genetic evidence that oxidative derivatives of retinoic acid are not involved in retinoid signaling during mouse development. Nat Genet 31:84–88 (2002).

Noy N: Physical-chemical properties and action of retinoids, in Nau H, Blaner WS (eds): Retinoids: The Biochemical and Molecular Basis of Vitamin A and Retinoid Action, pp 3–29 (Springer-Verlag, Berlin 1999).

Nurse PM: Nobel Lecture. Cyclin dependent kinases and cell cycle control. Biosci Rep 22:487–499 (2002).

Oakberg EF: A description of spermiogenesis in the mouse and its use in an analysis of the cycle of the seminiferous epithelium and germ cell renewal. Am J Anat 99:391–414 (1956).

Ohta H, Tohda A, Nishimune Y: Proliferation and differentiation of spermatogonial stem cells in the *W/Wv* mutant mouse testis. Biol Reprod 30:30 (2003).

Ong DE, Newcomer ME, Chytil F: Cellular retinoid-binding proteins, in Sporn MB, Roberts AB, Goodman DS (eds): The Retinoid, Biology, Chemistry, and Medicine, 2nd ed, pp 283–317 (Raven Press, New York 1994).

Orth JM: Proliferation of Sertoli cells in fetal and postnatal rats: a quantitative autoradiographic study. Anat Rec 203:485–492 (1982).

Packer AI, Wolgemuth DJ: Genetic and molecular approaches to understanding the role of retinoids in mammalian spermatogenesis, in Nau H, Blaner WS (eds): Retinoids: The Biochemical and Molecular Basis of Vitamin A and Retinoid Action, pp 347–368 (Springer-Verlag, Berlin 1999).

Pestell RG, Albanese C, Reutens AT, Segall JE, Lee RJ, Arnold A: The cyclins and cyclin-dependent kinase inhibitors in hormonal regulation of proliferation and differentiation. Endocr Rev 20:501–534 (1999).

Piedrafita FJ, Pfahl M: Nuclear retinoid receptors and mechanisms of action, in Nau H, Blaner WS (eds): Retinoids: The Biochemical and Molecular Basis of Vitamin A and Retinoid Action, pp 153–184 (Springer-Verlag, Berlin 1999).

Pitts JD, Hamilton AE, Kam E, Burk RR, Murphy JP: Retinoic acid inhibits junctional communication between animal cells. Carcinogenesis 7:1003–1010 (1986).

Quadro L, Blaner WS, Salchow DJ, Vogel S, Piantedosi R, Gouras P, Freeman S, Cosma MP, Colantuoni V, Gottesman ME: Impaired retinal function and vitamin A availability in mice lacking retinol-binding protein. EMBO J 18:4633–4644 (1999).

Rajan N, Kidd GL, Talmage DA, Blaner WS, Suhara A, Goodman DS: Cellular retinoic acid-binding protein messenger RNA: levels in rat tissues and localization in rat testis. J Lipid Res 32:1195–1204 (1991).

Rajguru SU, Kang YH, Ahluwalia BS: Localization of retinol (vitamin A) in rat testes. J Nutr 112:1881–1891 (1982).

Reed SI: Cell cycling? Check your brakes. Nat Cell Biol 4:E199–201 (2002).

Roberts JM: Evolving ideas about cyclins. Cell 98:129–132 (1999).

Rogers MB: Life-and-death decisions influenced by retinoids. Curr Top Dev Biol 35:1–46 (1997).

Ross AC: Retinoid production and catabolism: role of diet in regulating retinol esterification and retinoic acid oxidation. J Nutr 133:291S–296S (2003).

Russell LD, Ettlin RA, SinhaHikim AP, Clegg ED: Histological and histopathological evaluation of the testis (Cache River Press, Clearwater 1990).

Sakai Y, Meno C, Fujii H, Nishino J, Shiratori H, Saijoh Y, Rossant J, Hamada H: The retinoic acid-inactivating enzyme CYP26 is essential for establishing an uneven distribution of retinoic acid along the anterio-posterior axis within the mouse embryo. Genes Dev 15:213–225 (2001).

Satoh H, Zhong Y, Isomura H, Saitoh M, Enomoto K, Sawada N, Mori M: Localization of 7H6 tight junction-associated antigen along the cell border of vascular endothelial cells correlates with paracellular barrier function against ions, large molecules, and cancer cells. Exp Cell Res 222:269–274 (1996).

Schrans-Stassen BH, Saunders PT, Cooke HJ, de Rooij DG: Nature of the spermatogenic arrest in *Dazl–/–* mice. Biol Reprod 65:771–776 (2001).

Sherr CJ, Roberts JM: CDK inhibitors: positive and negative regulators of G1-phase progression. Genes Dev 13:1501–1512 (1999).

Shetty J, Marathe GK, Dighe RR: Specific immunoneutralization of FSH leads to apoptotic cell death of the pachytene spermatocytes and spermatogonial cells in the rat. Endocrinology 137:2179–2182 (1996).

Smith SM, Levy NS, Hayes CE: Impaired immunity in vitamin A-deficient mice. J Nutr 117:857–865 (1987).

Sobhon P, Mitranond V, Tosukhowong P, Chindaduangrat W: Cytological changes in the testes of vitamin-A-deficient rats. II. Ultrastructural study of the seminiferous tubules. Acta Anat 103:169–183 (1979).

Sueoka N, Lee HY, Walsh GL, Hong WK, Kurie JM: Posttranslational mechanisms contribute to the suppression of specific cyclin:CDK complexes by all-trans retinoic acid in human bronchial epithelial cells. Cancer Res 59:3838–3844 (1999).

Suzuki R, Shintani T, Sakuta H, Kato A, Ohkawara T, Osumi N, Noda M: Identification of RALDH-3, a novel retinaldehyde dehydrogenase, expressed in the ventral region of the retina. Mech Dev 98:37–50 (2000).

Tahayato A, Dolle P, Petkovich M: *Cyp26C1* encodes a novel retinoic acid-metabolizing enzyme expressed in the hindbrain, inner ear, first branchial arch and tooth buds during murine development. Gene Expr Patterns 3:449–454 (2003).

Taimi M, Helvig C, Wisniewski J, Ramshaw H, White J, Amad M, Korczak B, Petkovich M: A novel human cytochrome P450, CYP26C1 involved in metabolism of 9-cis and all-trans, isomers of retinoic acid. J Biol Chem 7:77–85 (2003).

Takase S, Ong DE, Chytil F: Transfer of retinoic acid from its complex with cellular retinoic acid-binding protein to the nucleus. Arch Biochem Biophys 247:328–334 (1986).

Thang SH, Kobayashi M, Matsuoka I: Regulation of glial cell line-derived neurotrophic factor responsiveness in developing rat sympathetic neurons by retinoic acid and bone morphogenetic protein-2. J Neurosci 20:2917–2925 (2000).

Thompson JN, Howell JM, Pitt GA: Vitamin A and reproduction in rats. Proc R Soc Lond B Biol Sci 159:510–535 (1964).

Tohda A, Matsumiya K, Tadokoro Y, Yomogida K, Miyagawa Y, Dohmae K, Okuyama A, Nishimune Y: Testosterone suppresses spermatogenesis in juvenile spermatogonial depletion (jsd) mice. Biol Reprod 65:532–537 (2001).

Tsutsui T, Hesabi B, Moons DS, Pandolfi PP, Hansel KS, Koff A, Kiyokawa H: Targeted disruption of CDK4 delays cell cycle entry with enhanced p27(Kip1) activity. Mol Cell Biol 19:7011–7019 (1999).

Van Pelt AM, de Rooij DG: The origin of the synchronization of the seminiferous epithelium in vitamin A-deficient rats after vitamin A replacement. Biol Reprod 42:677–682 (1990a).

van Pelt AM, de Rooij DG: Synchronization of the seminiferous epithelium after vitamin A replacement in vitamin A-deficient mice. Biol Reprod 43:363–367 (1990b).

van Pelt AM, de Rooij DG: Retinoic acid is able to reinitiate spermatogenesis in vitamin A-deficient rats and high replicate doses support the full development of spermatogenic cells. Endocrinology 128:697–704 (1991).

van Pelt AM, van den Brink CE, de Rooij DG, van der Saag PT: Changes in retinoic acid receptor messenger ribonucleic acid levels in the vitamin A-deficient rat testis after administration of retinoids. Endocrinology 131:344–350 (1992).

Vogel S, Gamble MV, Blaner WS: Biosynthesis, absorption, metabolism and transport of retinoids, in Nau H Blaner WS (eds): Retinoids: The Biochemical and Molecular Basis of Vitamin A and Retinoid Action, pp 31–95 (Springer-Verlag, Berlin 1999).

Walder L, Lutzelschwab R: Effects of 12-O-tetradecanoylphorbol-13-acetate (TPA), retinoic acid and diazepam on intercellular communication in a monolayer of rat liver epithelial cells. Exp Cell Res 152:66–76 (1984).

Walker WH: Molecular mechanisms controlling Sertoli cell proliferation and differentiation. Endocrinology 144:3719–3721 (2003).

Weil M, Jacobson MD, Raff MC: Are caspases involved in the death of cells with a transcriptionally inactive nucleus? Sperm and chicken erythrocytes. J Cell Sci 111:2707–2715 (1998).

Weiler R, He S, Vaney DI: Retinoic acid modulates gap junctional permeability between horizontal cells of the mammalian retina. Eur J Neurosci 11:3346–3350 (1999).

White JA, Beckett-Jones B, Guo YD, Dilworth FJ, Bonasoro J, Jones G, Petkovich M: cDNA cloning of human retinoic acid-metabolizing enzyme (hP450RAI) identifies a novel family of cytochromes P450. J Biol Chem 272:18538–18541 (1997).

Wing TY, Christensen AK: Morphometric studies on rat seminiferous tubules. Am J Anat 165:13–25 (1982).

Wolbach SB, Howe PR: Tissue changes following deprivation of fat-soluble A vitamin. J Exp Med 42:753–777 (1925).

Wolgemuth DJ, Watrin F: List of cloned mouse genes with unique expression patterns during spermatogenesis. Mamm Genome 1:283–288 (1991).

Wolgemuth DJ, Rhee K, Wu S, Ravnik SE: Genetic control of mitosis, meiosis and cellular differentiation during mammalian spermatogenesis. Reprod Fertil Dev 7:669–683 (1995).

Wolgemuth DJ, Laurion E, Lele KM: Regulation of the mitotic and meiotic cell cycles in the male germ line. Recent Prog Horm Res 57:75–101 (2002).

Yang ZW, Wreford NG, de Kretser DM: A quantitative study of spermatogenesis in the developing rat testis. Biol Reprod 43:629–635 (1990).

Zgombic-Knight M, Ang HL, Foglio MH, Duester G: Cloning of the mouse class IV alcohol dehydrogenase (retinol dehydrogenase) cDNA and tissue-specific expression patterns of the murine ADH gene family. J Biol Chem 270:10868–10877 (1995).

Zhai Y, Sperkova Z, Napoli JL: Cellular expression of retinal dehydrogenase types 1 and 2: effects of vitamin A status on testis mRNA. J Cell Physiol 186:220–232 (2001).

Zhang Q, Wang X, Wolgemuth DJ: Developmentally regulated expression of cyclin D3 and its potential in vivo interacting proteins during murine gametogenesis. Endocrinology 140:2790–2800 (1999).

Zheng WL, Bucco RA, Schmitt MC, Wardlaw SA, Ong DE: Localization of cellular retinoic acid-binding protein (CRABP) II and CRABP in developing rat testis. Endocrinology 137:5028–5035 (1996).

Zhong Y, Enomoto K, Tobioka H, Konishi Y, Satoh M, Mori M: Sequential decrease in tight junctions as revealed by 7H6 tight junction-associated protein during rat hepatocarcinogenesis. Jpn J Cancer Res 85:351–356 (1994).

Zhou Q, Stetler-Stevenson M, Steeg PS: Inhibition of cyclin D expression in human breast carcinoma cells by retinoids in vitro. Oncogene 15:107–115 (1997).

Cytogenet Genome Res 105:203–214 (2004)
DOI: 10.1159/000078190

Male germline-specific histones in mouse and man

D. Churikov, I.A. Zalenskaya, and A.O. Zalensky

The Jones Institute for Reproductive Medicine, Eastern Virginia Medical School, Norfolk, VA (USA)

Abstract. In mice and humans, the production of male gametes is a result of a complex multistep process of stem cell differentiation. The final product, the mature spermatozoon, is designed for the safe delivery of a haploid copy of the paternal genetic information to the oocyte in a structural state suitable for zygote formation and embryogenesis. A remarkable structural reorganization of chromosomes in germline cells during mammalian spermatogenesis has been characterized. The most important steps are connected with the recombination events during meiosis and the final packaging of the haploid genome in the genetically inert, compacted nucleus of the sperm. Underlying the changes in chromatin organization is the appearance of testis-specific histones. Although the existence of such histones has been known for decades, their exact functions still are not established. Deciphering of the mouse and human genomes has allowed a more detailed description of the organization and regulation of the testis-specific histone genes. In addition, it has facilitated the discovery of previously unknown proteins. This review summarizes contemporary information on these germline-specific/enriched histones in both the mouse and human and outlines early achievements in the identification of their functions.

Copyright © 2004 S. Karger AG, Basel

Histones are basic proteins responsible for the nucleosomal organization of eukaryotic DNA. Two molecules of each histone protein, H2A, H2B, H3, and H4 (core histones), form an octamer that interacts with ~ 146 base pairs of DNA to make up a nucleosome core (McGhee and Felsenfeld, 1980). The histones of the H1 class (linker histones) seal two turns of nucleosomal DNA at the surface of the nucleosome core and interact with the internucleosomal linker DNA. Each class of the histone proteins, except for H4, includes several subtypes that differ to variable extents in their amino acid sequences and are encoded by different genes. Albig and Doenecke (1997) proposed to classify histone genes with respect to their expression,

into three groups: (1) replication-dependent genes transcribed during the S phase of the cell cycle which produce the bulk of the somatic histones; (2) replication-independent genes transcribed at a low constant level throughout the cell cycle and in nondividing cells which supply the so-called "replacement histones", and (3) genes transcribed only in certain types of the differentiated tissues such as the testes. The tissue-specific genes can be expressed either in a replication-dependent or -independent mode. The male germline-specific/enriched histones from human and mouse, which belong to the latter two groups, will be in the focus of this review.

General overview

The compacted chromatin of the haploid genome within a mammalian sperm head has a highly specific organization that is fundamentally different from the nucleosomal architecture characteristic of somatic cells (Balhorn, 1982). DNA compaction is achieved largely by protamines (small highly basic proteins), which become the major protein component at the late stages of spermatogenesis. The replacement of histones by prot-

This work has been supported by NIH grant HD39830 to A.O.Z.

Received 17 November 2003; manuscript accepted 10 December 2003.

Request reprints from Andrei O. Zalensky
  The Jones Institute for Reproductive Medicine
  Eastern Virginia Medical School, Norfolk, VA 23507 (USA)
  telephone: 757-446-5002; fax: 757-446-8998
  e-mail: zalensao@evms.edu

amines is not direct and occurs in several steps. The histones, most of which are either testis-specific or testis-enriched variants, are first replaced by transitional proteins. These proteins are replaced later by protamines (reviewed in Wouters-Tyrou et al., 1998; Dadoune, 2003; Lewis et al., 2003a, 2003b; Meistrich et al., 2003). The testis-specific histones are a diverse group of proteins associated with different stages of differentiation from spermatogonia to spermatozoa. They are involved not only in a general reorganization and in remodeling of chromatin structure but also may have some novel functions specific for spermatogenesis. Phosphorylation of Ser139 in histone H2AX is associated with the XY body formation in spermatocytes (Fernandez-Capetillo et al., 2003b), and the human sperm histone H2B-like protein involved in telomere organization (Gineitis et al., 2000) are examples of the non-canonical functions for histones in the germline cells. While the histone variants present in the germ cells of testis and the time course of their appearance during spermatogenesis have been described in detail for the mouse (Meistrich et al., 1985; Meistrich, 1989), there is much less data for humans. Comparative analysis of testis-specific histone gene organization and their chromosomal locations that has become possible with the completion of human and mouse genome sequencing projects is proving useful in establishing true orthology between testis-specific histones in mouse and human. Studies using mouse transgenic and gene knockout models may lead to deciphering the functions of these testis-specific histones during mouse spermatogenesis and allow us to extrapolate this knowledge to their orthologs in humans.

The first step towards understanding the function(s) of the germline-specific histones is the identification of the testis/sperm-specific histone complement. Conventional methods involving the biochemical fractionation of proteins are difficult given the limited tissue amounts and complex cellular composition of the testis. However, new genome mining approaches, such as in silico subtraction and digital differential display, have been successful in identifying a number of germ cell-specific genes (Yan et al., 2002a, b) including the gene for the histone H1-like protein (Yan et al., 2003) and two human H2B-like proteins (Zalensky et al., 2002; Churikov et al., submitted).

Nucleosomal core histones

The histone (H3-H4)₂ tetramer and the two H2A-H2B heterodimers form the protein component of a nucleosome, which provides the primary level of DNA compaction preserved in all eukaryotes. Variations in each core histone may provide functional modulation in chromatin organization. In murine and human spermatogenic cells, testis-specific or enriched core histones, with the exception of H4, appear in chromatin at different stages during spermatogenesis (Brock et al., 1980; Meistrich et al., 1985; Doenecke et al., 1997b).

Testis-specific TH2B
The first testis-specific core histone, corresponding to H2B, was discovered in rat (Shires et al., 1975). It was observed that rat testis H2B (rTH2B) formed dimers after oxidation and, therefore, the presence of cysteine, a very unusual residue for somatic histones, was predicted (Shires et al., 1976). Later, Kim et al. (1987) cloned and compared cDNAs of rTH2B and its somatic counterpart from rat testes. Analysis of the deduced testis and somatic proteins (126 and 124 residues, respectively) revealed extensive divergence between the two H2B variants in the N-termini, with the most significant difference located in a stretch of nine amino acids starting at the third position (Kim et al., 1987). Sequencing confirmed the presence of a single cysteine residue at the 33rd position in the testis protein. In situ hybridization with probes corresponding to the divergent N-terminal regions of the two proteins revealed a reciprocal expression pattern, i.e., somatic H2B is transcribed in proliferating spermatogonia and preleptotene spermatocytes, whereas the rTH2B gene is transcribed exclusively in rat pachytene spermatocytes, where no significant DNA replication is observed (Kim et al., 1987). Lim and Chae (1992) demonstrated that the transcription of the rTH2B gene in spermatogonia is inhibited by the presence of a repressor protein; however, the up-regulation of the rTH2B gene in meiotic spermatocytes remains poorly understood.

Analysis of the histones from fractionated spermatogenic cells showed that as much as 90% of the H2B is replaced by TH2B by the mid pachytene stage (Meistrich et al., 1985). The extensive sequence changes in the TH2B N-terminal region, which is known to interact with DNA, could be responsible for the increased DNase I sensitivity of the mononucleosomes from rat pachytene spermatocytes compared to liver mononucleosomes (Rao and Rao, 1987). A loosened nucleosome structure might be advantageous for meiotic recombination and may facilitate replacement of histones with more basic transitional proteins and protamines later during spermatogenesis.

Choi and Chae (1991) showed that all the CpG sites in the promoter region of rTH2B gene are methylated in somatic tissues but not in testis, suggesting that methylation may play a casual role in the repression of rTH2B gene in somatic cells. This finding was exploited to clone the mouse homolog of the rTH2B gene. Thus, Choi et al. (1996) found that among 10–20 copies of mouse H2B histone genes, three copies are methylated in somatic tissues, but not in testis. One of these methylated H2B genes, mTH2B, showed a strikingly similar sequence to the TH2B gene of rat, and contained a cysteine residue at the 33rd position. During spermatogenesis, the mTH2B gene, similar to rat TH2B, is expressed predominantly in pachytene spermatocytes (Choi et al., 1996). mTH2B gene maps to mouse chromosome 13 within a major histone gene cluster. It corresponds to the independently identified mouse replication-dependent gene *Hist1h2ba* by Marzluff et al. (2002).

A monoclonal antibody raised against rat tyrosine hydrolase (TyH) was shown to cross-react with TH2B, but not with somatic H2B (Unni et al., 1995a). This cross-reactivity was due to a common epitope (a stretch of eight amino acids) present in TyH and the N-terminus of the murine TH2B (Unni et al., 1995a). Later, van Roijen et al. (1998) showed that the anti-TyH antibody recognized a 17-kDa protein in human testis and spermatozoa, which they suggested, is histone TH2B.

More recently, Zalensky et al. (2002) demonstrated that several human sperm proteins separated by two-dimensional AUT/SDS PAGE reacted with antibodies to histone H2B; two of these H2B-related polypeptides cross-react with antibodies to TyH. Using the sequence of the anti-TyH-reactive peptide as bait in BLAST searches of the human genome databases, Zalensky et al. (2002) identified a unique gene on human chromosome 6. Testis-specific expression of this gene was confirmed by RT-PCR. Analysis of the deduced amino acid sequence of the human testis/sperm-specific H2B (hTSH2B) (Zalensky et al., 2002) revealed high sequence identities with both rTH2B (93% identical) and mTH2B (95% identical). Noteworthy, the human protein lacks cysteine at the 33rd position, which is characteristic of both the rat and mouse proteins.

Human TSH2B gene, localized on 6p22.1, is intronless and encodes an mRNA that is not polyadenylated (Zalensky et al., 2002). The same gene was independently identified as a replication-dependent gene in the human histone 1 cluster and termed HIST1H2BA by Marzluff et al. (2002). Human TSH2B and mouse TH2B genes have been identified as true orthologs according to their position in the mouse and human histone gene clusters (Churikov, unpublished). It remains uncertain whether their protein products are functional homologs since the patterns of expression during spermatogenesis differ remarkably between human and mouse as discussed below.

In mouse, TH2B is synthesized and deposited onto the chromatin beginning from preleptotene stage with maximal levels of synthesis occurring in early primary spermatocytes (Meistrich et al., 1985; Unni et al., 1995b). Immunolocalization of TH2B in mouse and rat testicular sections revealed increased accessibility of the N-terminus of TH2B in elongating spermatids while the amount of TH2B detected by immunoblotting decreased (Unni et al., 1995a). This finding indicated that in elongating spermatids the N-terminus of TH2B may be less tightly bound to DNA, possibly in preparation for its removal from the chromatin. At step 12, staining became restricted to the basal area of nuclei and TH2B was not detected in murine sperm nuclei. However, a different pattern of TH2B localization was observed in human spermatogenesis. TH2B was detected throughout spermatogenesis, from spermatogonia to mature sperm (van Roijen et al., 1998; Zalensky et al., 2002; Zalensky et al., in preparation). Interestingly, hTH2B was found only in 30% of mature human sperm (van Roijen et al., 1998; Zalensky et al., 2002), and in these cells protein was localized in foci at the basal nuclear area (Zalensky et al., 2002).

Murine spermatid-specific H2B

Another H2B variant was found in round spermatids of the mouse (Moss et al., 1989) and rat (Unni et al., 1995b). Moss et al. (1989) isolated a cDNA clone representing one of the polyadenylated spermatid-specific histone transcripts and showed that it encodes an unusual H2B protein (mouse spermatid-specific H2B – mssH2B). MssH2B is very similar in sequence to other H2B variants but has an additional 12 amino acids at the C-terminus. Immunocytochemistry using antibodies specific to this C-terminus conducted on both light and electron microscopic levels showed that ssH2B is distributed uniformly throughout the round spermatid nucleus (Moss and Orth, 1993). Expression of the murine ssH2B is restricted to a narrow time window (Moss et al., 1989; Unni et al., 1995b); in rat, it appears postmeiotically, being maximal in step 7–8 spermatids, and decreasing during steps 9–11.

We identified a gene encoding histone ssH2B on the mouse chromosome 13 (cytogenetic location 13A3.1) in the major histone gene cluster. Interestingly, the mouse ssH2B gene is interrupted by a single intron, which separates exon 2 encoding the entire C-terminal hydrophobic extension from exon 1 encoding the first 127 amino acids.

We were unable to identify an orthologous gene in the human genome either by BLAST searches, or by location in the homologous region on human chromosome 6. Results of the immunoblotting of human testis and sperm proteins using antibodies against mssH2B developed by Moss and Orth (1993) also were negative (Zalensky, unpublished).

Novel H2B-like proteins in human germline cells

In search of an H2B-like protein component of the human sperm telomere-binding complex (Gineitis et al., 2000), we identified a novel cluster of histone H2B-related genes on the long arm of human chromosome Xq22 (Churikov et al., submitted). Analysis of the EST and reference mRNA databases revealed that at least four genes from this cluster are transcribed. All four genes encode divergent histone H2B-like proteins, which share only a moderate homology (50–52% sequence identity) within a conserved H2B domain with the conventional histone H2B. The tissue origin of the EST and cDNA clones corresponded to the novel genes indicating that two genes express in testis. These two H2B-like genes are in a head-to-head configuration and are separated by approximately 26 kb. Both genes consist of three exons with the third exon being noncoding. The deduced protein sequences are substantially longer (175 and 191 amino acid residues) than the conventional H2B protein sequence (126 amino acid residues); hence we designate them hTH2B-175 and hTH2B-191. The larger size of both proteins is due to an extended unique N-terminus. These novel H2B-like proteins share 70% sequence identity, whereas each protein is only 52% identical with the consensus sequence of somatic H2B. We confirmed testis-specific expression of the TH2B-175 gene (accession no. AY283369) by RT-RCR and demonstrated that the transcript contains a poly(A) tract.

Immunofluorescent localization of hTH2B-175 protein on testicular sections using antibodies raised against the unique 11-amino-acid-long peptide at the C-terminus demonstrated that the protein is present in the nuclei of round spermatids (Zalensky et al., in preparation). In these cells, hTH2B-175 is localized at the nuclear periphery, apparently associated with nuclear membrane. Human TH2B-175 had a similar localization pattern in mature spermatozoa (Zalensky et al., in preparation).

A search of the mouse genome does not reveal a gene homologous to the hTH2B-175 gene. PCR analysis of genomic DNA of several mammalian species including mouse and Rhesus monkey showed that this gene is primate-specific.

Preliminary analysis of the TH2B-175 promoter region revealed the absence of the conservative elements typically found in the replication-dependent genes. Instead, 115 bp upstream of the translation start codon is a cAMP-responsive element (CRE), which is capable of binding CRE modulator (CREM), a transcription factor that is present at high levels in testis and plays a critical role in activating the post-meiotic transcriptional cascade (Fimia et al., 2001). The presence of a CREM-binding site in TH2B-175 promoter suggests transcriptional activation of TH2B-175 during spermiogenesis, which is in a good agreement with our detection of the TH2B-175 protein in round spermatids.

Testis-specific H2As

Cloning and sequencing of rat genomic DNA containing the rTH2B gene revealed a new H2A gene located upstream of the rTH2B gene (Huh et al., 1991). It was shown that this new H2A gene, rTH2A, is also transcribed only in testis, similar to rTH2B (Huh et al., 1991). Earlier, Trostle-Weige et al. (1982) identified a testis-specific histone TH2A in rat and showed that it is synthesized primarily in pachytene spermatocytes. However, the amino acid composition of the histone TH2A purified by Trostle-Weige et al. (1982) differs from that deduced from the genomic DNA cloned by Huh et al. (1991), suggesting that these two histones are not the same protein.

Comparison of the deduced amino acid sequence of rTH2A and its somatic counterpart showed that they are much more similar compared to the TH2B/H2B pair. Because it was difficult to prepare gene-specific probes due to the extensive sequence similarity between the TH2A and H2A genes, in situ localization of their mRNAs has not been performed. S1 mapping analysis conducted on mRNAs isolated from rat testis of different ages indicated that TH2A mRNA appears in rat spermatogonia and its level dramatically increases as the proportion of primary spermatocytes increases (Huh et al., 1991).

The testis-specific rTH2A and rTH2B genes cloned by Huh et al. (1991) are transcribed in the opposite directions from the shared promoter containing an S-phase-specific transcription regulatory element. The presence of this S-phase-dependent regulatory element in the promoter of a pair of testis-specific genes transcribed in pachytene spermatocytes in the absence of DNA replication is surprising (Hwang et al., 1990).

Our analysis of the genomic organization of the hTSH2B gene showed that, similar to murine TH2Bs, human TSH2B also is coupled with an H2A gene, which is located 382 bp upstream and is transcribed in the opposite direction. These two human genes share a common promoter, which includes an S-phase-specific transcription regulatory element, and both genes have the 16-bp stem-loop at their 3′ ends that serve as a transcriptional terminator in replication-dependent histone genes (Dominski and Marzluff, 1999).

Testis-enriched histone H2AX

Although histone H2AX is not testis-specific, the content of H2AX in the total H2A protein is highest in the testes (Trostle-Weige et al., 1982). The H2AX variant contains a unique C-terminus of 22 amino acids when compared to the major H2A histones. Amino acid sequences of the mouse (Nagata et al., 1991) and human (Mannironi et al., 1989) H2AX histones differ only in four positions (97% identical). A unique feature of H2AX is the phosphorylation of Ser139 in response to DNA double-strand breaks (DSBs) (Rogakou et al., 1998). This phosphorylated form gamma H2AX (γ-H2AX) accumulates at the sites of DSBs including those produced during meiotic recombination (Mahadevaiah et al., 2001). During spermatogenesis, γ-H2AX peaks in leptotene spermatocytes that correlate with initiation of meiotic DSBs by Spo11, a topoisomerase II-like protein (Mahadevaiah et al., 2001).

By the time of synaptonemal complex formation, immunofluorescent staining for γ-H2AX disappears from the autosomal chromosomes, but the γ-H2AX protein remains associated with the XY sex body in pachytene spermatocytes (Mahadevaiah et al., 2001). Fernandez-Capetillo et al. (2003b) showed that X and Y chromosomes of H2AX-deficient spermatocytes did not condense, failed to form the sex body, and exhibited a severe abnormality in meiotic pairing. Moreover, other sex body-associated proteins, such as macroH2A1.2 and meiosis-specific XMR protein (Calenda et al., 1994) were not deposited on the sex chromosomes in the absence of H2AX. Thus, in male germ cells, H2AX plays a role in meiotic inactivation of the X and Y chromosomes. Apparently, H2AX exerts this function indirectly, via recruitment of other proteins associated with heterochromatin.

Fernandez-Capetillo et al. (2003a) recently demonstrated another novel function of histone H2AX, i.e., in H2AX-deficient mice telomere movement during meiosis was impaired.

Finally, immunolocalization of the γ-H2AX in mouse testis (Hamer et al., 2003) demonstrated that this protein is highly enriched in spermatogonial cells. γ-H2AX was found in intermediate and B spermatogonia where uniform nuclear γ-H2AX staining is observed. Similarly, the presence of γ-H2AX was observed in the mouse spermatogonial cell line GS-2spd by immunofluorescence (Zalenskaya, unpublished). We recently have shown that immunolocalization of γ-H2AX in sections of human testis follows the same pattern that was observed in mouse (Zalensky, in preparation).

This data indicates that the potential functions of H2AX phosphorylation during spermatogenesis are not restricted to the formation of γ-H2AX foci at DNA double-strand breaks.

MacroH2A and XY body formation

A macroH2A that is nearly three times the molecular size of conventional H2A histones was discovered as a 42-kDa protein that co-purified with rat liver nucleosomes (Pehrson and Fried, 1992). The N-terminal third of macro H2A is ~65% identical to the conventional H2A. The non-histone portion is unique and contains a short highly basic region, capable of binding nucleic acids, and a putative leucine zipper, a helical structure involved in protein-protein interactions (Pehrson and Fried, 1992). Several isoforms of this histone variant are known both in human and mouse: macroH2A1.1, macroH2A1.2, and macroH2A2.

Although macroH2A is not male germline-specific, it may have a specific function during spermatogenesis. MacroH2A1.2 localizes to the developing XY body during early pachytene (Hoyer-Fender et al., 2000a; Richler et al., 2000), similar to its accumulation in the inactive X chromosome in female somatic nuclei (Costanzi and Pehrson, 1998). However, meiotic sex chromosome inactivation is *Xist*-independent and is regulated by a different mechanism compared to X chromosome inactivation in female somatic cells (Turner et al., 2002). Interestingly, during meiosis macroH2A1.2 localizes to centromeric heterochromatin (Hoyer-Fender et al., 2000b) and to a focus within a portion of the pseudoautosomal region, the site of XY pairing during male meiosis (Turner et al., 2001). The localization of macroH2A1.2 is similar to M31, a protein involved in the trans-heterochromatization of the Y chromosome (Motzkus et al., 1999). Thus, macroH2A1.2 may be involved in the transferring the heterochromatization process from the X to the Y chromosome (Turner et al., 2001).

In early mouse spermatids, macroH2A1.2 have been immunolocalized to the chromocenter (Hoyer-Fender et al., 2000a), i.e. similar to localization of protein M31 (Hoyer-Fender et al., 2000b), suggesting that macroH2A1.2 participates in the higher order organization of chromosomes. Western-blotting data apparently showed presence of macro H2A1.2 in mouse spermatozoa (Hoyer-Fender et al., 2000a)

Testis-specific TH3, H3t and replacement variants H3.3A and H3.3B

The male germline-specific histone H3, TH3, was identified and purified from rat testes by Trostle-Weige et al. (1984). The synthesis of this protein takes place in rat spermatogonia, but the protein persists throughout the meiotic stages and is still present in round spermatids.

Albig et al. (1996) isolated a human solitary H3 histone gene, H3t (HIST3H3), mapped it to the chromosome 1q42, and showed that it is transcribed only in testis (Witt et al., 1996). Although human H3t gene is located outside the major histone gene clusters, it shows the consensus promoter and 3′ flanking regions that are typical for replication-dependent genes (Albig et al., 1996). This is rather surprising because it is transcribed in pachytene spermatocytes, where no significant DNA replication occurs.

In situ hybridization of sections from human testis with an H3t-specific probe revealed that expression of this gene is confined to primary spermatocytes (Witt et al., 1996). Because the deduced H3t amino acid sequence differs from the consensus mammalian H3 protein only at four positions (Albig et al., 1996), it is not possible to target H3t with specific antibodies. Consequently, the extent to which H3t replaces the somatic H3 histones during spermatogenesis is unknown.

Amino acid composition comparison between rat TH3 and human H3t suggests that they are not orthologous proteins. Furthermore, the H3t gene was not found in the mouse genome (Marzluff et al., 2002).

The replacement variant of histone H3, termed H3.3 (Wu and Bonner, 1981), also was detected in rat spermatogenic cells (Trostle-Weige et al., 1984). H3.3 represents between 13 and 20% of the total H3 protein in the various germinal cell types (Meistrich et al., 1985). Both in human and mouse, this replacement variant is encoded by two different genes, H3.3A (H3F3A) and H3.3B (H3F3B), located on different chromosomes (Albig et al., 1995; Bramlage et al., 1997). While the nucleotide sequences of H3.3A and H3.3B genes vary substantially in both the mouse and human (Albig et al., 1995), both genes encode an identical protein which differs at three positions from the conventional H3 (Franklin and Zweidler, 1977). In both species, the H3.3A and H3.3B genes are interrupted by three introns and contain polyadenylation signals in the 3′ untranslated regions (Wells et al., 1987; Albig et al., 1995; Mouse Genome Data).

Using H3.3A and H3.3B gene-specific probes, Bramlage et al. (1997) showed by RNase protection analysis and in situ hybridization that mouse H3.3A mRNA is present from pre- to post-meiotic cells, whereas expression of the H3.3B gene is essentially restricted to the cells of meiotic prophase. Therefore, it appears that the H3.3A gene exhibits basal expression throughout spermatogenesis, whereas transcription of the H3.3B gene is stage-specific. Interestingly, a cAMP/phorbol ester response element is involved in the transcriptional regulation of the human H3.3B gene (Witt et al., 1997, 1998), but it is absent from the H3.3A promoter (Wells et al., 1987). Again, a functional role for the differentially expressed H3 variants remains obscure.

Histone H4

Histone H4 is essential for nucleosome formation in all eukaryotes and the amino acid sequence of this protein is highly conserved. Not surprisingly, a testis-specific histone H4 protein variant has not been observed in any species studied. However, a histone H4 gene is uniquely transcribed during mammalian spermatogenesis and may be the only H4 gene that is expressed during meiotic prophase. This gene, H4t *(Hist1h4c)*, was first discovered in the rat genome in close proximity to the testicularly expressed H1t *(Hist1h1t)* (Grimes et al., 1987). Later, orthologous genes were isolated from both the mouse and human genomes (Drabent et al., 1995b). H4t mRNA level peaks in pachytene spermatocytes of all species studied (Wolfe et al., 1989; Drabent et al., 1995b), whereas somatic H4 gene transcripts are not detected (Wolfe et al., 1989). H4t gene expression is down-regulated in early spermatids (Grimes et al., 1987). In contrast to the closely associated testis-specific H1t gene, the H4t gene is not completely turned off in nongerminal cells (Wolfe et al., 1989; Drabent et al., 1995b). The function of this gene is unknown, especially because the primary structure of its protein product is 100% identical to the evolutionarily conserved somatic H4.

The histone H1 family

The H1 histone family is the most divergent class of histones. At least eight members are expressed in mammals (Wang et al., 1997; Khochbin, 2001). Histone H1 is associated with the linker DNA of the nucleosome, therefore it is called a linker histone. H1 plays a major role in the higher-order structure of chromatin (Thoma et al., 1979; Carter and van Holde, 1998). A tripartite structure typical for all H1 proteins (Bradbury et al., 1975) includes a central globular domain flanked by a short N- and a relatively longer C-terminal domain. The central globular domain is conserved among H1 subtypes, while both tail domains vary considerably in their lengths and sequences and constitute distinctive features of each subtype (Doenecke et al., 1997a). A recent study of transgenic expression in H1 knockout mice supports the idea that different linker histone subtypes introduce subtle modifications in the higher-order chromatin structure and differentially affect gene expression (Alami et al., 2003).

All known histone H1 subtypes, except the oocyte-specific variant H1oo (Tanaka et al., 2001), are detected in different testicular cell types, although in different relative proportions. Analysis of the stage-specific expression patterns revealed that the H1 subtypes appear in a particular sequence during spermatogenesis. Thus, in the type A and B spermatogonia, the testis-enriched H1a and somatic H1c variants comprise a majority of the linker histone complement (Meistrich et al., 1985; Franke et al., 1998). The testis-specific variant H1t is not detectable until the early spermatocyte stage, but it then accumulates rapidly and becomes the most abundant type by the mid-pachytene stage (Meistrich et al., 1985; Drabent et al., 1998). Apparently, in pachytene spermatocytes, the histone H1t partially replaces the DNA replication-dependent histone H1a which is enriched in dividing spermatogonia. As shown by immunostaining with an anti-H1a antibody (Drabent et al., 1996), the amount of H1a protein decreases gradually during meiosis, and the protein is virtually absent in post-meiotic cells. H1t persists at high levels through the two meiotic cell divisions and is detectable in early spermatids (Grimes et al., 1997). The protein disappears from elongating spermatids by the onset of HILS1 protein expression and just prior to the start of transitional protein and protamine expression (Yan et al., 2003). Finally, HILS1 is exclusively detected in the nuclei of elongating and elongated spermatids and is absent in spermatozoa.

The testis-enriched H1a and especially the testis-specific H1t variants have received the most attention of researchers (reviewed in Grimes et al., 2003). Transcriptional regulation of the histone H1t gene has become a tissue-specific model for studying transcriptional control during cellular differentiation (Wolfe and Grimes, 1993).

Histone H1a (H1.1 gene product)

Histone H1a (H1.1 gene product) is a predominant H1 subtype in the testis of mice (Lennox and Cohen, 1984) and rats (Meistrich et al., 1985). The H1.1 gene was isolated from both human (Eick et al., 1989) and mouse (Drabent et al., 1995a) genomes and mapped to syntenic regions of human chromosome 6 (Albig et al., 1993) and murine chromosome 13 (Drabent et al., 1995a). Both the human and mouse genes exhibit features of the DNA replication-dependent class of histones.

Southern blot analysis using human H1.1 as a probe showed that the H1.1 gene is highly conserved in higher primates (chimpanzee, orangutan, gorilla and rhesus monkey), while no cross-hybridization could be detected with DNA from other mammalian species (mouse, rat, hamster, and bull) demonstrating significant divergence in nucleotide sequences (Burfeind et al., 1994).

Franke et al. (1998) studied the expression of the H1.1 gene at both the mRNA and protein levels in mouse testis. They showed by in situ hybridization that H1.1 mRNA accumulates only in the cells aligned at the basal lamina of the seminiferous tubules, which corresponds to proliferating spermatogonia. In contrast to the H1.1 mRNA, the H1a protein was found in almost every layer of cells, although a gradual decrease of H1.1 antibody staining was seen from the basal lamina to the lumen, suggesting a decrease of H1.1 protein during spermatogenesis (Franke et al., 1998). Thus, the H1.1 gene is apparently transcribed exclusively in mouse spermatogonia (and possibly in preleptotene spermatocytes), but the protein persists in the germ cell chromatin until post-meiotic stages. In murine prepachytene spermatocytes, the H1a protein represents about 70% of the H1 complement, whereas in somatic tissues it is only a minor component of chromatin (Lennox and Cohen, 1984).

In contrast to the murine H1.1 gene, transcription of the human H1.1 gene occurs post-meiotically (Burfeind et al., 1994). Expression of the DNA replication-dependant H1.1 histone gene in nondividing post-meiotic cells is highly surprising. Based on the differences in expression patterns and primary structures, Burfeind et al. (1994) concluded that rodent and human H1.1 gene products are not functional homologues.

The mouse histone H1a shares only ~60% sequence similarity with the other subtypes (Drabent et al., 1995a). The structural divergence and cell-type-specific expression of H1a suggests it has a function different from the other H1 variants. Rabini et al. (2000) inactivated the H1.1 locus by homologous recombination and analyzed spermatogenesis in H1.1-deficient mice. Neither changes in testicular morphology nor any irregularity in spermatogenesis were observed. However, enhanced expression of histones H1.2 through H1.4 in the testis of H1.1-deficient mice was observed, suggesting that other replication-dependent histones could compensate for H1a (Rabini et al., 2000).

Spermatocyte-specific linker histone H1t

The H1t protein is highly divergent compared to the other linker histones in mammals. Mouse and human H1t share only 46 and 50% identity with the major somatic H1 variants in mice and humans, respectively (Drabent et al., 1991, 1993). The main sequence differences are in the internal part of the N-terminal domain and in the C-terminal domain. Compared to somatic histone H1 variants, the arginine-to-lysine ratio is increased in H1t (Doenecke et al., 1997). The replacement of lysine with arginine is most prevalent in the C-terminal domain, which is also shorter compared to somatic H1 variants (Cole et al., 1986). Changes in the C-terminal sequence also lead to a disappearance of potential phosphorylation sites.

Circular dichroism study of the H1t-containing chromatin reconstituted in vitro showed that H1t is a poor condenser of chromatin as compared to the other H1 subtypes (De Lucia et al., 1994; Khadake and Rao, 1995). Therefore, it was hypothesized that H1t-associated chromatin might have a more open conformation, facilitating meiotic recombination in pachytene spermatocytes and/or replacement of histones by transition proteins in early spermatids (Oko et al., 1996).

The H1t gene was isolated from several mammalian species including rat (Cole et al., 1986), human (Drabent et al., 1991), mouse (Drabent et al., 1993), and monkey (Koppel et al., 1994). In mouse and human the H1t genes are located in the major histone gene clusters on mouse chromosome 13 (Drabent et al., 1995a; Wang et al., 1996) and on human chromosome 6 (Albig et al., 1993; Koppel et al., 1994). Both mouse and human genes are intronless and encode poly(A)-mRNAs, with a conserved stem-loop structure in the 3′ ends (Drabent et al., 1991, 1993). These structures are necessary for mRNA processing and half-life regulation (Dominski and Marzluff, 1999). Analysis of the H1t promoter region (Wolfe et al., 1995) revealed a number of conserved elements found in other H1 genes as well as specific elements required for tissue-specific activation and repression of the H1t gene (Wolfe and Grimes, 1993). Grimes et al. (2003) recently reviewed the mechanisms of the complex transcriptional regulation of the H1t gene.

In mice, the testis-specific linker histone H1t gene is expressed in primary spermatocytes and transcription ceases by the late primary spermatocyte stage (Grimes et al., 1977, 1987, 1992, 1997; Wolfe and Grimes, 1993). More recently, evidence of low-level transcription in spermatogonia has been shown (Drabent et al., 1998). Furthermore, H1t mRNA is subject to rapid turnover (Drabent et al., 1996, 1998).

Histone H1t protein synthesis has been detected exclusively in pachytene primary spermatocytes, where H1t protein level reaches 55–60% of the total H1 histone complement. However, the protein is retained until the mid-spermatid stage of spermiogenesis when most of the histones including the germ line-specific types are replaced by transition proteins (Bucci et al., 1982; Meistrich et al., 1985; Oko et al., 1996).

The function of H1t in spermatogenesis has been examined using H1t-deficient mice created by homologous recombination (Drabent et al., 2000; Lin et al., 2000; Fantz et al., 2001). Surprisingly, mice lacking histone H1t were fertile and did not show any detectable abnormalities in spermatogenesis. Analogous to the H1.1 knockout, enhanced expression of the certain H1 subtypes was observed, suggesting other H1 subtypes may compensate for H1t.

Another possibility is that the major function for H1t is to provide a chromatin conformation that facilitates recombination. A decreased recombination rate resulting from the inactivation of the H1t gene would not produce any obvious phenotype nor would it affect fertility in F1 and in the next few generations. However, the lack of H1t function in knockouts may have a remote effect on population.

Spermatid-specific histone H1-like protein HILS1
Discovery of the HILS1 protein in mice and humans is an example of genome database mining to identify new members of the male germline-specific histones. The *Hils1* transcript initially was identified in the mouse using in silico subtraction of the testis cDNA libraries against the remaining libraries available from the mouse UniGene database (Yan et al., 2003). Genomic locations of the mouse *Hils1* gene and its human ortholog also were identified "electronically" by taking advantage of the mouse and human genome databases. Mouse *Hils1* is an intronless gene located on chromosome 11, whereas human HILS1 is interrupted by a small intron and is located in a syntenic region of chromosome 17 (Yan et al., 2003). The HILS1 protein is much more divergent from the histone H1 family members when compared to the H1t protein. The globular domains of both the mouse and human HILS1 share only about 40% amino acid identity with the globular domains of the somatic histones (Yan et al., 2003).

Hils1 mRNA expression in testis is confined to round spermatids; the RNA appears in step 4 spermatids, peaks in step 6–8 spermatids, and decreases in step 9 spermatids (Yan et al., 2003). The protein is detected exclusively in the nuclei of elongating and elongated spermatids. It first appears in step 9 spermatids, persists at high levels in step 10–13 spermatids, and then decreases abruptly in step 14 spermatids (Yan et al., 2003). The appearance of the Hils1 protein is slightly earlier than the transition nuclear proteins (TPs). Hils1 colocalizes with the TPs and protamines in the nuclei of elongating and elongated spermatids (steps 10–14). Immunofluorescent microscopy examination of the Hils1 subcellular localization showed that in step 12 spermatid nuclei Hils1 displayed a pattern of patchy spots in the chromatin, which was not identical to that displayed by TP1, but very similar to that of TP2 and protamine 1 (Yan et al., 2003).

The timing of Hils1 expression and its colocalization with TP2 and protamine 1 in the condensing chromatin of elongating and elongated spermatids indicates that it may participate in the nuclear protein replacement process, although its exact role is unclear. In addition, the globular domain of the HILS1 contains a winged helix, a highly conserved DNA-binding motif found in a large number of topologically related proteins with diverse biological functions (Gajiwala and Burley, 2000). This finding suggests that HILS1 also may be involved in gene regulation, DNA repair, and other specialized chromosome processes.

Recently, another group of investigators characterized expression of TISP64 (synonym of HILS1) during mouse spermiogenesis (Iguchi et al., 2003). Their results are in a good agreement with those published by Yan et al. (2003). The most exciting news though is on the *Hils1* mutants study in progress (Iguchi et al., 2003).

The histone complement in human sperm

In the human, unlike other mammals including the mouse, core histones are not displaced completely during spermiogenesis, and account for approximately 15% of basic chromosomal proteins in mature sperm (Tanphaichitr et al., 1978; Gusse et al., 1986; Gatewood et al., 1987). It was proposed that human sperm histone variants may contribute to the sequence-specific restructuring of chromatin and would be essential for epigenetic reprogramming of the male genome (Gatewood et al., 1987, 1990; Ward and Zalensky, 1996). Although human sperm histones were partially fractionated more than a decade ago (Gatewood et al., 1990), their identities and biological role remain, for the most part, unknown.

Recent studies (Zalensky et al., 2002; Churikov et al., submitted) characterized members of the human sperm histone complement. Several studies of sperm histones provided early evidence for their non-canonical functions. A pool of core histones is associated with sperm telomeres, which retain a nucleosomal organization unlike the majority of sperm chromatin packaged by protamines (Zalenskaya et al., 2000; Wykes and Krawets, 2003). The sperm-specific H2B-like protein participates in binding to double-stranded telomeric DNA (Zalensky

Table 1. Comparison of the male germline-specific/-enriched histone gene orthologs in mouse and man

Mouse

Protein	% identity[a] with major somatic variants	Gene[b]	Location	mRNA 3' end	Reference
H1a (H1.1)	62–68%	*Hist1h1a* intronless	13A3.1	stem-loop	Franke et al., 1998 Rabini et al., 2000
H1t	50–54%	*Hist1h1t* intronless	13A3.1	stem-loop	Drabent et al., 1993, 1996, 1998, 2000 Lin et al., 2000
HILS1	37–41% within globular domain	*Hils1* intronless	11D	poly(A)+	Yan et al., 2003 Iguchi et al., 2003
TH2B	88%	MTH2B (*Hist1h2ba*) intronless	13A3.1	stem-loop	Choi et al., 1996
ssH2B	97% within 119 N-terminal amino acids and unique C-terminal 12 amino acids	LOC328201 interrupted, two exons	13A3.1	poly(A)+	Moss et al., 1989 Moss and Orth, 1993 Unni et al., 1995
TH2B-175		absent			
H2AX	96–97% within 120 N-terminal and unique 22 C- terminal residues	*H2afx* intronless	9A5.2	0.6-kb stem-loop and 1.4-kb poly(A)+	Nagata et al., 1991
macroH2A1.2	64% within N-terminal third and 2/3 unique	*H2afy* interrupted 10 exons	13B1	poly(A)+	Pehrson and Fried, 1992
H3t		absent			
H3.3	96%	*H3f3a, H3f3b* interrupted 4 exons	13D2.3, 11E2	poly(A)+ poly(A)+	Bramlage et al., 1997
H4	100%	*Hist1h4c* intronless	13A2-A3	stem-loop	Drabent et al., 1995

[a] Protein sequence homologies were compared using NCBI Blast 2 Sequences Program using Blosum62 matrix and without filtering for low complexity.
[b] HGMW-approved symbol or locus/accession number.

Human

% identity[a] with major somatic variants	Gene[b]	Location	mRNA 3' end	Reference	Proposed function[c]
66–69%	HIST1H1A intronless	6p21.3	stem-loop	Burfeind et al., 1994	--
48–53%	HIST1H1T intronless	6p21.3	stem-loop	Steger et al., 1998	open chromatin, facilitates meiotic recombination and protein transitions, but dispensable
37–41% within globular domain	AY286318 two exons	17q21.33	poly(A)+	Yan et al., 2003	--
85%	hTSH2B (HIST1H2BA) intronless	6p22.1	stem-loop	van Roijen et al., 1998 Zalensky et al., 2002	--
	absent				chromatin- nuclear envelope interaction unconfirmed
52% within conserved H2B domain	AY283369 interrupted, three exons	Xq22.2	poly(A)+	Churikov and Zalensky, unpublished	telomere association unconfirmed
96–97% within 120 N-terminal and unique 22 C-terminal residues	H2AFX intronless	11q23.2–q23.3	0.6-kb stem- loop and 1.6-kb poly(A)+	Mannironi et al., 1989	recruitment of repair factors to recombin. DSBs; XY body formation; telomere dynamics
65% within N-terminal third and 2/3 unique	H2AFY interrupted 10 exons	5q31.3–q32	poly(A)+	Chadwick and Willard, 2001	transcriptional inactivation of sex chromosomes
97%	HIST3H3 (H3FT) intronless	1q42		Witt et al., 1996	--
96%	H3F3A, H3F3B interrupted 4 exons	1q41, 17q25	poly(A)+ poly(A)+	Wells et al., 1987 Albig et al., 1995	--
100%	HIST1H4C intronless	6p21.3	stem-loop	Drabent et al., 1995	--

[a] Protein sequence homologies were compared using NCBI Blast 2 Sequences Program using Blosum62 matrix and without filtering for low complexity.
[b] HGMW-approved symbol or locus/accession number.
[c] "--" no data available.

et al., 1997; Gineitis et al., 2000). Another pool of human sperm histones may organize centromeric chromatin since centromeric protein A (a variant of histone H3) is preserved in mature spermatozoa (Zalensky et al., 1993). Hypothetically, telomere and centromere domains of human sperm chromosomes could maintain a nucleosomal organization, which may be needed immediately for proper pronuclei formation after fertilization. It remains to be seen whether differences in the sperm histone content between humans and mice are important in fertilization and subsequent embryonic development.

Histone modifications during spermatogenesis

In addition to the histone variants, post-translational modification of histones is another potential mechanism to remodel chromatin during spermatogenesis. All histones in differentiating germline cells are subject to a number of modifications including acetylation, phosphorylation, methylation, ubiquitination, and ADP-rybosilation. For detailed information concerning histone modifications during spermatogenesis the reader is referred to recent papers (Hazzouri et al., 2000; Jason et al., 2002; Lewis et al., 2003; Pivot-Pajot et al., 2003; Sutovsky, 2003). Overall, each of these modifications can change the association between a core histone and DNA, other core histones, linker histones, and other chromosomal proteins. Most of these modifications occur within the histone tails and play an important role in the regulation of gene transcription and silencing, in meiotic reorganization of chromatin, and in histone replacement during spermiogenesis.

Conclusions

Histones of male germ cells and human spermatozoa have been the focus of studies during the last 30 years. Nevertheless, it appears that full histone complement in these cells is yet to be established. Furthermore, functions for most of these proteins are unknown and therefore described in very general terms such as "histone XX participates in remodeling of chromatin during spermatogenesis". Deciphering of the mouse and human genomes offers innovative possibilities for the identification, and assists in molecular and functional characterization of additional members of testis-/sperm-specific histone family.

Table 1 provides a summary of the data accumulated for germline-specific histones in mouse and man. Three variants of linker histone H1 and several (six in mouse and eight in human) variants of core histones have been characterized. For most proteins, the gene structure has been established. Most of the genes were found to be solitary, not belonging to any known cluster of somatic histone genes. Three genes, coding for proteins H1a, H1t, H4, are located within, and TH2A/TH2B gene pair is located at the very edge, of the major histone gene clusters in both species. Noteworthy, transcripts from these genes lack a poly(A) tract and, in this respect, are similar to mRNAs transcribed from somatic, replication-dependent histone genes. This group of genes is transcribed relatively early during spermatogenesis (from spermatogonia to spermatocytes). The remaining germline-specific genes have genomic locations outside the major histone gene clusters, often on different chromosomes. Most of them (but not all) have introns and their mRNAs are polyadenylated. The latter group of genes is transcribed relatively late during spermatogenesis (in spermatids) and probably is regulated by post-meiotic transcription factors, such as CREM or others yet to be identified.

Overall, the germline-specific histone complements are similar in both organisms with two exceptions; spermatid-specific H2B is unique for mice, while H2B-175 is found only in humans. In spite of the overall similarity in the testis-specific histone in these two species, several histones have different pat-

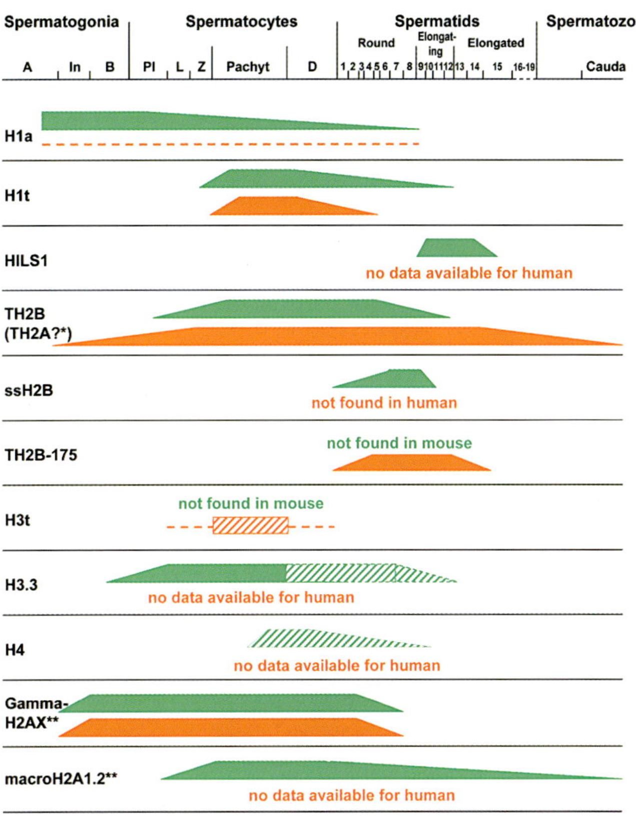

Fig. 1. Germline-specific/enriched histones during spermatogenesis. Filled shapes indicate kinetics of protein accumulation during murine (green) and human (orange) spermatogenesis. Dashed shapes indicate mRNA accumulation, when no protein data are available. The dashed lines indicate uncertainty in the timing of protein expression. * Information on histone TH2A accumulation is limited. Data available suggest that its pattern of accumulation during spermatogenesis may be similar to that of TH2B. ** Intranuclear localization is stage-specific.

terns of expression and chromatin accumulation during spermatogenesis (Fig. 1).

Functions for these male germline-specific/enriched histones remain unknown with the exception of macroH2A that participates in the inactivation of sex chromosomes, and of H2AX, which is involved in recombination, XY body formation and telomere dynamics. Unfortunately, males null for some histones, e.g. H1t, did not give an obvious phenotype, making it difficult to elucidate gene function (Lin et al., 2000; Fantz et al., 2001). The gene knockout experiments have demonstrated that linker histone subtypes are dispensable for mouse spermatogenesis. Even the absence of two to three linker histone subtypes can be, at least partially, compensated by the remaining others (Fan et al., 2001; Nayernia et al., 2003).

Knowledge of the mouse genome in combination with the recently developed immortalized spermatogonial cell line capable of differentiation to spermatid stage (Feng et al., 2002) (these cells do not make sperm) will be a powerful tool for examining the regulation and function of some germline histone genes.

A noticeable difference between human and mouse spermatozoa is the presence of core histones in human sperm nuclei. An initial indication for the unique role(s) played by human sperm histones was provided by the identification of the telomere-associated H2B-like protein (Gineitis et al., 2000). Still, the most important question concerning the importance of histones in human sperm for fertilization and/or early development remains open.

References

Alami R, Fan Y, Pack S, Sonbuchner TM, Besse A, Lin Q, Greally JM, Skoultchi AI, Bouhassira EE: Mammalian linker-histone subtypes differentially affect gene expression in vivo. Proc Natl Acad Sci USA 100:5920–5925 (2003).

Albig W, Doenecke D: The human histone gene cluster at the D6S105 locus. Hum Genet 101:284–294 (1997).

Albig W, Drabent B, Kunz J, Kalff-Suske M, Grzeschik KH, Doenecke D: All known human H1 histone genes except the H1(0) gene are clustered on chromosome 6. Genomics 16:649–654 (1993).

Albig W, Bramlage B, Gruber K, Klobeck HG, Kunz J, Doenecke D: The human replacement histone H3.3B gene (H3F3B). Genomics 30:264–272 (1995).

Albig W, Ebentheuer J, Klobeck G, Kunz J, Doenecke D: A solitary human H3 histone gene on chromosome 1. Hum Genet 97:486–491 (1996).

Balhorn R: A model for the structure of chromatin in mammalian sperm. J Cell Biol 93:298–305 (1982).

Bradbury EM, Chapman GE, Danby SE, Hartman PG, Riches PL: Studies on the role and mode of operation of the very-lysine-rich histone H1 (F1) in eukaryote chromatin. The properties of the N-terminal and C-terminal halves of histone H1. Eur J Biochem 57:521–528 (1975).

Bramlage B, Kosciessa U, Doenecke D: Differential expression of the murine histone genes H3.3A and H3.3B. Differentiation 62:13–20 (1997).

Brock WA, Trostle PK, Meistrich ML: Meiotic synthesis of testis histones in the rat. Proc Natl Acad Sci USA 77:371–375 (1980).

Bucci LR, Brock WA, Meistrich ML: Distribution and synthesis of histone 1 subfractions during spermatogenesis in the rat. Exp Cell Res 140:111–118 (1982).

Burfeind P, Hoyer-Fender S, Doenecke D, Tsaousidou S, Engel W: Expression of a histone H1 gene (H1.1) in human testis and Hassall's corpuscles of the thymus. Expression of a histone H1 gene (H1.1). Thymus 19:245–251 (1992).

Burfeind P, Hoyer-Fender S, Doenecke D, Hochhuth C, Engel W: Expression and chromosomal mapping of the gene encoding the human histone H1.1. Hum Genet 94:633–639 (1994).

Calenda A, Allenet B, Escalier D, Bach JF, Garchon HJ: The meiosis-specific *Xmr* gene product is homologous to the lymphocyte Xlr protein and is a component of the XY body. EMBO J 13:100–109 (1994).

Carter GJ, van Holde K: Self-association of linker histone H5 and of its globular domain: evidence for specific self-contacts. Biochemistry 37:12477–12488 (1998).

Chadwick BP, Willard HF: Histone H2A variants and the inactive X chromosome: identification of a second macroH2A variant. Hum Mol Genet 10:1101–1113 (2001).

Choi YC, Chae CB: DNA hypomethylation and germ cell-specific expression of testis-specific H2B histone gene. J Biol Chem 266:20504–20511 (1991).

Choi YC, Gu W, Hecht NB, Feinberg AP, Chae CB: Molecular cloning of mouse somatic and testis-specific H2B histone genes containing a methylated CpG island. DNA Cell Biol 15:495–504 (1996).

Cole KD, Kandala JC, Kistler WS: Isolation of the gene for the testis-specific H1 histone variant H1t. J Biol Chem 261:7178–7183 (1986).

Costanzi C, Pehrson JR: Histone macroH2A1 is concentrated in the inactive X chromosome of female mammals. Nature 393:599–601 (1998).

Dadoune JP: Expression of mammalian spermatozoal nucleoproteins. Microsc Res Tech 61:56–75 (2003).

De Lucia F, Faraone-Mennella MR, D'Erme M, Quesada P, Caiafa P, Farina B: Histone-induced condensation of rat testis chromatin: testis-specific H1t versus somatic H1 variants. Biochem Biophys Res Commun 198:32–39 (1994).

Doenecke D, Albig W, Bode C, Drabent B, Franke K, Gavenis K, Witt O: Histones: genetic diversity and tissue-specific gene expression. Histochem Cell Biol 107:1–10 (1997a).

Doenecke D, Drabent B, Bode C, Bramlage B, Franke K, Gavenis K, Kosciessa U, Witt O: Histone gene expression and chromatin structure during spermatogenesis. Adv Exp Med Biol 424:37–48 (1997b).

Dominski Z, Marzluff WF: Formation of the 3′ end of histone mRNA. Gene 239:1–14 (1999).

Drabent B, Kardalinou E, Doenecke D: Structure and expression of the human gene encoding testicular H1 histone (H1t). Gene 103:263–268 (1991).

Drabent B, Bode C, Doenecke D: Structure and expression of the mouse testicular H1 histone gene (H1t). Biochim Biophys Acta 1216:311–313 (1993).

Drabent B, Franke K, Bode C, Kosciessa U, Bouterfa H, Hameister H, Doenecke D: Isolation of two murine H1 histone genes and chromosomal mapping of the H1 gene complement. Mammal Genome 6:505–511 (1995a).

Drabent B, Kardalinou E, Bode C, Doenecke D: Association of histone H4 genes with the mammalian testis-specific H1t histone gene. DNA Cell Biol 14:591–597 (1995b).

Drabent B, Bode C, Bramlage B, Doenecke D: Expression of the mouse testicular histone gene H1t during spermatogenesis. Histochem Cell Biol 106:247–251 (1996).

Drabent B, Bode C, Miosge N, Herken R, Doenecke D: Expression of the mouse histone gene H1t begins at premeiotic stages of spermatogenesis. Cell Tissue Res 291:127–132 (1998).

Drabent B, Saftig P, Bode C, Doenecke D: Spermatogenesis proceeds normally in mice without linker histone H1t. Histochem Cell Biol 113:433–442 (2000).

Eick S, Nicolai M, Mumberg D, Doenecke D: Human H1 histones: conserved and varied sequence elements in two H1 subtype genes. Eur J Cell Biol 49:110–115 (1989).

Fan Y, Sirotkin A, Russell RG, Ayala J, Skoultchi AI: Individual somatic H1 subtypes are dispensable for mouse development even in mice lacking the H1(0) replacement subtype. Mol Cell Biol 21:7933–7943 (2001).

Fantz DA, Hatfield WR, Horvath G, Kistler MK, Kistler WS: Mice with a targeted disruption of the H1t gene are fertile and undergo normal changes in structural chromosomal proteins during spermiogenesis. Biol Reprod 64:425–431 (2001).

Feng LX, Chen Y, Dettin L, Pera RA, Herr JC, Goldberg E, Dym M: Generation and in vitro differentiation of a spermatogonial cell line. Science 297:392–395 (2002).

Fernandez-Capetillo O, Liebe B, Scherthan H, Nussenzweig A: H2AX regulates meiotic telomere clustering. J Cell Biol 163:15–20 (2003a).

Fernandez-Capetillo O, Mahadevaiah SK, Celeste A, Romanienko PJ, Camerini-Otero RD, Bonner WM, Manova K, Burgoyne P, Nussenzweig A: H2AX is required for chromatin remodeling and inactivation of sex chromosomes in male mouse meiosis. Dev Cell 4:497–508 (2003b).

Fimia GM, Morlon A, Macho B, De Cesare D, Sassone-Corsi P: Transcriptional cascades during spermatogenesis: pivotal role of CREM and ACT. Mol Cell Endocrinol 179:17–23 (2001).

Franke K, Drabent B, Doenecke D: Testicular expression of the mouse histone H1.1 gene. Histochem Cell Biol 109:383–390 (1998).

Franklin SG, Zweidler A: Non-allelic variants of histones 2a, 2b and 3 in mammals. Nature 266:273–275 (1977).

Gajiwala KS, Burley SK: Winged helix proteins. Curr Opin Struct Biol 10:110–116 (2000).

Gatewood JM, Cook GR, Balhorn R, Bradbury EM, Schmid CW: Sequence-specific packaging of DNA in human sperm chromatin. Science 236:962–964 (1987).

Gatewood JM, Cook GR, Balhorn R, Schmid CW, Bradbury EM: Isolation of four core histones from human sperm chromatin representing a minor subset of somatic histones. J Biol Chem 265:20662–20666 (1990).

Gineitis AA, Zalenskaya IA, Yau PM, Bradbury EM, Zalensky AO: Human sperm telomere-binding complex involves histone H2B and secures telomere membrane attachment. J Cell Biol 151:1591–1598 (2000).

Grimes SR Jr, Meistrich ML, Platz RD, Hnilica LS: Nuclear protein transitions in rat testis spermatids. Exp Cell Res 110:31–39 (1977).

Grimes S, Weisz-Carrington P, Daum H 3rd, Smith J, Green L, Wright K, Stein G, Stein J: A rat histone H4 gene closely associated with the testis-specific H1t gene. Exp Cell Res 173:534–545 (1987).

Grimes SR, Wolfe SA, Koppel DA: Tissue-specific binding of testis nuclear proteins to a sequence element within the promoter of the testis-specific histone H1t gene. Arch Biochem Biophys 296:402–409 (1992).

Grimes SR Jr, van Wert J, Wolfe SA: Regulation of transcription of the testis-specific histone H1t gene by multiple promoter elements. Mol Biol Rep 24:175–184 (1997).

Grimes SR, Wilkerson DC, Noss KR, Wolfe SA: Transcriptional control of the testis-specific histone H1t gene. Gene 304:13–21 (2003).

Gusse M, Sautiere P, Belaiche D, Martinage A, Roux C, Dadoune JP, Chevaillier P: Purification and characterization of nuclear basic proteins of human sperm. Biochim Biophys Acta 884:124–134 (1986).

Hamer G, Roepers-Gajadien HL, van Duyn-Goedhart A, Gademan IS, Kal HB, vanBuul PP, de Rooij DG: DNA double-strand breaks and gamma-H2AX signaling in the testis. Biol Reprod 68:628–634 (2003).

Hazzouri M, Pivot-Pajot C, Faure AK, Usson Y, Pelletier R, Sele B, Khochbin S, Rousseaux S: Regulated hyperacetylation of core histones during mouse spermatogenesis: involvement of histone deacetylases. Eur J Cell Biol 79:950–960 (2000).

Hoyer-Fender S, Costanzi C, Pehrson JR: Histone macroH2A1.2 is concentrated in the XY-body by the early pachytene stage of spermatogenesis. Exp Cell Res 258:254–260 (2000a).

Hoyer-Fender S, Singh PB, Motzkus D: The murine heterochromatin protein M31 is associated with the chromocenter in round spermatids and is a component of mature spermatozoa. Exp Cell Res 254:72–79 (2000b).

Huh NE, Hwang IW, Lim K, You KH, Chae CB: Presence of a bi-directional S phase-specific transcription regulatory element in the promoter shared by testis-specific TH2A and TH2B histone genes. Nucl Acids Res 19:93–98 (1991).

Hwang IW, Lim K, Chae CB: Characterization of the S-phase-specific transcription regulatory elements in a DNA replication-independent testis-specific H2B (TH2B) histone gene. Mol Cell Biol 10:585–592 (1990).

Iguchi N, Tanaka H, Yomogida K, Nishimune Y: Isolation and characterization of a novel cDNA encoding a DNA-binding protein (Hils1) specifically expressed in testicular haploid germ cells. Int J Androl 26:354–365 (2003).

Jason LJ, Moore SC, Lewis JD, Lindsey G, Ausio J: Histone ubiquitination: a tagging tail unfolds? Bioessays 24:166–174 (2002).

Khadake JR, Rao MR: DNA- and chromatin-condensing properties of rat testes H1a and H1t compared to those of rat liver H1bdec; H1t is a poor condenser of chromatin. Biochemistry 34:15792–15801 (1995).

Khochbin S: Histone H1 diversity: bridging regulatory signals to linker histone function. Gene 271:1–12 (2001).

Kim YJ, Hwang I, Tres LL, Kierszenbaum AL, Chae CB: Molecular cloning and differential expression of somatic and testis-specific H2B histone genes during rat spermatogenesis. Dev Biol 124:23–34 (1987).

Koppel DA, Wolfe SA, Fogelfeld LA, Merchant PS, Prouty L, Grimes SR: Primate testicular histone H1t genes are highly conserved and the human H1t gene is located on chromosome 6. J Cell Biochem 54:219–230 (1994).

Lennox RW, Cohen LH: The alterations in H1 histone complement during mouse spermatogenesis and their significance for H1 subtype function. Dev Biol 103:80–84 (1984).

Lewis JD, Abbott DW, Ausio J: A haploid affair: core histone transitions during spermatogenesis. Biochem Cell Biol 81:131–140 (2003a).

Lewis JD, Song Y, de Jong ME, Bagha SM, Ausio J: A walk through vertebrate and invertebrate protamines. Chromosoma 111:473–482 (2003b).

Lim K, Chae CB: Presence of a repressor protein for testis-specific H2B (TH2B) histone gene in early stages of spermatogenesis. J Biol Chem 267:15271–15273 (1992).

Lin Q, Sirotkin A, Skoultchi AI: Normal spermatogenesis in mice lacking the testis-specific linker histone H1t. Mol Cell Biol 20:2122–2128 (2000).

Mahadevaiah SK, Turner JM, Baudat F, Rogakou EP, de Boer P, Blanco-Rodriguez J, Jasin M, Keeney S, Bonner WM, Burgoyne PS: Recombinational DNA double-strand breaks in mice precede synapsis. Nat Genet 27:271–276 (2001).

Mannironi C, Bonner WM, Hatch CL: H2A.X a histone isoprotein with a conserved C-terminal sequence, is encoded by a novel mRNA with both DNA replication type and polyA 3′ processing signals. Nucl Acids Res 17:9113–9126 (1989).

Marret C, Avallet O, Perrard-Sapori MH, Durand P: Localization and quantitative expression of mRNAs encoding the testis-specific histone TH2B, the phosphoprotein p19, the transition proteins 1 and 2 during pubertal development and throughout the spermatogenic cycle of the rat. Mol Reprod Dev 51:22–35 (1998).

Marzluff WF, Gongidi P, Woods KR, Jin J, Maltais LJ: The human and mouse replication-dependent histone genes. Genomics 80:487–498 (2002).

McGhee JD, Felsenfeld G: Nucleosome structure. Annu Rev Biochem 49:1115–1156 (1980).

Meistrich ML: Histones and Other Basic Nuclear Proteins. Hnilica G, Stein G, Stein J (eds), pp165–182 (CRC Press, Orlando 1989).

Meistrich ML, Bucci LR, Trostle-Weige PK, Brock WA: Histone variants in rat spermatogonia and primary spermatocytes. Dev Biol 112:230–240 (1985).

Meistrich ML, Mohapatra B, Shirley CR, Zhao M: Roles of transition nuclear proteins in spermiogenesis. Chromosoma 111:483–488 (2003).

Moss SB, Orth JM: Localization of a spermatid-specific histone 2B protein in mouse spermiogenic cells. Biol Reprod 48:1047–1056 (1993).

Moss SB, Challoner PB, Groudine M: Expression of a novel histone 2B during mouse spermiogenesis. Dev Biol 133:83–92 (1989).

Motzkus D, Singh PB, Hoyer-Fender S: M31, a murine homolog of Drosophila HP1, is concentrated in the XY body during spermatogenesis. Cytogenet Cell Genet 86:83–88 (1999).

Nagata T, Kato T, Morita T, Nozaki M, Kubota H, Yagi H, Matsuhiro A: Polyadenylated and 3′ processed mRNAs are transcribed from the mouse histone H2A.X gene. Nucl Acids Res 19:2441–2447 (1991).

Nayernia K, Drabent B, Adham IM, Moschner M, Wolf S, Meinhardt A, Engel W: Male mice lacking three germ cell expressed genes are fertile. Biol Reprod 69:1973–1978 (2003).

Oko RJ, Jando V, Wagner CL, Kistler WS, Hermo LS: Chromatin reorganization in rat spermatids during the disappearance of testis-specific histone, H1t, and the appearance of transition proteins TP1 and TP2. Biol Reprod 54:1141–1157 (1996).

Pehrson JR, Fried VA: MacroH2A, a core histone containing a large nonhistone region. Science 257:1398–1400 (1992).

Pivot-Pajot C, Caron C, Govin J, Vion A, Rousseaux S, Khochbin S: Acetylation-dependent chromatin reorganization by BRDT, a testis-specific bromodomain-containing protein. Mol Cell Biol 23:5354–5365 (2003).

Rabini S, Franke K, Saftig P, Bode C, Doenecke D, Drabent B: Spermatogenesis in mice is not affected by histone H1.1 deficiency. Exp Cell Res 255:114–124 (2000).

Rao BJ, Rao MR: DNase I site mapping and micrococcal nuclease digestion of pachytene chromatin reveal novel structural features. J Biol Chem 262:4472–4476 (1987).

Richler C, Dhara SK, Wahrman J: Histone macroH2A1.2 is concentrated in the XY compartment of mammalian male meiotic nuclei. Cytogenet Cell Genet 89:118–120 (2000).

Rogakou EP, Pilch DR, Orr AH, Ivanova VS, Bonner WM: DNA double-stranded breaks induce histone H2AX phosphorylation on serine 139. J Biol Chem 273:5858–5868 (1998).

Shires A, Carpenter MP, Chalkley R: New histones found in mature mammalian testes. Proc Natl Acad Sci USA 72:2714–2718 (1975).

Shires A, Carpenter MP, Chalkley R: A cysteine-containing H2B-like histone found in mature mammalian testis. J Biol Chem 251:4155–4158 (1976).

Steger K, Klonisch T, Gavenis K, Drabent B, Doenecke D, Bergmann M: Expression of mRNA and protein of nucleoproteins during human spermiogenesis. Mol Hum Reprod 4:939–945 (1998).

Sutovsky P: Ubiquitin-dependent proteolysis in mammalian spermatogenesis, fertilization, and sperm quality control: killing three birds with one stone. Microsc Res Tech 61:88–102 (2003).

Tanaka M, Hennebold JD, Macfarlane J, Adashi EY: A mammalian oocyte-specific linker histone gene H1oo: homology with the genes for the oocyte-specific cleavage stage histone (cs-H1) of sea urchin and the B4/H1M histone of the frog. Development 128:655–664 (2001).

Tanphaichitr N, Sobhon P, Taluppeth N, Chalermisarachai P: Basic nuclear proteins in testicular cells and ejaculated spermatozoa in man. Exp Cell Res 117:347–356 (1978).

Thoma F, Koller T, Klug A: Involvement of histone H1 in the organization of the nucleosome and of the salt-dependent superstructures of chromatin. J Cell Biol 83:403–427 (1979).

Trostle-Weige PK, Meistrich ML, Brock WA, Nishioka K, Bremer JW: Isolation and characterization of TH2A, a germ cell-specific variant of histone 2A in rat testis. J Biol Chem 257:5560–5567 (1982).

Trostle-Weige PK, Meistrich ML, Brock WA, Nishioka K: Isolation and characterization of TH3, a germ cell-specific variant of histone 3 in rat testis. J Biol Chem 259:8769–8776 (1984).

Turner JM, Burgoyne PS, Singh PB: M31 and macroH2A1.2 colocalise at the pseudoautosomal region during mouse meiosis. J Cell Sci 114:3367–3375 (2001).

Turner JM, Mahadevaiah SK, Elliott DJ, Garchon HJ, Pehrson JR, Jaenisch R, Burgoyne PS: Meiotic sex chromosome inactivation in male mice with targeted disruptions of Xist. J Cell Sci 115:4097–4105 (2002).

Unni E, Mayerhofer A, Zhang Y, Bhatnagar YM, Russell LD, Meistrich ML: Increased accessibility of the N-terminus of testis-specific histone TH2B to antibodies in elongating spermatids. Mol Reprod Dev 42:210–219 (1995a).

Unni E, Zhang Y, Kangasniemi M, Saperstein W, Moss SB, Meistrich ML: Stage-specific distribution of the spermatid-specific histone 2B in the rat testis. Biol Reprod 53:820–826 (1995b).

van Roijen HJ, Ooms MP, Spaargaren MC, Baarends WM, Weber RF, Grootegoed JA, Vreeburg JT: Immunoexpression of testis-specific histone 2B in human spermatozoa and testis tissue. Hum Reprod 13:1559–1566 (1998).

Wang ZF, Krasikov T, Frey MR, Wang J, Matera AG, Marzluff WF: Characterization of the mouse histone gene cluster on chromosome 13: 45 histone genes in three patches spread over 1 Mb. Genome Res 6:688–701 (1996).

Wang ZF, Sirotkin AM, Buchold GM, Skoultchi AI, Marzluff WF: The mouse histone H1 genes: gene organization and differential regulation. J Mol Biol 271:124–138 (1997).

Ward WS, Zalensky AO: The unique, complex organization of the transcriptionally silent sperm chromatin. Crit Rev Eukaryot Gene Expr 6:139–147 (1996).

Wells D, Hoffman D, Kedes L: Unusual structure, evolutionary conservation of non-coding sequences and numerous pseudogenes characterize the human H3.3 histone multigene family. Nucl Acids Res 15:2871–2889 (1987).

Witt O, Albig W, Doenecke D: Testis-specific expression of a novel human H3 histone gene. Exp Cell Res 229:301–306 (1996).

Witt O, Albig W, Doenecke D: Transcriptional regulation of the human replacement histone gene H3.3B. FEBS Lett 408:255–260 (1997).

Witt O, Albig W, Doenecke D: cAMP/phorbol ester response element is involved in transcriptional regulation of the human replacement histone gene H3.3B. Biochem J 329:609–613 (1998).

Wolfe SA, Grimes SR: Histone H1t: a tissue-specific model used to study transcriptional control and nuclear function during cellular differentiation. J Cell Biochem 53:156–160 (1993).

Wolfe SA, Anderson JV, Grimes SR, Stein GS, Stein JS: Comparison of the structural organization and expression of germinal and somatic rat histone H4 genes. Biochim Biophys Acta 1007:140–150 (1989).

Wolfe SA, van Wert JM, Grimes SR: Expression of the testis-specific histone H1t gene: evidence for involvement of multiple *cis*-acting promoter elements. Biochemistry 34:12461–12469 (1995).

Wouters-Tyrou D, Martinage A, Chevaillier P, Sautiere P: Nuclear basic proteins in spermiogenesis. Biochimie 80:117–128 (1998).

Wykes SM, Krawetz SA: The structural organization of sperm chromatin. J Biol Chem 278:29471–29477 (2003).

Wu RS, Bonner WM: Separation of basal histone synthesis from S-phase histone synthesis in dividing cells. Cell 27:321–330 (1981).

Yan W, Burns KH, Ma L, Matzuk MM: Identification of Zfp393, a germ cell-specific gene encoding a novel zinc finger protein. Mech Dev 118:233–239 (2002a).

Yan W, Rajkovic A, Viveiros MM, Burns KH, Eppig JJ, Matzuk MM: Identification of *Gasz,* an evolutionarily conserved gene expressed exclusively in germ cells and encoding a protein with four ankyrin repeats, a sterile-alpha motif, and a basic leucine zipper. Mol Endocrinol 16:1168–1184 (2002b).

Yan W, Ma L, Burns KH, Matzuk MM: HILS1 is a spermatid-specific linker histone H1-like protein implicated in chromatin remodeling during mammalian spermiogenesis. Proc Natl Acad Sci USA 100:10546–10551 (2003).

Zalenskaya IA, Bradbury EM, Zalensky AO: Chromatin structure of telomere domain in human sperm. Biochem Biophys Res Commun 279:213–218 (2000).

Zalensky AO, Breneman JW, Zalenskaya IA, Brinkley BR, Bradbury EM: Organization of centromeres in the decondensed nuclei of mature human sperm. Chromosoma 102:509–518 (1993).

Zalensky AO, Tomilin NV, Zalenskaya IA, Teplitz RL, Bradbury EM: Telomere-telomere interactions and candidate telomere binding protein(s) in mammalian sperm cells. Exp Cell Res 232:29–41 (1997).

Zalensky AO, Siino JS, Gineitis AA, Zalenskaya IA, Tomilin NV, Yau P, Bradbury EM: Human testis/sperm-specific histone H2B (hTSH2B). Molecular cloning and characterization. J Biol Chem 277:43474–43480 (2002).

Cytogenet Genome Res 105:215–221 (2004)
DOI: 10.1159/000078191

Single-cell quantitative RT-PCR analysis of *Cpt1b* and *Cpt2* gene expression in mouse antral oocytes and in preimplantation embryos

L. Gentile,[a] M. Monti,[a] V. Sebastiano,[a] V. Merico,[a] R. Nicolai,[b] M. Calvani,[b] S. Garagna,[a] C.A. Redi,[a] and M. Zuccotti[c]

[a] Laboratorio di Biologia dello Sviluppo e Centro di Eccellenza in Biologia Applicata, Dipartimento di Biologia Animale, Università degli Studi di Pavia, Pavia;
[b] Sigma-Tau S.p.A., Pomezia;
[c] Dipartimento di Medicina Sperimentale, Sezione di Istologia ed Embriologia, Universita' degli Studi di Parma, Parma (Italy)

Abstract. Fatty acids represent an important energy source for preimplantation embryos. Fatty acids oxidation is correlated with the embryo oxygen consumption which remains relatively constant up to the 8-cell stage, but suddenly increases between the 8-cell and morula stages. The degradation of fatty acids occurs in mitochondria and is catalyzed by several carnitine acyl transferases, including two carnitine palmitoyl transferases, CPT-I and CPT-II. We have carried out a study to determine the relative number of transcripts of *Cpt1b* and *Cpt2* genes encoding for *m*-CPT-I and CPT-II enzymes, during mouse preimplantation development. Here we show that *Cpt1b* transcripts are first and temporally detected at the 2-cell stage and reappear at the morula and blastocyst stage. *Cpt2* transcripts decrease following fertilization to undetectable levels and are present again later at the morula stage. These results show that transcription of both *Cpt1b* and *Cpt2* is triggered at the morula stage, concomitantly with known increasing profiles of oxygen uptake and fatty acids oxidation. Based on the number of *Cpt2* transcripts detected, we could discriminate the presence of two groups of embryos with high and low number of transcripts, from the zygote throughout preimplantation development. To further investigate if the establishment of these two groups of embryos occurs prior to fertilization, we have analyzed the relative number of transcripts of both genes in antral and ovulated MII oocytes. As for preimplantation embryos, MII oocytes show two groups of *Cpt2* expression. Antral oocytes, classified according to their chromatin configuration in SN (surrounded nucleolus, in which the nucleolus is surrounded by a rim of Hoechst-positive chromatin) and NSN (not surrounded nucleolus, in which this rim is absent), show three groups with different numbers of *Cpt2* transcripts. All NSN oocytes have a number of *Cpt2* transcripts doubled compared to that of the group of MII oocytes with high expression. Instead, SN oocytes could be singled out into two groups with high and low numbers of *Cpt2* transcripts, similar to those found in MII oocytes. The results of this study point out a correlation between the timing of fatty acids oxidation during preimplantation development and the expression of two genes encoding two enzymes involved in the oxidative pathway. Furthermore, although the biological meaning for the presence of two groups of oocytes/embryos with different levels of *Cpt2* transcripts remains unclear, the data obtained suggest a possible correlation between the levels of *Cpt2* expression and embryo developmental competence.

Copyright © 2004 S. Karger AG, Basel

Supported by Sigma-Tau S.p.A. Pomezia, Italy, by MIUR-COFIN 2002-2003 and Ministero della Salute - Ricerca Ministeriale Finalizzata 2002.

Received 25 November 2003; manuscript accepted 5 December 2003.

Request reprints from Maurizio Zuccotti, Laboratorio di Biologia dello Sviluppo Dipartimento di Biologia Animale, Piazza Botta 9, 27100 Pavia (Italy)
telephone: 390 382 506 323; fax: 390 382 506 270
e-mail: maurizio.zuccotti@unipv.it

Very little is known about the metabolism of fatty acids during preimplantation development, although they represent an important source of energy for the embryo (Flynn and Hillman, 1978; Kane, 1979). Fatty acids are incorporated into lipids and are used for oxidation to carbon dioxide (Cholewa and Whitten, 1970). Fatty acid β-oxidation is correlated with the embryo oxygen consumption which remains relatively constant up to the 8-cell stage, but suddenly increases between the 8-cell and morula stages (Brinster, 1967; Flynn and Hillman, 1978; Houghton et al., 1996).

The degradation of fatty acids occurs in mitochondria (Lehninger, 1970) and carnitine is required for the transport of activated acyls (i.e., acyl-CoA) across the inner mitochondrial membrane (Bieber, 1988; McGarry and Brown, 1997). The reversible exchange of acyl groups between CoA and carnitine is catalyzed by several carnitine acyl transferases (CAT). CAT enzymes are mainly subdivided into two subgroups, the carnitine octanoyl transferases (COT), which are extra-mitochondrial proteins, and the carnitine palmitoyl transferases (CPT) that are mitochondrial enzymes. When preimplantation embryos reach the morula and blastocyst stage, glycolysis is the major metabolic pathway for energy production, and palmitic acid is the major substrate among fatty acids (Harvey et al., 2002). The activity of CPT results from the integrated action of two different proteins: a 68-kDa oligomer (Finocchiaro et al., 1990) located in the inner mitochondrial membrane (CPT-II) and a second subunit (CPT-I) present on either the outer side of the inner mitochondrial membrane or on the outer mitochondrial membrane (Esser et al., 1993; Di Lisa et al., 1995). CPT-I has two isoforms, l-CPT-I, found in the liver, and m-CPT-I, found in skeletal muscles (Taggart et al., 1999). These isoforms are encoded by two genomic sequences, named *Cpt1a*, located on mouse chromosome 19, and *Cpt1b*, located on chromosome 15 (Cox et al., 1998). CPT-II is encoded by a single copy gene located on mouse chromosome 4 (Yang et al., 1998).

Changes in the rate of fatty acids β-oxidation occurring during preimplantation development are indicative of a change in mitochondrial oxidative function. As a first step towards the understanding of fatty acids metabolism in preimplantation development, we have carried out a study to determine the relative number of transcripts of *Cpt1b* and *Cpt2* genes encoding for m-CPT-I and CPT-II enzymes. Making use of a single-cell sensitive semi-quantitative reverse transcriptase polymerase chain reaction (RT-PCR), we analyzed the expression of these two genes in antral and metaphase II (MII) oocytes and in preimplantation embryos.

Materials and methods

Animals

Adult 5-month-old male and 4-week-old female B6C3F1 mice were purchased from Charles River (Como, Italy). The animals were maintained in a temperature- and humidity-controlled room with 12-hour light and 12-hour dark phases.

Oocyte collection

Antral oocytes were isolated from the ovaries of unprimed females or from females injected with 7.5 IU PMSG (pregnant mare serum gondadotrophin) and sacrificed after 48 h, as previously described (Zuccotti et al., 1995; Longo et al., 2003). Following isolation, follicle-cell free antral oocytes were classified, on the basis of their chromatin organization, in SN (surrounded nucleolus) oocytes or NSN (not surrounded nucleolus) oocytes (Zuccotti et al., 1995). Single SN, NSN and MII oocytes were transferred to the bottom of a 0.2-ml Eppendorf tube containing 1.5 μl lysis buffer (Zuccotti et al., 2002) and processed as for RT-PCR.

MII oocytes were isolated from the oviducts of females following natural ovulation or treated with a first intraperitoneal injection of 7.5 IU PMSG followed 48 h later by an injection of 7.5 IU hCG (human chorionic gonadotrophin) and briefly exposed to M2 medium (Fulton and Whittingham, 1978) containing 500 U/ml hyaluronidase to remove cumulus cells.

Sperm isolation and capacitation

Sperm were isolated by puncturing cauda epididymes of 5-month-old male mice of proven fertility. Punctured cauda epididymes were transferred to the bottom of a 5-ml tube and gently covered with 1.5 ml of Whittingham medium (Whittingham, 1971) pre-warmed at 37 °C in 5 % CO_2 in air. Sperm were allowed to swim up for 20 min; the upper 400 μl suspension was transferred to a 30-mm Petri dish and covered with mineral oil. The drop was incubated at 37 °C in 5 % CO_2 atmosphere for 40 min for sperm capacitation.

IVF and embryo collection

MII oocytes isolated as described above were transferred to a 100-μl drop of Whittingham medium containing 1.8×10^6 sperm/ml. After 2 h, oocytes were transferred to a 40-μl drop of M16 medium (Whittingham, 1971) for embryonic development.

Single preimplantation embryos at different stages of development were transferred to the bottom of a 0.2-ml Eppendorf tube containing 1.5 μl lysis buffer as described above. Stages of preimplantation embryonic development analyzed were as follows: 1-cell, 6 h post insemination (pi); 2-cell, 32 h pi; 4-cell, 48 h pi; 8-cell, 56 h pi; morula, 72 h pi and blastocyst, 96 h pi.

RT-PCR

Relative amounts of *Cpt1b* and *Cpt2* gene transcripts were determined by using a semi-quantitative RT-PCR assay. At least 15–20 single oocytes/ embryos were analyzed for each developmental stage. Eighteen microliters of the following reaction mixture were added to each 2-μl oocyte/embryo sample: 1 μl (2,500 molecules) of exogenous *pAW109* RNA (used as control for normalization of results), 1× PCR Buffer II, 5 mM $MgCl_2$, 4 mM each dNTP, 2.5 μM oligo $d(T)_{16}$, 20 U of RNase inhibitor, 50 U of MuLV reverse transcriptase (Applera, Monza, Italy). Both RT and PCR reactions were performed on an Applied Biosystems GeneAmp 9700 thermocycler. The amplification program for the reverse transcription step was as follows: 25 °C for 10 min, 42 °C for 15 min, 99 °C for 5 min. After the reverse transcriptase reaction, samples were kept at 4 °C overnight; then each sample was split into four 5-μl parts and to each of these, 20 μl of the following PCR mixture (Applera, Monza, Italy) were added: 1× PCR Buffer I, 200 μM each dNTP, 250 nM of each specific primer, 1.25 IU of AmpliTaq polymerase. The exogenous control *(pAW109)*, the endogenous control *(Hprt)* and the two investigated genes (i.e., *Cpt1b* and *Cpt2*) were co-amplified with the following PCR cycle program: 95 °C for 5 min followed by 18 cycles of a two-step touchdown PCR with a first step at 95 °C for 15 s and a second step with a starting annealing/extension temperature of 63.5 °C, which decreases by 0.5 °C every cycle, for 90 s, and a final elongation step at 72 °C for 7 min. A second round of PCR amplification was performed for each gene sequence separately (nested PCR), by diluting 1 μl of the first-round reaction into 24 μl of a new reaction mixture prepared as described above, with primers designed internal to the first PCR sequence. The PCR cycle parameters were as follows: 95 °C for 5 min, followed by 23 cycles of 95 °C for 15 s, 55 °C for 60 s, 72 °C for 60 s and a final elongation step at 72 °C for 7 min. Ten microliters of PCR product were mixed with 2 μl loading buffer and were electrophoresed on a 2.5 % agarose gel in 0.5 % TBE containing 0.5 μg/ml ethidium bromide at 6 V/cm for 85 min. The products were visualized under short-wave length UV on a Bio-Rad Gel Doc system and densitometric analysis was performed with the Bio-Rad Quantity-One software. The relative number of transcripts (rnt) of the genes under study was obtained after normalization of the data with those of the exogenous control *pAW109*.

Of the four 5-μl aliquots obtained following reverse transcription, three were used to study the expression of *Hprt*, *Cpt1b* and *Cpt2*; one was employed for embryo sexing (Greenlee et al., 1998).

PCR primers

Table 1 shows the list of primers used for the PCR amplification of the investigated gene sequences. To avoid genomic amplification, one of the external primers was chosen to overlap exon/intron boundaries.

Table 1. List of the primer sequences used for the quantitative RT-PCR

Primer name	Sequence	Amplicon length	Reference
Cpt1b A	5'-CAAGTTCAGAGACGAACGCC-3'	262	–
Cpt1b B	5'-TCAAGAGCTGTTCTCCGAACTG-3'		–
Cpt1b C	5'-TTTGGGAACCACATCCGCCAA-3'	194	–
Cpt1b D	5'-TTATGCCTGTGAGCTGGCCAC-3'		–
Cpt2 A	5'-TCTGCCCAGCTTCCATCTTT-3'	263	–
Cpt2 B	5'-GGTGGACAGGATGTTGTGGTTT-3'		–
Cpt2 C	5'-GCCCAGCTTCCATCTTTACT-3'	254	–
Cpt2 D	5'-CAGGATGTTGTGGTTTATCCGC-3'		–
Hprt A	5'-CCTGCTGGATTACATTAAAGCACT-3'	358	Kay et al., 1993
Hprt B	5'-GTCAAGGGCATATCCAACAACAAAC-3'		Kay et al., 1993
Hprt C	5'-TCAGTCAACGGGGGACATAA-3'	218	Kay et al., 1993
Hprt D	5'-ATCCAACAAAGTCTGGCCTG-3'		Kay et al., 1993
pAW109 A	5'-AAACAGATGAAGTGCTCCTTCCAGG-3'	306	Wang et al., 1989
pAW109 B	5'-TGGAGAACACCACTTGTTGCTCCA-3'		Wang et al., 1989
pAW109 C	5'-CAGATGAAGTGCTCCTTCCA-3'	302	–
pAW109 D	5'-AGAACACCACTTGTTGCTCC-3'		–
Zfy A	5'-AAGATAAGCTTACATAATCACATGGA-3'	599–617	Kunieda et al., 1992
Zfy B	5'-CCTATGAAATCCTTTGCTGCACATGT-3'		Kunieda et al., 1992
Zfy C	5'-GTAGGAAGAATCTTTCTCATGCTGG-3'	199–217	Kunieda et al., 1992
Zfy D	5'-TTTTTGAGTGCTGATGGGTGACGG-3'		Kunieda et al., 1992
DXNds3 A	5'-GAGTGCCTCATCTATACTTACAG-3'	244	Kunieda et al., 1992
DXNds3 B	5'-TCTAGTTCATTGTTGATTAGTTGC-3'		Kunieda et al., 1992
DXNds3 C	5'-ATGCTTGGCCAGTGTACATAG-3'	111	Kunieda et al., 1992
DXNds3 D	5'-TCCGGAAAGCAGCCATTGGAGA-3'		Kunieda et al., 1992

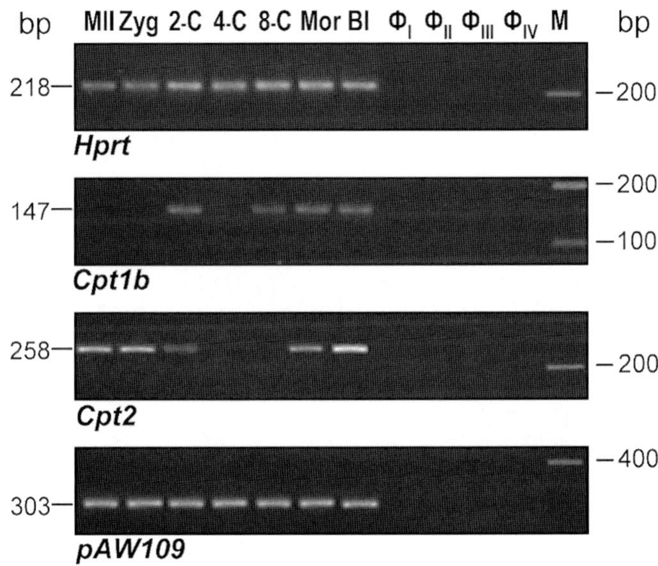

Fig. 1. An example of the results obtained after RT-PCR amplification of *Cpt1b*, *Cpt2* and *Hprt* sequences in metaphase II oocytes (MII) and in preimplantation embryos, from the zygote (Zyg) to the blastocyst (Bl) stage. *pAW109* is the exogenous reference standard for quantification of the number of transcripts of the genes under study (see Materials and methods).

Statistical analysis

Fifteen to 20 single oocytes or preimplantation embryos at the zygote, 2-cell, 4-cell, 8-cell, morula and blastocyst stage were analyzed through 3–4 experiments. The two-tail test was used to analyze the distribution of the data that were found to have a gaussian distribution. Means and standard deviations were compared with the Student t test, differences were considered when $P < 0.05$.

Results and discussion

Cpt1b, Cpt2 and Hprt expression during preimplantation development

Figure 1 shows an example of the results obtained after RT-PCR amplification of *Cpt1b*, *Cpt2* and *Hprt* sequences in preimplantation embryos. The relative number of transcripts (rnt) detected at each stage of oocyte maturation or embryonic development represents the steady-state level of transcripts present, which is the summation of transcription and transcript degradation.

The number of transcripts of the endogenous control gene, *Hprt*, increased significantly ($P < 0.05$) during the first cell cycle, from 3,236 ± 395 (mean ± standard deviation) rnt at the zygote stage to 3,969 ± 792 rnt at the 2-cell stage. Then, transcripts remained constant until the 8-cell stage and increased again, up to 4,588 ± 332 rnt, during the following cell divisions (Fig. 2).

Cpt1b and *Cpt2* genes showed different profiles of expression. *Cpt1b* transcripts were not detected in 1-cell embryos, they were temporally found (537 ± 147 rnt) at the S-phase of 2-cell embryos (32 h pi) and detected again at lower levels in few (22%) 8-cell embryos (162 ± 64 rnt); then they increased ($P < 0.05$) at the morula (503 ± 253 rnt) and blastocyst (629 ± 253 rnt) stages.

Cpt2 transcripts were present throughout preimplantation development (Fig. 2). From the 1-cell stage (2,967 ± 1,132 rnt) the number of transcripts decreased and almost disappeared at the 4-cell (121 ± 180 rnt) and 8-cell (153 ± 218 rnt) stage. Transcription of *Cpt2* began at the morula stage (1,763 ± 1,100 rnt) and increased abruptly, reaching its relative maximum at the blastocyst stage (7,727 ± 2,697 rnt). Interestingly, the coincident raise of both the number of *Cpt1b* and *Cpt2* transcripts

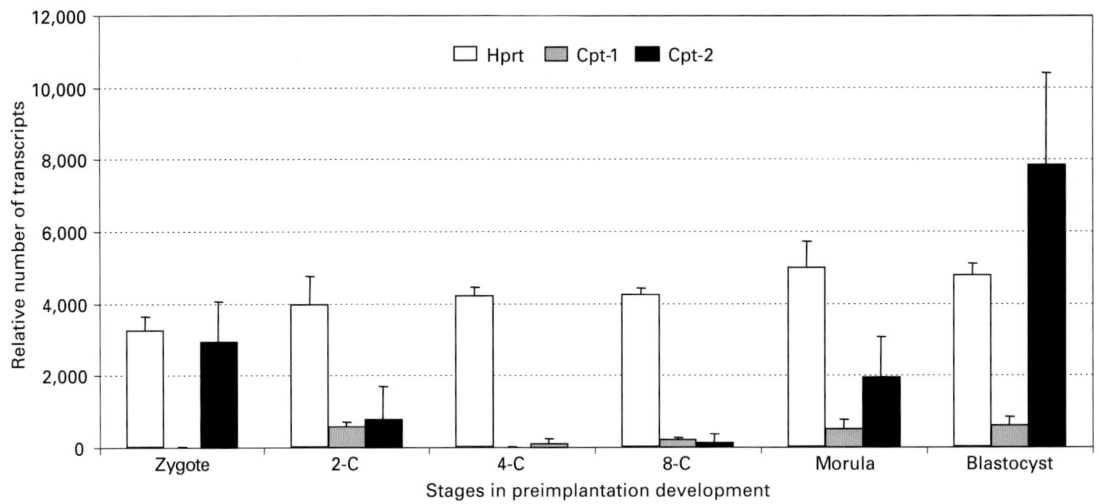

Fig. 2. Relative number of transcripts throughout all stages of preimplantation development. *Hprt* is expressed at the same level in all the embryos analyzed; *Cpt1b* is transiently expressed at 2-cell stage and is expressed again from the stage of 8-cell embryo. The number of transcripts of *Cpt2* is high at the zygote stage, decreases at 2-, 4-, 8-cell embryos and then abruptly increases from the morula stage.

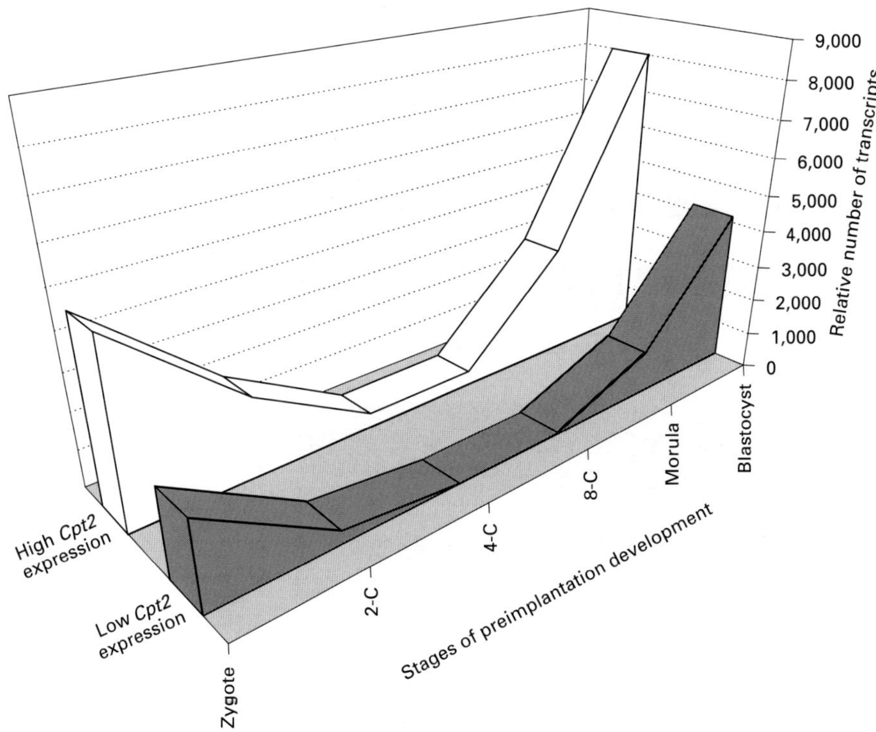

Fig. 3. The quantitative analysis of the number of *Cpt2* transcripts revealed the presence of two groups of embryos from 1-cell to the blastocyst stage of development: one with low and the other with high expression.

at the morula stage is correlated with biochemical, physiological and epigenetic modifications occurring at this stage of development: a) increasing profiles of oxygen uptake (Houghton et al., 1996) and fatty acids oxidation (Hillman and Flynn, 1980); b) morphological modifications of mitochondria (Stern et al., 1971); c) reorganization of chromatin structure (Clarke et al., 1992; Thompson, 1996) and DNA methylation (Santos et al., 2002).

Quantitative analysis of Cpt2 expression evidences the presence of two groups of embryos throughout preimplantation development

Figure 2 clearly shows the high degree of variability of *Cpt2* expression existing among embryos of the same developmental stage. Statistical analysis of the number of *Cpt2* transcripts at each stage of development evidences a bimodal gaussian distribution, in which it was possible to differentiate two distinct groups of embryos (M1–M2 ≥ 2σ) (Fig. 3). Interestingly, 30%

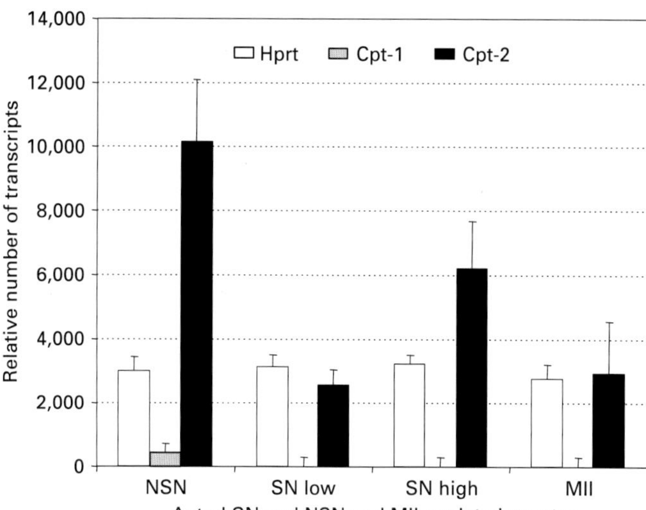

Fig. 4. The relative number of *Hprt* transcripts was not significantly different in NSN and SN oocytes and decreased in MII oocytes. *Cpt1b* was expressed only in antral NSN oocytes. NSN oocytes showed the highest number of *Cpt2* transcripts compared to SN and MII oocytes.

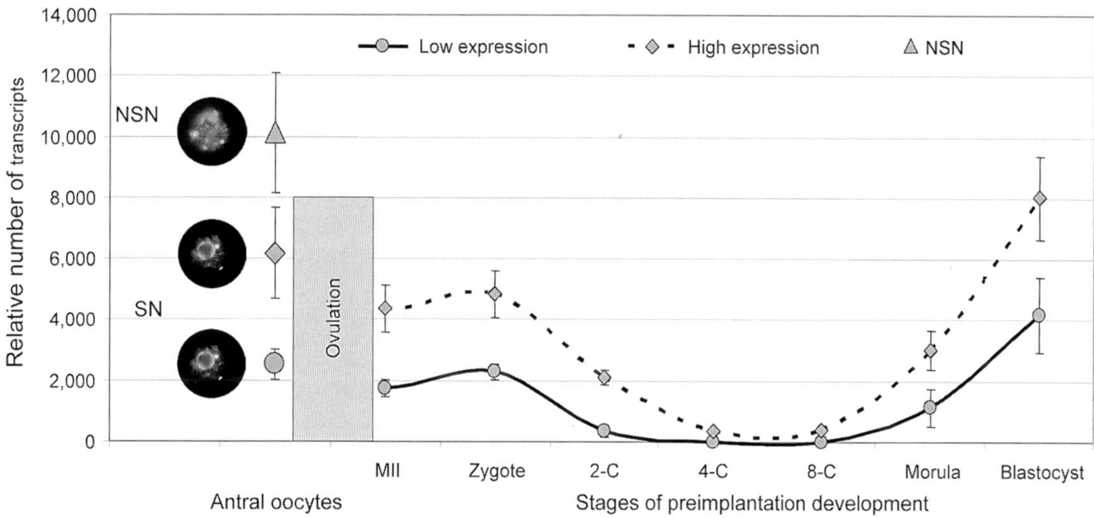

Fig. 5. The existence of two groups of *Cpt2* level of expression was found in antral SN and MII oocytes and in embryos throughout the preimplantation development. In NSN oocytes the highest number of transcripts was detected. It is likely that NSN are not ovulated while both groups of SN oocytes are.

of 1-cell embryos show a high number of *Cpt2* transcripts; this percentage increases to 70% in blastocysts.

Earlier studies have demonstrated a differential developmental rate between male and female embryos (Valdivia et al., 1993), which could be attributed to a different metabolism. We have sexed each of the preimplantation embryos analyzed and found no existing correlation between *Cpt2* expression and embryo gender (data not shown).

To further investigate if the establishment of these two groups of embryos occurs prior to fertilization, we have analyzed the expression profiles of *Cpt1b* and *Cpt2* genes in antral and in ovulated MII oocytes.

Cpt1b, Cpt2 and Hprt expression in antral and ovulated oocytes

Mouse antral oocytes are characterized by one of two nuclear morphologies (Mattson and Albertini, 1990; Debey et al., 1993; Zuccotti et al., 1995): SN (surrounded nucleolus), in which the nucleolus is surrounded by a rim of Hoechst-positive chromatin and NSN (not surrounded nucleolus), in which this rim is essentially absent. This morphological difference has a biological relevance as NSN oocytes are incapable of development beyond the 2-cell stage; whereas SN oocytes are capable of development to the blastocyst stage (Zuccotti et al., 1998, 2002).

As shown in Fig. 4, the relative number of *Hprt* transcripts was not significantly different in NSN and SN oocytes (3,069 ± 408 rnt) and decreased in MII oocytes (2,747 ± 463 rnt). *Cpt1b* was expressed only in antral NSN oocytes (428 ± 298 rnt).

In MII oocytes, the number of *Cpt2* transcripts was attested at 2,925 ± 1,625 rnt (Fig. 4). The statistical analysis demonstrated the existence, as for preimplantation embryos, of two groups of oocytes: 50% of which with a low (1,885 ± 198) and 50% with a high (4,302 ± 485) number of transcripts (Fig. 5).

When we analyzed antral oocytes, they showed three groups of *Cpt2* expression. NSN oocytes showed a number of *Cpt2* transcripts doubled (10,125 ± 1,973) compared to that of the group of MII oocytes with high expression. In SN oocytes we could distinguish two groups with a profile of *Cpt2* expression similar to that found in MII oocytes: 50% of SN oocytes had a high (6,183 ± 1,508) and 50% a low (2,538 ± 491) number of transcripts.

NSN oocytes are considered as a transcriptionally active, immature form, whereas SN oocytes are transcriptionally inactive (Parfenov et al., 1989; Longo et al., 2003; Zatsepina et al., 2003) and are believed to be ovulated. In Fig. 5, we hypothesize that both SN oocytes with high and low *Cpt2* expression are ovulated, since we found two groups with different *Cpt2* expression in MII oocytes. Interestingly, when compared to antral SN oocytes, the two groups of MII oocytes show a 30% decrease in the number of *Cpt2* transcripts, which is in accordance with the expected overall decrease of stored RNA occurring during the transition from antral to ovulated MII oocytes (Wassarman, 1988).

To make certain that the presence of two groups of oocytes/ embryos with high and low numbers of *Cpt2* transcripts was not artificial and due to the in vitro condition or the hormonal treatment, we isolated antral oocytes from unprimed females, MII oocytes and preimplantation embryos following natural ovulation or mating, respectively. The pattern of expression of *Hprt*, *Cpt1b* and *Cpt2* in antral and MII oocytes and in preimplantation embryos obtained in vivo was not significantly different from that described for the in vitro condition, with the exception of *Cpt1b* transcripts that were not detected in the 2-cell embryos.

Also for oocytes/embryos obtained in vivo, we were able to distinguish the presence of two groups with high and low numbers of *Cpt2* transcripts, thus confirming that this phenomenon occurs naturally and is not induced by the experimental procedures.

The equal presence of two groups of oocytes with low and high numbers of *Cpt2* transcripts and the increase of the latter up to 70% at the blastocyst stage may correlate with the embryo's developmental competence. To test this working hypothesis we are planning to isolate microbiopsies from oocytes/embryos, and compare the quantitative profile of *Cpt2* expression with the oocyte developmental competence following in vitro fertilization and further preimplantation embryo culture. In order to carry out this latter experiment, we have already set up a quantitative RT-PCR capable of detecting the presence of *Hprt* transcripts in a single microbiopsy of about 10–12 μm in diameter.

The results of this study point out a correlation between the timing of fatty acids oxidation during preimplantation development and the expression of two genes coding for two enzymes involved in the oxidative pathway. Furthermore, although the biological meaning of the presence of two groups of oocytes/embryos with different levels of *Cpt2* expression remains unclear, the data obtained suggest a possible correlation between the levels of *Cpt2* expression and the embryo's developmental competence. An ideal way to study the functions of this gene could be the use of the RNA interference methods, recently described in mouse oocytes (Svoboda et al., 2000) and preimplantation embryos (Wianny and Zernicka-Goetz, 2000; see also Svoboda, this issue).

References

Bieber LL: Carnitine. Annu Rev Biochem 57:261–283 (1988).

Brinster RL: Carbon dioxide production from lactate and pyruvate by the preimplantation mouse embryo. Exp Cell Res 47:634–637 (1967).

Cholewa JA, Whitten WK: Development of two-cell mouse embryos in the absence of a fixed nitrogen source. J Reprod Fertil 22:553–555 (1970).

Clarke HJ, Oblin C, Bustin M: Developmental regulation of chromatin composition during mouse embryogenesis: somatic histone H1 is first detectable at the 4-cell stage. Development 115:791–799 (1992).

Cox KB, Johnson KR, Wood PA: Chromosomal locations of the mouse fatty acid oxidation genes *Cpt1a*, *Cpt1b*, *Cpt2*, *Acadvl*, and metabolically related *Crat* gene. Mamm Genome 9:608–610 (1998).

Debey P, Szollosi MS, Szollosi D, Vautier D, Girousse A, Besombes D: Competent mouse oocytes isolated from antral follicles exhibit different chromatin organization and follow different maturation dynamics. Mol Reprod Dev 36:59–74 (1993).

Di Lisa F, Blank PS, Colonna R, Gambassi G, Silverman HS, Stern MD, Hansford RG: Mitochondrial membrane potential in single living adult rat cardiac myocytes exposed to anoxia or metabolic inhibition. J Physiol 486 (Pt 1):1–13 (1995).

Esser V, Britton CH, Weis BC, Foster DW, McGarry JD: Cloning, sequencing, and expression of a cDNA encoding rat liver carnitine palmitoyltransferase I. Direct evidence that a single polypeptide is involved in inhibitor interaction and catalytic function. J Biol Chem 268:5817–5822 (1993).

Finocchiaro G, Colombo I, DiDonato S: Purification, characterization and partial amino acid sequences of carnitine palmitoyl-transferase from human liver. FEBS Lett 274:163–166 (1990).

Flynn TJ, Hillman N: Lipid synthesis from [U14C]glucose in preimplantation mouse embryos in culture. Biol Reprod 19:922–926 (1978).

Fulton BP, Whittingham DG: Activation of mammalian oocytes by intracellular injection of calcium. Nature 273:149–151 (1978).

Greenlee AR, Krisher RL, Plotka ED: Rapid sexing of murine preimplantation embryos using a nested, multiplex polymerase chain reaction (PCR). Mol Reprod Dev 49:261–267 (1998).

Harvey AJ, Kind KL, Thompson JG: REDOX regulation of early embryo development. Reproduction 123:479–486 (2002).

Hillman N, Flynn TJ: The metabolism of exogenous fatty acids by preimplantation mouse embryos developing in vitro. J Embryol Exp Morphol 56:157–168 (1980).

Houghton FD, Thompson JG, Kennedy CJ, Leese HJ: Oxygen consumption and energy metabolism of the early mouse embryo. Mol Reprod Dev 44:476–485 (1996).

Kane MT: Fatty acids as energy sources for culture of one-cell rabbit ova to viable morulae. Biol Reprod 20:323–332 (1979).

Kay GF, Penny JD, Patel D, Ashworth A, Brockdorf N, Rastan S: Expression of *Xist* during mouse development suggests a role in the initiation of X chromosome inactivation. Cell 72:171–182 (1993).

Kunieda T, Xian M, Kobayashi E, Imamichi T, Moriwaki K, Toyoda Y: Sexing of mouse preimplantation embryos by detection of Y chromosome-specific sequences using polymerase chain reaction. Biol Reprod 46:692–697 (1992).

Lehninger AL: Mitochondria and calcium ion transport. Biochem J 119:129–138 (1970).

Longo F, Garagna S, Merico V, Orlandini G, Gatti R, Scandroglio R, Redi CA, Zuccotti M: Nuclear localization of NORs and centromeres in mouse oocytes during folliculogenesis. Mol Reprod Dev 66:279–290 (2003).

Mattson BA, Albertini DF: Oogenesis: chromatin and microtubule dynamics during meiotic prophase. Mol Reprod Dev 25:374–383 (1990).

McGarry JD, Brown NF: The mitochondrial carnitine palmitoyltransferase system. From concept to molecular analysis. Eur J Biochem 244:1–14 (1997).

Parfenov V, Potchukalina G, Dudina L, Kostyuchek D, Gruzova M: Human antral follicles: oocyte nucleus and the karyosphere formation (electron microscopic and autoradiographic data). Gamete Res 22:219–231 (1989).

Santos F, Hendrich B, Reik W, Dean W: Dynamic reprogramming of DNA methylation in the early mouse embryo. Dev Biol 241:172–182 (2002).

Stern S, Biggers JD, Anderson E: Mitochondria and early development of the mouse. J Exp Zool 176:179–191 (1971).

Svoboda P: Long dsRNA and silent genes strike back: RNAi in mouse oocytes and early embryos. Cytogenet Genome Res 105:422–434 (2004).

Svoboda P, Stein P, Hayashi H, Schultz RM: Selective reduction of dormant maternal mRNAs in mouse oocytes by RNA interference. Development 127:4147–4156 (2000).

Taggart RT, Smail D, Apolito C, Vladutiu GD: Novel mutations associated with carnitine palmitoyltransferase II deficiency. Hum Mutat 13:210–220 (1999).

Thompson EM: Chromatin structure and gene expression in the preimplantation mammalian embryo. Reprod Nutr Dev 36:619–635 (1996).

Valdivia RP, Kunieda T, Azuma S, Toyoda Y: PCR sexing and developmental rate differences in preimplantation mouse embryos fertilized and cultured in vitro. Mol Reprod Dev 35:121–126 (1993).

Wang AM, Doyle MV, Mark DF: Quantitation of mRNA by the polymerase chain reaction. Proc Natl Acad Sci USA 86:9717–9721 (1989).

Wassarman PM: The Mammalian Ovum, in Knobil E, Neill JD (eds): The Physiology of Reproduction, pp 69–102 (Raven Press, New York 1988).

Whittingham DG: Culture of mouse ova. J Reprod Fertil Suppl 14:7–21 (1971).

Wianny F, Zernicka-Goetz M: Specific interference with gene function by double-stranded RNA in early mouse development. Nat Cell Biol 2:70–75 (2000).

Yang BZ, Ding JH, Dewese T, Roe D, He G, Wilkinson J, Day DW, Demaugre F, Rabier D, Brivet M, Roe C: Identification of four novel mutations in patients with carnitine palmitoyltransferase II (CPT II) deficiency. Mol Genet Metab 64:229–236 (1998).

Zatsepina O, Baly C, Chebrout M, Debey P: The stepwise assembly of a functional nucleolus in preimplantation mouse embryos involves the Cajal (coiled) body. Dev Biol 253:66–83 (2003).

Zuccotti M, Piccinelli A, Rossi PG, Garagna S, Redi CA: Chromatin organization during mouse oocyte growth. Mol Reprod Dev 41:479–485 (1995).

Zuccotti M, Giorgi Rossi P, Martinez A, Garagna S, Forabosco A, Redi CA: Meiotic and developmental competence of mouse antral oocytes. Biol Reprod 58:700–704 (1998).

Zuccotti M, Boiani M, Ponce R, Guizzardi S, Scandroglio R, Garagna S, Redi CA: Mouse *Xist* expression begins at zygotic genome activation and is timed by a zygotic clock. Mol Reprod Dev 61:14–20 (2002).

Cytogenet Genome Res 105:222–227 (2004)
DOI: 10.1159/000078192

New mouse genetic models for human contraceptive development

C. Lessard, J.K. Pendola, S.A. Hartford, J.C. Schimenti, M.A. Handel, and J.J. Eppig

The Jackson Laboratory, Bar Harbor, ME (USA)

Abstract. Genetic strategies for the post-genomic sequence age will be designed to provide information about gene function in a myriad of physiological processes. Here an ENU mutagenesis program (http://reprogenomics.jax.org) is described that is generating a large resource of mutant mouse models of infertility; male and female mutants with defects in a wide range of reproductive processes are being recovered. Identification of the genes responsible for these defects, and the pathways in which these genes function, will advance the fields of reproduction research and medicine. Importantly, this program has potential to reveal novel human contraceptive targets.

The promise of the "post-sequence" genomic age lies in understanding the function of every gene and the application of this knowledge to the betterment of the human condition, including the regulation of fertility. Rapid and unplanned expansion of human populations has serious environmental, societal and medical consequences, particularly in underdeveloped countries. In order to provide safe, economical, and efficient fertility control for the world's burgeoning population, many researchers are attempting to develop novel contraceptive methods (Aitken, 2002; Anderson and Baird, 2002; Delves et al., 2002). To provide options other than hormonal manipulation, physical barriers, and surgical procedures, recent studies have focused on developing ways to manipulate reproductive processes using chemical analogs, protein repressors and vaccines. Although these strategies have been employed for many years, their success has been highly variable, and while some

show promise for future use, they must be made more reliable before they will be acceptable for public use. Importantly, valid as these approaches are, of necessity they focus on what we know about reproduction. There is much that we do not yet know and this new knowledge will be derived from genomic and proteomic strategies. For example, micro-array analyses have revealed hundreds of transcripts expressed after the initiation of meiosis in the male. They appear to be male germ cell specific and are potential targets for contraception (Schultz et al., 2003).

Our limited knowledge of the mechanisms and pathways of the processes that contribute to human fertility has restricted the development of novel contraceptive strategies. Despite increasing knowledge of genes acting in reproductive pathways, ideal targets for contraceptive agents remain elusive. The vast majority of genetic analyses of reproduction to date have focused targeted inactivation of known genes ("knockouts") to test their function (Matzuk and Lamb, 2002). However, there is good reason to suspect that many genes acting in specific reproductive functions are as yet unknown, because, first, there are many genes in the genome whose functions have not been discovered; and second, many gene knockouts had unexpected reproductive consequences. Thus what is needed is a totally unbiased approach to gene discovery, one that relies on no a priori assumptions about reproductive processes. Here we describe a program for mutagenesis and phenotype screening

Supported by a grant from the NIH (HD42137) to J.J.E, M.A.H. and J.C.S.

Received 8 October 2003; manuscript accepted 19 November 2003.

Request reprints from John J. Eppig, The Jackson Laboratory
   600 Main Street, Bar Harbor, ME 04609 (USA)
   telephone: 207-288-6349; fax: 207-288-6073; e-mail: jje@jax.org

C.L. and J.K.P. contributed equally to this paper.

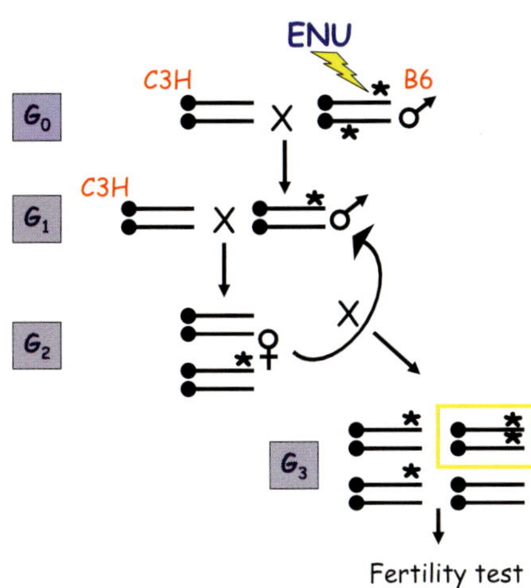

Fig. 1. ENU mutagenesis breeding scheme. Asterisk = ENU-induced mutation.

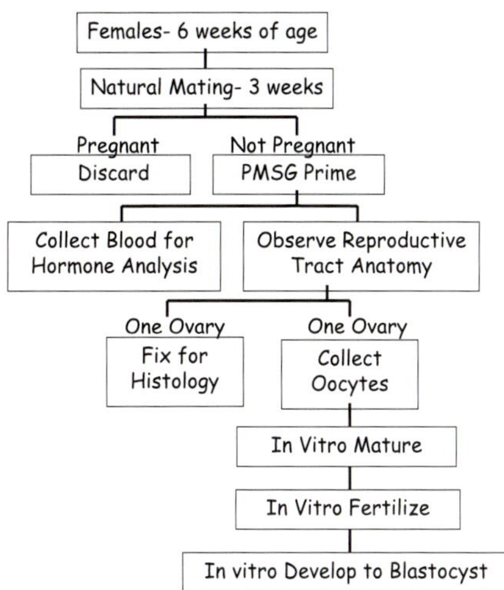

Fig. 2. Protocol for the female infertility clinic.

designed to identify mutant mice that fail to reproduce, which is the goal of contraception. This program will generate a large-scale resource of mutant mouse models of human infertility; it can reveal genes involved in key reproductive processes and in doing so unveil novel targets for new contraceptive agents. In this report we describe the categories of abnormal reproductive phenotypes obtained and consider their potential for leading to novel approaches to contraception. Ultimately, the utility of these mutations for enhancing our knowledge of reproductive pathways and suggesting new approaches to contraception relies on the ability to clone the relevant gene. In this "post sequence" era of the genome information explosion, an abundance of new tools is available that make this strategy possible (Venter et al., 2001, 2003; Waterston et al., 2002; Botstein and Risch, 2003). Thus, once an abnormal reproductive phenotype is identified, it is possible to determine the gene responsible and to analyze its normal biological function.

The mutagenesis and phenotype screening program at the Jackson Laboratory

The mutant mouse models resource we describe constitutes a program funded by the NIH for the purpose of generating models of human infertility. The potent mutagen ethylnitrosourea (ENU) is used to induce random, genome wide, single base-pair mutations (Shibuya and Morimoto, 1993; Justice, 1999; Noveroske et al., 2000). We screen for infertility of test animals to detect those mutations that disrupt the physiological pathways necessary for male and female reproduction. ENU effectively generates not only null mutations, which results in complete loss of the function of a single gene, but also hypomorphic mutations causing partial loss of gene function (Noveroske et al., 2000). These types of defects allow the generation of a full spectrum of mutant mouse genotypes and phenotypes, a key feature in modeling the many causes of human infertility and identifying multiple new contraceptive targets. Important-

ly, the primary screen for phenotypes is mating to test fertility (see below for details); thus it is not based on any prior assumptions about the nature of reproductive pathways, and thus it leads to unbiased gene discovery.

This project uses a 3-generation breeding screen designed to select recessive mutations affecting reproductive functions. Initially, male C57BL6/J (B6) mice (G_0 generation) are treated with ENU at doses known to induce a high frequency of mutations in the DNA of spermatogonial stem cells. Each of these males is mated to several females of a different inbred strain, C3HeB/FeJ (C3H), to generate G_1 offspring (Fig. 1). Each G_1 male carries a different and unique array of recessive DNA mutations acquired from his G_0 father. Each of these G_1 males is used to found a "family" that will segregate the unique set of mutations, some of which may affect reproduction. G_1 males are mated to C3H females to produce G_2 offspring. Females from the G_2 population, each having a 50% chance of carrying any particular mutation from her G_1 father, are mated back to the G_1 father to generate the G_3 population. In families segregating mutant genes, each G_3 animal has a 25% chance of being homozygous for a given mutant allele. Dominant mutations leading to infertility would be notable in either the G_1 male or G_2 females, but no attempt is made to recover these mutant lines. This mutagenesis and screening scheme is widely used and has several advantages: it targets the entire genome (except for the Y chromosome, which is not recovered if it renders G_1 males infertile, and the X chromosome, which is not recovered in the G_1 males), and it facilitates mapping of detected mutations by mixing two genetic backgrounds (B6 and C3H). Similar whole-genome screens are being used to detect mutations affecting a number of physiological parameters, as well as behavior (Balling, 2001).

The mutations detected in such a mutagenesis screen depend on the phenotype analysis that is conducted. The goal of this resource is to generate mutant models of infertility in order to understand further reproductive pathways and infertility as well as to generate new targets for contraception. Thus, to select

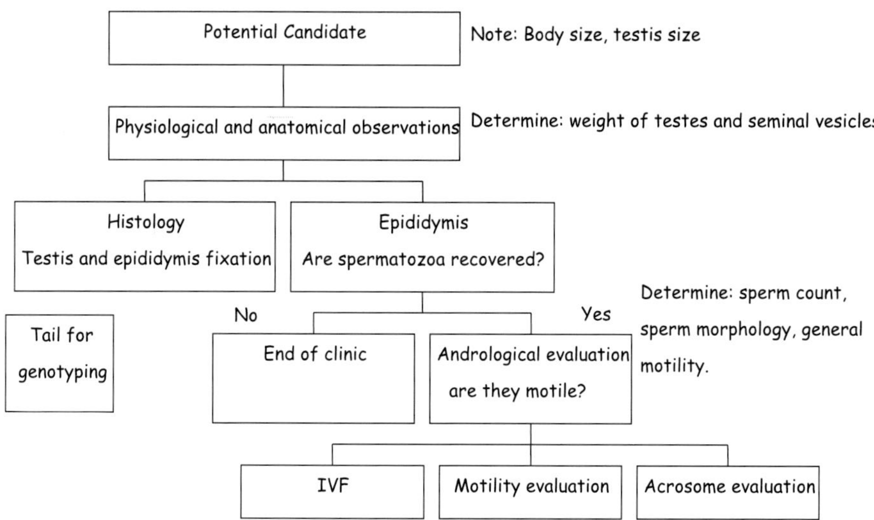

Fig. 3. Protocol for the male infertility clinic.

reproductive mutants, the primary phenotype screen of the G_3 mice is natural mating to wild-type animals. Animals that fail to reproduce are candidates for bearing a homozygous mutation affecting fertility. Subsequent secondary phenotype screens are designed to provide information about the specific aspect of reproduction that is affected. This is accomplished by screening the potentially mutant G_3 mice in female and male infertility "clinics" to assess reproductive tissues, gamete structure, and gamete function in fertilization and activation of development. This two-tiered phenotype screen, first identifies animals that fail to reproduce, and second, describes the phenotype.

In the female infertility "clinic" (Fig. 2), females that failed to become pregnant in the primary mating screen are euthanized and blood is collected for hormone analysis. Hormone levels regulate reproductive cycling and alterations in their serum levels can be a valuable indicator of abnormal conditions caused by or leading to reproductive disorders. For example, decreased levels of follicle-stimulating hormone could indicate a mutation affecting pituitary gonadotrophs, hypothalamic function, or hyper-secretion of steroid hormones by the ovary. In addition to serum analysis, the clinical examination includes gross observation of the reproductive tract for abnormalities and removal of ovaries for further study. One ovary is taken for histological analysis to assess follicle and oocyte morphology and the other is mechanically disrupted with needles to release oocytes that are collected for in vitro maturation (Eppig and O'Brien, 1996; O'Brien et al., 2003). In reproductively fit female mice, all stages of follicle development are represented within the ovary. This includes the reserve pool of primordial follicles, primary and secondary follicles that have been recruited into the growing pool, and various stages of antral follicle development. Oocytes in primordial follicles are arrested at prophase of the first meiotic division and remain arrested until after the preovulatory surge of luteinizing hormone. As oocytes grow, they accumulate various factors including the products of maternal effect genes that are necessary for the resumption of meiosis, fertilization, and preimplantation development.

Although it is highly strain dependent, the average hormonally primed female mouse will, upon dissection, yield approximately 15–25 germinal vesicle stage oocytes (per ovary) that are competent to resume meiosis in vitro, a process called oocyte maturation. Oocyte maturation is scored in vitro by noting the breakdown (dissolution) of the germinal vesicle, this being the primary indicator of meiotic resumption. Oocytes are then fertilized and cultured to assess fertilization and pre-implantation embryonic development success (Ho et al., 1995; Eppig and O'Brien, 1996; O'Brien et al., 2003). Defects in any of the processes of oocyte or embryo development could lead to infertility and could be targets for contraceptive development.

In the male infertility "clinic" (Fig. 3), G_3 males that mated and plugged but failed to impregnate wild-type females in the primary fertility screen are analyzed for a spectrum of reproductive parameters. Initially, the gross anatomy of the male is observed and reproductive organ (testis, epididymis, seminal vesicles) weights are determined; these weights can provide initial estimations of the extent of spermatogenesis and endocrine status. Tissues are preserved for subsequent histological analysis of germ cell development and morphology of somatic cells (Sertoli and Leydig cells). The cauda portion of one epididymis is dissected to release sperm. If sperm are recovered, sperm count, morphology, acrosomal status and motility are assessed. The sperm are used for in vitro fertilization of wild-type oocytes after which embryo development to the 2-cell and blastocyst stages is assessed. Taken together, these analyses assess the capacity of sperm to function in fertilization and egg activation; defects in these steps cause infertility and are thus attractive targets for contraception.

When multiple infertile animals with the same phenotype are found in a single family, they are deemed potential mutants. A breeding line is established with these families to confirm heritability and to map the chromosomal region containing the mutant allele. Polymorphisms between the strain on which the mutations were induced (C57BL6/J) and the background strain of the females to which mutagenized mice were crossed (C3HeB/FeJ) are used to facilitate gene mapping. This step is often the bottleneck, but must be accomplished as rapidly as possible to enable colony management (identification of the heterozygotes to be used to propagate the line) by PCR genotyping rather than by progeny testing. Once a defective gene responsible for infertility is identified, its normal function(s) and the pathway(s) in which it acts can be thoroughly investigated through molecular and biological studies.

Mutant models for investigating reproductive function

This mutagenesis program has been ongoing for only one year at the time of writing, but already it is clear that the aberrant mouse phenotypes reaped in the program will provide valuable resources for study of human reproduction and infertility.

For example, female infertile mice have been identified with abnormalities in gonad formation, oocyte/follicle development, ovulation, and pre-implantation embryonic development. Infertile females that do not show any in vitro clinic or histology abnormalities are also of interest and are categorized as unexplained infertility. Some of the aberrant phenotypes that have been identified thus far include poor ovary development, lack of oocytes, intra-follicular oocyte death, failed ovulation, inability of oocytes to complete meiotic maturation, and embryonic arrest prior to blastocyst formation.

Phenotypes of potential male reproductive mutants identified thus far fall into three categories: spermatogenic abnormalities, post-spermatogenic abnormalities and unexplained infertility. Spermatogenic abnormalities include spermatocyte arrest (possibly indicating meiotic failure) and defects of postmeiotic, spermiogenic differentiation. These defects are generally associated with low testis weight and a lack of mature spermatozoa in the epididymis. Mice with post-spermatogenic abnormalities have mature spermatozoa, but manifest a low fertilization rate, sometimes accompanied by low sperm motility. Some mice, whose sperm yield low fertilization rates, also exhibit low weight of accessory organs, suggesting a hormonal defect. Interestingly, some males that fail to produce offspring in natural mating do not show abnormalities in any of the clinical tests, including in vitro fertilization, thus modeling unexplained human male infertility.

Surprisingly, significantly more infertile males than infertile females have been recovered from this phenotype screen. The reason for this skewed distribution is unknown, but it may lead to interesting discoveries on the genetic complexity of the male versus female reproductive requirements. For example, it is estimated that nearly 4% of the genome is expressed in meiotic and postmeiotic male germ cells (Schultz et al., 2003), but the relative number of genes that must be expressed for successful development of the female gamete is not yet known.

From mutant models to contraception

A mutagenesis program such as the one described here has the potential to reveal novel and unsuspected genes that control reproductive processes in males and females. However, there are at least two major steps that must be accomplished between selecting a mutant model and applying its lessons to contraceptive development: first, gene cloning to delineate reproductive genomics or "transcriptomics," revealing site of gene expression; and second, reproductive "proteomics" to understand protein function and interactions in reproductive pathways and, importantly, in other physiological pathways. Thus extensive molecular biology and biochemistry studies lie ahead. Nonetheless, infertility phenotypes can target and focus research in promising directions.

The ideal contraceptive target would fulfill a number of criteria. First, the target should be a protein (or its encoding gene) that functions after gonadogenesis is complete. This would allow the option of using contraceptive control at any time during the adult reproductive life span. Second, its function should be tightly limited spatially and temporally such that its inhibition will lead to infertility without any other adverse biological repercussions. Third, inhibition of the gene product should lead to complete infertility so contraceptive action is dependable. Lastly, the infertility should be reversible once the inhibitory influence is removed, thus returning the individual to a normal fertile state. Because of differences in anatomy, tempo and hormonal regulation, strategies and targets for male and female contraceptives will differ greatly.

Importantly, from the perspective of contraceptive targets, our screen begins with the phenotype of interest, namely infertility when naturally mated. This strategy makes the screen totally unbiased with respect to the reproductive function being targeted. Subsequent clinical examinations were designed to discern the nature of the defects with regard to the key processes required for normal fertility: hormone production, gonadogenesis, gametogenesis, fertilization and subsequent pre-implantation development. For identification of unique contraceptive targets, we are particularly interested in mice that have normal hormone production and normal gonads but are abnormal with regard to some aspect of gametogenesis or gamete interaction.

Examination of ovaries from infertile females allows determination of whether all stages of follicle development are represented in a normal pattern, thus assessing hormonal function as well as folliculogenesis and gametogenesis. This evaluation includes, for example, assessment of the abundance of primordial follicles that constitute the reserved pool of oocytes needed for a full-length reproductive lifespan, as well as the presence of fully expanded antral follicle on the verge of ovulation. Absence of or abnormalities in any follicular stage would indicate defects in key genes regulating follicle growth and development. Artificially inhibiting oocyte or follicle development at specific stages could be contraceptive while still preserving the limited oocyte reserves for later use. Histological assessment of mutant ovaries also allows assessment of oocyte growth and whether oocytes are released by ovulation or, alternatively, trapped and degenerating in their surrounding follicles. In vitro maturation of fully-grown oocytes recovered from mice allows for functional assessment of whether they have achieved normal cytoplasmic and nuclear development. If normal development has taken place, fully-grown oocytes will be readily capable of undergoing germinal vesicle breakdown and meiotic resumption as noted by the loss of the germinal vesicle during in vitro maturation. These analyses can reveal factors controlling maturation. For example, the phosphatase CDC25B has the exclusive function of activating maturation promoting factor (MPF) in fully-grown oocytes, thus allowing breakdown of the germinal vesicle and resumption of meiosis (Lincoln et al., 2002). *Cdc25b* knockout mice are normal except that females are sterile due to a block in oocyte maturation. Little is known about the other factors that are responsible for nuclear and cytoplasmic maturation, thus additional key factors will likely be discovered in the current screen. Effectively suppressing the function of positive regulators such as *Cdc25b* or activating negative regulators of key steps in oocyte development could be an effective method of contraception.

Oocytes that undergo maturation during the clinical evaluation are subsequently fertilized and allowed to complete pre-implantation development to the blastocyst stage. The fertilization and development assay allows us to identify mutations that could be the result of defects in, for instance, egg-sperm binding, egg activation, and maternal messages necessary for early embryo development. These processes are complex and likely involve multiple pathways, and little is known about the maternally encoded genetic factors involved in assuring embryonic activation and development. One of the few genes that have been found to have a critical role in embryonic development is the maternal effect gene *Mater*, which, when disrupted, produces a sterile female phenotype due to early cleavage stage embryo arrest (Tong et al., 2000).

On the male side, evaluation of testis histology can reveal mutations that affect spermatogenesis as well as somatic cells. However, mutations causing testicular spermatogenic defects are unlikely to provide promising contraceptive targets. This is because the blood-testis barrier (formed by Sertoli cell junctions) prevents direct access of reagents to germ cells. Furthermore, destruction of germ cells can lead to auto-immune orchitis (Tung and Teuscher, 1995). Additionally, hormonal manipulations have not thus far been successful in controlling human spermatogenesis (Anderson and Baird, 2002). However, it is always possible that mutations in this category would reveal unexpected aspects of the hormonal control of spermatogenesis. Examples of such unanticipated effects are the consequences of targeted mutations of the oestrogen receptor 1(alpha) (Eddy et al., 1996) and the aromatase enzyme *(Cyp19a1)* (Fisher et al., 1998) genes, demonstrating an indirect role of estrogen on spermatogenesis (Hess et al., 1997; Robertson et al., 1999).

Two male mutant phenotypes that may represent the best potential for male contraception are abnormal sperm motility and impaired fertilization success, detected by screening andrological parameters and IVF function. Motility is necessary for a spermatozoon to achieve oocyte fertilization (Yanagimachi, 1994) and mutants with low sperm motility could reveal genes important for this function. An example of the kind of gene that could be detected is that encoding CATSPER1 (Ren et al., 2001), an ion channel expressed only in the testis. Deletion of the mouse *Catsper1* gene causes a reduction of motility so that sperm are unable to penetrate the zona pellucida. A second CATSPER protein, CATSPER2, has been detected in mouse

sperm (Quill et al., 2001), and partial deletion of the homologue in human males is associated with nonsyndromic male infertility (Avidan et al., 2003). Mutants with low in vitro fertilization success can reveal proteins important in the process of fertilization. This process includes sperm capacitation, primary zona pellucida binding, acrosomal reaction, sperm penetration and sperm-egg fusion (Yanagimachi, 1994). Proteins involved in these steps represent highly desirable targets for contraceptive agents (Brewis and Wong, 1999; McLeskey et al., 1998), but our knowledge of these processes is limited with respect to the proteins involved.

Finally, the secondary screening "clinic" reveals the important phenotype of unexplained infertility, and cloning the responsible genes may reveal new and unsuspected aspects of male reproduction. Mutants with this phenotype exhibit normal sperm morphology, motility, and function in IVF (encompassing both fertilization and embryo development). Nonetheless, these mutant males fail to produce viable pregnancies when mated with fertile females. Ultimately cloning of the genes and analysis of the encoded proteins will reveal their function, perhaps totally unanticipated. For example, surprisingly, sperm have odorant receptors (Spehr et al., 2003), which are hypothesized to mediate a chemotactic signal from the egg that attracts sperm. Disruption of this receptor could impair sperm transport in the uterus and oviduct, but not impair fertilization in vitro, thus reflecting an "unexplained infertility" phenotype. Unexplained infertility could result from mutation of a protein that affects sperm-oviduct binding. During the transit in the oviduct, the sperm has interactions with oviductal epithelial cells (Suarez, 2001, 2002). If an important protein involved in this interaction were mutated, it could stop the progression of the sperm inside the oviduct or impair the temporal coordination of fertilization of the oocyte. As an example, mice that are null for the *Adam2* gene encoding the protein fertilin β, exhibit impaired oviductal transport, as well as other fertilization-related phenotypes (Cho et al., 1998).

The occurrence of mutations affecting both male and female fertility without any obvious cause – the unexplained infertility mutants – may in the end provide the most novel and unexpected contraceptive targets. These mutant animals exhibit no health defects except for infertility, suggesting that the infertility derives from genes specific for reproductive processes. In addition, effects on gonadal development and function, if any, must be so minimal that there is negligible impact on physiological and psychological processes dependent upon reproductive hormone function.

Conclusions and perspective

The initial success of this program has already been demonstrated (Ward et al., 2003). At this writing, less than one year from initiation of the program, approximately 12 male and/or female infertility mutants have been selected for distribution and mapping analysis. Information about these mutants and how to obtain them can be found at http://reprogenomics.jax.org. The power of this program lies not only in its efficiency but also in the unbiased nature of the screen. Unlike gene targeting, random mutagenesis and selection based on phenotype can result in discovery of genes and pathways that would never have been suspected to have a reproductive role. Thus ENU mutagenesis can reveal the genetic factors that are indispensable for fertility. These valuable reproductive mutant resources can significantly advance the field of reproductive biology. These resources can lead to therapies for couples afflicted by infertility problems, estimated at one in six couples attempting to have children (Shah et al., 2003). Much of what is currently known about the genetic bases of human infertility comes from studies of genes on the Y chromosome (Cooke and Saunders, 2002; Shah et al., 2003). However, the mutant resources currently being generated will allow genetic identifi-cation of autosomal genes responsible for human fertility. Taken together, exploitation of these genetic resources can lead to the development of new contraceptives, one of the many missions undertaken by the World Health Organization to help control the world's growing population problem (Waites, 2003).

Acknowledgements

We are grateful for thoughtful comments on the manuscript from Drs. Wesley G. Beamer, Gregory A. Cox and Beverly Richards-Smith.

References

Aitken RJ: Immunocontraceptive vaccines for human use. J Reprod Immunol 57:273–287 (2002).

Anderson RA, Baird DT: Male contraception. Endocr Rev 23:735–762 (2002).

Avidan N, Tamary H, Dgany O, Cattan D, Pariente A, Thulliez M, Borot N, Moati L, Barthelme A, Shalmon L, Krasnov T, Ben-Asher E, Olender T, Khen M, Yaniv I, Zaizov R, Shalev H, Delaunay J, Fellous M, Lancet D, Beckmann JS: CATSPER2, a human autosomal nonsyndromic male infertility gene. Eur J Hum Genet 11:497–502 (2003).

Balling R: ENU mutagenesis: analyzing gene function in mice. A Rev Genomics Hum Genet 2:463–492 (2001).

Botstein D, Risch N: Discovering genotypes underlying human phenotypes: past successes for mendelian disease, future approaches for complex disease. Nature Genet 33(suppl):228–237 (2003).

Brewis IA, Wong CH: Gamete recognition: sperm proteins that interact with the egg zona pellucida. Rev Reprod 4:135–142 (1999).

Cho C, Bunch DO, Faure JE, Goulding EH, Eddy EM, Primakoff P, Myles DG: Fertilization defects in sperm from mice lacking fertilin beta. Science 281:1857–1859 (1998).

Cooke HJ, Saunders PT: Mouse models of male infertility. Nature Rev Genet 3:790–801 (2002).

Delves PJ, Lund T, Roitt IM: Antifertility vaccines. Trends Immunol 23:213–219 (2002).

Eddy EM, Washburn TF, Bunch DO, Goulding EH, Gladen BC, Lubahn DB, Korach KS: Targeted disruption of the estrogen receptor gene in male mice causes alteration of spermatogenesis and infertility. Endocrinology 137:4796–4805 (1996).

Eppig JJ, O'Brien MJ: Development in vitro of mouse oocytes from primordial follicles. Biol Reprod 54:197–207 (1996).

Fisher CR, Graves KH, Parlow AF, Simpson ER: Characterization of mice deficient in aromatase (ArKO) because of targeted disruption of the cyp19 gene. Proc Natl Acad Sci USA 95:6965–6970 (1998).

Hess RA, Bunick D, Lee KH, Bahr J, Taylor JA, Korach KS, Lubahn DB: A role for oestrogens in the male reproductive system. Nature 390:509–512 (1997).

Ho Y, Wigglesworth K, Eppig JJ, Schultz RM: Preimplantation development of mouse embryos in KSOM: Augmentation by amino acids and analysis of gene expression. Mol Reprod Dev 41:232–238 (1995).

Justice MJ: Mutagenesis of the mouse germline, in Jackson JJ, Abbott CM (eds): Mouse Genetics and Transgenics: A Practical Approach, (Oxford University Press, Oxford 1999).

Lincoln JA, Wickramasinghe D, Stein P, Schultz RM, Palko ME, De Miguel MP, Tessarollo L, Donavan PJ: Cdc25b phosphatase is required for resumption of meiosis during oocyte maturation. Nature Genetics 30:446–449 (2002).

Matzuk MM, Lamb DJ: Genetic dissection of mammalian fertility pathways. Nature Cell Biol 4 (suppl): s41–s49 (2002).

McLeskey SB, Dowds C, Carballada R, White RR, Saling PM: Molecules involved in mammalian sperm-egg interaction. Int Rev Cytol 177:57–113 (1998).

Noveroske JK, Weber JS, Justice MJ: The mutagenic action of N-ethyl-N-nitrosourea in the mouse. Mammal Genome 11:478–483 (2000).

O'Brien MJ, Pendola JK, Eppig JJ: A revised protocol for in vitro development of mouse oocytes from primordial follicles dramatically improves their developmental competence. Biol Reprod 68:1682–1686 (2003).

Quill TA, Ren D, Clapham DE, Garbers DL: A voltage-gated ion channel expressed specifically in spermatozoa. Proc Natl Acad Sci USA 98:12527–12531 (2001).

Ren D, Navarro B, Perez G, Jackson AC, Hsu S, Shi Q, Tilly JL, Clapham DE: A sperm ion channel required for sperm motility and male fertility. Nature 413:603–609 (2001).

Robertson KM, O'Donnell L, Jones ME, Meachem SJ, Boon WC, Fisher CR, Graves KH, McLachlan RI, Simpson ER: Impairment of spermatogenesis in mice lacking a functional aromatase (cyp 19) gene. Proc Natl Acad Sci USA 96:7986–7991 (1999).

Schultz N, Hamra FK, Garbers DL: A multitude of genes expressed solely in meiotic or postmeiotic spermatogenic cells offers a myriad of contraceptive targets. Proc Natl Acad Sci USA 100:12201–12206 (2003).

Shah K, Sivapalan G, Gibbons N, Tempest H, Griffin DK: The genetic basis of infertility. Reproduction 126:13–25 (2003).

Shibuya T, Morimoto K: A review of the genotoxicity of 1-ethyl-1-nitrosourea. Mutat Res 297:3–38 (1993).

Spehr M, Gisselmann G, Poplawski A, Riffell JA, Wetzel CH, Zimmer RK, Hatt H: Identification of a testicular odorant receptor mediating human sperm chemotaxis. Science 299:2054–2058 (2003).

Suarez SS: Carbohydrate-mediated formation of the oviductal sperm reservoir in mammals. Cells Tissues Organs 168:105–112 (2001).

Suarez SS: Formation of a reservoir of sperm in the oviduct. Reprod Domest Anim 37:140–143 (2002).

Tong ZB, Gold L, Pfeifer KE, Dorward H, Lee E, Bondy CA, Dean J, Nelson LM: Mater, a maternal effect gene required for early embryonic development in mice. Nat Genet 26:267–268 (2000).

Tung KS, Teuscher C: Mechanisms of autoimmune disease in the testis and ovary. Hum Reprod Update 1:35–50 (1995).

Venter JC, Adams MD, Myers EW, Li PW, Mural RJ, Sutton GG et al: The sequence of the human genome. Science 291:1304–1351 (2001).

Venter JC, Levy S, Stockwell T, Remington K, Halpern A: Massive parallelism, randomness and genomic advances. Nature Genet 33(suppl):219–227 (2003).

Waites GM: Development of methods of male contraception: impact of the World Health Organization Task Force. Fertil Steril 80:1–15 (2003).

Ward JO, Reinholdt LG, Hartford SA, Wilson LA, Munroe RJ, Schimenti KJ, Libby BJ, O'Brien M, Pendola JK, Eppig J, Schimenti JC: Towards the genetics of mammalian reproduction: induction and mapping of gametogenesis mutants in mice. Biol Reprod 69:1615–1625 (2003).

Waterston RH, Lindblad-Toh K, Birney E, Rogers J, Abril JF et al: Initial sequencing and comparative analysis of the mouse genome. Nature 420:520–562 (2002).

Yanagimachi R: Mammalian fertilization, in Knobil E, Neill J (eds): The Physiology of Reproduction, pp 189–317, 2nd ed (Raven Press, New York 1994).

Cytogenet Genome Res 105:228–234 (2004)
DOI: 10.1159/000078193

Mouse zona pellucida genes and glycoproteins

P.M. Wassarman, L. Jovine and E.S. Litscher

Brookdale Department of Molecular, Cell and Developmental Biology, Mount Sinai School of Medicine, New York, NY (USA)

Abstract. The zona pellucida (ZP) is a thick extracellular coat that surrounds all mammalian eggs. The ZP plays important roles during oogenesis, fertilization, and preimplantation development. The mouse ZP consists of only three glycoproteins, called ZP1, ZP2, and ZP3. All three glycoproteins are essential structural components of the ZP. Additionally, ZP3 serves as a primary sperm receptor and acrosome reaction-inducer, and ZP2 serves as a secondary sperm receptor during fertilization. ZP1, ZP2, and ZP3 are encoded by single-copy genes present on three different chromosomes. The genes are expressed exclusively by mouse oocytes as they grow and the cellular specificity can be ascribed to *cis*-acting sequences close to the site of transcription initiation and to certain *trans*-acting factors. Concomitantly, ZP polypeptides are synthesized, modified with N- and O-linked oligosaccharides, secreted, and assembled into crosslinked filaments that exhibit a structural repeat. Nascent ZP glycoproteins are incorporated into large secretory vesicles that fuse with the oocyte plasma membrane and deposit nascent ZP glycoproteins into the innermost layer of the thickening ZP. Each ZP polypeptide possesses several characteristic features, including an N-terminal signal sequence, a ZP domain, a consensus furin cleavage site, and a C-terminal transmembrane domain. The latter is required for assembly of nascent ZP polypeptides into a ZP, cleavage at the consensus furin cleavage site is required for secretion, and the ZP domain supports protein:protein interactions during ZP assembly. At ovulation, when meiotic maturation of oocytes occurs and chromosomes condense into bivalents, expression of the three ZP genes ceases. Using "knockout mice", in the absence of either ZP2 or ZP3 expression, a ZP fails to assemble around growing oocytes and females are infertile. There is no effect on males. In the absence of ZP1 expression, a disorganized ZP assembles around growing oocytes and females exhibit reduced fertility. These observations are consistent with the current model for ZP structure in which ZP2 and ZP3 form long Z filaments crosslinked by ZP1.

All mammalian eggs are surrounded by an extracellular coat called the *zona pellucida* (ZP). For fusion of sperm and egg ("fertilization") to occur, sperm must first bind to and then penetrate the thick ZP. Results of in vitro experiments suggest that sperm bind to the ZP in a relatively species-specific manner and that, following binding, undergo a form of cellular exocytosis, the acrosome reaction, that allows bound sperm to penetrate the ZP and reach the egg plasma membrane. In response to fertilization, the ZP becomes refractory to binding of free-swimming sperm and impenetrable to sperm already bound; these changes in the ZP constitute the so-called "slow block to polyspermy". During preimplantation development, the growing embryo is protected by the ZP and only just prior to implantation does the expanded blastocyst "hatch" from the ZP (Gwatkin, 1977; Yanagimachi, 1994).

During the past twenty years or so, most functions of the ZP during fertilization (e.g., primary and secondary binding and acrosome reaction-inducing activities) have been ascribed to specific ZP glycoproteins. This is especially true for mice. Mouse ZP glycoproteins, called ZP1, ZP2, and ZP3, and the genes encoding them have been isolated and characterized and, as a result, considerable progress has been made toward understanding structure-function relationships for these glycoproteins (Wassarman, 1999; Wassarman et al., 2001; Jovine et al., 2002a). It has also become clear that ZP glycoproteins from mice and all other mammals, including humans, are very similar. For example, the polypeptide primary structures of ZP2

Research from the authors' laboratory was supported in part by the National Institutes of Health, most recently by HD-35105. Luca Jovine is a postdoctoral fellow supported by the Human Frontier Science Program Organization.

Received 30 June 2003; manuscript accepted 14 July 2003.

Request reprints from Paul M. Wassarman, Brookdale Department of
Molecular Cell and Developmental Biology, Mount Sinai School of Medicine
One Gustave L. Levy Place, New York, NY 10029-6574 (USA)
telephone: 212-241-8616; fax: 212-860-9279
e-mail: paul.wassarman@mssm.edu

and ZP3 from mice and human beings, species separated by more than 10^8 years of evolution, are approximately 60% identical (McLeskey et al., 1998). Furthermore, human ZP glycoproteins can replace their mouse counterparts, at least as structural components of the mouse egg ZP (Rankin et al., 1998).

Here, we review some biochemical, molecular genetic, and structural aspects of mouse ZP genes and glycoproteins that have been investigated extensively during the past couple of decades. As the reader will see, a great deal has been learned about mouse ZP genes and glycoproteins during a relatively short period. Hopefully, the knowledge will be put to good use.

Characteristics of the mouse ZP

The plasma membrane of all mammalian eggs is surrounded by a ZP that, depending on the species, varies in thickness from ~ 2 to ~ 20 μm (Gwatkin, 1977; Wassarman, 1988a; Yanagimachi, 1994). The mouse egg ZP is a thick (average ~ 6.5 μm) extracellular coat containing, on average, ~ 3.5 ng of glycoprotein (~ 15% of total egg protein) (Fig. 1). Interestingly, the thickness of the mouse egg ZP varies from ~ 4.3 to ~ 8.1 μm, reflecting incorporation of different amounts of ZP glycoproteins into the ZP of different oocytes. The ZP can be completely dissolved at low pH (pH 2.5–3.5) or elevated temperatures indicating that ZP glycoproteins are held together by noncovalent interactions. The ZP can also be dissolved in the presence of reducing agents, suggesting that disulfides play an important role in maintaining ZP structure. The ZP is an extremely porous coat that can be penetrated by large macromolecules, such as immunoglobulins and enzymes, as well as by small viruses. The physical properties of the ZP suggest that it is quite elastic and, to a certain extent, resembles mucins. The ZP may represent a combination or fusion of the vitelline envelope and jelly coat that frequently surround eggs from nonmammalian species.

Characteristics of mouse ZP glycoproteins

The mouse ZP consists of only three glycoproteins, called ZP1, ZP2, and ZP3, that have average apparent M_rs of 200, 120, and 83 kDa, respectively (Bleil and Wassarman, 1980a; Wassarman, 1988a). Although these glycoproteins display extensive heterogeneity on high-resolution two-dimensional gels, the heterogeneity is attributable to oligosaccharides attached to ZP polypeptides and not to the polypeptides themselves (see below). ZP2 and ZP3 are present in approximately equimolar amounts in the ZP and each is significantly more abundant than ZP1. All three glycoproteins are acidic, having average pIs between 4.1 and 5.2, which is attributable to their asparagine– (N–) and serine/threonine– (O–) linked oligosaccharides, not to their polypeptides (pIs 6.5–6.7). There is evidence that oligosaccharides attached to ZP glycoproteins are sialylated and sulfated. ZP1 is a dimer of identical polypeptide chains held together by intermolecular disulfides; under reducing conditions, the ZP1 monomer has an average apparent M_r of 120 kDa.

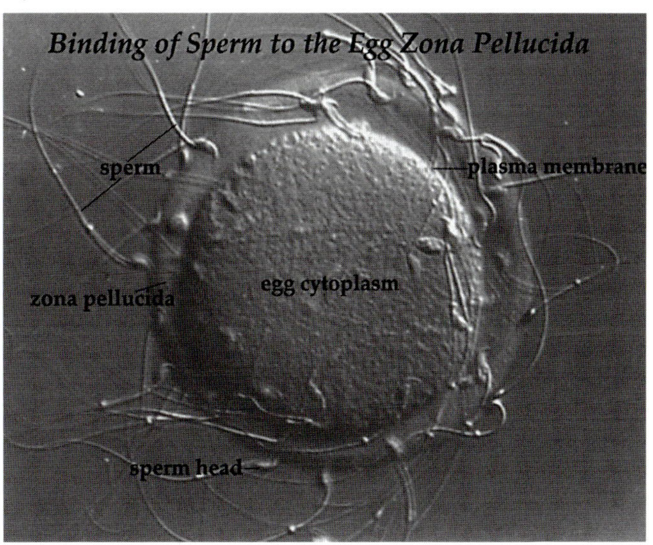

Fig. 1. Binding of mouse sperm to the unfertilized mouse egg ZP. Light micrograph (Nomarski differential interference contrast) of sperm bound to the ZP of an unfertilized egg in vitro.

Organization of Mouse ZP Polypeptides

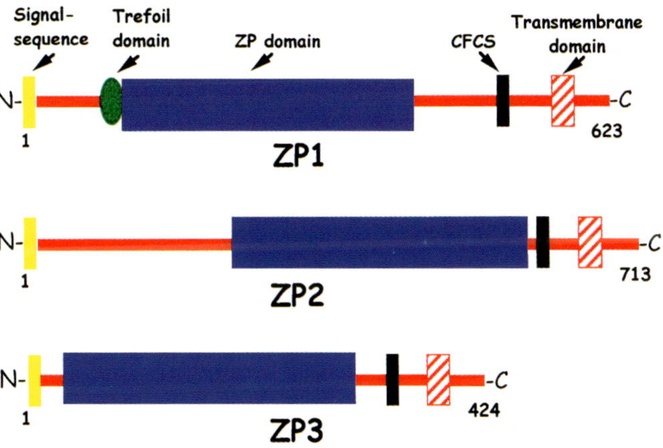

Fig. 2. Schematic representation of mouse ZP polypeptides. Depicted are several elements that comprise the polypeptide chains of mouse ZP1, ZP2, and ZP3 (623, 713, and 424 amino acids, respectively). These elements include an N-terminal signal-sequence (yellow), ZP domain (blue), consensus furin cleavage-site (CFCS; black), and transmembrane domain (red hatched). ZP1 also contains a trefoil domain (green) just N-terminal to the ZP domain. For further details see Jovine et al. (2002a).

Based on gene sequences (see below), the predicted M_rs of ZP1, ZP2, and ZP3 polypeptides are ~ 69 kDa (monomer; 623 amino acids), ~ 80 kDa (713 amino acids), and ~ 46 kDa (424 amino acids), respectively (Wassarman et al., 2001; Jovine et al., 2002a). Each glycoprotein possesses several (5–7) potential N-linked glycosylation sites (–Asn–X–Ser/Thr–), an undetermined number of O-linked oligosaccharides, an N-terminal signal sequence, a ZP domain, a consensus furin cleavage site (CFCS), and a predicted transmembrane domain located close to the C-terminus (Fig. 2). Additionally, ZP1 possesses a trefoil domain just N-terminal to its ZP domain. ZP glycoproteins are predicted to be relatively rich in β-structure (e.g., ZP1 ~ 21%, ZP2 ~ 32%, ZP3 ~ 28%) and have little, if any, α-helix content (<1.5%). It is likely that ZP glycoproteins possess novel structural folds. As is the case for other secreted proteins, apparently the Cys residues of ZP2 and ZP3 are present as intramolecular disulfides; as indicated above, ZP1 also has intermolecular disulfides.

The "ZP domain" (Bork and Sander, 1992) is a key feature of all ZP glycoproteins and has been found in a large number of proteins from a wide variety of organisms, including flies and worms (Wassarman et al., 2001; Jovine et al., 2002a). Most of these proteins act as receptors and/or have mechanical functions. For ZP1, ZP2, and ZP3 polypeptides, the ZP domain represents amino acids 271–542, 364–630, and 45–304, respectively; for ZP3, the ZP domain accounts for ~ 80% of the mature glycoprotein's polypeptide. The domain consists of ~ 260 amino acids, with eight conserved Cys residues, together with a relatively large number of amino acid positions with conserved character (e.g., hydrophobic or polar). It has been reported recently that the ZP domain is a module that allows polymerization of ZP glycoproteins (as well as other extracellular proteins) into long filaments (Jovine et al., 2002b). Furthermore, specific single mutations in the ZP domain can prevent incorporation of nascent ZP glycoproteins into the ZP. It is clear that X-ray crystallographic and NMR studies of ZP domain-containing proteins, such as ZP1, ZP2, and ZP3, would have a significant impact on our understanding of the biological activities of this functionally diverse group of proteins.

Functions of mouse ZP glycoproteins

A great deal of evidence has accumulated as to the functions of mouse ZP glycoproteins, especially during fertilization (McLeskey et al., 1998; Wassarman, 1999, 2003; Wassarman et al., 2001). In general, all three glycoproteins play structural roles in assembly of the ZP during oogenesis (discussed below). Furthermore, ZP3 serves as the primary sperm receptor to which free-swimming sperm bind in a relatively species-specific manner and also as the acrosome reaction-inducer following binding of sperm to the ZP. Apparently, acrosome-intact sperm recognize and bind to specific O-linked oligosaccharides present on ZP3; although the nature of the sperm protein that binds to ZP3 remains problematic. ZP2 serves as the secondary sperm receptor to which ZP-associated, acrosome-reacted sperm bind. There is evidence to suggest that binding of acrosome-reacted sperm to ZP2 is mediated by the sperm protease, acrosin, and/or its zymogen form, proacrosin (Howes and Jones, 2003). Following fertilization, changes in ZP2 and ZP3 (i.e., "zona reaction"), probably induced by the contents of egg cortical granules that are deposited into the ZP at the time of sperm-egg fusion, appear to account for the altered behavior of the ZP toward sperm.

Organization and expression of mouse ZP genes

ZP glycoproteins are encoded by single-copy genes located on different chromosomes. Mouse *Zp1, Zp2,* and *Zp3* are located on chromosomes 19, 7, and 5, respectively (Lunsford et al., 1990; Epifano et al., 1995a; Rankin and Dean, 2000). Genes encoding ZP glycoproteins exhibit conserved organization. For example, in mice and humans ZP1 and ZP3 consist of 12 and 8 exons, respectively, whereas ZP2 genes in mice and humans consist of 18 and 19 exons, respectively (Kinloch et al., 1988; Chamberlin and Dean, 1990; Liang and Dean, 1993; McLeskey et al., 1998). This reflects the presence of exon/intron boundaries that define the limits of distinct domains in ZP glycoproteins. ZP genes share TATAA boxes approximately 30 base pairs upstream of their transcription start-sites, as well as E-box sequences (CANNTG) at approximately –200 base pairs. It has been reported that E-boxes are involved in oocyte-specific expression of ZP genes that takes place coordinately upon binding of E12/FIGα heterodimers (Liang et al., 1997; Soyal et al., 2000). FIGα (factor in the germline α), is a germ cell specific, basic helix-loop-helix transcription factor, with a M_r of ~ 21 kDa (194 amino acids). Female mice that are nulls for FIGα are sterile, due to massive depletion of oocytes, and do not express *Zp1, Zp2,* or *Zp3*.

Transgenic mice also have been used to identify *cis*-acting sequences of the ZP3 gene that are responsible for targeting expression of the gene to growing oocytes (Lira et al., 1990, 1993). Expression of transgenes in which regions of ZP3 5′-flanking sequence were fused to the firefly luciferase gene ("reporter") was examined in transgenic mice. Such experiments revealed that as little as 153 nucleotides of ZP3 5′-flanking sequence was sufficient to target expression of the reporter to growing oocytes; although it was clear that enhancer elements were present between –153 and –470 nucleotides of the ZP3 5′-flanking sequence (Fig. 3). This ZP3 promoter has been used for cre-mediated recombination of loxP-flanked target genes in mouse oocytes (Lewandoski et al., 1997). Furthermore, an ovary-specific, DNA-binding protein was identified (~ 60 kDa) that bound to the sequence 5′-GATAA-3′ within the first 100 base pairs of the ZP3 promoter (Schickler et al., 1992). There also is evidence to suggest that promoters of human ZP genes can utilize the heterologous transcription machinery (transcription factors) in mouse oocytes (Liang and Dean, 1993). Overall, these observations indicate that DNA elements responsible for targeting expression of ZP genes to growing oocytes are located close to the transcription start-sites of the genes.

During growth of mouse oocytes, the absolute rate of total protein synthesis increases ~ 40-fold and ZP3 synthesis represents ~ 1.5–2.5% of the total. The number of copies of *Zp3* messenger RNA increases from undetectable levels in nongrowing (~ 12 μm) oocytes, to ~ 300,000 copies/oocyte in 60–70 μm oocytes, to ~ 240,000 copies/oocyte in fully-grown (~ 80 μm) oocytes, and to almost undetectable levels (~ 5,000 copies/egg) in unfertilized eggs following ovulation (Roller et al., 1989). Comparisons of the levels of *Zp1, Zp2,* and *Zp3* transcripts present in growing (~ 50–60 μm) oocytes suggest that they are present at a ratio of 1:4:4, respectively (Epifano et al., 1995b). ZP glycoprotein messenger RNA is undetectable in

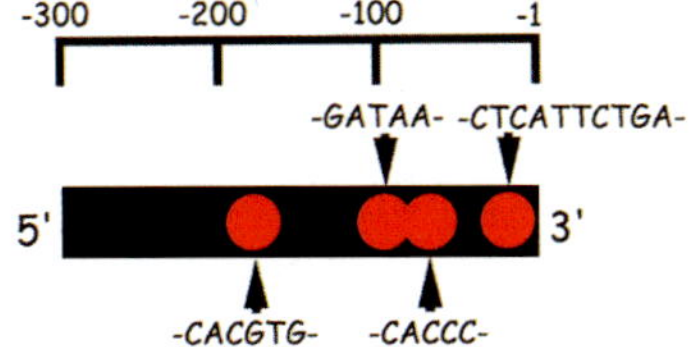

Fig. 3. Diagrammatic representation of the mouse *Zp3* 5'-flanking region and luciferase activities in oocytes from transgenic mice. Several putative regulatory elements that may contribute to *Zp3* expression include the following: A consensus sequence –CACGTG– (–186 to –181 nt) for binding of helix-loop-helix proteins; a consensus sequence –CACCC– (–85 to –81 nt) frequently found in close proximity to GATA sequences; a consensus sequence –GATAA– (–97 to –93 nt) for binding of specific transcription factors; a consensus sequence –CTCATTCTGA– (–16 to –7 nt) for initiator binding proteins. Also shown are mean luciferase activities in light units (LU) for individual oocytes from transgenic mice. These mice harbor a transgene containing 153 (153-ZP3/LUC), 470 (470-ZP3/LUC), or 6,500 (6,500-ZP3/LUC) nt of *Zp3* 5'-flanking sequence fused to the firefly luciferase gene. The background activity in these experiments was 99 LU/oocyte that was subtracted from the values shown. In each case, a minimum of 15 individual oocytes were assayed for luciferase activity. For further details see Lira et al., (1990, 1993).

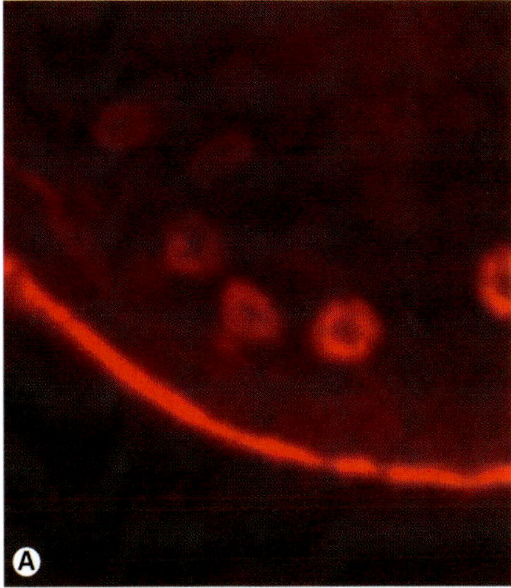

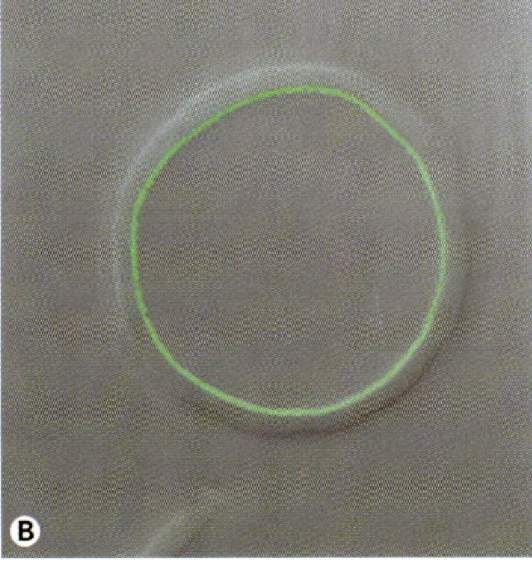

Fig. 4. Images of nascent ZP glycoproteins in mouse oocyte secretory vesicles and incorporated into the innermost layer of the oocyte ZP. (**A**) Shown is a fluorescent confocal image of secretory vesicles containing nascent ZP2 in a mouse growing oocyte microinjected with a *myc-Zp2* cDNA and immunolabeled with anti-Myc and an FITC-conjugated second antibody. The secretory vesicles are doughnut-shaped and colored red. For further details see Qi et al. (2002). (**B**) Shown is a composite of light and fluorescent confocal images of a ZP isolated from a mouse growing oocyte microinjected with a *Flag-Zp3* cDNA and immunolabeled with anti-Flag and an FITC-conjugated second antibody. The nascent ZP3 in the innermost layer of the ZP is colored green. For further details see Qi et al. (2002).

preimplantation embryos. Therefore, expression of ZP genes and synthesis of ZP glycoproteins in mice is restricted to growing oocytes. In this context, it is of interest to note that for some other mammalian species there are reports that ZP glycoproteins are synthesized by ovarian follicle cells, as well as by growing oocytes. For some fish and birds homologues of ZP glycoproteins are synthesized by follicle cells, as well as by the liver and transported through the blood to the ovary (Wassarman et al., 2001; Jovine et al., 2002a).

Synthesis and secretion of mouse ZP glycoproteins

As indicated above, ZP1, ZP2, and ZP3 are synthesized and secreted concomitantly by growing mouse oocytes for a period of 2–3 weeks (Bleil and Wassarman, 1980b; Wassarman et al., 1985; Wassarman, 1988a). The nascent polypeptides (described above) have their N-terminal signal-sequences removed and are N-glycosylated in the endoplasmic reticulum, giving rise to ZP1, ZP2, and ZP3 intermediates that have M_rs of 160, 91, and 53/56 kDa, respectively. All of these intermediates possess high-mannose-type, N-linked oligosaccharides. ZP2 has six N-linked and ZP3 either three or four N-linked oligosaccharides per molecule that are sensitive to digestion by either Endo H or α-mannosidase (Greve et al., 1982; Salzmann et al., 1983). The N-linked oligosaccharides of nascent ZP glycoproteins are processed to complex-type in the Golgi and become insensitive to Endo H, but sensitive to digestion by N-glycanase. At the same time, in the Golgi, O-linked oligosaccharides are added to each ZP glycoprotein. The increased biochemical activities of the Golgi are reflected in changes in its ultrastructure during oocyte growth (Wassarman and Josefowicz, 1978). The oocyte Golgi changes from flattened stacks of lamellae, with few, if any, vacuoles or granuoles, in the early stages of oocyte growth, to extensive arrays of swollen, stacked lamellae with many large vacuoles in the late stages of growth.

From the *trans* Golgi, nascent ZP glycoproteins are packaged into large secretory vesicles (~ 2 μm in diameter; approximately 10-times larger than somatic cell secretory vesicles), apparently with their transmembrane domains anchored in vesicle membrane (Qi et al., 2002) (Fig. 4). These vesicles then move to the plasma membrane of oocytes where they fuse and release ZP glycoproteins into the extracellular space. At some point during the latter stages of secretion, perhaps at the oocyte plasma membrane, ZP glycoproteins are cleaved at their CFCS (Litscher et al., 1999; Williams and Wassarman, 2001; Qi et al., 2002). Cleavage results in the removal of the C-terminal 78 and 71 amino acids, which includes their transmembrane domains, from ZP2 and ZP3, respectively. As a result of proteolytic cleavages (at the N-terminal signal-sequence and CFCS) and post-translational modifications with N- and O-linked oligosaccharides, mature forms of ZP1, ZP2, and ZP3 (apparent average M_rs 200, 120, and 83 kDa) are assembled into a ZP. In this context, recently it has been reported that mutant forms of ZP2 and ZP3, truncated before their transmembrane domains, were secreted as efficiently as wild-type ZP glycoproteins (Jovine et al., 2002b). However, the mutant glycoproteins were not incorporated into the ZP, suggesting that the presence of a transmembrane domain is required for ZP assembly.

It should be noted that mouse ZP glycoproteins can also be synthesized and secreted by many kinds of transfected cells, including embryonal carcinoma (EC) cells, 293 cells, Chinese hamster ovary (CHO) cells, and *Xenopus* oocytes (Kinloch et al., 1991; Beebe et al., 1992; Doren et al., 1999; Williams and Wasserman, 2001). However, the type and extent of glycosylation of ZP polypeptides vary considerably from one kind of transfected cell to another. These results strongly suggest that each ZP glycoprotein is synthesized and secreted independently (see below).

Assembly and structure of the mouse ZP

Non-growing oocytes contained in ovarian primordial follicles lack a ZP. As oocytes initiate growth and increase in diameter during 2–3 weeks, a thin ZP appears and then increases in thickness. Recently, epitope-tagged cDNAs for either mouse ZP2 or ZP3 were microinjected into the germinal vesicles (nuclei) of growing oocytes and nascent ZP glycoproteins were detected in the ZP by confocal microscopy (Qi et al., 2002). In each case, nascent ZP glycoprotein was found exclusively within a very thin layer at the inner surface of the ZP, close to the plasma membrane of the oocyte (Fig. 4). Therefore, it is likely that the ZP thickens from the inside (i.e., the region closest to the oocyte) and that the outermost layer of the ZP consists of glycoproteins synthesized very early in oocyte growth.

Perhaps, the most relevant information about mouse ZP assembly has come from experiments in which synthesis of individual ZP glycoproteins was ablated in growing oocytes, either by injection of antisense oligonucleotides into oocytes or by homologous recombination in mouse embryonic stem (ES) cells. In the latter case, homozygous null female mice were produced for each of the three ZP glycoproteins.

To target the degradation of ZP2 and ZP3 messenger RNA, a large excess of complimentary oligonucleotide was injected into the cytoplasm of isolated growing mouse oocytes (Tong et al., 1995). Within 16 h of injection, the targeted ZP glycoprotein was no longer synthesized by oocytes, whereas non-targeted ZP glycoprotein, tagged with radiolabeled amino acids,

continued to be synthesized. Interestingly, little, if any, radiolabeled, nascent ZP2 or ZP3 was found in ZP isolated from oocytes injected with either ZP2 or ZP3 antisense oligonucleotides. That is, the absence of synthesis of either glycoprotein, ZP2 or ZP3, prevented incorporation of the other glycoprotein into the ZP. These results strongly suggest that both nascent ZP2 and ZP3 must be present for incorporation into the thickening ZP.

Female mice that are homozygous nulls for ZP1 *(Zp1$^{-/-}$)*, ZP2 *(Zp2$^{-/-}$)*, or ZP3 *(Zp3$^{-/-}$)* have been produced by targeted mutagenesis and their phenotypes determined. *Zp2$^{-/-}$* (Rankin et al., 2001) and *Zp3$^{-/-}$* (Liu et al., 1996; Rankin et al., 1996) females proved to be infertile and this is attributable to the absence of a ZP around growing oocytes and unfertilized eggs. As was the case for the messenger-RNA degradation experiments described above, ZP3 and ZP2 were synthesized by *Zp2$^{-/-}$* and *Zp3$^{-/-}$* oocytes, respectively, but failed to be incorporated into a ZP. Again, these results strongly suggest that both glycoproteins must be present for incorporation into a ZP and that synthesis of each ZP glycoprotein occurs independently of the others. Interestingly, female mice that are homozygous nulls for ZP1 are fertile, although not at wild-type levels, and their oocytes possess a very loosely organized ZP (Rankin et al., 1999). Finally, female mice that are heterozygous nulls for ZP3 *(Zp3$^{+/-}$)* are as fertile as wild-type mice and their oocytes have a ZP that is about one-half the thickness of the ZP of oocytes recovered from wild-type mice (Wassarman et al., 1997). These observations with "knockout mice" are completely consistent with the current model for ZP structure.

There is evidence to suggest that the three mouse ZP glycoproteins are organized in a very specific manner in the ZP. In the current model for ZP structure, ZP2 and ZP3 constitute long filaments that make up the extracellular coat and ZP1 serves as the crosslinker between filaments (Greve and Wassarman, 1985; Wassarman and Mortillo, 1991) (Fig. 5). ZP2 and ZP3 are present every 140 Å or so along ZP filaments, as revealed in electron micrographs of antibody stained and dissolved ZP, and impart a visible structural periodicity to the filaments. While each ZP filament may be a single linear polymer of ZP2 and ZP3, some recent experiments raise the possibility that each filament may be composed of two protofilaments (Jovine et al., 2002b). If this is the case, each protofilament could be a homopolymer composed of either ZP2 or ZP3 or a heteropolymer composed of both ZP2 and ZP3. Further experiments need to be performed to determine the precise arrangement of glycoproteins in the ZP.

Final comments

Here, we have summarized some of the major findings of investigations on mouse ZP genes and glycoproteins during the past twenty years or so. Perhaps, most surprising at the time was finding that the mouse ZP, a structure with several biological functions, is composed of only three glycoproteins (Bleil and Wassarman, 1980a). This is accepted now to be the case for most, if not all, egg ZP, including the ZP of human eggs. Since 1980, contemporary biotechnology, including molecular clon-

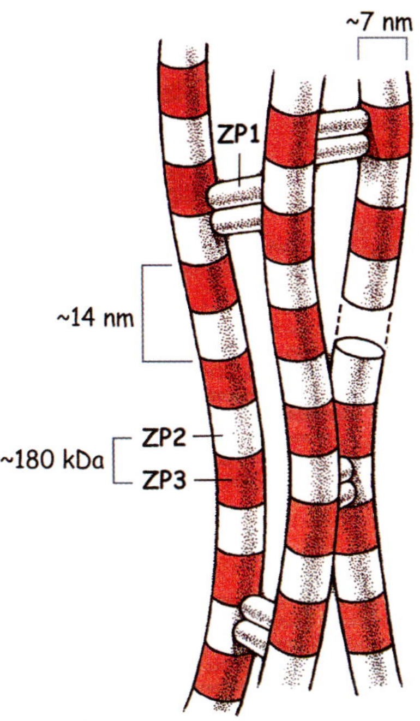

Fig. 5. Schematic representation of filaments that constitute the mouse egg ZP. Depicted are filaments composed of ZP2 and ZP3 that are cross-linked by ZP1 to form an extracellular coat (ZP) around growing oocytes and eggs. ZP2 and ZP3 are present every 14 nanometers or so along the filaments. For further details see Wassarman (1988b) and Wassarman and Mortillo (1991).

ing, transgenesis, and targeted mutagenesis, has been applied to mouse ZP genes and glycoproteins and has led to enormous progress in the field. Most recently it has been reported that oocytes have been derived from mouse embryonic stem (ES) cells and that the oocytes were enclosed in a coat resembling the ZP (Hubner et al., 2003). The ability to derive oocytes from ES cells has important implications for the field.

We hope that this brief review reflects the tremendous progress that has been made in understanding many aspects of mouse ZP genes and glycoproteins. It should be apparent that some aspects of what we have learned extend to many other kinds of secreted proteins in both vertebrates and invertebrates. While the story is by no means complete, what we have learned so far suggests in what directions we should go in our future research. Many questions remain unanswered. However, in this era of *"mouse genetics after the mouse genome"*, it is likely that results of clever in vivo and in vitro experimental approaches will provide answers to these questions in the very near future.

Acknowledgements

We thank Silvia Garagna for her kind invitation to contribute to this volume. We thank past and present members of our laboratory for their contributions to our research on the mouse zona pellucida. Due to page restrictions, we regret that in many instances reviews rather than primary publications are listed.

References

Beebe SJ, Leyton L, Burks B, Ishikawa M, Fuerst T, Dean J, Saling P: Recombinant mouse ZP3 inhibits sperm binding and induces the acrosome reaction. Dev Biol 151:48–54 (1992).

Bleil JD, Wassarman PM: Structure and function of the zona pellucida: Identification and characterization of the proteins of the mouse oocyte's zona pellucida. Dev Biol 76:185–202 (1980a).

Bleil JD, Wassarman PM: Synthesis of zona pellucida proteins by denuded and follicle-enclosed mouse oocytes. Proc Natl Acad Sci USA 77:1029–1033 (1980b).

Bork P, Sander C: A large domain common to sperm receptors (ZP2 and ZP3) and TGF-β type III receptor. FEBS Lett 300:237–240 (1992).

Chamberlin ME, Dean J: Human homolog of the mouse sperm receptor. Proc Natl Acad Sci USA 87:6014–6018 (1990).

Doren S, Landsberger N, Dwyer N, Gold L, Blanchette-Mackie J, Dean J: Incorporation of mouse zona pellucida proteins into the envelope of *Xenopus laevis* oocytes. Dev Genes Evol 209:330–339 (1999).

Epifano O, Liang L-F, Dean J: Mouse ZP1 encodes a zona pellucida protein homologous to egg envelope proteins in mammals and fish. J Biol Chem 270:27254–27258 (1995a).

Epifano O, Liang L-F, Familiari M, Moos MC, Dean J: Coordinate expression of the three zona pellucida genes during mouse oogenesis. Development 121:1947–1956 (1995b).

Greve JM, Wassarman PM: Mouse egg extracellular coat is a matrix of interconnected filaments possessing a structural repeat. J Mol Biol 181:253–264 (1985).

Greve JM, Salzman GS, Roller RJ, Wassarman PM: Biosynthesis of the major zona pellucida glycoprotein secreted by oocytes during mammalian oogenesis. Cell 31:749–759 (1982).

Gwatkin RBL: Fertilization Mechanisms in Man and Mammals (Plenum Press, New York 1977).

Howes EA, Jones R: Secondary binding of mammalian sperm to the egg zona pellucida. ChemTracts Biochem Mol Biol 16:134–141 (2003).

Hubner K, Fuhrmann G, Christenson LK, Kehler J, Reinbold R, De La Fuente R, Wood J, Strauss JF, Bolani M, Scholer HR: Derivation of oocytes from mouse embryonic stem cells. Science 300:1251–1256 (2003).

Jovine L, Litscher ES, Wassarman PM: Egg zona pellucida egg vitelline envelope and related extracellular glycoproteins. Adv Dev Biol Biochem 12:31–54 (2002a).

Jovine L, Qi H, Williams Z, Litscher E, Wassarman PM: The ZP domain is a conserved module for polymerization of extracellular proteins. Nature Cell Biol 4:457–461 (2002b).

Kinloch RA, Roller RJ, Fimiani CM, Wassarman DA, Wassarman PM: Primary structure of the mouse sperm receptor's polypeptide chain determined by genomic cloning. Proc Natl Acad Sci USA 85:6409–6413 (1988).

Kinloch RA, Mortillo S, Stewart CL, Wassarman PM: Embryonal carcinoma cells transfected with ZP3 genes differentially glycosylate similar polypeptides and secrete active mouse sperm receptor. J Cell Biol 115:655–664 (1991).

Lewandoski M, Wassarman KM, Martin GR: ZP3-cre, a transgenic mouse line for the activation of loxP-flanked target genes specifically in the female germ line. Curr Biol 7:148–151 (1997).

Liang L-F, Dean J: Conservation of mammalian secondary sperm receptor genes enables the promoter of the human gene to function in mouse oocytes. Dev Biol 156:399–408 (1993).

Liang L-F, Soyal SM, Dean J: FIGα a germ cell specific transcription factor involved in the coordinate expression of the zona pellucida genes. Development 124:4939–4947 (1997).

Lira SA, Kinloch RA, Mortillo S, Wassarman PM: An upstream region of the mouse ZP3 gene directs expression of firefly luciferase specifically to growing oocytes in transgenic mice. Proc Natl Acad Sci USA 87:7215–7219 (1990).

Lira SA, Schickler M, Wassarman PM: Cis-acting DNA elements involved in oocyte-specific expression of mouse sperm receptor gene mZP3 are located close to the gene's transcription start-site. Mol Reprod Dev 36:494–499 (1993).

Litscher ES, Qi H, Wassarman PM: Mouse zona pellucida glycoproteins mZP2, and mZP3 undergo carboxy-terminal proteolytic processing in growing oocytes. Biochemistry 38:12280–12287 (1999).

Liu C, Litscher ES, Mortillo S, Sakai Y, Kinloch RA, Stewart CL, Wassarman PM: Targeted disruption of the mZP3 gene results in production of eggs lacking a zona pellucida and infertility in female mice. Proc Natl Acad Sci USA 93:5431–5436 (1996).

Lunsford RD, Jenkins NA, Kozak CA, Liang L-F, Silan CM, Copeland NG, Dean J: Genomic mapping of murine ZP2 and ZP3 two oocyte-specific loci encoding zona pellucida proteins. Genomics 6:184–187 (1990).

McLeskey SB, Dowds C, Carballada R, White RR, Saling PM: Molecules involved in mammalian sperm-egg interaction. Intl Rev Cytol 177:57–113 (1998).

Qi H, Williams Z, Wassarman PM: Secretion and assembly of zona pellucida glycoproteins by growing mouse oocytes microinjected with epitope-tagged cDNAs for mZP2 and mZP3. Mol Biol Cell 13:530–541 (2002).

Rankin T, Dean J: The zona pellucida: Using molecular genetics to study the mammalian egg coat. Rev Reprod 5:114–121 (2000).

Rankin T, Familiari M, Lee E, Ginsberg AM, Dwyer N, Blanchette-Mackie J, Darago J, Dean J: Mice homozygous for an intentional mutation in the ZP3 gene lack a zona pellucida and are infertile. Development 122:2903–2910 (1996).

Rankin T, Tong Z-B, Castle PE, Lee E, Gore-Langton R, Nelson LM, Dean J: Human ZP3 restores fertility in ZP3 null mice without affecting order-specific sperm binding. Development 125:2415–2424 (1998).

Rankin T, Talbot P, Lee E, Dean J: Abnormal zonae pellucidae in mice lacking ZP1 results in early embryonic loss. Development 126:3847–3855 (1999).

Rankin TL, O'Brien M, Lee E, Wigglesworth K, Eppig J, Dean J: Defective zonae pellucidae in ZP2-null mice disrupt folliculogenesis fertility and development. Development 128:1119–1126 (2001).

Roller RJ, Kinloch RA, Hiraoka BY, Li SS-L, Wassarman PM: Gene expression during mammalian oogenesis and early embryogenesis: Quantification of three messenger-RNAs abundant in fully grown mouse oocytes. Development 106:251–261 (1989).

Salzmann GS, Greve JM, Roller RJ, Wassarman PM: Biosynthesis of the sperm receptor during oogenesis in the mouse. EMBO J 2:1451–1456 (1983).

Schickler M, Lira SA, Kinloch RA, Wassarman PM: A mouse oocyte-specific protein that binds to a region of mZP3 promoter responsible for oocyte-specific mZP3 gene expression. Mol Cell Biol 12:120–127 (1992).

Soyal SM, Amleh A, Dean J: FIGα a germ cell-specific transcription factor required for ovarian follicle formation. Development 127:4645–4654 (2000).

Tong Z, Nelson LM, Dean J: Inhibition of zona pellucida gene expression by antisense oligonucleotides injected into mouse oocytes. J Biol Chem 270:849–853 (1995).

Wassarman PM: Zona pellucida glycoproteins. A Rev Biochem 57:415–442 (1988a).

Wassarman PM: Fertilization in mammals. Sci Am 255:78–84 (1988b).

Wassarman PM: Mammalian fertilization: Molecular aspects of gamete adhesion exocytosis and fusion. Cell 96:175–183 (1999).

Wassarman PM: The mammalian egg zona pellucida: Structure and function during fertilization. Chem-Tracts Biochem Mol Biol 16:117–125 (2003).

Wassarman PM, Josefowicz WJ: Oocyte development in the mouse: An ultrastructural comparison of oocytes isolated at various stages of growth and meiotic competence. J Morphol 156:209–236 (1978).

Wassarman PM, Mortillo S: Structure of the mouse egg extracellular coat the zona pellucida. Intl Rev Cytol 130:85–109 (1991).

Wassarman PM, Bleil JD, Florman HM, Greve JM, Roller RJ, Salzmann GS, Samuels FG: The mouse egg's sperm receptor: What is it and how does it work? Cold Spring Harbor Symp Quant Biol 50:11–18 (1985).

Wassarman PM, Qi H, Litscher ES: Mutant female mice carrying a single mZP3 allele produce eggs with a thin zona pellucida but reproduce normally. Proc Roy Soc Lond B 26:323–328 (1997).

Wassarman PM, Jovine L, Litscher ES: A profile of fertilization in mammals. Nature Cell Biol 3:E59–E64 (2001).

Williams Z, Wassarman PM: Secretion of mouse ZP3 the sperm receptor requires cleavage of its polypeptide at a consensus furin cleavage-site. Biochemistry 40:929–937 (2001).

Yanagimachi R: Mammalian fertilization, in Knobil E, Neill J (eds): The Physiology of Reproduction, vol 1 pp 189–317 (Raven Press, New York 1994).

Cytogenet Genome Res 105:235–239 (2004)
DOI: 10.1159/000078194

Notch-1, c-kit and GFRα-1 are developmentally regulated markers for premeiotic germ cells

V. von Schönfeldt,[a] J. Wistuba,[a] and S. Schlatt[b]

[a] Institute of Reproductive Medicine of the University of Münster, Münster (Germany);
[b] Center for Research in Reproductive Physiology, Department of Cell Biology and Physiology,
University of Pittsburgh School of Medicine, Pittsburgh, PA (USA)

Abstract. Culture, transfection and immortalization of mouse germ line stem cells, germ cell transplantation and grafting of testicular tissue are milestones of recent biotechnological breakthroughs. Alone and in combination they offer new pathways to explore the cellular mechanisms responsible for pluripotency and the requirements of cells to enter the germ line. Efficient markers, isolation and culture systems as well as transfection approaches are developed to elucidate the molecular and cellular mechanisms leading to the development of male germ line cells. Here, we describe the localization pattern of c-kit, Notch-1 and GFRα-1 in postnatal, immature and adult testes. All three proteins are potentially useful markers for spermatogonial characterization and enrichment. First attempts and various future perspectives to use spermatogonial stem cells as pathway for the introduction of transgenes are discussed.

Spermatogonia have earned increasing interest of reproductive scientists during recent years since several scientific discoveries and methodological breakthroughs opened promising perspectives for novel applications on male germ line stem cells in basic research as well as clinical research areas (de Rooij, 1998; de Rooij and Grootegoed, 1998; Meachem et al., 2001; Brinster, 2002; Wistuba and Schlatt, 2002). Germ cell transplantation is now an established functional assay for the assessment of the biological activity of spermatogonial stem cells (Shinohara et al., 2001; Brinster, 2002; Orwig et al., 2002a; McLean et al., 2002; Nagano et al., 2003). Culture of these cells came into focus to address regulation and functional aspects of stem cell biology under defined conditions (Yoshinaga et al., 1991; Ogawa et al., 2000; Ohta et al., 2000; Orwig et al., 2002a; Izadyar et al., 2003; Kanatsu-Shinohara et al., 2003; Nagano et al., 2003). Nagano et al. (2003) demonstrated effects of supplementation with various growth factors and of co-cultures with different feeder cell lines on the stem cell fate decision (self renewal/differentiation) of cultured mouse spermatogonial stem cells. Exposure of spermatogonia to glial cell line-derived neurotrophic factor (GDNF), which was suggested to suppress differentiation of stem cells in vivo (Meng et al., 2002), evoked an increase of the number of spermatogonial stem cells in vitro. It became obvious from numerous experiments that spermatogonial stem cells – similar to embryonic, neural or hematopoietic stem cells – respond to environmental conditions and factors (Nagano et al., 2003).

Transfection of spermatogonia in vitro is a novel tool for manipulation of the germ line which is extremely powerful as it can be combined with germ cell transplantation to produce transgenic male gametes from transfected stem cells. This strategy has already been applied successfully to produce transgenic mice (Nagano et al., 2001) and rats (Hamra et al., 2002; Orwig et al., 2002b). Thus, it may become a most valuable tool for embryonic stem cell independent production of transgenic progeny and represents an alternative to the transformation of embryonic stem cells into germ cells in vitro as previously published (Toyooka et al., 2003). It eliminates all problems associated with isolation, maintenance, transfection and transformation of embryonic stem cells and production of chimeric offspring. In vitro transfection of spermatogonia has been shown to be feasible and clonal outgrowth of gonocytes in vitro (Hasthorpe et al., 1999, 2000) has recently been described.

Supported by grants and fellowships from the Deutsche Forschungsgemeinschaft (Schl 394/3-2, Schl 394/6-1).

Received 1 October 2003; revision accepted 15 October 2003.

Request reprints from Stefan Schlatt
  Center for Research in Reproductive Physiology
  Department of Cell Biology and Physiology
  University of Pittsburgh School of Medicine
  W952 Biomedical Science Tower, 3500 Terrace Street
  Pittsburgh, PA 15261 (USA); telephone: +1-412-648-8316
  fax: +1-412-648- 8315; e-mail: schlatt@pitt.edu

However, the target for approaches utilizing these new technologies as possible pathways for transgenesis depend on the spermatogonial stem cells. The testis contains many different types of spermatogonia of which only a small subfraction presents the spermatogonial stem cells. It is a prerequisite for the establishment of culture and transfection systems to obtain highly enriched and well characterized fractions of spermatogonial stem cells. In recent years, several new strategies using specific markers and/or specific features of spermatogonia have been described for enrichment and culture of gonocytes and spermatogonia (Morena et al., 1996; van den Ham, 1997; von Schönfeldt et al., 1999; de Miguel and Donovan, 2000; Shinohara and Brinster, 2000; van der Wee et al., 2001).

As an important step towards this goal this manuscript describes the testicular localization pattern of c-kit, Notch-1 and GFRα-1 at different developmental stages. All three factors have been shown to play important roles for male germ cell differentiation and expansion and for the cellular crosstalk inside the seminiferous epithelium (Yoshinaga et al., 1991; Dirami et al., 2001; Tadokoro et al., 2002). We show that the localization pattern of these three markers indicates that they are useful to identify different subpopulations of male germ line stem cells.

Materials and methods

Animals
The testes derive from outbred neonatal, immature and adult mice (strain: CD-1) derived from our institutional colony. The animals were maintained and all procedures were performed in accordance with the federal German law on the handling of experimental animals.

Immunohistochemistry
Testes were removed and routinely fixed in Bouin's solution for up to 12 h before transfer into 70 % ethanol. The tissue was routinely embedded in paraffin using an automated processor. Tissue sections of 5 μm were immunohistochemically stained for c-kit, Notch-1 and GFRα-1 using an indirect method with peroxidase-labeled secondary antibodies according to the manufacturer's instructions. Briefly, sections were deparaffinized in paraclear and rehydrated in a graded series of ethanol. For antigen retrieval, sections were heated in a microwave oven in Glycin/HCl buffer (50 mM, pH 3.5) for 10 min at 100 °C. Endogenous peroxidase activity was quenched by treatment with hydrogen peroxide (3 % for 5 min) followed by blocking of non-specific antibody-binding with 5 % normal goat serum supplemented with BSA (0.1 %) for 20 min at room temperature. All antibodies were diluted in TBS/BSA (0.1 %). The slides were incubated with primary antibody (rabbit anti-c-kit antibody (1:50), rabbit anti-Notch-1 antibody (1:100) or rabbit anti-GFRα-1 (1:50); Santa Cruz, CA, USA) at room temperature in a humidified chamber for 1 h and rinsed in TBS (10 mM TBS, 150 mM NaCl, pH 7.6) for 3 × 5 min between each of the following incubations. Sections incubated in TBS/BSA without primary antibody served as negative control. Adult testicular tissues were immunostained with HRP-conjugated goat anti-rabbit IgGs (1:50; Dako, Glostrup, Denmark) for 30 min. The label was visualized by incubation in 3,3′diaminobenzidine tetrahydrochloride in urea buffer for 5 to 12 min (Sigma-Aldrich, Taufkirchen, Germany). Positive staining appeared as a brown precipitate in the cells. Juvenile testicular tissues were immunostained using an LSAB-kit (DAKO Diagnostika, Hamburg, Germany) according to the manufacturer's instructions: Washes following incubation with primary antibody were carried out in TBS (2 × 5 min). Then the sections were incubated with biotinylated swine-anti-rabbit IgGs (10 min), washed and covered with streptavidin-AP-solution (10 min) and staining was finally visualized using new fuchsin (5 to 10 min).

All sections were counterstained in hematoxylin, mounted and analyzed by light microscopy.

Results

The results of the immunohistochemical localization of Notch-1 and GFRα-1 in adult testicular tissue of the mouse are shown in Fig. 1.

Notch-1-positive cells are located at the base of the germinal epithelium. The majority of labeled cells appears as groups of 2 or 4 cells in close proximity to each other and in contact with the basement membrane. They are embedded in between Sertoli cells. Higher magnification photomicrographs reveal cytoplasmic bridges connecting the labeled cells (Fig. 1b). Identical nuclear sizes and chromatin patterns of the immunopositive cells suggest that they are in a synchronous phase of the cell cycle, while unlabeled germ cells differ in these parameters. Taking cell size and localization into account, the Notch-1-positive cells can be addressed as undifferentiated A-spermatogonia with A_{pr} as the most common stained cell type followed by A_{al} (4 cells) and single spermatogonia, that appear to be A_s. All other cell types within the tubular compartment are unlabeled. All sections including the control show immunopositive label of extracellular spaces in the interstitial compartment which is due to unspecific binding of the secondary antibodies.

Specific staining of GFRα-1-positive cells is confined to few and mostly isolated germ cells in close proximity to the basement membrane. The expression pattern differs from Notch-1 since the label is restricted to single cells; no GFRα-1-positive cells connected by cytoplasmic bridges could be observed. Whenever immunopositive cells are located in close proximity, they are not connected through cytoplasmic bridges. Taken together, the morphology and the localization of the GFRα-1-positive cells suggest that they represent A_s-spermatogonia.

Representative micrographs of the immunohistochemical localization of Notch-1, c-kit and GFRα-1 in immature stages of the mouse testis are shown in Fig. 2.

All sections display faint unspecific cytoplasmic background staining due to the high sensitivity of the labeling approach. Intense specific staining for c-kit is localized in the cytoplasm of gonocytes from day 1 through day 6 post partum (dpp). The staining is not uniformly distributed throughout the cytoplasm of the immunopositive cells. Specific c-kit immunoreactivity is found both in gonocytes that are still located centrally within the tubules as well as in gonocytes that have completed migration to the periphery of the tubules (Fig. 2a). In other subpopulations of gonocytes, the staining is weak or absent. Sertoli cells do not show any specific staining. Within the interstitial compartment, the capillary endothelia display intense staining, which is also observed in control sections and therefore considered to be unspecific. From day 6 pp onwards, most of the gonocytes have completed migration and have matured into spermatogonia. At this point of development, only few gonocytes remain without contact to the basement membrane. As for the gonocyte population, the staining pattern of spermatogonia for c-kit is inconsistent, with some cells showing intense and other cells absence of c-kit staining.

Expression of GFRα-1 is observed predominantly in clones of 4 to 8 gonocytes around day 3 of postnatal development (Fig. 2b). These gonocytes seem to originate from clones undergoing synchronous cell divisions. The close proximity and

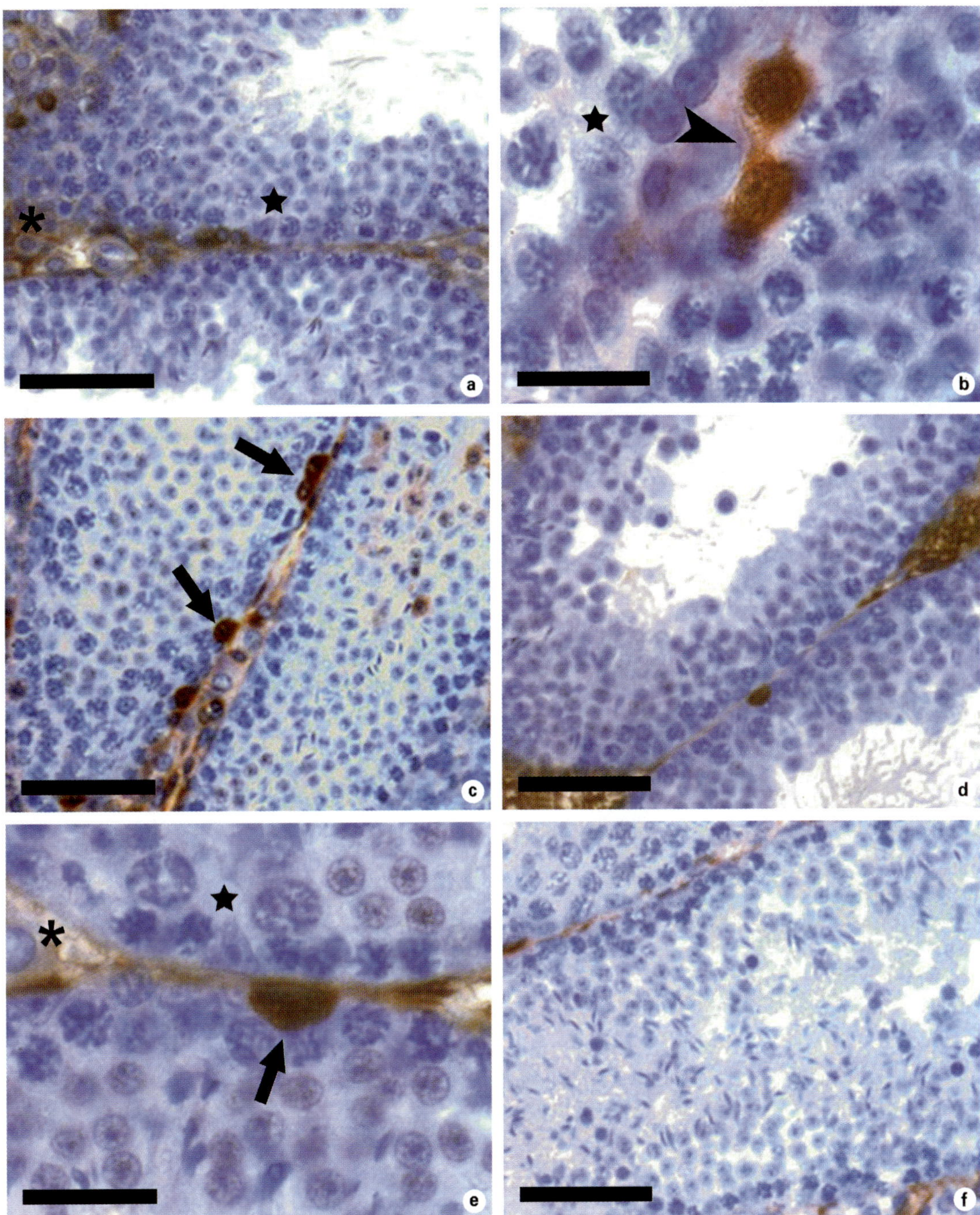

Fig. 1. Expression of Notch-1 (**a**, **b**) and GFRα-1 (**c–e**) in the adult mouse testis. The staining is confined to spermatogonia (arrows). Notch-1-positive spermatogonia often display a connection with cytoplasmic bridges (arrowheads). Sertoli cells (stars) and Leydig cells (asterisks) remain unstained. A representative control of method specificity shows no specific staining within the tubular compartment (**f**). Scale bars = 20 μm (**b**, **e**) or = 50 μm (**a**, **c**, **d**, **f**), respectively.

the identical nuclear size and chromatin pattern indicate that these cells are connected by cytoplasmic bridges comparable to the situation in A$_{al}$-spermatogonia observed in adult testes. Most often these immunopositive groups of gonocytes are located centrally within the tubules. Isolated gonocytes are GFRα-1-negative during this developmental period. There is no staining observed within Sertoli cells.

Notch-1 expression is seen in germ cells earliest from day 6 pp (Fig. 2c). The staining is confined to cells that have completed their migration and begin to flatten towards the basement membrane. Immunopositive spermatogonia in close contact to each other are frequently observed, while groups of gonocytes seem to be unlabeled. Sertoli cells, peritubular cells and interstitial cells are unlabeled.

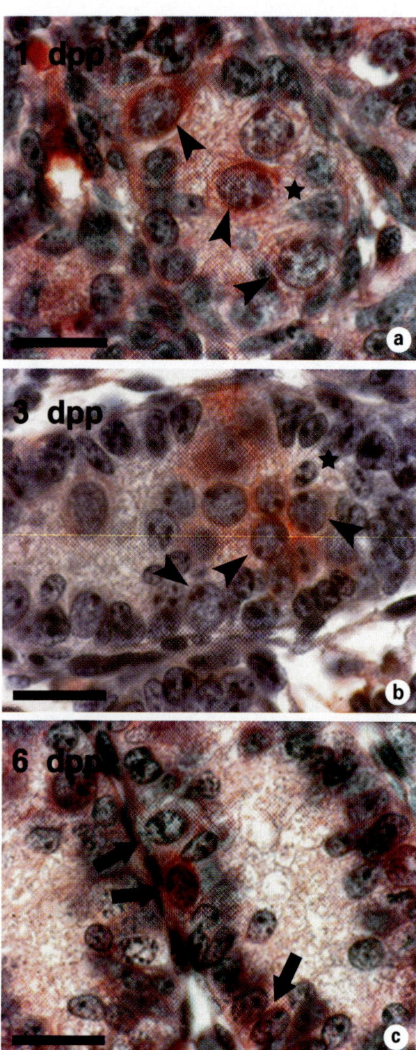

Fig. 2. Expression of c-kit (**a**), GFRα-1 (**b**) and Notch-1 (**c**) in postnatal testicular tissue of the mouse. Arrowheads indicate gonocytes and arrows indicate spermatogonia. Sertoli cells remain unstained (stars). Scale bars = 20 µm.

Discussion

Defining the mechanisms regulating self renewal and differentiation of spermatogonia is necessary to understand the premeiotic events of spermatogenesis and to select and characterize the male germ line stem cell fraction from the various subfractions of diploid germ cells. Here, we analyzed the cellular expression of several factors which are suspected to be involved in cell fate decision of undifferentiated spermatogonia in the adult mouse testis and during the first wave of spermatogenesis. Notch-1 is known to govern cell fate decision in a range of different tissues and stem cell systems of vertebrate and invertebrate species (Varnum-Finney et al., 1998; Artavanis-Tsakonas et al., 1999). Even though the importance of *Notch* has been demonstrated for the development of different organ systems, the role of *Notch* during spermatogenesis remains unclear – possibly due to the fact that the null mutation in the *Notch* gene is embryonically lethal (Swiatek et al., 1994; Hamada et al., 1999). *Notch* signaling either occurs between a group of equal cells – termed homotypic interaction – or is part of the communication of unequivalent cells (heterotypic interaction). Dirami et al. (2001) have identified further components of the *Notch* signaling pathway within the germinal epithelium of the mouse: The Notch-1 receptor protein is localized to spermatogonia and the Notch ligands Delta-1 and Jagged-1 are expressed exclusively in spermatogonia and Sertoli cells, respectively. We have also observed the Notch-1 receptor protein in spermatogonia of the adult and postpubertal mouse testis. From our data we hypothesize that the initiation of spermatogenesis is regulated by delta-mediated homotypic interaction between spermatogonia. The observation that mostly pairs or small clones of spermatogonia demonstrate positive Notch-1 staining suggests a role in cell interaction. The timing of earliest Notch expression around day 6 of postnatal development coincides with the onset of spermatogonial differentiation which strengthens our viewpoint.

While the general roles of c-kit/SCF was described a long time ago (Mintz, 1957; Mintz and Russel, 1957; Russel, 1979), some controversy remains about the exact function during premeiotic phases of spermatogenesis. Different kit- and kit-1-deficient mouse models display a variety of phenotypes ranging from complete absence of germ cells from the testis with some mutated alleles while others only display a diminished efficiency of spermatogenesis. In Sl^{17H}-deficient mice the transition from A_{al}- to A_1-spermatogonia is inhibited (de Rooij et al., 1999) indicating a role of c-kit receptor signaling during this transition. In vivo administration of anti-c-kit antibodies is followed by a depletion of A_1- to A_4-spermatogonia (Yoshinaga et al., 1991) while the same population of germ cells is maintained in tubular fragments if cultured in the presence of SCF (Yan et al., 2000) suggesting an involvement of c-kit also in regulating the subsequent steps of spermatogenesis. The expression of functional c-kit protein is necessary for the differentiation of spermatogonia as has been demonstrated by transplanting Sl-deficient germ cells into $W^{-/-}$ testes where they reinitiate full spermatogenesis to produce mature sperm (Ogawa et al., 2000). Before and during the first wave of spermatogenesis we found the expression of c-kit in germ cells to be inconsistent: At day 6 pp c-kit-immunonegative and immunopositive gonocytes are observed. Both populations can be located centrally or basally within the developing tubules. We speculate that not all gonocytes of this developmental stage share stem cell properties. Kinetic studies suggest that gonocytes not only transform into A_s-spermatogonia but some of them directly differentiate into A_2-spermatogonia (de Rooij, 1998). While a subpopulation of gonocytes is responsible for establishing the stem cell reserve, others act as developmentally more advanced A_{al}-spermatogonia and directly matured into A_1-cells. Since the c-kit receptor is considered to be a marker for differentiated spermatogonia, it can be speculated that c-kit-positive gonocytes are the latter ones with c-kit-negative gonocytes establishing the A_s population. This hypothesis could also account for the observation that some gonocytes are interconnected by cytoplasmic bridges similar to those shared by A_{pr}- or A_{al}-spermatogonia. Furthermore, the proliferation of A_s cells is independent of c-kit expression (Yoshinaga et al., 1991).

GFRα-1 is the receptor component of GDNF-mediated signal transduction. In the germinal epithelia of adult and juvenile mice, GFRα-1 expression is restricted to single spermatogonia which can be identified as A_s-spermatogonia due to their localization within the tubules and their nuclear morphology. A_s-spermatogonia are considered the stem cell population of the adult testis. They are low in number and proliferate at a slow rate. This is consistent with our observation, that GFRα-1-positive spermatogonia are rare and are almost never located in close proximity to other GFRα-1-positive spermatogonia of similar nuclear morphology. Confirming these findings, Dettin et al. (2003) demonstrated the GFRα-1 expression pattern in neonatal mouse germ cells to indicate the presence of A_s, A_{pr} and possibly A_{al} spermatogonia. Furthermore, a number of studies using different approaches have revealed that the GDNF-mediated pathway is responsible for cell fate decision of spermatogonial stem cells (Meng et al., 2002) and stem cell proliferation (Tadokoro et al., 2002).

In summary, we have described differential distribution patterns of three markers for male gonocytes and spermatogonia which are valuable for the discrimination and characterization of specific spermatogonial subtypes. The further exploration of spermatogonial stem cells and the development of new strategies for culture, transfection and transplantation opens new horizons to transfer the present knowledge of the mouse genome into generating new strains of transgenic mice and probably other transgenic mammals via the male germ line.

Acknowledgement

The authors are indebted to Prof. Dr. E. Nieschlag, Director of the Institute of Reproductive Medicine of the University of Münster for providing of the facilities of the Institute to perform the studies on the early postnatal development of male mouse germ cells.

References

Artavanis-Tsakonas S, Rand MD, Lake RJ: Notch signaling: cell fate control and signal integration in development. Science 284:770–776 (1999).

Brinster RL: Germline stem cell transplantation and transgenesis. Science 296:2174–2176 (2002).

de Miguel MP, Donovan PJ: Isolation and culture of mouse germ cells. Methods Mol Biol 137:403–408 (2000).

de Rooij DG: Stem cells in the testis. Int J Exp Pathol 79:67–80 (1998).

de Rooij DG, Grootegoed JA: Spermatogonial stem cells. Curr Opin Cell Biol 10:694–701 (1998).

de Rooij DG, Okabe M, Nishimune Y: Arrest of spermatogonial differentiation in jsd/jsd, Sl17H/Sl17H, and cryptorchid mice. Biol Reprod 61:842–847 (1999).

Dettin L, Ravindranath N, Hofmann MC, Dym M: Morphological characterization of the spermatogonial subtypes in the neonatal mouse testis. Biol Reprod 69:1565–1571 (2003).

Dirami G, Ravindranath N, Achi MV, Dym M: Expression of Notch pathway components in spermatogonia and Sertoli cells of neonatal mice. J Androl 22:944–952 (2001).

Hamada Y, Kadokawa Y, Okabe M, Ikawa M, Coleman JR, Tsujimoto Y: Mutation in ankyrin repeats of the mouse Notch2 gene induces early embryonic lethality. Development 126:3415–3424 (1999).

Hamra FK, Gatlin J, Chapman KM, Grellhesl DM, Garcia JV, Hammer RE, Garbers DL: Production of transgenic rats by lentiviral transduction of male germ-line stem cells. Proc Natl Acad Sci USA 99:14931–14936 (2002).

Hasthorpe S, Barbic S, Farmer PJ, Hutson JM: Neonatal mouse gonocyte proliferation assayed by an in vitro clonogenic method. J Reprod Fertil 116:335–344 (1999).

Hasthorpe S, Barbic S, Farmer PJ, Hutson JM: Growth factor and somatic cell regulation of mouse gonocyte-derived colony formation in vitro. J Reprod Fertil 119:85–91 (2000).

Izadyar F, den Ouden K, Creemers LB, Posthuma G, Parvinen M, de Rooij DG: Proliferation and differentiation of bovine type A spermatogonia during long term culture. Biol Reprod 68:272–281 (2003).

Kanatsu-Shinohara M, Ogonuki N, Inoue K, Miki H, Ogura A, Toyokuni S, Shinohara T: Long-term proliferation in culture and germline transmission of mouse male germline stem cells. Biol Reprod 69:612–616 (2003).

McLean DJ, Russell LD, Griswold MD: Biological activity and enrichment of spermatogonial stem cells in vitamin A-deficient and hyperthermia-exposed testes from mice based on colonization following germ cell transplantation. Biol Reprod 66:1374–1379 (2002).

Meachem S, von Schönfeldt V, Schlatt S: Spermatogonia: stem cells with a great perspective. Reproduction 121:825–834 (2001).

Meng X, Lindahl M, Hyvonen ME, Parvinen M, de Rooij DG, Hess MW, Raatikainen-Akohas A, Sainio K, Rauvala H, Lakso M, Pichel JG, Westphal H, Saarma M, Sariola H: Regulation of cell fate decision of undifferentiated spermatogonia by GDNF. Science 287:1489–1493 (2002).

Mintz B: Embryological development of primordial germ cells in the mouse: influence of a new mutation, Wj. J Embryol Exp Morphol 5:396–404 (1957).

Mintz B, Russell ES: Gene-induced embryological modifications of primordial germ cells in the mouse. J Exp Zool 134:207–237 (1957).

Morena AR, Boitani C, Pesce M, de Felici M, Stefanini M: Isolation of highly purified type A-spermatogonia from prepubertal rat testis. J Androl 17:708–717 (1996).

Nagano M, Brinster CJ, Orwig KE, Ryu BY, Avarbock MR, Brinster RL: Transgenic mice produced by retroviral transduction of male germ-line stem cells. Proc Natl Acad Sci USA 98:13090–13095 (2001).

Nagano M, Ryu BY, Brinster CJ, Avarbock MR, Brinster RL: Maintenance of mouse male germ line stem cells in vitro. Biol Reprod 68:2207–2214 (2003).

Ogawa T, Dobrinski I, Avarbock MR, Brinster RL: Transplantation of male germ line stem cells restores fertility in infertile mice. Nat Med 6:29–34 (2000).

Ohta H, Yomogida K, Dohmae K, Nishimune Y: Regulation of proliferation and differentiation in spermatogonial stem cells: the role of c-kit and its ligand SCF. Development 127:2125–2131 (2000).

Orwig KE, Shinohara T, Avarbock MR, Brinster RL: Functional analysis of stem cells in the adult rat testis. Biol Reprod 66:944–949 (2002a).

Orwig KE, Avarbock MR, Brinster RL: Retrovirus-mediated modification of male germline stem cells in rats. Biol Reprod 67:874–879 (2002b).

Russell ES: Hereditary anemias of the mouse: a review for geneticists. Adv Genet 20:357–459 (1979).

Shinohara T, Brinster RL: Enrichment and transplantation of spermatogonial stem cells. Int J Androl 23:89 (2000).

Shinohara T, Orwig KE, Avarbock MR, Brinster RL: Remodeling of the postnatal mouse testis is accompanied by dramatic changes in stem cell number and niche accessibility. Proc Natl Acad Sci USA 98:6186–6191 (2001).

Swiatek PJ, Lindsell CE, del Amo FF, Weinmaster G, Gridley T: Notch1 is essential for postimplantation development in mice. Genes Dev 8:707–719 (1994).

Tadokoro Y, Yomogida K, Ohta H, Tohda A, Nishimune Y: Homeostatic regulation of germinal stem cell proliferation by the GDNF/FSH pathway. Mech Dev 113:29–39 (2002).

Toyooka Y, Tsunekawa N, Akasu R, Noce T: Embryonic stem cells can form germ cells in vitro. Proc Natl Acad Sci USA 100:11457–11462 (2003).

van den Ham R, Van Pelt AM, de Miguel MP, van Kooten PJ, Walther N, van Dissel-Emiliani FM: Immunomagnetic cell isolation of fetal rat gonocytes. Am J Reprod Immunol 38:39–45 (1997).

van der Wee KS, Johnson EW, Dirami G, Dym TM, Hofmann MC: Immunomagnetic isolation and long term culture of mouse type A-spermatogonia. J Androl 22:696–704 (2001).

Varnum-Finney B, Purton LE, Yu M, Brashem-Stein C, Flowers D, Staats S, Moore KA, Le Roux I, Mann R, Gray G, Artavanis-Tsakonas S, Bernstein ID: The Notch ligand, Jagged-1, influences the development of primitive hematopoietic precursor cells. Blood 91:4084–4091 (1998).

von Schönfeldt V, Krishnamurthy H, Foppiani L, Schlatt S: Magnetic cell sorting is a fast and effective method of enriching viable spermatogonia from Djungarian hamster, mouse, and marmoset monkey testes. Biol Reprod 61:582–589 (1999).

Wistuba J, Schlatt S: Transgenic mouse models and germ cell transplantation: two excellent tools for the analysis of genes regulating male fertility. Mol Genet Metab 77:61–67 (2002).

Yan W, Suominen J, Toppari J: Stem cell factor protects germ cells from apoptosis in vitro. J Cell Sci 113:161–168 (2000).

Yoshinaga K, Nishikawa S, Ogawa M, Hayashi S, Kunisada T, Fujimoto T: Role of c-kit in mouse spermatogenesis: identification of spermatogonia as a specific site of c-kit expression and function. Development 113:689–699 (1991).

Cytogenet Genome Res 105:240–250 (2004)
DOI: 10.1159/000078195

Systems biology of the 2-cell mouse embryo

A.V. Evsikov,[a] W.N. de Vries,[a] A.E. Peaston,[a] E.E. Radford,[a] K.S. Fancher,[a] F.H. Chen,[b] J.A. Blake,[a] C.J. Bult,[a] K.E. Latham,[c] D. Solter,[a,b] and B.B. Knowles[a]

[a] The Jackson Laboratory, Bar Harbor, ME (USA); [b] Max-Planck-Institut für Immunbiologie, Freiburg (Germany); [c] The Fels Institute for Cancer Research and Molecular Biology, Temple University School of Medicine, Philadelphia, PA (USA)

Abstract. The transcriptome of the 2-cell mouse embryo was analyzed to provide insight into the molecular networks at play during nuclear reprogramming and embryonic genome activation. Analysis of ESTs from a 2-cell cDNA library identified nearly 4,000 genes, over half of which have not been previously studied. Transcripts of mobile elements, especially those of LTR retrotransposons, are abundantly represented in 2-cell embryos, suggesting their possible role in introducing genomic variation, and epigenetic restructuring of the embryonic genome. Analysis of Gene Ontology of the 2-cell-stage expressed genes outlines the major biological processes that guide the oocyte-to-embryo transition. These results provide a foundation for understanding molecular control at the onset of mammalian development.

Copyright © 2004 S. Karger AG, Basel

In metazoans, maternal factors accumulated in the ooplasm during oocyte growth are solely responsible for the control of maturation, fertilization and initial development of the newly formed embryo (Davidson, 1986). Accordingly, the transcriptionally silent fully grown oocyte is a striking example of a self-sufficient, well regulated cell that follows a strict route of completion of meiosis (maturation), activation upon fertilization, and progression through the first embryonic cell cycle(s). In this cell a unique environment is created, in which the genetic material of the gametes (or even that of a somatic cell) is transformed, or reprogrammed, and development can proceed until the newly formed embryonic genome is activated and transcription starts again. It is logical that the factors acting at the moment when maturation starts differ from those functioning when the embryonic genome is activated, but how is this change achieved? In the absence of transcription, such progressive change may be realized by posttranslational modification of proteins, time-dependent translational recruitment of stored maternal RNAs, and timely degradation of obsolete molecules (Solter et al., 2001).

In most non-mammalian species that have been studied, the first nuclear divisions proceed rapidly. The ooplasm is programmed to progress through the first morphogenetic events without the need for embryonic genome activation. In contrast, the first cell cycles of a newly formed mammalian embryo are very extended, with no differentiation events occurring prior to major activation of the embryonic genome (Telford et al., 1990). In the mouse, the ooplasmic control of developmental processes, from the start of oocyte maturation to embryonic genome activation at the 2-cell stage, occupies 40–44 h, or about 8% of prenatal development. The recent cloning of several mammals shows that factors present in the oocyte cytoplasm are pivotal for remodeling the introduced nuclei (Solter, 2000). In mice, nuclear reprogramming is essentially completed by the end of the second cell cycle, the time when the mouse embryonic genome is considered completely activated (Latham and Schultz, 2001).

What are the genes that act during this prolonged period of transcriptional silence and coordinate the transition from the specialized germ cells to the totipotent embryo? Classically, candidate genes were selected for study on the basis of evidence suggesting their involvement in the oocyte-to-embryo transi-

This work was supported by The National Institutes of Health grants HD37102 (B.B.K.), RR015253 (K.E.L.), HG00330 (J.A.B., C.J.B.), HG02273 (J.A.B.) and P30 CA34196, Department of Energy grant DE-FG02-99ER62850 (C.J.B.), The Lalor Foundation (A.V.E.), and the NHMRC of Australia (A.E.P.).

Received 25 November 2003; manuscript accepted 5 December 2003.

Request reprints from Barbara B. Knowles, The Jackson Laboratory
Bar Harbor, ME 04609 (USA); telephone: 1-207-288-6361
fax: 1-207-288-6071; e-mail: bbk@jax.org

tion and/or early development, and their function during this period was queried. Another approach started from a phenotype (e.g. developmental arrest at the preimplantation stage) with a search for the gene whose mutation might cause that phenotype.

The emergence of cDNA library technology and high throughput automated sequencing introduced a third way to study genes expressed during this period and offered the possibility of addressing far more powerful and precise questions about the molecular control of the oocyte-to-embryo transition. A subtractive cDNA library approach led us to the discovery of several novel genes, expression of which is stage specific during preimplantation development (Solter et al., 2001). Large-scale sequencing of cDNA libraries (e.g. Mammalian Gene Collection (MGC), RIKEN, WashU/HHMI mouse EST projects) provides another dimension, permitting inquiry as to whether a particular transcript is present in oocytes and/or preimplantation-stage embryos by sequence comparisons with public mouse EST datasets in GenBank. Other in silico mining approaches, for example the Digital Differential Display (DDD) tool of UniGene, may be used for identification of oocyte- and preimplantation stage-specific genes (Rajkovic et al., 2001).

Using these approaches, extensive lists of genes expressed in oocyte- and preimplantation embryo cDNA libraries have been published previously (Sasaki et al., 1998; Ko et al., 2000; Stanton and Green, 2001). Lists of genes, however, may be compared with a list of towns in a state. In the same sense as the latter is incomplete without a map, knowledge of genes expressed at a particular stage is of limited value. Functional annotation of genes, together with the understanding of the interactions of their products, would help to position them in the broad picture of biological processes involved in the oocyte-to-embryo transition.

To achieve this goal, we now describe a systems biology effort to functionally characterize the 2-cell mouse embryo transcriptome using the strength of bioinformatics tools, as well as in vitro validation of the in silico research. Knowledge of the components of this biological system provides access to the molecular networks controlling the onset of embryonic development.

Materials and methods

In silico analysis

14,813 ESTs of the Knowles Solter mouse 2-cell library (GenBank dbEST library ID.862) were downloaded in FASTA format from NCBI. EST sequences were assembled in consensus sequences (clusters) using the CAP3 assembly program (Huang and Madan, 1999). CAP3 produced 3,040 clusters and 3,165 single ESTs (singletons). CAP3-assembled clusters were individually checked for consistency of assembly. The Consed program (Gordon et al., 1998) was used to display the alignment of all ESTs in a cluster for visual checking. We identified 450 clusters with inconsistencies (i.e., only partially overlapping ESTs were assembled into a cluster, mainly because of the presence of unmasked repetitive elements). These were reclustered manually, by removing the partially overlapping ESTs from the consensus, and setting them as individual singletons, or by using stricter values for the CAP3 program, and resolved into 438 clusters and 369 singletons. The resulting 6,562 consensus sequences (3,028 clusters and 3,534 singletons) were each assigned a unique reference number, and BLAST searches of public databases were performed for every sequence. Batch searches against the public databases were performed using the BCM SearchLauncher (available from ftp://ftp.hgsc.bcm.tmc.edu/sl/software/search-launcher/). BLASTN, BLASTX, MEGABLAST searches were done using the NCBI BLAST engine (http://www.ncbi.nlm.nih.gov/BLAST/). For the identification and classification of repetitive elements we used RepeatMasker (http://ftp.genome.washington.edu/cgi-bin/RepeatMasker). Identification of mouse genes expressed only in somatic tissues was performed using the online digital differential display tool of UniGene (http://www.ncbi.nlm.nih.gov/UniGene/info_ddd.shtml). Searches against the mouse and human genome assemblies were performed using online search tools of the ENSEMBL project (http://www.ensembl.org/).

For the current study, the file of mouse genes determined to be expressed in the 2-cell library was indexed to the mouse genes in the Mouse Genome Informatics (MGI) database (http://informatics.jax.org/). The MGI accession ID was then associated with each gene. Mouse genes are annotated to the Gene Ontology (GO) (The Gene Ontology Consortium, 2001) by the MGI scientific curators using a combination of computational and manual methods. A set of high-level terms to cluster GO annotation results (GO bins) were determined from the three GO vocabularies. The GO annotations for these mouse genes were obtained from the MGI database and then clustered to the selected high-level term. The same queries and clustering were performed for all mouse genes in the MGI database. Because the GO vocabularies are structured as directed acyclic graphs, a given term may be a subterm of more than one parent. The practical result of this for GO binning purposes is that a gene may be grouped in more than one set.

The dataset of library-identified genes (version 1.0), along with the appropriate documentation, are currently available in Microsoft Excel format. The dataset and future updates are free for download from MGI (ftp://ftp.informatics.jax.org/pub/informatics/datasets/index.html), or available from authors upon request.

Oocyte and embryo isolation

All mice used in this study are B6D2F1/J. Full-grown oocytes, ovulated oocytes, zygotes and 2- to 8-cell embryos were collected as described (Oh et al., 2000); morulae and blastocysts were isolated at 72 and 84 h after mating, respectively. For RT-PCR and RT activity assays, batches of 55 oocytes or embryos were treated with 0.5 % pronase in M2 medium to remove zonae pellucidae, washed twice in M2 and 3 times in PBS supplemented with 0.4 % PVP (PBS/PVP), transferred in a minimal amount of PBS/PVP in sterile Eppendorf microcentrifuge tubes, and snap frozen until use. For RT assay, subconfluent cultured mouse embryo fibroblasts (CByB6F1×C57BL/6J-TgN(pPGKneobpA)3Ems) were harvested by trypsinization, washed twice in PBS, and snap frozen in aliquots until use.

Whole-mount immunostaining

Oocytes and embryos were treated with 0.5 % pronase in M2 medium (37 °C, 5 min) to remove the zona pellucida, washed twice in M2 and twice in PBS, PVP, and fixed in 2 % paraformaldehyde in PBS overnight at 4 °C. Embryos were permeabilized with 0.5 % Triton X-100 in PBS for 5–10 min at room temperature, transferred to PBS supplemented with 1 % BSA (Jackson ImmunoResearch) for 30 min, and blocked for 16 h at 4 °C in PBS containing 5 % normal donkey serum (Jackson ImmunoResearch). The embryos were washed three times in PBS, BSA incubated with the primary antibodies diluted in PBS, BSA at 37 °C for 1 h, washed in three changes of PBS, BSA (20 min, room temperature), incubated with the secondary antibodies in PBS, BSA at 37 °C for 1 h, washed in three changes of PBS, BSA, washed in PBS, and mounted on freshly made poly-L-lysine-coated slides. SlowFade Light kit (Molecular Probes) was used to prevent photobleaching. Oocytes and embryos were examined using a Leica TCS NT confocal microscope.

To obtain the optimal signal-to-noise ratio, each primary antibody was first tried at different dilutions (1:50 to 1:500). Primary antibodies were: anti-ELAVL1, anti-ELAVL2, and anti-UBE2I, goat polyclonal IgGs (SantaCruz Biotechnology, sc-5484, sc-5982, and sc-5231); anti-20S proteasome (α/β subunits), anti-PSMC4, anti-CSN8, and anti-XPO1: rabbit polyclonal IgGs (Affiniti, PW8155, PW8145, PW8290, and SantaCruz Biotechnology, sc5595). Secondary antibodies were Cy™ 3-conjugated donkey anti-rabbit and anti-goat IgGs (H+L) (Jackson ImmunoResearch). Secondary antibodies (0.65 mg/ml in 50 % glycerol) were used at 1:300 dilutions. To estimate the level of background staining, primary antibodies were omitted from the first incubation.

Isolation of probes, RT-PCR, QADB

IMAGE clones from the 2-cell library corresponding to the genes of interest were ordered from Incyte Genomics or ATCC, sequenced, and verified using GeneWorks 2.5.1 against the cluster sequence. Gene-specific primers with a projected annealing temperature of 60°C were designed using Gene-Works 2.5.1. Optimal buffers and annealing temperatures for PCR were determined empirically for each primer pair using the corresponding plasmid as a template.

RNA from oocytes and embryos was isolated as previously described (Oh et al., 2000). Oligo(dT) primers and Gibco BRL Superscript II kit (Life Technologies) were used for cDNA synthesis, and two oocyte or embryo equivalents were used in each reaction. PCR conditions for *Cpeb1, Tacc2, Tacc3, Elavl1, Elavl2, Mbtd1, Ing1, Psmc4* and *Xpo1* were: 97°C 30 s; followed by 40 cycles of 94°C, 45 s; (optimal temperature for each primer pair) for 60 s; 72°C for 90 s + 1 s per cycle. Primers (temperatures) were:

Cpeb1: TCCTAAGACATTTCCCAGTGG (F),
TGGGACAACCAAGAAGTTCC (R) (60°C);
Tacc2: GATTCCGTGAACAGCGTCTC (F),
CCGATGCATATGTTCACGTC (R) (60°C);
Tacc3: CGGCAGATCATCTGTTCTTTG (F),
TGGAAGCTGAAAGGCTTGAG (R) (60°C);
Elavl1: CCAAACGTCAAGGCAGTTATG (F),
GGGCAGCTCCAGTATATTCC (R) (60°C);
Elavl2: CAATAACACAGCCAATGGTCC (F),
GACCTGGTCGACAAGAATCC (R) (64°C);
Mbtd1: CAGCTGGCTGATTCCAAATC (F),
GCAGCAGATGGATCTGACTG (R) (60°C);
Ing1: ATGACATCACCTCAGGAACG (F),
GCTTACTTCCAAGTCACTCCC (R) (58°C);
Psmc4: CAAGGACGAGCAGAAGAACC (F),
TTCTGAACAAACTCCGAGCC (R) (62°C);
Xpo1: CGTGATTGGTGCACAGACAG (F),
GCATCCAACAAAGGAGGAAC (R) (60°C).

PCR conditions for MT transposon (MTT), ERVL and mitochondrial ATP synthase (*mt-Atp6*) were: 97°C 30 s; and (optimal number of cycles) of 94°C for 30 s; 60°C for 30 s; 72°C for 30 s. Primers (number of cycles) were:

MTT: TTCTGAGAGCACAGTGGCTG (F),
TCTTGGTGGTCATCTCATGG (R) (25 cycles);
ERVL: AATGCCCTTAAACCAACAGG (F),
TTTCCTTCACCTTCAGCCAG (R) (35 cycles);
mt-Atp6: TTCCACTATGAGCTGGAGCC (F),
GGTAGCTGTTGGTGGGCTAA (R) (35 cycles).

For the preparation of the probes, plasmid DNA was isolated and purified using the Plasmid Midi Kit (Qiagen) and digested with appropriate restriction enzymes. After being separated on a 1% agarose gel, the DNA bands were excised from the gel and extracted using the QIA Quick Gel Extraction Kit (Qiagen). These probes were used for <u>q</u>uantitative <u>a</u>mplification and <u>d</u>ot <u>b</u>lotting (QADB), as described previously (Wang et al., 2001).

Reverse transcriptase activity assays

The assay was conducted as described (Voisset et al., 2001), using the three described primers, with minor modifications. Briefly, aliquots of embryos or cells were lysed by repeat cycles of freeze-thawing in 20 μl of lysis buffer (20 mM Tris·HCl pH 8.4, 50 mM KCl, 2 mM Pefabloc (Roche Molecular Biologicals), 0.025% Triton X-100) before serial 10-fold dilutions in same buffer. 10 μl of each lysate dilution, or of serial dilutions of purified RT (MMLV-derived Superscript II; Invitrogen), were used as the RT source in an RT reaction (0.3 pmol MS2 RNA (Roche Molecular Biologicals), 0.4 mM dNTPs, 3 pmol gene-specific first primer, 4 mM MgCl₂, 20 mM Tris-HCl pH 8.4, 50 mM KCl, 8 mM dithiothreitol, 40 units RNasin (Promega), final volume 25 μl). Negative controls for each experiment were a reaction lacking RNA and a reaction lacking a source of RT.

2.5 μl of each RT reaction were used as the template in a PCR reaction (10 mM Tris-HCl pH 8.3, 50 mM KCl, 10 mM MgCl₂, 0.2 mM dNTPs, 3 pmol second primer; final volume 25 μl). 2.5 μl of this reaction were then used as template in a nested PCR identical to the previous reaction except for the addition of the third gene-specific primer to the reaction. For both PCRs, the cycling conditions were 94°C 30 s, and 35 rounds of 94°C for 30 s; 45°C for 30 s; 72°C for 30 s. PCR products were separated on 3% agarose gels, stained with ethidium bromide, and visualized by UV transillumination.

To calculate the RT activity per egg, embryo or cell (*x*), the following formula was used: $x = n \div y$, where n was the lowest RT activity detected in the Superscript II RT serial dilution, and y is the lowest egg, embryo, or cell number from which RT activity could be detected.

Results and discussion

The tool: cDNA library to ESTs to transcriptome

The 2-cell cDNA library (Knowles Solter 2-cell mouse; Rothstein et al., 1992, 1993) was made from 13,500 embryos isolated at 40–42 h post-hCG injection from oviducts of B6D2F1/J females mated to B6D2F1/J males for 1.5 h. It is known that by this time, more than 80% of embryos complete the S-phase and reside in G2 of the second cell cycle (Sawicki et al., 1978). Embryos were immediately processed for RNA isolation, so that the exposure to in vitro environment was reduced to minimum. Timed caging of females with males at 13 to 14.5 h post-hCG resulted in fertilization of all oocytes at approximately the same time, and minimized the heterogeneity within the embryo population. The large sample size excluded the need for PCR amplification of cDNA, consequently eliminating PCR-related artifacts and bias in transcript abundance (Rothstein et al., 1992; Solter et al., 2001), a clear advantage over the cDNA libraries prepared using PCR methods (Sasaki et al., 1998; Ko et al., 2000). Previous analysis showed that this 2-cell-embryo cDNA library is representative and has a >99% probability of including very rare transcripts (Rothstein et al., 1992).

14,813 ESTs from this library were produced by the WashU/HHMI mouse EST project. As a first step we identified the transcripts these ESTs represent. To facilitate the analysis, all overlapping EST sequences were assembled in consensus sequences or clusters, decreasing the number of individual sequences for analysis. BLASTN searches of GenBank were performed for each sequence. Results of this search were used to sort the sequences into major categories (Fig. 1, and supplementary material at http://karger.com/doi/10.1159/000078195, I): *i)* known nuclear genes (i.e., known genes of non-mitochondrial or retroviral/transposon origin that have Mouse Genome Informatics [MGI, http://www.informatics.jax.org/] and/or Locus-Link entries), *ii)* unknown genes presumed to be nuclear, *iii)* unknown sequences containing genomic repeats, *iv)* retroviral genes and genomic repeats, *v)* mitochondrial transcripts, and *vi)* obvious contaminants (i.e., bacterial and cloning vector sequences). The known nuclear genes *(i)* were classified, on the basis of sequence similarity to sequences in GenBank, into *a)* mouse genes curated by MGI, *b)* putative genes identified from NCBI's annotation of the public mouse genome assembly, and *c)* putative orthologs of genes from other species. An official MGI gene name was assigned for each sequence from group *a*, Locus-Link IDs for sequences in group *b*, and official name of an ortholog for sequences in group *c*. The assignments further reduced the seemingly independent clusters and singletons to representatives of distinct genes, and enabled accurate determination of the number of known genes represented in the library. Relatively small transcripts were generally represented by one consensus sequence. However, larger transcripts, or alternative transcripts of the same gene, were often represented by several clusters and

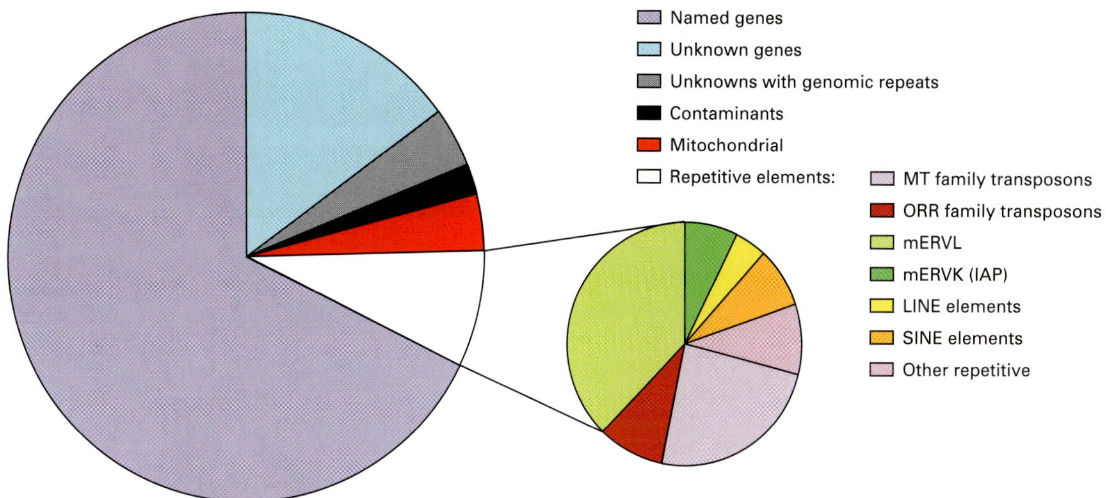

Fig. 1. The 2-cell mouse embryo transcriptome. Distribution of 14,813 2-cell library ESTs among six broad categories (see text). The "repetitive elements" segment is further subdivided into seven major classes in a smaller side chart.

singletons, and these were classified as corresponding to a single gene. Analysis of the 4,120 consensus sequences in category *(i)* (known genes) suggested that they actually represent 2,771 genes. If the same ratio holds for the 1,717 sequences from categories *ii* and *iii* (i.e., the original 4,120 clusters and ESTs represent 2,771 known genes), then the unknown sequences may actually represent $1,717 \times (2,771 \pm 4,120) = 1,155$ additional genes. This suggests that the sequenced ESTs represent approximately 4,000 different genes expressed at the 2-cell stage.

The estimated number of genes can be compared with other measures of the number of genes represented in the 2-cell mouse embryo library. The result is close to the number of Uni-Gene (http://www.ncbi.nlm.nih.gov/UniGene/) clusters that contain at least one EST from this library (3,499 "UniGenes", build 113), but is much lower than the respective number of AllGenes clusters (5,676, DoTS ver. 4; http://www.all-genes.org). The lesser number of UniGenes may result from the fact that only 71% of ESTs from the 2-cell library were used in assembly, because of the stringent criteria for EST trace sequence quality in UniGene. The AllGenes build procedure, unlike UniGene and our procedure, is transcript-oriented, and quite often two or more clusters correspond to alternative transcripts of a same gene; this would explain the "redundancy" of the DoTS assemblies. Our results indicate that at least one-tenth of the estimated number of genes in the mouse genome is represented in this analysis of the 2-cell library, although transcripts of low abundance may not be included in our results.

For the 250 most abundant genes in the library, a MEGA-BLAST search against the GenBank mouse EST database was performed. The "virtual Northern blot" helps to determine whether a gene is expressed in other mouse tissues and whether it can be found in other libraries from the same developmental stages. Ten of these 250 genes (4%) were found *only* in the cDNA libraries from oocytes or embryos up to the 8-cell stage, suggesting these are maternal mRNAs of genes that have oocyte-restricted expression. This implies that the physiology of the 2-cell embryo is controlled by a combination of many known genes and a few stage-specific genes.

Expression of mobile elements: regulatory aspects

Basal levels of transcription of repetitive elements have been reported for most cells (Kazazian, 1998), including preimplantation embryos (Piko et al., 1984; Taylor and Piko, 1987). In the 2-cell embryo library, 8% of ESTs represent transcripts of repetitive elements. Investigation of the spectrum of repetitive elements expressed at the two-cell stage revealed that 80% of these transcripts represent retrotransposons or related sequences (Fig. 1). Expression of the MT transposon-like element (Heinlein et al., 1986) is ubiquitous in oocytes and throughout preimplantation development (Fig. 2A). Although ESTs of ORR1A and ORR1B transposons (Smit, 1993) are present in our 2-cell library, as well as other libraries of 2- to 16-cell embryos and blastocysts, cDNA libraries prepared from oocytes and zygotes do not contain such ESTs, suggesting these are expressed at the initiation of embryonic genome activation. Our analysis also revealed a high expression of endogenous retroviruses (Fig. 1). Expression of intracisternal A-particles (IAP, murine ERVK) has been reported previously in oocytes and preimplantation embryos of all stages (Piko et al., 1984). On the other hand, the extraordinarily high expression of another endogenous retrovirus, murine ERVL (Benit et al., 1997), is transient (Fig. 2A) and coincident with embryonic genome activation at the 2-cell stage (Wang et al., 2001).

To determine whether the retroviral transcripts are translated in oocytes and preimplantation-stage embryos, a PCR-based assay for reverse transcriptase (RT) activity (Voisset et al., 2001) was performed. From the oocyte to the early 2-cell stage, there was less than 50 nU of RT activity per egg or embryo. At the late 2-cell stage, RT activity increased to about 500 nU per embryo, then fell to approximately 50 nU per embryo at later stages (Fig. 2B). This approach could not distin-

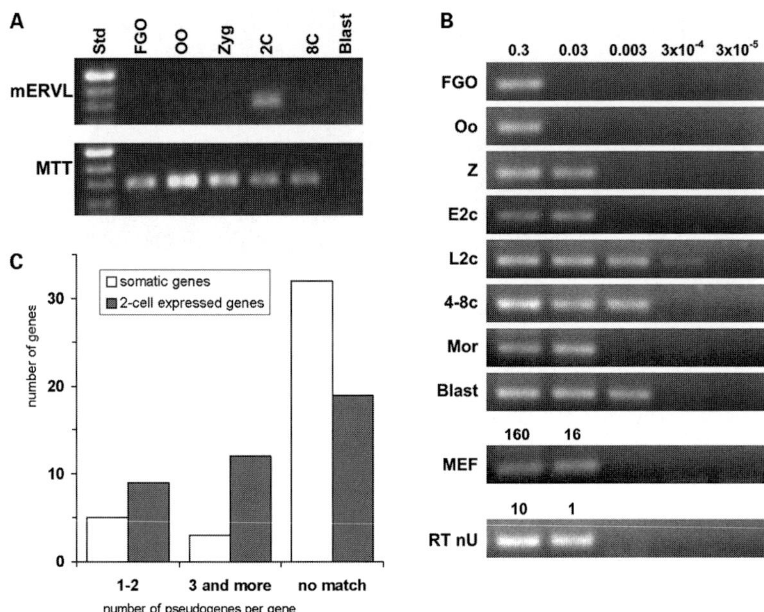

Fig. 2. Expression of mobile elements in mouse oocytes and preimplantation embryos. (**A**). Expression of mERVL and MT transposon. Std: DNA MW markers; FGO, full grown oocytes; OO, ovulated (metaphase II arrested) oocytes; Zyg, zygotes; 2C, 2-cell embryos; 8C, 8-cell embryos; Blast, blastocysts. (**B**) Reverse transcriptase activity in eggs and preimplantation embryos. The figure shows a representative example of the nested PCR performed on cDNA constructed using serial dilutions of egg, embryo or cell lysates as the source of the RT, and MS2 RNA as the template, with gene-specific primer. The numbers indicate the number of eggs, embryos or cells from whose lysates the RT activity was derived. FGO, full grown oocyte; Oo, ovulated oocyte; Z, zygote; E2c, early 2-cell embryo; L2c, late 2-cell embryo; 4-8c, 4- to 8-cell-stage embryos; Mor, morula; Blast, blastocyst; MEF, mouse embryo fibroblast. Serial dilutions of commercial reverse transcriptase (RT nU) were used as a control. (**C**) 2-cell-embryo expressed genes are retrotransposed more frequently than somatic genes. Of the 80 genes tested (supplementary material, II), more 2-cell-stage expressed genes have 1–2, and ≥ 3, putative pseudogenes, than the somatically expressed genes. Among the 2-cell-embryo expressed genes, there are fewer genes not having pseudogenes ("no match" group).

guish between the sources of reverse transcriptase (i.e., *pol* genes of ERVL, IAP and LINEs), but the results suggest a high level of RT activity at this time. The difference in cell volume between a mouse oocyte and a cultured mouse embryo fibroblast is approximately 30-fold. Therefore, these results indicate that the concentration of RT activity in oocytes and preimplantation embryos is 180-fold, and in the late 2-cell embryos 1,800-fold higher than that in this somatic cell.

These results pose the intriguing possibility that retrotransposition events which occur in the germline or in early embryos may become fixed in the next generation. To test the hypothesis that genes expressed in the 2-cell-stage embryo would have a greater chance of being retrotransposed than genes expressed exclusively in somatic tissues, we randomly selected and compared a group of 40 somatically expressed genes with a group of 40 genes found in the 2-cell library. Somatic genes, represented in cDNA libraries from normal somatic tissues, but not in cDNA libraries from oocytes/preimplantation embryos, ovaries or testes, were selected using the DDD tool of UniGene. BLASTN analysis of an open reading frame sequence of each gene against the mouse genome assembly (Mouse Genome Consortium, 2002) was performed, and every BLAST hit was scored as *i)* a match to the gene itself, or *ii)* a match to a paralogous gene, or *iii)* a contiguous sequence match, ≥ 80% aligned, or *iv)* contiguous sequence match, <80% aligned. Assuming that evolutionarily recent retrotransposition events would fall into group *iii)*, the "somatic" and "2-cell-stage expressed" groups of genes were compared (Fig. 2C, and supplementary material at http://karger.com/doi/10.1159/000078195, II). A significant difference between the two groups was found (P = 0.0047, Mann-Whitney U test), lending credence to the notion that this developmental period is a hot spot for retrovirus-induced evolutionary changes in genome organization. A case study of LINE1 retrotransposition during oogenesis in human (Brouha et al., 2002), the observation that genes expressed during spermatogenesis may be reverse-transcribed in vivo (Zhong and Kleene, 1999), and studies of expressed pseudogenes (Kleene et al., 1998; Chen et al., 2002a) all support the notion that retrotransposed genes and pseudogenes may derive from genes expressed in gametes and early-cleavage-stage embryos.

The abundance of transcripts that contain a genomic repeat in the 5′ UTR, most commonly an LTR of the MT transposon (MTA) or ERVL (LTR_ERVL), is an intriguing feature of 2-cell library identified genes. For example, transcripts of four novel genes, putative paralogs of *Uba52*, are extremely abundant in the library: LOC216814, LOC237780, LOC235733 and LOC216818 are represented by 52, 8, 8 and 5 ESTs, respectively. The genomic organization of these genes is very similar in that each contains two exons, the first exon is an LTR_ERVL or MTA, while the second exon contains the rest of the 5′ UTR, the coding sequence, and the 3′ UTR. In the cases of LOC216814, LOC235733 and LOC216818, the 3′ UTR is immediately followed by a poly(A) stretch 10–20 nucleotides long, suggesting that these transcribed genes originated by retrotransposition.

In silico analysis suggests these genes are transiently expressed at the 2-cell stage and are silenced later in development and in adults (the only exception is that one EST for LOC216814 is found in the mouse early blastocyst cDNA library, GenBank dbEST library ID.1021). A similar genomic organization, that is a transcription start site and the first exon residing within a repetitive element, has been reported for two other 2-cell-stage-expressed genes, *Rnf33* and *Rnf35* (Chen et al., 2002b).

LTRs contain multiple transcription factor binding sites, but in most cells and tissues repetitive elements are hyper-methylated and silenced (Yoder et al., 1999). We suggest that the loci described above are expressed in 2-cell embryos because genome-wide silencing of the mobile and repetitive elements has yet to be established. Thus, repetitive and mobile elements may contribute specific regulatory variability to the expression of such loci in the early-cleavage-stage embryos by targeting genes for developmentally regulated expression and subsequent silencing by epigenetic mechanisms during chromosomal restructuring. It has been suggested that genes which share common regulation may form functionally linked networks (Davidson, 1986). We do not know if the loci above code for functional proteins, but it is clear that they represent a pool of gene expression and/or function variants, which may impact embryonic phenotype and act as substrates for natural selection.

Experimental evidence suggests that RNA interference (RNAi) is functional at this stage (Svoboda et al., 2000); *eIF2C2* and *Dicer1*, encoding two of the few known mammalian elements of RNAi machinery (Hutvagner and Zamore, 2002), are found in the 2-cell embryo cDNA library. It has been suggested that RNAi is required for posttranslational silencing of repetitive elements in animal cells (Hammond et al., 2001). Recently, the initiation of targeted heterochromatin formation and transcriptional repression at sites of repetitive element transcription in the fission yeast *Schizosaccharomyces pombe* was shown to require the RNA interference machinery (Jenuwein, 2002). This raises the possibility that in early embryos RNAi is involved in the epigenetic restructuring of the mammalian embryonic genome at the time of its activation.

Annotation of the 2-cell embryo transcriptome

To gain insight into the overall physiology of the 2-cell embryo, we used the existing GO annotations (The Gene Ontology Consortium, 2001) for library-identified genes. Gene Ontology is based on three major structured controlled vocabularies for describing genes: molecular function, biological process and cellular component. To visualize and facilitate the interpretation of the trends in gene expression at this stage, 9–10 high-level terms (GO bins) to cluster GO annotation results were used. For comparative purposes, the dataset of genes that have elevated expression in three types of stem cells (Ramalho-Santos et al., 2002) and the whole dataset of mouse genes in the MGI database were also sorted using the same GO bins (Fig. 3, and supplementary material at http://karger.com/doi/10.1159/

000078195, III *C*). Approximately one-half of MGI-indexed, library-identified genes have GO annotations; the rest are either novel or not yet annotated. This analysis allowed us to build a snapshot of the molecular processes controlling the 2-cell embryo.

Translational regulation: Many oocyte-transcribed mRNAs are deadenylated and stored in the ooplasm for later translation. The abundance of gene transcripts whose products are translational regulators and RNA binding proteins (Fig. 3A) underlines the expected tight translational control of maternal transcripts during the oocyte-to-embryo transition. An overwhelming body of evidence ties the processes involved in translational control and mRNA stability to the presence of certain regulatory motifs in the 3′UTR of mRNAs, which are recognized by respective binding proteins. We have shown previously that one such motif, Cytoplasmic Polyadenylation Element (CPE), was present in 30% (75 out of 249) of polyadenylation signal-containing 2-cell EST clusters (Oh et al., 2000). Two proteins are responsible for delayed translation of CPE-containing transcripts, a regulatory CPE-binding protein (CPEB), characterized in *Xenopus laevis*, mouse, and other species (Mendez and Richter, 2001), and an essential partner of CPEB Maskin, identified and studied in *Xenopus* only (Stebbins-Boaz et al., 1999). As expected, ESTs for *Cpeb1* are present in the 2-cell library, apparently representing the residual maternal transcripts laid down in the oocyte (Fig. 4A). To our knowledge, the functional homolog of Maskin has not been identified in mammals. TBLASTN search of the mouse genome assembly with the protein sequence of *Xenopus* Maskin produced four significant hits, to LOC245600, *Tacc3*, *Tacc2* and *Tacc1* gene loci. Reciprocal BLASTP analysis of the amino acid sequences revealed that the protein products of these genes share a 29–46% identity and 44–61% similarity to Maskin. Of these four candidates, *Tacc2* and *Tacc3* ESTs are present in the 2-cell library, and we confirmed their expression during this time by RT-PCR (Fig. 4A). Moreover, TACC3 protein is abundant in mouse oocytes (Hao et al., 2002). Additional searches of the mouse EST database with the mRNA sequences of the four TACC genes confirmed a high level of representation of *Tacc3* and *Tacc2*, but not *Tacc1* or LOC245600 ESTs in other oocyte and preimplantation embryo libraries. This result strongly suggests that *Tacc2* and *Tacc3* are functional homologs of *Maskin*; accordingly these genes may play a role in translational regulation in mammalian oocytes and preimplantation embryos.

Other regulatory motifs in the 3′UTRs, AU-rich elements (AREs) were originally identified as elements responsible for destabilization and rapid degradation of mRNAs in somatic cells (Chen and Shyu, 1995). However, in oocytes and early embryos of *Xenopus* these motifs seem to be responsible for fast deadenylation, but not necessarily degradation, of mRNAs (Voeltz and Steitz, 1998). One of the factors that stabilize the ARE-containing mRNAs is the RNA-binding protein HuR (ELAVL1), of the ELAV family (Brennan and Steitz, 2001). In the mouse, there are four ELAV-like genes – *Elavl1*, *Elavl2*, *Elavl3* and *Elavl4*. The first two are present in the 2-cell library. RT-PCR analysis, and whole-mount immunostaining of oocytes and preimplantation embryos confirm the presence of both proteins, and their mRNAs, in the egg and in cleavage-

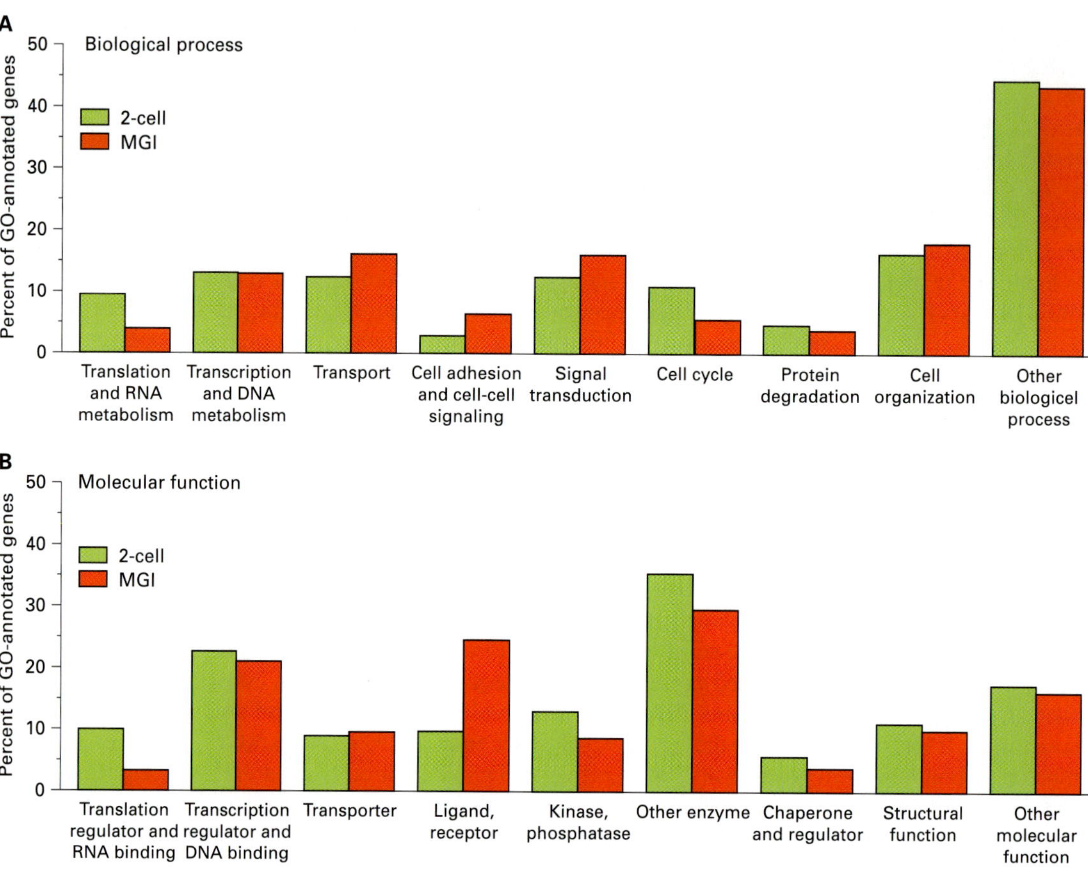

Fig. 3. Functional annotation of the 2-cell mouse embryo transcriptome. Distribution of the library-identified genes (2-cell), and all genes in the MGI database (MGI), within Gene Ontology bins. (**A**) Biological process (GO:0008150) the gene product is involved in. (**B**) Molecular function (GO:0003674) of the gene product.

stage embryos (Fig. 4A, B). This, together with an observation that in *Xenopus* eggs mRNAs bound by a member of the ELAV family, elrA, are polyadenylated and translated only after fertilization (Wu et al., 1997), suggests these proteins play an important role in translational regulation of stored maternal mRNAs during the oocyte-to-embryo transition.

Protein degradation machinery: Timely protein modifications and degradation of obsolete proteins are a requirement for the changing physiology during the oocyte-to-embryo transition. A large proportion (13%) of the annotated 2-cell-stage identified genes are kinases or phosphatases (Fig. 3B), and 5% are involved in protein degradation (Fig. 3A). The multitude and abundance of the genes of the ubiquitination and related pathways of targeted protein modification (supplementary material at http://karger.com/doi/10.1159/000078195, I) indicates their importance during this time. UBE2I (UBC9), a major sentrin (SUMO-1, UBL1) ligase found in the library, localizes to the nucleus at the 2-cell stage (Fig. 5), in agreement with the known role of UBE2I as a mediator of function of many nuclear proteins (Buschmann et al., 2001; Kang et al., 2001).

Expression of proteasome components during this developmental period is particularly interesting because of the many processes this organelle may regulate, such as progression of

meiosis in mammalian oocytes (Josefsberg et al., 2000) and translational regulation by degradation of CPEB in *Xenopus* (Mendez et al., 2002). The assembled (26S) proteasome is a complex of the barrel-shaped proteolytic (20S) particle, and the regulatory (19S) particle, which in turn consists of two multiprotein complexes, the "lid" and the "base" (Ferrell et al., 2000). Thirty-nine ESTs, or 0.26% of the 2-cell library, represent different subunits of the proteasome. Indeed, in vitro culture of zygotes and 2-cell embryos in M16 medium supplemented with MG132, a reversible inhibitor of proteasome proteolytic activity, blocks embryonic development (Evsikov et al., unpublished data). The 20S particle preferentially localizes to the nuclei of the Full-Grown Oocytes (FGO) and 1- to 16-cell embryos (Fig. 5). However, in blastocysts the 20S proteasome appears more evenly distributed between the blastomere nuclei and cytoplasm, as found in somatic cells (Reits et al., 1997).

PSMC4 protein (TBP7, RPT3) is one of the six ATPases of the 19S particle "base". The function of *Psmc4* is essential for preimplantation development (Sakao et al., 2000), and transcripts of this gene are present in all stages tested (Fig. 6A). The 19S particle is dynamic, and its subunits may play multiple roles in the cell (Ferrell et al., 2000). Interestingly, PSMC4 does not colocalize with the 20S complex until the 2-cell stage

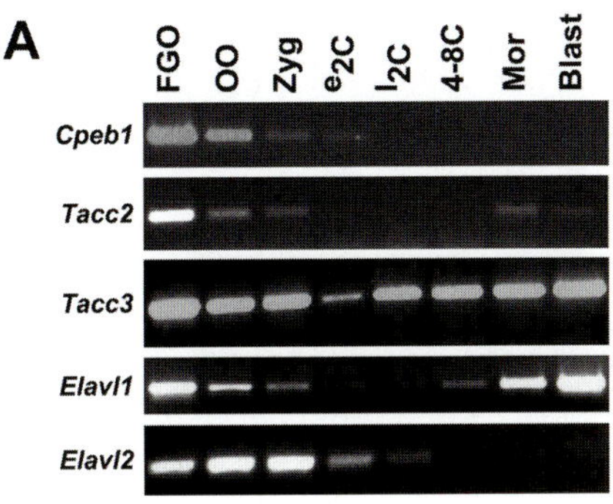

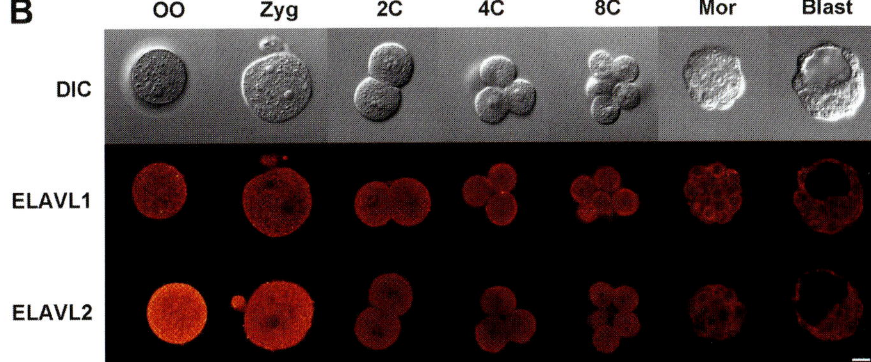

Fig. 4. Expression of translational control genes in oocytes and preimplantation embryos. (**A**) RT-PCR detection of *Cpeb1*, *Tacc2*, *Tacc3*, *Elavl1* and *Elavl2* transcripts. All abbreviations as in Fig. 2. (**B**) Whole-mount immunostaining for ELAVL1 (Hu antigen R; Hua) and ELAVL2 (Hu antigen B; mel-N1) proteins of oocytes and embryos. DIC, differential interference contrast images of respective stages. Note a marked decrease in the immunostaining for ELAVL2 after the zygote stage. Scale bar: 20 µm.

(Fig. 5), suggesting that *Psmc4* may also have non-proteasomal functions in preimplantation mouse embryos.

Regulation of the cell cycle: The notion that a zygotic clock regulates the oocyte-to-embryo transition from completion of meiosis until embryonic genome activation (Latham and Schultz, 2001) draws attention to genes controlling the cell cycle. Indeed, analysis of the GO annotations of the library-identified genes revealed that 11 % are involved in controlling the cell cycle (Fig. 3A). Given that the first, second and third cell cycles of the mouse embryo are very prolonged (18–20, 15–17 and 11–13 h, respectively), one might expect to find an abundance of negative cell cycle regulators. Components of the COP9 signalosome, a multisubunit particle structurally related to the 19S lid (Wei et al., 1998), represent 0.15 % of the ESTs in the 2-cell cDNA library. COP9 negatively regulates G1/S transition, modulating

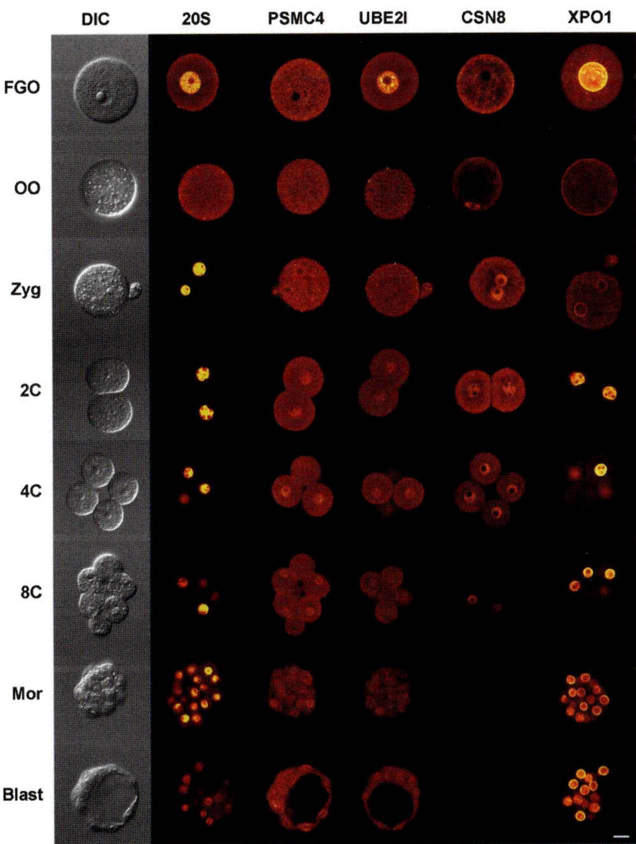

Fig. 5. Protein products of 2-cell library-identified genes in oocytes and embryos. 20S, 20S (α/β) complex of the proteasome; PSMC4 (TBP7, RPT3), proteasome 26S ATPase subunit 4; UBE2I (UBC9), ubiquitin-conjugating enzyme E2I; CSN8, COP9 signalosome subunit 8, "expressed sequence AA408242"; XPO1, exportin, CRM1 (yeast) homolog; FGO, full-grown oocyte; OO, ovulated oocyte; Zyg, zygote; 2C, 2-cell embryo; 8C, 8-cell embryo; Mor, morula; Blast, blastocyst; DIC, differential interference contrast. Scale bar: 20 µm.

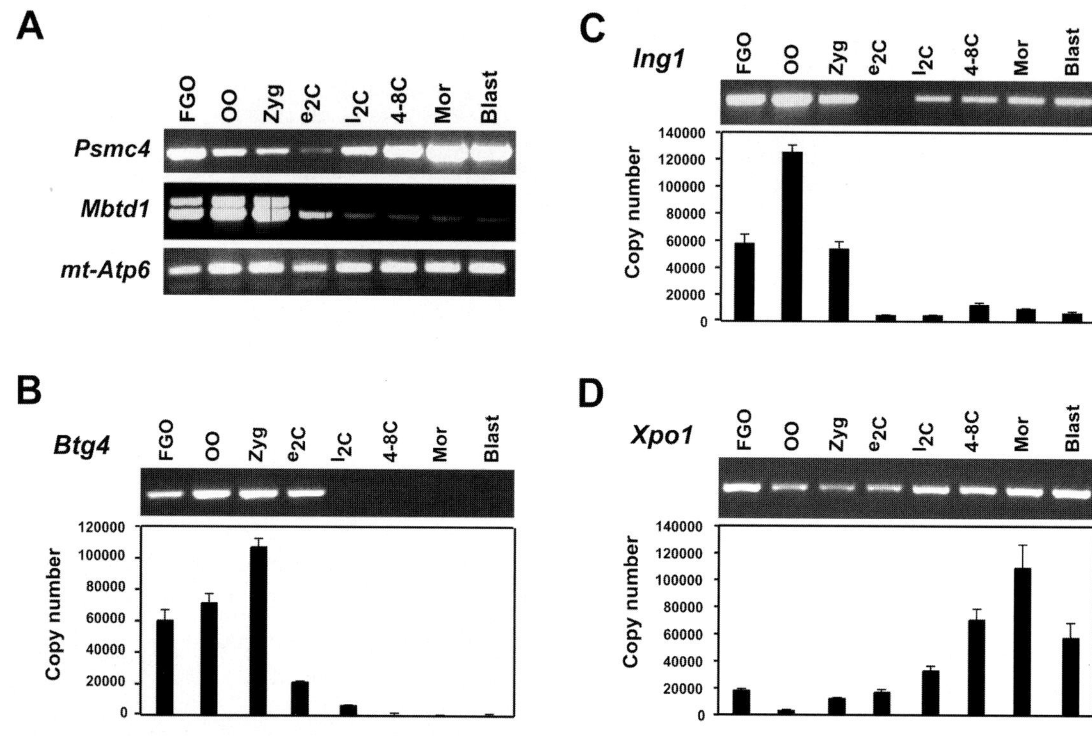

Fig. 6. RT-PCR and QADB detection of library-identified transcripts. (**A**) RT-PCR of *Psmc4* and *Mbtd1*. *mt-Atp6* (mitochondrial transcript, control) shown. RT-PCR and QADB (Wang et al., 2001) analysis of (**B**) *Btg4*, (**C**) *Ing1* and (**D**) *Xpo1* expression in oocytes and embryos. All abbreviations as in Fig. 2.

Skp1-Cullin-F-box protein (SCF) ubiquitin ligase activity by deneddylation of cullins (Yang et al., 2002). We followed CSN8, a structural component of the COP9 signalosome, in eggs and embryos (Fig. 5). When the cell cycle length of blastomeres shortens after the third cleavage, CSN8 protein becomes undetectable by immunostaining. Similarly, the abundance of the *Btg4* transcripts, a negative regulator of G1/S transition (Buanne et al., 2000) whose ESTs comprise 0.26% of the 2-cell library, dramatically decreases after the 2-cell stage (Fig. 6B), concurrent with the shortening of cell cycles.

Transcriptional regulation and chromatin remodeling: The oocyte-to-embryo transition is characterized by extensive remodeling of chromatin organization at all levels (Latham and Schultz, 2001). About 13% of GO annotated, library-identified genes encode transcriptional regulators and DNA binding proteins (Fig. 3A). Studies on the distribution of acetylated core histones in preimplantation mouse embryos and experiments with inhibition of histone deacetylase activity reveal that the degree of histone acetylation is pivotal for embryonic genome activation and establishing the program of zygotic transcription (Latham and Schultz, 2001; Solter et al., 2001). However, nothing is known about the mode of regulation of histone acetylation during this developmental period. Two alternative products of the PHD-finger gene *Ing1*, p47 (ING1a) and p33 (ING1b), differentially regulate histone H3 and H4 acetylation via their interactions with histone deacetylase and histone acetyltransferase, respectively (Vieyra et al., 2002). Twenty-two ESTs for *Ing1* are found in the library. *Ing1* mRNA is abundant

in oocytes and zygotes, absent in the early 2-cell embryos, and reappears after embryonic genome activation (Fig. 6C). *Ing1* is thus a candidate gene for master regulation of histone acetylation in oocytes and preimplantation embryos.

We have previously suggested that specific negative regulators of transcription may control embryonic genome activation, and described an Id-like gene *Maid*, capable of dominant-negative transcriptional repression (Hwang et al., 1997). In silico analysis revealed that two other novel potential transcriptional repressors, *Mbtd1* and *D11Ertd530e*, are abundantly expressed in the 2-cell embryo. *Mbtd1* ("mbt domain containing 1") is represented by 28 ESTs (0.19% of the library). RT-PCR revealed that oocytes and zygotes contain an alternative maternal transcript of *Mbtd1* which is no longer detectable by the early 2-cell stage (Fig. 6A), and the abundance of the second transcript also decreases at this time. Its putative protein product contains four "malignant brain tumor" domains (MBT), present in some members of the Polycomb group of transcriptional repressors (Bornemann et al., 1998). A few mammalian genes containing this domain have been described (Tomotsune et al., 1999; Usui et al., 2000), but the function of the MBT protein family remains largely unknown. Another novel gene, *D11Ertd530e* ("DNA segment, Chr 11, ERATO Doi 530, expressed"), C2H2 zinc finger transcriptional repressor, is very abundant, represented by 27 ESTs (0.18% of the library). The putative *D11Ertd530e* protein product is highly similar (40% identity, 60% similarity) to Drosophila Suppressor of zeste 12 *(Su(z)12)*, a Polycomb group gene essential for germ cell development (Birve et al., 2001), suggesting an ancient yet unknown role of these genes in oogenesis and/or the first stages of embryonic development.

Nucleo-cytoplasmic transport: Transport of different molecules is an important component of the 2-cell embryo physiology (Fig. 3A, B). Nucleo-cytoplasmic transport might be of particular importance because of the remodeling which chromatin undergoes at this stage. Library analysis reveals that the major elements of the nucleo-cytoplasmic transport machinery of the 2-cell embryo are essentially identical to that of every cell. *Xpo1* (exportin, CRM1), which is involved in nuclear export signal-mediated transport of proteins from the nucleus (Fornerod and Ohno, 2002), is highly abundant in the library (29 ESTs) and at all preimplantation stages (Fig. 6D). Indeed, XPO1 is localized to the nuclear envelope in mouse FGO and preimplantation embryos (Fig. 5). XPO1-dependent nuclear export is quite universal and regulates a multitude of cellular processes (Fornerod and Ohno, 2002). Inhibition of XPO1 function by leptomycin B in Xenopus embryos leads to developmental arrest at the gastrula-neurula transition (Callanan et al., 2000), while treatment with the same drug arrests swine embryos at the 4-cell stage (Cabot et al., 2002), just prior to embryonic genome activation in this species. These results suggest that *Xpo1* has a pivotal role in establishing the program of gene expression; therefore, in the 2-cell mouse embryo exportin may regulate the pattern of gene expression during embryonic genome activation, possibly by controlling the pool of transcription factors present in the nucleus.

Concluding remarks

The transcriptome of the 2-cell embryo is a reflection of the physiological state of the mammalian embryo as it emerges from ooplasmic control of development to the complexities of conditional cell fate specifications at later stages (Davidson, 1990). By the 2-cell stage, only the most stable maternal RNAs remain in the cytoplasm, implying their functional importance.

Simultaneously, the first transcripts from the new embryonic genome become apparent. Our analysis of this "transitional" transcriptome revealed elevated levels of transcripts whose products regulate translation and the cell cycle, as well as reduced number of transcripts representing ligands, receptors and other molecules involved in cell-cell signaling. Indeed, elimination of maternal pool of *Cdh1* transcripts, encoding the cell adhesion protein E-cadherin, and *Catnb1*, which encodes β-catenin, a central player in both cell adhesion and Wnt signaling pathways, had little effect on the oocyte-to-embryo transition and preimplantation development (de Vries et al., submitted). This provides a molecular foundation for a long-standing hypothesis that profound changes occurring during the oocyte-to-embryo transition in mammals, including nuclear reprogramming, are cell-autonomous processes independent of external signaling.

The breadth of data obtained here presents a starting point for other large-scale transcriptome analyses, such as microarray experiments or chromosomal expression studies. In one such application, we analyzed whether the genes, recently found to have elevated expression levels in embryonic, hematopoietic and neuronal stem cells (Ivanova et al., 2002; Ramalho-Santos et al., 2002) were expressed in this totipotent cell. Forty percent of such genes are indeed represented in the 2-cell cDNA library (supplementary material at http://karger.com/doi/10.1159/000078195, III *A*, III *B*; but also see Evsikov and Solter, 2003). Finally, a curated, publicly accessible database of the genes expressed at the 2-cell stage will be a significant resource for research into the molecules and biological processes involved in embryonic genome activation and the initiation of mammalian embryogenesis.

Acknowledgements

The authors thank Dr. John Eppig and Dr. Rick Woychik for critical reading of the manuscript. We also thank B.G. Patel for technical assistance with the QADB analysis.

References

Benit L, De Parseval N, Casella JF, Callebaut I, Cordonnier A, Heidmann T: Cloning of a new murine endogenous retrovirus, MuERV-L, with strong similarity to the human HERV-L element and with a gag coding sequence closely related to the Fv1 restriction gene. J Virol 71:5652–5657 (1997).

Birve A, Sengupta AK, Beuchle D, Larsson J, Kennison JA, Rasmuson-Lestander A, Muller J: *Su(z)12*, a novel *Drosophila* Polycomb group gene that is conserved in vertebrates and plants. Development 128:3371–3379 (2001).

Bornemann D, Miller E, Simon J: Expression and properties of wild-type and mutant forms of the *Drosophila* sex comb on midleg (SCM) repressor protein. Genetics 150:675–686 (1998).

Brennan CM, Steitz JA: HuR and mRNA stability. Cell Mol Life Sci 58:266–277 (2001).

Brouha B, Meischl C, Ostertag E, de Boer M, Zhang Y, Neijens H, Roos D, Kazazian HH Jr: Evidence consistent with human L1 retrotransposition in maternal meiosis I. Am J Hum Genet 71:327–336 (2002).

Buanne P, Corrente G, Micheli L, Palena A, Lavia P, Spadafora C, Lakshmana MK, Rinaldi A, Banfi S, Quarto M, Bulfone A, Tirone F: Cloning of PC3B, a novel member of the PC3/BTG/TOB family of growth inhibitory genes, highly expressed in the olfactory epithelium. Genomics 68:253–263 (2000).

Buschmann T, Lerner D, Lee CG, Ronai Z: The Mdm-2 amino terminus is required for Mdm2 binding and SUMO-1 conjugation by the E2 SUMO-1 conjugating enzyme Ubc9. J Biol Chem 276:40389–40395 (2001).

Cabot RA, Hannink M, Prather RS: CRM1-mediated nuclear export is present during porcine embryogenesis, but is not required for early cleavage. Biol Reprod 67:814–819 (2002).

Callanan M, Kudo N, Gout S, Brocard M, Yoshida M, Dimitrov S, Khochbin S: Developmentally regulated activity of CRM1/XPO1 during early *Xenopus* embryogenesis. J Cell Sci 113:451–459 (2000).

Chen CY, Shyu AB: AU-rich elements: characterization and importance in mRNA degradation. Trends Biochem Sci 20:465–470 (1995).

Chen HH, Liu TY, Huang CJ, Choo KB: Generation of two homologous and intronless zinc-finger protein genes, zfp352 and zfp353, with different expression patterns by retrotransposition. Genomics 79:18–23 (2002a).

Chen HH, Liu TY, Li H, Choo KB: Use of a common promoter by two juxtaposed and intronless mouse early embryonic genes, rnf33 and rnf35: implications in zygotic gene expression. Genomics 80:140–143 (2002b).

Davidson EH: Gene Activity in Early Development. (Academic Press, New York 1986).

Davidson EH: How embryos work: a comparative view of diverse modes of cell fate specification. Development 108:365–389 (1990).

Evsikov AV, Solter D: Comment on "Stemness": "transcriptional profiling of embryonic and adult stem cells" and "A stem cell molecular signature" (II). Science 302:393 (2003).

Ferrell K, Wilkinson CR, Dubiel W, Gordon C: Regulatory subunit interactions of the 26S proteasome, a complex problem. Trends Biochem Sci 25:83–88 (2000).

Fornerod M, Ohno M: Exportin-mediated nuclear export of proteins and ribonucleoproteins. Results Probl Cell Differ 35:67–91 (2002).

The Gene Ontology Consortium: Creating the gene ontology resource: design and implementation. Genome Res 11:1425–1433 (2001).

Gordon D, Abajian C, Green P: Consed: a graphical tool for sequence finishing. Genome Res 8:195–202 (1998).

Hammond SM, Caudy AA, Hannon GJ: Post-transcriptional gene silencing by double-stranded RNA. Nat Rev Genet 2:110–119 (2001).

Hao Z, Stoler MH, Sen B, Shore A, Westbrook A, Flickinger CJ, Herr JC, Coonrod SA: TACC3 expression and localization in the murine egg and ovary. Mol Reprod Dev 63:291–299 (2002).

Heinlein UA, Lange-Sablitzky R, Schaal H, Wille W: Molecular characterization of the MT-family of dispersed middle-repetitive DNA in rodent genomes. Nucl Acids Res 14:6403–6416 (1986).

Huang X, Madan A: CAP3: a DNA sequence assembly program. Genome Res 9:868–877 (1999).

Hutvagner G, Zamore PD: A microRNA in a multiple-turnover RNAi enzyme complex. Science 297:2056–2060 (2002).

Hwang SY, Oh B, Fuchtbauer A, Fuchtbauer EM, Johnson KR, Solter D, Knowles BB: Maid: a maternally transcribed novel gene encoding a potential negative regulator of bHLH proteins in the mouse egg and zygote. Dev Dyn 209:217–226 (1997).

Ivanova NB, Dimos JT, Schaniel C, Hackney JA, Moore KA, Lemischka IR: A stem cell molecular signature. Science 298:601–604 (2002).

Jenuwein T: An RNA-guided pathway for the epigenome. Science 297:2215–2218 (2002).

Josefsberg LB, Galiani D, Dantes A, Amsterdam A, Dekel N: The proteasome is involved in the first metaphase-to-anaphase transition of meiosis in rat oocytes. Biol Reprod 62:1270–1277 (2000).

Kang ES, Park CW, Chung JH: Dnmt3b, de novo DNA methyltransferase, interacts with SUMO-1 and Ubc9 through its N-terminal region and is subject to modification by SUMO-1. Biochem Biophys Res Commun 289:862–868 (2001).

Kazazian HH Jr: Mobile elements and disease. Curr Opin Genet Dev 8:343–350 (1998).

Kleene KC, Mulligan E, Steiger D, Donohue K, Mastrangelo MA: The mouse gene encoding the testis-specific isoform of Poly(A) binding protein (Pabp2) is an expressed retroposon: intimations that gene expression in spermatogenic cells facilitates the creation of new genes. J Mol Evol 47:275–281 (1998).

Ko MS, Kitchen JR, Wang X, Threat TA, Wang X, Hasegawa A, Sun T, Grahovac MJ, Kargul GJ, Lim MK, Cui Y, Sano Y, Tanaka T, Liang Y, Mason S, Paonessa PD, Sauls AD, DePalma GE, Sharara R, Rowe LB, Eppig J, Morrell C, Doi H: Large-scale cDNA analysis reveals phased gene expression patterns during preimplantation mouse development. Development 127:1737–1749 (2000).

Latham KE, Schultz RM: Embryonic genome activation. Front Biosci 6:D748–D759 (2001).

Mendez R, Richter JD: Translational control by CPEB: a means to the end. Nat Rev Mol Cell Biol 2:521–529 (2001).

Mendez R, Barnard D, Richter JD: Differential mRNA translation and meiotic progression require Cdc2-mediated CPEB destruction. EMBO J 21:1833–1844 (2002).

Oh B, Hwang S, McLaughlin J, Solter D, Knowles BB: Timely translation during the mouse oocyte-to-embryo transition. Development 127:3795–3803 (2000).

Piko L, Hammons MD, Taylor KD: Amounts, synthesis, and some properties of intracisternal A particle-related RNA in early mouse embryos. Proc Natl Acad Sci USA 81:488–492 (1984).

Rajkovic A, Yan MSC, Klysik M, Matzuk M: Discovery of germ cell-specific transcripts by expressed sequence tag database analysis. Fertil Steril 76:550–554 (2001).

Ramalho-Santos M, Yoon S, Matsuzaki Y, Mulligan RC, Melton DA: "Stemness": transcriptional profiling of embryonic and adult stem cells. Science 298:597–600 (2002).

Reits EA, Benham AM, Plougastel B, Neefjes J, Trowsdale J: Dynamics of proteasome distribution in living cells. EMBO J 16:6087–6094 (1997).

Rothstein JL, Johnson D, DeLoia JA, Skowronski J, Solter D, Knowles BB: Gene expression during preimplantation mouse development. Genes Dev 6:1190–1201 (1992).

Rothstein JL, Johnson D, Jessee J, Skowronski J, DeLoia JA, Solter D, Knowles BB: Construction of primary and subtracted cDNA libraries from early embryos. Methods Enzymol 225:587–610 (1993).

Sakao Y, Kawai T, Takeuchi O, Copeland NG, Gilbert DJ, Jenkins NA, Taked K, Akira S: Mouse proteasomal ATPases Psmc3 and Psmc4: genomic organization and gene targeting. Genomics 67:1–7 (2000).

Sasaki N, Nagaoka S, Itoh M, Izawa M, Konno H, Carninci P, Yoshiki A, Kusakabe M, Moriuchi T, Muramatsu M, Okazaki Y, Hayashizaki Y: Characterization of gene expression in mouse blastocyst using single-pass sequencing of 3995 clones. Genomics 49:167–179 (1998).

Sawicki W, Abramczuk J, Blaton O: DNA synthesis in the second and third cell cycles of mouse preimplantation development. A cytophotometric study Exp Cell Res 112:199–205 (1978).

Smit AF: Identification of a new, abundant superfamily of mammalian LTR-transposons. Nucl Acids Res 21:1863–1872 (1993).

Solter D: Mammalian cloning: advances and limitations. Nat Rev Genet 1:199–207 (2000).

Solter D, de Vries WN, Evsikov AV, Peaston AE, Chen FH, Knowles BB: Fertilization and activation of the embryonic genome, in Tamm P, Rossant J (eds): Mouse Development: Patterning and Organogenesis, pp 5–19 (Academic Press, New York 2001).

Stanton JL, Green DP: Meta-analysis of gene expression in mouse preimplantation embryo development. Mol Hum Reprod 7:545–552 (2001).

Stebbins-Boaz B, Cao Q, de Moor CH, Mendez R, Richter JD: Maskin is a CPEB-associated factor that transiently interacts with eIF-4E. Mol Cell 4:1017–1027 (1999).

Svoboda P, Stein P, Hayashi H, Schultz RM: Selective reduction of dormant maternal mRNAs in mouse oocytes by RNA interference. Development 127:4147–4156 (2000).

Taylor KD, Piko L: Patterns of mRNA prevalence and expression of B1 and B2 transcripts in early mouse embryos. Development 101:877–892 (1987).

Telford NA, Watson AJ, Schultz GA: Transition from maternal to embryonic control in early mammalian development: a comparison of several species. Mol Reprod Dev 26:90–100 (1990).

Tomotsune D, Takihara Y, Berger J, Duhl D, Joo S, Kyba M, Shirai M, Ohta H, Matsuda Y, Honda BM, Simon J, Shimada K, Brock HW, Randazzo F: A novel member of murine Polycomb-group proteins, Sex comb on midleg homolog protein, is highly conserved, and interacts with RAE28/mph1 in vitro. Differentiation 65:229–239 (1999).

Usui H, Ichikawa T, Kobayashi K, Kumanishi T: Cloning of a novel murine gene Sfmbt, Scm-related gene containing four mbt domains, structurally belonging to the Polycomb group of genes. Gene 248:127–135 (2000).

Vieyra D, Loewith R, Scott M, Bonnefin P, Boisvert FM, Cheema P, Pastyryeva S, Meijer M, Johnston RN, Bazett-Jones DP, McMahon S, Cole MD, Young D, Riabowol K: Human ING1 proteins differentially regulate histone acetylation. J Biol Chem 277:29832–29839 (2002).

Voeltz GK, Steitz JA: AUUUA sequences direct mRNA deadenylation uncoupled from decay during Xenopus early development. Mol Cell Biol 18:7537–7545 (1998).

Voisset C, Tonjes RR, Breyton P, Mandrand B, Paranhos-Baccala G: Specific detection of RT activity in culture supernatants of retrovirus-producing cells, using synthetic DNA as competitor in polymerase enhanced reverse trancriptase assay. J Virol Methods 94:187–193 (2001).

Wang Q, Chung YG, de Vries WN, Struwe M, Latham KE: Role of protein synthesis in the development of a transcriptionally permissive state in one-cell stage mouse embryos. Biol Reprod 65:748–754 (2001).

Wei N, Tsuge T, Serino G, Dohmae N, Takio K, Matsui M, Deng XW: The COP9 complex is conserved between plants and mammals and is related to the 26S proteasome regulatory complex. Curr Biol 8:919–922 (1998).

Wu L, Good PJ, Richter JD: The 36-kilodalton embryonic-type cytoplasmic polyadenylation element-binding protein in Xenopus laevis is ElrA, a member of the ELAV family of RNA-binding proteins. Mol Cell Biol 17:6402–6409 (1997).

Yang X, Menon S, Lykke-Andersen K, Tsuge T, Xiao D, Wang X, Rodriguez-Suarez RJ, Zhang H, Wei N: The COP9 signalosome inhibits p27(kip1) degradation and impedes G1-S phase progression via deneddylation of SCF Cul1. Curr Biol 12:667–672 (2002).

Yoder JA, Walsh CP, Bestor TH: Cytosine methylation and the ecology of intragenomic parasites. Trends Genet 13:335–340 (1999).

Zhong X, Kleene KC: cDNA copies of the testis-specific lactate dehydrogenase (LDH-C) mRNA are present in spermatogenic cells in mice, but processed pseudogenes are not derived from mRNAs that are expressed in haploid and late meiotic spermatogenic cells. Mammal Genome 10:6–12 (1999).

Cytogenet Genome Res 105:251–256 (2004)
DOI: 10.1159/000078196

Expression of the proliferation marker Ki-67 during early mouse development

H. Winking,[a] J. Gerdes,[b] and W. Traut[a]

[a] Institut für Biologie, Universität Lübeck, Lübeck;
[b] Abteilung Immunologie und Zellbiologie, LG Tumorbiologie, Forschungszentrum Borstel, Borstel (Germany)

Abstract. In somatic tissues, the mouse Ki-67 protein (pKi-67) is expressed in proliferating cells only. Depending on the stage of the cell cycle, pKi-67 is associated with different nuclear domains: with euchromatin as part of the perichromosomal layer, with centromeric heterochromatin, and with the nucleolus. In gametes, sex-specific expression is evident. Mature MII oocytes contain pKi-67, whereas pKi-67 is not detectable in mature sperm. We investigated the re-establishment of the cell cycle-dependent distribution of pKi-67 during early mouse development. After fertilization, male and female pronuclei exhibited very little or no pKi-67, while polar bodies were pKi-67 positive. Towards the end of the first cell cycle, prophase chromosomes of male and female pronuclei simulta-neously got decorated with pKi-67. In 2-cell embryos, the distribution pattern changed, presumably depending on the progress of development of the embryo, from a distribution all over the nucleus to a preferential location in the nucleolus precursor bodies (NPBs). From the 4-cell stage onwards, pKi-67 showed the regular nuclear relocations known from somatic tissues: during mitosis the protein was found covering the chromosome arms as a constituent of the perichromosomal layer, in early G1 it was distributed in the whole nucleus, and for the rest of the cell cycle it was associated with NPBs or with the nucleolus.

Copyright © 2004 S. Karger AG, Basel

The Ki-67 protein (pKi-67) is a large nuclear protein of about 360 kDa. The protein and its monoclonal antibodies have been known for 20 years (Gerdes et al., 1983). The locations of the genes encoding the human and the mouse Ki-67 mRNAs have been mapped (Fonatsch et al., 1991; Traut et al., 1988), and the gene has been sequenced (Schlüter et al., 1993). In the human, two different cDNA species of 9,768 and 8,688 bp have been found. These two isoforms are generated by alternative splicing of exon 7 of the Ki-67 gene (Duchrow et al., 1996).

Although many interesting facts have been gathered since the discovery of pKi-67, the function of the protein has not yet been elucidated (Scholzen and Gerdes, 2000; Traut et al., 2002b). One observation has made the study of pKi-67 very attractive for many scientists, namely the strong association with cell proliferation. Because of this property, pKi-67 and the corresponding antibodies have been widely used as tools to detect proliferating cells in human neoplasms (Gerdes et al., 1987).

As shown by cell cycle analysis, pKi-67 is present in all phases of the cell cycle (G1, S, G2, and mitosis) of proliferating cells, but is absent in resting cells (G0 and the resting stage of oocyte maturation) (Endl et al., 1997; Traut et al., 2002b). The location of pKi-67 is confined to the nucleus during interphase and to chromosomes during mitosis. The protein shows a striking spatial and temporal regulation of its location during the progression of the cell cycle (Gerdes et al., 1984; Kill, 1996; Starborg et al., 1996; Bridger et al., 1998; Traut et al., 2002a). In early G1 of somatic cells, pKi-67 co-locates with heterochromatic regions and relocates later in G1 to the nucleoli where it remains until the end of G2. At the G2/M transition, pKi-67 relocates again. During mitosis it is found as a constituent of the perichromosomal layer (Starborg et al., 1996; Traut et al., 2002a). In male and female meiocytes, similar nuclear redistributions of pKi-67 were found, but at comparably different stages of the meiotic divisions. While co-localization with heterochromatin is seen in early G1 of somatic cells, meiocytes display this type of distribution in zygotene and pachytene, stages that are regarded as belonging to the prophase of the first meiotic divison (Starborg et al., 1996; Traut et al., 2002b).

Received 8 October 2003; manuscript accepted 3 November 2003.

Request reprints from: Dr. Heinz Winking
  Institut für Biologie, Universität Lübeck
  Ratzeburger Allee 160, DE–23538 Lübeck (Germany)
  telephone: +49-451-5004117; fax: +49-451-5004815
  e-mail: winking@molbio.mu-luebeck.de

In this paper we have analyzed the expression and localization of pKi-67 in early mouse development because of two main reasons. First, the analysis of gametogenesis revealed that the mature gametes differ with respect to the presence of pKi-67. In mature oocytes, which are in the MII stage, pKi-67 is present as part of the perichromosomal layer. During spermiogenesis, in contrast, the protein is present only in very early spermatid stages, but absent in all subsequent stages including mature sperm (Traut et al., 2002b). Thus, some pKi-67 is maternally transmitted and the question to be answered is when during embryonic development do the parental genomes become equally associated with pKi-67. Second, during most of the cell cycle of proliferating cells, pKi-67 is part of the nucleolus. But it is known that at very early embryonic stages a functional nucleolus is missing. Instead, nucleolus precursor bodies (NPBs) are present which are subsequently replaced by functional, transcriptionally active nucleoli (Flechon and Kopecny, 1998).

Materials and methods

Animals

Outbred NMRI females with an all-acrocentric karyotype were mated with CD males with multiple Robertsonian translocations (Johannisson and Winking, 1994). CD males were used in order to label paternal chromosomes and pronuclei. Mouse pre-implantation embryos were recovered by flushing the uterus or oviduct of pregnant females. The flushing medium was an isotonic saline (2.2 % sodium citrate). The embryos were collected with the aid of a micropipette and individually transferred with a small amount of the saline to grease-free slides. The embryos were fixed by pouring a few drops of fixative (methanol:acetic acid 3:1) onto the embryos. The fixation procedure was monitored under a dissecting microscope. After spreading of the fixative, the embryonic cells were air dried by blowing. The position of the embryo was marked with a diamond pencil underneath the slides. The slides were stored at –20 °C until further use.

The presence of a vaginal plug in the morning indicated a successful copulation. Embryonic age was estimated by assuming a midnight fertilization. Embryos from 20 females were analyzed. The age of the embryos was between 9 and 81 h. The developmental stages within this sample ranged from the pronucleus stage up to 64-cell embryos.

Silver staining

After immunostaining and DAPI staining, a silver staining method (Ag staining) (Albini et al., 1984) was applied to visualize the argyrophilic nucleolar proteins and to study their co-localization with pKi-67.

Immunocytochemistry

The preparations were additionally fixed for 10 min in 4 % buffered paraformaldehyde, washed twice in TBS (0.01 M Tris(hydroxymethyl)aminomethane, 0.15 M NaCl, pH 7.5), briefly rinsed in distilled water and submerged in citrate buffer (pH 6.0) at 95 °C for about 45 min. The slides were washed once in distilled water and twice in TBS at room temperature. They were incubated for 30 min at room temperature in a moist chamber with S24, a polyclonal rabbit antiserum against the murine Ki-67 protein (Kosco-Vilbois et al., 1997). After extensive washes in TBS, signals were detected with goat anti-rabbit IgG conjugated with Alexa Fluor 568 (Molecular Probes, Eugene, Oreg., USA). The slides were washed twice in TBS, treated again with 4 % buffered paraformaldehyde, counterstained with DAPI (4′ 6-diamidino-2-phenylindole) and mounted in Antifade (0.233 g 1,4-diazobicyclo(2.2.2)-octane; 1 ml 0.2 M Tris-HCl, pH 8.0; 9 ml glycerol).

Images

Fluorescence images of AlexaFluor568 and DAPI signals were taken with a cooled CCD camera and the Zeiss fluorescence filter combinations 2 and 15. Bright field images of the silver stained preparations were taken at the same magnification. The fluorescence images were false-coloured and merged using the Adobe Photoshop (Version 4) program package.

Results

Pronucleus stage

Zygotes with a developmental age of 9, 9.5, 12, 17, and 21 h contained pronuclei only. Male and female pronuclei were distinguished by their size difference, the bigger ones were recorded as male and the smaller ones as female pronuclei verified by the marker chromosomes. In 21 hour-old zygotes some pronuclei had reached the prophase stage of the first cell cycle. After immunostaining for the Ki-67 protein, little or no pKi-67 was found in male and female pronuclei (Fig. 1a, b) prior to the prophase stage. At prophase, the surface of the chromosomes in male and female pronuclei was lightly decorated with pKi-67 (Fig. 2a, b). As in normal somatic mitoses, the strongly DAPI-positive centromeric regions were free of pKi-67. Consecutive Ag-staining revealed some Ag-positive areas in the vicinity of centromeres (Fig. 2b), the chromosomal sites of nucleolus organizing regions (NORs). The NORs were also free of pKi-67 (Traut et al., 2002a). The polar bodies, however, showed an intense pKi-67 reaction. The background, probably the cytoplasm of the zygote, showed some immunofluorescence staining (Fig. 1b). The cytoplasmic pKi-67 reaction may be artifactual but was regularly observed at this stage.

2-cell embryos

Embryos of a developmental age from 37 to 40 h were at the 2-cell stage. After immunostaining, the nuclei displayed an overall distribution of pKi-67, with some spots having a higher concentration. Ag staining revealed a few dark spots of argyrophilic proteins, some of them co-localized with the pKi-67 spots (Fig. 3a, b).

Among the age group of 43–44 h, mainly 2-cell and a few 4-cell embryos were present. Thus, this age group included the transition from the 2- to the 4-cell stage. The Ag staining of 2-cell embryos revealed nuclei with a homogeneous network of chromatin and round bodies, the nucleolus precursor bodies (NPBs) (Flechon and Kopecny, 1998), with darker stained subdomains (Fig. 4b). After immunostaining, a preferential location of pKi-67 in NPBs was found (Fig. 4a). Regions of higher pKi-67 concentrations at the periphery and within the NPBs mostly coincided with the stronger Ag-positive NPB subdomains (Fig. 4b).

Figs. 1–5. Zygote and 2-cell stages. pKi-67 was visualized by immunostaining with S24 (red), chromatin by DAPI staining (blue; merged with pKi-67 signals in **1a–5a**), argyrophilic proteins by Ag staining (black and white images). f – female pronucleus; m – male pronucleus; pb – nucleus of the polar body. Arrows point to some pKi-67- and Ag-positive sites; arrowheads point to NORs in prophase chromosomes. Bar represents 50 μm. (**1a, b**) Zygote (9 h), pronuclei in interphase. (**2a, b**) Zygote (21 h), pronuclei in prophase. (**3a, b**) 2-cell stage (37 h). (**4a, b**) 2-cell stage (40 h); the insert in **4a** shows the pKi-67 signal of the polar body (not merged with DAPI signals). (**5a, b**) 2-cell stage (43 h). Nuclei at the G2/M transition; the nucleus on the right is the more advanced one, it has smaller pKi-67/Ag patches and part of the pKi-67 has already moved to the emerging chromatids. The insert in **5a** shows the pKi-67 signal of the polar body alone (not merged with DAPI signals).

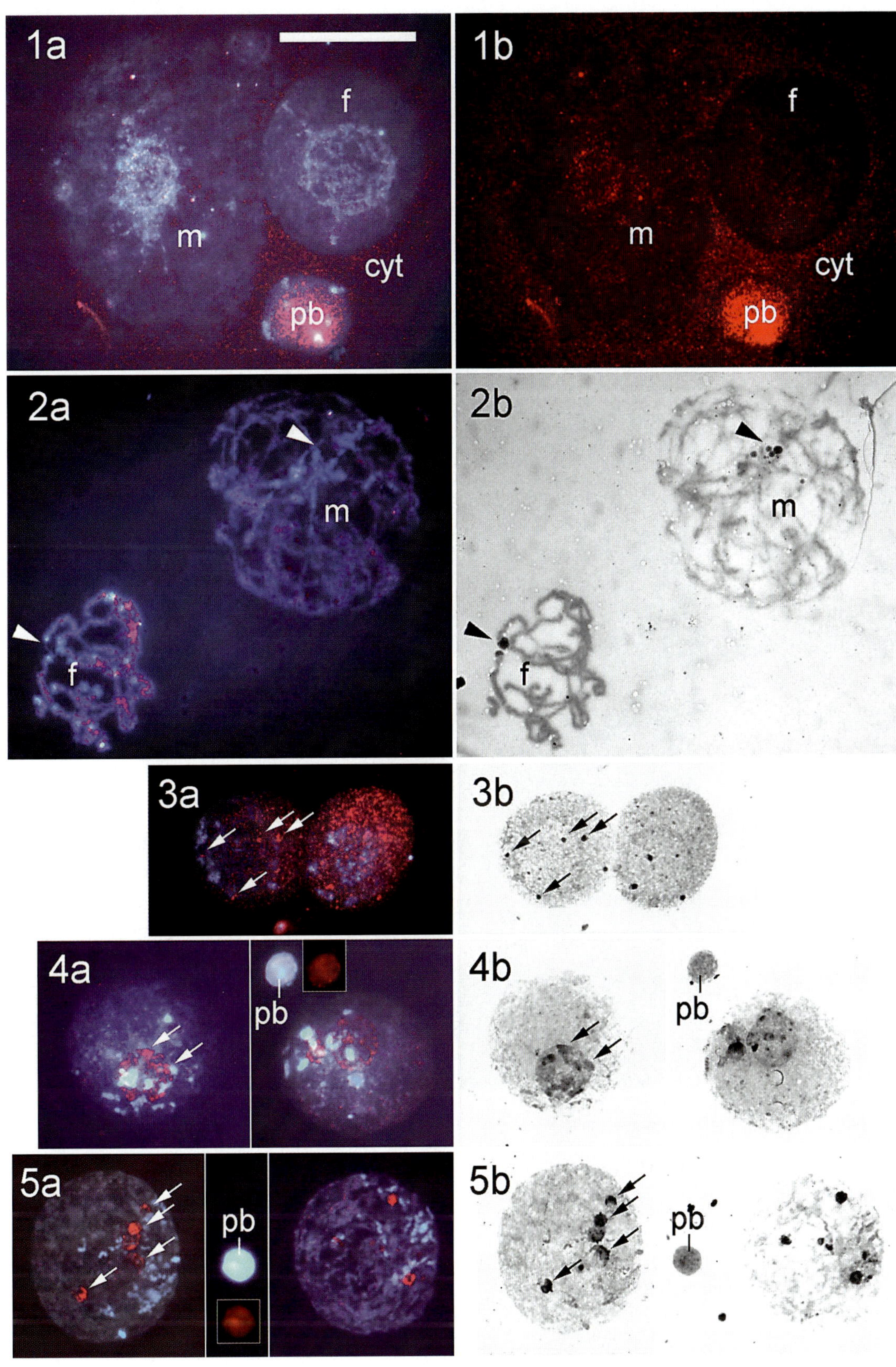

1a
f
m
cyt
pb
1b
f
m
cyt
pb
2a
m
f
2b
m
f
3a
3b
4a
pb
4b
pb
5a
pb
5b
pb

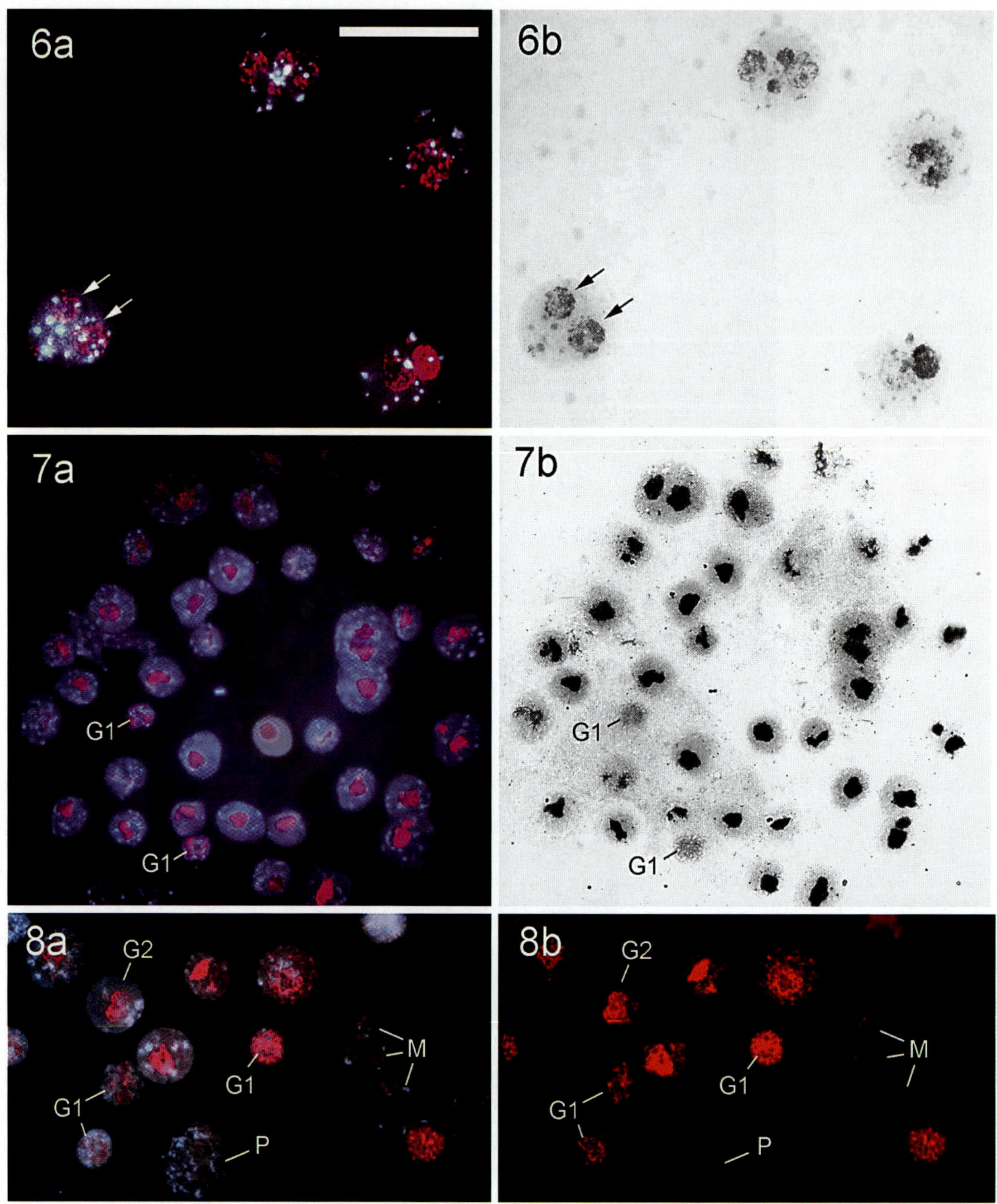

Figs. 6–8. Embryos with 4 to 37 cells. G1: G1 stage; G2: G2 stage; P: prophase; M: mitosis. Other explanations as in Figs. 1–5. (**6a, b**) 4-cell stage (60 h). (**7a, b**) Embryo with 37 nuclei (81 h). (**8a, b**) 32-cell stage (83 h). Merged image (**a**), and pKi-67 image alone (**b**).

Discernible threads of chromatin characterized the nuclei of some embryos as being in transition to the prophase of the second cell cycle. In such Ag stained nuclei, the large roundish NPBs were no longer present and appeared to have disassembled. Instead, Ag staining detected a few smaller spots of aryrophilic proteins (Fig. 5b, arrows). Immunostaining revealed pKi-67 at these sites (Fig. 5a). We consider this distribution pattern of pKi-67 and argyrophilic proteins as representative for cells in the G2/M transition. In more advanced prophase

nuclei, pKi-67 was found evenly covering the threads of chromatin (data not shown). Polar bodies were still intensely pKi-67 positive during the 2-cell stage (Fig. 4a, insert).

4-cell embryos

In nuclei of 4-cell embryos, Ag staining revealed roundish bodies similar in appearance to those present in 2-cell embryos (Fig. 6b). Hence, they were also regarded as NPBs. Immunostaining detected a concentration of pKi-67 in the NPBs

(Fig. 6a). We found pKi-67-positive and pKi-67-negative polar bodies in the 4-cell stage.

8- to 64-cell embryos

In interphase nuclei of 8- to 64-cell stages, Ag staining detected patches of irregular shape. The morphology and the strong and even reaction to Ag staining of these patches resembled that of nucleoli in somatic proliferating cells. Early G1 nuclei, which can be recognised by their small size, had an almost homogeneous distribution of pKi-67 (Fig. 7a, G1). Ag staining indicated the absence of functional nucleoli at this stage of the cell cycle (Fig. 7b, G1). With progression of the cell cycle, pKi-67 moved to the reorganized nucleoli as seen by the co-localization with argyrophilic nucleolar proteins (Fig. 7a, b). Relocation of pKi-67 from the nucleolus to the chromosome periphery was apparent in prophase and metaphase chromosomes (Fig. 8, P, M). The euchromatic chromosome arms were decorated with pKi-67 whereas the centromeric heterochromatic regions remained free (Fig. 8a, b).

Discussion

Female and male gametes are different in their supply of pKi-67. While mature sperm nuclei do not carry the protein, mature oocyte nuclei do. Therefore, with respect to pKi-67, mouse embryonic development starts with differently equipped parental nuclei. In the mouse, the parental genomes are organized in separate pronuclei up to the first cell division. Using immunocytochemistry we detected little or no pKi-67 in male as well as female pronuclei of 9-hour embryos, while the nuclei of the polar bodies were markedly pKi-67 positive. Thus, within a few hours after fertilization, the two parental nuclei gained similarity. The similarity continued when at prophase of the first cell division safely detectable amounts of pKi-67 were recorded in both pronuclei. Simultaneously, the male and the female pronucleus acquired pKi-67. It was then part of the perichromosomal layer of the chromosomes. This location is typical for non-embryonic mitoses as well. After the first cell division, all embryonic nuclei are diploid.

The presence of pKi-67 in polar bodies for about 40 h needs some explanation in the light of the very short half life of pKi-67 which was estimated be around 1 to 1.5 h (Bruno and Darzynkiewicz, 1992; Heidebrecht et al., 1996). Since synthesis of pKi-67 by the polar body is unlikely, the long persistence of pKi-67 is probably due to the absence of the degradation machinery.

With respect to the unknown function, it is interesting to note that pKi-67 makes its first appearance in the embryonic nuclei at the first mitosis. In this stage as well as in all subsequent embryonic and non-embryonic mitoses, it is a component of the perichromosomal layer. Its presence in the perichromosomal layer may, therefore, not only be a means of transport as thought before (Traut et al., 2002a) but include a functional necessity.

It is generally assumed that embryonic development starts with inactivated parental genomes. In G2 of the second cell cycle, the major wave of embryonic genome activation takes place (Zuccotti et al., 2002; Zatsepina et al., 2003). Until this time, the embryo is dependent on maternally derived transcripts and proteins stored in the ooplasm. For the proper translation of successfully activated embryonic genes, the embryo needs the establishment of a functional, transcriptionally active nucleolus with a variety of proteins. Around the time of genome activation up to the 4-cell stage, we found NPBs instead of nucleoli present in embryonic cells.

From the regular association of pKi-67 with the nucleolus in proliferating somatic cells, a role as an intensifier of ribosome biogenesis has been postulated (MacCallum and Hall, 2000). During interphase of the second zygotic cell cycle, pKi-67 showed two main distribution patterns: dispersed throughout the nucleus, and restricted to NPBs. Taking the age of the 2-cell embryos into account, the relocation of pKi-67 to the NPBs takes place late in the second cell cycle. Thus, at about the time of genome activation, pKi-67 has entered the NPBs. The migration of the nucleolar proteins fibrillarin and nucleolin into NPBs takes place at approximately the same time (Baran et al., 1995). Thus the timing fits well with the postulated role in ribosome biogenesis postulated by MacCallum and Hall (2000).

From the third to the seventh embryonic cell cycle, pKi-67 is present for the major part of the cycle within NPBs or nucleoli, which is quite similar to observations made in adult proliferating cells (Traut et al., 2002b). Therefore, with respect to the presence and location of pKi-67, embryonic cells attain the profile of non-embryonic proliferating cells already in the third cell cycle. This profile includes a cycling spatial redistribution from the nucleoli to the periphery of mitotic chromosomes during mitosis. A similar redistribution is known from other nucleolar proteins and is not unique to pKi-67 (Fair et al., 2001).

Acknowledgements

The technical assistance of Elzbietha Manthey and Constanze Reuter is gratefully acknowledged.

References

Albini S, Jones G, Wallace B: A method for preparing two-dimensional surface-spreads of synaptonemal complexes from plant meiocytes for light and electron microscopy. Exp Cell Res 152:280–285 (1984).

Baran V, Vesela J, Rehak P, Koppel J, Flechon J: Localization of fibrillarin and nucleolin in nucleoli of mouse preimplantation embryos. Mol Reprod Dev 40:305–310 (1995).

Bridger JM, Kill IR, Lichter P: Association of pKi-67 with satellite DNA of the human genome in early G1 cells. Chromosome Research 6:13–24 (1998).

Bruno S, Darzynkiewicz Z: Cell cycle dependent expression and stability of the nuclear protein detected by Ki-67 antibody in HL-60 cells. Cell Prolif 25:31–40 (1992).

Duchrow M, Schlüter C, Wohlenberg C, Flad H-D, Gerdes J: Molecular characterization of the gene locus of the human cell proliferation-associated nuclear protein defined by monoclonal antibody Ki-67. Cell Prolif 29:1–12 (1996).

Endl E, Steinbach P, Knuchel R, Hofstaedter F: Analysis of cell cycle-related Ki-67 and p120 expression by flow cytometric BrdUrd-Hoechst/7AAD and immunolabeling technique. Cytometry 29:233–241 (1997).

Fair T, Hyttel P, Lonergan P, Boland MP: Immunolocalization of nucleolar proteins during bovine oocyte growth, meiotic maturation, and fertilization. Biol Reprod 64: 1516–1525 (2001).

Flechon J, Kopecny V: The nature of the "nucleolus precursor body" in early preimplantation embryos: a review of fine structure cytochemical, immunocytochemical and autoradiographic data related to nucleolar function. Zygote 6: 183–191 (1998).

Fonatsch C, Duchrow M, Rieder H, Schlüter C, Gerdes J: Assignment of the human Ki-67 gene (MKI67) to 10q25-qter. Genomics 11:476–477 (1991).

Gerdes J, Lemke H, Baisch H, Wacker H-H, Schwab U, Stein H: Cell cycle analysis of a cell proliferation-associated human nuclear antigen defined by the monoclonal antibody Ki-67. J Immunol 133:1710–1715 (1984).

Gerdes J, Schwab U, Lemke H, Stein H: Production of a mouse monoclonal antibody reactive with a human nuclear antigen associated with cell proliferation. Int J Cancer 31:13–20 (1983).

Gerdes J, Stein H, Pileri S, Rivano M, Gobbi M, Ralfkiaer E, Nielsen K, Pallesen G, Bartels H, Palestro G, Delsol G: Prognostic relevance of tumour-cell growth fraction in malignant non-Hodgin's lymphomas. Lancet II: 448–449 (1987).

Heidebrecht HJ, Buck F, Haas K, Wacker HH, Parwaresch R: Monoclonal antibodies Ki-S3 and Ki-S5 yield new data on the "Ki-67" proteins. Cell Prolif 29:413–425 (1996).

Johannisson R, Winking H: Synaptonemal complexes of chains and rings in mice heterozygous for multiple Robertsonian translocations. Chromosome Research 2:137–145 (1994).

Kill IR: Localization of the Ki-67 antigen within the nucleolus. J Cell Sci 109:1253–1263 (1996).

Kosco-Vilbois MH, Zentgraf H, Gerdes J, Bonnefoy JY: To "B" or not to "B" a germinal center. Immunology Today 18: 225–230 (1997).

MacCallum DE, Hall PA: The location of pKi-67 in the outer dense fibrillary compartment of the nucleolus points to a role in ribosome biogenesis during the cell division cycle. J Pathol 190:537–44 (2000).

Schlüter C, Duchrow M, Wohlenberg C, Becker MH, Key G, Flad HD, Gerdes J: The cell proliferation-associated antigen of antibody Ki-67: a very large ubiquitous nuclear protein with numerous repeated elements representing a new kind of cell cycle-maintaining proteins. J Cell Biol 123:513–522 (1993).

Scholzen T, Gerdes J: The Ki-67 protein: from the known and the unknown. J Cell Physiol 182:311–322 (2000).

Starborg M, Gell K, Brundell E, Höög C: The murine Ki-67 proliferation antigen accumulates in the nucleolar and heterochromatic regions of interphase cells and at the periphery of the mitotic chromosomes in a process essential for cell cycle progression. J Cell Sci 109:143–153 (1996).

Traut W, Endl E, Garagna S, Scholzen T, Schwinger E, Gerdes J, Winking H: Chromatin preferences of the perichromosomal layer constituent pKi-67. Chromosome Res 10:685–694 (2002a).

Traut W, Endl E, Scholzen T, Gerdes J, Winking H: The temporal and spatial distribution of the proliferation associated Ki-67 protein during female and male meiosis. Chromosoma 111:156–164 (2002b).

Traut W, Scholzen T, Winking H, Kubbutat MHG, Gerdes J: Assignment of the murine Ki-67 gene (Mki67) to chromosome band 7F3-F5 by in situ hybridization. Cytogenet Cell Genet 83:12–13 (1998).

Zatsepina O, Baly C, Chebrout M, Debey P: The stepwise assembly of a functional nucleolus in preimplantation mouse embryos involves the Cajal (coiled) body. Dev Biol 253:66–83 (2003).

Zuccotti M, Boiani M, Ponce R, Guizzardi S, Scandroglio R, Garagna S, Redi C: Mouse Xist expression begins at zygotic genome activation and is timed by a zygotic clock. Mol Reprod Dev 61:14–20 (2002).

Cytogenet Genome Res 105:257–269 (2004)
DOI: 10.1159/000078197

Genetic and genomic approaches to study placental development

M. Hemberger[a] and U. Zechner[b]

[a] Genes and Development Research Group, Department of Biochemistry and Molecular Biology, University of Calgary, Calgary, Alberta (Canada);
[b] Institute for Human Genetics, Johannes Gutenberg University Mainz, Mainz (Germany)

Abstract. Recent technological advances in genetic manipulation and expression profiling offer excellent opportunities to elucidate the molecular mechanisms controlling developmental processes during embryogenesis. Thus, this revolution also strongly benefits studies of the molecular genetics of placental development. Here we review the findings of several expression profiling analyses in extraembryonic tissues and assess how this work can contribute to the identification of essential components governing placental development. We further discuss the relevance of these components in the context of genetic manipulation experiments. In conclusion, the intelligent combination of genetic and genomic approaches will substantially accelerate the progress in identifying the key molecular pathways of placental development.

Copyright © 2004 S. Karger AG, Basel

Placental development and function

Extraembryonic development starts with differentiation of the outer layer of the blastocyst, the trophectoderm. Trophectodermal cells contribute exclusively to the formation of extraembryonic tissues, most importantly the placenta. The major function of the placenta is that of an exchange organ, transporting nutrients and oxygen to – and waste products from – the embryo. In the early mouse embryo before the definitive placenta is formed, nutrient and gas exchange between maternal and fetal blood circulations is a diffusion process that takes place across the yolk sac membranes. The cell type responsible for attracting maternal blood to the conceptus is the tropho-

blast giant cell. They surround the entire implantation site and represent the outermost layer of fetal cells that are in direct contact with the (maternal) uterine tissue (Muntener and Hsu, 1977; Mossman, 1987). Giant cells synthesize a number of angiogenic and vasodilatory hormones to increase blood flow towards the embryo and form an anastomosing network of blood sinuses lining the implantation site (Muntener and Hsu, 1977). In the first half of gestation, passive diffusion of nutrients between the maternal blood spaces and fetal blood vessels is sufficient to ensure growth and survival of the embryo (Muntener and Hsu, 1977; Welsh and Enders, 1991; Cross et al., 1994; Cross, 1998; Rossant and Cross, 2001).

Around mid-gestation (embryonic day [E] 10) of mouse development, however, an active transport mechanism is required to meet the increasing demands of the growing fetus. This function is accomplished by the labyrinth layer of the placenta that forms at this stage of development and is essential for embryonic survival. Essential for labyrinth formation is a process during which the allantois, the embryonic "waste" organ derived from the mesodermal cell lineage, fuses with the trophectoderm-derived chorion, a layer that harbors trophoblast stem cells (Uy et al., 2002). Allantoic cells are responsible for embryonic blood vessel outgrowth into the forming labyrinth, whereas the surrounding trophoblasts fuse to form a syncytio-

M.H. was a recipient of a Human Frontier Science Program Organization long-term fellowship.

Received 28 October 2003; manuscript accepted 23 December 2003.

Request reprints from Dr. Ulrich Zechner
Institute for Human Genetics, Johannes Gutenberg University Mainz
Langenbeckstrasse 1, D–55101 Mainz (Germany)
telephone: +49 6131 175850; fax: +49 6131 175689
e-mail: zechner@humgen.klinik.uni-mainz.de

trophoblast layer that is in direct contact with the maternal blood circulation. There is a close interdependence between both cell types to develop normally, resulting in a tightly intermingled network of fetal blood vessels and maternal blood spaces that provide an enormous surface area for nutrient and gas exchange (Amoroso, 1952; Hernandez-Verdun, 1974). Many mutant mouse strains that die at around mid-gestation exhibit labyrinth defects. It can be speculated that placental dysfunction is the underlying cause of death for many, if not most, mid-gestational lethal mutants because malformations of the embryo proper are much better tolerated up to this stage of development (Cross et al., 1994; Cross, 1998; Rossant and Cross, 2001).

In addition to the labyrinth, the mature mouse placenta consists of the spongiotrophoblast and a thin layer of trophoblast giant cells that are again located at the border to the maternal component of the placenta, the decidua. In the second half of gestation, sufficient maternal blood supply to the labyrinth layer is ensured by two factors. Firstly, the vascular endothelium of maternal arteries entering the implantation site is completely replaced by trophoblast resulting in loss of vasoconstrictive control. Secondly, placental glycogen cells which differentiate within the spongiotrophoblast and undergo an interstitial invasion into the decidua most likely function to further increase blood flow into the placenta (Adamson et al., 2002).

The physiological requirements of extraembryonic tissues demonstrate the importance of their normal differentiation and function in order to support embryonic development and survival. This is particularly the case for trophoblast giant cells, glycogen cells and cells contributing to formation of the labyrinth, namely syncytiotrophoblast and endothelial cells of the fetal blood vessels. Because of the absolute necessity of the placenta for mammalian development and reproduction, it is extremely important to elucidate all factors involved in differentiation and function of these cell types. Many insights have been gained by analysis of mouse mutants that exhibit placental defects (Copp, 1995; Rossant and Cross, 2001). From these data, at least some major molecular pathways have been elucidated that are critical for development of various trophoblast subtypes. However, these pathways are often incomplete and the full set of individual players still has to be determined. In this context, recent large-scale expression analyses are extremely beneficial as they provide a powerful tool to reveal potential members of these pathways and to identify genes of similar function that are expressed in the placenta and therefore represent candidates potentially required for trophoblast differentiation.

Expression profiling approaches

In recent years, gene expression in mouse extraembryonic tissues has been analyzed several times on a genome-wide scale by cDNA array-based technology. The underlying aims of these studies were different and served to describe the divergence of gene expression profiles between embryonic and placental tissues (Tanaka et al., 2000) and their derived stem cell populations (Tanaka et al., 2002), between extraembryonic tissues at various stages of development (Hemberger et al., 2001), and between placentae of normal fetuses and those derived by nuclear transfer (NT) (Humpherys et al., 2002; Suemizu et al., 2003). Thus, Tanaka et al. (2000) identified 720 genes that displayed statistically significant differences in expression between placenta and embryo at E12.5 of development. Among the 289 genes more highly expressed in the placenta, 61 genes were categorized as placenta-specific. In the study by Tanaka et al. (2002), expression profiles unique for pluripotent embryonic stem (ES) cells, extraembryonic-restricted trophoblast stem (TS) cells (derived from E3.5 blastocysts), and terminally-differentiated mouse embryo fibroblast (MEF) cells from E12.5 embryos as well as genes common only to ES and TS cells were identified. Among 2,150 genes differentially expressed between ES and TS cells, 1,526 genes were more highly expressed in ES cells and 624 genes were more highly expressed in TS cells. Hemberger et al. (2001) used cDNA subtraction between extraembryonic tissues of early- (E7.5) and late-stage (E17.5) embryos to generate stage-specific cDNA pools that were used for screening of high-density mouse UniGene cDNA arrays. A total of 638 cDNA clones were identified, 488 with the E7.5-specific probe and 150 with the E17.5-specific probe. Humpherys et al. (2002) analyzed abnormal gene expression in neonatal cloned mice and found the expression of 286 genes to be changed at least twofold in NT placentae derived from cumulus cell-derived clones and the expression of 221 genes to be changed at least twofold in NT placentae derived from ES cell-derived cell clones, each compared with placentae from normally fertilized controls. With the same motivation, Suemizu et al. (2003) compared the expression profiles of E19.5 mouse NT placentae derived from ES cells with the expression profiles of control E19.5 mouse NT placentae derived from one-cell embryos. Two ES-derived NT placentae were compared with control placentae in separate experiments and showed differential expression of 1,807 and 1,964 genes, respectively. As a nature of array technology, these reports can only concentrate on overall expression changes and closer analysis of one or a few genes at best. Yet, these studies are extremely beneficial as they provide valuable information on individual gene expression in extraembryonic tissues, irrespective of the initial aim of the study. The same benefit is provided by large-scale sequencing projects of expressed sequence tags (ESTs) derived from extraembryonic tissues and trophoblast stem (TS) cells. Thus, Ko et al. (1998) sequenced 2,103 cDNAs from an E7.5 extraembryonic tissue library and identified 3,186 ESTs, ~40% of which were novel to the sequence database. The lists of genes identified in these analyses can therefore be reexamined for any group of genes of interest, in particular those emerging to be essential for extraembryonic/placental development. In this review, we re-evaluate data from various expression profiling approaches to emphasize their potential in identifying components of critical cellular pathways. We focus on four major gene groups, i.e. growth factors, growth factor receptors, signal transducers and transcription factors, discuss their importance for extraembryonic development as revealed by loss-of-function experiments, and examine reported cDNA array data for further members of these groups that can be speculated to have a crucial role in placental development and function.

Growth factors and growth factor receptors

Growth factors are proteins that bind to receptors on the cell surface, with the primary result of inducing cellular proliferation and/or differentiation. Many growth factors are quite versatile, stimulating cellular division in numerous different cell types, while others are specific to a particular cell type. In the following, the latest findings about the growth factor signaling pathways known to play essential roles during placental development are summarized.

Recent molecular and genetic data indicate that fibroblast growth factor (Fgf) signaling is one of the key components of early trophoblast development. A member of the fibroblast growth factor family, Fgf4, is required for the maintenance of trophoblast stem cells (Tanaka et al., 1998). Fgf4 is expressed in the inner cell mass (ICM) of the embryo at the blastocyst stage and subsequently in the early postimplantation epiblast (Niswander and Martin., 1992; Rappolee et al., 1994). One of the four known Fgf receptors, Fgfr2, is strongly expressed in trophectoderm cells and is therefore considered to be the main trophoblast stem cell-specific Fgf receptor (Haffner-Krausz et al., 1999). Further support for this assumption comes from the phenotypes of two different mutations in the *Fgfr2* gene: a null mutation leads to peri-implantation death whereas a hypomorphic mutation compromises chorioallantoic attachment (Arman et al., 1998; Xu et al., 1998). In addition, null alleles of *Cdx2* or *Eomes*, which are both transcription factors considered to be putative downstream targets of Fgf signaling, also result in peri-implantation death (Chawengsaksophak et al., 1997; Russ et al., 2000).

Another system known to be involved in placental development is the transforming growth factor β (Tgf-β) signaling pathway. One protein in the Tgf-β superfamily, Nodal, is expressed specifically in the spongiotrophoblasts of the mouse placenta (Ma et al., 2001). Conceptuses with an insertional null mutation in the *Nodal* gene completely lack the spongiotrophoblast and labyrinthine region and exhibit increased giant cell formation (Ma et al., 2001). Moreover, in a hypomorphic *Nodal* mutant not only the giant cell layer but also the spongiotrophoblast layer is increased whereas a reduced size of the labyrinth is observed (Ma et al., 2001). Another subgroup of the Tgf-β superfamily constitutes the bone morphogenetic (BMP) proteins. Two members of the BMP family, Bmp5 and Bmp7, are coexpressed in the allantois from gastrulation onwards. A null mutation of both *Bmp5* and *Bmp7* causes lethality at E10.5 displaying allantoic defects and placental failure (Solloway and Robertson, 1999). Members of the Tgf-β superfamily signal through a heteromeric complex consisting of two types of transmembrane serine/threonine kinases known as type I and type II receptors. Conceptuses lacking Tgf-β type I receptor (TβRI, *Tgfbr1*), also termed activin receptor-like kinase 5 (ALK-5) and closely related to the nodal type I receptor ALK-7/Acvr1c, die at midgestation, exhibiting severe defects in vascular development of the yolk sac and placenta, with embryonic vessels apparently unable to sprout properly into the labyrinthine layer (Larsson et al., 2001). Furthermore, the importance of the Tgf-β signaling pathway in placental development is also stressed by the phenotypes of null mutants of different Smad family

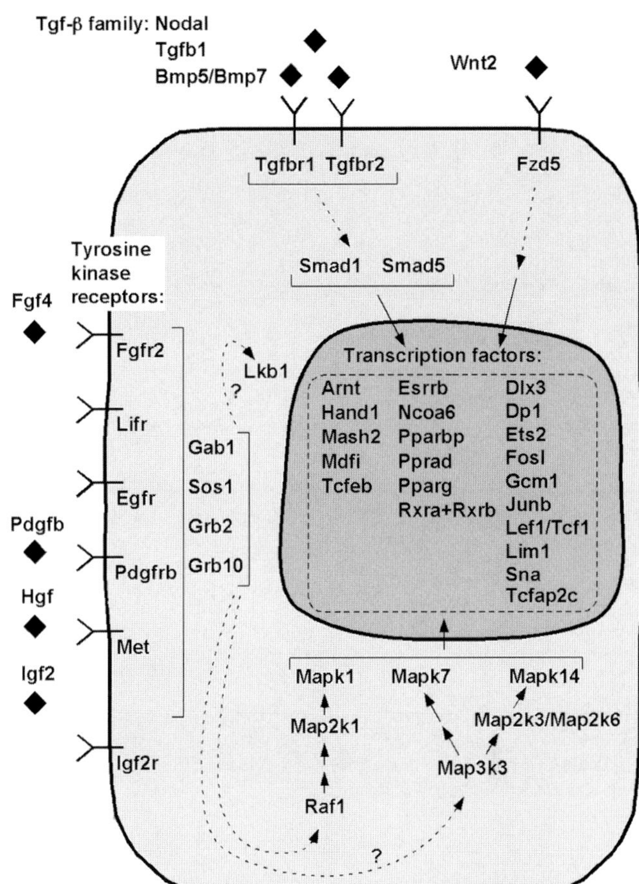

Fig. 1. Schematic representation of growth factors, growth factor receptors, intracellular signaling molecules and transcription factors that have been demonstrated to be important for placental development by genetic ablation. Only those genes for which an abnormal development of extraembryonic tissues was observed are shown. Dashed arrows indicate a (potentially) indirect transduction of the signal via intermediate signaling molecules. The transcription factors are grouped into bHLH transcription factors and bHLH interacting proteins (1st column), nuclear hormone receptors and associated proteins (2nd column) and transcription factors belonging to various other groups (3rd column). Please note that the Igf2 signal is transmitted via the Igf1 receptor tyrosine kinase (Igf1r), whereas Igf2r acts as a sink for Igf2 leading to its degradation.

members, which represent the intracellular mediators of Tgf-β signaling (see below).

Mutations in several other growth factor signaling cascades have also been reported to interfere with labyrinthine development (Fig. 1). First, mice deficient for the genes encoding hepatocyte growth factor *(Hgf)* or its receptor *Met* show markedly reduced numbers of labyrinthine cells and die before birth between E14.5 and E16.5 (Bladt et al., 1995; Uehara et al., 1995). Second, a similar phenotype leading to perinatal lethality is associated with a null mutation of the platelet-derived growth factor β *(Pdgfb)* or the platelet-derived growth factor receptor β *(Pdgfrb)* gene (Ohlsson et al., 1999). Third, labyrinthine together with spongiotrophoblast deficiency is observed in mutants for the epidermal growth factor receptor *(Egfr)* gene (Sibilia and Wagner, 1995; Threadgill et al., 1995). Fourth, leukemia inhibitory factor receptor *(Lifr)* mutant mice succumb

Table 1. Growth factors and related molecules identified in large-scale expression profiling approaches

Gene	Symbol	Accession no.
placental lactogen I	*Pl1 (Csh1)*	M35662
placental lactogen II	*Pl2 (Csh2)*	NM_008865
proliferin	*Plf*	NM_031191
proliferin-related protein	*Plfr*	X02594
prolactin-like protein A	*Prlpa*	AF011383
Prolactin-like protein I	*Prlpi*	NM_013766
prolactin-like protein E	*Prlpe*	BC051671
Prolactin-like protein I	*Prlpi*	NM_013766
fibroblast growth factor binding protein 1	*Fgfbp1*	NM_008009
latent transforming growth factor beta-binding protein 4	*Ltbp4*	NM_175641
transforming growth factor beta 1	*Tgfb1*	BC013738
Platelet derived growth factor alpha	*Pdgfa*	NM_008808
adrenomedullin	*Adm*	NM_009627
Vascular endothelial growth factor	*Vegf*	NM_009505
Activin betaC	*Inhbc*	U95962
glypican 1	*Gpc1*	NM_016696
preadipocyte growth factor (glutathione S-transferase pi)	*Gstp2*	AI286503
pre-B-cell colony-enhancing factor	*Pbef*	BC018358
stem cell growth factor	*Scgf*	NM_009131
alpha fetoprotein mRNA	*Afp*	NM_007423
milk fat globule-EGF factor 8 protein	*Mfge8*	NM_008594
Matrilin 2	*Matn2*	NM_016762

Table 2. Growth factor receptors and related molecules identified in large-scale expression profiling approaches

Gene	Symbol	Accession no.
fibroblast growth factor regulated protein 2	*Fgfrp2 (Fn14)*	NM_013749
FGF receptor activating protein	*Frag1*	U57715
transforming growth factor beta receptor II	*Tgfbr2*	BC016262
transforming growth factor beta receptor III	*Tgfbr3*	NM_011578
Insulin-like growth factor II receptor	*Igf2r*	NM_010515
integrin alpha 3A	*Itga3*	NM_013565
integrin beta 4	*Itgb4*	NT_039521
Colony stimulating factor 1 receptor	*Csf1r*	NM_007779
endothelial differentiation, sphingolipid G-protein-coupled receptor 3	*Edg3*	NM_010101
growth arrest specific 1	*Gas1*	X65128
growth arrest specific 5	*Gas5*	NM_013525
Parathyroid hormone receptor	*Pthr1*	NM_011199
Interleukin 1 receptor, type I	*Il1r1*	NM_008362
Folate receptor 1	*Folr1*	NM_008034
Scavenger receptor class B1	*Srb1 (Scarb1)*	NM_016741
toll/interleukin-1 receptor 8	TIR8	AF113795
adrenocorticotropin hormone receptor	ACTHR (*Mc2r*)	D31952

to peri- or postnatal death, lacking well-organized labyrinthine and spongiotrophoblast layers (Ware et al., 1995). Finally, knockout data for one member of the wingless-related (Wnt) signaling pathway, *Wnt2*, and the Wnt receptor gene *Fzd5* show that both genes are indispensable for correct labyrinthine development (Monkley et al., 1996; Ishikawa et al., 2001). In addition, the essential role of all these signaling pathways is underlined by the labyrinthine defects occurring in conceptuses deficient for their specific downstream mediators and targets which are discussed below.

One of the major players in the genetic control of placental growth is the insulin-like growth factor II *(Igf2)* gene. *Igf2* is expressed in both the labyrinth and spongiotrophoblast layer of the placenta (Redline et al., 1993). Already over a decade ago, it was demonstrated that a targeted disruption of the *Igf2* gene results in growth reduction of both fetus and placenta during the last two thirds of gestation (DeChiara et al., 1990, 1991; Baker et al., 1993). A more detailed phenotypic analysis of *Igf2* null placentae showed a reduction in the number of the glycogen trophoblast cells (Lopez et al., 1996). A recent study by Constancia et al. (2002) described the placental phenotype of conceptuses lacking specifically the labyrinth-specific P0 *Igf2* transcript. Interestingly, the absence of this transcript, which makes up only about 10% of the total *Igf2* mRNA, causes the same placental size reduction at the end of gestation as the complete lack of *Igf2* mRNA. Further, the size reduction is accompanied by a decrease in passive and an increase in active transport of nutrients across the placenta. These results support a role of Igf2 in both the placental supply of, and the genetic demand for, maternal nutrients to the mammalian fetus. Most of the biological effects of Igf2 are mediated by the type 1 receptor, whereas the type II receptor, Igf2r, is involved in Igf2 degradation. As expected from the placental hypoplasia of *Igf2* null conceptuses, an *Igf2r* null mutation results in an excess of Igf2 levels and, as a consequence, in placental hyperplasia as one of the major outcomes (Eggenschwiler et al., 1997).

When studying the list of growth factors and other extracellular molecules identified in the various large-scale expression profiling approaches, the first most striking observation is the frequent occurrence of genes coding for members of the prolactin family of hormones *(Pl1, Pl2, Prlpa, Prlpe, Plf, Plfr)* (Table 1). In mice and rats, this gene family comprises the pituitary hormones prolactin (Prl) and growth hormone (Gh) as well as more than twenty other placental-specific hormones that are closely related in sequence to Prl and Gh. In humans only one placenta-specific member of this family, placental lactogen I (PL-I), has been identified up to now. In general, the placenta-specific members of this family appear to fulfill important regulatory tasks in angiogenesis (fetal, uterine and placental), hematopoiesis and lymphocyte function during pregnancy (Linzer and Fisher, 1999). Particular opposing functions in angiogenesis and vascularization of the implantation site have been attributed to Plf, which is secreted specifically by the trophoblast giant cells, and Prp/Plfr, which is expressed in giant cells as well as the spongiotrophoblasts. Thus the angiogenic activity of Plf seems to stimulate vascularization of the implantation site whereas the antiangiogenic activity of Prp is postulated to inhibit it (Jackson et al., 1994). This is supported by the markedly reduced vascularization of the implantation site observed in mice carrying a null mutation of the *Gata2* transcription factor, which is known to be essential for Plf expression (Ma et al., 1997, see also below). Mice with a functional disruption of the various rodent Prl-like placental hormones have not been described up to now. It is assumed, that the analysis of such knockout mice will significantly accelerate the identification of novel functions of this gene family.

The second most obvious feature in the list of growth factor and growth factor receptor genes detected in the expression profiling approaches is the presence of several factors implicated in Fgf- and Tgf-β signaling (Tables 1 and 2). For Fgf signaling, these genes include *Fgfbp1*, *Frag1* and *Fgfrp2*. *Fgfbp1* codes for fibroblast growth factor-binding protein (Fgf-bp) 1, a secreted protein that can bind Fgf 1 and 2. Fgf-bp1 enhances Fgf1- and Fgf2-dependent proliferation of NIH-3T3 fibroblasts and Fgf2-induced extracellular signal-regulated kinase 2 phosphorylation (Tassi et al., 2001). *Frag1* encodes Fgf receptor activating protein, a membrane-spanning protein that potently activates Fgfr2 by C-terminal fusion through chromosomal rearrangement (Lorenzi et al., 1999). *Fgfrp2* is synonymous with *Fn14,* a gene encoding an Fgf1 inducible membrane-spanning protein that may play a role in cell-matrix interactions (Meighan-Mantha et al., 1999). Given the pivotal role of Fgf signaling in early trophoblast development it is tempting to speculate about the functional involvement of these three molecules in placental development. Genes implicated in Tgf-β signalling are *Tgfb1*, Activin betaC *(Inhbc)*, *Tgfbr2*, *Tgfbr3* and *Ltbp4*. *Tgfb1* and *Tgfbr2* are known to play essential roles in the development of extraembryonic tissues. 50% of *Tgfb1* null conceptuses and all *Tgfbr2* null conceptuses die at midgestation due to defects in yolk sac vasculogenesis and hematopoiesis (Dickson et al., 1995; Oshima et al., 1996). Mice with a null mutation in Activin betaC are viable and survive to adulthood suggesting that Activin betaC is not essential for embryonic and extraembryonic development (Lau et al., 2000). *Tgfbr3* deficiency results in late gestation embryonic lethality due to proliferative defects in heart and apoptosis in liver (Stenvers et al., 2003). It may be worthwhile to perform an additional detailed analysis of extraembryonic tissues in these embryos. *Ltbp4* encodes latent transforming growth factor-beta binding protein 4, a member of the family of fibrillin/Ltbp glycoproteins. The secretion and activation of Tgf-betas is regulated by their association with latency-associated proteins and Ltbps. In vitro studies predict a dual role for Ltbps as structural components of the extracellular matrix (ECM) and as local regulators of Tgf-β signaling (Taipale et al., 1998). Recently, it was shown that a homozygous hypomorphic mutation of *Ltbp4* in the mouse causes abnormal lung development, cardiomyopathy, and colorectal cancer (Sterner-Kock et al., 2002). The generation of *Ltbp4* null mice may provide an insight into the functions of this gene in placental development.

Expression profiling assays identified two further important growth factor and growth factor receptor genes: *Pdgfa* (platelet-derived growth factor α) and *Igf2r* (Tables 1 and 2). Both genes were found to be trophoblast stem cell-specific (Tanaka et al., 2000). A mouse platelet-derived growth factor A chain *(Pdgfa)* null allele was shown to be homozygous lethal, with two distinct restriction points, one prenatally before E10 and one postnatally (Boström et al., 1996). It will be interesting to determine if prenatal lethality of *Pdgfa* null mice is associated with defects in extraembryonic tissues. The identification of Igf2r further supports the pivotal role of Igf2 signaling in the genetic control of placental growth.

Members of the signal transduction cascade

The binding of extracellular signaling molecules to their specific receptors activates a series of intracytosolic mediators that are potent to transfer the signal into the nucleus, ultimately to induce gene expression changes. Several different general signaling pathways exist that are, however, likely to be intertwined at many levels.

One of these general modes of signal transduction is the Tgf-β/Smad pathway. Smad proteins are intracytosolic mediators of Tgf-β signaling that can translocate into the nucleus after activation and association with a growing number of interacting proteins that are potent modulators of the signal (Shi and Massague, 2003; Zwijsen et al., 2003). The phenotypes of some Smad mutations provide further evidence for the importance of Tgf-β signaling in extraembryonic development. This is demonstrated by the absence of Smad1 and Smad5 that lead to mid-gestational lethality due to a failure of chorio-allantoic fusion (Chang et al., 1999; Yang et al., 1999; Lechleider et al., 2001; Tremblay et al., 2001).

Another general mechanism of signal transduction involves growth factor receptors with tyrosine kinase activity. It has become clear that receptor tyrosine kinase signaling leading to mitogen-activated protein kinase (MAPK) activation is a major determinant of placental labyrinth formation. Upon activation by ligand binding, growth factor receptors with tyrosine kinase activity associate with a few adaptor molecules and activate Ras that in turn transduces the signal onto a series of cytosolic serine/threonine kinases, which ultimately leads to activation of MAPKs. Several groups of MAPK signal transduction pathways have been identified in mammals, with the MAPKs including extracellular signal-regulated kinase (ERK), c-Jun N-terminal kinase (JNK) and p38. MAPKs have the ability to translocate into the nucleus and activate transcription factors, which then leads to gene expression changes in response to the receptor-bound growth factor.

Mutations have been introduced at many levels of this signaling cascade (Fig. 1) and cause strikingly similar placental defects. Typically, mutations in individual signal transducers result in a small, underdeveloped labyrinth albeit the extent of this phenotype may vary. Outgrowth of fetal blood vessels into the labyrinth is impaired and often accompanied by a failure of secondary branching of vessels. This placental defect is observed in mutants of the signaling adaptor molecules Gab1 (Itoh et al., 2000) and Grb2 (Saxton et al., 2001) that bind to activated growth factor receptors and associate with the guanine nucleotide exchange factor Sos. Very similar to the *Gab1*- and *Grb2*-deficient phenotype, labyrinth formation is blocked in the absence of Sos1 (Qian et al., 2000). Placental labyrinth development also fails to proceed normally in mutants of the Ras-activated kinase Raf1 (Wojnowski et al., 1998; Mikula et al., 2001) and in the absence of the subsequently activated downstream MAPK kinase Mek1 *(Map2k1)* (Giroux et al., 1999). Mek1 is known to be the activating kinase of Erk1 *(Mapk3)* and Erk2 *(Mapk1)*. Erk1-mutant mice have been generated but do not exhibit an obviously altered placental phenotype (Pages et al., 1999). Only most recently, Erk2 deficiency has been described to result in failure of ectoplacental cone and extraembryonic ectoderm formation. This early onset phenotype may reflect a direct requirement of Erk2 activation in transduction of the Fgf4/Fgfr2 signal to control proliferation of the polar trophectoderm (Saba-El-Leil et al., 2003).

Table 3. Intracellular signaling molecules identified in large-scale expression profiling approaches

Gene	Symbol	Accession no.
Growth factor receptor bound protein 2-associated protein 1	*Gab1*	XM_134516
Growth factor receptor bound protein 7	*Grb7*	BC003295
Growth factor receptor bound protein 10	*Grb10*	NM_010345
G protein-coupled receptor 97	*Gpr97*	NM_173036
G protein-coupled receptor kinase 5	*Gprk5*	AF040756.1
G protein B subunit homolog guanine nucleotide binding protein, alpha 11	*Gna11*	W12402
guanine nucleotide binding protein, alpha 12	*Gna12*	AA259670
guanine nucleotide binding protein beta 1	*Gnb1*	NM_030987
guanine nucleotide binding protein, beta-2, related sequence 1	*Gnb2-rs1*	AI115190
ral guanine nucleotide dissociation stimulator,-like 2	*Rgl2 (Rab21)*	AA544542
cAMP-regulated guanine nucleotide exchange factor I	*Epac (Rapgef4)*	AA475255
developmentally regulated GTP-binding protein 1	*Drg1*	NM_007879.1
SH3-domain GRB2-like B1 (endophilin)	*Sh3glb1*	BC024362
SH3 and PX domain-containing protein SH3PX1	*Snx9 (Sh3px1)*	AA754673
SH3-containing protein p4015	*Argbp2*	AA538339
SH3 domain protein 2A	*Sh3gl2 (Sh3d2a)*	NM_019535.1
SH3-containing protein SH3P2	*Sh3d3 (Ostf1)*	U58888.1
(rat) SH3 domain binding protein	*CR16*	U25281.1
v-ra1 homolog B (ras-related)	*Ralb*	BC006907
RAS p21 protein activator 3	*Rasa3*	AA756830
member RAS oncogene family	*Rab3a*	AA509606
RAB/Rip protein	*Hrb*	AF057287.1
similar to H. sapiens ras GTPase-activating-like protein	*Iqgap1*	NM_016721
ras-GTPase-activating protein/ SH3-domain-binding protein 2 similiar to R. norvegicus RhoB gene	*G3bp2*	NM_011816.1
binder of Rho GTPase 4	*Cdc42ep4 (Borg4)*	AA544302
Janus kinase 1	*Jak1*	NM_146145
v-crk-associated tyrosine kinase substrate	*Bcar1 (Crkas)*	NM_009954
p21 (Cdkn1a)-activated kinase 2	*Pak2*	XM_148586
mitogen-activated protein kinase kinase kinase kinase 4	*Map4k4/Nik*	XM_129778
Nik related kinase	*Nrk*	NM_013724
Mapkkk7 interacting protein 2	*Map3k7ip2*	NM_138667
mitogen activated protein kinase 1	*Mapk1/Erk2*	BC006708
mitogen activated protein kinase 14	*Mapk14*	NM_011951.1
sim. To H.sapiens serine/threonine kinase 24	*Stk24*	XM_053212
(human) FUSED serine/threonine kinase	*STK36*	AF200815.1
catenin alpha 2	*Catna2*	NM_145732
putative intracellular signaling protein	*Trip6*	AA682123
A kinase anchor protein 2	*Akap2*	AI119368
embryonal Fyn-associated substrate	*Efs*	BC005438.1
pituitary tumor-transforming 1	*Pttg1*	NM_013917.1
fibroblast growth factor inducible 13	*Ppm1g (Fin13)*	NM_008014.1
diacylglycerol kinase alpha	*Dgka (Dagk1)*	BC006713.1

In contrast to the Gab1/Grb2/Sos1/Raf1/Mek1/Erk pathway, the JNK pathway does not appear to be important for extraembryonic development. Mutations have been introduced into all three Jnk genes, but single as well as double mutants do not have altered placental phenotypes (Kuan et al., 1999).

However, the p38 MAPK pathway is again critical for placental development and mutations in its transducing molecules cause labyrinthine phenotypes very similar to those observed in Erk1/2 pathway mutants. In p38α *(Mapk14)*-deficient mice, it has also been demonstrated that the failure of labyrinth forma-tion and blood vessel outgrowth is due to trophoblast defects and does not require p38α function in vascular endothelial cells that are derived from a different cell lineage, the extraembryonic mesoderm (Adams et al., 2000; Mudgett et al., 2000). As would be expected, mutations in upstream regulators of p38α result in mid-gestational lethality due to placental labyrinth failure as well (Fig. 1). This has been described for the p38α-activating MAPK kinase kinase Mekk3 *(Map3k3)* (Yang et al., 2000) and most recently for double mutants in the intermediate MAPK kinase genes Mkk3 *(Map2k3)* and Mkk6 *(Map2k6)* (Brancho et al., 2003).

In addition to direct Mkk3/6 and subsequent p38α activation, Mekk3 is also an upstream regulator of Erk5 *(Mapk7)* via Mek5 *(Map2k5)* activation (Fig. 1). Whereas the impact of Mek5 on placental development is not known from mouse mutants, Erk5 deficiency again causes a small labyrinth phenotype with a lack of embryonic blood vessel outgrowth into this region (Regan et al., 2002).

These sets of serine/threonine kinase signal transduction cascades are likely to be intertwined on many levels and cannot be seen as separate linear pathways. There is also evidence that there are further factors involved in signal transduction that have not been placed into any of the known cascades yet. This has become clear from mutation of the *Lkb1* gene encoding a serine/threonine kinase of unknown function with no identified substrates in vivo. *Lkb1* deficiency is embryonic lethal as the result of placental insufficiency, and the phenotype is again characterized by a defect in labyrinth development and a failure of embryonic blood vessel invasion into the placenta (Ylikorkala et al., 2001).

The identification of *Lkb1* as being essential for placental and therefore embryonic development has demonstrated the likely possibility for the presence of additional crucial factors in signal transduction pathways during placenta formation. The evaluation of gene expression in extraembryonic tissues provides a useful tool to identify further candidates belonging to this functional group of genes. There is also the need for other, yet unidentified signal transduction molecules arising from the fact that most of the factors analyzed to date result in placental labyrinth defects. However, growth factor signaling is essential much earlier during embryogenesis and governs trophoblast stem cell proliferation and differentiation immediately after implantation. It therefore can be speculated that these early signals are transmitted by another set of molecules and/or by specific combinations of known members of the signal transduction cascades.

Table 3 summarizes genes that were identified in the various large-scale expression profiling approaches as being expressed in trophoblast derivatives and that play a role in transmitting growth factor signals within the intracellular environment. These include several growth factor receptor bound adapter proteins such as Gab1, Grb7, and Grb10. Gab1 has been described above to be important for placental labyrinth development and was shown to be upregulated in placentae from cloned embryos (Humpherys et al., 2002). Grb10 is also known to affect placental (and fetal) development and growth (Charalambous et al., 2003). Since Grb7 belongs to the same family of adapter proteins as Grb10 and Grb14, it will be inter-

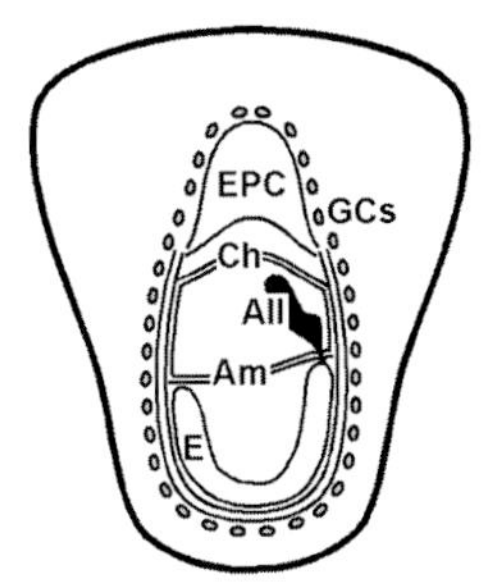

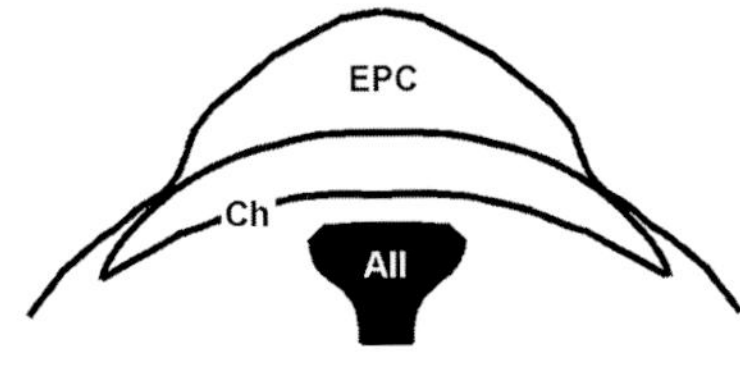

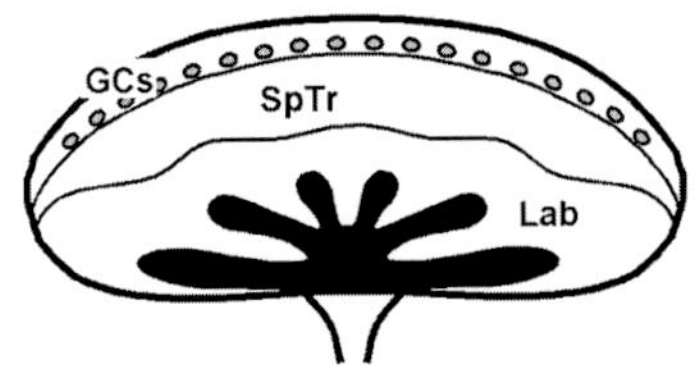

Giant cell formation and function:

Gene	Reference
Arnt	[Abbott and Buckalew, 2000; Adelman et al., 2000; Kozak et al., 1997]
Dp1	[Kohn et al., 2003]
Esrrb	[Luo et al., 1997]
Gata2/3	[Ma et al., 1997]
Hand1	[Riley et al., 1998; Firulli et al., 1998]
Mash2	[Guillemot et al., 1994]
Mdfi	[Kraut et al., 1998]
Sna	[Nakayama et al., 1998]

Chorioallantoic fusion:

Gene	Reference
Esrrb	[Luo et al., 1997]
Lef1/Tcf1	[Galceran et al., 1999]
Lim1	[Shawlot and Behringer, 1995]

Labyrinth developent and vascularization:

Gene	Reference
Arnt	[Abbott and Buckalew, 2000; Adelman et al., 2000; Kozak et al., 1997]
Dlx3	[Morasso et al., 1999]
Fosl	[Schreiber et al., 2000]
Gcm1	[Anson-Cartwright et al., 2000]
Junb	[Schorpp-Kistner et al., 1999]
Ncoa6	[Antonson et al., 2003; Kuang et al., 2002; Zhu et al., 2003]
Pparbp	[Crawford et al., 2002; Zhu et al., 2000]
Ppard	[Barak et al., 2002]
Pparg	[Barak et al., 1999]
Rxra + Rxrb	[Sapin et al., 1997 Wendling et al., 1999]
Tcfap2c	[Auman et al., 2002; Werling and Schorle, 2002]
Tcfeb	[Steingrimsson et al., 1998]

Fig. 2. Summary of transcription factors that are important for extraembryonic development as known from gene targeting approaches. The developmental stages focussed upon are (left) the early post-implantation stage (~ E7.5) during which giant cell formation and function is most important; (middle) the phase just prior to midgestation when chorioallantoic fusion occurs (~ E8.5–E9), and (right) post-midgestation (E10 onwards) where labyrinth formation and vascularization is a major process of placental development. Genes interfering with developmental processes during these stages are listed underneath the schematic representations.

esting to investigate its role in trophoblast differentiation. Furthermore, a strikingly large group of guanine nucleotide binding, Ras-related and Ras-activating GTPases are expressed in the placenta (Table 3), and it can be speculated that at least some of these factors are important for its formation. A double mutation in two guanine nucleotide binding proteins, G alpha(q) and G alpha(11), has been reported to lead to embryonic lethality at E11 due to cardiomyocyte hypoplasia (Offermanns et al., 1998). However, early embryonic cardiac defects are often secondary to placental dysfunction (Adams et al., 2000; Hemberger and Cross, 2001) and it will be important to investigate the development of extraembryonic tissues in these mice.

Many of the genes identified in the large-scale expression analyses have protein serine/threonine kinase activity *(Jak1,* STK36, *Stk24, Pak2, Map4k4, Nrk, Mapk1)* or are directly implicated in the protein serine/threonine kinase signaling pathway *(Akap2, Ppm1g).* Given the known impact of other serine/threonine kinases in particular of the MAP signaling cascade on placental development, these factors may also play a vital role in trophoblast differentiation. As for *Map4k4,* genetic inactivation causes mid-gestational lethality between E9.5 and E10.5 (Xue et al., 2001). The most obvious defect described is a mesodermal patterning defect. Because intrauterine death at this stage is most often due to defects in extraembryonic development, notably failure of chorioallantoic fusion or labyrinth formation, it will again be extremely interesting to re-analyze the *Map4k4*-deficient embryos for development of extraembryonic tissues. In addition, Map4k4 binds to Mekk1 and is able to activate the JNK pathway; however, the *Map4k4*-mutant phenotype is not observed in the absence of *Jnk1/ Mapk8* (Dong et al., 1998), *Jnk2/Mapk9* (Sabapathy et al., 1999) or *Jnk3/Mapk10* (Yang et al., 1997). These findings imply that other *Map4k4* target molecules exist that are important for *Map4k4*-mediated signaling during early embryonic development.

Transcription factors

As the ultimate mediators of cell differentiation, the analysis of transcription factor mutants has been vital to understand trophoblast development and function. For each important phase during placenta formation, transcription factors have been identified that govern differentiation, morphogenesis and function. Several members of the basic helix-loop-helix (bHLH) family of transcription factors are crucial for early post-implantation development (Fig. 2). Well-known in their function governing giant cell differentiation are the bHLH factors Hand1 and Mash2. Hand1 functions to promote giant cell formation and Hand1-mutants die at around E8.5–E9.5 exhib-

Table 4. Transcription factors identified in large-scale expression profiling approaches

Gene	Symbol	Accession no.
Placenta and embryonic expression gene	*Pem*	NM_008818
Placenta-specific homeobox 1	*Psx1*	NM_008955
Myeloid ecotropic viral integration site 1, PBX/exd homeobox family	*Meis1*	NM_010789.1
POU domain, class 2, transcription factor 1	*Pou2f1*	AA711419
POU domain, class 5, transcription factor 1	*Pou5f1*	AA549756
SCIP, myelin-specific POU-domain transcriptional repressor	*Pou3f1*	M72711.1
heart and neural crest derivatives expressed transcript 1	*Hand1*	S79216
upstream transcription factor 1	*Usf1*	NM_009480.1
similar to zinc finger protein 220	*Myst3*	BC024786
zinc finger protein		AA607860
Putative zinc finger protein		AA616844
vascular endothelial zinc finger 1/Notch4-like protein	*Vezf1*	AA097970
zinc-finger protein 100 mRNA	*Zfp100*	W20703
zinc finger protein 76		AA471918
Putative zinc finger protein 136		
pleomorphic adenoma gene like 2, zinc finger transcription factor	*Plagl2*	NM_018807.1
Putative zinc finger protein		BC019219
PHD zinc finger transcription factor 12	*Phf12*	XM_031423.1
transcription factor 20	*Tcf20*	NM_013836
GATA-binding protein 1	*Gata1*	NM_008089.1
GATA-binding protein 2	*Gata2*	NM_008090.1
GATA-binding protein 3	*Gata3*	NM_008091.1
transcription factor AP-2, gamma	*Tcfap2c*	NM_009335.1
RAR-related orphan receptor alpha	*Rora*	NM_013646
nuclear receptor co-repressor 2	*Ncor2*	NM_011424
E26 avian leukemia oncogene 2	*Ets2*	BC005486
ets-related transcription factor	*Etv5*	AY004174.1
Ets-family transcription factor E74-like factor 1	*Elf1*	NM_007920
Krüppel-like factor 5, basic transcription element binding protein 2	*Klf5*	NM_009769
T box transcription factor 1	*Tbx1*	AF171100
TBX1/ LPS-induced TNF-alpha factor	*Litaf*	NM_019980.1
High mobility group box transcription factor 1 (rat)	*Hbp1*	NM_013221
LIM protein FHL2 (four and a half LIM domains 2)	*Fhl2*	AA419967
(human) Tim50, nuclear LIM interactor		XM_046104.1
Paired box gene 8	*Pax8*	AA597048
Myocyte enhancer factor 2B (MADS box transcription enhancer factor 2, polypeptide B)	*Mef2b*	AA516926
transforming growth factor beta 1 induced transcript 4	*Tgfb1i4*	AI114976
Tripartite motif protein	*Trim33*	AF220138.1
nuclear factor of activated T-cells, cytoplasmic 3	*Nfatc3*	XM_134391
transcription factor UBF		
transcription factor (p38 interacting protein)		XM_130880
cellular repressor of E1A-stim. genes	*Creg*	AF084524
Cbp/p300-interacting transactivator, melanocyte-specific gene 1	*Cited1 (Msg1)*	U65091
transcriptional repressor SIN3B	*Sin3b*	L38622
transcriptional co-activator p52	*Psip2/Ledgf*	AA177775
Cofactor required for Sp1 transcriptional activation subunit 2	*Crsp2*	NM_004229.1
Period 3, circadian oscillators	*Per3*	NM_011067.1
Putative human transcription intermediary factor 1-alpha		
CCAAT binding factor 1	*Cbf (Cebpa-rs1)*	NM_009882

iting only a thin shell of considerably smaller "giant" cells surrounding their implantation sites (Firulli et al., 1998; Riley et al., 1998). Mash2 by contrast acts to inhibit trophoblast differentiation into giant cells, either by supporting cell proliferation or by inhibiting their differentiation (Guillemot et al., 1994).

Hand1 affects Mash2 activity indirectly, whereas another regulator of this system, I-mfa, inhibits Mash2 directly by preventing the nuclear import and DNA-binding ability of Mash2 (Kraut et al., 1998).

Since differentiation of giant cells requires a switch from the mitotic to an endoreduplicative cell cycle, an important control point in this process is regulation of cell cycle progression. In addition to several cell cycle regulators that are important for giant cell differentiation, endoreduplication is enhanced – directly or indirectly – by the transcription factors Hand1 and Dp1 as giant cells are not only reduced in number but also in ploidy in both mutants (Firulli et al., 1998; Riley et al., 1998; Kohn et al., 2003).

Transcription factors are also critically involved in controlling extraembryonic development during later post-implantation stages. Several are known to be involved in the processes of chorioallantoic fusion, in determination of the invagination sites where allantoic mesoderm starts to grow into the forming labyrinth, and in the branching morphogenetic remodelling that establishes the vascular network of the placental labyrinth layer (Fig. 2). As can be expected from the complexity of labyrinth formation and function, many transcription factors are important for proper vascularization of this placental layer and for the concurrent development of syncytiotrophoblast-lined maternal blood spaces. A "small labyrinth" phenotype similar to that seen in many mutants of components of the signaling cascade is observed in the absence of several bHLH factors and in the absence of both components of the AP1 transcription factor, Junb (Schorpp-Kistner et al., 1999) and Fosl (Schreiber et al., 2000). The lack of labyrinthine vascularization caused by AP1 deficiency is particularly interesting as AP1 is a direct target of ERK MAPK activation. This phenocopy therefore suggests that during placental development, Junb and Fosl are the most relevant transcription factor subunits activated by the ERK1/2 pathway.

Labyrinth formation is also affected by several nuclear hormone receptors that function to transmit steroid hormone signals directly into the nucleus (Fig. 2). One of the most prominent steroid hormone effectors is retinoic acid, and mutants in many retinoic acid receptor components are characterized by labyrinth malformations (Fig. 2). Interestingly, mutants in a retinoic acid-responsive transcription factor, AP2-γ *(Tcfap2c)*, are at least as severely affected showing no signs of labyrinth formation, reduced tropoblast proliferation and embryonic death between E7.5 and E9.5 (Auman et al., 2002; Werling and Schorle, 2002).

From these examples it is obvious that many transcription factors belonging to different classes play an important role in extraembryonic development. As expected, the expression profiling approaches identified several of them that were known to be critical for placental development, such as *Gata2* and *Gata3* (Ma et al., 1997), *Tcfap2c* (Auman et al., 2002; Werling and Schorle, 2002), *Ets2* (Yamamoto et al., 1998) and *Hand1* (Firulli et al., 1998; Riley et al., 1998). The array analyses revealed expression of many other transcription factors in the trophoblast cell lineage (Table 4) most of which were not known to be expressed in trophoblast derivatives before. Some of them are of particular interest as several lines of evidence suggest their importance for extraembryonic development. The Ets-related gene *Etv5* for example has been found to be expressed specifically in tissues undergoing branching morphogenesis such as the lung, salivary gland, kidney, and mammary gland (Chotteau-Lelievre et al., 2003). Branching morphogenesis also occurs during the initial phases of labyrinth development and, therefore, expression of *Etv5* in the placenta supports this previously found correlation. Because of the highly restricted gene expression pattern it can be speculated that *Etv5*, like *Gcm1*, is critical for the tissue remodeling process that occurs in the developing labyrinth. Similarly, *Tgfb1i4* is dynamically expressed at sites of epithelial-mesenchymal interactions (Dohrmann et al., 1999), a morphological feature again found in the labyrinthine environment of trophoblast in direct contact with endothelial cells of the ingrowing blood vessels. Tgfb1i4 is a putative leucine-zipper transcription factor induced by TGFβ (Shibanuma et al., 1992), and at least some of the importance of this signaling pathway might be executed by *Tgfb1i4* in extraembryonic tissues.

Several transcription factors identified by the array analyses are involved in retinoic acid signaling. Because of the importance of this pathway in trophoblast differentiation these genes are excellent candidates to be essential for placental development. *Ncor2* is a silencing mediator of retinoid hormone receptors (Park et al., 1999). Since retinoic acid promotes trophoblast differentiation into a giant cell fate (Yan et al., 2001), *Ncor2* may be required for maintenance of the trophoblast stem cell population. NM_145732 is a nuclear steroid receptor related to the retinoic acid receptor RAR (Giguere et al., 1994). Interactions with RAR have been shown for the Krüppel-like zinc finger transcription factor Klf5 which activates TGFβ expression (Shindo et al., 2002). Interestingly, a homozygous loss of *Klf5* is embryonic lethal before E8.5 (Shindo et al., 2002), a time point highly suggestive of an extraembryonic defect. A role in placental development could also be speculated and has most recently been demonstrated for *Msg1/Cited1*, a transactivator enhancing Smad-mediated transcription (Rodriguez et al., 2004). Cited1 also binds to the heat shock cognate Hsc70 (Yahata et al., 2000) that has been found to be most abundantly expressed in the ectoplacental cone and invasive trophoblast giant cells (Hemberger et al., 2001).

Epigenetic control of placental development

Several findings support a strong influence of epigenetic factors and especially genomic imprinting on extraembryonic development. Genomic imprinting is a characteristic of mammals in which an imbalance of maternal and paternal genomes is deleterious to development. The non-equivalence of maternally and paternally derived genomes was first identified in mice by nuclear transfer studies of embryos with one of the two parental genomes (uniparental embryos) (Surani et al., 1984). Neither androgenetic nor gynogenetic pregnancies were viable: the androgenetic products showed exclusively extraembryonic development, thus indicating that the expression of specific genes on the paternal genome is crucial for placental development; the gynogenetic products had poorly developed placen-

tae but visible embryos thus indicating that maternally expressed genes are essential for the development of the embryo itself. The subsequent analysis of mouse embryos carrying both copies of a specific homologous chromosome or a specific chromosomal region from one parent revealed several regions that harbor imprinted genes responsible for abnormal extraembryonic phenotypes (Fundele et al., 1997). Additional evidence for the essential role of imprinted genes during placental development came from the analysis of mouse mutants. Thus, placental defects were observed in knockout conceptuses of several paternally expressed (e.g. *Igf2*, *Peg1* and *Peg3*) and maternally expressed (e.g. *H19*, *Igf2r*, *Cdkn1c* (alias *p57Kip2*), *Ipl*, *Mash2*, *Grb10* and *Slc22a3*) genes (DeChiara et al., 1991; Guillemot et al., 1994; Eggenschwiler et al., 1997; Lefebvre et al., 1998; Li et al., 1999; Takahashi et al., 2000; Zwart et al., 2001; Constancia et al., 2002; Frank et al., 2002; Charalambous et al., 2003). Interestingly, two of these genes were also identified in the large-scale expression profiling approaches. The maternally expressed *Igf2r* gene was found to be more highly expressed in TS cells compared to ES cells (Tanaka et al., 2002). As already mentioned above, an *Igf2r* null mutation results in an excess of Igf2 levels and as a consequence, in placental hyperplasia as one of the major outcomes (Eggenschwiler et al., 1997). Further, the *Grb10* gene was found to be significantly downregulated in hyperplastic placentae of both cumulus cell-derived and ES cell-derived NT placentae compared to controls (Humpherys et al., 2002). Consistent with these results, disruption of the maternal allele of *Grb10* results in overgrowth of both the embryo and placenta such that mutant mice are at birth approximately 30 % larger than normal (Charalambous et al., 2003).

Evolutionary and comparative aspects of placentogenesis

The data presented also must be considered in the context of placental evolution and human placentation. Extraembryonic membranes are found in all amniotes, which include reptiles, birds, and mammals as well as in sharks, bony fishes, and some amphibia. The formation of a placenta occurs in mammals (except monotremes) and in numerous species of squamate reptiles and cartilaginous fishes. Placental morphology is characterized by a unique diversity between or even within mammalian orders making the placenta more variable in structure than any other mammalian organ (Mossman, 1937). Thus, at one extreme, extraembryonic and uterine epithelia are closely apposed, but there is little or no destruction of maternal tissues (epitheliochorial placentae occurring in pigs, horses, whales and lemurs). At the other extremes, extraembryonic tissues breach the walls of maternal blood vessels, and the placenta gains direct access to circulating maternal blood (haemochorial placentae occurring in rodents, lagomorphs, insectivores, bats and, among primates, in tarsiers, monkeys, apes and humans). Other categories are found in ruminants (syndesmochorial) and carnivores (endotheliochorial) (Haig, 1993). Further, rodents and humans have different subtypes of haemochorial placentae. Rodents have three trophoblast cell layers (haemotrichorial), whereas humans have a single syncytial layer (haemomonochorial). These dramatic differences between mammalian placentae question whether genetic data obtained for the mouse placenta can be generalized for other mammalian and especially human placentae. However, this obvious objection can be met by the fact that considerable similarities exist as well. Thus, the identification of both genes specifying trophoblast cell structures and the expression sites of these genes made it possible to define functionally homologous cell types in placentae of mice and humans. First, both trophoblast giant cells in rodents and extravillous cytotrophoblast cells in humans compose the outer layer of the placenta and mediate implantation and invasion of the uterus. Second, at least some of the spongiotrophoblast cells in rodents are analogous to column cytotrophoblasts in humans. Third, both the murine labyrinthine trophoblasts and the floating chorionic villi in humans are analogous sites where fetal and maternal blood circulate in close association for physiological exchange. Genes, which are both essential for murine trophoblast development and expressed in analogous cell types in humans, include homologues of *Mash2, Hand1, Gcm1, Hgf* and *Met* (Alders et al., 1997; Somerset et al., 1998; Nait-Oumesmar et al., 2000; Knofler et al., 2002). Further, the increasing knowledge about genes governing placental development also provides some clues for the molecular evolution of the placenta. Contrary to the initial prediction that the development of an evolutionary relatively new organ like the placenta is regulated by newly established placental-specific genes, it rather appears that either pre-existing genes necessary for embryonic development acquired new functions in placentation (e.g. *Hand1*) or paralogues of pre-existing genes are generated by gene duplications and recruited for placenta-specific functions (e.g. *Mash2*) (Guillemot et al., 1994; Riley et al., 1998).

Conclusion

We have shown here that expression profiling approaches, irrespective of their initial aim, are suitable to detect genes already known to play essential roles during placental development. These data validate expression profiling approaches as an important tool for the identification of new genes that are likely to be important for trophoblast differentiation and/or function. In the future, the application of this technology in the analysis of mouse mutants will further extend our knowledge about the complex genetic basis of placentogenesis. Because homologous cell types have been identified in human and mouse extraembryonic tissues, the comparison of gene expression profiles between mouse and human trophoblast, both normal and abnormal, will give insights into the molecular defects underlying the different clinical conditions of human pregnancy.

References

Abbott BD, Buckalew AR: Placental defects in ARNT-knockout conceptus correlate with localized decreases in VEGF-R2, Ang-1, and Tie-2. Dev Dyn 219:526–538 (2000).

Adams RH, Porras A, Alonso G, Jones M, Vintersten K, Panelli S, Valladares A, Perez L, Klein R, Nebreda AR: Essential role of p38alpha MAP kinase in placental but not embryonic cardiovascular development. Mol Cell 6:109–116 (2000).

Adamson SL, Lu Y, Whiteley KJ, Holmyard D, Hemberger M, Pfarrer C, Cross JC: Interactions between trophoblast cells and the maternal and fetal circulation in the mouse placenta. Dev Biol 250:358–373 (2002).

Adelman DM, Gertsenstein M, Nagy A, Simon MC, Maltepe E: Placental cell fates are regulated in vivo by HIF-mediated hypoxia responses. Genes Dev 14:3191–3203 (2000).

Alders M, Hodges M, Hadjantonakis AK, Postmus J, van Wijk I, Bliek J, de Meulemeester M, Westerveld A, Guillemot F, Oudejans C, Little P, Mannens M: The human Achaete-Scute homologue 2 (ASCL2, HASH2) maps to chromosome 11p15.5, close to IGF2 and is expressed in extravillus trophoblasts. Hum Mol Genet 6:859–867 (1997).

Amoroso EC: Placentation, in Parkes AS (ed): Marshall's Physiology of Reproduction, 3rd edition, pp 127–311 (Longmans, London 1952).

Anson-Cartwright L, Dawson K, Holmyard D, Fisher SJ, Lazzarini RA, Cross JC: The glial cells missing-1 protein is essential for branching morphogenesis in the chorioallantoic placenta. Nat Genet 25:311–314 (2000).

Antonson P, Schuster GU, Wang L, Rozell B, Holter E, Flodby P, Treuter E, Holmgren L, Gustafsson JA: Inactivation of the nuclear receptor coactivator RAP250 in mice results in placental vascular dysfunction. Mol Cell Biol 23:1260–1268 (2003).

Arman E, Haffner-Krausz R, Chen Y, Heath JK, Lonai P: Targeted disruption of fibroblast growth factor (FGF) receptor 2 suggests a role for FGF signaling in pregastrulation mammalian development. Proc Natl Acad Sci USA 95:5082–5087 (1998).

Auman HJ, Nottoli T, Lakiza O, Winger Q, Donaldson S, Williams T: Transcription factor AP-2gamma is essential in the extra-embryonic lineages for early postimplantation development. Development 129:2733–2747 (2002).

Baker J, Liu JP, Robertson EJ, Efstratiadis A: Role of insulin-like growth factors in embryonic and postnatal growth. Cell 75:73–82 (1993).

Barak Y, Nelson MC, Ong ES, Jones YZ, Ruiz-Lozano P, Chien KR, Koder A, Evans RM: PPAR gamma is required for placental, cardiac, and adipose tissue development. Mol Cell 4:585–595 (1999).

Barak Y, Liao D, He W, Ong ES, Nelson MC, Olefsky JM, Boland R, Evans RM: Effects of peroxisome proliferator-activated receptor delta on placentation, adiposity, and colorectal cancer. Proc Natl Acad Sci USA 99:303–308 (2002).

Bladt F, Riethmacher D, Isenmann S, Aguzzi A, Birchmeier C: Essential role for the c-met receptor in the migration of myogenic precursor cells into the limb bud. Nature 376:768–771 (1995).

Boström H, Willetts K, Pekny M, Leveen P, Lindahl P, Hedstrand H, Pekna M, Hellstrom M, Gebre-Medhin S, Schalling M, Nilsson M, Kurland S, Tornell J, Heath JK, Betsholtz C: PDGF-A signaling is a critical event in lung alveolar myofibroblast development and alveogenesis. Cell 85:863–873 (1996).

Brancho D, Tanaka N, Jaeschke A, Ventura JJ, Kelkar N, Tanaka Y, Kyuuma M, Takeshita T, Flavell RA, Davis RJ: Mechanism of p38 MAP kinase activation in vivo. Genes Dev 17:1969–1978 (2003).

Chang H, Huylebroeck D, Verschueren K, Guo Q, Matzuk MM, Zwijsen A: *Smad5* knockout mice die at mid-gestation due to multiple embryonic and extraembryonic defects. Development 126:1631–1642 (1999).

Charalambous M, Smith FM, Bennett WR, Crew TE, Mackenzie F, Ward A: Disruption of the imprinted *Grb10* gene leads to disproportionate overgrowth by an Igf2-independent mechanism. Proc Natl Acad Sci USA 100:8292–8297 (2003).

Chawengsaksophak K, James R, Hammond VE, Kontgen F, Beck F: Homeosis and intestinal tumours in *Cdx2* mutant mice. Nature 386:84–87 (1997).

Chotteau-Lelievre A, Montesano R, Soriano J, Soulie P, Desbiens X, de Launoit Y: PEA3 transcription factors are expressed in tissues undergoing branching morphogenesis and promote formation of ductlike structures by mammary epithelial cells in vitro. Dev Biol 259:241–257 (2003).

Constancia M, Hemberger M, Hughes J, Dean W, Ferguson-Smith A, Fundele R, Stewart F, Kelsey G, Fowden A, Sibley C, Reik W: Placental-specific IGF-II is a major modulator of placental and fetal growth. Nature 417:945–948 (2002).

Copp AJ: Death before birth: clues from gene knockouts and mutations. Trends Genet 11:87–93 (1995).

Crawford SE, Qi C, Misra P, Stellmach V, Rao MS, Engel JD, Zhu Y, Reddy JK: Defects of the heart, eye, and megakaryocytes in peroxisome proliferator activator receptor-binding protein (PBP) null embryos implicate GATA family of transcription factors. J Biol Chem 277:3585–3592 (2002).

Cross JC: Formation of the placenta and extraembryonic membranes. Ann NY Acad Sci 857:23–32 (1998).

Cross JC, Werb Z, Fisher SJ: Implantation and the placenta: key pieces of the development puzzle. Science 266:1508–1518 (1994).

Cross JC, Flannery ML, Blanar MA, Steingrimsson E, Jenkins NA, Copeland NG, Rutter WJ, Werb Z: *Hxt* encodes a basic helix-loop-helix transcription factor that regulates trophoblast cell development. Development 121:2513–2523 (1995).

DeChiara TM, Efstratiadis A, Robertson EJ: A growth-deficiency phenotype in heterozygous mice carrying an insulin-like growth factor II gene disrupted by targeting. Nature 345:78–80 (1990).

DeChiara TM, Robertson EJ, Efstratiadis A: Parental imprinting of the mouse insulin-like growth factor II gene. Cell 64:849–859 (1991).

Dickson MC, Martin JS, Cousins FM, Kulkarni AB, Karlsson S, Akhurst RJ: Defective haematopoiesis and vasculogenesis in transforming growth factor-beta 1 knock out mice. Development 121:1845–1854 (1995).

Dohrmann CE, Belaoussoff M, Raftery LA: Dynamic expression of TSC-22 at sites of epithelial-mesenchymal interactions during mouse development. Mech Dev 84:147–151 (1999).

Dong C, Yang DD, Wysk M, Whitmarsh AJ, Davis RJ, Flavell RA: Defective T cell differentiation in the absence of Jnk1. Science 282:2092–2095 (1998).

Eggenschwiler J, Ludwig T, Fisher P, Leighton PA, Tilghman SM, Efstratiadis A: Mouse mutant embryos overexpressing IGF-II exhibit phenotypic features of the Beckwith-Wiedemann and Simpson-Golabi-Behmel syndromes. Genes Dev 11:3128–3142 (1997).

Firulli AB, McFadden DG, Lin Q, Srivastava D, Olson EN: Heart and extra-embryonic mesodermal defects in mouse embryos lacking the bHLH transcription factor Hand1. Nat Genet 18:266–270 (1998).

Frank D, Fortino W, Clark L, Musalo R, Wang W, Saxena A, Li CM, Reik W, Ludwig T, Tycko B: Placental overgrowth in mice lacking the imprinted gene Ipl. Proc Natl Acad Sci USA 99:7490–7495 (2002).

Fundele R, Surani MA, Allen ND: Consequences of genomic imprinting for fetal development, in Reik W, Surani MA (eds): Genomic Imprinting, pp 98–117 (IRL, Oxford 1997).

Galceran J, Farinas I, Depew MJ, Clevers H, Grosschedl R: Wnt3a−/−-like phenotype and limb deficiency in Lef1(−/−)Tcf1(−/−) mice. Genes Dev 13:709–717 (1999).

Giguere V, Tini M, Flock G, Ong E, Evans RM, Otulakowski G: Isoform-specific amino-terminal domains dictate DNA-binding properties of ROR alpha, a novel family of orphan hormone nuclear receptors. Genes Dev 8:538–553 (1994).

Giroux S, Tremblay M, Bernard D, Cardin-Girard JF, Aubry S, Larouche L, Rousseau S, Huot J, Landry J, Jeannotte L, Charron J: Embryonic death of Mek1-deficient mice reveals a role for this kinase in angiogenesis in the labyrinthine region of the placenta. Curr Biol 9:369–372 (1999).

Guillemot F, Nagy A, Auerbach A, Rossant J, Joyner AL: Essential role of Mash-2 in extraembryonic development. Nature 371:333–336 (1994).

Haffner-Krausz R, Gorivodsky M, Chen Y, Lonai P: Expression of Fgfr2 in the early mouse embryo indicates its involvement in preimplantation development. Mech Dev 85:167–172 (1999).

Haig D: Genetic conflicts in human pregnancy. Quart Rev Biol 68: 495–532 (1993).

Hemberger M, Cross JC: Genes governing placental development. Trends Endocrinol Metab 12:162–168 (2001).

Hemberger M, Cross JC, Ropers HH, Lehrach H, Fundele R, Himmelbauer H: UniGene cDNA array-based monitoring of transcriptome changes during mouse placental development. Proc Natl Acad Sci USA 98:13126–13131 (2001).

Hernandez-Verdun D: Morphogenesis of the syncytium in the mouse placenta. Ultrastructural study. Cell Tissue Res 148:381–396 (1974).

Humpherys D, Eggan K, Akutsu H, Friedman A, Hochedlinger K, Yanagimachi R, Lander ES, Golub TR, Jaenisch R: Abnormal gene expression in cloned mice derived from embryonic stem cell and cumulus cell nuclei. Proc Natl Acad Sci USA 99:12889–12894 (2002).

Ishikawa T, Tamai Y, Zorn AM, Yoshida H, Seldin MF, Nishikawa S, Taketo MM: Mouse Wnt receptor gene *Fzd5* is essential for yolk sac and placental angiogenesis. Development 128:25–33 (2001).

Itoh M, Yoshida Y, Nishida K, Narimatsu M, Hibi M, Hirano T: Role of Gab1 in heart, placenta, and skin development and growth factor- and cytokine-induced extracellular signal-regulated kinase mitogen-activated protein kinase activation. Mol Cell Biol 20:3695–3704 (2000).

Jackson D, Volpert OV, Bouck N, Linzer DI: Stimulation and inhibition of angiogenesis by placental proliferin and proliferin-related protein. Science 266:1581–1584 (1994).

Knofler M, Meinhardt G, Bauer S, Loregger T, Vasicek R, Bloor DJ, Kimber SJ, Husslein P: Human Hand1 basic helix-loop-helix (bHLH) protein: extra-embryonic expression pattern, interaction partners and identification of its transcriptional repressor domains. Biochem J 361:641–651 (2002).

Ko MS, Threat TA, Wang X, Horton JH, Cui Y, Pryor E, Paris J, Wells-Smith J, Kitchen JR, Rowe LB, Eppig J, Satoh T, Brant L, Fujiwara H, Yotsumoto S, Nakashima H: Genome-wide mapping of unselected transcripts from extraembryonic tissue of 7.5-day mouse embryos reveals enrichment in the t-complex and under-representation on the X chromosome. Hum Mol Genet 7:1967–1978 (1998).

Kohn MJ, Bronson RT, Harlow E, Dyson NJ, Yamasaki L: Dp1 is required for extra-embryonic development. Development 130:1295–1305 (2003).

Kozak KR, Abbott B, Hankinson O: ARNT-deficient mice and placental differentiation. Dev Biol 191:297–305 (1997).

Kraut N, Snider L, Chen C, Tapscott SJ, Groudine M: Requirement of the mouse I-mfa gene for placental development and skeletal patterning. EMBO J 17:6276–6288 (1998).

Kuan CY, Yang DD, Samanta Roy DR, Davis RJ, Rakic P, Flavell RA: The Jnk1 and Jnk2 protein kinases are required for regional specific apoptosis during early brain development. Neuron 22:667–676 (1999).

Kuang SQ, Liao L, Zhang H, Pereira FA, Yuan Y, DeMayo FJ, Ko L, Xu J: Deletion of the cancer-amplified coactivator AIB3 results in defective placentation and embryonic lethality. J Biol Chem 277:45356–45360 (2002).

Larsson J, Goumans MJ, Sjostrand LJ, van Rooijen MA, Ward D, Leveen P, Xu X, ten Dijke P, Mummery CL, Karlsson S: Abnormal angiogenesis but intact hematopoietic potential in TGF-beta type I receptor-deficient mice. EMBO J 20:1663–1673 (2001).

Lau AL, Kumar TR, Nishimori K, Bonadio J, Matzuk MM: Activin betaC and betaE genes are not essential for mouse liver growth, differentiation, and regeneration. Mol Cell Biol 20:6127–6137 (2000).

Lechleider RJ, Ryan JL, Garrett L, Eng C, Deng C, Wynshaw-Boris A, Roberts AB: Targeted mutagenesis of *Smad1* reveals an essential role in chorioallantoic fusion. Dev Biol 240:157–167 (2001).

Lefebvre L, Viville S, Barton SC, Ishino F, Keverne EB, Surani MA: Abnormal maternal behaviour and growth retardation associated with loss of the imprinted gene *Mest*. Nat Genet 20:163–169 (1998).

Li L, Keverne EB, Aparicio SA, Ishino F, Barton SC, Surani MA: Regulation of maternal behavior and offspring growth by paternally expressed Peg3. Science 284:330–333 (1999).

Linzer DI, Fisher SJ: The placenta and the prolactin family of hormones: regulation of the physiology of pregnancy. Mol Endocrinol 13:837–840 (1999).

Lopez MF, Dikkes P, Zurakowski D, Villa-Komaroff L: Insulin-like growth factor II affects the appearance and glycogen content of glycogen cells in the murine placenta. Endocrinology 137:2100–2108 (1996).

Lorenzi MV, Castagnino P, Aaronson DC, Lieb DC, Lee CC, Keck CL, Popescu NC, Miki T: Human FRAG1 encodes a novel membrane-spanning protein that localizes to chromosome 11p15.5, a region of frequent loss of heterozygosity in cancer. Genomics 62:59–66 (1999).

Luo J, Sladek R, Bader J-A, Matthyssen A, Rossant J, Giguere V: Placental abnormalities in mouse embryos lacking the orphan nuclear receptor ERR-b. Nature 388:778–782 (1997).

Ma GT, Roth ME, Groskopf JC, Tsai FY, Orkin SH, Grosveld F, Engel JD, Linzer DI: GATA-2 and GATA-3 regulate trophoblast-specific gene expression in vivo. Development 124:907–914 (1997).

Ma GT, Soloveva V, Tzeng SJ, Lowe LA, Pfendler KC, Iannaccone PM, Kuehn MR, Linzer DI: Nodal regulates trophoblast differentiation and placental development. Dev Biol 236:124–135 (2001).

Meighan-Mantha RL, Hsu DK, Guo Y, Brown SA, Feng SL, Peifley KA, Alberts GF, Copeland NG, Gilbert DJ, Jenkins NA, Richards CM, Winkles JA: The mitogen-inducible Fn14 gene encodes a type I transmembrane protein that modulates fibroblast adhesion and migration. J Biol Chem 274:33166–33176 (1999).

Mikula M, Schreiber M, Husak Z, Kucerova L, Ruth J, Wieser R, Zatloukal K, Beug H, Wagner EF, Baccarini M: Embryonic lethality and fetal liver apoptosis in mice lacking the c-raf-1 gene. EMBO J 20:1952–1962 (2001).

Monkley SJ, Delaney SJ, Pennisi DJ, Christiansen JH, Wainwright BJ: Targeted disruption of the *Wnt2* gene results in placentation defects. Development 122:3343–3353 (1996).

Morasso MI, Grinberg A, Robinson G, Sargent TD, Mahon KA: Placental failure in mice lacking the homeobox gene *Dlx3*. Proc Natl Acad Sci USA 96:162–167 (1999).

Mossman HW: Comparative morphogenesis of the foetal membranes and accessory uterine structures. Contr Embryol Carnegie Instn 26:127–146 (1937).

Mossman HW: Vertebrate Fetal Membranes: Comparative Ontogeny And Morphology, Evolution, Phylogenetic Significance, Basic Functions, Research Opportunities (Rutgers University Press, New Brunswick 1987).

Mudgett JS, Ding J, Guh-Siesel L, Chartrain NA, Yang L, Gopal S, Shen MM: Essential role for p38alpha mitogen-activated protein kinase in placental angiogenesis. Proc Natl Acad Sci USA 97:10454–10459 (2000).

Muntener M, Hsu YC: Development of trophoblast and placenta of the mouse. A reinvestigation with regard to the in vitro culture of mouse trophoblast and placenta. Acta Anat 98:241–252 (1977).

Nait-Oumesmar B, Copperman AB, Lazzarini RA: Placental expression and chromosomal localization of the human Gcm 1 gene. J Histochem Cytochem 48:915–922 (2000).

Nakayama H, Scott IC, Cross JC: The transition to endoreduplication in trophoblast giant cells is regulated by the mSNA zinc-finger transcription factor. Dev Biol 199:150–163 (1998).

Niswander L, Martin GR: Fgf-4 expression during gastrulation, myogenesis, limb and tooth development in the mouse. Development 114:755–68 (1992).

Offermanns S, Zhao LP, Gohla A, Sarosi I, Simon MI, Wilkie TM: Embryonic cardiomyocyte hypoplasia and craniofacial defects in G alpha q/G alpha 11-mutant mice. EMBO J 17:4304–4312 (1998).

Ohlsson R, Falck P, Hellstrom M, Lindahl P, Bostrom H, Franklin G, Ahrlund-Richter L, Pollard J, Soriano P, Betsholtz C: PDGFB regulates the development of the labyrinthine layer of the mouse fetal placenta. Dev Biol 212:124–136 (1999).

Oshima M, Oshima O, Taketo MM: TGF-β receptor type II deficiency results in defects of yolk sac hematopoiesis and vasculogenesis. Dev Biol 179:297–302 (1996).

Pages G, Guerin S, Grall D, Bonino F, Smith A, Anjuere F, Auberger P, Pouyssegur J: Defective thymocyte maturation in p44 MAP kinase (Erk 1) knockout mice. Science 286:1374–1377 (1999).

Park EJ, Schroen DJ, Yang M, Li H, Li L, Chen JD: SMRTe, a silencing mediator for retinoid and thyroid hormone receptors-extended isoform that is more related to the nuclear receptor corepressor. Proc Natl Acad Sci USA 96:3519–3524 (1999).

Qian X, Esteban L, Vass WC, Upadhyaya C, Papageorge AG, Yienger K, Ward JM, Lowy DR, Santos E: The Sos1 and Sos2 Ras-specific exchange factors: differences in placental expression and signaling properties. EMBO J 19:642–654 (2000).

Rappolee DA, Basilico C, Patel Y, Werb Z: Expression and function of FGF-4 in peri-implantation development in mouse embryos. Development 120:2259–2269 (1994).

Redline RW, Chernicky CL, Tan HQ, Ilan J, Ilan J: Differential expression of insulin-like growth factor-II in specific regions of the late (post day 9.5) murine placenta. Mol Reprod Dev 36:121–129 (1993).

Regan CP, Li W, Boucher DM, Spatz S, Su MS, Kuida K: Erk5 null mice display multiple extraembryonic vascular and embryonic cardiovascular defects. Proc Natl Acad Sci USA 99:9248–9253 (2002).

Riley P, Anson-Cartwright L, Cross JC: The Hand1 bHLH transcription factor is essential for placentation and cardiac morphogenesis. Nat Genet 18:271–275 (1998).

Rodriguez TA, Sparrow DB, Scott AN, Withington SL, Preis JI, Michalicek J, Clements M, Tsang TE, Shioda T, Beddington RSP, Dunwoodie SL: *Cited1* is required in trophoblasts for placental development and for embryo growth and survival. Mol Cell Biol 24:228–244 (2004).

Rossant J, Cross JC: Placental development: lessons from mouse mutants. Nat Rev Genet 2:538–548 (2001).

Russ AP, Wattler S, Colledge WH, Aparicio SA, Carlton MB, Pearce JJ, Barton SC, Surani MA, Ryan K, Nehls MC, Wilson V, Evans MJ: Eomesodermin is required for mouse trophoblast development and mesoderm formation. Nature 404:95–99 (2000).

Saba-El-Leil MK, Vella FDJ, Vernay B, Voisin L, Chen L, Labrecque N, Ang S-L, Meloche S: An essential function of the mitogen-activated protein kinase Erk2 in mouse trophoblast development. EMBO Reports 4:964–968 (2003).

Sabapathy K, Hu Y, Kallunki T, Schreiber M, David JP, Jochum W, Wagner EF, Karin M: JNK2 is required for efficient T-cell activation and apoptosis but not for normal lymphocyte development. Curr Biol 9:116–125 (1999).

Sapin V, Dolle P, Hindelang C, Kastner P, Chambon P: Defects of the chorioallantoic placenta in mouse RXRa null fetuses. Dev Biol 191:29–41 (1997).

Saxton TM, Cheng AM, Ong SH, Lu Y, Sakai R, Cross JC, Pawson T: Gene dosage-dependent functions for phosphotyrosine-Grb2 signaling during mammalian tissue morphogenesis. Curr Biol 11:662–670 (2001).

Schorpp-Kistner M, Wang ZQ, Angel P, Wagner EF: JunB is essential for mammalian placentation. EMBO J 18:934–948 (1999).

Schreiber M, Wang Z, Jochum W, Fetka I, Elliott C, Wagner EF: Placental vascularisation requires the AP-1 component Fra1. Development 127:4937–4948 (2000).

Scott IC, Anson-Cartwright L, Riley P, Reda D, Cross JC: The Hand1 basic helix-loop-helix transcription factor regulates trophoblast giant cell differentiation via multiple mechanisms. Mol Cell Biol 20:530–541 (2000).

Shawlot W, Behringer RR: Requirement for Lim1 in head-organizer function. Nature 374:425–430 (1995).

Shi Y, Massague J: Mechanisms of TGF-beta signaling from cell membrane to the nucleus. Cell 113:685–700 (2003).

Shibanuma M, Kuroki T, Nose K: Isolation of a gene encoding a putative leucine zipper structure that is induced by transforming growth factor beta 1 and other growth factors. J Biol Chem 267:10219–10224 (1992).

Shindo T, Manabe I, Fukushima Y, Tobe K, Aizawa K, Miyamoto S, Kawai-Kowase K, Moriyama N, Imai Y, Kawakami H, Nishimatsu H, Ishikawa T, Suzuki T, Morita H, Maemura K, Sata M, Hirata Y, Komukai M, Kagechika H, Kadowaki T, Kurabayashi M, Nagai R: Kruppel-like zinc-finger transcription factor KLF5/BTEB2 is a target for angiotensin II signaling and an essential regulator of cardiovascular remodeling. Nat Med 8:856–863 (2002).

Sibilia M, Wagner EF: Strain-dependent epithelial defects in mice lacking the EGF receptor. Science 269:234–238 (1995).

Solloway MJ, Robertson EJ: Early embryonic lethality in Bmp5; Bmp7 double mutant mice suggests functional redundancy within the 60A subgroup. Development 126:1753–1768 (1999).

Somerset DA, Li XF, Afford S, Strain AJ, Ahmed A, Sangha RK, Whittle MJ, Kilby MD: Ontogeny of hepatocyte growth factor (HGF) and its receptor (c-met) in human placenta: reduced HGF expression in intrauterine growth restriction. Am J Pathol 153:1139–1147 (1998).

Steingrimsson E, Tessarollo L, Reid SW, Jenkins NA, Copeland NG: The bHLH-Zip transcription factor Tfeb is essential for placental vascularization. Development 125:4607–4616 (1998).

Stenvers KL, Tursky ML, Harder KW, Kountouri N, Amatayakul-Chantler S, Grail D, Small C, Weinberg RA, Sizeland AM, Zhu HJ: Heart and liver defects and reduced transforming growth factor beta2 sensitivity in transforming growth factor beta type III receptor-deficient embryos. Mol Cell Biol 23:4371–4385 (2003).

Sterner-Kock A, Thorey IS, Koli K, Wempe F, Otte J, Bangsow T, Kuhlmeier K, Kirchner T, Jin S, Keski-Oja J, von Melchner H: Disruption of the gene encoding the latent transforming growth factor-beta binding protein 4 (LTBP-4) causes abnormal lung development, cardiomyopathy, and colorectal cancer. Genes Dev 16:2264–2273 (2002).

Suemizu H, Aiba K, Yoshikawa T, Sharov AA, Shimozawa N, Tamaoki N, Ko MS: Expression profiling of placentomegaly associated with nuclear transplantation of mouse ES cells. Dev Biol 253:36–53 (2003).

Surani MA, Barton SC, Norris ML: Development of reconstituted mouse eggs suggests imprinting of the genome during gametogenesis. Nature 308:548–550 (1984).

Taipale J, Saharinen J, Keski-Oja J: Extracellular matrix-associated transforming growth factor: role in cancer cell growth and invasion. Adv Cancer Res 75:87–134 (1998).

Takahashi K, Kobayashi T, Kanayama N: p57(Kip2) regulates the proper development of labyrinthine and spongiotrophoblasts. Mol Hum Reprod 6:1019–1025 (2000).

Tanaka S, Kunath T, Hadjantonakis AK, Nagy A, Rossant J: Promotion of trophoblast stem cell proliferation by FGF4. Science 282:2072–2075 (1998).

Tanaka TS, Jaradat SA, Lim MK, Kargul GJ, Wang X, Grahovac MJ, Pantano S, Sano Y, Piao Y, Nagaraja R, Doi H, Wood WH 3rd, Becker KG, Ko MS: Genome-wide expression profiling of mid-gestation placenta and embryo using a 15,000 mouse developmental cDNA microarray. Proc Natl Acad Sci USA 97:9127–9132 (2000).

Tanaka TS, Kunath T, Kimber WL, Jaradat SA, Stagg CA, Usuda M, Yokota T, Niwa H, Rossant J, Ko MS: Gene expression profiling of embryo-derived stem cells reveals candidate genes associated with pluripotency and lineage specificity. Genome Res 12:1921–1928 (2002).

Tassi E, Al-Attar A, Aigner A, Swift MR, McDonnell K, Karavanov A, Wellstein A: Enhancement of fibroblast growth factor (FGF) activity by an FGF-binding protein. J Biol Chem 276:40247–40253 (2001).

Threadgill DW, Dlugosz AA, Hansen LA, Tennenbaum T, Lichti U, Yee D, LaMantia C, Mourton T, Herrup K, Harris RC, et al: Targeted disruption of mouse EGF receptor: effect of genetic background on mutant phenotype. Science 269:230–234 (1995).

Tremblay KD, Dunn NR, Robertson EJ: Mouse embryos lacking Smad1 signals display defects in extra-embryonic tissues and germ cell formation. Development 128:3609–3621 (2001).

Uehara Y, Minowa O, Mori C, Shiota K, Kuno J, Noda T, Kitamura N: Placental defect and embryonic lethality in mice lacking hepatocyte growth factor/scatter factor. Nature 373:702–705 (1995).

Uy GD, Downs KM, Gardner RL: Inhibition of trophoblast stem cell potential in chorionic ectoderm coincides with occlusion of the ectoplacental cavity in the mouse. Development 129:3913–3924 (2002).

Ware CB, Horowitz MC, Renshaw BR, Hunt JS, Liggitt D, Koblar SA, Gliniak BC, McKenna HJ, Papayannopoulou T, Thoma B, et al: Targeted disruption of the low-affinity leukemia inhibitory factor receptor gene causes placental, skeletal, neural and metabolic defects and results in perinatal death. Development 121:1283–1299 (1995).

Welsh AO, Enders AC: Chorioallantoic placenta formation in the rat: II. Angiogenesis and maternal blood circulation in the mesometrial region of the implantation chamber prior to placenta formation. Am J Anat 192:347–365 (1991).

Wendling O, Chambon P, Mark M: Retinoid X receptors are essential for early mouse development and placentogenesis. Proc Natl Acad Sci USA 96:547–551 (1999).

Werling U, Schorle H: Transcription factor gene AP-2 gamma essential for early murine development. Mol Cell Biol 22:3149–3156 (2002).

Wojnowski L, Stancato LF, Zimmer AM, Hahn H, Beck TW, Larner AC, Rapp UR, Zimmer A: Craf-1 protein kinase is essential for mouse development. Mech Dev 76:141–149 (1998).

Xu X, Weinstein M, Li C, Naski M, Cohen RI, Ornitz DM, Leder P, Deng C: Fibroblast growth factor receptor 2 (FGFR2)-mediated reciprocal regulation loop between FGF8 and FGF10 is essential for limb induction. Development 125:753–765 (1998).

Xue Y, Wang X, Li Z, Gotoh N, Chapman D, Skolnik EY: Mesodermal patterning defect in mice lacking the Ste20 NCK interacting kinase (NIK). Development 128:1559–1572 (2001).

Yahata T, de Caestecker MP, Lechleider RJ, Andriole S, Roberts AB, Isselbacher KJ, Shioda T: The MSG1 non-DNA-binding transactivator binds to the p300/CBP coactivators, enhancing their functional link to the Smad transcription factors. J Biol Chem 275:8825–8834 (2000).

Yamamoto H, Flannery ML, Kupriyanov S, Pearce J, McKercher SR, Henkel GW, Maki RA, Werb Z, Oshima RG: Defective trophoblast function in mice with a targeted mutation of Ets2. Genes Dev 12:1315–1326 (1998).

Yan J, Tanaka S, Oda M, Makino T, Ohgane J, Shiota K: Retinoic acid promotes differentiation of trophoblast stem cells to a giant cell fate. Dev Biol 235:422–432 (2001).

Yang DD, Kuan CY, Whitmarsh AJ, Rincon M, Zheng TS, Davis RJ, Rakic P, Flavell RA: Absence of excitotoxicity-induced apoptosis in the hippocampus of mice lacking the Jnk3 gene. Nature 389:865–870 (1997).

Yang J, Boerm M, McCarty M, Bucana C, Fidler IJ, Zhuang Y, Su B: Mekk3 is essential for early embryonic cardiovascular development. Nat Genet 24:309–313 (2000).

Yang X, Castilla LH, Xu X, Li C, Gotay J, Weinstein M, Liu PP, Deng CX: Angiogenesis defects and mesenchymal apoptosis in mice lacking SMAD5. Development 126:1571–1580 (1999).

Ylikorkala A, Rossi DJ, Korsisaari N, Luukko K, Alitalo K, Henkemeyer M, Makela TP: Vascular abnormalities and deregulation of VEGF in Lkb1-deficient mice. Science 293:1323–1326 (2001).

Zhu Y, Qi C, Jia Y, Nye JS, Rao MS, Reddy JK: Deletion of PBP/PPARBP, the gene for nuclear receptor coactivator peroxisome proliferator-activated receptor-binding protein, results in embryonic lethality. J Biol Chem 275:14779–14782 (2000).

Zhu YJ, Crawford SE, Stellmach V, Dwivedi RS, Rao MS, Gonzalez FJ, Qi C, Reddy JK: Coactivator PRIP, the peroxisome proliferator-activated receptor-interacting protein, is a modulator of placental, cardiac, hepatic, and embryonic development. J Biol Chem 278:1986–1990 (2003).

Zwart R, Verhaagh S, Buitelaar M, Popp-Snijders C, Barlow DP: Impaired activity of the extraneuronal monoamine transporter system known as uptake-2 in *Orct3/Slc22a3*-deficient mice. Mol Cell Biol 21:4188–4196 (2001).

Zwijsen A, Verschueren K, Huylebroeck D: New intracellular components of bone morphogenetic protein/Smad signaling cascades. FEBS Lett 546:133–139 (2003).

Cytogenet Genome Res 105:270–278 (2004)
DOI: 10.1159/000078198

ATP levels in clone mouse embryos

M. Boiani, V. Gambles, and H.R. Schöler

Germline Development Group, Center for Animal Transgenesis and Germ Cell Research, New Bolton Center, University of Pennsylvania, Kennett Square, PA (USA)

Abstract. The transfer of a somatic cell nucleus into an oocyte results in a dramatic change of the transcriptional profile. Although the underlying process of chromatin remodeling after nuclear transfer so far has not been defined at the molecular level, adenosine triphosphate (ATP) is considered a crucial component. In our study, clones were cultured in six media (M16, CZB, KSOM(aa), α-MEM, D-MEM, G1/G2) and energy levels compared to fertilized embryos. Our data indicate that nuclear transfer into an oocyte does not cause a significant change of ATP content subsequent to chromatin remodeling and pronucleus-like nucleus formation. During cleavage, while ATP levels of normal embryos were kept within narrow limits, they were less restricted in clones. ATP variability was eminent until the clones reached the blastocyst stage. This profile is consistent with the emerging representation of cleavage-stage clone embryos as mosaics, whereby intra-embryo variations compound inter-embryo variations. The issue of energy in embryos remains a complex one. The existence of multiple and integrated pathways governing the physiology, metabolism and consequently the ATP level of a cell contrasts the single gene readout situation exemplified by *Oct4*. Such principal differences may explain why Oct4 is a prognostic marker for clones whereas ATP is not.

Copyright © 2004 S. Karger AG, Basel

Nuclear functions, including DNA replication and gene transcription, critically depend on efficient transport of molecules to and from the nucleus. Mitochondrial adenosine triphosphate (ATP) production is required to support energy-consuming processes at the nuclear envelope (Dzeja et al., 2002). Similarly, the remodeling of a somatic cell nucleus into a pronucleus-like nucleus after transfer into the ooplasm is a critical event. Unlike that of a sperm nucleus, the remodeling of a somatic cell nucleus is developmentally unscheduled, and there may not be a homeostatic mechanism in the oocyte to compensate for its demands. One requisite for nuclear remodeling within an oocyte must therefore be the provision of cellular energy.

Changes of nuclear structure and activity have been associated with the relocation of mitochondria and ATP production in the cytoplasm of growing oocytes (Calarco, 1995; Fulka et al., 1998). At fertilization, cytoplasm of sea urchin oocytes contains an ATP-dependent activity that remodels sperm chromatin (Medina et al., 2001). In vitro, remodeling of mammalian somatic nuclei in *Xenopus* egg extracts depends on external ATP supply (Kikyo et al., 2000). In development, the ATP content of oocytes and embryos is critical for nucleic acid and protein synthesis (Quinn and Wales, 1971, 1973) and it has been suggested as an indicator for the developmental potential of human (Slotte et al., 1990; Van Blerkom et al., 1995), mouse (Quinn and Wales, 1971, 1973; Leese et al., 1984), and bovine (Stojkovic et al., 2001) oocytes and embryos. Thus, ATP is widely essential for biochemical reactions including processes that result in nuclear remodeling.

No indicator of energy content such as ATP has been firmly established and characterized for clone embryos. Recent studies revealed that the physiology and substrate metabolism of clones are unlike those of normal embryos (Chung et al., 2002; Gao et al., 2003). Physiology and metabolism entail ATP con-

This work was supported by the Marion Dilley and David George Jones Funds and the Commonwealth and General Assembly of Pennsylvania (M.B., V.G., H.S.) and by grant NIH 1RO1HD42011-01 to H.S.

Received 13 October 2003; manuscript accepted 10 December 2003.

Request reprints from: Dr. Hans R. Schöler, New Bolton Center
382 W. Street Road, Kennett Square, PA 19348 (USA)
telephone: +1-610-444-5800; fax: +1-610-925-8121
e-mail: scholer@vet.upenn.edu

sumption (for nucleus remodeling), production (donor nucleus-driven metabolism) and maintenance (supply of donor nucleus-encoded gene products to mitochondria). Due to either of these processes, donor nuclei reprogrammed either fully, partially, or not at all would be expected to produce different levels of ATP in blastomeres. These considerations prompted us to investigate whether mouse oocytes subjected to meiotic spindle removal ("enucleation") and nuclear transfer from ovarian cumulus cells, would indeed experience changes of ATP content and the residue is enough for development as measured by rates of blastocyst formation. Since the energy status of a cell may be altered by culture conditions, the reconstructed oocytes were compared in M16, CZB, KSOM(aa), α-MEM, D-MEM and G1/G2. This comparison will help define whether the abnormal expression of genes in clone embryos has specific causes that could be pursued at the molecular level or, it is part of a global perturbation associated with culture conditions.

Our findings indicate that the direct energy demand for cumulus cell nucleus remodeling within the ooplasm, as measured by ATP content after nuclear transfer and after pronucleus-like nucleus formation, is minor. Nevertheless, disturbances of the ATP profile become apparent during cleavage, and fade at the blastocyst stage, without a prominent effect of the culture medium. This raises the question such as whether previous differences in ATP within/between embryos were corrected and disappeared, or if clones went through a selection whereby only those with certain features (e.g. metabolism, energy status, ATP content, gene expression, pluripotency, or a combination) formed blastocysts. The ATP variation in clones may correspond to a "latent" cost for nuclear remodeling continuing in blastomeres during preimplantation. Possible factors responsible for the altered ATP profiles in clones are discussed.

Materials and methods

Recipient oocyte and donor nucleus collection
Eight- to ten-week-old C57Bl/6J X C3H/HeN female mice (henceforth referred to as B6C3F1; Taconic, Germantown, NY) were superovulated with 7.5 U of pregnant mare's serum gonadotropin (PMSG, Calbiochem) followed 48 h later by 7.5 U of human chorionic gonadotropin (hCG, Calbiochem). Cumulus-oocyte complexes were collected 14–15 h post-hCG and cumulus cells enzymatically removed. Hyaluronidase (ICN, activity >5,000 IU/mg) was used at 50 U/ml in albumin-free, HEPES-buffered CZB medium (Chatot et al., 1989) containing 5.56 mM glucose, 5 mM HEPES, 20 mM sodium bicarbonate, 0.1 % w/v polyvinylpyrrolidone (PVP, 40 kDa, Calbiochem). After incubation in a 100-μl drop of hyaluronidase solution for 15–20 min at 28 °C, the cumulus-free oocytes were removed, washed three times in HEPES-buffered CZB medium and then placed in α-MEM culture medium (see below). The cumulus cells dispersed in hyaluronidase solution were left in the drop, added with 1-to-1 volume of α-MEM and stored at 4 °C until use as nucleus donors for cloning. Such handling of cumulus cells was shown to result in full development (Boiani et al., 2002, 2003). Mice used in this study were maintained and used for experimentation according to the guidelines of the Institutional Animal Care and Use Committee (IACUC) of the University of Pennsylvania.

Micromanipulation and culture of clone mouse embryos
Micromanipulations were performed as follows. In a 6-cm glass-bottomed dish under silicon oil (Sigma) B6C3F1 recipient oocytes were enucleated in batches of 20 in HEPES-buffered CZB medium (PVP 0.1 % w/v; albumin-free) in the presence of 1 μg/ml cytochalasin B (ICN, dissolved at

1 mg/ml in dimethyl sulfoxide, DMSO) at 28 °C. After enucleation was completed, the batch of oocytes concerned was washed and left to recover for 1 to 2 h in drops of α-MEM medium at 37 °C. Nuclear transplantation from B6C3F1 cumulus cells was done in hypertonic (110 %) HEPES-buffered CZB medium (as above, except with PVP 1 % w/v). Oocytes were processed for injection in batches of 20 within a 10-min time span, using piezo-driven (PMM 150 FU; PrimeTech, Japan) borosilicate needles (Clark Instruments, UK) loaded with mercury and operated with DIC optics (Nikon, Japan) at 28 °C. After micromanipulation, oocytes were left to recover in 1:1 mixture of HEPES-buffered CZB and α-MEM medium for 1 to 2 h on the bench at 28 °C prior to incubation in α-MEM medium at 37 °C. The reconstructed oocytes were then activated for 6 h in Ca-free α-MEM supplemented with 10 mM $SrCl_2$ (ICN), 5 μg/ml cytochalasin B (final carrier DMSO 0.5 % v/v) and 0.4 % w/v bovine serum albumin (ICN). This treatment was also used to induce parthenogenetic activation of metaphase II oocytes when required. Presence of cytochalasin B was required to prevent extrusion of a pseudo polar body upon oocyte activation. Prior to transfer to development media, the 1-cell clones were extensively washed in serial drops of α-MEM medium with 3 % bovine serum albumin for 30 min at 37 °C. Control embryos were fertilized embryos obtained by mating of B6C3F1 gonadotropin-primed females in the same strain background. The 1-cell embryos were randomly assigned to six groups (culture media M16, CZB, KSOM(aa), α-MEM, D-MEM, or G1/G2). The culture media were purchased (α-MEM from Sigma; D-MEM from Invitrogen) or prepared in the laboratory according to recipe (see Nagy et al., 2003 and references therein) using milliQ water (Millipore) and chemicals from Calbiochem. CZB was increased from 0 to 5.56 mM glucose and G1 switched to G2 at the 4-cell stage. All six groups of embryos were cultured in parallel in 25-μl drops overlaid with mineral oil (ICN) extracted with water, in a 35-mm Petri dish (Corning).

Determination of total ATP content in single preimplantation embryos
Embryos were collected from the culture dish at the following stages measured from the start of the activation period: 1-cell (8 h post-activation, hpa), early 2-cell (17 hpa), late 2-cell (29 hpa), 4-cell (42 hpa), 8-cell/morula (54 hpa), compacted morula (72 hpa), and blastocyst (96 hpa), to be processed for the ATP assay.

The quantitative content of ATP was measured by the luminescence generated from the ATP-dependent luciferin-luciferase reaction (Spielmann et al., 1981):

$$ATP + d\text{-luciferin} + O_2 \rightarrow Oxyluciferin + AMP + pyrophosphate + CO_2 + Light\ (560\ nm)$$

An ATP standard solution was prepared by dissolving powdered ATP (Calbiochem, purity >95 %) in milliQ water to a final concentration of 0.1 M. The pH was adjusted to 7.4 and the actual concentration was verified with a spectrophotometer at 259 nm (ATP molar extinction coefficient = 15,400). The 0.1-M ATP stock solution was stored at –20 °C until use. To obtain a standard curve (Fig. 1), a series of known quantities (picomoles) of ATP were measured concurrently with the extracted sample ATP (see below). To account for the presence of ADP and AMP in the cell extracts, equal amount of ADP (Calbiochem, purity >99 %) and AMP (Calbiochem, purity >95 %) was added to the ATP standard. An 11-point standard curve (ATP, ADP, AMP, 0.0 through 1.0 picomoles each) was included in each assay.

Embryos were prepared as follows. They were picked from the culture drop, and washed twice in 35-mm plastic dishes filled with isotonic solution (NaCl 0.9 % w/v, PVP 0.1 % w/v; sterile-filtered) for 5 min each wash. They were individually placed in 200 μl milliQ water in PCR tubes using a flame-polished, mouth-controlled micropipette, and the tubes were maintained on ice until further processing. To allow for nucleotide release and prevent ATP degradation by enzymatic activities, the tubes were heated (Yang et al., 2002) at 95 °C for 40 min on a PCR thermocycler. The samples were stored at –20 °C until analysis within two weeks, otherwise at –80 °C. Our pre-trials indicated that the ATP content is stable at –20 °C for one month.

Each single extracted embryo (200 μl) was split into three aliquots of 65 μl each, to allow for either replicates of ATP measurement, or measurement of ADP and AMP. The aliquots were transferred to a white 96-well plate (FluoroNunc), which was placed in the chamber of a luminometer (Mediators PhL, Austria) and left for 30 min prior to processing. This allows for the autoluminescence of the white plate to dissipate. Fifty microliters of reconstituted luciferase/luciferin reagent (Promega) was automatically dispensed to each well. Following a 4-second lag time, light emission was inte-

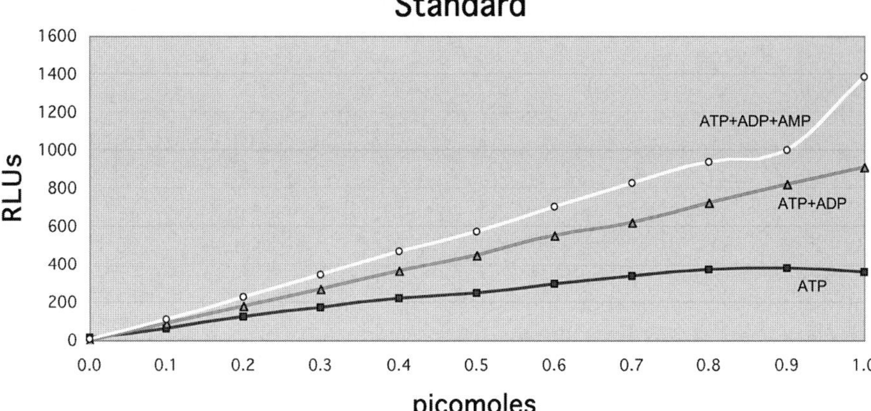

Fig. 1. Standard curves used for determination of energy content of embryos. The lower standard curve (ATP) was used as the reference; the intermediate (ATP+ADP) and upper (ATP+ADP+AMP) curves were used to validate the linearity of the method and were obtained after conversion of ADP and AMP to ATP.

grated for 15 s to give the Relative Light Units (RLUs). After conversion of the RLUs to picomoles of ATP, based on the standard curve (Fig. 1), the ATP value of the aliquot was multiplied by 3 to give that of the sample.

Differential labeling of the inner cell mass (ICM) and trophectoderm (TE) of blastocysts

The nuclei of blastocyst ICM and TE cells were differentially labeled with polynucleotide-specific fluorochromes and counted as described by Boiani et al. (2003) based on the established methods (Solter and Knowles, 1975; Handyside and Hunter, 1984).

Statistics

All experiments were repeated at least three times, except for the ATP profiles, in which clone and fertilized embryos were assayed side-by-side in two complete sets of measures. Data are presented as mean values, and variation is indicated with the standard deviation (SD). The SD of percentages was calculated according to Bailey (1959).

Results

Effect of oocyte micromanipulation, reconstruction and activation on ATP content

Nucleus remodeling is one reason for energy consumption, but there are several other possible reasons. We were interested to determine if micromanipulation per se diminishes the ATP content of mouse oocytes. For example, as established in *Xenopus* oocytes and other cells subject to mild mechanical deformation, mechanical stimuli can trigger release of ATP in the extracellular medium (Nakamura and Strittmatter, 1996; Maroto and Hamill, 2001).

Manipulated mouse oocytes were measured for ATP content against non-manipulated oocytes from the same pool. Just prior to activation, enucleated oocytes (n = 10) had an average ATP content of 0.34 ± 0.01 picomoles, cumulus cell nucleus-transplanted oocytes (n = 10) had 0.39 ± 0.01 picomoles, and metaphase II oocytes (n = 10) had 0.27 ± 0.02 picomoles. At the end of the 6-hour activation period, enucleated, nucleus-transplanted and metaphase II oocytes had average ATP contents of 0.38 ± 0.01, 0.46 ± 0.04 and 0.40 ± 0.01 picomoles, respectively. The estimated ATP content of a cumulus cell was 0.018 ± 0.007 picomoles (n = 280 cumulus cells total, in samples of 5, 10, 100 cells), contributing marginally to the increase of ATP content observed after nuclear transfer. A typical ATP

standard calibration is shown in Fig. 1. ATP was measured further in the activated oocytes during culture in α-MEM (supplemented with cytochalasin B in order to prevent fragmentation of enucleated oocytes upon release from the activation medium). At the time equivalent to the 4-cell stage, average ATP contents were 0.18 ± 0.02, 0.17 ± 0.02 and 0.12 ± 0.02 picomoles ATP in the enucleated, nucleus-transplanted and parthenogenetic 1-cell embryos, respectively (n = 10).

These results indicate that (1) oocyte micromanipulation and reconstruction per se do not cause a detectable reduction of ATP in clones within 48 h, and (2) nucleus remodeling defined as formation of a pronucleus-like nucleus bears a minor direct/immediate cost in terms of ATP, as evident from the similarity between enucleated oocytes with/without a nucleus transplant. Whether this energy state is transient or stable can be tested by follow up of the ATP content during cleavage of activated oocytes. This took place in vitro to avoid the inadequacy of the mouse oviduct to support cleavage of 1- and 2-cell clone embryos (Boiani et al., 2002; Teruhiko Wakayama, personal communication).

Influence of culture medium on the ATP profile of clone embryos

ATP levels were measured in single embryos cultured in M16, CZB, KSOM(aa), α-MEM, D-MEM, G1/G2 from the 1-cell to the blastocyst stage (Fig. 2). In each trial we had about 200 oocytes reconstructed from cumulus cell nuclei, activated simultaneously and cultured in parallel. Although blastocysts were obtained in all six media, a 2- and 4-cell stage block of development was eminent for most clones cultured in D-MEM and in G1/G2, and blastocyst rates were consistently higher in α-MEM than in the other media (Fig. 3). Clone and fertilized embryos had distinct profiles of ATP during development in M16, CZB, KSOM(aa) and α-MEM. ATP profiles in fertilized embryos were more dynamic and consistent with the shifts in ATP synthesis previously reported for mouse preimplantation development (Ginsberg and Hillman, 1973, 1975) (Fig. 4, left panel). Although the report did not analyze the 1-cell stage, and fertilized embryos were not cultured, the data indicated a progressive decrease of ATP at each successive cleavage stage, from 1.4 to 0.4 picomoles. ATP values in this study were lower,

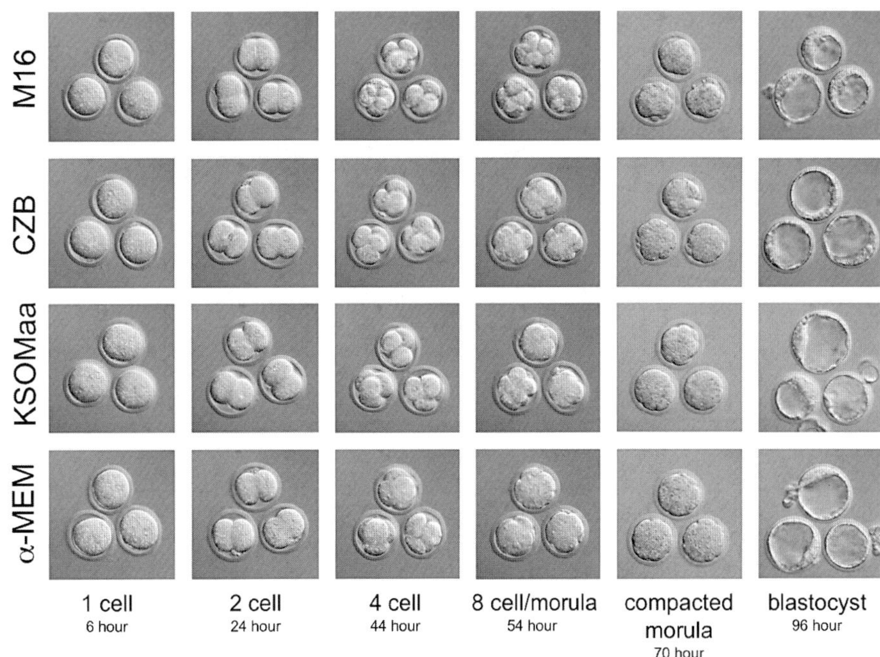

Fig. 2. Morphology of clone embryos and stages used for the ATP analysis, from the 1-cell to the blastocyst stage.

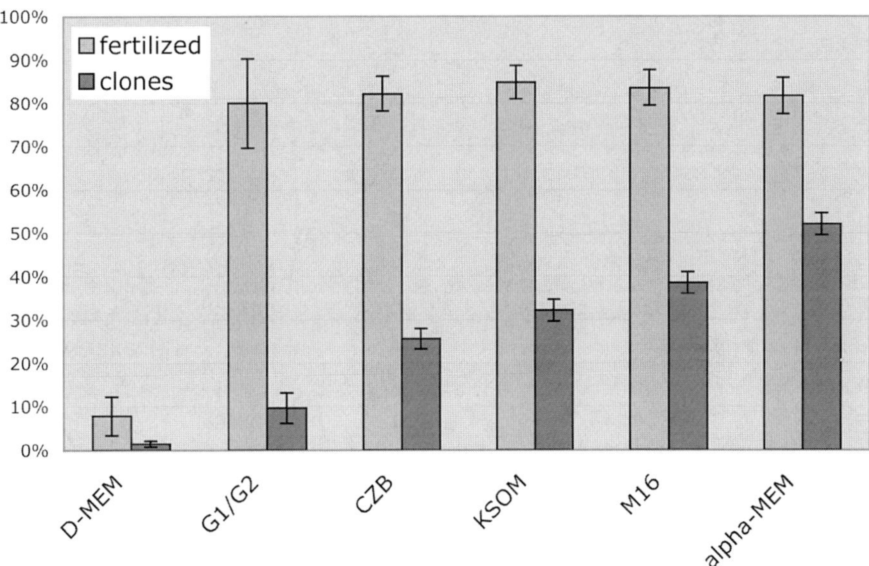

Fig. 3. Percentage of blastocysts formed in different media (clones, n = 148, 96, 117, 197, 1 and 7 blastocysts in M16, CZB, KSOM(aa), α-MEM, D-MEM, G1/G2, respectively; fertilized, n = 66, 74, 73, 67, 3 and 12 blastocysts in M16, CZB, KSOM(aa), α-MEM, D-MEM, G1/G2, respectively). Error bars indicate standard deviations.

from 0.4 to 0.1 picomoles, which, in part may be due to the different procedure for nucleotide extraction (physical versus chemical), and in line with those of other reports based on later technologies (Spielmann and Erickson, 1983). As opposed to fertilized embryos, the ATP content of clone embryos did not present a reproducible profile. Discrepancies were observed between media in the same experiment and also between replicates of the same culture medium. The ATP curves of clones cultured in CZB were consistently below those of other media (Fig. 4, right panel).

In 2-cell embryos of the NMRI strain, a decrease in ATP is characteristic for this stage in vivo but not found in embryos that do not go further in vitro (Spielmann et al., 1984). It is possible that the ATP drop at the regular 2-cell stage embryos is due to stress related to the culture condition. Clones did not have a decrease of ATP at the 2-cell stage as observed in fertilized embryos. We cannot exclude that the "missing" ATP decrease in the clones was due to the culture conditions and the particular mouse strain used, or that the decrease was shifted and thus between two time points. Regardless of the culture medium used, the inconsistencies observed in the ATP content during cleavage dissipated as the clones reached the blastocyst stage.

These results indicate that (1) α-MEM is a consistently successful medium for clone mouse embryo culture; (2) clones do not respond to culture conditions the same way as fertilized

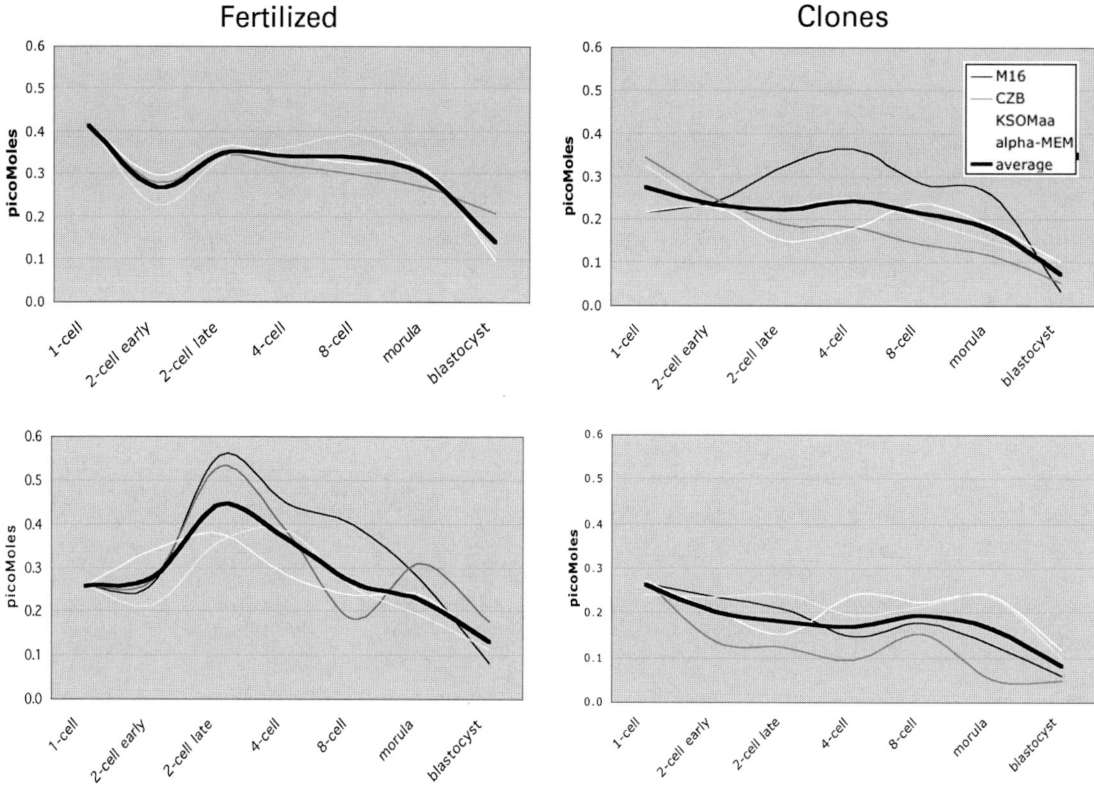

Fig. 4. ATP content in clone and fertilized embryos cultured in M16, CZB, KSOM(aa), α-MEM media. For each panel, three embryos were measured per stage and medium.

embryos, resulting in inconsistent ATP levels, and (3) pre-blastocyst stage clones are more heterogeneous with respect to the ATP levels than blastocysts. This suggests that only a subset of embryos with certain features may have formed blastocysts.

Relationship between cell number and ATP content of clone blastocysts

At the blastocyst stage the ATP content of clones is similar when different media are used. To resolve whether this reflects a more general similarity between blastocysts or, rather, the variability of ATP content during cleavage is allowing only embryos with certain characteristics to form blastocysts, clone blastocysts were analyzed for cell numbers and lineage allocation. Morphologically, clone blastocysts formed in M16, CZB, KSOM(aa) and α-MEM were similar (Fig. 2). However, total cell numbers (M16 = 45 ± 14, CZB = 36 ± 15, KSOM(aa) = 44 ± 19, α-MEM = 45 ± 18) and ICM ratio (M16 = 0.33, CZB = 0.33, KSOM(aa) = 0.23, α-MEM = 0.37) were lower compared to controls (87 ± 17, 64 ± 19, 67 ± 19, 77 ± 22 total cells; ICM ratio 0.32, 0.48. 0.48, 0.57 in M16, CZB, KSOM(aa) and α-MEM, respectively) (n>10 blastocysts in each culture group). Thus, most clone blastocysts had an ICM as defined by overall morphology, but comprised of only about 15 cells versus 30–40 in the control embryos. Future gene expression analyses will need to show if the forming ICMs transcribe genes required to establish pluripotency.

Since clone blastocysts present a broad distribution of total cell numbers around the mean (Boiani et al., 2003), and yet the ATP values of clones cultured in different media converge at the blastocyst stage, the cell number is unlikely to be a causal factor in the distortion of the ATP profile.

Discussion

Prior to focusing on specific aspects of nuclear function involving ATP in clone embryos, we ruled out that mechanical intervention on the oocyte during micromanipulations could cause a "mechanosensitive" ATP release (see hypothesis 1, below). Excluding this possibility was important, because it would have invalidated the study of nucleus remodeling and reprogramming under an energy perspective, and clone viability might have been compromised even before nucleus remodeling could begin. The question was then addressed if ATP could be used or produced differently in clone compared to fertilized embryos.

Several distinct aspects were noticed under our experimental conditions. Preimplantation failure of clone mouse embryos did not correlate with distinct deviations of the ATP profile associated with culture conditions. In this respect it was important to compare as many as six culture media. Because chromatin-remodeling activities are ATP-dependent (Varga-Weisz,

2001) and changes of chromatin configuration anticipate those in gene expression (Barton and Emerson, 1994), we surmised that differences in ATP content could reflect differences in remodeling efficiency, which in return result in differences in gene expression. This led us to consider hypothesis 2 (see below) whereby ATP could be directly consumed by chromatin remodeling activities. We also considered other possible causes as to why the ATP content of the nucleus-transplanted oocyte and embryo was different from that of fertilized counterparts. ATP could fail to be regenerated at the rate needed due to insufficient export of transcripts from the transplanted nucleus (hypothesis 3, see below) or, the transplanted nucleus may have a "memory" of its donor cell and confer the donor cell character to the embryo (including ATP metabolism; hypothesis 4, see below).

Hereafter we outline these four hypotheses, discuss them in the context of our experimental results, and finally draw our conclusions.

Hypothesis 1: Reduction of ATP in clones by a "mechanosensitive" release process

It has been shown that even mild mechanical deformation of *Xenopus* oocytes and neurons evokes ATP release so-called "mechanosensitive" (Nakamura and Strittmatter, 1996; Maroto and Hamill, 2001). For instance, a *Xenopus* oocyte releases ATP at a basal rate of 0.01 fmol/s, increasing to 50 fmol/s upon gentle stimulation. If the same rate of ATP release applied to mouse oocytes undergoing piezo-assisted micromanipulation, the ATP content of one oocyte (known to be in the picomole range) would be depleted in about 20 s; this being approximately the combined time of enucleation and nuclear transfer for one mouse oocyte. Since the developmental ability of clone embryos is so low, whereas sham-manipulated oocytes are not impaired (Heindryckx et al., 2001), we addressed this question by measuring the ATP content of manipulated and non-manipulated mouse oocytes. Our data indicate that, under the experimental conditions, the average ATP content of an enucleated mouse oocyte is not lower than that of an intact oocyte. Moreover, the ATP content of a nucleus-transplanted oocyte is not lower than that of an intact or enucleated mouse oocyte. Although the mechanosensitive release of ATP is an interesting event, we consider its contribution to cloning-associated defects to be marginal. Since the ATP content does not seem to be altered by micromanipulation, the long-term effects of the process remain to be determined.

Hypothesis 2: Consumption of the ATP content in clones by chromatin remodeling activities

The oocyte possesses a limited capacity to decondense and remodel sperm into functional pronuclei (Perreault and Zirkin, 1982). In mouse oocytes fertilized with up to four sperm the developmentally scheduled swelling, protamine/histone exchange, and demethylation of all male pronuclei occurs normally; however, these processes occur only partially as more sperm are added (Santos et al., 2002). These facts raise the question if the oocyte capacity to remodel transplanted somatic nuclei is limited and due to other causes. Given its distinct protein-DNA packaging, the somatic nucleus contains a double or quadruple DNA content compared to sperm, and in addition, the somatic nucleus unlike sperm is an experimental artifact that does not have its correspondence in natural development. To address if the nucleus remodeling capacity of the ooplasm is limited under our experimental conditions, we used the content in ATP as an indicator. Specifically, we looked for differences in ATP profile during development of cumulus cell nucleus-transplanted oocytes compared to oocytes that were only "enucleated" (no nuclear transfer) and oocytes that were not prepared at all (no enucleation, no nuclear transfer). Our data indicate that, under the experimental conditions, enucleated and nucleus-transplanted activated oocytes have similar ATP contents up to the chronological 4-cell stage, and both are higher than that of parthenogenetic embryos. Thus, the presence of a cumulus cell nucleus in the ooplasm does not appear to deplete ATP from the ooplasm before development commences; rather, removing chromosomes from the metaphase II oocyte may prevent an inhibitory effect of the maternal genome on ATP metabolism.

During cleavage, the ATP profile of clone embryos displays more variation than that of fertilized embryos. The inferior ATP content of clones cultured in CZB cannot be attributed to a higher utilization towards more advanced reprogramming, since the match for low ATP levels and ICM-restricted *Oct4* gene expression is actually poor (manuscript in preparation). We speculate that while ATP enables chromatin-remodeling activities, subsequent ATP homeostasis depends on how such remodeling was effected, in terms of gene expression resulting from it and necessary to maintain nucleus-cytoplasmic interactions. The functional arrangement of mitochondria in the cytoplasm, and the supply of nucleus-encoded polypeptides to mitochondria, need to be set properly after nuclear transplantation (see hypothesis 3). These considerations lead to the possible conclusion that ATP reflects, rather than govern, the extent of reprogramming observed in cumulus cell clones.

Hypothesis 3: Failure to (re)generate ATP at sufficient levels

As the transplanted nucleus is remodeled in the ooplasm in an energy-dependent process, transcription in the nucleus is required to supply mitochondria with RNAs and proteins. Mitochondria are not self-sufficient organelles, but the electron transport and oxidative phosphorylation system require contributions from both the nuclear and the mitochondrial genetic systems (Chinnery, 2003). It has been shown that targeted genetic disruption of a single, nucleus-encoded mitochondrial transcription factor, NRF-1, results in a peri-implantation lethal phenotype in mice (Huo and Scarpulla, 2001). This demonstrates that nucleus activity is required already at preimplantation stages, although the bulk of mitochondrial DNA replication and transcription occurs around gastrulation. Arguably, defects in this nucleus-cytoplasm dialogue may set the stage for the irregular ATP profiles observed in clones.

In this study, disturbances in the ATP profile were not observed in freshly reconstituted cumulus cell-derived clone embryos, but instead occurred during cleavage. Considering that a cumulus cell contains very few mitochondria compared to a zygote (10^3–10^4 are probably based on mitochondrial DNA copy number; Steuerwald et al., 2000; Huo and Scarpulla,

2001), the cumulus cell nucleus is likely to express lower levels of nuclear genes encoding mitochondrial proteins, which would predict lower ATP amounts to be measured in clones. In fact, ATP levels in clones were not diminished, rather they were not controlled as per normal development, suggestive of nucleus-cytoplasm misinteraction.

Besides the level of activity, another important issue is the transport of information from the nucleus to the mitochondria and vice versa. The approximately 300-fold increase in cell volume associated with somatic cell (10 μm) nuclear transfer into an oocyte (70 μm) argues that even if remodeling/reprogramming was complete, a somatic nucleus is still inadequate to govern metabolic reactions on such a greater scale than the original donor cell. The new-established trafficking between the somatic nucleus and the ooplasm may be too inefficient to take over ATP production to maintain a vital level. Conversely, the unstable ATP level may impact the intracellular signaling network based on the second messenger c-AMP, whose participation in genome reprogramming has been surmised (Burnside and Collas, 2002). It cannot be excluded that clones form such different blastocyst rates in CZB and α-MEM, because one medium allows for a more effective cytoplasmic organization than the other (Bavister and Squirrell, 2000). The cytoplasmic distribution of mitochondria in relation to the level of glucose and phosphate in the medium, and its impact on hamster embryo viability, have been described (Barnett et al., 1997; Ludwig et al., 2001). In fertilized cat embryos, oscillations of ATP level similar to those described in this study were attributed to changes of mitochondrial structure and activity, different metabolic pathways, and to the energy source available/preferred during the different stages of cleavage (Freistedt et al., 2001).

Hypothesis 4: Regulation of ATP content in clones on the basis of the somatic character of the nucleus ("memory effect")

It has been proposed that a transplanted donor nucleus alters the physiology or metabolism of mouse embryos during preimplantation development, accounting in part for culture medium preferences and, therefore, the abnormal activation of embryonic genes or inactivation of somatic genes. This is consistent with the poor developmental rates observed in vivo, such that 1- and 2-cell stage clone embryos develop significantly better in vitro than in a mouse oviduct (Boiani et al., 2002), indicative of different nutritional needs. Examples of abnormal gene expression include *Oct4* in cumulus cell clones cultured in M16 (Boiani et al., 2002) and the glucose transporter *glut4* in myoblast clones cultured in HAMF10/DMEM (Gao et al., 2003), respectively.

The choice of the culture medium may enhance or retard the expression of somatic cell features by the clone. Providing various concentrations of glucose (0–5.56 mM), a typical substrate for glycolysis in somatic cells but not in early embryos, did not result in altered levels of ATP in cumulus cell clones. This contrasts with the expectation that a high rate of glycolysis would cause inhibition of mitochondrial respiration, known as the "Crabtree effect" (Koobs, 1972), in the embryo. A high rate of glucose internalization and utilization in mouse clones was surmised by Chung and colleagues (Chung et al., 2002) following observations on cumulus cell clones; and by Gao and colleagues (Gao et al., 2003) following observations on myoblast clones. One important technical aspect to consider is that cytochalasin B, used to prevent pseudo polar body extrusion in the nucleus-transplanted oocytes (regardless of the nucleus donor), is known to interfere with glucose transport (Dan-Goor et al., 1997). Secondly, the nuclear transfer technique used with the myoblasts (electrofusion) is such that the entire donor cell, not only the nucleus, became part of the reconstructed oocyte. This may account for the glucose transporter 4 (GLUT4) making its prompt appearance on the membrane of the 1- and 2-cell myoblast-derived clone embryos. It would be interesting to test if the substrate preference for glucose of myoblast-derived clones is also observed in clones obtained by nucleus injection rather than electrofusion. Although the uptake of glucose by clones was not measured in this study, the clone ATP profile may not be simply related to a role of this substrate in energy metabolism or to the presence of more or less glucose in the culture medium. It is safe to assume that even if clones would absorb more glucose than the fertilized counterparts, this may not necessarily translate in an energy response. In fact, glucose can be metabolized through the pentose phosphate pathway that produces ribose 5-phosphate, a precursor for de novo purine synthesis used as building block for DNA. Interestingly, the pentose phosphate pathway also leads to production of GTP and ATP, which is involved in the production of the second messenger cyclic adenosine monophosphate (c-AMP). Thus, signaling between blastomeres might change in response to glucose, while the energy content does not. The possibility of blastomere-blastomere coupling leading to epigenetic complementation has been put forward (Boiani et al., 2003).

Conclusion

Previously, the expression profile of *Oct4* during development had been discussed as a prognostic marker in mouse cloning (Boiani et al., 2002, 2003). However, as it emerges from the present study, such a correlation between gene expression and development cannot simply be extended to ATP, a key substrate for biochemical reactions including those for nuclear remodeling. Although cumulus cell-derived clone mouse embryos present more variability than fertilized counterparts, they do not exhibit dramatic change of ATP content and this cannot be accounted for the demise of clones. The existence of multiple pathways governing the physiology, metabolism and ultimately ATP level of a cell, as opposed to the binary on-off situation for a single gene, e.g. *Oct4,* may contribute to the discrepancy between ATP and Oct4 in their prognostic value. Also contributing is the great enrichment in oocytes of ATP and mitochondria relative to any other cell – a key feature for oocytes ability to support meiotic maturation and early development. Such a large endowment would buffer energy variations that may actually have occurred during the chromatin remodeling process in the ooplasm or be continuing in blastomeres. The natural variation in the number of mitochondria between oocytes would downgrade the prognostic value of ATP for clones, although fertilized embryos subject to the same natural variation did not present comparable fluctuations in ATP.

Thus, the inconsistent ATP levels appear to be specific to the cloning procedure.

It is interesting to note that clones cultured in α-MEM outperformed those cultured in all other media in terms of blastocyst rates. This medium contains inositol and it has been shown that the chromatin remodeling machinery can be regulated by the inositol signaling pathway (Shen et al., 2003; Steger et al., 2003). Together with the fact that the activity of phosphatidyl inositol 3 (PI3)-kinase is necessary for proliferation and differentiation potentials of mouse ES cells (Shen et al., 2003; Jirmanova et al., 2002), which have the ability to reprogram somatic nuclei upon cell fusion reminiscent of nuclear reprogramming by oocyte (Tada et al., 2001), there might be a possibility that this novel pathway of inositol is involved in the remodeling process after nuclear transfer into oocytes. Since the inositol signaling pathway is linked to the cell membrane, there might be also a possibility that manipulation of the oocyte membrane during spindle removal and nuclear transfer leaves marks, which would be passed on to the blastomere membrane upon cleavage, and set the stage for functional differences among blastomeres. Mosaicism of gene expression has been reported in clone embryos (porcine, Park et al., 2002; mouse, Boiani et al., 2002). It remains to be seen if mosaicism in clone embryos applies to the distribution and function of mitochondria, thereby accounting for the variable and inconsistent ATP profiles observed.

Acknowledgements

The authors would like to thank K. John McLaughlin and Nina Hillman for much helpful comments on the manuscript, as well as Sigrid Eckardt and Fatima Cavaleri for critical reading the manuscript. We also thank Areti Malapetsa for editing the manuscript in its final version.

References

Bailey NTJ: Statistical Methods in Biology, p 200 (UNIBOOKS, The English Universities Press, London, 1959).

Barnett DK, Clayton MK, Kimura J, Bavister BD: Glucose and phosphate toxicity in hamster preimplantation embryos involves disruption of cellular organization, including distribution of active mitochondria. Mol Reprod Dev 48:227–237 (1997).

Barton MC, Emerson BM: Regulated expression of the beta-globin gene locus in synthetic nuclei. Genes Dev 8:2453–2465 (1994).

Bavister BD, Squirrell JM: Mitochondrial distribution and function in oocytes and early embryos. Hum Reprod 15:189–198 (2000).

Boiani M, Eckardt S, Schöler HR, McLaughlin KJ: Oct4 distribution and level in mouse clones: consequences for pluripotency. Genes Dev 16:1209–1219 (2002).

Boiani M, Eckardt S, Leu AN, Schöler HR, McLaughlin KJ: Pluripotency deficit in mouse clones overcome by clone-clone aggregation: epigenetic complementation? EMBO J 22:5304–5312 (2003).

Burnside AS, Collas P: Induction of Oct-3/4 expression in somatic cells by gap junction-mediated cAMP signaling from blastomeres. Eur J Cell Biol 81:585–591 (2002).

Calarco PG: Polarization of mitochondria in the unfertilized mouse oocyte. Dev Genet 16:36–43 (1995).

Chatot CL, Ziomek CA, Bavister BD, Lewis JL, Torres I: An improved culture medium supports development of random-bred 1-cell mouse embryos in vitro. J Reprod Fertil 86:679–688 (1989).

Chinnery PF: Searching for nuclear-mitochondrial genes. Trends Genetics 19:60–62 (2003).

Chung YG, Mann MR, Bartolomei MS, Latham KE: Nuclear-cytoplasmic "tug of war" during cloning: effects of somatic cell nuclei on culture medium preferences of preimplantation cloned mouse embryos. Biol Reprod 66:1178–1184 (2002).

Dan-Goor M, Sasson S, Davarashvili A, Almagor M: Expression of glucose transporter and glucose uptake in human oocytes and preimplantation embryos. Human Reprod 12:2508–2510 (1997).

Dzeja PP, Bortolon R, Perez-Terzic C, Holmuhamedov EL, Terzic A: Energetic communication between mitochondria and nucleus directed by catalyzed phosphotransfer. Proc Natl Acad Sci USA 99:10156–10161 (2002).

Freistedt P, Stojkovic P, Wolf E, Stojkovic M: Energy status of nonmatured and in vitro-matured domestic cat oocytes and of different stages of in vitro-produced embryos: enzymatic removal of the zona pellucida increases adenosine triphosphate content and total cell number of blastocysts. Biol Reprod 65:793–798 (2001).

Fulka J Jr, Karnikova L, Moor RM: Oocyte polarity: ICSI, cloning and related techniques. Hum Reprod 13:3303–3305 (1998).

Gao S, Chung YG, Williams JW, Riley J, Moley K, Latham KE: Somatic cell-like features of cloned mouse embryos prepared with cultured myoblast nuclei. Biol Reprod 69:48–56 (2003).

Ginsberg L, Hillman N: ATP metabolism in cleavage-staged mouse embryos. J Embryol Exp Morphol 30:267–282 (1973).

Ginsberg L, Hillman N: Shifts in ATP synthesis during preimplantation stages of mouse embryos. J Reprod Fertil 43:83–90 (1975).

Handyside AH, Hunter S: A rapid procedure for visualising the inner cell mass and trophectoderm nuclei of mouse blastocysts in situ using polynucleotide-specific fluorochromes. J Exp Zool 231:429–434 (1984).

Heindryckx B, Rybouchkin A, Van Der Elst J, Dhont M: Effect of culture media on in vitro development of cloned mouse embryos. Cloning 3:41–50 (2001).

Huo L, Scarpulla RC: Mitochondrial DNA instability and peri-implantation lethality associated with targeted disruption of nuclear respiratory factor 1 in mice. Mol Cell Biol 21:544–654 (2001).

Jirmanova L, Afanassieff M, Gobert-Gosse S, Markossian S, Savatier P: Differential contributions of ERK and PI3-kinase to the regulation of cyclin D1 expression and to control of the G1/S transition in mouse embryonic stem cells. Oncogene 15:5515–5528 (2002).

Kikyo N, Wade PA, Guschin D, Ge H, Wolffe AP: Active remodeling of somatic nuclei in egg cytoplasm by the nucleosomal ATPase ISWI. Science 289:2360–2362 (2000).

Koobs DH: Phosphate mediation of the Crabtree and Pasteur effects. Science 178:127–133 (1972).

Leese HJ, Biggers JD, Mroz EA, Lechene C: Nucleotides in a single mammalian ovum or preimplantation embryo. Anal Biochem 140:443–448 (1984).

Ludwig TE, Squirrell JM, Palmenberg AC, Bavister BD: Relationship between development, metabolism, and mitochondrial organization in 2-cell hamster embryos in the presence of low levels of phosphate. Biol Reprod 65:1648–1654 (2001).

Maroto R, Hamill OP: Brefeldin A block of integrin-dependent mechanosensitive ATP release from Xenopus oocytes reveals a novel mechanism of mechanotransduction. J Biol Chem 276:23867–23872 (2001).

Medina R, Gutierrez J, Puchi M, Imschenetzky M, Montecino M: Cytoplasm of sea urchin unfertilized eggs contains a nucleosome remodeling activity. J Cell Biochem 83:554–562 (2001).

Nagy A, Gertenstein M, Vinterstein K, Behringer R: Manipulating the Mouse Embryo, p 164 (Cold Spring Harbor Laboratory Press, Cold Spring Harbor 2003).

Nakamura F, Strittmatter SM: P2Y1 purinergic receptors in sensory neurons: contribution to touch-induced impulse generation. Proc Natl Acad Sci USA 93:10465–10470 (1996).

Park KW, Lai L, Cheong HT, Cabot R, Sun QY, Wu G, Rucker EB, Durtshi D, Bonk A, Samuel M, Rieke A, Day BN, Murphy CN, Carter DB, Prather RS: Mosaic gene expression in nuclear transfer-derived embryos and the production of cloned transgenic pigs from ear-derived fibroblasts. Biol Reprod 66:1001–1005 (2002).

Perreault SD, Zirkin BR: Sperm nuclear decondensation in mammals: role of sperm-associated proteinase in vivo. J Exp Zool 224:253–257 (1982).

Quinn P, Wales RG: Adenosine triphosphate content of preimplantation mouse embryos. J Reprod Fertil 25:133–135 (1971).

Quinn P, Wales RG: The relationships between the ATP content of preimplantation mouse embryos and their development in vitro during culture. J Reprod Fertil 35:301–309 (1973).

Santos F, Hendrich B, Reik W, Dean W: Dynamic reprogramming of DNA methylation in the early mouse embryo. Dev Biol 241:172–182 (2002).

Shen X, Xiao H Ranallo R, Wu WH, Wu C: Modulation of ATP-dependent chromatin-remodeling complexes by inositol polyphosphates. Science 299:112–114 (2003).

Slotte H, Gustafson O, Nylund L, Pousette A: ATP and ADP in human pre-embryos. Hum Reprod 5:319–322 (1990).

Solter D, Knowles BB: Immunosurgery of mouse blastocyst. Proc Natl Acad Sci USA 72:5099–5102 (1975).

Spielmann H, Erickson RP: Normal adenylate ribonucleotide content in mouse embryos homozygous for the t12 mutation. J Embryol Exp Morphol 78:43–51 (1983).

Spielmann H, Jacob-Muller U, Schulz P: Simple assay of 0.1–1.0 pmol of ATP, ADP, and AMP in single somatic cells using purified luciferin luciferase. Anal Biochem 113:172–178 (1981).

Spielmann H, Jacob-Muller U, Schulz P, Schimmel A: Changes of the adenine ribonucleotide content during preimplantation development of mouse embryos in vivo and in vitro. J Reprod Fertil 71:467–473 (1984).

Steger DJ, Haswell ES, Miller AL, Wente SR, O'Shea EK: Regulation of chromatin remodeling by inositol polyphosphates. Science 299:114–116 (2003).

Steuerwald N, Barritt JA, Adler R, Malter H, Schimmel T, Cohen J, Brenner CA: Quantification of mtDNA in single oocytes, polar bodies and subcellular components by real-time rapid cycle fluorescence monitored PCR. Zygote 8:209–215 (2000).

Stojkovic M, Machado SA, Stojkovic P, Zakhartchenko V, Hutzler P, Goncalves PB, Wolf E: Mitochondrial distribution and adenosine triphosphate content of bovine oocytes before and after in vitro maturation: correlation with morphological criteria and developmental capacity after in vitro fertilization and culture. Biol Reprod 64:904–909 (2001).

Tada M, Takahama Y, Abe K, Nakatsuji N, Tada T: Nuclear reprogramming of somatic cells by in vitro hybridization with ES cells. Curr Biol 11:1553–1558 (2001).

Van Blerkom J, Davis PW, Lee J: ATP content of human oocytes and developmental potential and outcome after in-vitro fertilization and embryo transfer. Hum Reprod 10:415–424 (1995).

Varga-Weisz P: ATP-dependent chromatin remodeling factors: nucleosome shufflers with many missions. Oncogene 20:3076–3085 (2001).

Yang NC, Ho WM, Chen YH, Hu ML: A convenient one-step extraction of cellular ATP using boiling water for the luciferin-luciferase assay of ATP. Anal Biochem 306:323–327 (2002).

Cytogenet Genome Res 105:279–284 (2004)
DOI: 10.1159/000078199

Maternal and environmental factors in early cloned embryo development

S. Gao and K.E. Latham

The Fels Institute for Cancer Research and Molecular Biology and Department of Biochemistry, Temple University
School of Medicine, Philadelphia, PA (USA)

Abstract. Cloning by somatic cell nuclear transfer (SCNT) in mammals has revealed the remarkable ability of an oocyte to reprogram somatic cell nuclei and induce them to recapitulate the developmental program. Despite the success, cloning remains very inefficient. This review summarizes recent observations from cloning in mice that reveal some of the likely causes for the present inefficiency. One cause appears to be the slow pace of reprogramming combined with the early onset of genome transcription, which together cause cloned embryos to elaborate many somatic cell characteristics even before the first cleavage division. The altered phenotypes of cloned embryos render standard embryo culture conditions grossly sub-optimum. Another cause appears to be a hitherto unappreciated contribution of spindle-associated factors to early embryo development. As current procedures remove the spindle and associated factors, cloned embryos lack these factors. These observations are providing new insight into basic mammalian embryology. They also reveal possible changes to protocols that could improve the overall success of cloning.

Copyright © 2004 S. Karger AG, Basel

A prevailing view of cloning by somatic cell nuclear transfer (SCNT) is that reprogramming happens during the first few hours after introduction of the donor cell nucleus into the oocyte. Arrested metaphase II (MII) stage oocyte recipients clearly support the development of nuclear transfer embryos much more effectively than zygote stage recipients (Latham et al., 1994; Wakayama et al., 1998). Moreover, successful cloning is facilitated by allowing the nucleus to reside in the MII ooplasm for a minimal period (e.g., 1 h) before oocyte activation (Wakayama et al., 1998). Potentially critical processes occur in the MII stage ooplasm that do not occur in zygote ooplasm, including changes in chromatin structure related to chromosome condensation (Szollosi et al., 1988; Bordignon et al., 1999, 2001; Gao et al., 2003b). Additionally, the zygote cytoplasm may exert negative effects on introduced nuclei (Latham et al., 1994). Although such global changes in chromatin structure occur, and although the MII ooplasm clearly possesses the ability to initiate the cloning process, the overall timing and pace of actual nuclear reprogramming (i.e., silencing of the donor gene expression pattern coupled to the initiation of the embryonic gene expression pattern) remain poorly understood.

Our recent studies of cloning in mice have provided data to suggest that reprogramming is either very limited during the preimplantation period, or occurs progressively during this time and continues into the early post-implantation period. Specifically, we have found that even before the first cleavage division cloned embryos express many somatic cell characteristics and corresponding alterations in culture requirements and molecular properties (Chung et al., 2002; Gao et al., 2003a). Other laboratories have also observed altered culture requirements in cloned embryos (Heindryckx et al., 2001; Inoue et al., 2003). We have also shown repeatedly that these altered characteristics in cloned embryos are largely abrogated if the oocyte spindle-chromosome complex is not removed. Taken together, these observations indicate two fundamental limitations in the

Supported in part by grants from the NIH/NICHD (HD38381 and HD43092).

Received 12 August 2003; revision accepted 7 October 2003.

Request reprints from Keith E. Latham
　　The Fels Institute for Cancer Research and
　　Molecular Biology and Department of Biochemistry
　　Temple University School of Medicine
　　3307 North Broad Street, Philadelphia, PA 19140 (USA)
　　telephone: 215-707-7577; fax: 215-707-1454
　　e-mail: klatham@temple.edu

most widely applied cloning protocols that likely underlie much of the observed developmental restriction. These are, firstly, a failure to provide an appropriate culture environment for the cloned embryos, and, secondly, the depletion of critical spindle-associated factors from the cloned constructs. These observations are summarized below.

Alterations in cloned embryo culture requirements

The protocols initially employed for cloning in mice incorporated the use of dimethylsulfoxide (DMSO) as a solvent for cytochalasin B, a microfilament inhibitor that is added during the 5- to 6-hour oocyte activation step to prevent polar body extrusion and loss of chromosomes from the diploid donor genome (Wakayama et al., 2001). More recent studies have revealed that the use of DMSO as a solvent in this manner can affect cloned embryo phenotype, and thus obscure important characteristics of the clones (Chung et al., 2002), or in some cases conceal disadvantages of some protocols and advantages of others (Gao et al., 2003a). It thus has become apparent that the inclusion of DMSO in the procedure can complicate the experimental outcome, and interpretations of the results. Our laboratory has therefore sought to modify the cloning protocol in order to eliminate DMSO from the procedure. Chief among the modifications that we have made has been the development of alternative culture conditions that support enhanced preimplantation and term development in clones as compared to standard embryo culture media. These accomplishments are relevant beyond the immediate concern of improving cloning efficiency. Indeed, these results reveal that the cloned embryos have dramatically altered phenotypes as compared to normal fertilized embryos.

One of the earliest differences in culture requirement manifested by cloned embryos is an early preference for glucose-containing media. Cloned embryos progress from the 1-cell to the 2-cell stage at a nearly 2-fold greater rate in CZB medium supplemented with 5.5 mM glucose as compared with unsupplemented CZB medium (Chung et al., 2002). This preference for glucose marks cloned embryos as being substantially different from normal mouse embryos, as the latter do not require glucose during the preimplantation period, and under some circumstances glucose can be detrimental to normal embryos (Biggers and McGinnis, 2001). More importantly, the elaboration of this altered phenotype even before the first cleavage division indicates that the donor cell genome likely begins to modify embryo metabolism and physiology as soon as gene transcription begins during the second half of the 1-cell stage (Latham et al., 1992; Bouniol et al., 1995; Christians et al., 1995; Aoki et al., 1997; Latham, 1999; Latham and Schultz, 2001).

A second striking difference in cloned embryos is their preference for media with higher osmolarities and more somatic cell-like formulations. For example, better development to the 8-cell stage was achieved in Whitten's medium than in CZB, CZB + glucose, or KSOM (Chung et al., 2002; Gao et al., 2003a). This indicates that those aspects of normal embryo physiology and homeostasis for which media like KSOM are optimized no longer predominate in the cloned embryos.

A third striking difference observed for cloned embryo culture is that cloned embryos have altered responses to amino acids in the culture media. Specifically, cloned embryos display a high rate of developmental arrest in KSOM supplemented with amino acids, as compared to KSOM without amino acids (Heindryckx et al., 2001; Chung et al., 2002; Gao et al., 2003a). This negative effect of amino acid supplementation is not typically seen with fertilized embryos. Recent studies have revealed that amino acid supplementation, particularly with glutamine, can lead to increased ammonium concentrations in embryo culture medium, and that this can adversely affect long-term development (Lane and Gardner, 2003). The response of cloned embryos to amino acid supplementation could reflect either an increased production of ammonium or other disruptions in intracellular pH or osmotic regulation.

For cumulus cell donor nuclei, we have had the greatest success in preimplantation development using the somatic cell culture medium MEMα. Using this medium, we have increased the number of blastocysts and the total number of cells within blastocysts obtained in comparison to using standard embryo culture media such as Whitten's medium or KSOM. The MEMα culture medium also supports development to term at a rate slightly above 2.0 % (Fig. 1). Fertilized embryos develop in this medium to the blastocyst stage, but the overall rate of development is reduced. Thus, cumulus-cell-cloned embryos clearly thrive in a more somatic cell-like environment, whereas fertilized embryos, as expected, perform better in media optimized for normal embryos.

Cloned embryos produced with myoblast nuclei display even more profound alterations in culture requirements. These embryos failed to develop efficiently beyond the 2-cell stage in any standard embryo culture medium employed (Whitten's, CZB, KSOM), despite very efficient blastocyst formation among fertilized and parthenogenetic control embryos. Surprisingly, the same somatic cell culture medium preferred by the myoblasts themselves (a 1:1 mixture of Ham's F10 and DMEM) supported a 40 % rate of blastocyst formation (Gao et al., 2003a). Neither Ham's F10 nor DMEM alone produced as good a result, indicating that the cloned embryos actually prefer the mixture, as do their nuclear donor cells. The preference for this medium persisted beyond the morula stage, as replacement of the medium with embryo culture media was detrimental. These results provide a striking illustration of the persistence of somatic cell characteristics in cloned embryos.

Elaboration of somatic cell characteristics

Aside from altered culture medium requirements, cloned embryos display alterations at the functional and molecular levels indicative of incomplete reprogramming and unique phenotypes.

Glucose uptake and glucose transporter expression
The preference of cumulus-cell-cloned embryos for glucose-containing media indicated that clones may be able to take up and utilize glucose to a greater degree than normal embryos. We observed that glucose uptake is increased in clones made

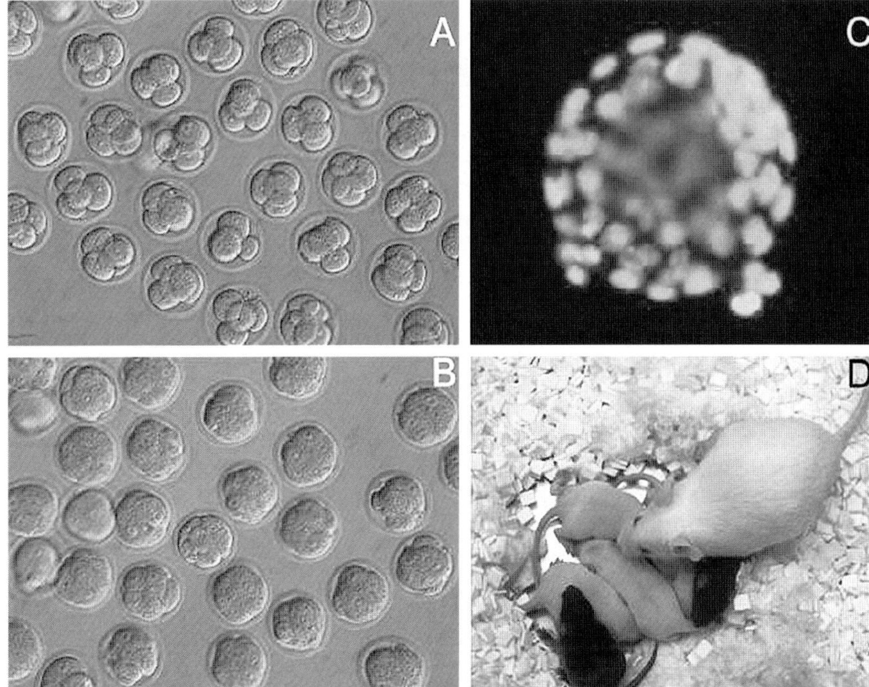

Fig. 1. Cloning in mice. Panels **A–C** show cloned mouse embryos produced using cumulus cell nuclei, and cultured in MEMα medium to the 4-cell, morula, and blastocyst stages. (**A**) 4-cell stage clones, (**B**) compacted 8-cell/morula stage clones, (**C**) blastocyst stage clone stained with DAPI to illustrate cell number. Panel **D** shows two cloned offspring (black) together with their albino nursing foster mother and her albino pups.

with either cumulus cell or myoblast nuclei (Gao et al., 2003a). Significantly, this increased glucose uptake was evident even by the late 1-cell stage when the preference for glucose is also manifested. The increased glucose uptake was also seen during the 2-cell stage.

Along with the increased glucose uptake, we observed changes in the expression of glucose transporters in clones. The GLUT1 transporter is widely expressed among diverse cell types and is expressed throughout preimplantation development. Normally, the GLUT1 (renamed SLC2A1) transporter becomes concentrated at the cell surface at the 8-cell stage. Although GLUT1 does not appear to be over-expressed in cloned embryos, it displays precocious localization to the cell surface at the 2-cell stage (Gao et al., 2003a). At later stages when normal embryos display localization of GLUT1 to the cell surface, clones fail to show this localization. Myoblast-cloned embryos also aberrantly express the GLUT4 transporter throughout preimplantation development. The GLUT4 (renamed SLC2A4) transporter is expressed in muscle but not in normal embryos. Thus, even after as many as 5 days in culture and the formation of expanded blastocysts with an average of 60 cells each, myoblast-cloned embryos continue to express the GLUT4 transporter. They cease to express MyoD, however, indicating that some degree of reprogramming occurs. Overall, these studies illustrate clearly that cloned embryos display a variety of somatic cell characteristics at the physiological, cellular, and molecular levels.

Alterations in mRNA regulation

We also found that cloned embryos aberrantly express the somatic form of the DNA methyl transferase DNMT1 (Chung et al., 2003). The *Dnmt1* gene produces an oocyte-specific, tes-tis-specific, and somatic form of protein. The oocyte form (Dnmt1o) differs from the somatic form due to alternative promoter and first exon usage during mRNA processing. Dnmt1o is normally the only form of protein during the preimplantation period, although the somatic form of the message is present from the 2-cell stage onward. Dnmt1o is expressed as an abundant maternal protein, and enters the nucleus only during the 8-cell stage when it is believed to be responsible for maintaining imprints for one round of replication (Carlson et al., 1992; Mertineit et al., 1998; Ratnam et al., 2002). In contrast to normal fertilized embryos, cloned embryos express the somatic form of Dnmt1 (Chung et al., 2003). Because the somatic form of the protein is not expressed in normal embryos, this indicates that the post-transcriptional regulation of Dnmt1 expression is disrupted. This may be due to an inability to maintain a translational block upon the somatic Dnmt1 mRNA. If this is the case, this could signify a broader defect affecting the regulation of maternal mRNA translation as well. Because maternal mRNA translation appears to be critical for regulating a variety of events in the early embryo (Latham and Schultz, 2001), such an effect could impede normal developmental progression in clones. This could compound the effects of disruptions in the post-transcriptional regulation of other mRNAs expressed embryonically.

Alterations in cytoplasm-nucleus protein trafficking

Along with the aberrant expression of the somatic form of Dnmt1, we also observed a greatly reduced uptake of the oocyte form of Dnmt1 into the nuclei at the 8-cell stage. Thus, although there was the normal, abundant supply of the oocyte form in the ooplasm of the wild-type oocytes used to make the cloned constructs, double labeling using one antibody specific for the somatic form and one that recognizes both forms failed to indicate abundant nuclear staining with the common antibody in excess of what was seen with the specific antibody. Additionally, although the somatic form was seen in the nuclei at the 8-cell stage, reminiscent of entry of the oocyte form into nuclei of normal 8-cell embryos, this nuclear staining was mosaic in nature. Thus, neither form of the Dnmt1 protein was able to enter the nuclei efficiently, and some nuclei failed to acquire either form of Dnmt1. This may contribute to defects in genomic imprinting (Mann et al., 2003). The failure of a key, developmentally regulated cytoplasmic-nuclear movement of a protein like Dnmt1 may indicate broader defects in cytoplasmic-nuclear protein trafficking. Such a defect could seriously inhibit the ability of the developing embryonic cytoplasm to reprogram the genome as development progresses, as well as disrupting basic gene regulation and developmental processes.

Effects of donor cell type

Several of the points made above reveal an important effect of donor cell type on cloned embryo phenotype. For example, the culture system preferred by cumulus cell clones (MEMα) is not the same as the one preferred by myoblast clones (F10:DMEM). Additionally, we have seen that the established myoblast cell line C2C12, which has comparatively relaxed growth requirements, produces cloned embryos that also have relaxed requirements (Gao et al., 2003a). Other unpublished studies from our laboratory reveal that other myoblast cell lines can produce different efficiencies of cloned embryo development to the blastocyst stage. These observations indicate that because characteristics of the donor somatic cell are elaborated in the cloned embryos, the exact phenotype of cloned embryos varies with donor cell type.

Realization of this direct relationship between donor cell phenotype and cloned embryo phenotype is relevant to the interpretation of experimental results. In a recent study, it was reported that clones made with different donor cell types display different disruptions in epigenetic histone modifications, specifically, histone H3 methylation and DNA methylation (Santos et al., 2003). These different classes of cloned embryos displayed different capacities to develop further. One can interpret such results in the context of different epigenetic states of the donor cells. Given the effects of the culture system, however, the possibility must also be considered that differences in the epigenotype of clones can also arise as a result of different responses to the culture environment, just as the culture environment can alter the epigenotype of normal fertilized embryos (Doherty et al., 2000; Khosla et al., 2001). Additionally, different donor cell types may express different levels of methyl transferases, and such a difference could persist in the cloned embryos.

Effects of the spindle-chromosome complex on phenotype

The first step undertaken in the microsurgical production of cloned embryos is the removal of the spindle-chromosome complex (SCC). This accomplishes the essential task of eliminating the oocyte genome. Simultaneously, this step also eliminates the spindle and a variety of factors that may be preferentially associated with it, as well as proteins associated with the chromosomes and kinetochores. These include such proteins as spindlin, centrosomal and motor proteins, protein kinases such as polo-like and aurora-A and aurora-B, protein phosphatases, and other microtubule-associated proteins. Although some of these proteins exist in ooplasmic reservoirs or may be replenished after removal of the SCC, it is possible that other proteins may be irretrievably depleted by SCC removal. This could affect not only chromosome segregation at the first mitosis, but also the correct regulation of the cell cycle.

Delayed SCC removal

Yin et al. (2002a) employed a strategy of removing the SCC 1–2 h after SCNT instead of before SCNT in a study using rabbit embryos. High-quality blastocysts were obtained and an apparently greater efficiency of implantation, but no term progeny were obtained. A similar strategy was used successfully to produce porcine clones, but without an apparent increase in efficiency (Yin et al., 2002b). Thus, delayed SCC removal may be beneficial, but perhaps insufficient by itself to produce a major improvement in clone development, suggesting that essential factors are still removed with the SCC.

Phenotype of tetraploid constructs

We have repeatedly observed that the presence of an oocytic SCC can alter the degree to which somatic cell characteristics are elaborated in nuclear transfer embryos. When the cloning procedure is undertaken without removal of the SC, the result is a tetraploid embryo possessing a diploid complement of oocyte-derived chromosomes as well as the diploid somatic cell nucleus. Tetraploid embryos do not exhibit a preference for glucose-containing media (Chung et al., 2002), and neither do they exhibit the same enhanced glucose uptake as seen in diploid clones (Gao et al., 2003a). The GLUT1 transporter is not precociously localized to the cell surface in tetraploids, and tetraploids show reduced expression of GLUT4 relative to diploid clones (Gao et al., 2003a). Tetraploid embryos do not display aberrant expression of the somatic form of Dnmt1 (Chung et al., 2003), and tetraploid bovine embryos show more complete demethylation of satellite sequence DNA than diploid clones (Kang et al., 2001). Tetraploid embryos also display less pronounced perturbations in the methylation and expression of imprinted genes than diploid clones (Mann et al., 2003).

It is possible that some of the differences between tetraploid embryos and diploid clones are due to the expression of genes that may be uniquely programmed for expression from an authentic embryonic (represented here by oocytic) genome. If such is the case, then diploid clones obviously would lack such expression unless their genome is suitably reprogrammed very early during development.

SCC effects on the loss of H1 linker histone exchange capacity

It is also possible that some of the effects of the maternal SCC are due to regulatory activities provided by SCC-associated proteins. Evidence that this occurs is seen in the effects of the SCC on histone H1 linker switching following SCNT. We observed that after SCNT, the oocyte-specific H1FOO rapidly assembles onto chromatin within 5 min of SCNT, and completely displaces somatic H1s by 60 min (Gao et al., 2003b). The ability of the oocyte to mediate this switch is developmentally regulated, being greatly diminished by 2 h post-activation and largely absent by the zygote stage. Interestingly, the loss of the ability to mediate H1 switching is accelerated in the absence of the SCC. Thus, it appears that some activity associated with the SCC may normally promote H1 transitions.

Effects in primates

Another demonstration of the importance of SCC-associated proteins is seen when cloning procedures are applied to the rhesus monkey (Simerly et al., 2003). Recent studies reveal that cloned rhesus constructs made with adult somatic cell nuclei are unable to assemble normal spindles at the first mitotic division, and thus tend to develop aneuploidy. The removal of the SCC may greatly deplete the oocyte of spindle-associated proteins, and the somatic cell centrosome may be unable to substitute.

These observations clearly illustrate the importance of spindle-associated proteins to the early embryo, and the potential problems associated with SCC removal during cloning. On the one hand, this may provide a valuable research model for examining the roles of spindle-associated factors during normal development. Unfortunately, this situation may also demand either that the cloning procedures be altered to avoid the loss of essential spindle-associated proteins, or that the oocytes to be used for cloning be supplemented with the relevant molecules once those factors become known.

Conclusions

The overall regulation of cellular metabolism, homeostasis, and mitosis appears to be disrupted in cloned embryos. This appears to be largely a result of the unexpected continuation of the donor somatic cell gene expression program as the embryo initiates gene transcription, combined with the apparent depletion of important spindle-associated factors. These deficiencies in cloned embryos raise a number of scientific questions, and suggest possible avenues by which the cloning procedures may eventually be improved.

One basic question that needs to be addressed is what determines which genes become reprogrammed the soonest, and which become reprogrammed later or not at all. Additionally, the question arises whether the ability of a given gene to be reprogrammed and the timing of that reprogramming differ with different donor cell types. The inherent difficulty in addressing these questions is that the overall pace and extent of reprogramming will likely be affected by cloned embryo health, which is in turn dependent upon the culture system employed. Within narrowly defined experimental parameters, however, it may be possible to establish some genes as being more easily reprogrammed than others. This would in turn permit detailed molecular studies to understand the basis for such differences.

The degree to which the transferred nucleus of a given donor cell type will alter embryonic metabolism, physiology, and homeostasis is likely to vary according to the exact array of genes that are mis-expressed, and thus is likely to vary with donor cell type. This creates additional limitations in the present ability to use cloning technology to address basic questions about nuclear potency. While it may at first appear that donor cell type-specific differences in cloning efficiency may reveal difference in "reprogrammability" of the genome of those donor cell types, this premise is flawed, because different donor cell types will be expected to yield embryos with different culture requirements. It is not our intention here to discount the idea that different donor cell types may have different epigenetic states that render them less suitable for cloning. For example, the differentiated state or cell cycle status could lead to epigenetic differences in chromatin structure that affect reprogramming. However, conclusions about differences in reprogrammability of some nuclei may only be valid within the context of a given culture system, and may in fact disappear as optimized culture systems are identified for different donor cell types. Thus an inability of a given type of donor nucleus to support clonal development can reflect technical deficiencies in the procedure rather than inherent epigenetic or genetic restrictions in nuclear potential. It is thus critical as we apply cloning technology to address basic questions of mammalian embryology that we bear in mind that this technology is very new and requires further refinement.

The effects of spindle-associated factors on cloned embryo development are both novel and exciting. These results reveal that such factors have larger parts to play in development than merely facilitating chromosome segregation during mitosis. Such factors may interact closely with the cell cycle, the cytoskeleton, and possibly with chromatin-associated factors, thereby affecting sub-cellular protein localization, cytoarchitecture, and gene expression in unexpected ways. Further studies of these effects using nuclear transfer technology and other approaches (e.g., genetic) should help to uncover the basis for such effects. It may also become possible to improve cloning success, and possibly the success of other approaches, by augmenting the oocyte with the relevant spindle-associated factors.

A variety of studies have reported on epigenetic defects in cloned embryos (e.g., Humpherys et al., 2001, 2002; Kang et al., 2001; Santos et al., 2003; Chung et al., 2003; Mann et al., 2003). Given the effects of culture on epigenotype of normal embryos, it is reasonable to suppose that similar effects would arise in cloned embryos. Such effects could be more severe for cloned embryos grown in standard embryo culture media, as such media are clearly poorly suited to such embryos. Further progress in discovering optimized media for cloned embryos could thus reduce the aberrant epigenotypes, and this could also lead to more successful term development.

References

Aoki F, Worrad DM, Schultz RM: Regulation of transcriptional activity during the first and second cell cycles in the preimplantation mouse embryo. Dev Biol 181:296–307 (1997).

Biggers JD, McGinnis LK: Evidence that glucose is not always an inhibitor of mouse preimplantation development in vitro. Hum Reprod 16:153–163 (2001).

Bordignon, V, Clarke HJ, Smith LC: Developmentally regulated loss and reappearance of immunoreactive somatic histone H1 on chromatin of bovine morula-stage nuclei following transplantation into oocytes. Biol Reprod 61:22–30 (1999).

Bordignon V, Clarke HJ, Smith LC: Factors controlling the loss of immunoreactive somatic histone H1 from blastomere nuclei in oocyte cytoplasm: a potential marker of nuclear reprogramming. Dev Biol 233:192–203 (2001).

Bouniol C Nguyen E, Debey P: Endogenous transcription occurs at the late 1-cell stage in the mouse embryo. Exp Cell Res 218:57–62 (1995).

Carlson LL, Page AW, Bestor TH: Properties and localization of DNA methyltransferase in preimplantation mouse embryos: implications for genomic imprinting. Genes Dev 6:2536–2541 (1992).

Christians E, Campion E, Thompson E, Renard JP: Expression of the *HSP 70.1* gene, a landmark of early zygotic gene activity in the mouse embryo, is restricted to the first burst of transcription. Development 112:113–122 (1995).

Chung YG, Mann MR, Bartolomei MS, Latham KE: Nuclear-cytoplasmic "tug of war" during cloning: effects of somatic cell nuclei on culture medium preferences of preimplantation cloned mouse embryos. Biol Reprod 66:1178–1184 (2002).

Chung YG, Ratnam S, Chaillet JR, Latham KE: Abnormal post-transcriptional gene regulation controlling DNA methyltransferase expression in cloned mouse embryos. Biol Reprod 69:146–153 (2003).

Doherty AS, Mann MRW, Tremblay KD, Bartolomei MS, Schultz RM: Differential effects of culture on imprinted *H19* expression in the preimplantation mouse embryo. Biol Reprod 62:1526–1535 (2000).

Gao S, Chung YG, Williams JW, Riley J, Moley K, Latham KE: Somatic cell-like features of cloned mouse embryos prepared with cultured myoblast nuclei. Biol Reprod 69:48–56 (2003a).

Gao S, Chung YG, Parseghian MH, King GJ, Adashi EY, Latham KE: Rapid H1 linker histone transitions following fertilization or somatic cell nuclear transfer: evidence for a uniform developmental program in mice. Dev Biol 266:62–75 (2003b).

Heindryckx B, Rybouchkin A, Van Der Elst J, Dhont M: Effect of culture media on in vitro development of cloned mouse embryos. Cloning 3:41–50 (2001).

Humpherys D, Eggan K, Akutsu H, Hochedlinger K, Rideout WM 3rd, Biniszkiewicz D, Yanagimachi R, Jaenisch R: Epigenetic instability in ES cells and cloned mice. Science 293:95–97 (2001).

Humpherys D, Eggan K, Akutsu H, Friedman A, Hochedlinger K, Yanagimachi R, Lander ES, Golub TR, Jaenisch R: Abnormal gene expression in cloned mice derived from embryonic stem cell and cumulus cell nuclei. Proc Natl Acad Sci USA 99:12889–12894 (2002).

Inoue K, Ogonuki N, Mochida K, Yamamoto Y, Takano K, Kohda T, Ishino F, Ogura A: Effects of donor cell type and genotype on the efficiency of mouse somatic cell cloning. Biol Reprod 69:1394–1400 (2003).

Kang YK, Koo DB, Park JS, Choi YH, Lee KK, Han YM: Influence of oocyte nuclei on demethylation of donor genome in cloned bovine embryos. FEBS Lett 499:55–58 (2001).

Khosla S, Dean W, Brown D, Reik W, Feil R: Culture of preimplantation mouse embryos affects fetal development and the expression of imprinted genes. Biol Reprod 64:918–26 (2001).

Lane M, Gardner DK: Ammonium induces aberrant blastocyst differentiation, metabolism, pH regulation, gene expression and subsequently alters fetal development in the mouse. Biol Reprod 69:1109–1117 (2003).

Latham KE: Mechanisms and control of embryonic genome activation in mammalian embryos. Int Rev Cytol 193:71–124 (1999).

Latham KE, Schultz RM: Embryonic genome activation. Front Biosci 6:D748–759 (2001).

Latham KE, Solter D, Schultz RM: Acquisition of a transcriptionally permissive state during the 1-cell stage of mouse embryogenesis. Dev Biol 149:457–462 (1992).

Latham KE, Garrels JI, Solter D: Alterations in protein synthesis following transplantation of mouse 8-cell stage nuclei to enucleated 1-cell embryos. Dev Biol 163:341–350 (1994).

Mann MR, Chung YG, Nolen LD, Verona RI, Latham KE, Bartolomei MS: Disruption of imprinted gene methylation and expression in cloned mouse embryos. Biol Reprod 69:902–914 (2003).

Mertineit C, Yoder JA, Taketo T, Laird DW, Trasler JM, Bestor TH: Sex-specific exons control DNA methyltransferase in mammalian germ cells. Development 125:889–897 (1998).

Ratnam S, Mertineit C, Ding F, Howell CY, Clarke HJ, Bestor TH, Chaillet JR, Trasler JM: Dynamics of Dnmt1 methyltransferase expression and intracellular localization during oogenesis and preimplantation development. Dev Biol 245:304–314 (2002).

Santos F, Zakhartchenko V, Stojkovic M, Peters A, Jenuwein T, Wolf E, Reik W, Dean W: Epigenetic marking correlates with developmental potential in cloned bovine preimplantation embryos. Curr Biol 13:1116–1121 (2003).

Simerly C, Dominko T, Navara C, Payne C, Capuano S, Gosman G, Chong KY, Takahashi D, Chace C, Compton D, Hewitson L, Schatten G: Molecular correlates of primate nuclear transfer failures. Science 300:297 (2003).

Szollosi D, Czolowska R, Szollosi MS, Tarkowski AK: Remodeling of mouse thymocyte nuclei depends on the time of their transfer into activated, homologous oocytes. J Cell Sci 91:603–613 (1988).

Wakayama T, Yanagimachi R: Effect of cytokinesis inhibitors, DMSO and the timing of oocyte activation on mouse cloning using cumulus cell nuclei. Reproduction 122:49–60 (2001).

Wakayama T, Perry ACF, Zuccotti M, Johnson KR, Yanagimachi R: Full-term development of mice from enucleated oocytes injected with cumulus cell nuclei. Nature 394:369–374 (1998).

Yin XJ, Kato Y, Tsunoda Y: Effect of delayed enucleation on the developmental potential of nuclear transferred oocytes receiving adult and fetal fibroblast cells. Zygote 10:217–222 (2002a).

Yin XJ, Tani T, Yonemura I, Kawakami M, Miyamoto K, Hasegawa R, Kato Y, Tsunoda Y: Production of cloned pigs from adult somatic cells by chemically assisted removal of maternal chromosomes. Biol Reprod 67:442–446 (2002b).

Cytogenet Genome Res 105:285–291 (2004)
DOI: 10.1159/000078200

Nuclear reprogramming in mammalian somatic cell nuclear cloning

H. Tamada and N. Kikyo

Stem Cell Institute, Division of Hematology, Oncology and Transplantation, Department of Medicine, University of Minnesota, Minneapolis, MN (USA)

Abstract. Nuclear cloning is still a developing technique used to create genetically identical animals by somatic cell nuclear transfer into unfertilized eggs. Despite an intensive effort in a number of laboratories, the success rate of obtaining viable offspring from this technique remains less than 5%. In the past few years many investigators reported the reprogramming of specific nuclear activities in cloned animals, such as genome-wide gene expression patterns, DNA methylation, genetic imprinting, histone modifications and telomere length regulation. The results highlight the tremendous difficulty the clones face to reprogram the original differentiation status of the donor nuclei. Nevertheless, nuclei prepared from terminally differentiated lymphocytes can overcome this barrier and produce apparently normal mice. Study of this striking nuclear reprogramming activity should significantly contribute to our understanding of cell differentiation in more physiological settings.

Copyright © 2004 S. Karger AG, Basel

Introduction

Given the enormous complexity of the gene regulatory pathways, it is remarkable that the entire cell differentiation program can be completely erased and properly re-established in somatic cell nuclear cloning (Hochedlinger and Jaenisch, 2002b; Oback and Wells, 2002). Nuclei taken from terminally differentiated B cells can produce the entire body of embryos with apparently normal functions as long as the extra-embryonic tissues are supplied externally by the tetraploid blastocysts (Hochedlinger and Jaenisch, 2002a). Simple nuclear injection was insufficient for the B cell nuclei to acquire pluripotency and the clones had to pass through ES cells to produce live pups in this experiment; nevertheless, this remarkable finding indicates that differentiated nuclei can be de-differentiated in the oocyte and embryonic environment. Nuclear cloning is arguably one of the most powerful experimental systems to study the reprogramming of cell differentiation. In *Xenopus* cloning, all active genes in the donor nuclei are shut off soon after nuclear transfer is completed. Several hours after nuclear transfer, embryonic nuclei start to express development specific genes that follow the normal time course of the zygotic gene activation (Chan and Gurdon, 1996; Byrne et al., 2002). One can argue that this genetic reprogramming is simply due to dilution of the donor nuclear components by the proteins and RNA stored in the large frog eggs. However, that argument is not convincing in the case of mouse cloning, where reprogrammed gene expression initiates as early as the 2-cell stage in tiny embryos. Although the success rate of animal cloning is still extremely low (Table 1), extensive reprogramming of differentiation at the cellular level is accomplished in aborted embryos that contain well differentiated tissues. Since such a drastic and rapid nuclear reprogramming is rare in living cells, nuclear cloning will provide a unique window to dissect the cell differentiation mechanisms.

The cloning of Dolly the sheep in 1997 triggered a wide interest in mammalian cloning (Wilmut et al., 1997) followed by a number of mammalian clones created from adult cell nuclei as shown in Table 1. The recent progress in mammalian cloning needs to be interpreted in the context of the long cloning history initiated by Briggs and King, who created swimming frog tadpoles by injecting blastomere nuclei into unfertilized eggs in 1952 (Briggs and King, 1952). In the pre-Dolly era, key

This work is supported by the NIH (GM068027), the American Cancer Society (IRG-58-001-43-IRG40), the Minnesota Medical Foundation, and the Graduate School of the University of Minnesota.

Received 6 October 2003; manuscript accepted 12 November 2003.

Request reprints from: Dr. Nobuaki Kikyo, Stem Cell Institute
Division of Hematology, Oncology and Transplantation
Department of Medicine, University of Minnesota, MMC 716
420 Delaware St. SE, Minneapolis, MN 55455 (USA)
telephone: +1-612-624-0498; fax: +1-612-624-2436
e-mail: kikyo001@tc.umn.edu

Table 1. List of cloned mammals

Reference	Species	Donor Cells	Cloning Efficiency (Live birth/Manipulated oocytes) (%)
Wilmut et al., 1997	Sheep	Mammary epithelial cells	1/277 (0.4)
Wakayama et al., 1998	Mouse	Cumulus cells	41/2468 (1.7)
Kato et al., 1998	Bovine	Cumulus cells	5/99 (5.0)
		Oviductal cells	3/150 (2.0)
Cibelli et al., 1998	Bovine	Fetal fibroblasts	4/276 (1.4)
Baguisi et al., 1999	Goat	Fetal fibroblasts	3/285 (1.1)
Onishi et al., 2000	Pig	Fetal fibroblasts	1/210 (0.5)
Polejaeva et al., 2000	Pig	Adult granulosa cells	5/183 (2.7)
Betthauser et al., 2000	Pig	Fetal cells	Not available
Chesne et al., 2002	Rabbit	Cumulus cells	6/1852 (0.3)
Shin et al., 2002	Cat	Cumulus cells	Not available
Woods et al., 2003	Mule	Fetal fiborblasts	1/334 (0.3)
Galli et al., 2003	Horse	Fibroblasts	1/841 (0.1)
Roh et al., 2003	Rat	2-cell stage embryos	6/139 (4.3)
Zhou et al., 2003	Rat	Fibroblasts	Not available

concepts, such as the importance of cell cycle compatibility between the donor nuclei and host eggs, and the progressive decrease of cloning efficiency related to the donor differentiation stage, were already established (Gurdon, 1986; Sun and Moor, 1995). One factor that has accelerated the cloning research is the successful mouse cloning by Wakayama and colleagues using adult cumulus cell nuclei (Wakayama et al., 1998). Until then, the blastocyst was believed to be the last stage compatible as a nuclear cloning donor (Solter, 2000). Supported by a wealth of background information, mouse cloning enabled us to investigate reprogramming of genetic imprinting, reactivation of the inactive X chromosome, and potential problems of ES cells as the source of the donor nuclei, none of which were possible with other species (see below for references).

In addition to low birth rate, live cloned animals demonstrate a variety of pathological conditions such as respiratory failure, placental dysfunctions and large offspring syndrome (Young et al., 1998; Rhind et al., 2003). It is usually difficult to trace the origin of these ailments to a few responsible genes. Probably, they reflect the cumulative effects of many faulty gene expressions. Because these abnormalities were not passed on to the offspring of the cloned mice, these phenotypes represent aberrant gene expression by deficient epigenetic reprogramming rather than genetic changes in cloned animals (Tamashiro et al., 2002). Even though nuclear reprogramming is a complicated process, by focusing on a certain aspect of the nuclear events, it is possible to dissect and understand the basic science behind the reprogramming as demonstrated by our recent finding of the nucleolar disassembly in egg cytoplasm (Gonda et al., 2003).

In this review, we will discuss the reprogramming of genome-wide gene expression, DNA methylation, histone modifications and telomere length regulation that occur during nuclear cloning. Reflecting the recent wide attention to the cloning field, numerous insightful reviews are available on various aspects of nuclear cloning. The readers are recommended to refer to the following papers: Campell (1999), Gurdon (1999), Wakayama and Yanagimachi (1999), Kikyo and Wolffe (2000), Hochedlinger and Jaenisch (2002b), Oback and Wells (2002), Mullins et al. (2003) for a more general account of cloning, McLaren (2000), Solter (2000), Gurdon and Byrne (2003) for the historical background, Jaenisch et al. (2002), Dean et al. (2003) for epigenetic reprogramming and Wade and Kikyo (2002) for biochemistry of the nuclear reprogramming.

Abnormal gene expression in cloned animals

Several groups have compared gene expression patterns in clones and control animals as summarized in Table 2. By analyzing expression of eight developmentally important genes in cloned blastocysts, Wrenzycki and colleagues reported that several genes were properly activated in the blastocysts, but with a marked difference in the gene expression levels (Wrenzycki et al., 2001). These differences were found to be dependent on parameters in the nuclear transfer procedure, including the activation protocol, the cell cycle of the donor cells and the passage number of the donor cells. Daniels and colleagues also reported a similar finding based on the study of a different set of genes specific to early embryonic development (Daniels et al., 2000, 2001).

To understand the genome-wide difference in the gene expression patterns between cloned mice and fertilization-derived controls, a DNA microarray was employed using RNA isolated from placentas and livers of these mice (Humpherys et al., 2002). The result showed that less than 3% of over 12,000 genes were expressed abnormally in the clone's placentas. Placentas tend to overgrow in clones, but there was no clear relationship between the additional growth and the aberrantly expressed genes. The livers of the clones showed a less conspicuous abnormality in gene expression than placentas, which may occur as liver is a more homogeneous tissue with smaller number of differentiated cell types than placenta. It is impressive that more than 97% of the genes could be properly silenced or activated in the cloned embryos in this comprehensive genome-wide analysis. However, it is important to note that this study examined RNA isolated from a whole tissue; and by doing so, an irregularity of the gene expression in each cell may have been averaged (see below for improper spatial distribution of Oct4 as an example).

Table 2. Aberrant gene expression patterns in cloned mammals

Reference	Donor	Cells	Number of the tissue	Aberrantly expressed genes[a]/total number of genes examined	Detection method
Daniels et al., 2000	Bovine	Granulosa cells	2-cell embryo to blastocyst (4–10)	3/7 (42.9 %)	RT-PCR
Daniels et al., 2001	Bovine	Fetal epithelial cells	Blastocyst (62)	1/4 (25 %)	RT-PCR
Wrenzycki et al., 2001	Bovine	Follicular cell line	Blastocyst	1/8 (12.5%) to 3/8 (37.5%)	RT-PCR
Humpherys et al., 2002	Mouse	ES cells	Placenta (12)	221/12,654 (1.7%)	Microarray
			Liver (13)	26/12,654 (0.2%)	
		Cumulus cells	Placenta (14)	286/12,654 (2.3%)	
Suemizu et al., 2003	Mouse	ES cells	Placenta (2)	Clone 1: 1,807/15,247 (11.9%)	Microarray
				Clone 2: 1,964/15,247 (12.9%)	

[a] Gene expression is defined as aberrant in the microassay analysis when the gene expression level in the cloned mice is 2-fold higher or lower than in the controls derived from fertilization.

The transcription factor Oct4 is essential to maintain pluripotency of early mouse blastomeres (Pesce and Scholer, 2001). *Oct4* is exclusively expressed in germ cells and early embryonic cells; therefore, it must be reactivated soon after nuclear transfer in the somatic cell clones. Indeed, more than 80% of the cumulus cell clones reactivated *Oct4* at the correct stage, but 54.7% of the clones showed aberrantly high level of the *Oct4* transcript in the trophectoderm at the blastocyst stage when *Oct4* expression is normally limited to the inner cell mass (Boiani et al., 2002). Recently, it was reported that *Oct4* expression could be specifically reactivated in mouse thymocyte nuclei and human lymphocyte nuclei injected into *Xenopus* oocytes, suggesting that the regulatory mechanisms for this pluripotency-specific gene are probably evolutionarily conserved (Byrne et al., 2003). Another pluripotency gene *Nanog* is a newly discovered homeoprotein specifically expressed in morulae, inner cell mass and ES cells (Chambers et al., 2003; Mitsui et al., 2003). Nanog is required to maintain these cells pluripotent, independently of the LIF/Stat3 pathway used by the Oct4 signaling system. It remains to be examined if *Nanog* demonstrates correct spatial and temporal profiles of reactivation in cloned embryos.

Reprogramming of DNA methylation and imprinting

DNA methylation of cytosine at the CpG dinucleotides plays vital roles in the regulation of gene expression in mammalian development (Bird, 2002; Li, 2002). DNA methylation suppresses gene expression by recruiting methyl-CpG binding proteins, such as MeCP2, MBD1, MBD2 and MBD3, as well as associated histone deacetylases, co-repressor proteins and chromatin remodeling machineries to the promoter of specific genes. At least three DNA methyltransferases are involved in the methylation of new CpG sites and maintenance of the already methylated CpG during DNA replication. Ubiquitously expressed DNMT1 functions primarily as a maintenance methylase that methylates CpG sites on the newly synthesized DNA strand copying the existing methylation pattern on the template DNA strand. Developmentally regulated DNMT3a and DNMT3b are responsible for methylation of new CpG sites to establish de novo CpG methylation patterns, especially in early development and germ cell development.

The DNA methylation pattern shows global changes during early mouse development (Dean et al., 2003). Upon fertilization a majority of the sperm-derived genomic DNA is rapidly demethylated before the onset of DNA replication by an uncharacterized active mechanism (Mayer et al., 2000; Santos et al., 2002). In contrast, oocyte-derived DNA is passively demethylated only after DNA replication initiates, by the nuclear exclusion of DNMT1. The global level of DNA methylation remains at the lowest level in the morula and blastocyst stages until implantation, when sudden genome-wide de novo methylation occurs by DNMT3a and DNMT3b. The genome-wide demethylation and remethylation in early embryos seems to be conserved across species as observed in cow, rat and pig, although their timing with respect to developmental stages is slightly different (Dean et al., 2001). Successfully cloned embryos have to follow these methylation dynamics to erase the tissue-specific DNA methylation pattern and establish a new embryo-specific DNA methylation pattern on numerous genes.

A majority of the cloned bovine embryos show a gross abnormality in the genome-wide DNA methylation level and DNA methylation patterns on various repetitive sequences when compared with fertilization-derived controls (Table 3). The DNA methylation level in clones can be higher or lower than that in the control embryos depending on the donor cell types, target DNA sequences, examined embryonic stages and detection methods. The abnormality of the DNA methylation level is also substantially variable among individual clones (Kang et al., 2001a) and extremely abnormal embryos may have died before the analysis was done. Indeed, DNA methylation was undetectable in six out of nine spontaneously aborted bovine clones, but the methylation level was normal in the clones that survived to adulthood (Cezar et al., 2003).

Bovine somatic nuclei are resistant to the erasure of DNA methylation in early embryogenesis described above and the clones have a tendency to preserve the DNA methylation patterns inherited from the donor cells (Bourc'his et al., 2001; Dean et al., 2001). Re-establishment of DNA methylation was also potentially deregulated by precocious de novo methylation in clones (Dean et al., 2001). This abnormal methylation transition in cloned embryos could be due to the specific features of the somatic chromatin structure and/or defective regulation of DNMTs. For example, cloned mouse embryos expressed the somatic form of DNMT1 at abnormally high level and showed

Table 3. DNA methylation status in cloned mammals

Reference	Species	Donor cells	DNA sequence	Degree of DNA methylation[a]	Detection method
Kang et al., 2001b	Pig	Fetal fibroblasts	Centromeric satellite PRE-1 (euchromatic repeat)	Embryos: C=A Embryos C<A	Bisulfite
Dean et al., 2001	Bovine	Fetal fibroblasts	Whole genome	Embryos: C>A	Immunofluorescence[b]
Bourc'his et al., 2001	Bovine	Adult fibroblasts	Whole genome	Embryos: C>A	Immunofluorescence[b]
Kang et al., 2001a	Bovine	Fetal fibroblasts	Satellite I, Satellite II, SINE and 18S rRNA	In all sequences, Embryos D=C>A	Bisulfite
Kang et al., 2002	Bovine	Fetal fibroblasts	Tissue specific promoters	Embryos: C=A	Bisulfite
Cezar et al., 2003	Bovine	Genital ridge cells	Whole genome	Summary of 4 donor cell types	Reverse phase HPLC, restriction enzyme
		Fetal skin cells		Aborted fetuses: C<<A	
		Adult skin cells		Live fetuses: C<A	
		Fetal and adult cumulus cells		Adults: C=A	

[a] A: age-matched controls derived from fertilization, C: cloned animals and D: the donor animals.
[b] Immunofluorescence with anti-5-methylcytosine antibody.

defective nucleo-cytoplasmic translocation of the oocyte form of DNMT1 (Chung et al., 2003). Culture conditions of the cloned embryos are also known to affect DNA methylation as shown by loss of methylation in the regulatory CpG site of the *H19* gene depending on the culture medium of the embryos (Doherty et al., 2000).

DNA methylation of imprinted genes is established during germ cell development and is protected from the genome-wide demethylation and re-methylation in early development by an unknown mechanism (Li, 2002). It is intriguing to understand whether methylation imprinting in the donor somatic nuclei is protected from the global changes of DNA methylation in the early embryos as effectively as that in the fertilized nuclei. While Inoue and colleagues found normal allele-specific expression of seven imprinted genes in mouse embryos obtained from Sertoli cells (Inoue et al., 2002), two other groups reported grossly disrupted imprinting in cumulus cell clones (Humpherys et al., 2002; Mann et al., 2003). This abnormality in the imprinting status may suggest susceptibility of the methylation imprinting in the somatic nuclei to the global methylation changes during early embryogenesis. Epigenetic markers for the inactive X chromosome can also be erased and re-established on either X chromosome in cloning (Eggan et al., 2000) with the exception of some X-linked genes (Xue et al., 2002). ES cell-derived mouse clones show a striking variation in the DNA methylation pattern and imprinted gene expression, perhaps reflecting the instability of DNA methylation during the ES cell culture. In spite of this, some ES cell-derived clones developed to term implying that the epigenetic noise paused by aberrant DNA methylation and imprinting can be compensated by other mechanisms (Humpherys et al., 2001; Jaenisch et al., 2002). This notion is consistent with the routine success in producing ES cell chimeras in transgenic experiments.

Histone modifications in cloned animals

Global release and uptake of linker histone H1 is another challenge for the donor nuclei during the nuclear reprogramming. The histone H1 exists at a very low level in mature mouse oocytes and gradually becomes abundant around the 4-cell stage (Clarke et al., 1997; Adenot et al., 2000). Following this temporal profile, blastomere nuclei lose histone H1 upon injection into oocytes and reacquire histone H1 during the subsequent development (Bordignon et al., 2001). This DNA replication-independent transition of the histone H1 level was also observed in bovine clones (Bordignon et al., 1999). In *Xenopus* somatic nuclei incubated in egg extract, the molecular chaperone nucleoplasmin is responsible for the exchange of the somatic linker histone with the egg type linker histone B4 (Dimitrov and Wolffe, 1996). It is likely that mammalian nucleoplasmin (Burns et al., 2003) is involved in the loss of histone H1 from the donor nuclei, although its physiological meaning is unknown.

Alteration of histone modifications is also an important aspect of chromatin remodeling in cloning. Histones receive a number of covalent modifications including acetylation, methylation, phosphorylation, ubiquitination and ADP-ribosylation at the amino termini protruding from the chromatin core. A specific combination of these histone modifications on a given gene provides a recognition site for interacting molecules and thus contributes to regulating the gene activity (histone code hypothesis) (Strahl and Allis, 2000; Jenuwein and Allis, 2001). Bovine oocytes and early embryos express several histone acetylases and deacetylases with some variability in the transcript levels depending on the developmental stages (McGraw et al., 2003). In mouse oocytes, histone H3 and H4 are globally deacetylated on several lysines at the metaphase II of the second meiosis, which was reproduced in somatic nuclei transferred into the same stage of oocytes (Kim et al., 2003). This genome-wide decrease of histone acetylation may contribute to the erasure of the previous gene expression patterns specific to the donor cell differentiation.

Methylation on histone H3 lysine 9 (H3-K9) is usually associated with gene inactivation and acetylation on H3-K9 is linked with gene activation (Fischle et al., 2003). Fertilized control mouse embryos become hypoacetylated on H3-K9 at the 4-cell stage and are gradually hyperacetylated after the 8-cell stage (Santos et al., 2003). In contrast, cloned embryos retain hyperacetylation on H3-K9 throughout these stages. At the blastocyst stage, the cloned embryos show hypermethylation on H3-K9 in the trophectoderm compared with the controls. The

Cytogenet Genome Res 105:285–291 (2004)

Table 4. Telomere length in cloned mammals

Reference	Species	Donor cells	Number of cloned animals	Telomere length in clones[a]	Telomere activity in clones
Shiels et al., 1999	Sheep	Mammary epithelial cells	3	D=C<A	Not tested
Wakayama et al., 2000	Mouse	Cumulus cells	35	D<C	Not tested
Tian et al., 2000	Bovine	Adult fibroblasts and cumulus cells	10	D<C=A	Detected
Kato et al., 1998	Bovine	Adult ear cells	3	Ear cells: D<C White blood cells: D>C=A	
Lanza et al., 2000	Bovine	Senescent fibroblasts	6	D<C>A	Detected
Betts et al., 2001	Bovine	Adult fibroblasts Fetal fibroblasts Granulosa cells	Total 21	D<C=A	Detected
Miyashita et al., 2002	Bovine	Oviductal epithelial cells	9	D>C<A	Not tested
		Mammary epithelial cells	1	D>C<A	Not tested
		Muscle cells	2	D=C<A	Not tested
		Skin fibroblasts	2	D=C<A	Not tested
		Blastomere cells	6	C>A	Not tested
Clark et al., 2003	Sheep	Fetal fibroblasts	2	D<C<A	Not tested

[a] A: age-matched control animals derived from fertilization, C: cloned animals and D: donor animals.

detailed enzymology responsible for these transitions of histone acetylation and methylation in early embryos is not yet available, but these aberrant histone modifications should almost certainly affect expression of a number of genes.

Telomere restoration in clones

Telomeres are DNA-protein complexes at the ends of eukaryotic chromosomes essential for chromosomal integrity and normal cell growth (McEachern et al., 2000; Blasco, 2002). Vertebrate telomere DNA is composed of tandem repeats of the sequence TTAGGG and a 3′-overhang that forms a t-loop with the double-stranded DNA protecting the 3′ end of telomeres. Because conventional DNA polymerases cannot replicate the lagging strand at the 5′ end, telomeric DNA is progressively lost with each round of cell division, 50–150 base pairs per cell division in human cells, unless the cells express the ribonucleoprotein complex telomerase. The enzymatic core of telomerase consists of the reverse transcriptase TERT (telomerase reverse transcriptase) and its template RNA TR (telomerase RNA). While TR is ubiquitously expressed, TERT expression is limited to germ cells and stem cells in the normal human body. When telomeres become shorter than the critical threshold in somatic cells due to a lack of TERT, p53- and Rb-regulated DNA damage responses trigger growth arrest (replicative senescence) (Maser and DePinho, 2002). If TP53 and RB are inactivated, these cells can bypass this growth arrest, but the cells will eventually die because of massive chromosome end fusions triggered by the cumulative telomere loss (crisis).

The telomerase activity is subjected to multiple levels of regulatory mechanisms (Blasco, 2002; Kyo and Inoue, 2002). Transcriptional regulation of TERT by c-Myc, Max and Sp1 is one of the most critical control mechanisms. Alternative splicing of TERT also regulates the telomerase activity by producing more than six forms of transcripts including truncated forms and dominant-negative forms of TERT. Subcellular localization of TERT adds another layer of regulation to the telomerase activity. For instance, telomerase activity in activated T cells is not dependent on the total TERT protein level, rather the activity is defined by nuclear translocation of TERT accompanied by phosphorylation (Liu et al., 2001). While transfected TERT is localized in the nucleoli in normal cells and released into nucleoplasm only in the S phase, it is always widely distributed in the cancer cell nucleoplasm (Wong et al., 2002). In addition to these direct modifications of TERT, a number of telomere binding proteins play essential roles in modulating the telomere length (Blasco, 2002). For example, TRF1 and TRF2 bind to the TTAGGG repeats and negatively regulate the telomere length through interactions with other proteins on telomeric DNA (de Lange, 2002).

Since telomere length is tightly linked with cellular senescence, it is intriguing to examine whether shortened telomeres in somatic nuclei can be restored in nuclear cloning as one aspect of re-juvenilization at the cellular level. A comparison of telomere length among donors, clones and age-matched controls was reported by several groups (Table 4). These results indicate that shortened adult cell telomeres can be restored during early development of cloned animals but the degree of telomere elongation is quite variable. Even in a single study, there was significant variation in telomere length among individual clones and among different tissues isolated from a single clone (Betts et al., 2001; Miyashita et al., 2002), underscoring the complexity and difficulty of telomere length control in clones. If aborted embryos, which potentially harbor telomeres with extremely abnormal length, are included, the efficiency of telomere restoration may be even less than what is being reported. It is not known whether the telomere elongation in clones is a secondary effect of non-specifically activated telomerase or a regulated telomere restoration reflecting the cell's effort to compensate for defective telomeres.

To understand the functional consequences of telomere restoration, two groups examined whether the replicative lifespan of senescent cells could be elongated by nuclear cloning. Lanza and colleagues found that clone-derived bovine fibroblast cells, which contained fully restored telomeres comparable to the age-matched controls, showed longer proliferative lifespan than the senescent donor fibroblast cells (Lanza et al., 2000). However, when Clark and colleagues tested cloned sheep fibroblast cells that harboured partially restored telomeres, the proliferative lifespan of the cells was not extended (Clark et al., 2003). Thus, it remains to be examined to what extent the restored telomeres can influence the proliferative lifespan of these cells. It is also still unclear whether the resetting of proliferative lifespan of isolated cells has something to do with the lifespan of cloned animals. When whole embryonic extract was tested, the temporal profile of the telomerase activity during development was similar in cloned bovine embryos and fertilized embryos (Xu and Yang, 2001). Because a number of factors contribute to define the telomerase activity and telomere length in vivo as described above, the next step will be to examine individual accessory factors of the telomerase complex and subnuclear localization of TERT in cloned animals.

Conclusion

To be successful, clones have to erase the previous differentiation memory and establish embryo-specific gene expression profiles within a short period of time. This is accomplished through large-scale reorganization of chromatin structure and functions, as exemplified by genome-wide DNA methylation and re-methylation, adjustment of expression level for imprinted genes, reactivation of inactive X chromosome genes, global changes of histone modifications and exchange of linker histones. They also have to repair shortened telomeres as an essential step to restore replicative competence. It is almost inconceivable that differentiated cell nuclei can achieve this daunting task, albeit with extremely low efficiency. Some of the successful clones may have been derived from nuclei of stem cells embedded within somatic tissues, requiring less extensive nuclear reprogramming. The rare occurrence of stem cells in these tissues may explain the low efficiency of cloning and this possibility needs to be carefully examined using strict criteria for the donor cell differentiation. If creation of perfectly normal animals is the goal of cloning, it may be quite difficult, if not impossible (Rhind et al., 2003), but if the goal of the cloning study is to understand how the embryonic environment is trying to reprogram the differentiated nuclei, then the study will be a precious source of insight into the normal cell differentiation mechanisms.

Acknowledgements

We thank Justin Wudel for his critical comments on the manuscript.

References

Adenot PG, Campion E, Legouy E, Allis CD, Dimitrov S, Renard J, Thompson EM: Somatic linker histone H1 is present throughout mouse embryogenesis and is not replaced by variant H1 degrees. J Cell Sci 113:2897–2907 (2000).

Baguisi A, Behboodi E, Melican DT, Pollock JS, Destrempes MM, Cammuso C, Williams J, Nims SD, Porter CA, Midura P, et al: Production of goats by somatic cell nuclear transfer. Nat Biotechnol 17:456–461 (1999).

Betthauser J, Forsberg E, Augenstein M, Childs L, Eilertsen K, Enos J, Forsythe T, Golueke P, Jurgella G, Koppang R, et al: Production of cloned pigs from in vitro systems. Nat Biotechnol 18:1055–1059 (2000).

Betts D, Bordignon V, Hill J, Winger Q, Westhusin M, Smith L, King W: Reprogramming of telomerase activity and rebuilding of telomere length in cloned cattle. Proc Natl Acad Sci USA 98:1077–1082 (2001).

Bird A: DNA methylation patterns and epigenetic memory. Genes Dev 16:6–21 (2002).

Blasco MA: Telomerase beyond telomeres. Nat Rev Cancer 2:627–633 (2002).

Boiani M, Eckardt S, Scholer HR, McLaughlin KJ. Oct4 distribution and level in mouse clones: consequences for pluripotency. Genes Dev 16:1209–1219 (2002).

Bordignon V, Clarke HJ, Smith LC: Developmentally regulated loss and reappearance of immunoreactive somatic histone H1 on chromatin of bovine morula-stage nuclei following transplantation into oocytes. Biol Reprod 61:22–30 (1999).

Bordignon V, Clarke HJ, Smith LC: Factors controlling the loss of immunoreactive somatic histone H1 from blastomere nuclei in oocyte cytoplasm: a potential marker of nuclear reprogramming. Dev Biol 233:192–203 (2001).

Bortvin A, Eggan K, Skaletsky H, Akutsu H, Berry DL, Yanagimachi R, Page DC, Jaenisch R: Incomplete reactivation of Oct4-related genes in mouse embryos cloned from somatic nuclei. Development 130:1673–1680 (2003).

Bourc'his D, Le Bourhis D, Patin D, Niveleau A, Comizzoli P, Renard JP, Viegas-Pequignot E: Delayed and incomplete reprogramming of chromosome methylation patterns in bovine cloned embryos. Curr Biol 11:1542–1546 (2001).

Briggs R, King TJ: Transplantation of living nuclei from blastula cells into enucleated frogs' eggs. Proc Natl Acad Sci USA 38:455–463 (1952).

Burns KH, Viveiros MM, Ren Y, Wang P, DeMayo FJ, Frail DE, Eppig JJ, Matzuk MM: Roles of NPM2 in chromatin and nucleolar organization in oocytes and embryos. Science 300:633–636 (2003).

Byrne JA, Simonsson S, Gurdon JB: From intestine to muscle: nuclear reprogramming through defective cloned embryos. Proc Natl Acad Sci USA 99:6059–6063 (2002).

Byrne JA, Simonsson S, Western PS, Gurdon JB: Nuclei of adult mammalian somatic cells are directly reprogrammed to Oct-4 stem cell gene expression by amphibian oocytes. Curr Biol 13:1206–1213 (2003).

Campbell KH: Nuclear transfer in farm animal species. Semin Cell Dev Biol 10:245–252 (1999).

Cezar GG, Bartolomei MS, Forsberg EJ, First NL, Bishop MD, Eilertsen KJ: Genome wide epigenetic alterations in cloned bovine fetuses. Biol Reprod 68:1009–1014 (2003).

Chambers I, Colby D, Robertson M, Nichols J, Lee S, Tweedie S, Smith A: Functional expression cloning of Nanog, a pluripotency sustaining factor in embryonic stem cells. Cell 113:643–655 (2003).

Chan AP, Gurdon JB: Nuclear transplantation from stably transfected cultured cells of *Xenopus*. Int J Dev Biol 40:441–451 (1996).

Chesne P, Adenot PG, Viglietta C, Baratte M, Boulanger L, Renard JP: Cloned rabbits produced by nuclear transfer from adult somatic cells. Nat Biotechnol 20:366–369 (2002).

Chung YG, Ratnam S, Chaillet JR, Latham KE: Abnormal regulation of DNA methyltransferase expression in cloned mouse embryos. Biol Reprod 69:146–153 (2003).

Cibelli JB, Stice SL, Golueke PJ, Kane JJ, Jerry J, Blackwell C, Ponce de Leon FA, Robl JM: Cloned transgenic calves produced from nonquiescent fetal fibroblasts. Science 280:1256–1258 (1998).

Clark AJ, Ferrier P, Aslam S, Burl S, Denning C, Wylie D, Ross A, de Sousa P, Wilmut I, Cui W: Proliferative lifespan is conserved after nuclear transfer. Nat Cell Biol 5:535–538 (2003).

Clarke HJ, Bustin M, Oblin C: Chromatin modifications during oogenesis in the mouse: removal of somatic subtypes of histone H1 from oocyte chromatin occurs post-natally through a post-transcriptional mechanism. J Cell Sci 110:477–487 (1997).

Daniels R, Hall V, Trounson AO: Analysis of gene transcription in bovine nuclear transfer embryos reconstructed with granulosa cell nuclei. Biol Reprod 63:1034–1040 (2000).

Daniels R, Hall VJ, French AJ, Korfiatis NA, Trounson AO: Comparison of gene transcription in cloned bovine embryos produced by different nuclear transfer techniques. Mol Reprod Dev 60:281–288 (2001).

de Lange T: Protection of mammalian telomeres. Oncogene 21:532–540 (2002).

Dean W, Santos F, Stojkovic M, Zakhartchenko V, Walter J, Wolf E, Reik W: Conservation of methylation reprogramming in mammalian development: aberrant reprogramming in cloned embryos. Proc Natl Acad Sci USA 98:13734–13738 (2001).

Dean W, Santos F, Reik W: Epigenetic reprogramming in early mammalian development and following somatic nuclear transfer. Semin Cell Dev Biol 14:93–100 (2003).

Dimitrov S, Wolffe AP: Remodeling somatic nuclei in *Xenopus laevis* egg extracts: molecular mechanisms for the selective release of histones H1 and H1(0) from chromatin and the acquisition of transcriptional competence. EMBO J 15:5897–5906 (1996).

Doherty AS, Mann MR, Tremblay KD, Bartolomei MS, Schultz RM: Differential effects of culture on imprinted H19 expression in the preimplantation mouse embryo. Biol Reprod 62:1526–1535 (2000).

Eggan K, Akutsu H, Hochedlinger K, Rideout WM 3rd, Yanagimachi R, Jaenisch R: X-Chromosome inactivation in cloned mouse embryos. Science 290:1578–1581 (2000).

Fischle W, Wang Y, Allis CD: Histone and chromatin cross-talk. Curr Opin Cell Biol 15:172–183 (2003).

Galli C, Lagutina I, Crotti G, Colleoni S, Turini P, Ponderato N, Duchi R, Lazzari G: Pregnancy: a cloned horse born to its dam twin. Nature 424:635 (2003).

Gonda K, Fowler J, Katoku-Kikyo N, Haroldson J, Wudel J, Kikyo N: Reversible disassembly of somatic nucleoli by the germ cell proteins FRGY2a and FRGY2b. Nat Cell Biol 5:205–210 (2003).

Gurdon JB: Nuclear transplantation in eggs and oocytes. J Cell Sci 4:287–318 (1986).

Gurdon JB: Genetic reprogramming following nuclear transplantation in Amphibia. Semin Cell Dev Biol 10:239–243 (1999).

Gurdon JB, Byrne JA: The first half-century of nuclear transplantation. Proc Natl Acad Sci USA 100:8048–8052 (2003).

Hochedlinger K, Jaenisch R: Monoclonal mice generated by nuclear transfer from mature B and T donor cells. Nature 415:1035–1038 (2002a).

Hochedlinger K, Jaenisch R: Nuclear transplantation: lessons from frogs and mice. Curr Opin Cell Biol 14:741–748 (2002b).

Humpherys D, Eggan K, Akutsu H, Hochedlinger K, Rideout WM 3rd, Biniszkiewicz D, Yanagimachi R, Jaenisch R: Epigenetic instability in ES cells and cloned mice. Science 293:95–97 (2001).

Humpherys D, Eggan K, Akutsu H, Friedman A, Hochedlinger K, Yanagimachi R, Lander ES, Golub TR, Jaenisch R: Abnormal gene expression in cloned mice derived from embryonic stem cell and cumulus cell nuclei. Proc Natl Acad Sci USA 99:12889–12894 (2002).

Inoue K, Kohda T, Lee J, Ogonuki N, Mochida K, Noguchi Y, Tanemura K, Kaneko-Ishino T, Ishino F, Ogura A: Faithful expression of imprinted genes in cloned mice. Science 295:297 (2002).

Jaenisch R, Eggan K, Humpherys D, Rideout WM 3rd, Hochedlinger K: Nuclear cloning, stem cells, and genomic reprogramming. Cloning Stem Cells 4:389–396 (2002).

Jenuwein T, Allis CD: Translating the histone code. Science 293:1074–1080 (2001).

Kang YK, Koo DB, Park JS, Choi YH, Chung AS, Lee KK, Han YM: Aberrant methylation of donor genome in cloned bovine embryos. Nat Genet 28:173–177 (2001a).

Kang YK, Koo DB, Park JS, Choi YH, Kim HN, Chang WK, Lee KK, Han YM: Typical demethylation events in cloned pig embryos. Clues on species-specific differences in epigenetic reprogramming of a cloned donor genome. J Biol Chem 276:39980–39984 (2001b).

Kang YK, Park JS, Koo DB, Choi YH, Kim SU, Lee KK, Han YM: Limited demethylation leaves mosaic-type methylation states in cloned bovine preimplantation embryos. EMBO J 21:1092–1100 (2002).

Kato Y, Tani T, Sotomaru Y, Kurokawa K, Kato J, Doguchi H, Yasue H, Tsunoda Y: Eight calves cloned from somatic cells of a single adult. Science 282:2095–2098 (1998).

Kikyo N, Wolffe AP: Reprogramming nuclei: insights from cloning, nuclear transfer and heterokaryons. J Cell Sci 113:11–20 (2000).

Kim JM, Liu H, Tazaki M, Nagata M, Aoki F: Changes in histone acetylation during mouse oocyte meiosis. J Cell Biol 162:37–46 (2003).

Kyo S, Inoue M: Complex regulatory mechanisms of telomerase activity in normal and cancer cells: how can we apply them for cancer therapy? Oncogene 21:688–697 (2002).

Lanza RP, Cibelli JB, Blackwell C, Cristofalo VJ, Francis MK, Baerlocher GM, Mak J, Schertzer M, Chavez EA, Sawyer N, et al: Extension of cell lifespan and telomere length in animals cloned from senescent somatic cells. Science 288:665–669 (2000).

Li E: Chromatin modification and epigenetic reprogramming in mammalian development. Nat Rev Genet 3:662–673 (2002).

Liu K, Hodes RJ, Weng N: Cutting edge: telomerase activation in human T lymphocytes does not require increase in telomerase reverse transcriptase (hTERT) protein but is associated with hTERT phosphorylation and nuclear translocation. J Immunol 166:4826–4830 (2001).

Mann MR, Chung YG, Nolen LD, Verona RI, Latham KE, Bartolomei MS: Disruption of imprinted gene methylation and expression in cloned preimplantation stage mouse embryos. Biol Reprod 69:902–914 (2003).

Maser RS, DePinho RA: Connecting chromosomes, crisis, and cancer. Science 297:565–569 (2002).

Mayer W, Niveleau A, Walter J, Fundele R, Haaf T: Demethylation of the zygotic paternal genome. Nature 403:501–502 (2000).

McEachern MJ, Krauskopf A, Blackburn EH: Telomeres and their control. Annu Rev Genet 34:331–358 (2000).

McGraw S, Robert C, Massicotte L, Sirard MA: Quantification of histone acetyltransferase and histone deacetylase transcripts during early bovine embryo development. Biol Reprod 68:383–389 (2003).

McLaren A: Cloning: pathways to a pluripotent future. Science 288:1775–1780 (2000).

Mitsui K, Tokuzawa Y, Itoh H, Segawa K, Murakami M, Takahashi K, Maruyama M, Maeda M, Yamanaka S: The homeoprotein Nanog is required for maintenance of pluripotency in mouse epiblast and ES cells. Cell 113:631–642 (2003).

Miyashita N, Shiga K, Yonai M, Kaneyama K, Kobayashi S, Kojima T, Goto Y, Kishi M, Aso H, Suzuki T, et al: Remarkable differences in telomere lengths among cloned cattle derived from different cell types. Biol Reprod 66:1649–1655 (2002).

Mullins LJ, Wilmut I, Mullins JJ: Nuclear Transfer in Rodents. J Physiol, published online on October 17, 2003 as 10.1113/jphysiol.2003.049742.

Oback B, Wells D: Donor cells for nuclear cloning: many are called, but few are chosen. Cloning Stem Cells 4:147–168 (2002).

Onishi A, Iwamoto M, Akita T, Mikawa S, Takeda K, Awata T, Hanada H, Perry AC: Pig cloning by microinjection of fetal fibroblast nuclei. Science 289:1188–1190 (2000).

Pesce M, Scholer HR: Oct-4: gatekeeper in the beginnings of mammalian development. Stem Cells 19:271–278 (2001).

Polejaeva IA, Chen SH, Vaught TD, Page RL, Mullins J, Ball S, Dai Y, Boone J, Walker S, Ayares DL, et al: Cloned pigs produced by nuclear transfer from adult somatic cells. Nature 407:86–90 (2000).

Rhind SM, King TJ, Harkness LM, Bellamy C, Wallace W, DeSousa P, Wilmut I: Cloned lambs-lessons from pathology. Nat Biotechnol 21:744–745 (2003).

Rideout WM 3rd, Eggan K, Jaenisch R: Nuclear cloning and epigenetic reprogramming of the genome. Science 293:1093–1098 (2001).

Roh S, Guo J, Malakooti N, Morrison JR, Trounson AO, Du ZT: Birth of rats by nuclear transplantation using 2-cell stage embryo as donor nucleus and recipient cytoplasm. Theriogenology 59:283 (2003).

Santos F, Hendrich B, Reik W, Dean W: Dynamic reprogramming of DNA methylation in the early mouse embryo. Dev Biol 241:172–182 (2002).

Santos F, Zakhartchenko V, Stojkovic M, Peters A, Jenuwein T, Wolf E, Reik W, Dean W: Epigenetic marking correlates with developmental potential in cloned bovine preimplantation embryos. Curr Biol 13:1116–1121 (2003).

Shiels PG, Kind AJ, Campbell KH, Waddington D, Wilmut I, Colman A, Schnieke AE: Analysis of telomere lengths in cloned sheep. Nature 399:316–317 (1999).

Shin T, Kraemer D, Pryor J, Liu L, Rugila J, Howe L, Buck S, Murphy K, Lyons L, Westhusin M: A cat cloned by nuclear transplantation. Nature 415:859 (2002).

Solter D: Mammalian cloning: advances and limitations. Nat Rev Genet 1:199–207 (2000).

Strahl BD, Allis CD: The language of covalent histone modifications. Nature 403:41–45 (2000).

Suemizu H, Aiba K, Yoshikawa T, Sharov AA, Shimozawa N, Tamaoki N, Ko MS: Expression profiling of placentomegaly associated with nuclear transplantation of mouse ES cells. Dev Biol 253:36–53 (2003).

Sun FZ, Moor RM: Nuclear transplantation in mammalian eggs and embryos. Curr Top Dev Biol 30:147–176 (1995).

Tamashiro KL, Wakayama T, Akutsu H, Yamazaki Y, Lachey JL, Wortman MD, Seeley RJ, D'Alessio DA, Woods SC, Yanagimachi R, Sakai RR: Cloned mice have an obese phenotype not transmitted to their offspring. Nat Med 8:262–267 (2002).

Tian XC, Xu J, Yang X: Normal telomere lengths found in cloned cattle. Nat Genet 26:272–273 (2000).

Wade PA, Kikyo N: Chromatin remodeling in nuclear cloning. Eur J Biochem 269:2284–2287 (2002).

Wakayama T, Yanagimachi R: Cloning the laboratory mouse. Semin Cell Dev Biol 10:253–258 (1999).

Wakayama T, Perry ACF, Zucotti M, Johnson KR, Yanagimachi R: Full term development of mice from enucleated oocytes injected with cumulus cell nuclei. Nature 394:369–374 (1998).

Wakayama T, Shinkai Y, Tamashiro KL, Niida H, Blanchard DC, Blanchard RJ, Ogura A, Tanemura K, Tachibana M, Perry AC, et al: Cloning of mice to six generations. Nature 407:318–319 (2000).

Wilmut I, Schnieke AE, McWhir J, Kind AJ, Campbell KH: Viable offspring derived from fetal and adult mammalian cells. Nature 385:810–813 (1997).

Wong JM, Kusdra L, Collins K: Subnuclear shuttling of human telomerase induced by transformation and DNA damage. Nat Cell Biol 4:731–736 (2002).

Woods GL, White KL, Vanderwall DK, Li GP, Aston KI, Bunch TD, Meerdo LN, Pate BJ: A mule cloned from fetal cells by nuclear transfer. Science 301:1063 (2003).

Wrenzycki C, Wells D, Herrmann D, Miller A, Oliver J, Tervit R, Niemann H: Nuclear transfer protocol affects messenger RNA expression patterns in cloned bovine blastocysts. Biol Reprod 65:309–317 (2001).

Xu J, Yang X: Telomerase activity in early bovine embryos derived from parthenogenetic activation and nuclear transfer. Biol Reprod 64:770–774 (2001).

Xue F, Tian XC, Du F, Kubota C, Taneja M, Dinnyes A, Dai Y, Levine H, Pereira LV, Yang X: Aberrant patterns of X chromosome inactivation in bovine clones. Nat Genet 31:216–220 (2002).

Young LE, Sinclair KD, Wilmut I: Large offspring syndrome in cattle and sheep. Rev Reprod 3:155–163 (1998).

Zhou Q, Renard J-P, Le Friec G, Brochard V, Beaujean N, Cherifi Y, Fraichard A, Cozzi J: Generation of fertile cloned rats by regulating oocytes activation. Science 302:1179 Epub (2003).

Cytogenet Genome Res 105:292–301 (2004)
DOI: 10.1159/000078201

Spatial genome organization during T-cell differentiation

S.H. Kim,[a] P.G. McQueen,[b] M.K. Lichtman,[c] E.M. Shevach,[c] L.A. Parada[a] and T. Misteli[a]

[a] National Cancer Institute, NIH; [b] Mathematical and Statistical Laboratory, DCB, CIT, NIH, and
[c] National Institute of Allergy and Infectious Diseases, Laboratory of Immunology, Bethesda, MD (USA)

Abstract. The spatial organization of genomes within the mammalian cell nucleus is non-random. The functional relevance of spatial genome organization might be in influencing gene expression programs as cells undergo changes during development and differentiation. To gain insight into the plasticity of genomes in space and time and to correlate the activity of specific genes with their nuclear position, we systematically analyzed the spatial genome organization in differentiating mouse T-cells. We find significant global reorganization of centromeres, chromosomes and gene loci during the differentiation process. Centromeres were repositioned from a preferentially internal distribution in undifferentiated cells to a preferentially peripheral position in differentiated CD4+ and CD8+ cells. Chromosome 6, containing the differentially expressed T-cell markers CD4 and CD8, underwent differential changes in position depending on whether cells differentiated into CD4+ or CD8+ thymocytes. Similarly, the two marker loci CD4 and CD8 showed distinct behavior in their position relative to the chromosome 6 centromere at various stages of differentiation. Our results demonstrate that significant spatial genome reorganization occurs during differentiation and indicate that the relationship between dynamic genome topology and single gene regulation is highly complex.

The mammalian cell nucleus is a highly compartmentalized organelle (Lamond and Earnshaw, 1998; Dundr and Misteli, 2001). In addition to the well-established compartmentalized nature of many intranuclear protein domains (Lamond and Earnshaw, 1998; Matera, 1999; Dundr and Misteli, 2001), the distribution of the genetic material within the three-dimensional space of the interphase nucleus is also heterogeneous. The genetic material that makes up each single chromosome is not dispersed throughout the nucleus, but is confined to a spatially limited nuclear subvolume, referred to as a chromosome territory (Cremer and Cremer, 2001; Parada and Misteli, 2002).

The position of chromosome territories in the nucleus is not random (Cremer and Cremer, 2001; Parada and Misteli, 2002).

In plants and *Drosophila*, chromosomes are often arranged in a Rabl configuration with their telomeres and centromeres clustered, typically on opposing sides of the nucleus (Hochstrasser et al., 1986; Abranches et al., 1998). In yeast, centromeres cluster in a Rabl-like configuration (Jin et al., 2000). In mammalian cells, chromosomes are either organized in radial patterns relative to the center of the nucleus or in clusters relative to each other (Parada and Misteli, 2002). Radial patterns have been characterized in lymphocytes and fibroblasts and have been correlated to a chromosome's gene density with gene-dense chromosomes positioned towards the nuclear center and gene-poor chromosomes positioned towards the periphery (Croft et al., 1999; Boyle et al., 2001). Radial positioning according to chromosome size has also been described (Sun et al., 2000). In addition to radial positioning, chromosomes may also be located in preferential positions relative to each other. In mouse lymphocytes a cluster of chromosomes 12, 14, and 15 has been characterized (Parada et al., 2002). The relative positioning pattern of these three chromosomes is conserved through mitosis and during cancerous transformation (Parada et al., 2002).

S.H.K. and M.K.L. are HHMI-NIH Research Scholars. T.M. is a Keith R. Porter Fellow.

Received 1 October 2003; manuscript accepted 17 October 2003.

Request reprints from: Dr. Tom Misteli, National Cancer Institute, NIH
41 Library Road, Bldg. 41, B610, Bethesda, MD 20892 (USA)
telephone: +1-301-402-3959; fax: +1-928-832-0970
e-mail: mistelit@mail.nih.gov

Although chromosome territories are distinct physical entities, they are not solid, impermeable masses of chromatin (Cremer and Cremer, 2001; Van Driel et al., 2003). In contrast, studies on the diffusional motion of proteins through the nucleus demonstrate relatively high access of regulatory factors even to highly condensed, centromeric heterochromatin regions (Phair and Misteli, 2000; Pederson, 2001; Cheutin et al., 2003; Festenstein et al., 2003). Based on these observations it appears that the nucleoplasm is an extensively branched compartment that reaches well into the interior of chromosome territories and fills the volume between chromatin fibers (Pederson, 2001; Van Driel et al., 2003). Presumably as a consequence of the open nature of chromatin domains and the free accessibility of regulators to genes, there appears to be no strict correlation between the activity of a gene and its position in the interior or at the surface of a chromosome territory (Kurz et al., 1996; Nogami et al., 2000; Mahy et al., 2002).

While these observations argue against a strict requirement for positioning in the control of gene expression, several lines of evidence indicate an important role for spatial positioning effects on gene regulation. Highly active genome regions can be expulsed from chromosome territories, potentially increasing the exposure to regulatory factors or to place sequences near nuclear compartments that supply regulatory factors (Volpi et al., 2000; Mahy et al., 2002; Williams et al., 2002). A functional role of positioning is also indicated by the fact that radial chromosome distribution patters are evolutionarily conserved (Tanabe et al., 2002). Relative positioning might be functionally important in the formation of chromosome aberrations since proximally positioned chromosomes are presumably more likely to undergo illegitimate re-joining than chromosomes distally positioned from each other (Roix et al., 2003). Indeed, preferential proximal positioning of frequent translocation partners has been demonstrated for several loci combinations, including BCR-ABL and myc-IGH (Lukasova et al., 1997; Neves et al., 1999; Roix et al., 2003).

If spatial genome organization contributes to gene regulation, one might expect that genes alter their spatial position when cells undergo dramatic changes in their gene expression programs, for example during differentiation. Consistent with this prediction, during lymphoblast differentiation, the position of several differentiation specific genes changes in correlation with their transcriptional activity (Brown et al., 1997; Francastel et al., 2000, 2001). As genes become activated they are moved out of heterochromatin blocks, whereas during their inactivation they are moved into heterochromatin regions (Brown et al., 1997; Francastel et al., 2001; Skok et al., 2001). Furthermore, immunoglobulin genes are repositioned from the periphery in hematopoetic progenitors towards the center in pro-B cells (Kosak et al., 2002).

Mouse T-cell differentiation is an ideal model system to systematically study the correlation between spatial genome organization and gene activity since the differentiation process is molecularly relatively well understood and highly purified fractions of cells at various stages of differentiation can be isolated. Development of T lymphocytes occurs in the thymus in an ordered multi-step process to eventually give rise to MHC class-II-restricted CD4+ helper cells and MHC class-I-restricted CD8+ cytotoxic T cells (Kioussis and Ellmeier, 2002). The differentiation process originates from bone marrow derived lymphoid progenitor cells, which do not express either of the T-cell marker loci CD4 or CD8. These double-negative cells (DN) differentiate into immature double-positive cells (DP), which express both CD4 and CD8 markers. At this stage the DP cells undergo selection and differentiate into CD4+CD8– T-helper cells (CD4+) and CD4–CD8+ cytotoxic T cells (CD8+). Here we report on the systematic characterization of the spatial reorganization of the CD4 and CD8 loci, of centromeres and of chromosome 6 during T-cell differentiation.

Materials and methods

Isolation of thymocyte populations

DN, DP, CD4+ and CD8+ thymocytes were isolated on a FACStar TM Cell Sorter as described (Piccirillo et al., 2002). In brief, fresh thymi were collected from 6–8-week-old C57BL/6 mice. Single cell suspensions were prepared by passing cells through a sterile wire mesh. For the purification of double-negative cells, single- and double-positive thymocytes were removed from the sample by negative selection via depletion of CD4+ and CD8+ thymocytes with magnetic beads on the AutoMACS magnetic separation system according to the manufacturer's protocol. The resulting cell preparations were incubated with FITC-anti-CD8 antibody, PE-B220 antibody, and Tricolor-CD4 antibody in PBS/2 % FCS for 15 min at 4 °C. The cells were then washed and resuspended in RPMI 1640/10 % FCS without phenol red for FACS sorting. B cells were excluded during sorting based on their positive signal for B220. The purity of the final preparations was typically >95 %.

Indirect immunofluorescence and fluorescence in situ hybridization

For all microscopy procedures, cells were fixed immediately after isolation in methanol:acetic acid (3:1) or in 3 % paraformaldehyde in PBS, pH 7.25, for 20 min at room temperature, followed by permeabilization with 0.5 % Triton/PBS for 5 min. Results were independent of fixation procedure used. Indirect immunofluorescence to detect centromeres was carried out as described (Misteli and Spector, 1996) using the ANA-C (Sigma) antibody at 1:100 in PBS. The primary antibody was detected with a Texas Red-anti-human IgG antibody (Vector Technologies). For chromosome painting, a whole chromosome painting probe was prepared from flow-sorted chromosome 6 by degenerate oligonucleotide-primed polymerase chain reaction to incorporate Texas Red-dUTP for labeling as previously described (Parada et al., 2002). Genomic DNA probes for detection of CD4, CD8 and centromere 6 were obtained from bacterial artificial chromosome clones RP23-130D15 (CD4), RP23-229N15 (CD8) and D6MIT83 (centromere 6) (Korenberg et al., 1999). Clones were chosen using Ensembl Genome Browser (http//www.ensembl.org), Locuslink (http://www.ncbi.nih.gov/LocusLink/) and Genome Mapviewer (http://www.ncbi.nlm.nih.gov/mapview). Detailed protocols for FISH and probe generation are described in Parada et al. (2002), and at http://rex.nci.nih.gov/RESEARCH/basic/lrbge/cbge.html. Briefly, probes were produced by nick translation with the use of digoxigenin- or biotin-conjugated dUTP (Roche). Cells were hybridized to probe combinations overnight in a moist chamber at 37 °C. Digoxigenin- and biotin-labeled probes were detected with anti-digoxigenin-FITC (Roche) and streptavidin-Cy5 (Amersham), respectively. Nuclear DNA was detected using DAPI.

Microscopy and quantitative analysis

Samples were examined on a Nikon Eclipse E800 microscope equipped with epi-fluorescence optics and a Photometrics MicroMax cooled CCD camera controlled by Metamorph 4.96. For quantitation of centromere positions, single equatorial planes were analyzed, for analysis of chromosome 6 or gene loci, 3-dimensional z-stacks were obtained and all nuclei in these images that displayed the expected number of distinct hybridization signals were collected into 2D maximum intensity projection image collages. For determination of the radial position of chromosome territories, the perimeter of each cell as well as chromosome 6 were drawn on the maximum projection and the center of mass for the nucleus and the chromosome territory determined using Metamorph software. These regions of interest were collated

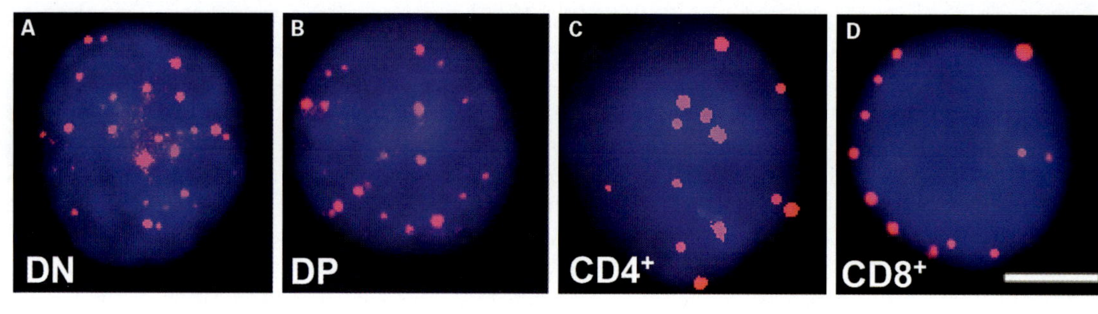

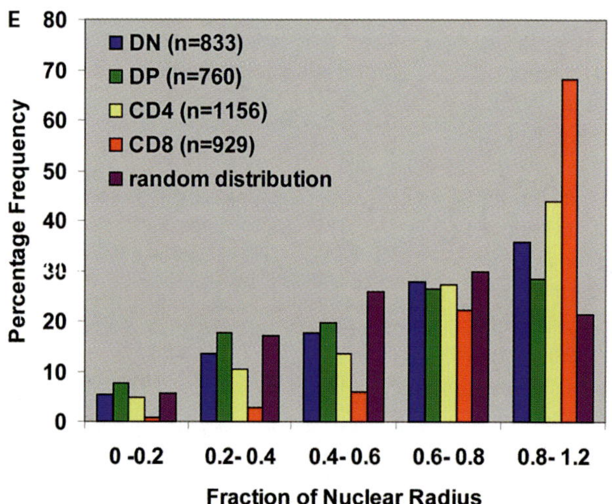

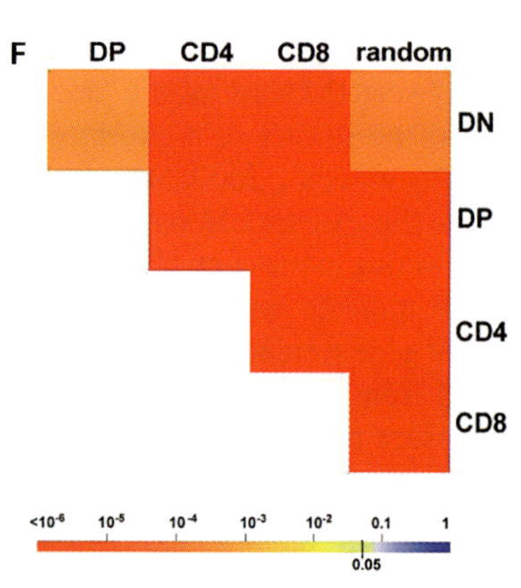

Fig. 1. Radial distribution of centromeres during T-cell differentiation. Centromeres of all chromosomes were detected by a pan-centromere antibody in (**A**) DN, (**B**) DP, (**C**) CD4+ or (**D**) CD8+ cells fixed in paraformaldehyde. Single equatorial planes are shown. Centromeres (red), DAPI (blue). (**E**) Quantitation of the radial distribution of centromeres. Data was binned into 5 concentric areas with equidistant radius from the center. (**F**) Pairwise comparison of cumulative radial distributions using the KS test. The random distribution represents a two-dimensional projection of points uniformly placed in a three-dimensional sphere. n indicates the number of analyzed centromeres. Bar: ∼ 5 µm.

with their corresponding regions through the focal planes to verify that the regions were representative of the entire chromosome. For measurement of loci positions, geometrical coordinate centers of loci and nuclei identified using manual intensity thresholds were measured in automated Metamorph analyses as described (Roix et al., 2003). Nuclear radii (R_n) were calculated from $R_n = (A/\pi)^{0.5}$, where (A) is the thresholded nuclear pixel area. Radial locus position was calculated as $(L:C)/R_n$, where (L:C) is the distance from the center of an individual locus to the nuclear center. The absolute spatial separations between loci or between loci and centromeres ($L_1:L_2$) were normalized as a fraction of nuclear radius ($L_1:L_2$) to account for natural variations in nuclear size, which may influence relative positioning. Radial positioning data was subdivided into bins as indicated in the figures. A random distribution was generated by 2D projection of points uniformly placed in a three-dimensional sphere (Roix et al., 2003). Calculations and graphs were made using Excel. For quantitative measurements between 100 and 500 nuclei from multiple experiments were analyzed.

Statistical analysis

For statistical analysis cumulative positioning and distance distributions were compared pairwise using the Kolmogorov- Smirnov test (KS-test) (Stevens, 1982; Press et al., 1992). The null hypothesis was that the experimentally measured distributions are produced by the same underlying random process. The smaller the p value, the less likely the two experimental distributions are from the same underlying process. We also conducted contingency table χ^2 analysis on the binned data points (Fienberg, 1982; Press et al., 1992). In general we found that the KS test produced the more sensitive

comparisons in the sense that *P* values were smaller. Both tests as well as the simulation of uniform distributions were performed using standard numeral algorithms coded in Java. The color-coded heat diagrams were generated in Mathematica 4.2.

Results

Peripherization of centromeres during T-cell differentiation

To test whether genomes are re-organized globally during T-cell differentiation, we first analyzed the nuclear distribution of all centromeres. DN, DP, CD4+ and CD8+ cell populations were isolated by FACS analysis as previously described using the combinatorial presence of CD4 and CD8 as markers (Piccirillo et al., 2002; see Materials and methods). The purity of each cell population was routinely above 95% (data not shown). To detect centromeres of all chromosomes, we performed immunofluorescence microscopy using a pan centromeric antibody (Fig. 1). By qualitative inspection of equatorial optical sections, centromeres in DN and DP cells were distributed throughout the nucleus with no apparent pattern (Fig. 1A, B). In contrast, in differentiated CD4+ cells and in CD8+ cells,

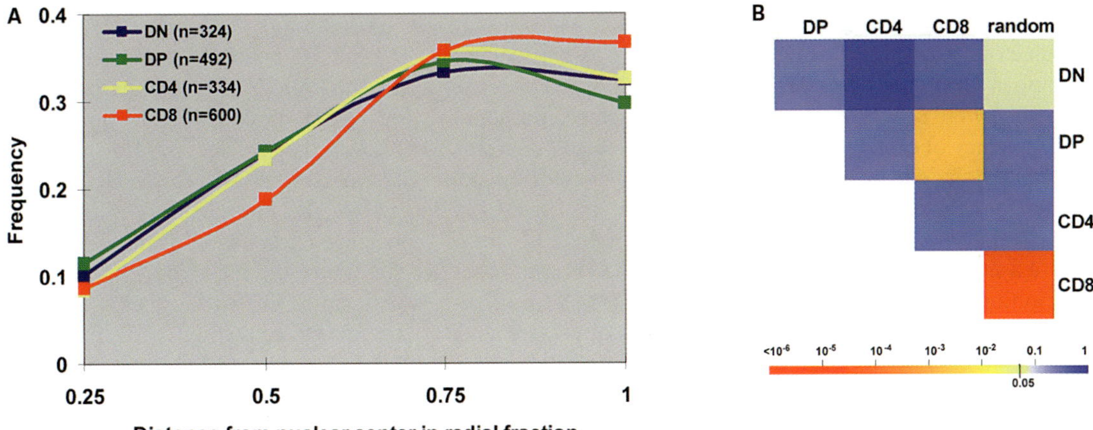

Fig. 2. Radial distribution of centromere 6 during T-cell differentiation. Centromere 6 was detected in paraformaldehyde-fixed cells using a specific FISH probe. (**A**) Quantitation of the radial distribution of centromere 6 on projected stacks of optical sections. Data was binned into 5 concentric areas with equidistant radius from the center. (**B**) Pairwise comparison of cumulative radial distributions using the KS test. n indicates the number of analyzed centromeres.

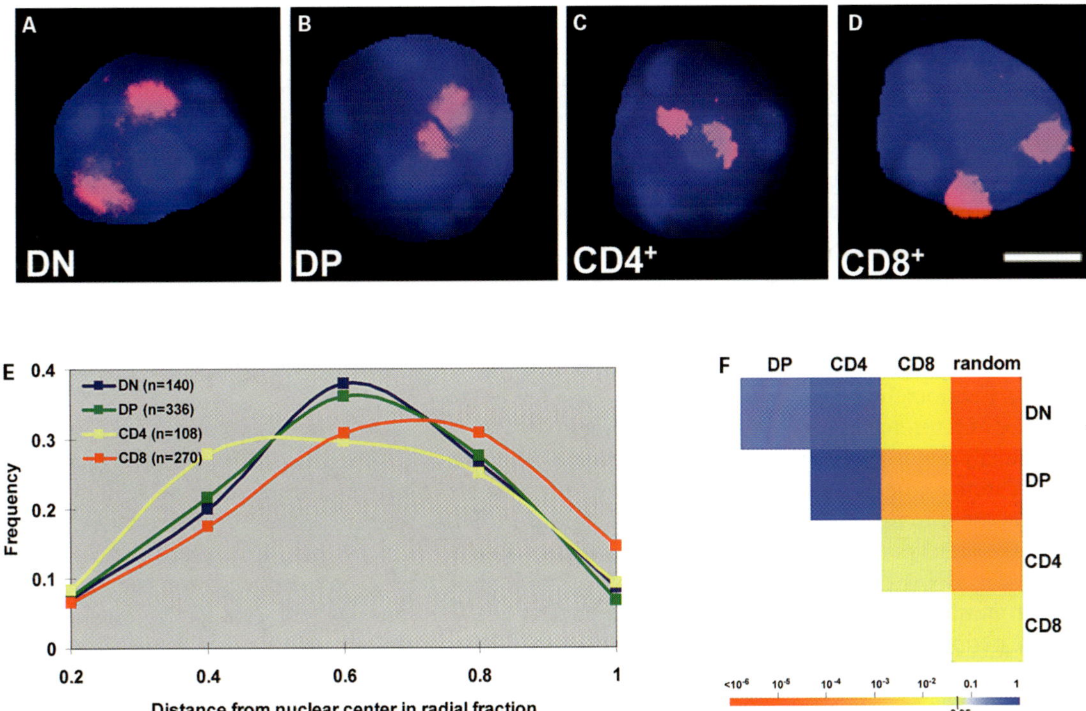

Fig. 3. Radial distribution of chromosome 6 during T-cell differentiation. Chromosome 6 was detected by a specific whole-chromosome painting probe in (**A**) DN, (**B**) DP, (**C**) CD4+ or (**D**) CD8+ cells fixed in paraformaldehyde. Projections of image stacks are shown. Chromosome 6 (red), DAPI (blue). (**E**) Quantitation of the radial distribution of chromosome 6. Data was binned into 5 concentric areas with equidistant radius from the center. (**F**) Pairwise comparison of cumulative radial distributions using the KS test. n indicates the number of analyzed chromosomes. Bar: ~ 5 µm.

centromeres appeared more frequently towards the nuclear periphery (Fig. 1C, D). The peripheralization was particularly striking in CD8+ cells (Fig. 1D). For quantitative analysis, the cell nuclei were divided into equidistant, concentric areas and the frequency of centromere positioning within each area was determined in equatorial image sections as previously described for chromosomes and gene loci (Croft et al., 1999; Roix et al., 2003) (Fig. 1E). Comparison of the centromere distribution to a simulated random distribution indicated non-random positioning of centromeres in all cell types (Fig. 1F). Confirming our qualitative observations, 37 and 28% of centromeres were found in the outermost area in DN and DP cells, respectively, but 44 and 68% of centromeres in CD4+ and CD8+ cells were found in the most peripheral region (Fig. 1E). Pairwise comparison using the KS-test indicated highly significant differences between the overall distributions of centromeres in the single positive cell types CD4+ and CD8+ compared to DN and DP cells (Fig. 1F). Comparable results were obtained in paraformaldehyde or methanol/acetic acid-fixed cells. These observations indicate the occurrence of large-scale genome reorganization during the differentiation of DP cells into CD4+ or CD8+ cells.

Positioning of chromosome 6

The genes for the two major differentiation markers CD4 and CD8 both map to mouse chromosome 6. We thus asked whether the centromeres of chromosome 6 behaved similarly as the general centromeres population or whether the presence of these differentially expressed markers affected centromere 6 positioning. To this end we visualized chromosome 6 centromeres by fluorescence in situ hybridization (FISH). As observed for the general population, centromeres 6 were essentially identically distributed in DN and DP cells (Fig. 2A). In CD4+ cells, the distribution was unchanged, but a significant reduction of the more internal centromeres and a corresponding increase in the frequency of the most peripheral population of centromeres were observed in CD8+ cells (Fig. 2A; $P = 0.0015$). Centromeres of chromosome 6 thus roughly follow the behavior of centromeres of the other chromosomes.

To analyze not only the position of centromere 6 but the entire chromosome 6, we used a specific whole-chromosome painting probe to detect chromosomes 6 by FISH (Fig. 3A–D). While the radial distribution of the chromosome was virtually identical in DN and DP cells with a preferential positioning of the chromosome roughly halfway between the nuclear center and the periphery (Fig. 3A, B), differentiated CD4+ and CD8+ cells showed distinct distribution patterns (Fig. 3C, D). In CD4+ cells, the position of chromosome 6 was shifted towards the center of the nucleus, whereas in CD8+ cells it was shifted significantly towards the periphery (Fig. 3C–F). Since in both differentiated cell populations one of the two markers is expressed and the other silenced, the differential positioning of the centromere and of chromosome 6 indicate that the radial position of the entire chromosome is not critical for expression of its genes. Furthermore, in combination with the results on centromere 6 positioning, we conclude that the position of the centromeres alone is not necessarily a faithful indicator of the position of the entire chromosome.

Positioning of CD4 and CD8 gene loci

The CD4 and CD8 marker loci are differentially expressed during the progression from DN precursor cells to single positive T-cells. To ask whether the activity of these genes correlates with their nuclear position, we analyzed the location of the CD4 and CD8 loci by FISH using specific BAC probes (Fig. 4A–D). The position of the CD4 locus was essentially unchanged between DN and DP cells (Fig. 4A, B, E). In CD4+ cells, where the locus is highly expressed, the CD4 gene was significantly shifted towards the interior of the nucleus compared to DP cells (Fig. 4C, E, F; $P = 8.9 \times 10^{-5}$). However, in CD8+ cells, where CD4 is not expressed, the preferential position of the locus was shifted towards the periphery (Fig. 4D, E; $P = 0.008$). While the relocation of the active CD4 in CD4+ cells towards the interior and the repositioning of the inactive CD4 in CD8+ cells towards the periphery might indicate a correlation between locus activity and radial positioning, the identical distribution of CD4 in DN and DP cells, where the locus is differentially expressed, indicates that radial positioning does not strictly correlate with gene activity.

To test whether the CD8 locus behaves in a similar fashion, we localized CD8 by FISH (Fig. 4A–D). Similar to the situation for CD4, the CD8 locus was virtually identically distributed in DN and DP cells (Fig. 4A, B). In CD4+ cells, the CD8 locus did not change its position significantly compared to DN and DP cells, further indicating that radial position does not correlate with gene activity (Fig. 4C, G, H). The lack of correlation was further supported by the absence of any positional shift in CD8+ cells, where the gene is highly expressed (Fig. 4D, G, H). As previously noted for several loci, the localization of both CD4 and CD8 in all cell types was distinct from a random distribution (Fig. 4F, H; Roix et al., 2003). Taken together, these observations clearly demonstrate that repositioning of the marker gene loci occurs during the differentiation process. However, it is also clear from the distinct movements of the two marker loci upon gene silencing that gene activity does not correlate with a particular, predictable type of repositioning.

Relative positioning of CD4 and CD8 loci

The CD4 locus maps to chromosome 6F2 and the CD8 locus to chromosome 6C3. Both loci are distal from the centromere and are separated by 58.3 Mb. To ask whether activation of these loci affects their positions relative to the other locus on the same chromosome, we simultaneously visualized the CD4 and CD8 loci by dual-color FISH and measured the relative interdistance between the most proximal locus pair in each cell (Fig. 5A). As expected, since the two loci are localized on the same chromosome, the vast majority of pairs were found separated by less than 40% of nuclear diameter (Fig. 5A). A statistically significant difference between the interdistances of the loci in DN and DP was detected (Fig. 5A, B). In DN cells a somewhat higher fraction of loci pairs were in close proximity compared to the DP cells, possibly indicative of decondensation of chromatin in DP cells compared to DN cells (Fig. 5A, B). No significant difference in relative positioning was found between DN or DP cells compared to single-positive cells. We conclude that the simultaneous activation of CD4 and CD8 results in a slight separation of the two loci.

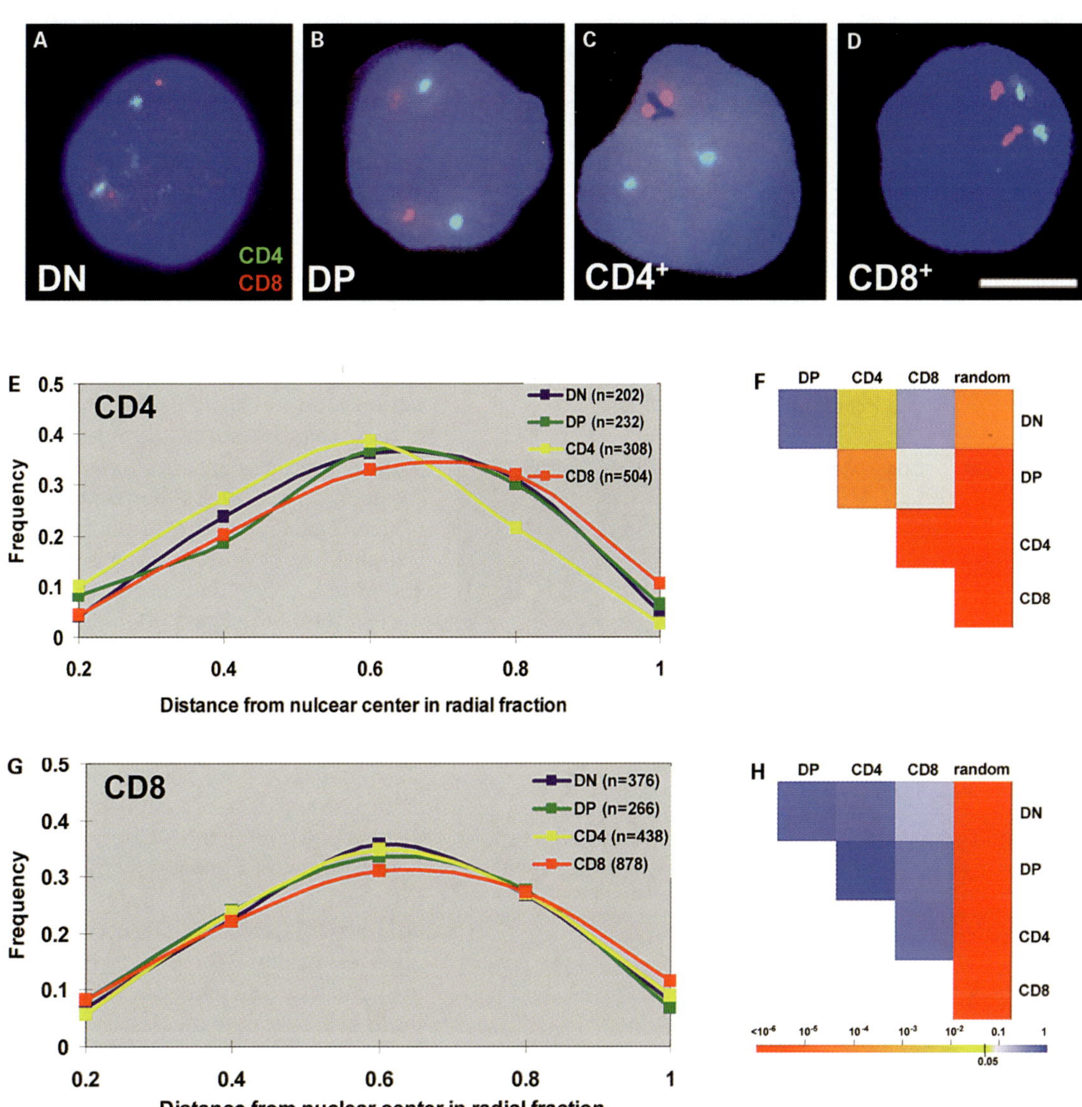

Fig. 4. Radial distribution of CD4 and CD8 loci during T-cell differentiation. CD4 (green) and CD8 (red) were detected by specific probes in (**A**) DN, (**B**) DP, (**C**) CD4+ or (**D**) CD8+ cells fixed in paraformaldehyde. DAPI is shown in blue. (**E, G**) Quantitation of the radial distribution of (**E**) CD4 and (**G**) CD8. Data was binned into 5 concentric areas with equidistant radius from the center. (**F, H**) Pairwise comparison of cumulative radial distributions using the KS test for (**F**) CD4 and (**H**) CD8. n indicates the number of analyzed loci. Bar: ~ 5 μm.

Highly transcribed genome regions have previously been reported to frequently localize outside of the main body of chromosome territories (Volpi et al., 2000; Mahy et al., 2002; Williams et al., 2002). Upon activation these regions are expelled from chromosome territories in the form of extended loops. Although CD4 and CD8 are not believed to be in large, highly transcribed chromosomal regions, we tested whether the active CD4 and CD8 loci were part of an extrachromosomal loop. To this end, we performed triple-color FISH on DP cells to detect CD4, CD8 and the chromosome territory (Fig. 5C). We find that both expressed loci are associated with the chromosome 6 territory. The same configuration was found in DN cells (data not shown). Based on this observation, we suggest that neither CD4 nor CD8 are part of an extended chromatin loop. This conclusion is consistent with the lack of change in their relative positioning, since expulsion of one locus on an extra-chromosomal loop would result in an increase in the relative distance to the other locus.

Centromeric proximity of CD4 and CD8

Centromeres and pericentromeric regions are generally transcriptionally silenced and several gene loci have been demonstrated to become associated with centromeric heterochromatin upon silencing (Brown et al., 1997; Francastel et al., 2000, 2001). Since CD4 and CD8 are silenced in CD8+ and CD4+ cells, respectively, and are both silenced in DN cells, we tested whether inactivation of these loci correlated with more proximal positioning to centromeric regions. To this end we

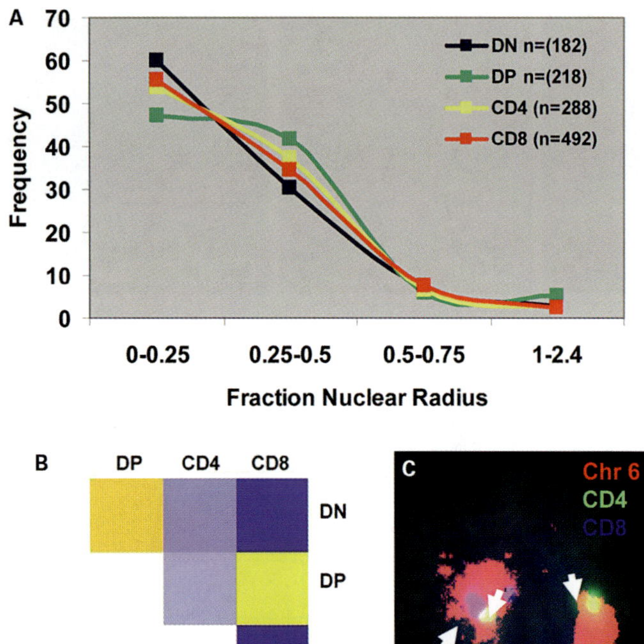

Fig. 5. Relative distribution of CD4 and CD8 loci during T-cell differentiation. (**A**) Quantitation of the relative distance between the most proximal CD4 and CD8 pair in each cell. Data was divided into 4 bins. (**B**) Pairwise comparison of cumulative radial distributions using the KS-test. (**C**) Simultaneous detection of chromosome 6 (red), CD4 locus (green) and CD8 locus (blue) in a DP cell fixed in paraformaldehyde to test whether the loci are present in extended loops outside of the chromosome territory. Both loci are found within the chromosome territory. Arrowheads indicate CD4 loci, arrows indicate CD8 loci. DAPI is in blue. n indicates the number of analyzed loci pairs. Bar: ~ 5 µm.

performed dual FISH for CD4 or CD8 and chromosome 6 centromeres and measured the relative interdistances between the closest pair (Fig. 6). We find a strong correlation between gene inactivity and increased proximity to chromosome 6 centromere for the CD8 locus (Fig. 6A–D, I, J). In DN cells, the inactive CD8 locus was found close to centromere 6, whereas in DP cells, the distance between centromere 6 and the active CD8 locus increased (Fig. 6A, B, I, J). Even more strikingly, in CD4+ cells, the inactive CD8 locus was found near centromeres (Fig. 6C, I, J), but in CD8+ cells, the now active locus had moved away significantly from the centromere (Fig. 6D, I, J). This shift was statistically significant ($P = 0.0023$). While these observations are consistent with a silencing role of centromeres, the CD4 locus behaved very differently (Fig. 6E–H). Both DP and CD4+ cells, where the CD4 locus is active, exhibited somewhat closer association of the CD4 locus with chromosome 6 centromere than in the DN and CD8+ cells, where CD4 is inactive (Fig. 6E–H, K, L). Notably, though, for both CD4 and CD8 their relative position to the centromere in the inactive state in single-positive cells was similar to that in DN cells and the relative centromere position in the active

state in single-positive cells was similar to that in DP cells. We conclude from these observations that the relative position of the loci to centromeres does not necessarily correlate with their gene activity.

Discussion

We provide here a characterization of the changes in spatial organization of the genome during a cellular differentiation process. Analysis of entire chromosomes, centromeres and the two key marker gene loci CD4 and CD8 during T-cell differentiation indicates significant changes in the spatial positioning of these elements during the differentiation process. Our observations are consistent with reports of global genome reorganization in several other differentiation systems. Peripherization of centromeres, as seen in the transition from DP to CD8+ cells, has been reported during myoblast differentiation (Chaly and Munro, 1996). In contrast, centromeres accumulate in the nuclear interior during Purkinje cell differentiation, suggesting that peripherization of centromeres is not a general feature of differentiated cells (Choh and De Boni, 1996; Martou and De Boni, 2000). Repositioning of entire chromosome territories, as seen for chromosome 6 during the progression of DP cells to CD4+ and CD8+ cells, has also been observed during cell-cycle exit of fibroblasts and lymphoblasts (Bridger et al., 2000) and distinct positions of the X chromosome have been found in cortical neurons of epileptic foci compared to normal neurons (Manuelidis, 1984). Changes in the position of genes, as here observed for CD4 and CD8, occur frequently during differentiation for example for BCR, ABL, PML and C-MYC during hematopoetic differentiation and for immunoglobulin loci during B-cell development (Lukasova et al., 1997; Neves et al., 1999; Bartova et al., 2000; Kosak et al., 2002). The sum of these and other observations clearly demonstrates that the spatial, dynamic reorganization of genomes is a common event during cellular differentiation.

Our observations allow us to draw a general, although crude, model of the genome reorganization that occurs during T-cell differentiation. The transition from DN to DP cells occurs with relatively moderate changes in large-scale organization. Neither centromeres in general, nor centromere 6 or chromosome 6 show dramatic repositioning during this time. As DP cells enter the CD4 pathway, however, several significant changes occur. The chromosome 6 territory appears to move towards the center of the nucleus, although at the same time the chromosome 6 centromere remains essentially stationary or is slightly moved towards the periphery. Both the CD4 and CD8 locus remain stationary compared to DP cells. And while the CD8 locus now becomes associated with centromeric heterochromatin, the CD4 locus does not change its relative position to the centromere. Differentiation from DP to CD8+ cells involves more dramatic changes in spatial organization. In this case, chromosome 6 becomes strongly peripheralized and the CD4 locus is dramatically shifted towards the periphery. The CD8 locus does not change its radial position, but remains distant from the centromere, whereas the CD4 locus appears to dissociate from the centromere.

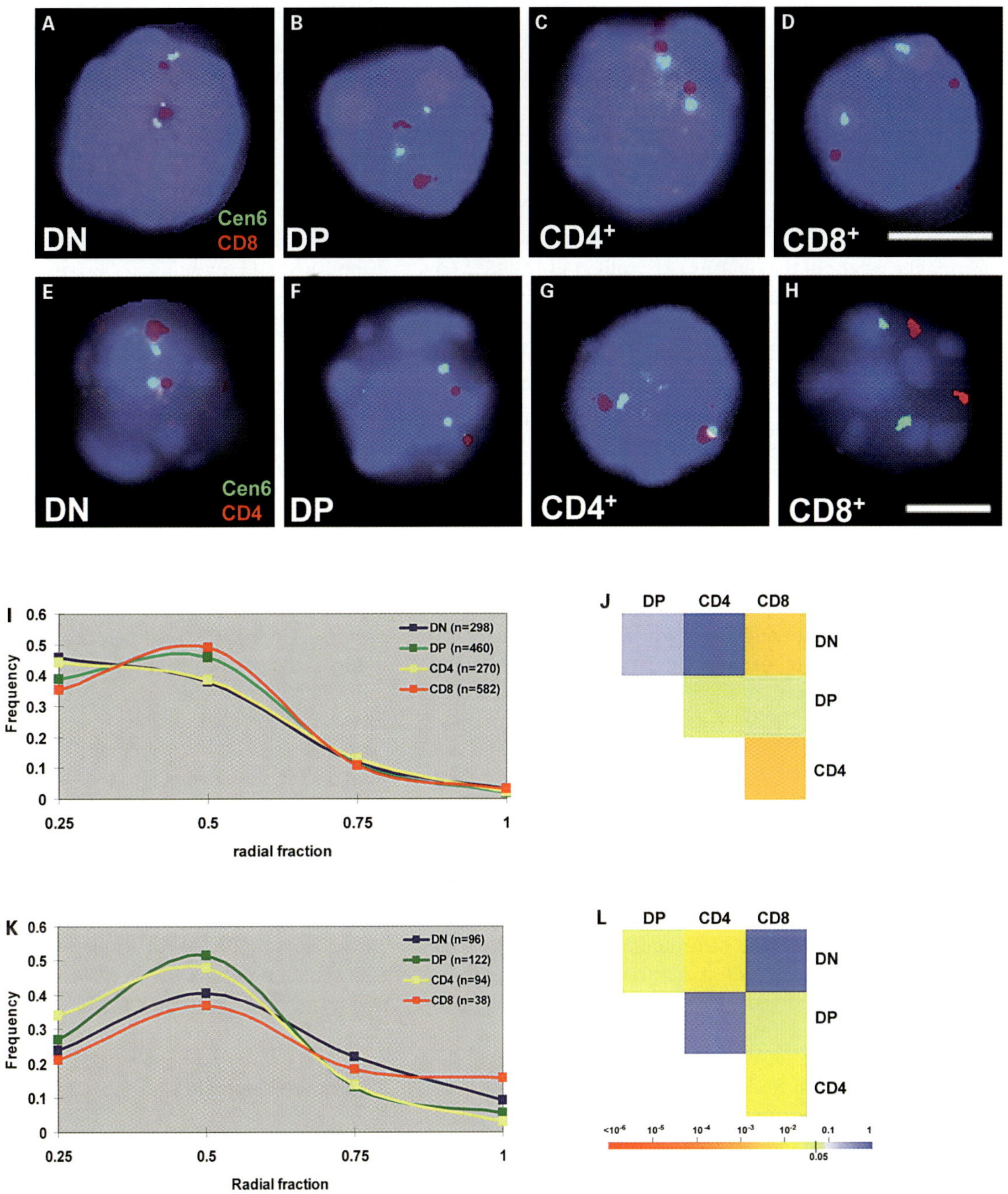

Fig. 6. Relative association of CD4 or CD8 loci with chromosome 6 centromere during T-cell differentiation. Centromere 6 (green) and CD4 or CD8 (red) were detected by specific probes in (**A, E**) DN, (**B, F**) DP, (**C, G**) CD4+ or (**D, H**) CD8+ cells fixed in paraformaldehyde. DAPI is shown in blue. (**I, K**) Quantitation of the relative distance between the closest pair of centromere 6 and (**I**) CD8 or (**K**) CD4 in each cell nucleus. Data was divided into 4 bins. (**J, L**) Pairwise comparison of cumulative radial distributions using the KS-test for (**J**) centromere 6-CD8 distances and (**H**) centromere 6-CD4 distances. n indicates the number of analyzed centromere-loci pairs. Bar: ∼ 5 μm.

Our finding that transition from DN to DP occurs in the absence of significant changes in genome organization, but the differentiation into single-positive cells occurs via large scale genome rearrangements, implies that the mechanisms of gene silencing in DN and in single-positive cells are distinct and that the silencing in DN cells is independent of the spatial organization. This difference might be related to the fact that while CD4 and CD8 in DN cells are only temporarily inactive and need to be reactivated as cells become DP, silencing of these loci in single-positive cells is permanent. This observation is consistent with a model in which CD4 and CD8 in DN cells are in a potentiated transcriptional state, but their expression is inhibited by cis-acting regulatory factors. In contrast, silencing of loci in single-positive cells may be mediated by epigenetic modifications that alter chromatin structure (Kioussis and Ellmeier, 2002).

The systematic analysis of chromosomes, centromeres and specific gene loci described here allows us to test the general validity of several correlations between gene activity and nuclear positioning. First, the radial position of genes has been linked to their gene activity. In many mammalian cells, the nuclear periphery is enriched in transcriptionally silent heterochromatin. Thus, active genes might preferentially be found towards the center of the nucleus (Andrulis et al., 1998; Mahy et al., 2002). This model is supported by the observation that early-replicating genome regions, which are indicative of active domains, localize towards the interior of the nucleus (Sadoni et al., 1999). Consistently with this notion we find internalization of the CD4 locus in CD4 cells. However, our observation that no significant change in positioning of active and inactive loci was found between DN and DP cells and the slight movement of the active CD8 locus towards the periphery argue against a general correlation between radial positioning and gene activity. This conclusion is consistent with the observation that transcription sites can be found throughout the nuclear volume (Wansink et al., 1993). Second, a number of loci, many of which involved in B- and T-cell differentiation have been reported to become repositioned away from the centromere upon activation (Brown et al., 1997, 1999; Francastel et al., 2001; Kosak et al., 2002). While we observe a correlation between the relative locus-centromere positioning and gene activity for CD4 and CD8, increased proximity did not necessarily result in transcriptional silencing. CD8 was found more distant from centromeres in its active state in DP and CD8+ cells, but CD4 was found closer to its centromere in its active state in DP and CD4+ cells. Thus, while the relative positioning of a locus to its centromere appears to correlate with its activity, distal positioning from the centromere does not necessarily indicate locus activity. Third, it is attractive to speculate that differentially regulated genes on the same chromosome might be spatially separated from each other to provide different nuclear environments each supportive of differential regulation for the genes (Lemon and Tjian, 2000). Our observation of unchanged relative positioning of CD4 and CD8 argues against this model. Similarly, our observation that both CD4 and CD8 were associated with the chromosome territory in DP cells indicates that these two marker genes are not looped out from the chromosome body as has been observed for several highly transcribed genomic regions (Volpi et al., 2000; Mahy et al., 2002; Williams et al., 2002). The conclusion from these observations must be that despite the clear changes in nuclear positioning, it is still unclear how positioning within the nucleus, relative to centromeres, and relative to other genes relates to gene activity.

A major question in understanding spatial genome organization is how patterns of chromosome or loci organization are established and maintained. Two extreme models can be envisioned. It can be argued that the transcriptional activity of a genome region determines the physical properties of the chromatin. It is well established that transcriptionally active chromatin is generally more decondensed than inactive chromatin. The physical properties of chromatin and the sum of interacting proteins, such as polymerases, might determine the distribution of chromosomes and genes in a self-organizing manner (Cook, 2002). For example, it is possible that the decondensed, active regions of chromatin physically push the inactive heterochromatin regions towards the periphery. Furthermore, the interaction of active regions with polymerases might provide anchor points, which determine the spatial organization of the genome (Cook, 2002). In contrast to such a self-organization model is the possibility of defined, molecular organizers, which might anchor particular genome regions to specific places within the nuclear architecture. Candidates for this function are sparse at present, but the identification of a matrix type protein, SATB1, in thymocytes that tethers specific gene loci and affects their function in a cell-type specific manner might be an indication that such genome organizers exist (Cai et al., 2003). These considerations clearly show that much work remains to be done to understand how the spatial organization of genomes relates to their function. The study of changes in positioning of chromosome and loci in differentiation systems as reported here should provide a useful tool to begin to address this fundamental aspect of genome function.

Acknowledgments

We thank Sotiria Sotiriou and Jeff Roix for critical comments.

References

Abranches R, Beven AF, Aragon-Alcaide L, Shaw PJ: Transcription sites are not correlated with chromosome territories in wheat nuclei. J Cell Biol 143:5–12 (1998).

Andrulis ED, Neiman AM, Zappulla DC, Sternglanz R: Perinuclear localization of chromatin facilitates transcriptional silencing. Nature 394:592–595 (1998).

Bartova E, Kozubek S, Kozubek M, Jirsova P, Lukasova E, Skalnikova M, Cafourkova A, Koutna I: Nuclear topography of the c-myc gene in human leukemic cells. Gene 244:1–11 (2000).

Boyle S, Gilchrist S, Bridger JM, Mahy NL, Ellis JA, Bickmore WA: The spatial organization of human chromosomes within the nuclei of normal and emerin-mutant cells. Hum Mol Genet 10:211–219 (2001).

Bridger JM, Boyle S, Kill IR, Bickmore WA: Re-modelling of nuclear architecture in quiescent and senescent human fibroblasts. Curr Biol 10:149–152 (2000).

Brown KE, Guest SS, Smale ST, Hahm K, Merkenschlager M, Fisher AG: Association of transcriptionally silent genes with Ikaros complexes at centromeric heterochromatin. Cell 91:845–854 (1997).

Brown KE, Baxter J, Graf D, Merkenschlager M, Fisher AG: Dynamic repositioning of genes in the nucleus of lymphocytes preparing for cell division. Mol Cell 3:207–217 (1999).

Cai S, Han HJ, Kohwi-Shigematsu T: Tissue-specific nuclear architecture and gene expression regulated by SATB1. Nat Genet 34:42–51 (2003).

Chaly N, Munro SB: Centromeres reposition to the nuclear periphery during L6E9 myogenesis in vitro. Exp Cell Res 223:274–278 (1996).

Cheutin T, McNairn AJ, Jenuwein T, Gilbert DM, Singh PB, Misteli T: Maintenance of stable heterochromatin domains by dynamic HP1 binding. Science 299:721–725 (2003).

Choh V, De Boni U: Spatial repositioning of centromeric domains during regrowth of axons in nuclei of murine dorsal root ganglion neurons in vitro. J Neurobiol 31:325–332 (1996).

Cook PR: Predicting three-dimensional genome structure from transcriptional activity. Nat Genet 32:347–352 (2002).

Cremer T, Cremer C: Chromosome territories, nuclear architecture and gene regulation in mammalian cells. Nat Rev Genet 2:292–301 (2001).

Croft JA, Bridger JM, Boyle S, Perry P, Teague P, Bickmore WA: Differences in the localization and morphology of chromosomes in the human nucleus. J Cell Biol 145:1119–1131 (1999).

Dundr M, Misteli T: Functional architecture in the cell nucleus. Biochem J 356:297–310 (2001).

Festenstein R, Pagakis SN, Hiragami K, Lyon D, Verreault A, Sekkali B, Kioussis D: Modulation of heterochromatin protein 1 dynamics in primary mammalian cells. Science 299:719–721 (2003).

Fienberg SE (1982) in Kotz S, Johnson NL (eds): Encyclopedia of Statistical Science, pp 161–171 (Wiley and Sons, New York, 1982).

Francastel C, Schubeler D, Martin DI, Groudine M: Nuclear compartmentalization and gene activity. Nat Rev mol Cell Biol 1:137–143 (2000).

Francastel C, Magis W, Groudine M: Nuclear relocation of a transactivator subunit precedes target gene activation. Proc Natl Acad Sci USA 98:12120–12125 (2001).

Hochstrasser M, Mathog D, Gruenbaum Y, Saumweber H, Sedat JW: Spatial organization of chromosomes in the salivary gland nuclei of *Drosophila melanogaster*. J Cell Biol 102:112–123 (1986).

Jin QW, Fuchs J, Loidl J: Centromere clustering is a major determinant of yeast interphase nuclear organization. J Cell Sci 113:1903–1912 (2000).

Kioussis D, Ellmeier W: Chromatin and CD4, CD8A and CD8B gene expression during thymic differentiation. Nat Rev Immunol 2:909–919 (2002).

Korenberg JR, Chen XN, Devon KL, Noya D, Oster-Granite ML, Birren BW: Mouse molecular cytogenetic resource: 157 BACs link the chromosomal and genetic maps. Genome Res 9:514–523 (1999).

Kosak ST, Skok JA, Medina KL, Riblet R, Le Beau MM, Fisher AG, Singh H: Subnuclear compartmentalization of immunoglobulin loci during lymphocyte development. Science 296:158–162 (2002).

Kurz A, Lampel S, Nickolenko JE, Bradl J, Benner A, Zirbel RM, Cremer T, Lichter P: Active and inactive genes localize preferentially in the periphery of chromosome territories. J Cell Biol 135:1195–1205 (1996).

Lamond AI, Earnshaw WC: Structure and function in the nucleus. Science 280:547–553 (1998).

Lemon B, Tjian R: Orchestrated response: a symphony of transcription factors for gene control. Genes Dev 14:2551–2569 (2000).

Lukasova E, Kozubek S, Kozubek M, Kjeronska J, Ryznar L, Horakova J, Krahulcova E, Horneck G: Localisation and distance between ABL and BCR genes in interphase nuclei of bone marrow cells of control donors and patients with chronic myeloid leukaemia. Hum Genet 100:525–535 (1997).

Mahy NL, Perry PE, Bickmore WA: Gene density and transcription influence the localization of chromatin outside of chromosome territories detectable by FISH. J Cell Biol 159:753–763 (2002).

Mahy NL, Perry PE, Gilchrist S, Baldock RA, Bickmore WA: Spatial organization of active and inactive genes and noncoding DNA within chromosome territories. J Cell Biol 157:579–589 (2002).

Manuelidis L: Different central nervous system cell types display distinct and nonrandom arrangements of satellite DNA sequences. Proc Natl Acad Sci USA 81:3123–3127 (1984).

Martou G, De Boni U: Nuclear topology of murine, cerebellar purkinje neurons: changes as a function of development. Exp Cell Res 256:131–139 (2000).

Matera AG: Nuclear bodies: multifaceted subdomains of the interchromatin space. Trends Cell Biol 9:302–309 (1999).

Misteli T, Spector DL: Serine/threonine phosphatase 1 modulates the subnuclear distribution of pre-mRNA splicing factors. Mol Biol Cell 7:1559–1572 (1996).

Neves H, Ramos C, da Silva MG, Parreira A, Parreira L: The nuclear topography of ABL, BCR, PML, RARalpha genes: evidence for gene proximity in specific phases of the cell cycle and stages of hematopoietic differentiation. Blood 93:1197–1207 (1999).

Nogami M, Kohda A, Taguchi H, Nakao M, Ikemura T, Okumura K: Relative locations of the centromere and imprinted SNRPN gene within chromosome 15 territories during the cell cycle in HL60 cells. J Cell Sci 113: 2157–2165 (2000).

Parada L, Misteli T: Chromosome positioning in the interphase nucleus. Trends Cell Biol 12:425–432 (2002).

Parada L, McQueen P, Munson P, Misteli T: Conservation of relative chromosome positioning in normal and cancer cells. Curr Biol 12: 1692–1697 (2002).

Pederson T: Protein mobility within the nucleus – what are the right moves? Cell 104:635–638 (2001).

Phair RD, Misteli T: High mobility of proteins in the mammalian cell nucleus. Nature 404:604–609 (2000).

Piccirillo CA, Letterio JJ, Thornton AM, McHugh RS, Mamura M, Mizuhara H, Shevach EM: CD4(+)CD25(+) regulatory T cells can mediate suppressor function in the absence of transforming growth factor beta1 production and responsiveness. J Exp Med 196:237–246 (2002).

Press WH, Teuolskky SA, Vettering WT, Flannery BP; in Press WH, Teuolskky SA, Vettering WT, Flannery BP (eds): Numerical Recipes in C, chapter 14 (Cambridge University Press, Cambridge UK, 1992).

Roix JJ, McQueen PG, Munson PJ, Parada LA, Misteli T: Spatial proximity of translocation-prone gene loci in human lymphomas. Nat Genet 34:287–291 (2003).

Sadoni N, Langer S, Fauth C, Bernardi G, Cremer T, Turner BM, Zink D: Nuclear organization of mammalian genomes. Polar chromosome territories build up functionally distinct higher order compartments. J Cell Biol 146:1211–1226 (1999).

Skok JA, Brown KE, Azuara V, Caparros ML, Baxter J, Takacs K, Dillon N, Gray D, Perry RP, Merkenschlager M, Fisher AG: Nonequivalent nuclear location of immunoglobulin alleles in B lymphocytes. Nat Immunol 2:848–854 (2001).

Stevens MA (1982) in Kotz S, Johnson N (eds): Encyclopedia of Statistical Science, pp 393–396 (Wiley and Sons, New York, 1982).

Sun HB, Shen J, Yokota H: Size-dependent positioning of human chromosomes in interphase nuclei. Biophys J 79:184–189 (2000).

Tanabe H, Muller S, Neusser M, von Hase J, Calcagno E, Cremer M, Solovei I, Cremer C, Cremer T: Evolutionary conservation of chromosome territory arrangements in cell nuclei from higher primates. Proc Natl Acad Sci USA 99:4424–4429 (2002).

Van Driel R, Fransz PF, Verschure PJ: The eukaryotic genome: a system regulated at different hierarchical levels. J Cell Sci 116:4067–4075 (2003).

Volpi EV, Chevret E, Jones T, Vatcheva R, Williamson J, Beck S, Campbell RD, Goldsworthy M, Powis SH, Ragoussis J, Trowsdale J, Sheer D: Large-scale chromatin organization of the major histocompatibility complex and other regions of human chromosome 6 and its response to interferon in interphase nuclei. J Cell Sci 113:1565–1576 (2000).

Wansink DG, Schul W, van der Kraan I, van Steensel B, van Driel R, de Jong L: Fluorescent labelling of nascent RNA reveals transcription by RNA polymerase II in domains scattered throughout the nucleus. J Cell Biol 122:282–293 (1993).

Williams RR, Broad S, Sheer D, Ragoussis J: Subchromosomal positioning of the epidermal differentiation complex (EDC) in keratinocyte and lymphoblast interphase nuclei. Exp Cell Res 272:163–175 (2002).

Cytogenet Genome Res 105:302–310 (2004)
DOI: 10.1159/000078202

Positional changes of pericentromeric heterochromatin and nucleoli in postmitotic Purkinje cells during murine cerebellum development

I. Solovei,[a] N. Grandi,[a] R. Knoth,[b] B. Volk,[b] and T. Cremer[a]

[a] Department of Biology II, Human Genetics, Ludwig Maximilians University (LMU), Munich;
[b] Department of Neuropathology, Neurocentre, University of Freiburg, Freiburg (Germany)

Abstract. Previous studies revealed changes of pericentromeric heterochromatin arrangements in postmitotic Purkinje cells (PCs) during postnatal development in the mouse cerebellum (Manuelidis, 1985; Martou and De Boni, 2000). Here, we performed vibratome sections of mouse cerebellum (vermis) at P0 (day of birth), at various stages of the postnatal development (P2–P21), as well as in very young (P28) and 17-months-old adults. FISH was carried out on these sections with major mouse satellite DNA in combination with immunostaining of the nucleolar protein B23 (nucleophosmin). Laser confocal microscopy, 3D reconstructions and quantitative image analysis were employed to describe changes in the number and topology of chromocenters and nucleoli. At all stages of postnatal PC development heterochromatin clusters were typically associated either with nucleoli or with the nuclear periphery, while non-associated clusters were rare (<1% at P0 to P21 and about 3% in adult stages). At P0, about 2–4 nucleoli and 7–8 pericentromeric heterochromatin clusters were variably located within PC nuclei. The relative volume of heterochromatin clusters associated with the nucleoli (about 50%) was rough-

ly equal to the volume of clusters associated with the nuclear periphery. Positional changes of both nucleoli and centromeres towards the nuclear center occurred between P0 and P6. At P6 the average number of chromocenters per PC nucleus had decreased to about five. In agreement with previous studies, one or occasionally two nucleoli were noted at the nuclear center surrounded by major perinucleolar heterochromatin clusters. The relative volume of these perinucleolar clusters increased to about 84%, while the volume of clusters in the nuclear periphery decreased to about 15%. At subsequent postnatal stages, the arrangement of most pericentromeric heterochromatin around a central nucleolus was maintained. In adult animals, however, we observed a partial redistribution of heterochromatin towards the nuclear periphery. The average total number of pericentromeric heterochromatin signals increased again to about ten. The volume of heterochromatin associated with the nuclear periphery roughly doubled (30%), while the volume of the perinucleolar heterochromatin decreased correspondingly.

Purkinje cells (PCs) are known as the efferent elements of the cerebellar cortex. They form a monolayer located at the boundary between the molecular layer at the cerebellar periphery and the granular layer towards the cerebellar interior. In adult animals PCs show a characteristic cellular and nuclear morphology. In pioneering non-isotopic in situ hybridization studies, Laura Manuelidis employed mouse satellite DNA as a probe to pericentromeric heterochromatin of different mouse neuronal cell types (Manuelidis, 1984a, b). In these studies Manuelidis described cell-type-specific patterns of satellite DNA clusters in neuronal cell nuclei, including nuclei from PCs. In newly born mice, PCs are already in a postmitotic state but not yet terminally differentiated. In detailed follow-up studies of PC nuclei during postnatal development starting a week after mouse birth, Manuelidis found that the nuclear morphology begins to assume its adult shape between day 7 to day 11: a central nucleolus became recognizable and nuclei showed 2 to 4 heterochromatin clusters, similar to nuclei of adult PCs.

Supported by a grant from the Deutsche Forschungsgemeinschaft to T.C. (Cr 59/20-3).

Received 14 October 2003; manuscript accepted 15 December 2003.

Request reprints from Irina Solovei, Department of Biology II
Humangenetik, Ludwig Maximilians University (LMU)
Richard Wagner Str. 10, 80333 München (Germany)
telephone: +49-89-2180-6713; fax: +49-89-2180-6719
e-mail: Irina.Solovei@lrz.uni-muenchen.de

At this stage, some clusters were adherent to the nucleolus, while others were still located at the nuclear periphery – in contrast to adult Purkinje neurons, which showed a rather exclusive positioning of heterochromatin around a single centrally located nucleolus (Manuelidis, 1985). These observations strongly indicate a migration of pericentromeric heterochromatin during terminal differentiation away from the nuclear membrane and towards the nuclear center. During the same period synapses of the dendrite tree commence. A similar architecture of adult PC nuclei with major clusters of constitutive heterochromatin around a large central nucleolus was also observed in other species, such as guinea pigs and hamsters (Manuelidis, 1984a, b). The correlation between these profound changes in the nuclear architecture and the formation of a functional dendrite tree has raised considerable interest in a possible role of a dynamic higher order nuclear architecture in the cell-type-specific functioning of PCs and other neuronal cell types. The evolutionary conservation of cell-type-specific patterns, despite differences in centromere DNA sequences, further emphasizes a functional role that is not yet understood.

In subsequent studies, Martou and De Boni (2000) confirmed a striking kinetochore movement in PC nuclei between the day of birth (P0) and day 29 after birth (P29) when the mouse becomes a young adult with a fully developed cerebellum. These authors performed immunostaining of cerebellar sections with CREST antiserum at different postnatal stages. Following light optical sectioning of PCs, they counted the total number of kinetochore signals per nucleus and determined the percentages of centromeres associated with the nucleolus and the nuclear periphery, respectively, or showing an intermediate position, remote from either nucleoli or the nuclear periphery. During postnatal differentiation of PCs, they found changes in the aggregational state of kinetochores, including a decrease of the signal numbers at early stages of postnatal development and an increase during later stages. The fraction of kinetochore signals found in the nuclear periphery decreased during terminal PC differentiation, while the fraction associated with the centrally located nucleolus increased.

In the present study we used state-of-the-art technology including laser confocal microscopy, 3D reconstructions, and quantitative image analysis of mouse PC nuclei to describe changes in the number and topology of centromere clusters and nucleoli at defined days of postnatal development (P0, P2, P3, P4, P6, P14, P21), in very young adults (P28), as well as in fully matured 17-month-old adults.

Materials and methods

Murine cerebellum sections

Adult C57Bl/6 mice (P28 and 17 months old) and mice at different postnatal developmental stages (P6, P14, P21) were anaesthetized with isoflurane (Abbott, Wiesbaden, Germany) and by intraperitoneal administration of a mixture of ketamine (100 mg/kg; Parke Davis, Berlin, Germany) and xylazine (Rompun, 5 mg/kg; Bayer, Leverkusen, Germany). Mice were transcardially perfused with Ringer's solution and then with 4 % formaldehyde in PBS (pH 7.2) for 10 min, followed by immersion fixation of the removed brains in the same fixative overnight. Brains of P0, P2, P3, and P4 mice were removed unfixed immediately after decapitation and fixed by immersion into 4 % formaldehyde overnight. Tissues were embedded in 4 % agarose and sections of 60 μm thickness through the vermis of the cerebellum were cut coronally on a Vibratome (VT1000S; Leica, Heidelberg, Germany). Sections were kept in PBS with 0.1 % sodium azide at 4 °C until use. Purkinje cells (PCs) from any part of the vermis were studied for stages P0 to P3; at older stages, the second and third folia of the vermis could be clearly recognized and PCs from only these two folia were analyzed.

Immunostaining

For immunostaining individual floating sections were permeabilized with 0.5 % Triton X-100 for 1 h and then sequentially incubated with primary and secondary antibodies diluted in PBS with 0.1 % Triton X-100 and 1 % BSA at 37 °C. As a PC marker, calbindin was detected with a mouse anti-calbindin antibody (1:1000; monoclonal, IgG; Swant, Bellinzona, Italy) for 12 h. Nucleoli were detected by protein B23 (nucleophosmin/NPM) staining using a mouse anti-B23 antibody (1:500, monoclonal; IgG, Sigma) for 12 h. In both cases, incubation with primary antibody was followed by incubation with goat anti-mouse antibody conjugated to Alexa 488 (1:500; IgG [H+L], F(ab')$_2$; Molecular Probes, Eugene, OR, USA) for 3 h. Nuclei were counterstained with DAPI (0.5 μg/ml) and TO-PRO-3 (1 μM) in PBS. Sections were mounted on microscopic slides with antifade Vectashield (Vector).

Fluorescence in situ hybridization (FISH)

As a marker of all centromere regions in mouse cells, a mouse major satellite probe repeat was generated by PCR with 5'-GCG AGA AAA CTG AAA ATC AC and 5'-TCA AGT CGT CAA GTG GAT G primers and murine genomic DNA as a template. The probe was labeled with TAMRA-dUTP (Perkin Elmer) or Cy3-dUTP (Amersham Biosciences) by nick translation and dissolved in hybridization mixture (50 % formamide, 10 % dextran sulfate, 1× SSC) at a concentration of 10 ng/μl.

Floating vibratome sections were used in experiments performed at P0, P6 and P28, where FISH was combined with immunostaining of the nucleolus marker B23 (nucleophosmin). Sections were permeabilized with 1 M NaSCN (at 37 °C for 45 min), digested with 0.06 mg/μl pepsin (in 0.01 N HCl at 37 °C for 15 min) and equilibrated in 50 % formamide/2× SSC (at 37 °C for 30 min). Thereafter a single section was placed on a microscope slide, covered with probe in hybridization mixture, mounted under a glass chamber, and sealed with rubber cement. The chamber consisted of a 12 × 12 mm coverslip with glued glass strips with a height of 0.7 mm. Typically, 20–40 μl of hybridization mixture was sufficient to fill the chamber. Tissue DNA and probe DNA were denatured simultaneously at 85 °C on a hot block for 5 min. Hybridization was performed at 37 °C for 3–5 days in humid dark boxes. After hybridization sections were washed in 2× SSC at 37 °C (3 × 10 min), blocked in PBS with 2 % BSA and 0.1 % Triton X-100 for 15 min, and stained with anti-B23 antibody (see above). Finally, sections were counterstained with DAPI and TO-PRO-3, and mounted in antifade (Vectashield).

It was difficult to handle floating mouse cerebellum sections of early postnatal stages due to their small size and fragility. To overcome this problem and ensure the comparability of quantitative evaluation for early and late postnatal stages, we performed FISH to dehydrated sections attached to SuperFrost slides in all experiments which did not require anti-B23 immunostaining. Sections were placed in a drop of water on the slide, air-dried at room temperature for 30 min and dehydrated through an ethanol series (30, 50, 70, 90, and 100%). Permeabilization and pepsin digestion of attached sections were performed as described above. Then sections were again dehydrated, air-dried, mounted in hybridization mixture under the glass chamber, and sealed with rubber cement. While the hybridization efficiency obtained with floating and attached sections was similar, the morphology of the floating sections, which were never dried during the whole procedure (in contrast to attached sections) was superior.

Microscopy

Sections were viewed using an epifluorescence microscope (Axioplan II; Zeiss, Germany) equipped with Plan Apo 20×, 40×, 63×, and 100×/1.4 oil immersion objectives and filter sets for DAPI, Alexa 488, Cy3, and TO-PRO-3. Digital images of metaphase spreads and DAPI-stained sections were acquired by a CCD-camera (CoolView; Photonic Science, UK) using CytoVision software (Applied Imaging, UK). The 8-bit grayscale single fluorochrome images were overlaid to an RGB image assigning a false color to each fluorochrome.

Series of optical sections through PC nuclei were collected using a Leica TCS SP confocal system equipped with a Plan Apo 63×/1.3 NA oil immer-

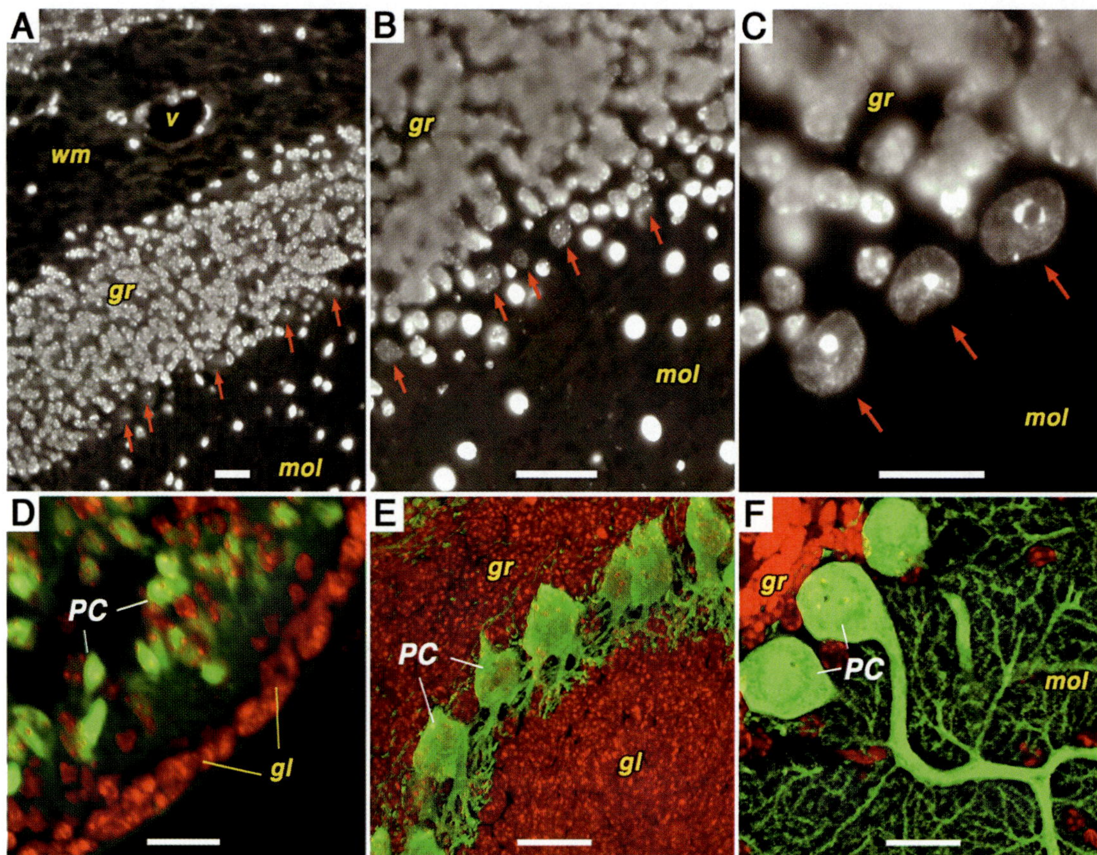

Fig. 1. Vibratom sections through vermis of cerebellum of an adult mouse. (**A–C**) Localization and morphology of Purkinje cell nuclei in DAPI-stained sections at increasing microscopic magnification. PC nuclei are the biggest in the tissue and reveal weakly stained nucleoplasm with brightly stained chromocenters. *gr*: granular layer; *mol*: molecular layer; *wm*: white matter, *v*: blood vessel. (**D–F**) Maximum intensity projections of confocal serial sections of PCs stained with anti-calbindin (green) and TO-PRO-3 (red). (**D**) PCs at P0 have a multilayer arrangement and are situated beneath the germinal layer *(gl)* from which granular cells originate. (**E**) PCs at P6 form a monolayer between granular cells *(gr)* and germinal layer *(gl)*. Note the beginning of the dendrite tree growth at this stage. (**F**) PCs at P28 with a completely developed dendrite tree protruding into the molecular layer *(mol)*. Scale bars are 20 µm in **C–F**, and 50 µm in **A** and **B**.

sion objective. Fluorochromes were visualized using an argon laser with excitation wavelengths of 488 nm (for Alexa 488) and 514 nm (for Cy3 and TAMRA), and a helium-neon laser with the excitation wavelength of 633 nm (for TO-PRO-3). For each optical section, images were collected sequentially for two or three fluorochromes. Stacks of 8-bit grayscale images were obtained with axial distances of ~ 300 nm between optical sections and pixel sizes of 80 nm. For a typical PC nucleus from adult mouse a complete image stack comprised about 35 sections.

Image processing and quantitative assessment of chromocenter arrangement in Purkinje cells

Galleries of RGB confocal images were assembled using ImageJ and Adobe Photoshop programs. Three-dimensional reconstruction of nuclei and their chromocenters was performed by surface rendering of image stacks using Amira 2.3 TGS (http://www.amiravis.com). This program was also used for volume measurements of nuclei after nuclear segmentation. ImageJ (http://rsb.info.nih.gov/ij) was employed for signal counting in pseudo-colored image stacks and volume measurements of chromocenters.

For a quantitative evaluation of the distribution of nucleoli and chromocenters on mid sections through Purkinje cell nuclei, the RRD (Relative Radius Distribution) computer program was used (for detailed description see Cremer et al., 2001). Briefly, the program works as follows: (1) The gravity center of a given nucleus and its borders are determined on the basis of the nuclear counterstain signal. (2) Borders of chromocenters or nucleoli are determined by thresholding of the fluorescence signals in the respective color channels. (3) The nuclear radius in any direction from the nuclear center (defined as the intensity gravity center of the DNA counterstain) to the segmented nuclear edge is normalized to 100 % and the nuclear space is divided into 25 shells of equal thickness (each covering 4 % of the total radius). In this way the distribution of pericentromeric heterochromatin or nucleoli (estimated as fluorescence signal intensity) was measured and expressed as a function of the relative distances of each shell from the nuclear center. For signals with a random radial distribution we expect the same distribution as for the counterstained nuclear DNA. Significant deviations from the counterstain curve indicate a non-random radial distribution. To compare relative positions of targeted structures, we calculated the average relative radius (ARR). ARR presents the mean value of the distribution of all distances from all voxels representing a signal to the gravity center of the counterstained nucleus.

Results

Figures 1A and B show the typical location of PCs in fluorescent images from sections through adult cerebellar vermis following nuclear DNA staining with DAPI. Higher magnification

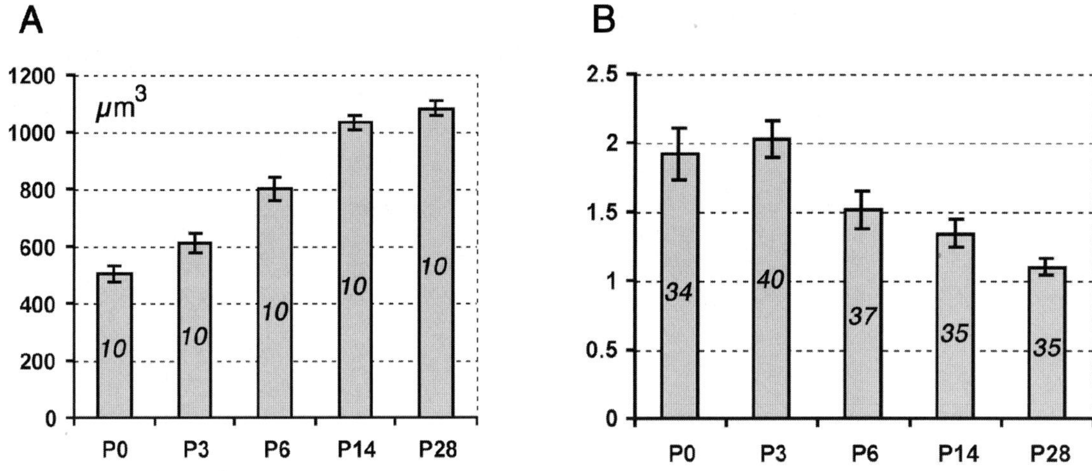

Fig. 2. Changes of the nuclear volume (**A**) and the number of nucleoli (**B**) in PC nuclei as a function of postnatal cerebellum development. Bars represent standard errors of the mean; numbers of analyzed nuclei per stage are shown inside the columns.

(Fig. 1C) reveals the large nuclei and the typical nuclear staining pattern of PCs in more detail. Most of the volume of PC nuclei stains weakly with DAPI (due to a strong chromatin decondensation), while heterochromatin forms 2–3 bright globules closely associated with the nucleoli situated in the center of the nucleus. Nucleoli could be readily identified in PC nuclei stained with DAPI and/or TO-PRO-3 as round non-stained areas of the nucleoplasm encircled by more intensely stained chromatin (see e.g. Fig. 3B–E). In preliminary experiments we tested whether PCs could be unequivocally identified based on their location and their nuclear staining pattern. For this purpose we combined DNA staining with DAPI and TO-PRO-3 with immunofluorescent staining of calbindin, a calcium-binding protein which provides a highly specific marker protein for the detection of PCs in the cerebellum (Legrand et al., 1983; Nordquist et al., 1988; Fenili and De Boni, 2003) (Fig. 1D–F). We first viewed cerebellar cortex sections in the DAPI or TO-PRO-3 channel and identified potential PCs; thereafter we switched to the calbindin channel and confirmed the presence of calbindin in all these cells. We conclude that it is sufficient to employ DNA staining alone for the identification of PCs. In contrast to DAPI, TO-PRO-3 has its emission in the far-red part of the spectrum and could be detected by the confocal microscope available for the present experiments. Accordingly, we employed DAPI to screen tissue sections by conventional epifluorescence microscopy and TO-PRO-3 to perform light optical sectioning. The positioning of the PC monolayer and the morphology of PC nuclei typical for adult mice were already manifested at P6 of postnatal mouse development. At the earlier stages, PCs were still represented by a multilayer (P0–P2, Fig. 1D). Granular cells, initially localized outward to PCs in the outer granular layer (also called germinal layer), had migrated inward and could be observed on either side of PCs (P3–P4). Nevertheless, additional immunostaining with calbindin revealed that DAPI/TO-PRO-3 stained PC nuclei could be distinguished from nuclei of other cell types solely on the basis of their location and size.

We measured the size of PC nuclei at five stages of the cerebellum development: P0, P3, P6, P14, and P28. Image stacks were obtained from ten TO-PRO-3 stained nuclei at each developmental stage, their 3D reconstructions were performed and volumes were calculated using Amira software. Intense growth of dendrite trees between stages P3 and P14 (Fig. 1D–F) was accompanied by a significant growth of nuclear volume, from 500 to 1000 μm^3 (Fig. 2A). Similar changes in volume of PC nuclei (from 400 to about 1000 μm^3) were described for developing rat cerebellum between P6 and adult stages (Lentz and Lapham, 1970). Two, three or even four small nucleoli were typically found in the cerebellum of newborn mice (see e.g. Fig. 5A). During postnatal development of the cerebellum the average number of nucleoli per nucleus decreased to 1–2 at P6 (Fig. 2B). PCs of adult animals usually had a single, very large central nucleolus (see e.g. Figs. 3B, E, 5C).

The mouse major satellite repeat was used to label the pericentromeric heterochromatin of all mouse chromosomes (Fig. 3A). PC nuclei in the second and the third folia of the cerebellar vermis were scanned using confocal microscope and light optical image stacks were arranged as galleries. Figure 3B exemplifies a typical distribution of pericentromeric heterochromatin in a PC nucleus. In murine interphase nuclei, centromere heterochromatin often fuses to chromocenters of variable sizes. Therefore, the observed number of major satellite signals in PC nuclei was always below the diploid chromosome number (2n = 40). Noteworthy, one or two heterochromatin globules stained with DAPI/TO-PRO-3 were not recognized by major satellite probe indicating that they did not contain major satellite but other DNA. These globules were typically observed at the nuclear periphery (not shown) or adjacent to the nucleolus (arrows on Fig. 3B, D).

To trace the dynamics of chromocenter formation, about 30 nuclei were scanned at each stage of postnatal development (P0, P2, P3, P4, P6, P14, P21) and for adult animals (P28 and 17 months old). Data for P28 and 17-month-old animals were pooled, since they were not significantly different (individual

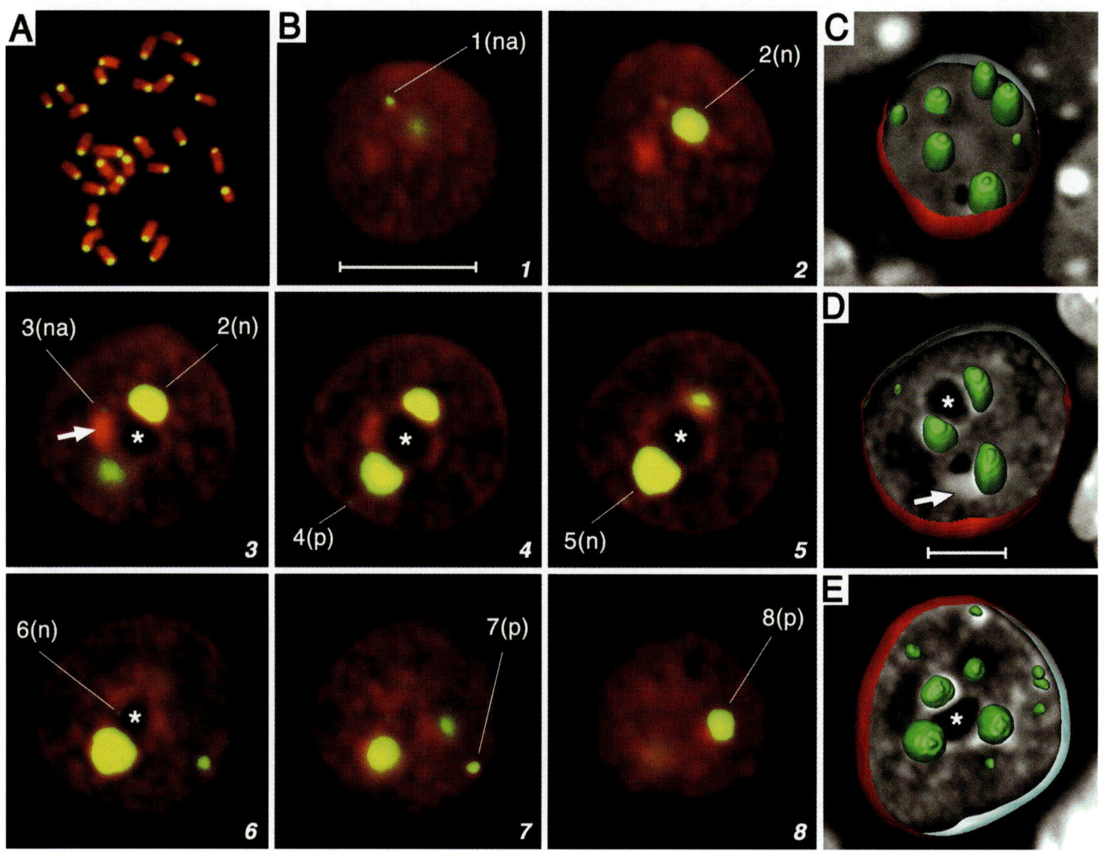

Fig. 3. (A) Metaphase spread from a mouse fibroblast after FISH with the mouse major satellite (m.s.) DNA probe and DAPI counterstain (red, false color) demonstrates staining of pericentromeric heterochromatin (green). Due to intense staining of pericentromeric heterochromatin with both DNA stain and m.s. probe, pericentromeric regions have yellow color. **(B)** Gallery of confocal sections (numbered from 1 to 8) through the mid-part of a PC nucleus at P21 stained with TO-PRO-3 (red) after 3D FISH with the m.s. probe (green). Every third section is shown. Eight heterochromatin signals are numbered in order of their appearance and marked with respect to their intranuclear location as adjacent to the nuclear periphery (p) or to the nucleolus (n), or to none of these structures (non-adjacent: na). The arrow in section 3 denotes a big heterochromatin cluster adjacent to the nucleolus (*), but not stained by the m.s. probe. Due to intense staining of pericentromeric heterochromatin with both DNA stain (red) and m.s. probe (green), pericentromeric regions on **A** and **B** appear yellow. **(C–E)** 3D reconstruction of PCs at P0 **(C)**, P6 **(D)**, and P28 **(E)**. Heterochromatin signals (green) are shown together with a partial reconstruction of the nuclear border (red). An optical mid nuclear section of TO-PRO-3 counterstain is shown in gray color. Scale bar in **B** is 10 μm, scale bar in **D** is 5 μm and applied to **C** and **E**.

data not shown). For the same reason, data for early stages of development (P2, P3, and P4) were also pooled into one category, P2–4. In newborn mice (P0), the average total number of chromocenters per PC nucleus was 7.6 ± 0.4. At P6, the number had decreased to about 5 ± 0.3 and increased again at the subsequent stages (P14 and P21). The highest average number of 10.4 ± 0.6 chromocenters per PC nucleus was observed in adults (Fig. 4A). With regard to their location in the nucleus, individual chromocenters were classified to three categories (Fig. 3B): adjacent to the nuclear periphery (p), adjacent to the nucleoli (n), and non-adjacent to either periphery or nucleoli (na). The peripheral signals (p) revealed the same developmental dynamics as described above for the total number of signals (Fig. 4B, black columns), i.e. the average number of peripherally located chromocenters was reduced from about five at P0 to two at P6 and then increased again from P6 to adults. The number of nucleoli-adjacent signals increased from P0 to P21 (Fig. 4B, gray columns). The number of the non-adja-

cent signals was very low at all stages, although a slight increase was apparent from P0 to adult stages (Fig. 4B, white columns).

Pericentromeric heterochromatin signals significantly varied in size and position depending on the developmental stage and number of chromosomes participating in the formation of chromocenters. Chromocenters in PC nuclei of P0–P4 animals were rather uniform in size (about 1.5 μm, Fig. 3C) and were observed both at the nuclear periphery and adjacent to nucleoli. In contrast, in P6 to adult animals, most pericentromeric heterochromatin was organized into 2–3 large chromocenters (on average about 3 μm in diameter) adjacent to 1 or 2 centrally located nucleoli. Peripherally located chromocenters in adult animals were much smaller (on average about 1 μm in diameter). The non-adjacent signals were typically small and apparently formed by pericentromeric regions of very few or single chromosomes (Fig. 3B).

Our counts of the numbers of chromocenters at different stages of postnatal development did not take into account their

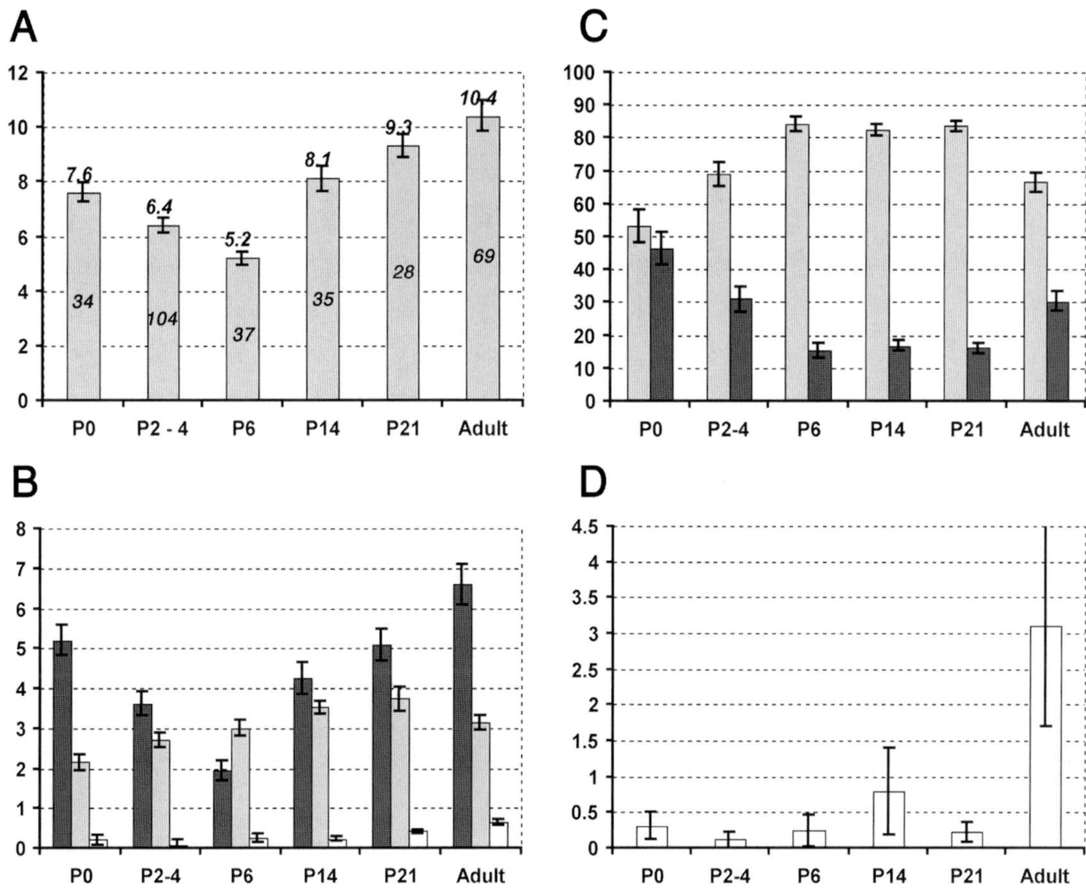

Fig. 4. Quantitative analysis of PC nuclei at different stages of postnatal development. (**A**) Average total number of pericentromeric signals. (**B**) Average number of signals adjacent to the nuclear periphery (black columns), to the nucleoli (gray columns), and of non-adjacent signals (white columns). (**C**) Average volumes of signals associated with the nuclear periphery (black columns) or with nucleoli (gray columns) given as per cent to the total signal volume. (**D**) Volumes of non-adjacent signals given as per cent to the total signal volume. Bars represent standard errors of the mean. The numbers of nuclei analyzed per stage are shown inside columns in **A** and applied to the rest of the graphs.

pronounced size variations. This prompted us to attempt volume measurements of individual chromocenters. Figures 4C and D show the relative volume of the three signal categories defined above (p, n, na) as percentage of the total volume of pericentromeric heterochromatin signals occupied in each nucleus. Taking into account that voxel numbers strongly depend on the chosen threshold that separates a signal from background, absolute volume measurements are unreliable, if not impossible, while relative volume measurements may provide more useful information. These data indicate that at P0, roughly half of the volume of pericentromeric heterochromatin was associated with nucleoli (Fig. 4C, gray columns), while the other half was associated with the nuclear periphery (Fig. 4C, black columns). At P2–4 the fraction of the volume of nucleolus associated chromocenters increased to 70% and at P6 to 80%, while only about 15% of the volume remained associated with the nuclear periphery. This distribution was stably retained at subsequent stages (P14 and P21). In PC nuclei from adult animals the fraction of the volume of nucleolus associated heterochromatin decreased slightly (to about 65%), while the amount of

peripheral heterochromatin increased from about 16 to 30%. Relative volumes of clusters which were neither associated with nucleoli nor with the nuclear envelope were very small at all developmental stages: less than 1% for P0–P21 and about 3% for adult animals (Fig. 4D).

The data described above support a correlation in space and time between the dynamics of nucleolus and chromocenter positioning during the postmitotic differentiation of PC nuclei. To study the distribution of pericentromeric heterochromatin and nucleoli in PC nuclei in more detail, we combined FISH with the major satellite probe with anti-B23 immunostaining for three developmental stages showing major differences in the distribution of pericentromeric heterochromatin, P0, P6, and P28. The dense packaging of cells in the cerebellum tissue made it difficult to resolve upper and lower borders of PC nuclei due to the limited Z-resolution of the laser confocal microscope, which under practical conditions is slightly less than 1 μm. Since borders of nuclei could be much better defined in the XY plane (due to the better resolution of about 300 nm), we used mid optical nuclear sections through PC

Fig. 5. (**A–C**) Representative mid sections through PC nuclei at P0 (**A**), P6 (**B**), P28 (**C**) after 3D FISH with mouse major satellite probe (red), immunostaining of the nucleolar marker B23 (green), and DNA counterstain with TO-PRO-3 (blue). Bar = 10 µm. (**D–F**) Radial distribution of B23 (green curves), pericentromeric heterochromatin (red curves) and TO-PRO-3-stained DNA (blue lines) in mid sections of PC nuclei at P0, P6 and P28. (**G**) Comparison of average relative radii (ARR) of nucleoli (green), chromocenters (red), and for the TO-PRO-3 counterstain (blue). (**H**) Scheme of major positional changes of the number and location of nucleoli (green) and pericentromeric heterochromatin clusters at P0, P6 and P28. Arrows in the nuclear scheme at P0 indicate the inward direction of movements and clustering of both nucleoli and heterochromatin that occurs between P0 and P6. As a result a large central nucleolus is typically present in PC nuclei at P6. Arrows indicate the outward movements of a part of the perinucleolar heterochromatin between P6 and P28. In addition to the de-clustering of the perinucleolar heterochromatin fraction traveling back to the nuclear periphery, it is possible that heterochromatin permanently located at the nuclear periphery splits and forms two or more smaller clusters. For simplicity, the scheme shows a nuclear projection where peripheral clusters are only drawn at the X,Y-nuclear rim, in reality peripheral clusters can be associated anywhere at the 3D nuclear envelope, most likely with the lamina.

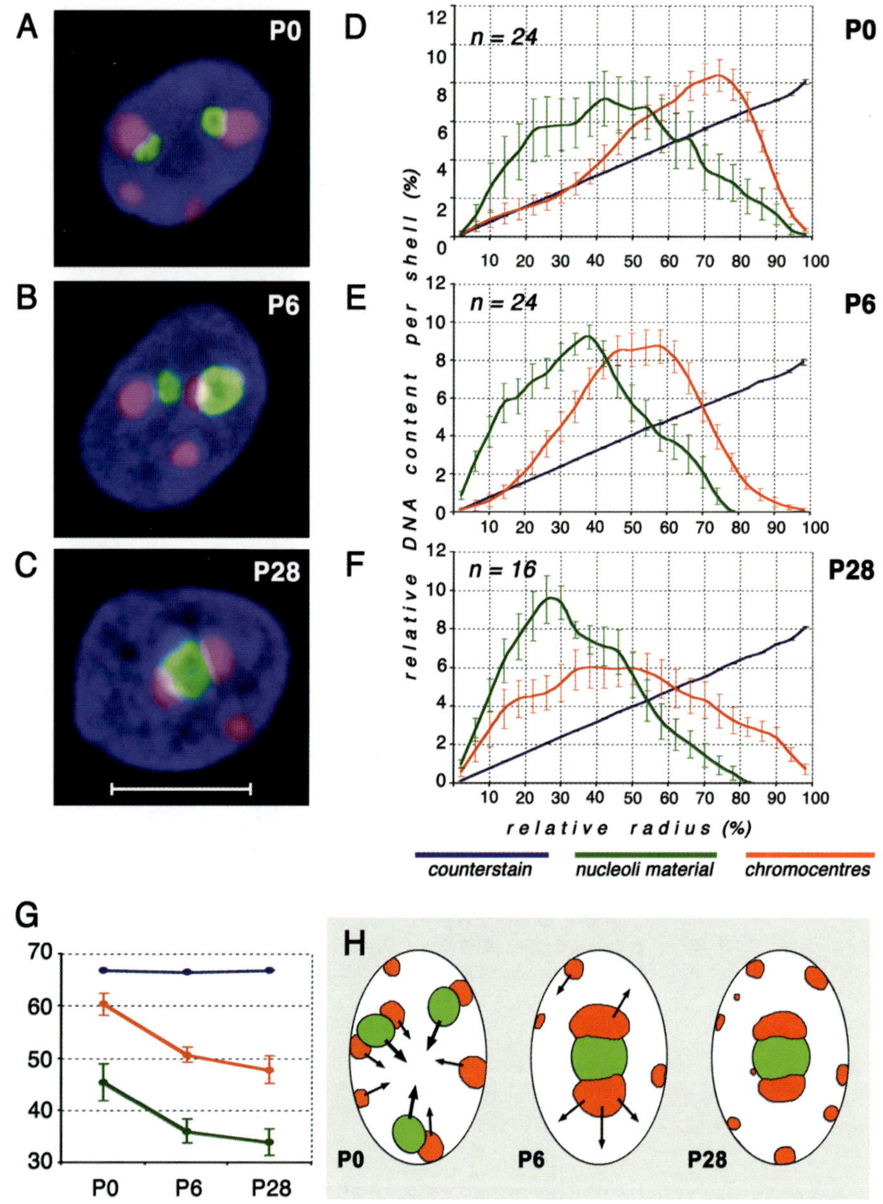

nuclei (Fig. 5A–C) to estimate the radial distribution of the nucleoli and chromocenters using the RRD (Relative Radius Distribution) computer program. Figures 5D and E demonstrate a significant shift of B23-stained nucleolar material and chromocenters from the nuclear periphery towards the nuclear center from P0 to P6. At P28 (Fig. 5F), nucleolar material showed an even more pronounced central location, while the distribution curve for the major satellite DNA signals had become broader reflecting the significant shift of pericentromeric heterochromatin back to the nuclear periphery. In accordance with the findings shown in Fig. 5D–F, the average relative radii (ARR) decreased significantly from P0 to P28 (Fig. 5G). In conclusion, compared to PC nuclei at P0, both nucleolar material and pericentromeric heterochromatin occupied more central positions at the later stages of postnatal PC development.

Discussion

Our study demonstrates changes in the number and spatial arrangement of pericentromeric heterochromatin in nuclei of postmitotic murine PCs during postnatal terminal differentiation. In the following discussion we will compare our results with the findings of previous studies and discuss biological implications.

Changes of nuclear architecture during terminal PC differentiation

Our results are consistent with the major findings of the pioneering studies performed by Manuelidis (1984b) and Martou and De Boni (2000). Martou and De Boni (2000) quantified the number of immunostained kinetochore clusters in PC nuclei and found that their average number decreased from P0 to P4, remained similar until P12, and increased again in stages from P15 to adults. These data correspond well with the similar decrease and increase that we observed for clusters of pericentromeric heterochromatin. It is noteworthy that the numbers for kinetochore signals reported by Martou and De Boni (2000) were higher for corresponding developmental stages than our counts of pericentromeric heterochromatin signals. This difference, apparently, reflects the fact that several clusters of kinetochores often contribute to a single chromocenter. The smaller size of immunologically detectable kinetochore signals resulted in a better spatial resolution compared to chromocenters. Accordingly, Martou and De Boni also counted higher average numbers of intermediately located kinetochore signals compared to non-adjacent chromocenters counted in the present study. An unexplained discrepancy between the results of Martou and De Boni and our study reflect the dynamic changes in the location of kinetochore clusters and chromocenters, respectively, at later stages of postnatal development. From P0 to P6 Martou and De Boni, as well as our group, consistently found a decrease of peripherally located kinetochores/chromocenters and a corresponding increase of nucleoli associated clusters. At subsequent stages (P7 to P29) and in adult mice, Martou and De Boni did not observe significant changes of the spatial cluster distribution already apparent at P5–P6. In contrast, we observed the redistribution of a fraction of pericentromeric heterochromatin clusters towards the nuclear periphery during later stages of development (P14, P21), as well as in an early adult (P28) and a 17-month-old adult.

In addition to previous studies, the present study included a quantitative assessment of dynamic changes in the position of nucleoli. For both, chromocenters and nucleoli, our data indicate movements from the nuclear periphery towards the nuclear center during the first week of postnatal development, from P0 to P6 (Fig. 5H). Centripetal movements of several nucleoli resulted in the formation of a big centrally located nucleolus (or occasionally two nucleoli), while centripetal movements of the pericentromeric heterochromatin yielded large heterochromatin clusters capping the central nucleolus.

Reduction of nucleoli number was also observed in differentiating PCs in rat and chicken (Lafarga et al., 1995). Most likely the formation of the large central nucleolus results from the fusion of several smaller nucleoli. However, the alternative possibility, i.e. that some nucleoli became inactive and disappeared, cannot be firmly excluded at present, since both silver staining or B23 staining only detect nucleoli with transcriptionally active NORs but not inactive NORs. While this question can only be settled unequivocally by FISH with NOR-specific DNA probes, we prefer the first explanation. In case that the alternative explanation were true, we should have observed the gradual disappearance of peripherally located nucleoli and the gradual growth of a centrally located nucleolus during P2, P3 and P4; however, this was not the case.

In addition to the obvious temporal correlation of nucleolar and heterochromatin movements, the observation that a fraction of pericentromeric heterochromatin already associated with nucleoli at P0 suggests that this fraction moves together with the nucleoli. It is not clear at present what is cause and what is effect in these coordinated centripetal movements of pericentromeric heterochromatin and nucleoli. We consider two possibilities: either heterochromatin is the primary moving structure and associated nucleoli follow or nucleoli move and associated heterochromatin follows. Heterochromatin movements may be of a primary importance since we observed that a fraction of chromocenters associated with the nuclear periphery but not with nucleoli also took part in these centripetal movements indicating that these movements occurred independently of the nucleoli. We cannot exclude, however, the possibility that other chromosomal structures than pericentromeric heterochromatin play a causal role for the observed movements. While the topology of the nucleolus almost did not change after P6, we observed a redistribution of some of the pericentromeric heterochromatin capping the nucleolus to the nuclear periphery in early (P28) and mature adults (17 months old). This movement is reflected by an increase of the average number of peripherally located chromocenters from 2 at P6 to 6.5 in adults.

Neither the mechanisms of these centripetal and centrifugal chromatin movements in postmitotic nuclei, nor the possible functional implications of such movements can be explained at present. Notably, the partial redistribution of pericentromeric heterochromatin coincides with the onset of the intense growth of the dendrite trees of PCs (Fig. 1E). Moreover, the interval between P6 and P21 is characterized by a maximum of synapse formation and synaptic activities, decreasing afterwards (due to reduction of number of spines) to the physiological level in adult animals (Weiss and Pysh, 1978). According to developmental studies of Hendelman and Aggerwal (1980), synaptic interconnections of differentiating PCs with afferents and local interneurons (basket cells and stellate cells) are greatly enhanced at P7–P10 by the formation of spines on the perisomatic region (Hendelman and Aggerwal, 1980). Enhanced synaptic input certainly means increasing physiological activity of the developing PCs involving active transcription of PC-specific genes. Martou et al. (2002) described an example of intranuclear sequence relocation correlated with onset of gene expression by showing radial relocation of the *Plcβ3* gene between P3 and P5 coincident with its de novo expression between P2 and P7. Whether changes in higher-order chromatin arrangements in postmitotic PCs during the postnatal development of the cerebellum are a causal necessity or a consequence of changes of the gene expression pattern remains an intriguing question. Expression profiling data for PCs at different differentiation stages would make it possible to establish sets of DNA probes for the 3D nuclear localization of genes that become active or silent during terminal PC differentiation. Even the unequivocal demonstration of a correlation between changes in the expression status of genes and changes in their spatial nuclear and chromatin domain topology will not be sufficient per se to prove a causal relationship. A decision between cause and consequence can only be based on new approaches that allow the analysis of gene expression and silencing, respectively, after experimental manipulation of gene positions and their chromatin environment in living cells. While it seems possible to develop living cell approaches for such a purpose – e.g., using organotypic slices of cerebellum in vitro (Fenili and De Boni, 2003) – they are presently restricted to cell cultures (e.g., Tsukamoto et al., 2000; Tumbar and Belmont, 2001). Such studies cannot, however, replace comprehensive studies of higher-order chromatin and gene arrangements in various cell types and in various species based on 3D multicolor FISH procedures. We anticipate that such studies will open a large field of investigations leading to an enhanced understanding of the contribution of higher-order chromatin architecture to the various epigenetic levels of gene regulation (Van Driel et al., 2003).

Acknowledgements

We thank Joachim Walter from our group for help with the quantitative evaluation of the data.

References

Cremer M, von Hase J, Volm T, Brero A, Kreth G, Walter J, Fischer C, Solovei I, Cremer C, Cremer T: Non-random radial higher-order chromatin arrangements in nuclei of diploid human cells. Chromosome Res 9:541–567 (2001).

Fenili D, De Boni U: Organotypic slices in vitro: repeated, same-cell, high-resolution tracking of nuclear and cytoplasmic fluorescent signals in live, transfected cerebellar neurons by confocal microscopy. Brain Res Brain Res Protoc 11:101–110 (2003).

Hendelman WJ, Aggerwal AS: The Purkinje neuron. I. A Golgi study of its development in the mouse and in culture. J Comp Neurol 193:1063–1079 (1980).

Lafarga M, Andres MA, Fernandez-Viadero C, Villegas J, Berciano MT: Number of nucleoli and coiled bodies and distribution of fibrillar centres in differentiating Purkinje neurons of chick and rat cerebellum. Anat Embryol 191:359–367 (1995).

Legrand C, Thomasset M, Parkes CO, Clavel MC, Rabie A: Calcium-binding protein in the developing rat cerebellum. An immunocytochemical study. Cell Tissue Res 233:389–402 (1983).

Lentz RD, Lapham LW: Postnatal development of tetraploid DNA content in rat Purkinje cells: a quantitative cytochemical study. J Neuropathol Exp Neurol 29:43–56 (1970).

Manuelidis L: Different central nervous system cell types display distinct and nonrandom arrangements of satellite DNA sequences. Proc Natl Acad Sci USA 81:3123–3127 (1984a).

Manuelidis L: Active nucleolus organizers are precisely positioned in adult central nervous system cells but not in neuroectodermal tumor cells. J Neuropathol Exp Neurol 43:225–241 (1984b).

Manuelidis L: Indications of centromere movement during interphase and differentiation. Ann NY Acad Sci 450:205–221 (1985).

Martou G, De Boni U: Nuclear topology of murine, cerebellar Purkinje neurons: changes as a function of development. Exp Cell Res 256:131–139 (2000).

Martou G, Park PC, De Boni U: Intranuclear relocation of the Plc beta3 sequence in cerebellar Purkinje neurons: temporal association with de novo expression during development. Chromosoma 110:542–549 (2002).

Nordquist DT, Kozak CA, Orr HT: cDNA cloning and characterization of three genes uniquely expressed in cerebellum by Purkinje neurons. J Neurosci 8:4780–4789 (1988).

Tsukamoto T, Hashiguchi N, Janicki SM, Tumbar T, Belmont AS, Spector DL: Visualization of gene activity in living cells. Nat Cell Biol 2:871–878 (2000).

Tumbar T, Belmont AS: Interphase movements of a DNA chromosome region modulated by VP16 transcriptional activator. Nat Cell Biol 3:134–139 (2001).

Van Driel R, Fransz PF, Verschure PJ: The eukaryotic genome: a system regulated at different hierarchical levels. J Cell Sci 116:4067–4075 (2003).

Weiss GM, Pysh JJ: Evidence of loss of Purkinje cell dendrites during late development: a morphometric Golgi analysis in the mouse. Brain Res 154:219–230 (1978).

Cytogenet Genome Res 105:311–315 (2004)
DOI: 10.1159/000098203

Murine metastable epialleles and transgenerational epigenetic inheritance

S. Chong and E. Whitelaw

School of Molecular and Microbial Biosciences, University of Sydney, New South Wales (Australia)

Abstract. For decades it has been recognized that a number of genes in the mouse genome do not behave according to Mendelian rules. Some of these genes exhibit unique patterns of expression, including variegation, variable expressivity in the context of isogenicity and transgenerational epigenetic inheritance. They are known as metastable epialleles. A correlation between expression and epigenetic state has been observed at these alleles, and information is emerging about the temporal course of erasure and re-establishment of these epigenetic marks during mouse development. Perhaps the most intriguing aspect of this process is the stochastic nature of the re-establishment of epigenetic state. This review aims to discuss the genomic structure and function of known and suspected metastable epialleles, to gain insight into the possible molecular mechanisms involved, and to aid in the future identification of other such alleles in the mouse and man.

Copyright © 2004 S. Karger AG, Basel

Metastable epialleles represent a distinct and novel group of epigenetically-sensitive genes that display variegation, variable expressivity in genetically identical individuals and transgenerational epigenetic inheritance. Although metastable epialleles have been well characterised in plants for some time, this phenomenon was not clearly demonstrated in mammals until recently. To date, we, and others have identified up to a dozen or so of these alleles in mice (Rakyan et al., 2002). The three best-characterised endogenous murine metastable epialleles are *agouti viable yellow*, *axin-fused* and *CDK5 activator-binding protein-IAP*.

This work was supported by a grant from the National Health and Medical Research Council of Australia to E.W.

Received 7 November 2003; manuscript accepted 2 December 2003.

Request reprints from: Dr. Emma Whitelaw
    School of Molecular and Microbial Biosciences, Biochemistry Building
    G08, University of Sydney, NSW 2006 (Australia)
    telephone: +61-2-9351-2549; fax: +61-2-9351-4726
    e-mail: e.whitelaw@mmb.usyd.edu.au

Agouti viable yellow

The *agouti viable yellow* (A^{vy}) allele is a dominant mutation of the murine *agouti (A)* locus, caused by the insertion of an intracisternal A-particle (IAP) retrotransposon approximately 100 kb upstream of the *agouti* coding exons (Duhl et al., 1994; Fig. 1A). Murine IAPs are endogenous transposable elements that are closely related to retroviruses. They are flanked by 5′ and 3′ long-terminal repeats (LTRs) that regulate transcription of the internal viral genes. At the A^{vy} locus, expression of the *agouti* coding exons is controlled by different upstream regions: use of wild-type *agouti* promoters results in hair-cycle-specific expression and an agouti coat colour (each hair is black with a subapical yellow band) whereas use of a cryptic LTR promoter at the 3′ end of the inserted IAP results in constitutive expression and a yellow coat (Duhl et al., 1994).

Variegation in mice heterozygous for A^{vy} is apparent as a mottled coat, which is a mosaic of yellow and agouti patches (Fig. 1B). Variable expressivity, that is, individual mice displaying different coat colour phenotypes ranging from completely yellow, to mottled, to completely agouti (referred to as pseudoagouti) is also seen among isogenic A^{vy}/a mice (Fig. 1B). A sire with a yellow coat is capable of producing offspring exhibiting the full range of coat colour phenotypes from yellow

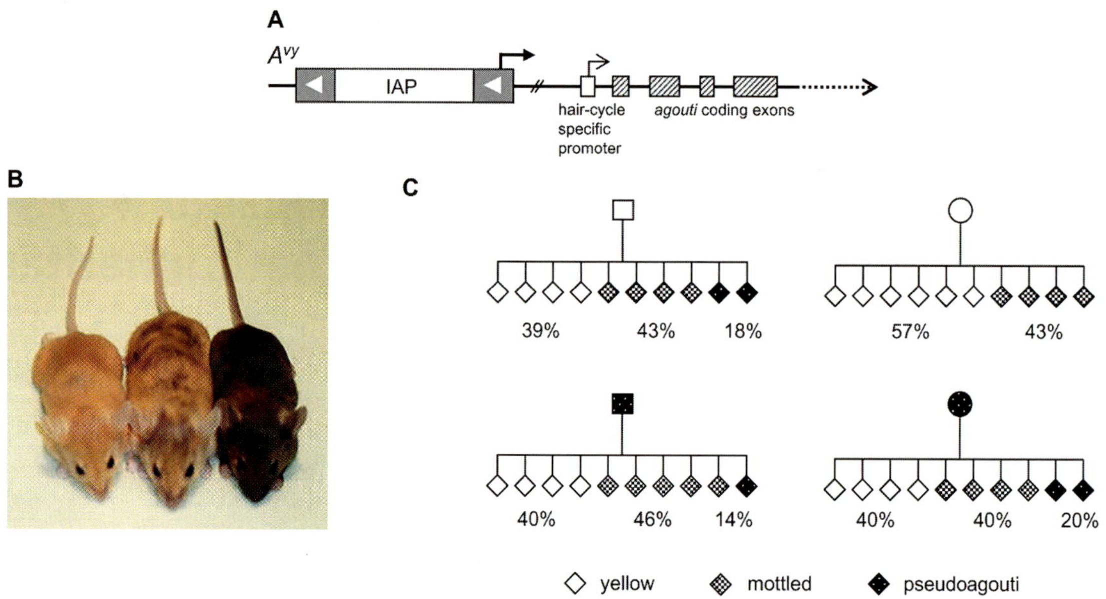

Fig. 1. *Agouti viable yellow.* (**A**) Map of the A^{vy} allele (not to scale) showing the location of the IAP insertion, *agouti* hair-cycle-specific promoter and coding exons. The position (grey boxes) and orientation of the IAP LTRs are as marked. A large black arrow indicates the direction of transcription from the cryptic IAP LTR promoter. (**B**) Isogenic C57BL/6J-A^{vy}/a mice showing a range of coat colour phenotypes. (**C**) A^{vy} inheritance. A^{vy}/a mice of the indicated phenotype were mated to congenic a/a mice and the offspring phenotype is indicated (%). a/a offspring have been omitted from the pedigrees. There is no significant difference in the range of offspring phenotypes arising from yellow and pseudoagouti sires. There is a significant difference in the proportion of offspring phenotypes arising from yellow and pseudoagouti dams.

to pseudoagouti, and a sire or dam with a pseudoagouti coat is similarly able to produce yellow, mottled and pseudoagouti offspring. Notably, yellow dams are more likely to produce yellow offspring, and pseudoagouti dams are more likely to produce pseudoagouti offspring (Wolff, 1978; Morgan et al., 1999; Fig. 1C). A number of additional experiments suggested that this was the result of transgenerational epigenetic inheritance (Morgan et al., 1999). Previous work in this laboratory has demonstrated that, in somatic tissues, the LTR at the 3′ end of the inserted IAP and adjacent A^{vy} sequence is hypermethylated in pseudoagouti mice and less methylated in yellow mice (Morgan et al., 1999). Mottled mice exhibit an intermediate methylation state. A^{vy} expression also correlates with a specific chromatin structure, in which the less methylated A^{vy} allele of yellow mice is more sensitive to DNaseI digestion than the hypermethylated A^{vy} allele of pseudoagouti mice (Nicola Vickaryous, personal communication). Interestingly, we have found that the somatic methylation patterns are retained in the gametes, inherited by the zygote and then erased sometime between fertilisation and blastocyst formation during the development of the mouse embryo (Marnie Champ, personal communication; Rakyan et al., 2003). This provides a possible molecular mechanism for transgenerational epigenetic inheritance. We have proposed that ultimately, epigenetic state and therefore phenotype is determined by the efficiency of clearing combined with the stochastic re-establishment of epigenetic marks in the developing embryo (Rakyan et al., 2003).

Axin-fused

Another endogenous murine allele that displays the hallmark characteristics of a metastable epiallele is *axin-fused* (*Axin^{Fu}*). *Axin* (<u>ax</u>is <u>in</u>hibition) encodes an inhibitor of the Wnt signalling pathway, which is involved in axis formation in the mouse embryo. The murine *fused* allele (*Axin^{Fu}*) is a mutation of *axin* that manifests most obviously as a kinked tail. Inbred 129P4/RrRk mice heterozygous for *Axin^{Fu}* exhibit variable expressivity as an array of tail phenotypes, ranging from kinked tails of varying degrees (penetrant) to a normal tail (silent). *Fused* is also associated with an IAP retrotransposon insertion, but unlike A^{vy} the IAP is located within intron 6 (Vasicek et al., 1997; Fig. 2A). We have found that the presence or absence of the kinky tail phenotype in mice heterozygous for *Axin^{Fu}* also correlates with differential DNA methylation (Rakyan et al., 2003). In particular, the LTR at the 3′ end of the inserted IAP and adjacent *Axin^{Fu}* intron 6 sequence is hypermethylated in silent animals and less methylated in penetrant animals. Moreover, transcripts were found to initiate immediately downstream of the IAP LTR, exclusively in kinky tailed mice (Rakyan et al., 2003). Again, the LTR appears to act as a cryptic promoter and/or enhancer, but this time producing shorter transcripts containing exons 7–10 only. The resulting protein would lack the amino-terminal end of wild-type Axin, and may cause kinky tails by interfering with wild-type Axin function. In contrast to A^{vy}, transgenerational epigenetic inheritance of *Axin^{Fu}* occurs regardless of the parental origin of the allele (Rakyan et al., 2003). That is, penetrant offspring occur at a higher

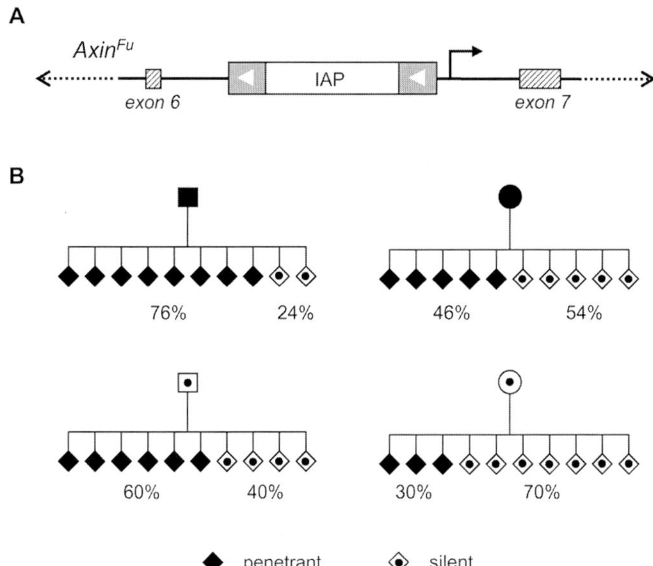

Fig. 2. *Axin-fused.* (**A**) Map of the *Axin^Fu* allele (not to scale) showing the IAP insertion into intron 6. The position (grey boxes) and orientation of the IAP LTRs are as marked. A large black arrow indicates transcription resulting from IAP LTR promoter/enhancer activity. (**B**) *Axin^Fu* inheritance. *Axin^Fu*/+ 129P4/RrRk mice of the indicated phenotype were mated to congenic +/+ mice and the offspring phenotype is indicated (%). +/+ offspring have been omitted from the pedigrees. There is a significant difference in the proportion of offspring phenotypes arising from penetrant and silent sires. There is also a significant difference in the proportion of offspring phenotypes arising from penetrant and silent dams.

frequency if the parent (either sire or dam) has a penetrant phenotype and vice versa (Fig. 2B). F_1 hybrid studies subsequently revealed that epigenetic inheritance in mice is susceptible to the strain background of the egg (Rakyan et al., 2003). In particular, the C57BL/6J egg is much more efficient at clearing paternally derived epigenetic marks than the 129P4/RrRk egg. Thus, when introduced into a C57BL/6J egg, the paternally inherited *Axin^Fu* marks are not retained. This may explain why the A^{vy} allele, which has been maintained in the C57BL/6J background, does not show epigenetic inheritance through the male germline.

How many metastable epialleles are there?

The frequency of metastable epialleles in the mouse genome is unclear, and ongoing work in our laboratory has been aimed at the further identification and characterisation of these alleles. It is possible, even likely, that this type of allele and consequently transgenerational epigenetic inheritance occurs in other mammals as well, however, there is an inherent difficulty in detecting metastable epialleles in animal systems that are not inbred. In the mouse, IAPs are present at approximately 1,000 copies per haploid genome (Kuff and Lueders, 1988). Given their transcriptional competence, their ability to influence the transcription of neighbouring genes and the relative abundance of IAPs in the mouse genome, one might expect more metastable epialleles to exist. A^{vy} and *Axin^Fu* were identified mainly because of their relatively dramatic effects on visible phenotypes; however, other murine metastable epialleles may not be so easily recognised. We have designed a strategy to aid in the identification of metastable epialleles at the molecular level. Both A^{vy} and *Axin^Fu* contain IAP retrotransposons of the IΔ1 subclass, inserted in an orientation such that IAP transcription is opposite to that of the adjacent epiallele. Cryptic LTR promoter/enhancer activity then results in the production of chimeric (A^{vy}) or truncated (*Axin^Fu*) transcripts. The wealth of information available following the release of the mouse genome sequence has allowed us to use a bioinformatic approach to identify murine metastable epialleles. Initially, mouse EST databases were scanned for chimeric transcripts containing sequences of the IAP IΔ1 subclass, and a third metastable epiallele, *Cabp^IAP*, was identified in this way (Riki Druker, manuscript in preparation). This approach, however, would not have identified alleles similar to *Axin^Fu*, whose truncated transcripts do not contain any IAP-derived sequences. Subsequent searches have focused on examining murine genomic DNA for IAP insertions of the appropriate subclass and orientation. The difficulty with this approach is that many candidates match these criteria. Differential methylation of the LTR at the 3′ end of the inserted IAP (corresponding to the IAP promoter region) correlates with both variable expressivity and epigenetic inheritance at A^{vy} and *Axin^Fu*, and we have attempted to use this correlation as an indicator of a metastable epiallele. Two genes analysed in this way were *Tomosyn* and *mitogen activated protein kinase kinase kinase 5* (*Map3k5*) both of which contain intronic IAP insertions in C57BL/6J mice. Southern blot analysis, with methylation sensitive restriction enzymes, of the LTR-*Map3k5* or LTR-*Tomosyn* junction in numerous mice showed a high degree of DNA methylation in all individuals and it was concluded that these genes were unlikely to display the complex and variable expression patterns typical of metastable epialleles (Michelle Holland, personal communication).

Stochastic re-establishment of epigenetic state

One of the most intriguing aspects of metastable epialleles is the stochastic establishment of epigenetic state, which occurs early in development. What elements or events determine whether such a mark will be set-up and maintained in a particular cell? Cytosine methylation-based silencing of IAPs has been proposed to function as a host-defence mechanism and IAP hypermethylation is observed in almost all murine tissues and stages of development (Bestor, 1998; Walsh et al., 1998; Lane et al., 2003). Moreover, IAPs appear to be mostly resistant to the epigenetic reprogramming that takes place during both gametogenesis and embryogenesis (Lane et al., 2003). There is also additional evidence that at least a proportion of IAPs are sequestered to heterochromatic regions of both the Syrian hamster and mouse genomes (Kuff and Lueders, 1988). So how is the activity and sustained expression of an IAP element established against such odds? Protection against the heterochro-

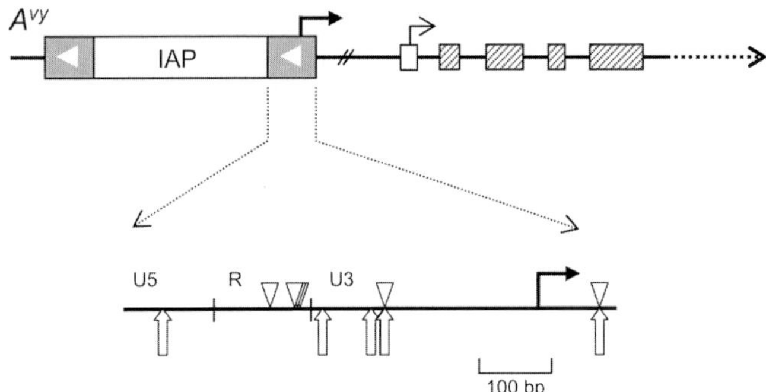

Fig. 3. Putative Sp1 and GAGA factor binding sites in the A^{vy} IAP LTR. A map (not to scale) of the A^{vy} allele is shown above, and an expanded map (to scale) of an IAP LTR is shown below. The three functional domains (U3–R–U5) of the LTR are shown, and the large black arrow indicates the initiation of transcription from the cryptic IAP LTR promoter. GAGA factor binding sites (up to 16 % mismatch) are indicated by inverted triangles above the line, and Sp1 binding sites (up to 18 % mismatch) are indicated by arrows below the line. Binding sites were identified using TRANSFAC v4.0.

matic repression of reporter genes in a yeast system can be conferred by the binding of mammalian factors such as Sp1, CTCF and GAGA factor (Ishii and Laemmli, 2003). Sequence analysis of the A^{vy} and $Axin^{Fu}$ IAP LTRs reveal multiple potential binding sites for Sp1 and GAGA factor (Fig. 3), but clearly occupation of these binding sites is hindered in some way at the majority of IAPs in the mouse genome.

It seems likely that the re-establishment of epigenetic marks at metastable epialleles occurs in a gene-specific manner, and is determined by the chromosomal environment. Barbot and colleagues (Barbot et al., 2002) have shown that genomic context can affect the epigenetic state and expression of a specific IAP of the IΔ1 subclass. That is, progressive demethylation of an IAP promoter, located in intron 1 of the *m.nocturnin* housekeeping gene, was observed upon repeated induction of *m.nocturnin* and IAP expression in a circadian manner in mouse liver (Barbot et al., 2002). Wild-type *agouti* is expressed in the middle stages of the hair growth cycle in adult skin (Millar et al., 1995). It is also detectable at E10.5 and by E14.5, expression is observed over the entire ventral head and trunk as well as the dorsal distal limbs (Millar et al., 1995). Wild-type *axin* is expressed ubiquitously in adults and E7.5–16.5 embryos (Zeng et al., 1997). It is possible that the stochastic establishment of epigenetic marks at metastable epialleles may result from competing molecular mechanisms: the need for the host genome to epigenetically silence potentially harmful retrotransposons versus the epigenetically active state necessary for appropriate expression of the adjacent endogenous gene early in development.

Subtle parent-of-origin effects

All three murine metastable epialleles identified to date are in chromosomal locations adjacent to clusters of imprinted genes. $Cabp^{IAP}$ and *agouti* are located approximately 640 kb

apart on distal chromosome 2, approximately 2,500 kb from *Nnat*, *Nespas*, *Nesp*, *Gnasx1* and *Gnas*. *Axin* is on proximal chromosome 17, roughly 13,000 kb from *Igf2r*, *Igf2ras*, *Slc22a2* and *Slc22a3*. On the other hand, only one imprinted gene, *Zac1* (3,300 kb away) has been identified near *Tomosyn* on proximal chromosome 10, and *Nocturnin* is on chromosome 3, which is currently thought to lack imprinted genes. This chromosomal positioning is tantalising, given that metastable epialleles do display subtle parent-of-origin effects. Classically imprinted genes display monoallelic expression based upon parental origin. That is, if the paternally inherited allele is expressed in all tissues, the maternally inherited allele will be correspondingly silent and vice versa. Although the expression of metastable epialleles does not strictly correspond to parental origin, there is a shift in the probability of expression based upon the parent-of-origin of the allele. At $Axin^{Fu}$, for example, there is 76 % chance of a kinky tail phenotype if the allele is inherited from a kinky tailed sire, but the likelihood of a kinky tail drops to 46 % when inherited from a kinky tailed dam (Rakyan et al., 2003; Fig. 2B). It is tempting to speculate that the proximity of metastable epialleles to clusters of imprinted genes explains their parental-dependent pattern of expression.

Mitotic and meiotic memory

We have previously proposed a model for transgenerational epigenetic inheritance in which some memory of the parental epigenetic state is transmitted to the next generation via the incomplete clearing of epigenetic marks post-fertilisation (Rakyan et al., 2003). Stable inheritance of a particular transcriptional state through mitosis and meiosis has been described in *Drosophila* (Cavalli and Paro, 1998). Interestingly, this inheritance is mediated by chromatin structure, which in turn, is defined by "cellular memory modules" that contain targets for Polycomb group and trithorax group proteins such as GAGA factor (Maurange and Paro, 2002). Moreover, Rank and colleagues (Rank et al., 2002) have proposed that transcription and the passage of the RNA polymerase II complex through a cellular memory module changes the default silent state to an active state by relieving repression and/or recruiting active chromatin remodelling complexes. This requirement for transcription in conferring heritable expression states is striking, and raises the possibility that a transcription-based mechanism can be invoked not only for variegation and variable expressivity but also for transgenerational epigenetic inheritance. The expression and epigenetic state of a metastable epiallele would be independently established in each cell, as either inactive, by the defensive silencing of IAPs, or active, due to the prevailing transcription of the endogenous gene during embryogenesis. The IAP LTR may then function as a cellular memory module, and the resulting epigenetic state would be heritable both mitotically and meiotically.

Significance

One compelling aspect of metastable epialleles is the possibility that they are present in humans and that the phenotypes involved may manifest as disease. Constitutive expression of murine A^{vy} results not only in a yellow coat, but also in obesity and insulin resistance (Miltenberger et al., 1997). Furthermore, these "yellow obese" mice can then produce a greater number of "yellow obese" offspring. Indirect support for the idea that metastable epialleles are likely to be present in humans can be found in reports of human monozygotic twins displaying discordance for schizophrenia and Beckwith-Wiedemann syndrome (Tsujita et al., 1998; Weksberg et al., 2002). This, combined with the observation that 8% of the human genome is of retroviral origin (International Human Genome Sequencing Consortium, 2001), suggests that metastable epialleles may indeed exist in humans and have significant phenotypic effects.

Acknowledgements

We thank Marnie Champ, Riki Druker, Michelle Holland and Nicola Vickaryous for use of unpublished data.

References

Barbot W, Dupressoir A, Lazar V, Heidmann T: Epigenetic regulation of an IAP retrotransposon in the aging mouse: progressive demethylation and desilencing of the element by its repetitive induction. Nucleic Acids Res 30:2365–2373 (2002).

Bestor TH: The host defence function of genomic methylation patterns. Novartis Found Symp 214:187–195 (1998).

Cavalli G, Paro R: The *Drosophila* Fab-7 chromosomal element conveys epigenetic inheritance during mitosis and meiosis. Cell 93:505–518 (1998).

Duhl DMJ, Vrieling H, Miller KA, Wolff GL, Barsh GS: Neomorphic agouti mutations in obese yellow mice. Nat Genet 8:59–65 (1994).

International Human Genome Sequencing Consortium: Initial sequencing and analysis of the human genome. Nature 409:860–921 (2001).

Ishii K, Laemmli UK: Structural and dynamic functions establish chromatin domains. Mol Cell 11:237–248 (2003).

Kuff EL, Lueders KK: The intracisternal A-particle gene family: structure and functional aspects. Adv Cancer Res 51:183–276 (1988).

Lane N, Dean W, Erhardt S, Hajkova P, Surani A, Walter J, Reik W: Resistance of IAPs to methylation reprogramming may provide a mechanism for epigenetic inheritance in the mouse. Genesis 35:88–93 (2003).

Maurange C, Paro R: A cellular memory module conveys epigenetic inheritance of hedgehog expression during *Drosophila* wing imaginal disc development. Genes Dev 16:2672–2683 (2002).

Millar SE, Miller MW, Stevens ME, Barsh GS: Expression and transgenic studies of the mouse agouti gene provide insight into the mechanisms by which mammalian coat color patterns are generated. Development 121:3223–3232 (1995).

Miltenberger RJ, Mynatt RL, Wilkinson JE, Woychik RP: The role of the agouti gene in the yellow obese syndrome. J Nutr 127:1902S–1907S (1997).

Morgan HD, Sutherland HGE, Martin DIK, Whitelaw E: Epigenetic inheritance at the agouti locus in the mouse. Nat Genet 23:314–318 (1999).

Rakyan VK, Blewitt ME, Druker R, Preis JI, Whitelaw E: Metastable epialleles in mammals. Trends Genet 18:348–351 (2002).

Rakyan VK, Chong S, Champ ME, Cuthbert PC, Morgan HD, Luu KVK, Whitelaw E: Transgenerational inheritance of epigenetic states at the murine Axin^{Fu} allele occurs after maternal and paternal transmission. Proc Natl Acad Sci USA 100:2538–2543 (2003).

Rank G, Prestel M, Paro R: Transcription through intergenic chromosomal memory elements of the *Drosophila* bithorax complex correlates with an epigenetic switch. Mol Cell Biol 22:8026–8034 (2002).

Tsujita T, Niikawa N, Yamashita H, Imamura A, Hamada A, Nakane Y, Okazaki Y: Genomic discordance between monozygotic twins discordant for schizophrenia. Am J Psychiatry 155:422–424 (1998).

Vasicek TJ, Zeng L, Guan X-J, Zhang T, Costantini F, Tilghman SM: Two dominant mutations in the mouse Fused gene are the result of transposon insertions. Genetics 147:777–786 (1997).

Walsh CP, Chaillet JR, Bestor TH: Transcription of IAP endogenous retroviruses is constrained by cytosine methylation. Nat Genet 20:116–117 (1998).

Weksberg R, Shuman C, Caluseriu O, Smith AC, Fei Y-L, Nishikawa J, Stockley TL, Best L, Chitayat D, Olney A, Ives E, Schneider A, Bestor TH, Li M, Sadowski P, Squire J: Discordant KCNQ1OT1 imprinting in sets of monozygotic twins discordant for Beckwith-Wiedemann syndrome. Hum Mol Genet 11:1317–1325 (2002).

Wolff GL: Influence of maternal phenotype on metabolic differentiation of agouti locus mutants in the mouse. Genetics 88:529–539 (1978).

Zeng L, Fagotto F, Zhang T, Hsu W, Vasicek TJ, Perry WL, Lee JJ, Tilghman SM, Gumbiner BM, Costantini F: The mouse fused locus encodes Axin, an inhibitor of the Wnt signalng pathway that regulates embryonic axis formation. Cell 90:181–192 (1997).

Cytogenet Genome Res 105:316–324 (2004)
DOI: 10.1159/000078204

DNA methylation in mouse gametogenesis

R. Marchal, A. Chicheportiche, B. Dutrillaux, and J. Bernardino-Sgherri

Laboratoire de radiosensibilité des cellules germinales, Département de Radiobiologie et Radiopathologie,
CEA/DSV/SEGG/LRCG Fontenay-aux-roses (France)

Abstract. DNA methylation is involved in many biological processes and is particularly important for both development and germ cell differentiation. Several waves of demethylation and de novo methylation occur during both male and female germ line development. This has been found at both the gene and all genome levels, but there is no demonstrated correlation between them. During the postnatal germ line development of spermatogenesis, we found very complex and drastic DNA methylation changes that we could correlate with chromatin structure changes. Thus, detailed studies focused on localization and expression pattern of the chromatin proteins involved in both DNA methylation, histone tails modification, condensin and cohesin complex formation, should help to gain insights into the mechanisms at the origin of the deep changes occurring during this particular period.

DNA methylation is the most studied epigenetic modification of the DNA. It is characterized by the covalent addition of a methyl group to the carbon 5 of the cytosine residue, particularly in CpG dinucleotides in mammals. This enzymatic reaction is driven by DNA methyltransferases. Mutations in the currently known DNA methyltransferases have shown that DNA methylation is essential for mouse development and germ cell differentiation (Li et al., 1992; Okano et al., 1999; Bourc'his et al., 2001; Maatouk and Resnick, 2003).

It is established that DNA methylation is implicated in gene silencing through chromatin modification and remodelling (Li, 2002) and also in chromosomal stability (Almeida et al., 1993; Eden et al., 2003; Gaudet et al., 2003; Lengauer, 2003). Histone modification is also an epigenetic factor tightly linked to DNA methylation (Fuks et al., 2000; Datta et al., 2003). Both seem to drive dynamic transcriptional silencing and activity during mouse development and somatic cell differentiation. These functions are for a part mediated by proteins, which specifically recognize methylated sequences and interact with histone deacetylase or other proteins implicated in chromatin remodelling (Li, 2002). DNA methyltransferases can also be part of transcription silencing, independent of their methylase activity (Fuks et al., 2001). Recently, it has been reported that methylated sequences may be linked to chromatin cohesion and compaction (Bernardino-Sgherri and Dutrillaux, 2001; Bernardino-Sgherri et al., 2002a, b; Flagiello et al., 2002; Hakimi et al., 2002).

Considering the involvement of DNA methylation in gametogenesis, most of the studies question its implication in parental imprinting which is established during both male and female mouse germ line differentiation (Kaneko-Ishino et al., 2003). Less is known about the roles of the overall or repeated sequences DNA methylation changes that have been reported so far. The relationship between genome-wide DNA methylation changes and those occurring at the gene level has not been elucidated.

The importance of DNA methylation for proper germ cell differentiation has been shown by the use of demethylating agents, such as 5-aza-deoxycytidine (5-aza-dC). For example, inhibition of spermatogonia differentiation into spermatocytes has been induced when 5-aza-dC was administrated to neonatal mice (Raman and Narayan, 1995). In adult rats and mice, abnormal gametogenesis and preimplantation development followed both mitotic and meiotic germ cell exposure (Doerksen and Trasler, 1996; Doerksen et al., 2000).

R.M. is a postdoctoral fellow, supported by the Program of Nuclear Toxicology of the CEA.

This work was supported by Electricité de France and by The Program of Nuclear Toxicology of the Comissariat à l'Energie Atomique (CEA).

Received 25 September 2003; accepted 17 October 2003.

Request reprints from: Dr. J. Bernardino-Sgherri
Laboratoire de radiosensibilité des cellules germinales
Département de Radiobiologie et Radiopathologie
CEA/DSV/SEGG/LRCG BP 6 92265 Fontenay-aux-roses cedex (France)
telephone: +33 (0) 1 46 54 80 04; fax: +33 (0) 1 46 54 99 06
e-mail: jacqueline.bernardino@cea.fr.

Because of this important role of DNA methylation for normal gametogenesis, it is of interest to carry on experiments enabling a better understanding of the mechanisms and identification of enzymes involved in DNA methylation changes during both spermatogenesis and oogenesis.

This report aims at summing up the findings in the field of DNA methylation in rodent gametogenesis and aims to point out the gaps that exist in the literature as well as the questions, which remain open, particularly about the role of widespread genomic DNA methylation changes during gametogenesis.

Gametogenesis, a complex maturation process

Gametogenesis is a complex process consting of successive periods of germ cell proliferation, cell cycle arrest and differentiation, with strong differences between female and male germ line establishment.

In the male mouse, gametogenesis can be divided into two main periods. The first one, also called the first round of spermatogenesis, corresponds to the establishment of gametogenesis leading to the production of the first spermatozoa in 35-day-old males. The second one refers to the continuous differentiation of some stem cell spermatogonia to mature spermatozoa, which occurs during all the adult life span (Kluin et al., 1982). In the mouse embryo, the first primordial germ cells (PGC) appear at about 7.5 days post coitum (dpc). Between 8.5 and 10 dpc, they rapidly proliferate during their migration towards the embryonic gonads (Molyneaux et al., 2001). At about 12.5 dpc, sex differentiation occurs. When germ cells are surrounded by fetal Sertoli cells, they are referred to gonocytes or prospermatogonia. Their proliferation ceases between 13.5 and 15 dpc and birth. It starts again after birth, giving rise to presumed spermatogonia stem cells among which some actively proliferate, differentiate and enter meiosis by 10 days post partum (dpp).

In the female mouse, oogenesis is almost achieved at the end of the fetal period. Oogonia actively divide, enter meiosis and undergo a protracted arrest at diplonema. After birth, oocytes increase in size during the growing follicular phase, while acquiring the competence to enter the final stages of meiosis. Full completion of gametogenesis occurs in the adult, leading to the formation of mature oocytes at each ovulation (once every 4 days in the mouse). Oocytes, which are then arrested at the metaphase II stage of meiosis, either degenerate or end meiosis upon fertilization.

Thus, whereas in the male mouse, there are several periods of germ cell proliferation and differentiation, in the female, the establishment of gametogenesis consists of one period of proliferation and several periods of maturation of non-replicating germ cells.

DNA methylation in Primordial Germ Cells (PGCs)

DNA methylation in PGCs has been studied in relation to its major role in genetic imprinting and X chromosome reactivation (Maatouk and Resnick, 2003). It has previously been shown that PGC overall DNA methylation was very low (Monk et al., 1987). These data correlate with our findings (Fig. 1A and 2A) using an immunocytological detection of methylated cytosine residues (Reynaud et al., 1992). This period of low DNA methylation, from 11.5 to 15.5 dpc, was interpreted as the period where the parental imprinted memories were erased (Lee et al., 2002). A recent study gave an accurate description of the ontogeny of DNA methylation changes in imprinted and non-imprinted genes as well as in repetitive DNA sequences of PGCs (Hajkova et al., 2002). This study highlighted a more complicated view of DNA methylation changes in these cells according to the type of sequences that were analysed. PGCs appear to undergo DNA demethylation at imprinted and non-imprinted loci in a small window of time corresponding to their migration into the genital ridges (from 10.5 to 12.5 dpc). However, some CpG sites within the imprinted gene *Igf2r* start to be demethylated in some PGCs before the colonization of the genital ridges (9.5 dpc) (Sato et al., 2003). By 12.5–13.5 dpc, the genomes of both XX and XY germ cells appear to be strongly hypomethylated at numerous loci, including those involved in parental imprinted alleles (Ueda et al., 2000; Lee et al., 2002). Promoters of several genes that may be involved in germ cell differentiation contain cytosine residues, which are methylated at 10.5 dpc, but unmethylated by 13.5 dpc (Maatouk and Resnick, 2003). This suggests that germ cell genome wide demethylation in PGCs may contribute to both parental imprint erasure and differentiation.

Because the majority of CpGs in the genome lie in repetitive sequences, some studies were performed to determine their methylation status in PGCs. Different classes of repetitive sequences have been studied. The minor and major satellite sequences, located in juxtacentromeric regions, consist of large arrays of simple repeats of a few base pairs (Hastie, 1989). The other repetitive sequences are interspersed at various loci throughout the genome. They comprise about 45 % of the total genome and derive from transposable elements. They form short interspersed nuclear elements (SINEs) represented by B1 and B2 elements in rodents and are mainly located in R-bands while long interspersed nuclear elements (LINEs) lie in G-bands (Boyle et al., 1990). Intracisternal A-particles (IAP) appear widely dispersed over the chromosome length (Kuff et al., 1986).

While single copy genes are demethylated in PGCs at 12.5–13.5 dpc, bulk DNA remains substantially methylated. This includes many classes of repeated sequences such as the minor satellite, IAP and LINE-1 (Hajkova et al., 2002; Lane et al., 2003; Lees-Murdock et al., 2003). This is in accordance with the lack of transcription of IAPs sequences during this period (Dupressoir and Heidmann, 1996) as methylation of these sequences has been linked to their transcription inhibition (Feenstra et al., 1986).

Very little data is available on the DNA methyltransferases expression pattern in PGCs. Using immunohistochemistry on testis cross-sections of 12.5-day-old embryos, when the overall DNA methylation of the germ cells is low, Sakai et al. (2001) have shown that Dnmt1 could be detected whereas the oocyte specific variant Dnmt1o was not. This was confirmed by Hajkova et al. (2002) who found that Dnmt1 was expressed in the

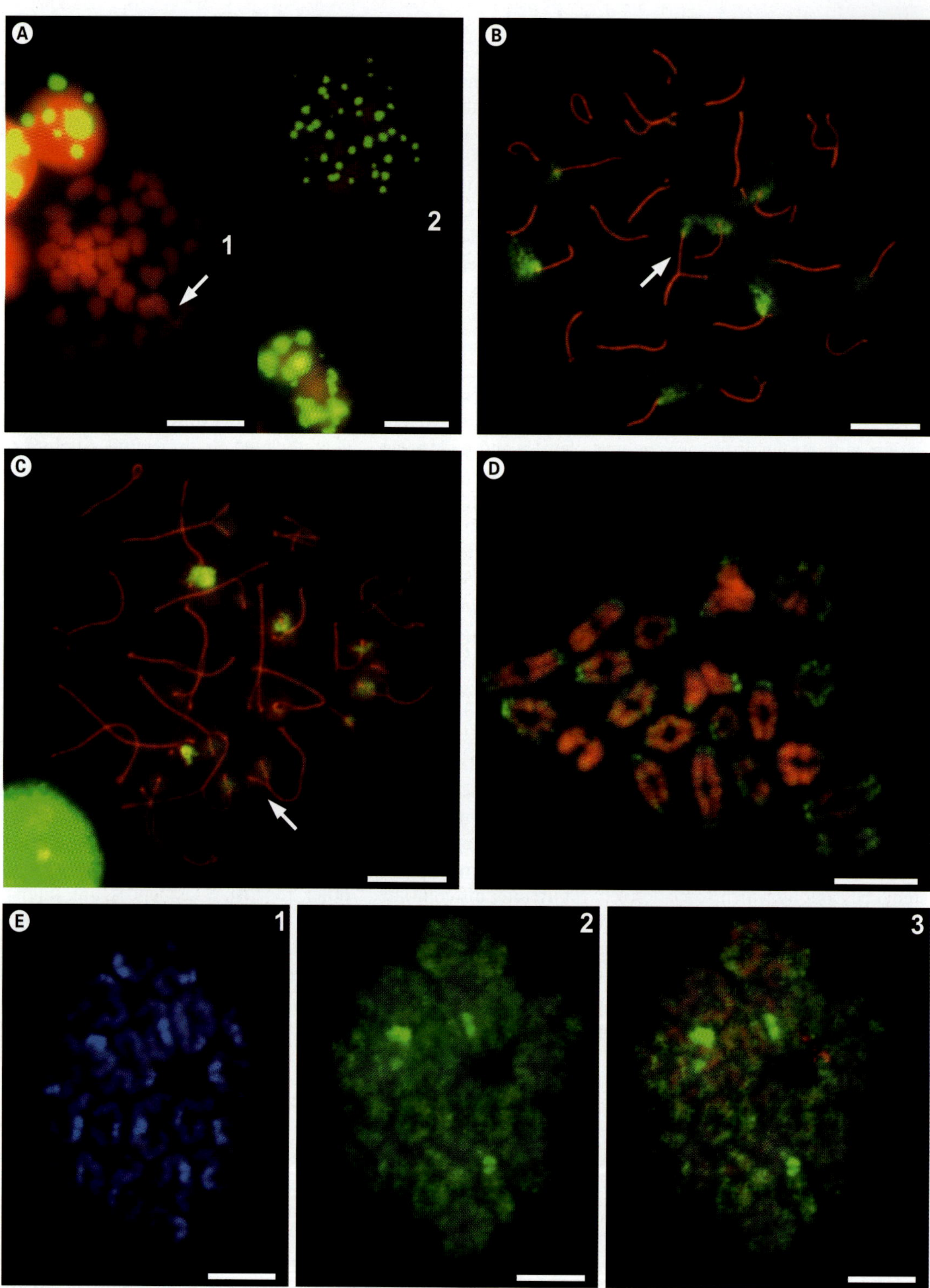

Fig. 1. DNA methylation patterns during female gametogenesis. (**A1**) Oogonium with poorly methylated DNA (red). A single centromeric region is methylated (green, arrow). (**A2**) Somatic cell from the same preparation. All centromeric regions are intensely labeled. (**B**) Late zygonema and (**C**) early diplonema. Intense labeling is restricted to some centromeric regions. Their asynapsis (**B**) or desynapsis (**C**) exhibit strong differences between homologous centromeric regions (arrows). DNA from loops surrounding synaptonemal complexes in red is slightly labeled, making a green background (not visible, but present in **C**). (**D**) Diakinesis exhibiting a regrouping of methylated DNA on bivalents. Asymmetrical labeling is due to that of centromeric heterochromatin. (**E**) Metaphase II oocyte, left panel (1): DAPI staining; middle panel (2): anti-5mC antibody staining; right panel (3): both anti-5mC antibody and propidium iodide stainings. Some centromeric regions are intensely labeled. Chromosome preparations and immunostainings were performed as described (Bernardino et al., 2000; Peters et al., 1997). In red: either DNA counterstaining with propidium iodide (**A, D, E3**) or synaptonemal complex staining with an anti-SCP3 antibody (**B, C**). Bar = 10 µm.

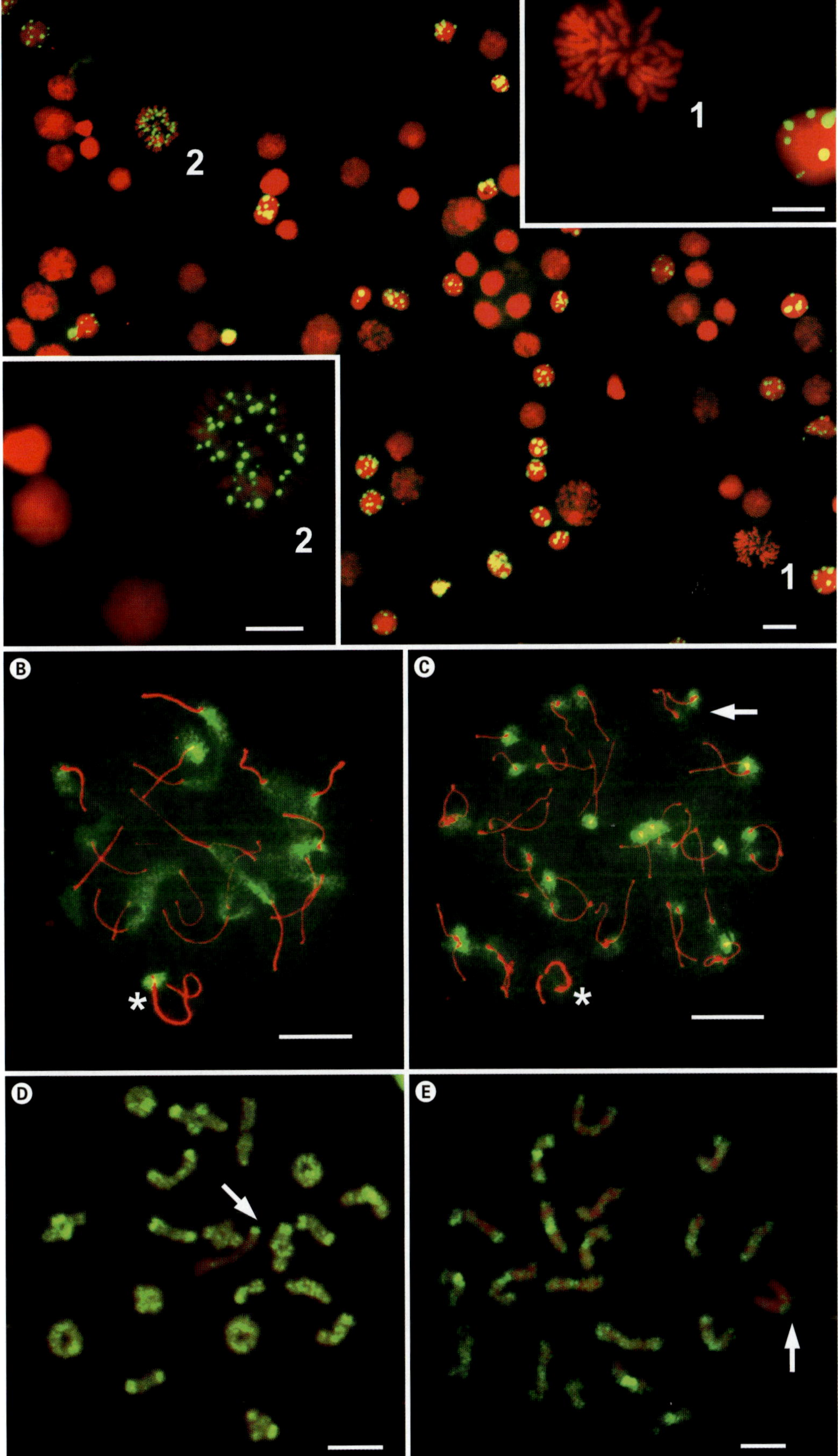

Fig. 2. Imunocytochemical detection of methylated cytosine in situ on chromosome spreads at different stages of male gametogenesis. (**A**) Gonocyte (1) and somatic (2) cell metaphase (13.5 dpc) on the same field. Higher magnification of the metaphases are shown in the inserts. (**B**) Pachynema. (**C**) Diplonema, exhibiting different methylation levels of asynapsed centromeric regions of homologs (arrow). The centromeric region (*) of the X chromosome is labeled in (**B**) but not in (**C**). (**D**) Metaphase I. (**E**) Metaphase II (20, X). In the sex chromosomes (arrows) strong DNA methylation is limited to the centromeric region of the X. Chromosome preparations and immunostainings were performed as for Fig. 1. Bar = 10 μm.

nucleus of almost all PGCs between 10.5 and 13.5 dpc, at higher levels than in somatic cells. Dnmt3a was not found in PGCs whereas Dnmt3b was highly expressed, but predominantly located in the cytoplasm (Hajkova et al., 2002), suggesting that it might not be active in these cells. Mechanisms of DNA demethylation involving cytoplasm sequestration of DNA methyltransferases and/or demethylases activities, not yet identified, may participate to the loss of DNA methylation, which suddenly occurs between 12.5 and 13.5 dpc. This demethylation is obviously not linked to the replication dependent passive demethylation described in the embryo (Rougier et al., 1998) and postnatal spermatogonia (Coffigny et al., 1999).

DNA methylation changes during oogenesis

There is little information about DNA methylation during oogenesis in female mice. Most studies aimed to understand the establishment of imprinting. The use of antibodies against methylated cytosines (5mC) showed low methylation levels in both nuclei and chromosomes of proliferating oogonia at 13.5 dpc (Fig. 1A1). Heterogeneous and low labeling could be detected at some juxtacentromeric regions. This is in contrast to their strong and homogeneous labeling in somatic cells (Fig. 1A2). In oocytes at both late zygonema (Fig. 1B) and late pachytene/diplotene stage of meiosis (16.5 dpc) (Fig. 1C), the overall DNA methylation remains rather low, except at some juxtacentromeric regions. Their labeling is often different between homologs (Fig. 1B, C). These findings fit with molecular data indicating extensive undermethylation at *Msp*I sites of interspersed repetitive sequences LINE-1, IAP and MUP (Murine Urinary Protein) at diplonema (Sanford et al., 1987). At later stages, including metaphase I (Fig. 1D) and II (Fig. 1E), chromosome arms were symmetrically stained. The staining of the centromeric regions was heterogeneous and weak as observed in oogonia, fitting the molecular data describing undermethylation of centromeric satellite DNA (Sanford et al., 1984). Molecular data showed that IAP, LINE-1 and MUP sequences are unmethylated in mature oocytes whereas *Alu* sequences are methylated (Rubin et al., 1994). Thus, during oogenesis, as spermatogenesis, the re-methylation of euchromatin seems to be symmetrical between homologous chromatids, whereas that of heterochromatin does not seem to follow any rule.

The expression pattern of some DNA methyltransferases, and particularly the oocyte-specific variant Dnmt1o was studied in oocytes from 17.5–18.5 dpc to 70 dpp mice (Ratnam et al., 2002). When female germ cells enter meiosis neither Dnmt1 nor Dnmt1o are expressed. After birth, when oocytes arrested at diplonema start growing, Dnmt1 mRNA is detected but not the protein. In contrast, Dnmt1o protein is expressed from 5 dpp to the adult age. This correlates with the progressive increase of the number of methylated sites within differentially methylated regions (DMR) of imprinted genes starting before 10 dpp and completed at metaphase II stage (Lucifero et al., 2002). Dnmt3L, which has been shown to have no DNA methyltransferase activity, seems to be important to establish methylation imprints in the female (Bourc'his et al., 2001), probably through its physical interaction with both Dnmt3a and Dnmt3b de novo methyltransferases (Hata et al., 2002). Disruption of both Dnmt3a and its variant Dnmt3a2 perturbs de novo methylation of maternally imprinted genes during oocyte maturation (Chen et al., 2002). Dnmt3L may target these DNA methyltransferases specifically to DMR of imprinted genes in the oocytes (Hata et al., 2002). The expression of the oocyte-specific DNA methyltransferase, Dnmt1o, during the oocyte growth (Ratnam et al., 2002) is consistent with the maintenance rather than the establishment of the maternal imprints. Effectively, imprinting establishment was found to be normal in Dnmt1o-deficient oocytes (Howell et al., 2001; Jaenisch and Bird, 2003). Increasing amounts of both Dnmt1o mRNA and protein have also been described between GVBD stage (Germinal Vesicle Break Down) and metaphase II stage oocytes (Ratnam et al., 2002).

DNA methylation changes during spermatogenesis

Late prenatal and neonatal periods

As mentioned above, male germ cells undergo a period of DNA demethylation between 10.5 to 13.5 dpc (Fig. 2A1). Between 15.5 and 16.5 dpc, we found a strong increase of overall DNA methylation in quiescent germ cell nuclei (Coffigny et al., 1999). This DNA hypermethylation status involved only euchromatin and was maintained until birth when germ cells resumed their cycle. Accordingly, some repeated sequences such as minor satellite, LINE-1 and IAP 5′ long terminal repeats (LTR) have been found to be re-methylated in quiescent germ cells (Walsh et al., 1998; Lees-Murdock et al., 2003). Methylation of IAP-LTR sequences has been associated with their transcription inhibition (Feenstra et al., 1986; Walsh et al., 1998). LTR activity has however been found in some gonocytes of fetal mice (Dupressoir and Heidmann, 1996). This suggests that IAP-LTR activity may either be regulated by additional factors to DNA methylation in some germ cells or that the methylation status of IAP-LTR may differ from one cell to another. Some imprinted and non-imprinted genes have also been shown to be methylated in quiescent germ cells (Kafri et al., 1992; Davis et al., 2000; Ueda et al., 2000).

After birth, we found that the strong overall DNA methylation of euchromatin decreased along with successive divisions corresponding to a passive mechanism of demethylation (Coffigny et al., 1999; Bernardino et al., 2000; Bernardino-Sgherri et al., 2002a). Thus, the methylation status of euchromatin and heterochromatin was completely different. Whereas euchromatin was strongly methylated in germ cells just leaving their cycle arrest, juxtacentromeric heterochromatin remained weakly methylated (Coffigny et al., 1999; Bernardino-Sgherri et al., 2002a). With successive divisions, whereas euchromatin was losing its strong DNA methylation, juxtacentromeric heterochromatin was progressively methylated, involving chromosomes at random (unpublished data). These findings suggest that different mechanisms are involved in the regulation of methylation according to the chromatin structure. During all prenatal and postnatal period, the DNA methylation pattern of

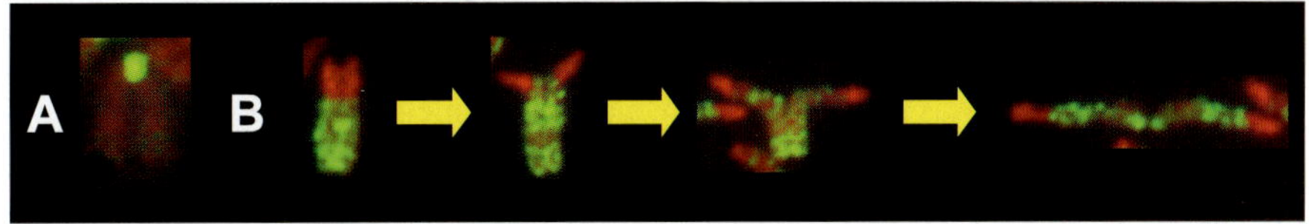

Fig. 3 . Picture assembly showing neonatal male germ cell chromosome characteristic segregation dynamic at first division after their cell cycle resumption (**B**), which is never observed in testicular somatic cells (**A**). In green: anti-5mC antibody staining. In red: DNA counterstained with propidium iodide. Weakly methylated germ cell juxtacentromeric region is less compacted than that of somatic cell. Strongly methylated germ cell chromatids are cohesive and compacted, whereas weakly methylated somatic cell chromatids are less cohesive and less compacted. Contrary to conventional chromatid segregation at metaphase/anaphase transition in somatic cells in postnatal germ cell, juxtacentromeric regions split up before chromatids, telomeric regions being the last to split.

Sertoli cell chromosomes did not change (Fig. 2A2). It was characterized by a low methylation of euchromatin and a strong methylation of juxtacentromeric regions. Putting these observations together with the compaction status of the chromosomes, we found that whatever the cell and structure type considered (euchromatin or heterochromatin), high DNA methylation levels correlated with strong compaction and low DNA methylation levels with weak compaction, giving chromosomes a fuzzy appearance. We thus hypothesized that a relationship exists between DNA methylation and chromatin compaction. In vitro studies on human lymphocytes aiming at artificially lowering DNA methylation levels using 5-azadC demonstrated intricate relationships between chromosome compaction and DNA methylation depending partially on the chromosome DNA methylation status before replication (Flagiello et al., 2002).

Another specific characteristic of germ cell chromosomes, as they undergo their first mitosis after birth, is that they exhibit a strong cohesion of their chromatids and a weak cohesion of their undermethylated centromeric regions (Bernardino-Sgherri et al., 2002a) (Fig. 3B). The latter evokes the so-called heterochromatin puffing described in the Roberts syndrome (Van Den Berg and Francke, 1993). The occurrence and extent of this feature in germ cells decreases at later stages, i.e. after several consecutive cell divisions, when their juxtacentromeric regions become more methylated. Consistent with the molecular data (Ponzetto-Zimmerman and Wolgemuth, 1984), juxtacentromeric regions remain undermethylated in spermatogonia compared to somatic tissue. In neonatal testicular somatic cells, such a low cohesion was never observed in their strongly methylated juxtacentromeric regions (Fig. 3A). In contrast, a low cohesion of their weakly methylated chromatids was a characteristic feature. This was also observed in chromosomes of germ cells having lost their strong DNA methylation after many divisions. Whether the concomitant changes in euchromatin DNA methylation and chromatin compaction are also involved in chromatid cohesion is not known. It is however tempting to anticipate a relationship between DNA methylation and both chromosome compaction and cohesion.

The recent development of knockout mice for genes such as *Atrx* (Alpha thalassemia/mental retardation syndrome X-linked) and *Lsh* (Lymphoid specific helicase) provided the first evidence that the deficiency in enzyme activities independent of DNA methyltransferases could lead to dramatic depletion of genomic DNA methylation levels (Gibbons et al., 2000; Dennis et al., 2001). Moreover, histone methyltransferase activities have been found to be necessary to target de novo DNA methyltransferase to juxtacentromeric heterochromatin DNA repeated sequences (Lehnertz et al., 2003). This suggests that chromatin proteins may drive the establishment of DNA methylation at least in these sequences. The characterization of the expression pattern and localization of proteins such as phosphorylated or methylated modified histones on germ cells chromosomes will be helpful to understand the interplay between DNA methylation and chromosome structure.

The expression pattern of the DNA methyltransferases at both mRNA and protein level during the neonatal period are not well known. The most recent study has focused on testing, by immunohistochemistry, whether Dnmt1 could be involved in the de novo DNA methylation, which occurs by day 15.5 pc. It was found that Dnmt1 expression was downregulated from day 15.5 pc until birth when only some postnatal gonocytes were found to be stained (Sakai et al., 2001) but the authors did not correlate this observation with the proliferation status of these cells. During the same period, Dnmt1o was never detected in germ cells. At later time points, using an antibody which recognizes Dnmt1, decreased expression and activity were found in total testis extracts from mice between 6 dpp and the adult age (Trasler et al., 1992; Benoit and Trasler, 1994). It would be interesting to analyse the expression pattern of the DNA methyltransferases when the strong DNA methylation pattern modifications occur. It can be expected that Dnmt3 family members are involved in both the strong hypermethylation in the fetal non-replicating germ cells and the de novo methylation of the juxtacentromeric regions after birth. Contrary to Dnmt1, Dnmt3a and Dnmt3b exhibit de novo methylation activities with some specificity to repeated sequences (Chen et al., 2003). It also remains to identify which of their isoforms (Okano et al., 1998; Chen et al., 2002) or if yet unknown testis-specific methylases could be involved.

Adults

The methylation pattern of many single copy and imprinted genes has been studied in the adult testis (Trasler et al., 1990; Ariel et al., 1991). Most of testis-specific genes are demethylated and expressed in the testis, and methylated in nonexpressing somatic tissues. This suggests that DNA methylation is involved in gene expression regulation (Hisano et al., 2003). In addition, it was found that both gene de novo methylation and demethylation events occur after completion of DNA replication, during meiotic prophase (Trasler et al., 1990). An accurate analysis of two imprinted regions of the *H19* gene throughout spermatogenesis has shown that the timing of methylation acquisition in both parental alleles was different in one of the two regions studied (Davis et al., 1999). Whereas paternal allele methylation took place before the onset of meiosis, methylation of the maternal allele was acquired during meiosis. Taken together these studies suggest DNA methylation targeting is tightly regulated during meiosis.

Using methylation-sensitive enzymes on testis cytological preparations del Mazo et al. (1994) found that the global DNA methylation level changes during spermatogenesis. Methylated cytosines on adult testis cross-sections have been detected, using a specific antibody, in all cell types but spermatocytes (Loukinov et al., 2002). However the authors did not study in detail the staining of the different spermatogonial cell types. Although these cell types can be distinguished by their chromatin structure on histological preparations (Chiarini-Garcia and Russell, 2001), fine DNA methylation changes and nuclear distribution are unlikely to be detected by this approach. For instance, on postnatal testis sections de novo methylation of juxtacentromeric regions could not be detected whereas this information could be obtained on cytological preparations (Coffigny et al., 1999; Bernardino et al., 2000). Using this antibody on isolated cell preparations, we found that methylated cytosines could be detected in every stage of spermatogenesis, including pachynema, where an enrichment of methylation involved some centromeric regions at random, as demonstrated by the different methylation status of homologous chromosomes (Bernardino et al., 2000). Consistent with the low levels of DNA methylation at pachynema, DNA methylation may be down regulated to allow recombination to take place (Bestor and Tycko, 1996). Indeed, DNA methylation has been shown to impede recombination events in some organisms, including V(D)J recombination in mammalian cells (Maloisel and Rossignol, 1998; Rassoulzadegan et al., 2002; Nakase et al., 2003). As shown in Fig. 2B and C, anti-5mC labeling occurs at centromeric regions and in DNA surrounding but not on synaptonemal complexes, where crossovers occur. Dnmt1 protein is detected only from spermatogonia to leptonema/zygonema (Mertineit et al., 1998). It is downregulated at the pachynema through the transient transcription of a pachytene-specific and untranslated mRNA (Mertineit et al., 1998). Neither localisation nor pattern expression of other known DNA methyltransferases have been addressed but some Dnmt3 family members are abundant in the adult testis (Chen et al., 2002).

Compared to somatic tissues, mouse major satellite DNA is undermethylated as early as in primitive type A spermatogonia and until in mature sperm DNA (Ponzetto-Zimmerman and Wolgemuth, 1984). Mouse minor satellite is also undermethylated in the male germ line (Sanford et al., 1984). However, repeated sequences such as LINE-1 and IAP are highly methylated in DNA from pachytene spermatocytes, in round spermatids, and epididymal sperm (Ponzetto-Zimmerman and Wolge-

muth, 1984; Sanford et al., 1987). De novo methylation also seems to occur late in germ cell development since some genes are remethylated only in epididymal spermatozoa (Ariel et al., 1994; Xie et al., 2002). This suggests that DNA methylation of some genes is needed for the completion of spermatozoa maturation. However, the underlying mechanisms are not known and the expression of the de novo methyltransferases has not been examined in epididymis.

Concluding remarks and perspectives

Most molecular studies on DNA methylation during male and female gametogenesis were performed at a limited number of CpG sites of some tissue-specific or imprinted genes. They were also focused on restricted periods of the germ line development, while the postnatal period was not studied. The mechanisms that drive the transition from gonocytes to undifferentiated spermatogonia, and later on to differentiated spermatogonia remains to be elucidated. Indeed this period is difficult to explore by molecular approaches. The undifferentiated spermatogonia are poorly represented in the seminiferous tubules. The morphological distinction between the different germ cell types is not well defined (Dettin et al., 2003). The availability of new methods of neonatal germ cells purification (Moore et al., 2002; Orwig et al., 2002) will be helpful to focus on this period. These approaches will hopefully allows us to identify the proteins involved in wide spread DNA methylation changes as well as to try to correlate these changes with those occurring at repeated, single-copy and imprinted DNA sequences. It will be interesting to better correlate data on DNA methylation at the molecular and cytogenetic scales. As indicated by the methylated status of the transcriptionally active pseudoautosomal regions of the X chromosomes (Bernardino et al., 2000), unmethylated genes may be harboured in methylated DNA at the scale of chromosome bands. Finally, molecular studies on purified cells and cytological approaches will help to answer crucial questions about the mechanisms involved in the relationship between overall DNA methylation status and chromatin structure, including the strong differences between heterochromatin and euchromatin compaction and cohesion.

Acknowledgements

We thank B. Malfoy for his critical reading of the manuscript and A. Niveleau for kindly providing us with anti-5methylcytosine antibodies.

References

Almeida A, Kokalj-Vokac N, Lefrancois D, Viegas-Péquignot E, Jeanpierre M, Dutrillaux B, Malfoy B: Hypomethylation of classical satellite DNA and chromosome instability in lymphoblastoid cell lines. Hum Genet 91:538–546 (1993).

Ariel M, McCarrey J, Cedar H: Methylation patterns of testis-specific genes. Proc Natl Acad Sci USA 88:2317–2321 (1991).

Ariel M, Cedar H, McCarrey J: Developmental changes in methylation of spermatogenesis-specific genes include reprogramming in the epididymis. Nat Genet 7:59–63 (1994).

Benoit G, Trasler JM: Developmental expression of DNA methyltransferase messenger ribonucleic acid, protein, and enzyme activity in the mouse testis. Biol Reprod 50:1312–1319 (1994).

Bernardino J, Lombard M, Niveleau A, Dutrillaux B: Common methylation characteristics of sex chromosomes in somatic and germ cells from mouse, lemur and human. Chromosome Res 8:513–525 (2000).

Bernardino-Sgherri J, Dutrillaux B: Compaction, stainability and methylation of the late replicating X chromosome in mouse female fibroblasts. Cytogenet Cell Genet 94:79–81 (2001).

Bernardino-Sgherri J, Chicheportiche A, Niveleau A, Dutrillaux B: Unusual chromosome cleavage dynamic in rodent neonatal germ cells. Chromosoma 111:341–347 (2002a).

Bernardino-Sgherri J, Flagiello D, Dutrillaux B: Overall DNA methylation and chromatin structure of normal and abnormal X chromosomes. Cytogenet Genome Res 99:85–91 (2002b).

Bestor TH, Tycko B: Creation of genomic methylation patterns. Nat Genet 12:363–367 (1996).

Bourc'his D, Xu GL, Lin CS, Bollman B, Bestor TH: Dnmt3L and the establishment of maternal genomic imprints. Science 294:2536–2539 (2001).

Boyle AL, Ballard SG, Ward DC: Differential distribution of long and short interspersed element sequences in the mouse genome: chromosome karyotyping by fluorescence in situ hybridization. Proc Natl Acad Sci USA 87:7757–7761 (1990).

Chen T, Ueda Y, Xie S, Li E: A novel Dnmt3a isoform produced from an alternative promoter localizes to euchromatin and its expression correlates with active de novo methylation. J Biol Chem 277:38746–38754 (2002).

Chen T, Ueda Y, Dodge JE, Wang Z, Li E: Establishment and maintenance of genomic methylation patterns in mouse embryonic stem cells by Dnmt3a and Dnmt3b. Mol Cell Biol 23:5594–5605 (2003).

Chiarini-Garcia H, Russell LD: High-resolution light microscopic characterization of mouse spermatogonia. Biol Reprod 65:1170–1178 (2001).

Coffigny H, Bourgeois C, Ricoul M, Bernardino J, Vilain A, Niveleau A, Malfoy B, Dutrillaux B: Alterations of DNA methylation patterns in germ cells and Sertoli cells from developing mouse testis. Cytogenet Cell Genet 87:175–181 (1999).

Datta J, Ghoshal K, Sharma SM, Tajima S, Jacob ST: Biochemical fractionation reveals association of DNA methyltransferase (Dnmt) 3b with Dnmt1 and that of Dnmt 3a with a histone H3 methyltransferase and Hdac1. J Cell Biochem 88:855–864 (2003).

Davis TL, Trasler JM, Moss SB, Yang GJ, Bartolomei MS: Acquisition of the H19 methylation imprint occurs differentially on the parental alleles during spermatogenesis. Genomics 58:18–28 (1999).

Davis TL, Yang GJ, McCarrey JR, Bartolomei MS: The H19 methylation imprint is erased and re-established differentially on the parental alleles during male germ cell development. Hum Mol Genet 9:2885–2894 (2000).

del Mazo J, Prantera G, Torres M, Ferraro M: DNA methylation changes during mouse spermatogenesis. Chromosome Res 2:147–52 (1994).

Dennis K, Fan T, Geiman T, Yan Q, Muegge K: Lsh, a member of the SNF2 family, is required for genome-wide methylation. Genes Dev 15:2940–2944 (2001).

Dettin L, Ravindranath N, Hofmann MC, Dym M: Morphological characterization of the spermatogonial subtypes in the neonatal mouse testis. Biol Reprod 69:1565–1571 (2003).

Doerksen T, Trasler JM: Developmental exposure of male germ cells to 5-azacytidine results in abnormal preimplantation development in rats. Biol Reprod 55:1155–1162 (1996).

Doerksen T, Benoit G, Trasler JM: Deoxyribonucleic acid hypomethylation of male germ cells by mitotic and meiotic exposure to 5-azacytidine is associated with altered testicular histology. Endocrinology 141:3235–3244 (2000).

Dupressoir A, Heidmann T: Germ line-specific expression of intracisternal A-particle retrotransposons in transgenic mice. Mol Cell Biol 16:4495–4503 (1996).

Eden A, Gaudet F, Waghmare A, Jaenisch R: Chromosomal instability and tumors promoted by DNA hypomethylation. Science 300:455 (2003).

Feenstra A, Fewell J, Lueders K, Kuff E: In vitro methylation inhibits the promotor activity of a cloned intracisternal A-particle LTR. Nucleic Acids Res 14:4343–4352 (1986).

Flagiello D, Bernardino-Sgherri J, Dutrillaux B: Complex relationships between 5-aza-dC induced DNA demethylation and chromosome compaction at mitosis. Chromosoma 111:37–44 (2002).

Fuks F, Burgers WA, Brehm A, Hughes-Davies L, Kouzarides T: DNA methyltransferase Dnmt1 associates with histone deacetylase activity. Nat Genet 24:88–91 (2000).

Fuks F, Burgers WA, Godin N, Kasai M, Kouzarides T: Dnmt3a binds deacetylases and is recruited by a sequence-specific repressor to silence transcription. EMBO J 20:2536–2544 (2001).

Gaudet F, Hodgson JG, Eden A, Jackson-Grusby L, Dausman J, Gray JW, Leonhardt H, Jaenisch R: Induction of tumors in mice by genomic hypomethylation. Science 300:489–492 (2003).

Gibbons RJ, McDowell TL, Raman S, O'Rourke DM, Garrick D, Ayyub H, Higgs DR: Mutations in ATRX, encoding a SWI/SNF-like protein, cause diverse changes in the pattern of DNA methylation. Nat Genet 24:368–371 (2000).

Hajkova P, Erhardt S, Lane N, Haaf T, El-Maarri O, Reik W, Walter J, Surani MA: Epigenetic reprogramming in mouse primordial germ cells. Mech Dev 117:15–23 (2002).

Hakimi MA, Bochar DA, Schmiesing JA, Dong Y, Barak OG, Speicher DW, Yokomori K, Shiekhattar R: A chromatin remodelling complex that loads cohesin onto human chromosomes. Nature 418:994–998 (2002).

Hastie ND: Highly repeated DNA families in the genome of *Mus musculus*, in Lyon MF, Searle AG (eds): Genetic Variants and Strains of the Laboratory Mouse, pp 559–573 (Oxford University Press, 1989).

Hata K, Okano M, Lei H, Li E: Dnmt3L cooperates with the Dnmt3 family of de novo DNA methyltransferases to establish maternal imprints in mice. Development 129:1983–1993 (2002).

Hisano M, Ohta H, Nishimune Y, Nozaki M: Methylation of CpG dinucleotides in the open reading frame of a testicular germ cell-specific intronless gene, *Tact1/Actl7b*, represses its expression in somatic cells. Nucleic Acids Res 31:4797–4804 (2003).

Howell CY, Bestor TH, Ding F, Latham KE, Mertineit C, Trasler JM, Chaillet JR: Genomic imprinting disrupted by a maternal effect mutation in the *Dnmt1* gene. Cell 104:829–838 (2001).

Jaenisch R, Bird A: Epigenetic regulation of gene expression: how the genome integrates intrinsic and environmental signals. Nat Genet 33:245–254 (2003).

Kafri T, Ariel M, Brandeis M, Shemer R, Urven L, McCarrey J, Cedar H, Razin A: Developmental pattern of gene-specific DNA methylation in the mouse embryo and germ line. Genes Dev 6:705–714 (1992).

Kaneko-Ishino T, Kohda T, Ishino F: The regulation and biological significance of genomic imprinting in mammals. J Biochem 133:699–711 (2003).

Kluin PM, Kramer MF, de Rooij DG: Spermatogenesis in the immature mouse proceeds faster than in the adult. Int J Androl 5:282–294 (1982).

Kuff EL, Fewell JE, Lueders KK, DiPaolo JA, Amsbaugh SC, Popescu NC: Chromosome distribution of intracisternal A-particle sequences in the Syrian hamster and mouse. Chromosoma 93:213–219 (1986).

Lane N, Dean W, Erhardt S, Hajkova P, Surani A, Walter J, Reik W: Resistance of IAPs to methylation reprogramming may provide a mechanism for epigenetic inheritance in the mouse. Genesis 35:88–93 (2003).

Lee J, Inoue K, Ono R, Ogonuki N, Kohda T, Kaneko-Ishino T, Ogura A, Ishino F: Erasing genomic imprinting memory in mouse clone embryos produced from day 11.5 primordial germ cells. Development 129:1807–1817 (2002).

Lees-Murdock DJ, De Felici M, Walsh CP: Methylation dynamics of repetitive DNA elements in the mouse germ cell lineage. Genomics 82:230–237 (2003).

Lehnertz B, Ueda Y, Derijck AA, Braunschweig U, Perez-Burgos L, Kubicek S, Chen T, Li E, Jenuwein T, Peters AH: Suv39h-mediated histone H3 lysine 9 methylation directs DNA methylation to major satellite repeats at pericentric heterochromatin. Curr Biol 13:1192–1200 (2003).

Lengauer C: Cancer. An unstable liaison. Science 300:442–443 (2003).

Li E: Chromatin modification and epigenetic reprogramming in mammalian development. Nat Rev Genet 3:662–673 (2002).

Li E, Bestor TH, Jaenisch R: Targeted mutation of the DNA methyltransferase gene results in embryonic lethality. Cell 69:915–926 (1992).

Loukinov DI, Pugacheva E, Vatolin S, Pack SD, Moon H, Chernukhin I, Mannan P, Larsson E, Kanduri C, Vostrov AA, Cui H, Niemitz EL, Rasko JE, Docquier FM, Kistler M, Breen JJ, Zhuang Z, Quitschke WW, Renkawitz R, Klenova EM, Feinberg AP, Ohlsson R, Morse HC, 3rd, Lobanenkov VV: BORIS, a novel male germ-line-specific protein associated with epigenetic reprogramming events, shares the same 11-zinc-finger domain with CTCF, the insulator protein involved in reading imprinting marks in the soma. Proc Natl Acad Sci USA 99:6806–6811 (2002).

Lucifero D, Mertineit C, Clarke HJ, Bestor TH, Trasler JM: Methylation dynamics of imprinted genes in mouse germ cells. Genomics 79:530–538 (2002).

Maatouk DM, Resnick JL: Continuing primordial germ cell differentiation in the mouse embryo is a cell-intrinsic program sensitive to DNA methylation. Dev Biol 258:201–208 (2003).

Maloisel L, Rossignol JL: Suppression of crossing-over by DNA methylation in *Ascobolus*. Genes Dev 12:1381–1389 (1998).

Mertineit C, Yoder JA, Taketo T, Laird DW, Trasler JM, Bestor TH: Sex-specific exons control DNA methyltransferase in mammalian germ cells. Development 125:889–897 (1998).

Molyneaux KA, Stallock J, Schaible K, Wylie C: Time-lapse analysis of living mouse germ cell migration. Dev Biol 240:488–498 (2001).

Monk M, Boubelik M, Lehnert S: Temporal and regional changes in DNA methylation in the embryonic, extraembryonic and germ cell lineages during mouse embryo development. Development 99:371–382 (1987).

Moore TJ, de Boer-Brouwer M, van Dissel-Emiliani FM: Purified gonocytes from the neonatal rat form foci of proliferating germ cells in vitro. Endocrinology 143:3171–3174 (2002).

Nakase H, Takahama Y, Akamatsu Y: Effect of CpG methylation on RAG1/RAG2 reactivity: Implications of direct and indirect mechanisms for controlling V(D)J cleavage. EMBO Rep 4:774–780 (2003).

Okano M, Xie S, Li E: Dnmt2 is not required for de novo and maintenance methylation of viral DNA in embryonic stem cells. Nucleic Acids Res 26:2536–2540 (1998).

Okano M, Bell DW, Haber DA, Li E: DNA methyltransferases Dnmt3a and Dnmt3b are essential for de novo methylation and mammalian development. Cell 99:247–257 (1999).

Orwig KE, Ryu BY, Avarbock MR, Brinster RL: Male germ-line stem cell potential is predicted by morphology of cells in neonatal rat testes. Proc Natl Acad Sci USA 99:11706–11711 (2002).

Peters AH, Plug AW, van Vugt MJ, de Boer P: A drying-down technique for the spreading of mammalian meiocytes from the male and female germline. Chromosome Res 5:66–68 (1997).

Ponzetto-Zimmerman C, Wolgemuth DJ: Methylation of satellite sequences in mouse spermatogenic and somatic DNAs. Nucleic Acids Res 12:2807–2822 (1984).

Raman R, Narayan G: 5-Aza deoxyCytidine-induced inhibition of differentiation of spermatogonia into spermatocytes in the mouse. Mol Reprod Dev 42:284–290 (1995).

Rassoulzadegan M, Magliano M, Cuzin F: Transvection effects involving DNA methylation during meiosis in the mouse. EMBO J 21:440–450 (2002).

Ratnam S, Mertineit C, Ding F, Howell CY, Clarke HJ, Bestor TH, Chaillet JR, Trasler JM: Dynamics of Dnmt1 methyltransferase expression and intracellular localization during oogenesis and preimplantation development. Dev Biol 245:304–314 (2002).

Reynaud C, Bruno C, Boullanger P, Grange J, Barbesti S, Niveleau A: Monitoring of urinary excretion of modified nucleosides in cancer patients using a set of six monoclonal antibodies. Cancer Lett 61:255–262 (1992).

Rougier N, Bourc'his D, Gomes DM, Niveleau A, Plachot M, Paldi A, Viegas-Pequignot E: Chromosome methylation patterns during mammalian preimplantation development. Genes Dev 12:2108–2113 (1998).

Rubin CM, VandeVoort CA, Teplitz RL, Schmid CW: Alu repeated DNAs are differentially methylated in primate germ cells. Nucleic Acids Res 22:5121–127 (1994).

Sakai Y, Suetake I, Itoh K, Mizugaki M, Tajima S, Yamashina S: Expression of DNA methyltransferase (Dnmt1) in testicular germ cells during development of mouse embryo. Cell Struct Funct 26:685–6891 (2001).

Sanford J, Forrester L, Chapman V, Chandley A, Hastie N: Methylation patterns of repetitive DNA sequences in germ cells of *Mus musculus*. Nucleic Acids Res 12:2823–2836 (1984).

Sanford JP, Clark HJ, Chapman VM, Rossant J: Differences in DNA methylation during oogenesis and spermatogenesis and their persistence during early embryogenesis in the mouse. Genes Dev 1:1039–1046 (1987).

Sato S, Yoshimizu T, Sato E, Matsui Y: Erasure of methylation imprinting of *Igf2r* during mouse primordial germ-cell development. Mol Reprod Dev 65:41–50 (2003).

Trasler JM, Hake LE, Johnson PA, Alcivar AA, Millette CF, Hecht NB: DNA methylation and demethylation events during meiotic prophase in the mouse testis. Mol Cell Biol 10:1828–1834 (1990).

Trasler JM, Alcivar AA, Hake LE, Bestor T, Hecht NB: DNA methyltransferase is developmentally expressed in replicating and non-replicating male germ cells. Nucleic Acids Res 20:2541–2545 (1992).

Ueda T, Abe K, Miura A, Yuzuriha M, Zubair M, Noguchi M, Niwa K, Kawase Y, Kono T, Matsuda Y, Fujimoto H, Shibata H, Hayashizaki Y, Sasaki H: The paternal methylation imprint of the mouse *H19* locus is acquired in the gonocyte stage during foetal testis development. Genes Cells 5:649–659 (2000).

Van Den Berg DJ, Francke U: Roberts syndrome: a review of 100 cases and a new rating system for severity. Am J Med Genet 47:1104–1123 (1993).

Walsh CP, Chaillet JR, Bestor TH: Transcription of IAP endogenous retroviruses is constrained by cytosine methylation. Nat Genet 20:116–1167 (1998).

Xie W, Han S, Khan M, DeJong J: Regulation of ALF gene expression in somatic and male germ line tissues involves partial and site-specific patterns of methylation. J Biol Chem 277:17765–17774 (2002).

Cytogenet Genome Res 105:325–334 (2004)
DOI: 10.1159/000078205

DNA methylation profiles of CpG islands for cellular differentiation and development in mammals

K. Shiota

Cellular Biochemistry, Animal Resource Sciences, Veterinary Medical Sciences, University of Tokyo, Tokyo (Japan)

Abstract. DNA methylation has been implicated in mammalian development. Transcription units contain CpG islands, but expression of CpG island associated genes in normal tissues was not believed to be controlled by DNA methylation. There are, however, numerous CpG islands containing tissue-dependent and differentially methylated regions (T-DMR), which are potential methylation sites in normal cells and tissues. Genomic scanning which focused on T-DMRs in CpG islands revealed that the DNA methylation profile of each cell/tissue is more complicated than previously considered. Differentiation of cells is associated with both methylation and demethylation, which occur at multiple loci. The epigenetic system characterized by DNA methylation requires cells to memorize gene expression patterns, thus, standardizing cellular phenotypes.

Copyright © 2004 S. Karger AG, Basel

Methylation of DNA is involved in various biological phenomena through gene silencing, stabilizing chromosomal structure and suppressing the mobility of retrotransposons (Bird and Wolffe, 1999; Walsh and Bestor, 1999; Bird, 2002; Shiota and Yanagimachi, 2002). A single fertilized egg gives rise to a complex multi-cellular organism consisting of at least 200 differentiated cell types. Most cells differentiate without changes in DNA sequence through activation of a particular set of genes and inactivation of others. The molecular basis for the memory of activated or inactivated gene sets, which is inherited by the next generation of cells, is critical for differentiation and development of multicellular organisms. Epigenetics is the study of heritable changes in gene activity without changes in DNA sequences (Russo et al., 1996). Methylation of the cytosine residue in a CpG dinucleotide sequence is a characteristic of the vertebrate genome. In vertebrates, methylation of DNA mainly occurs at the 5-position of cytosine in a CpG dinucleotide forming 5-methylcytosine (Bird, 1978; Gruenbaum et al., 1981) (Fig. 1). Recently we found that there are many CpG islands with tissue-dependent and differentially methylated regions (T-DMR) (Shiota et al., 2002). In this review, DNA methylation pattern implies the methylation status of each CpG, whereas DNA methylation profile indicates the array of methylation status of the T-DMR consisting of multiple CpGs. Here, I describe the genome-wide DNA methylation profiles specific to cells and tissues.

DNA methylation and gene function

Methylation of DNA plays a profound role in transcriptional repression of gene expression through several mechanisms (Bird and Wolffe, 1999; Bird, 2002). Generally, DNA of inactive genes is more heavily methylated than that of active ones; conversely demethylation of DNA reactivates gene expression in vivo and in vitro (Benvenisty et al., 1985; Shemer et al., 1996; Szyf, 1996; Kuramasu et al., 1998). Methylation of cytosine residues interrupts the recognition and binding of transcription factors (Hark et al., 2000; Bell and Felsenfeld, 2000; Takizawa et al., 2001; Maier et al., 2003). Another mechanism involves methyl CpG binding proteins (MBDs), which interact with transcription repressors and chromatin remodel-

Supported by PROBRAIN, and Grants in Aid for Scientific Research from the Ministry of Education, Culture, Sports, Science and Technology of Japan.

Received 30 September 2003; manuscript accepted 2 December 2003.

Request reprints from Kunio Shiota, Cellular Biochemistry
Animal Resource Sciences, Veterinary Medical Sciences
University of Tokyo, 1-1-1 Yayoi, Bunkyo-ku
Tokyo 113 113-8657 (Japan); telephone: +81-3-5841-5472
fax: +81-3-5841-8189; e-mail: ashiota@mail.ecc.u-tokyo.ac.jp

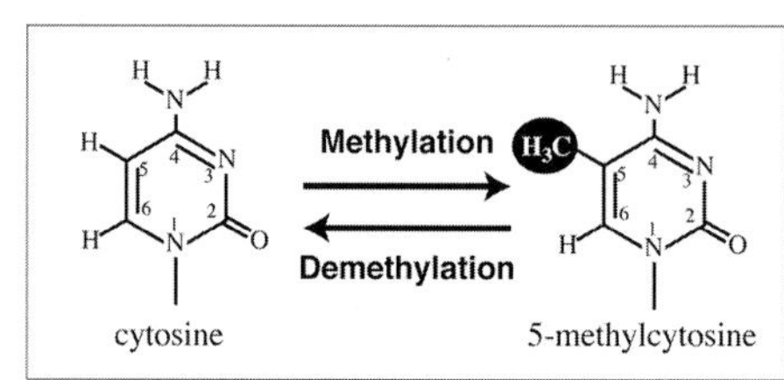

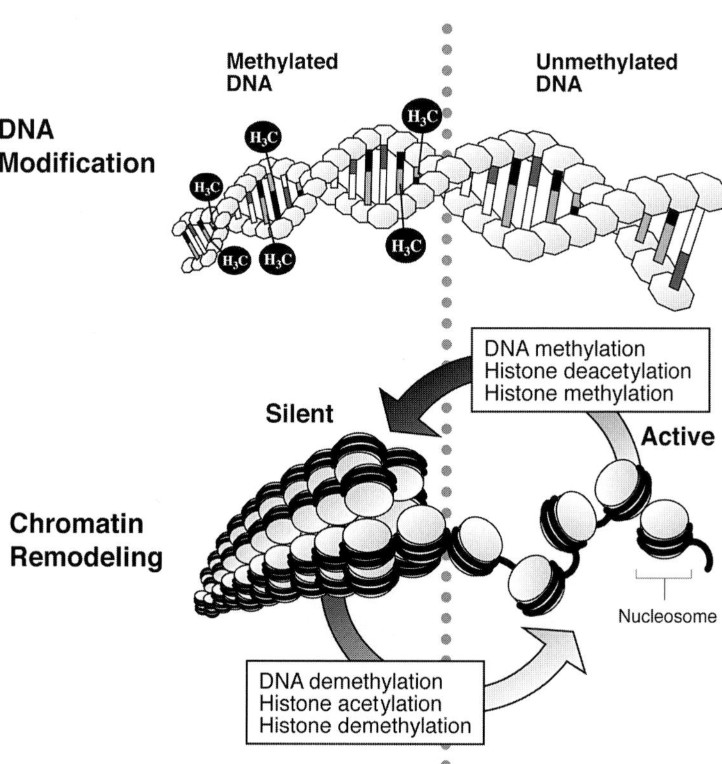

Fig. 1. DNA methylation and chromatin remodeling are the main epigenetic mechanisms for gene control. 5-Methylcytosine is the only methylated base in mammalian DNA. DNA methylation mainly occurs on cytosine residues of CpG dinucleotides. DNA methylation, histone deacetylation, and histone methylation (Lys 9) induce chromatin structure condensation and gene-silencing. Conversely, CpG demethylation, histone acetylation and histone demethylation induce the active state with relaxed chromatin form. The main epigenetic regulators involved in the process are DNA methyltransferases such as Dnmt1, 3a and 3b, several histone deacetylases, histone methyltransferases such as Suv39h and G9a, and histone deacetylases.

ing factors (Bird, 2002; Li, 2002). MBDs are involved in chromatin organization and gene silencing via recruitment of histone deacetylases (HDACs) (Wade, 2001). Individual MBDs have specific methylated DNA binding properties, and interact with transcriptional repressors and chromatin remodeling factors (Bird, 2002; Li, 2002). For example, MeCP2, an MBD, represses gene expression by recruiting mSin3A, which interacts with HDAC1 (Jones et al., 1998, Nan et al., 1998). Histone deacetylation and histone (H3, Lys9) methylation cause the chromatin configuration to condense (Tachibana et al., 2001, 2002). Changing the chromatin configuration also affects DNA methylation (Lachner and Jenuwein, 2002; Li, 2002). The relationship between DNA methylation and remodeling the chromatin structure is shown in Fig. 1. Through these mechanisms, genes become refractory regardless of the presence of transcription factors. Such reagents as 5-azacytidine (5-aza-C) and 5-aza-deoxycytidine (5-aza-dC) (Goffin and Eisenhauer, 2002), which inhibit DNA methylation, and Trichostatin A (TSA) (Yoshida et al., 1990), which inhibits histone deacetylase (HDAC), have been used to examine whether epigenetic regulation is involved in gene expression.

CpG islands with T-DMR in normal tissues and cells

Sequences of CpGs are not evenly distributed in the mammalian genome. They appear at a 10 to 20 times higher density in selected regions than in other regions, and regions enriched with CpGs are known as CpG islands (Bird, 1987; Gardiner-Garden and Frommer, 1987). These CpG islands are used as landmarks to find genomic regions in bulk DNA sequences, because CpG islands are generally found in transcription units. Data from the human and mouse genome projects confirmed that CpG islands are located in gene areas. Generally, it has been recognized that CpG islands are unmethylated in normal tissues (Yen et al., 1984; Bird et al., 1985; Cross and Bird, 1995; De Smet et al., 1999), except the CpG islands involved in X inactivation (Norris et al., 1991; Tribioli et al., 1992; Heard et al., 1997) and genomic imprinting (Ferguson-Smith et al., 1993; Razin and Cedar, 1994; Barlow, 1995). However, gene repression mediated by DNA methylation occurs in the genes associated with CpG islands as described below.

Sphingosine kinase (SPHK) 1 is a key enzyme catalyzing the production of sphingosine 1-phosphate, which functions in cel-

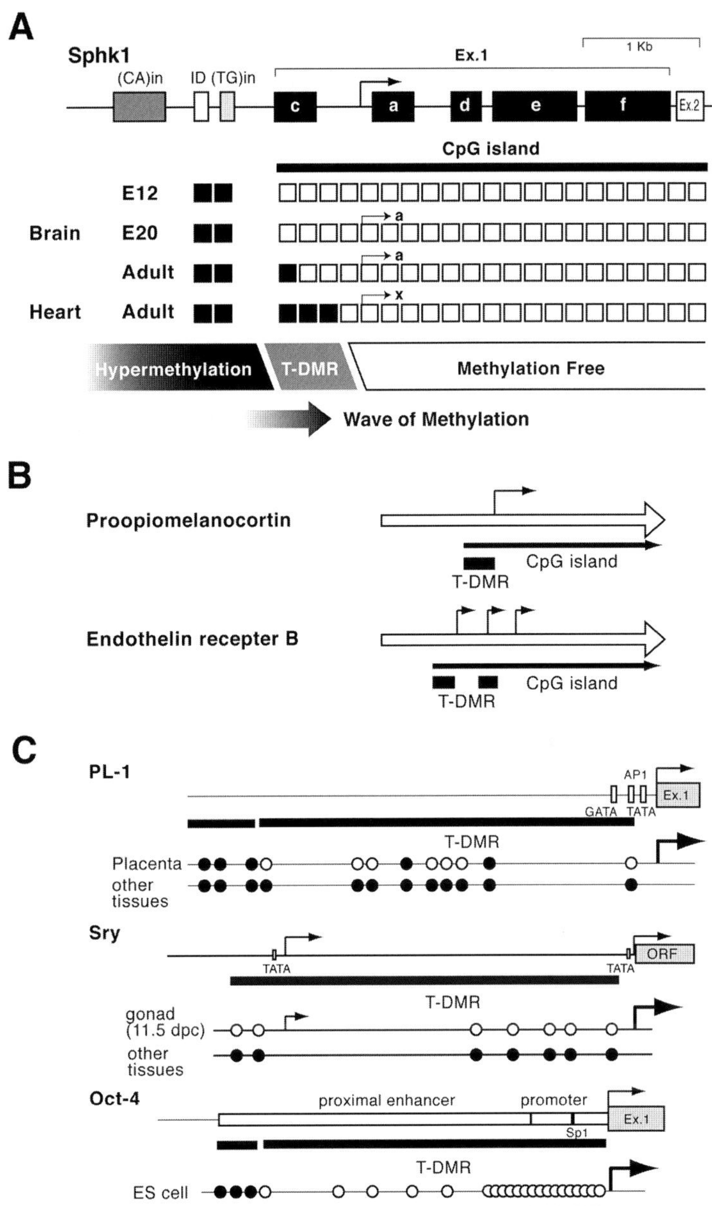

Fig. 2. Tissue-specific or developmentally dependent genes controlled by DNA methylation. (**A**) CpG islands with tissue-dependent and differentially methylated region (T-DMR). T-DMR is located in the CpG island of mouse SPHK1 gene. T-DMR is methylated in heart and unmethylated in brain in the adult. Methylation of T-DMR inhibits the transcription of a subtype of *Sphk1a*. Methylation status is different among tissues, and is regulated depending on development. (**B**) Several genes associated with CpG islands are regulated by DNA methylation at T-DMRs in normal tissues. (**C**) Tissue-specific genes with fewer CpG sequences. There are only a few CpGs in the 5′ regulatory regions of the PL-I and *Sry* genes, while *Oct4* has a relatively high density of CpGs. These are tissue- or developmentally dependent genes regulated by DNA methylation.

lular survival and differentiation (Postma et al., 1996; Edsall et al., 1997; Olivera et al., 1999). In the rat *Sphk1* gene, there are six alternative first exons for mRNA subtypes *(Sphk1a, b, c, d, e and f)* within a 3.7-kb CpG island (Imamura et al., 2001) (Fig. 2A). The CpG island contains a T-DMR (200 bp), which is located about 800 bp upstream of the first exon of *Sphk1a* (Imamura et al., 2001). The T-DMR is located in a limited area at the edge of the CpG island. The T-DMR is unmethylated in the brain and heart at embryonic day 12, and the methylation level gradually increases in the heart to birth, whereas it remains low in the brain. The T-DMR is hypomethylated in the adult rat brain where *Sphk1a* is expressed, whereas it is hypermethylated in the adult heart where gene expression is repressed. There are T-DMRs in the CpG islands of genes such as

endothelin receptor B (Pao et al., 2001), proopiomelanocortin (Newell-Price et al., 2001) and maspin (Futscher et al., 2002), and they are normally methylated depending on the tissues. Thus, there are CpG islands containing T-DMRs, which are methylation sites in normal tissues (Fig. 2B).

In general, repetitive sequences are targets of DNA methylation (Walsh and Bestor, 1999). In the rat *Sphk1* gene, there is a heavily methylated area with TG repeats, a microsatellite sequence of rat brain-specific identifier (ID) and CA repeats upstream of T-DMR and an area downstream free of methylation (Imamura et al., 2001) (Fig. 2A). Therefore, T-DMR is located at the boundary separating the hypermethylated and unmethylated areas in the *Sphk1* locus. If the methylation wave goes over the boundary, *Sphk1a* gene is repressed as seen in the heart while the gene is expressed in the brain when the wave is prevented. The *cis*-elements surrounding T-DMR are important in determining the tissue- or developmental stage-dependent methylation status. It was reported that clustered CG boxes with consensus Sp1 recognition sites were essential for protecting the CpG island from methylation (Brandeis et al., 1994; Macleod et al., 1994; Mummaneni et al., 1998). Repeated sequences of (ATAAA)$_n$ separate the hypermethylated area from the methylation-free area in human GSTP1 CpG island (Millar et al., 2000). Interestingly, there is a homologous sequence to an Sp1 recognition site in the 5′ end of the ID sequence of rat *Sphk1* gene and four clustered ATAAA sequences immediately downstream of the ID sequence. In this context, there may be functional sequences within or adjacent to the T-DMR.

Tissue-specific genes with fewer CG sequences that are regulated by DNA methylation

There exist other types of genes without CpG islands. Rat placental lactogen-I (rPL-I) gene contains few CpG sites with a TATA box (Fig. 2B). Expression of the rPL-I gene is strictly regulated so that the gene is silent in most tissues except trophoblast giant cells in the placenta. In the rPL-I gene promoter, there are only 17 CpG sites in the 3.4 kb of the 5′-flanking region (Fig. 2C). The flanking region of the rPL-I gene is hypomethylated specifically in the placenta (Cho et al., 2001). Treatment with 5-aza-dC or TSA, of a fibroblast cell line in which gene expression was silent, reactivated rPL-I gene expression. Furthermore, in vitro methylation of a rPL-I promoter-reporter construct repressed promoter activity. The density of CpG dinucleotides is lower than one CpG per 220 nucleotides, which was the minimally required density for repression by MeCP2 (Meehan et al., 1992; Nan et al., 1993). Indeed, over expression of MeCP2 caused further suppression of promoter activity. Similar to rPL-I gene, promoter regions of prolactin and growth hormone family member genes also contain few CpG sites with a TATA box, and methylation of the promoters inversely correlates to gene expression (Ngo et al., 1996). Thus, DNA methylation-mediated gene silencing is involved in various gene controls, regardless of richness of CpGs.

Master genes controlled by DNA methylation

In mammalian embryogenesis, the first differentiation event sets the trophectoderm aside from the inner cell mass (ICM) lineage at the blastocyst stage. After implantation, trophectoderm and trophoblast cells further give rise to a major part of the placenta and some extra-embryonic membranes (Cross et al., 1994), while the ICM alone become the fetus proper and germ cells (Nagy et al., 1993). Embryonic stem cells (ES cells) established from ICM have been studied for more than two decades. Trophoblast stem cells (TS cells) also have been established from explanted blastocysts (Tanaka et al., 1998). The TS cells contributed exclusively to the tissue of trophoblast cell lineage, but not to the fetus. Embryonic germ cells (EG cells),

established from primordial germ cells (Matsui et al., 1992; Resnick et al., 1992), showed properties similar to ES cells.

The *Oct4* gene does not have a CpG island, but has a CG-rich and TATA-less promoter (Okazawa et al., 1991; Sylvester and Schöler, 1994). In mice, *Oct4* is expressed in the oocyte and preimplantation embryo, and is later restricted to only the ICM of the blastocyst (Okamoto et al., 1990; Rosner et al., 1990; Schöler et al., 1990); therefore, expression is restricted to totipotent and pluripotent cells. In *Oct4* deficient embryos, the ICM loses pluripotency, and trophoblast cells no longer proliferate and differentiate into developing placenta (Nichols et al., 1998). The *Oct4* gene is expressed in ES cells but not TS cells (Tanaka et al., 1998). Reduction of *Oct4* gene expression leads to differentiation of ES cells to TS cells under adequate culture conditions (Niwa et al., 2000). Therefore, *Oct4* is the master gene critical for differentiation of these stem cells. The enhancer/promoter regions are hypomethylated in ES cells, but heavily methylated in TS cells (Fig. 2C) (Hattori et al., 2004). In the placenta of DNA methyltransferase 1 *(Dnmt1)*-deficient mice, most CpGs in the enhancer/promoter region were unmethylated, and the *Oct4* gene was aberrantly expressed. Chromatin immunoprecipitation assay revealed that *Oct4* enhancer/promoter regions are hyper-acetylated in ES cells compared to TS cells, thus demonstrating that DNA methylation status is linked to the chromatin structure of the *Oct4* gene.

The *Sry* gene is the master gene for testis differentiation in mammals (Koopman et al., 1991). There are very few CpG dinucleotides in the promoter regions; eight CpGs are located 2 kb upstream of the linear form of mouse *Sry* (Fig. 2C). In vitro DNA methylation of the mouse *Sry* 5′-flanking region represses promoter activity. Furthermore, the 5′-flanking region of the *Sry* gene is hypomethylated in the male gonad in vivo, and this occurs only when the *Sry* gene is expressed in the gonad, indicating that sex determination is regulated by *Sry* gene expression mediated by temporal DNA demethylation (Nishino et al., 2004).

Methylation of DNA has long been speculated to be involved in the establishment of cell type-specific gene expression (Riggs, 1975; Holliday and Pugh, 1975). In addition to the authentic genes, there are many unknown genes having T-DMRs. Such genes may be overlooked during mRNA screening due to lower level expression in a limited number of cells, and are potentially master genes controlling mammalian development and cellular differentiation.

Genome-wide DNA methylation profiles focusing on T-DMR of CpG islands

Information is still limited concerning DNA methylation and gene area. Also, quantity and type of genes regulated by DNA methylation remain to be clarified. It is unfortunate that CpG islands were believed to be unmethylated regions in normal tissues, leading to the misunderstanding that tissue-specific genes with CpG islands are not the targets of DNA methylation.

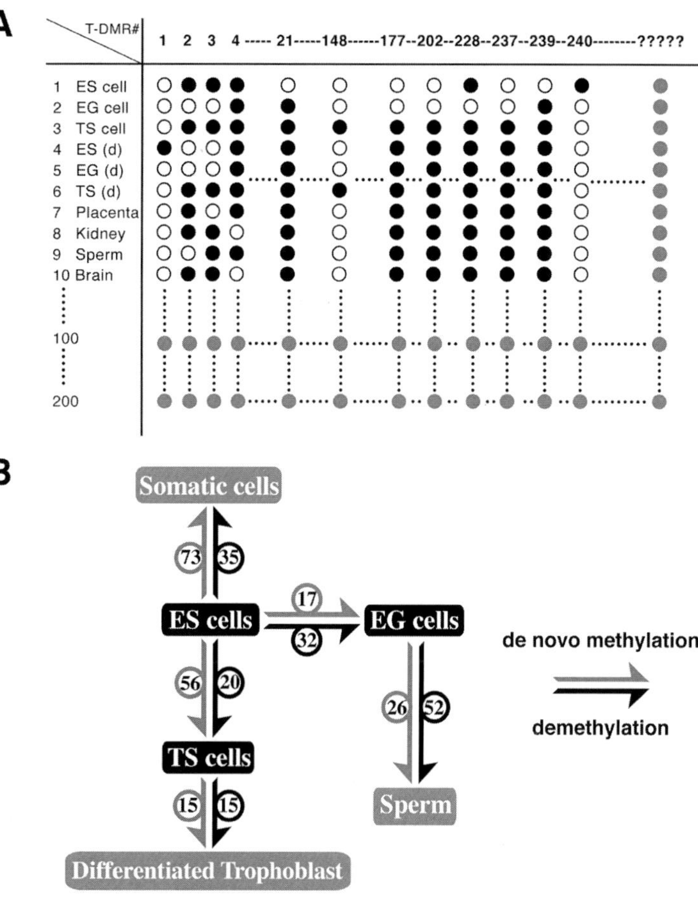

C DNA methylation level of CpG islands containing T-DMR

Cells/Tissues	% (total 247spots)
ES cell	51.4
EG cell	45.3
TS cell	66.4
TS cell (diff)	66.4
Placenta	52.6
Kidney	51.8
Sperm	33.6
Brain	47.0

Fig. 3. DNA methylation profiles specific to cell types and tissues. T-DMR panels prepared based on RLGS profiles. (**A**) Epigenetic marks by DNA methylation specific to cell and tissue types. T-DMR methylation status of 10 different genomes from mouse stem cells (ES, TS, EG before and after differentiation), germ cells and tissues are shown. There are at least 200 different cell types in the body, indicating that genomic loci with altered methylation status are numerous and widespread. Open circles represent unmethylated T-DMR at corresponding cells, while closed circles indicate methylated ones. (**B**) Epigenetic differences and similarities between cell types. Circled number indicates the sum of the T-DMRs, which are methylated or demethylated in association with differentiation of cells (ex. from TS to trophoblast) or inter-conversion between ES and TS cells. (**C**) DNA methylation level of CpG islands containing T-DMR.

A systematic survey of DNA methylation patterns regarding CpG islands is most interesting from the aspect of epigenetic participation in the differentiation and development of mammals. The human genome project identified 30,000–40,000 protein-coding genes, and there are approximately 29,000 CpG islands. There are 30,000 genes and 15,000 CpG islands in the mouse genome (International Human Genome Sequencing Consortium, 2001; Venter et al., 2001; Mouse Genome Sequencing Consortium, 2002). Tissue-specific promoters revealed that 50% of CpG islands are linked to tissue-specific genes (Suzuki et al., 2001). The remaining tissue-specific promoters do not associate with CpG islands. Consequently, it is clear that genes associated with CpG islands should be included in the list of genes investigated for DNA methylation-mediated gene silencing.

Several genomic scanning methods have been developed (Kaneda et al., 2003). Restriction Landmark Genomic Scanning (RLGS) can perform rapid analysis of methylation profiles of thousands of CpG islands associated with genes in parallel (Okazaki et al., 1995; Ohgane et al., 1998). Scanning of 1,500 CpG islands of genomes from ten different cell types and tissues, including ES, EG, TS cells before and after differentiation revealed 247 T-DMRs (Shiota et al., 2002) (Fig. 3A). Considering that there are 15,000 CpG islands in the mouse genome and there are, at least, 200 different cell types in the mammalian body, the total number of CpG islands with T-DMR will be much greater. It is clear that CpG islands having T-DMR were numerous and widespread in the genome. A recent study by restriction enzyme-based library cloning identified normally methylated CpG islands in the human genome (Strichman-Almashanu et al., 2002).

Comparison of ES, EG and TS cells revealed that they have distinct and unique DNA methylation profiles. Thus, it was possible to distinguish similar stem cells by analyzing the DNA methylation profiles. The DNA methylation profile of TS cells greatly contrasts with those from ES and EG cells. Similarly, germ cells and somatic cells had very different DNA methylation profiles. The T-DMR panel clearly indicates that DNA methylation profile is cell-type specific. Thus the T-DMR panel is useful to distinguish cell types.

Differentiation of cells associates with changing the DNA methylation pattern

Forced demethylation, using 5-aza-dC, induced mouse fibroblastic cell lines, 10T1/2 and 3T3 cells, to differentiate into myoblasts (Taylor and Jones, 1979). Similarly, treatment of bone marrow stromal cells with 5-aza-C or 5-aza-dC induced differentiation into cardiomyocytes (Makino et al., 1999). Therefore, one may consider that differentiation involves overall demethylation of genomic DNA. However, genome-wide DNA methylation status has not been analyzed in these studies. In the in vitro differentiation model of Rcho-1 cells, differentiation into trophoblast giant cells was accompanied by a change in the methylation profile of genomic DNA as a result of both methylation and demethylation (Ohgane et al., 2002). Thus, it is conceivable that such reagents may shock the cells to initiate rearrangement of the epigenetic status.

Information concerning T-DMRs is useful for evaluating epigenetic changes of cells during differentiation (Fig. 3B). Epigenetic distance, measured by counting the number of T-DMRs with different methylation status, also reveals epigenetic similarities among given cell types (Shiota et al., 2002). The ES cells are pluripotent cells that differentiate into all somatic cell types and germ cells (Nagy et al., 1993; Brook and Gardner, 1997), while TS cells are committed to differentiate into trophoblast lineage. During differentiation of ES cells, methylation status was changed at a total of 108 T-DMRs; 35 loci were demethylated and 73 loci were methylated. In contrast, differentiation of TS cells was accompanied by a change in methylation status at only 30 loci, with demethylation at 15 loci and methylation at 15 loci. The number of differentially methylated T-DMRs is larger in ES than TS cells. A comparison of methylation profiles between EG cells and sperm indicating demethylation at 52 T-DMRs and methylation at 26 T-DMRs also suggests that differentiation involves large epigenetic changes. Thus, differentiation of cells is associated with both methylation and demethylation, and the dual change of epigenetic marks occurs at multiple loci. Therefore, cell differentiation is linked to a new DNA methylation profile, indicating a more complex process than has been considered previously. In this context, the epigenetic system may be functioning to prevent reverse differentiation.

Changes in DNA methylation pattern in development

It is clear that active demethylation as well as de novo methylation are not restricted to the fetal developmental period, but both processes occur concomitantly during differentiation of various cell types. Therefore, the change of DNA methylation status during development should not be simplified. Regarding T-DMR, DNA methylation level was highest in TS cells and placenta, while the level was lowest in sperm (Fig. 3C). This contradicts the previously held belief that there is a genome-wide de-methylation process that occurs before the implantation period until blastocyst formation, and then a genome-wide de novo methylation at and after the gastrulation stage (Monk et al., 1987; Kafri et al., 1992). We must, therefore, carefully consider that the previous concept was established from data using mainly repetitive sequences as well as a limited number of genes. Repetitive sequences such as satellite DNA and endogenous retroviruses occupy a large part of the mammalian genome, while the area for protein-coding genes is very limited. Most CpGs are methylated in the mammalian genome; the value varies between 60–90% depending on the report (Gruenbaum et al., 1981; Razin et al., 1984; Cross and Bird, 1995). Therefore, overall level of DNA methylation reflects the hypermethylation status of gene-poor regions such as repetitive sequences rather than gene regions.

Abnormal DNA methylation status in cloned animals

The rate of cloned animal production is generally quite low; average 2–3% or less of reconstituted eggs develop into live offspring (Solter, 2000). It is most likely that incomplete DNA methylation does not allow cells of the embryo to correctly express genes required for development and survival of the embryo and fetus. Cloned animals that survived beyond birth have nearly correct DNA methylation profiles as previously reported (Ohgane et al., 2001). In cloned mice, the DNA methylation profile of T-DMR was 99% identical to the naturally produced control. Therefore, the cloned animals that developed to full term have almost normal DNA methylation profiles and are fairly good copies of nuclear donor animals. So, the epigenetic system is more flexible than previously thought. However, cloned animals have a variety of abnormal symptoms at and after birth (Wakayama and Yanagimachi, 1999; Lanza et al., 2000; Tamashiro et al., 2000; Tanaka et al., 2001). It is important to note that all six cloned mice, we analyzed, had imperfect DNA methylation profiles without exception (unpublished data). Each cloned animal had different DNA methylation aberrations, and the extent of abnormality and loci varied among different individuals.

Epigenetic change during development initiated by somatic cell nuclear transfer is markedly different from that in the animal produced by the normal reproductive process using germ cells. In cloned animals, the DNA methylation profile must be changed from that of the donor somatic cell to that of different cell types through that of zygote and early embryo. Therefore, there may be several loci susceptible to epigenetic error during nuclear transfer. The placentas of cloned mice have expanded spongiotrophoblast layers with increased incidence of glycogen cell differentiation (Tanaka et al., 2001). Placentomegaly is a phenomenon commonly observed in cloned mice regardless of sex, strain and cell type of somatic donor nuclei (Wakayama et al., 1998; Wakayama and Yanagimachi, 1999; Ogura et al., 2000). Recently, we observed that T-DMR #148 was an epigenetic hot spot with frequent epigenetic errors in cloned mice placenta (Ohgane et al., 2004). The T-DMR #148 was methylated only in TS cells but was unmethylated in ES cells, EG cells and somatic cells. More importantly, T-DMR #148 was located in the CpG island of Spalt-like 3 gene *(Sall3)* locus, at the telomeric E3 subregion of mouse chromosome 18. This locus has been suggested to be involved in the human disease 18q deletion syndrome (Kohlhase et al., 1999; Schulz et al., 2002).

Mechanism for formation and maintenance of DNA methylation patterns

Methylation patterns of DNA are generated by complex interplay between de novo DNA methylation and demethylation (Fig. 4A). There are de novo DNA methyltransferases such as Dnmt3a and Dnmt3b (Okano et al., 1998, 1999). Demethylation occurs through passive mechanisms such as the absence of maintenance Dnmt activity or through active enzymatic reactions. Demethylation of DNA by active enzymatic activity has been observed in chick embryos (Jost et al., 1995, 1997, 1999a; Fremont et al., 1997), mouse myoblast (Jost and Jost, 1994; Jost et al., 1999b) and a rat myogenic cell line (Weiss et al., 1996), although the molecules responsible for the enzymatic activity have not been isolated.

The methylated or unmethylated status of a CpG site is copied from parental DNA to daughter DNA. The DNA methylation pattern in the genome is maintained by the maintenance DNA methyltransferase activity of Dnmt1 (Bestor, 1992; Li et al., 1992), allowing inheritance of appropriate DNA methylation patterns. The carboxy-terminal region of Dnmt1 is a catalytic domain transferring the methyl groups from S-adenosyl-L-methionine (SAM) to cytosine in CpG dinucleotides. The N-terminal region of Dnmt1 regulates Dnmt1 function by protein-protein interactions with proteins such as PCNA (Chuang et al., 1997), DMAP1 (Rountree et al., 2000), HDAC1/2 (Rountree et al., 2000; Fuks et al., 2000) and Rb (Robertson et al., 2000), which are associated with DNA replication, histone acetylation, and gene expression during the cell cycle. Recently, we found a direct interaction between Dnmt1 and MeCP2 (Kimura and Shiota, 2003). The MeCP2 protein binds to single CpG dinucleotides that are symmetrically methylated (Lewis et al., 1992), whereas other MBDs have little or no binding activity to single symmetrically methylated CpG dinucleotides (Hendrich and Bird, 1998). The interactive domain was identified in the N-terminal of Dnmt1. The region of MeCP2 that interacts with Dnmt1 corresponds to the transcription repressor domain which recruits HDACs via corepressor mSin3A. The genomic DNA methylation pattern may be maintained by the Dnmt1-MeCP2 complex, bound to hemimethylated DNA (Fig. 4B).

Regarding Dnmt1-MeCP2 interaction, it still remains unclear if there is a specific target for the genome or other mole-

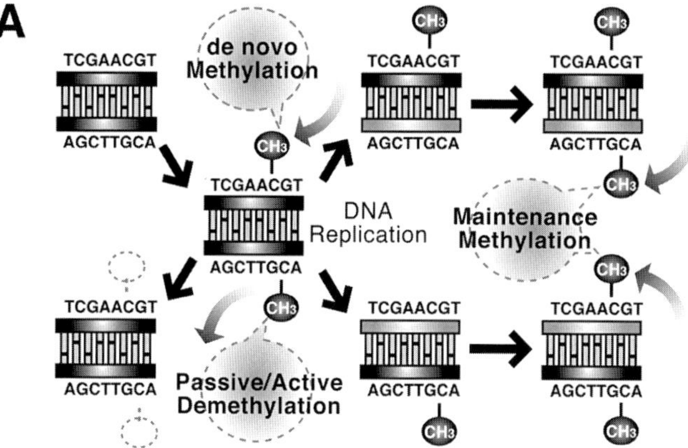

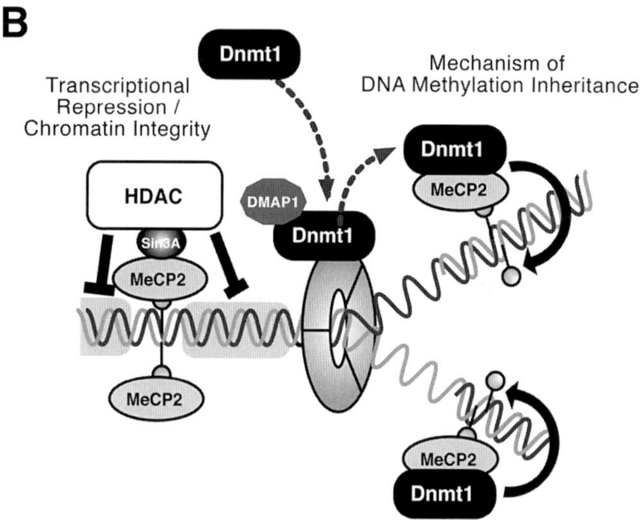

Fig. 4. Model for inheritance of DNA methylation pattern. **(A)** DNA methylation is heritable to the next cell generation. Methylation of DNA mainly occurs in CpG sequences, which are base paired to the same sequence on the other strand of DNA helix. The DNA methylation pattern on the parental DNA strand serves as template for the methylation of the daughter DNA strand, causing the pattern to be inherited by maintenance methyltransferase following DNA replication. Differentiation of cells is associated with changing the pattern of DNA methylation. De novo methyltransferase and the demethylation mechanism are involved in new pattern formation. **(B)** Model for the molecular mechanism of DNA methylation inheritance and transcriptional repression. During DNA replication, Dnmt1-MeCP2 complexes bind to hemimethylated CpG dinucleotides in double-stranded DNA and methylate cytosine residues in the daughter DNA strand. Additionally, Dnmt1 forms complexes with HDACs. The MeCP2-Dnmt1 complex does not contain the histone deacetylase, HDAC1. Alternately, the complex MeCp2-mSin3A-HDAC bound to fully methylated, double-stranded DNA, represses gene activity.

cules involved in DNA methylation. Mutations of MeCP2 have been implicated in several disorders including Rett syndrome (Amir et al., 1999; Guy et al., 2001; Chen et al., 2001). However, the function of MeCP2 and its precise mechanism of action are not known. In addition to the authentic function, MeCP2 may participate in maintaining the DNA methylation pattern by association with Dnmt1 in the developing brain or in the process of DNA repair. Although mutations of the Dnmt and MBD family genes caused loss of DNA methylation (Lei et al., 1996) and abnormal development (Guy et al., 2001; Chen et al., 2001), little is understood regarding which protein complexes are involved in this process.

To keep the methylation pattern, maintenance of both methylated and unmethylated status of CpG sites during DNA replication is necessary. Normally, unmethylated regions may show different fidelities from normally methylated regions. Interestingly, errors in DNA methylation pattern were mainly due to de novo methylation in unmethylated regions (Ushijima et al., 2003). In addition to the participation of Dnmt1, proper recognition of de novo Dnmts seems to be involved in maintaining the DNA methylation pattern (Liang et al., 2002; Chen et al., 2003).

Factors affecting the formation of DNA methylation profiles

DNA methylation profile changes during differentiation of cells, or changing the DNA methylation profile causes differentiation of cells. Removal of LIF and FGF from ES and TS cell cultures, respectively, causes differentiation of the cells and changes the DNA methylation profile (Shiota et al., 2002; Kremenskoy et al., 2003). Factors affecting differentiation should directly or indirectly contribute to changing DNA methylation. For example, steroids and vitamins including retinoic acid (Yan et al., 2001) are well known to be involved in cellular differentiation. In addition, several other factors including nutrition affect DNA methylation. Reduced quantity of folate, which affects activity of enzymes supplying SAM, causes genomic instability (Blount et al., 1997; Jacob, 1999; Friso and Choi, 2002; Van Den Veyver, 2002) and genomic hypomethylation (Friso et al., 2002). A methyl-deficient diet has been shown to induce liver cancer associated with hypomethylation (Dizik et al., 1991; Wainfan and Poirier, 1992).

Conclusions and future directions

There are numerous CpG islands containing tissue-dependent and differentially methylated regions (T-DMR) which are potential methylation sites in normal tissues. In contrast with the stable DNA sequence and rapidly changing fate of mRNA, DNA methylation is both stable and changeable. Currently, epigenetics is a new paradigm for the biology of health and disease. Investigation of aberrant methylation in CpG islands has been an approach to identify candidate oncogenes (Antequera et al., 1990; Baylin et al., 1998; Delgado et al., 1998; Jones and Laird, 1999). Epigenetic errors cause other diseases as suggested in cloned animals. In addition to information about DNA methylation status, how pattern and profile formation is controlled remains to be studied. Interplay of extra-/intracellular signaling, sequence-specific DNA-binding proteins, Dnmts, DNA demethylase, non-coding RNAs, and proteins affecting chromatin structure will be the keys of the new era.

Acknowledgements

I thank Drs. Naka Hattori, Koichiro Nishino, Mayumi Oda, Masako Suzuki, Hideki Sakamoto, and Kazuhiko Imakawa for communication and help in preparing the manuscript.

References

Amir RE, Van den Veyver IB, Wan M, Tran CQ, Francke U, Zoghbi HY: Rett syndrome is caused by mutations in X-linked MECP2, encoding methyl-CpG-binding protein 2. Nat Genet 23:185–188 (1999).

Antequera F, Boyes J, Bird A: High levels of de novo methylation and altered chromatin structure at CpG islands in cell lines. Cell 62:503–514 (1990).

Barlow DP: Gametic imprinting in mammals. Science 270:1610–1613 (1995).

Baylin SB, Herman JG, Graff JR, Vertino PM, Issa JP: Alterations in DNA methylation: a fundamental aspect of neoplasia. Adv Cancer Res 72:141–196 (1998).

Bell AC, Felsenfeld G: Methylation of a CTCF-dependent boundary controls imprinted expression of the *Igf2* gene. Nature 405:482–485 (2000).

Benvenisty N, Mencher D, Meyuhas O, Razin A, Reshef L: Sequential changes in DNA methylation patterns of the rat phosphoenolpyruvate carboxykinase gene during development. Proc Natl Acad Sci USA 82:267–271 (1985).

Bestor TH: Activation of mammalian DNA methyltransferase by cleavage of a Zn binding regulatory domain. EMBO J 11:2611–2617 (1992).

Bird AP: Use of restriction enzymes to study eukaryotic DNA methylation. II. The symmetry of methylated sites supports semi-conservative copying of the methylation pattern. J Mol Biol 118:49–60 (1978).

Bird AP: CpG islands as gene markers in the vertebrate nucleus. Trends Genet 3:342–347 (1987).

Bird A: DNA methylation patterns and epigenetic memory. Genes Dev 16:6–21 (2002).

Bird AP, Wolffe AP: Methylation-induced repression – belts, braces, and chromatin. Cell 99:451–454 (1999).

Bird A, Taggart M, Frommer M, Miller OJ, Macleod D: A fraction of the mouse genome that is derived from islands of nonmethylated, CpG-rich DNA. Cell 40:91–99 (1985).

Blount BC, Mack MM, Wehr CM, MacGregor JT, Hiatt RA, Wang G, Wickramasinghe SN, Everson RB, Ames BN: Folate deficiency causes uracil misincorporation into human DNA and chromosome breakage: implications for cancer and neuronal damage. Proc Natl Acad Sci USA 94:3290–3295 (1997).

Brandeis M, Frank D, Keshet I, Siegfried Z, Mendelsohn M, Nemes A, Temper V, Razin A, Cedar H: Sp1 elements protect a CpG island from de novo methylation. Nature 371:435–438 (1994).

Brook FA, Gardner RL: The origin and efficient derivation of embryonic stem cells in the mouse. Proc Natl Acad Sci USA 94:5709–5712 (1997).

Chen RZ, Akbarian S, Tudor M, Jaenisch R: Deficiency of methyl-CpG binding protein-2 in CNS neurons results in a Rett-like phenotype in mice. Nat Genet 27:327–331 (2001).

Chen T, Ueda Y, Dodge JE, Wang Z, Li E: Establishment and maintenance of genomic methylation patterns in mouse embryonic stem cells by Dnmt3a and Dnmt3b. Mol Cell Biol 23:5594–605 (2003).

Cho JH, Kimura H, Minami T, Ohgane J, Hattori N, Tanaka S, Shiota K: DNA methylation regulates placental lactogen I gene expression. Endocrinology 142:3389–3396 (2001).

Chuang LS, Ian HI, Koh TW, Ng HH, Xu G, Li BF: Human DNA-(cytosine-5) methyltransferase-PCNA complex as a target for p21WAF1. Science 277:1996 2000 (1997).

Cross JC, Werb Z, Fisher SJ: Implantation and the placenta: key pieces of the development puzzle. Science 266:1508–1518 (1994).

Cross SH, Bird AP: CpG islands and genes. Curr Opin Genet Dev 5:309–314 (1995).

Delgado S, Gomez M, Bird A, Antequera F: Initiation of DNA replication at CpG islands in mammalian chromosomes. EMBO J 17:2426–2435 (1998).

De Smet C, Lurquin C, Lethe B, Martelange V, Boon T: DNA methylation is the primary silencing mechanism for a set of germ line- and tumor-specific genes with a CpG-rich promoter. Mol Cell Biol 19:7327–7335 (1999).

Dizik M, Christman JK, Wainfan E: Alterations in expression and methylation of specific genes in livers of rats fed a cancer promoting methyl-deficient diet. Carcinogenesis 12:1307–1312 (1991).

Edsall LC, Pirianov GG, Spiegel S: Involvement of sphingosine 1-phosphate in nerve growth factor-mediated neuronal survival and differentiation. J Neurosci 17:6952–6960 (1997).

Ferguson-Smith AC, Sasaki H, Cattanach BM, Surani MA: Parental-origin-specific epigenetic modification of the mouse *H19* gene. Nature 362:751–755 (1993).

Fremont M, Siegmann M, Gaulis S, Matthies R, Hess D, Jost JP: Demethylation of DNA by purified chick embryo 5-methylcytosine-DNA glycosylase requires both protein and RNA. Nucl Acids Res 25:2375–2380 (1997).

Friso S, Choi SW: Gene-nutrient interactions and DNA methylation. J Nutr 132:2382S–2387S (2002).

Friso S, Choi SW, Girelli D, Mason JB, Dolnikowski GG, Bagley PJ, Olivieri O, Jacques PF, Rosenberg IH, Corrocher R, Selhub J: A common mutation in the 5,10-methylenetetrahydrofolate reductase gene affects genomic DNA methylation through an interaction with folate status. Proc Natl Acad Sci USA 99:5606–5611 (2002).

Fuks F, Burgers WA, Brehm A, Hughes-Davies L, Kouzarides T: DNA methyltransferase Dnmt1 associates with histone deacetylase activity. Nat Genet 24:88–91 (2000).

Futscher BW, Oshiro MM, Wozniak RJ, Holtan N, Hanigan CL, Duan H, Domann FE: Role for DNA methylation in the control of cell type specific maspin expression. Nat Genet 31:175–179 (2002).

Gardiner-Garden M, Frommer M: CpG islands in vertebrate genomes. J Mol Biol 196:261–282 (1987).

Goffin J, Eisenhauer E: DNA methyltransferase inhibitors – state of the art. Ann Oncol 13:1699–1716 (2002).

Gruenbaum Y, Stein R, Cedar H, Razin A: Methylation of CpG sequences in eukaryotic DNA. FEBS Lett 124:67–71 (1981).

Guy J, Hendrich B, Holmes M, Martin JE, Bird A: A mouse *Mecp2*-null mutation causes neurological symptoms that mimic Rett syndrome. Nat Genet 27:322–326 (2001).

Hark AT, Schoenherr CJ, Katz DJ, Ingram RS, Levorse JM, Tilghman SM: CTCF mediates methylation-sensitive enhancer-blocking activity at the *H19/Igf2* locus. Nature 405:486–489 (2000).

Hattori N, Nishino K, Ko Y, Hattori N, Ohgane J, Tanaka S, Shiota K: Epigenetic control of mouse *Oct4* gene expression in embryonic stem cells and trophoblast stem cells. J Biol Chem 279:17063–17069 (2004).

Heard E, Clerc P, Avner P: X-chromosome inactivation in mammals. Annu Rev Genet 31:571–610 (1997).

Hendrich B, Bird A: Identification and characterization of a family of mammalian methyl-CpG binding proteins. Mol Cell Biol 18:6538–6547 (1998).

Holliday R, Pugh JE: DNA modification mechanisms and gene activity during development. Science 187:226–232 (1975).

Imamura T, Ohgane J, Ito S, Ogawa T, Hattori N, Tanaka S, Shiota K: CpG island of rat sphingosine kinase-1 gene: tissue-dependent DNA methylation status and multiple alternative first exons. Genomics 76:117–125 (2001).

International Human Genome Sequencing Consortium: Initial sequencing and analysis of the human genome. Nature 409:860–921 (2001).

Jacob RA: The role of micronutrients in DNA synthesis and maintenance. Adv Exp Med Biol 472:101–113 (1999).

Jones PA, Laird PW: Cancer epigenetics comes of age. Nat Genet 21:163–167 (1999).

Jones PL, Veenstra GJ, Wade PA, Vermaak D, Kass SU, Landsberger N, Strouboulis J, Wolffe AP: Methylated DNA and MeCP2 recruit histone deacetylase to repress transcription. Nat Genet 19:187–191 (1998).

Jost JP, Jost YC: Transient DNA demethylation in differentiating mouse myoblasts correlates with higher activity of 5-methyldeoxycytidine excision repair. J Biol Chem 269:10040–10043 (1994).

Jost JP, Siegmann M, Sun L, Leung R: Mechanisms of DNA demethylation in chicken embryos. Purification and properties of a 5-methylcytosine-DNA glycosylase. J Biol Chem 270:9734–9739 (1995).

Jost JP, Fremont M, Siegmann M, Hofsteenge J: The RNA moiety of chick embryo 5-methylcytosine-DNA glycosylase targets DNA demethylation. Nucl Acids Res 25:4545–4550 (1997).

Jost JP, Schwarz S, Hess D, Angliker H, Fuller-Pace FV, Stahl H, Thiry S, Siegmann M: A chicken embryo protein related to the mammalian DEAD box protein p68 is tightly associated with the highly purified protein-RNA complex of 5-MeC-DNA glycosylase. Nucl Acids Res 27:3245–3252 (1999a).

Jost JP, Siegmann M, Thiry S, Jost YC, Benjamin D, Schwarz S: A re-investigation of the ribonuclease sensitivity of a DNA demethylation reaction in chicken embryo and G8 mouse myoblasts. FEBS Lett 449:251–254 (1999b).

Kafri T, Ariel M, Brandeis M, Shemer R, Urven L, McCarrey J, Cedar H, Razin A: Developmental pattern of gene-specific DNA methylation in the mouse embryo and germ line. Genes Dev 6:705–714 (1992).

Kaneda A, Takai D, Kaminishi M, Okochi E, Ushijima T: Methylation-sensitive representational difference analysis and its application to cancer research. Ann NY Acad Sci 983:131–141 (2003).

Kimura H, Shiota K: Methyl-CpG-binding protein, MeCP2, is a target molecule for maintenance DNA methyltransferase, Dnmt1. J Biol Chem 278:4806–4812 (2003).

Kohlhase J, Hausmann S, Stojmenovic G, Dixkens C, Bink K, Schulz-Schaeffer W, Altmann M, Engel W: SALL3, a new member of the human spalt-like gene family, maps to 18q23. Genomics 62:216–222 (1999).

Koopman P, Gubbay J, Vivian N, Goodfellow P, Lovell-Badge R: Male development of chromosomally female mice transgenic for *Sry*. Nature 351:117–121 (1991).

Kremenskoy M, Kremenska Y, Ohgane J, Hattori N, Tanaka S, Hashizume K, Shiota K: Genome-wide analysis of DNA methylation status of CpG islands in embryoid bodies, teratomas, and fetuses. Biochem Biophys Res Commun 311:884–890 (2003).

Kuramasu A, Saito H, Suzuki S, Watanabe T, Ohtsu H: Mast cell-/basophil-specific transcriptional regulation of human L-histidine decarboxylase gene by CpG methylation in the promoter region. J Biol Chem 273:31607–31614 (1998).

Lachner M, Jenuwein T: The many faces of histone lysine methylation. Curr Opin Cell Biol 14:286–298 (2002).

Lanza RP, Cibelli JB, Blackwell C, Cristofalo VJ, Francis MK, Baerlocher GM, Mak J, Schertzer M, Chavez EA, Sawyer N, Lansdorp PM, West MD: Extension of cell life-span and telomere length in animals cloned from senescent somatic cells. Science 288:665–669 (2000).

Lei H, Oh SP, Okano M, Juttermann R, Goss KA, Jaenisch R, Li E: De novo DNA cytosine methyltransferase activities in mouse embryonic stem cells. Development 122:3195–3205 (1996).

Lewis JD, Meehan RR, Henzel WJ, Maurer-Fogy I, Jeppesen P, Klein F, Bird A: Purification, sequence, and cellular localization of a novel chromosomal protein that binds to methylated DNA. Cell 69:905–914 (1992).

Li E: Chromatin modification and epigenetic reprogramming in mammalian development. Nat Rev Genet 3:662–673 (2002).

Li E, Bestor TH, Jaenisch R: Targeted mutation of the DNA methyltransferase gene results in embryonic lethality. Cell 69:915–926 (1992).

Liang G, Chan MF, Tomigahara Y, Tsai YC, Gonzales FA, Li E, Laird PW, Jones PA: Cooperativity between DNA methyltransferases in the maintenance methylation of repetitive elements. Mol Cell Biol 22:480–491 (2002).

Macleod D, Charlton J, Mullins J, Bird AP: Sp1 sites in the mouse *Aprt* gene promoter are required to prevent methylation of the CpG island. Genes Dev 8:2282–2292 (1994).

Maier H, Colbert J, Fitzsimmons D, Clark DR, Hagman J: Activation of the early B-cell-specific mb-1 (Ig-alpha) gene by Pax-5 is dependent on an unmethylated Ets binding site. Mol Cell Biol 23:1946–1960 (2003).

Makino S, Fukuda K, Miyoshi S, Konishi F, Kodama H, Pan J, Sano M, Takahashi T, Hori S, Abe H, Hata J, Umezawa A, Ogawa S: Cardiomyocytes can be generated from marrow stromal cells in vitro. J Clin Invest 103:697–705 (1999).

Matsui Y, Zsebo K, Hogan BL: Derivation of pluripotential embryonic stem cells from murine primordial germ cells in culture. Cell 70:841–847 (1992).

Meehan RR, Lewis JD, Bird AP: Characterization of MeCP2, a vertebrate DNA binding protein with affinity for methylated DNA. Nucl Acids Res 20:5085–5092 (1992).

Millar DS, Paul CL, Molloy PL, Clark SJ: A distinct sequence (ATAAA)$_n$ separates methylated and unmethylated domains at the 5′-end of the GSTP1 CpG island. J Biol Chem 275:24893–24899 (2000).

Monk M, Boubelik M, Lehnert S: Temporal and regional changes in DNA methylation in the embryonic, extraembryonic and germ cell lineages during mouse embryo development. Development 99:371–382 (1987).

Motiwala T, Ghoshal K, Das A, Majumder S, Weichenhan D, Wu YZ, Holman K, James SJ, Jacob ST, Plass C: Suppression of the protein tyrosine phosphatase receptor type O gene (PTPRO) by methylation in hepatocellular carcinomas. Oncogene 22:6319–6331 (2003).

Mouse Genome Sequencing Consortium: Initial sequencing and comparative analysis of the mouse genome. Nature 420:520–562 (2002).

Mummaneni P, Yates P, Simpson J, Rose J, Turker MS: The primary function of a redundant Sp1 binding site in the mouse Aprt gene promoter is to block epigenetic gene inactivation. Nucl Acids Res 26:5163–5169 (1998).

Nagy A, Rossant J, Nagy R, Abramow-Newerly W, Roder JC: Derivation of completely cell culture-derived mice from early-passage embryonic stem cells. Proc Natl Acad Sci USA 90:8424–8428 (1993).

Nan X, Meehan RR, Bird A: Dissection of the methyl-CpG binding domain from the chromosomal protein MeCP2. Nucl Acids Res 21:4886–4892 (1993).

Nan X, Ng HH, Johnson CA, Laherty CD, Turner BM, Eisenman RN, Bird A: Transcriptional repression by the methyl-CpG-binding protein MeCP2 involves a histone deacetylase complex. Nature 393:386–389 (1998).

Newell-Price J, King P, Clark AJ: The CpG island promoter of the human proopiomelanocortin gene is methylated in nonexpressing normal tissue and tumors and represses expression. Mol Endocrinol 15:338–348 (2001).

Ngo V, Gourdji D, Laverriere JN: Site-specific methylation of the rat prolactin and growth hormone promoters correlates with gene expression. Mol Cell Biol 16:3245–3254 (1996).

Nichols J, Zevnik B, Anastassiadis K, Niwa H, Klewe-Nebenius D, Chambers I, Schöler H, Smith A: Formation of pluripotent stem cells in the mammalian embryo depends on the POU transcription factor Oct4. Cell 95:379–391 (1998).

Nishino K, Hattori N, Tanaka S, Shiota K: DNA methylation-mediated control of *Sry* gene expression in mouse gonadal development. J Biol Chem 279:22306–22313 (2004).

Niwa H, Miyazaki J, Smith AG: Quantitative expression of Oct-3/4 defines differentiation, dedifferentiation or self-renewal of ES cells. Nat Genet 24:372–376 (2000).

Norris DP, Brockdorff N, Rastan S: Methylation status of CpG-rich islands on active and inactive mouse X chromosomes. Mamm Genome 1:78–83 (1991).

Ogura A, Inoue K, Ogonuki N, Noguchi A, Takano K, Nagano R, Suzuki O, Lee J, Ishino F, Matsuda J: Production of male cloned mice from fresh, cultured, and cryopreserved immature Sertoli cells. Biol Reprod 62:1579–1584 (2000).

Ohgane J, Aikawa J, Ogura A, Hattori N, Ogawa T, Shiota K: Analysis of CpG islands of trophoblast giant cells by restriction landmark genomic scanning. Dev Genet 22:132–140 (1998).

Ohgane J, Wakayama T, Kogo Y, Senda S, Hattori N, Tanaka S, Yanagimachi R, Shiota K: DNA methylation variation in cloned mice. Genesis 30:45–50 (2001).

Ohgane J, Hattori N, Oda M, Tanaka S, Shiota K: Differentiation of trophoblast lineage is associated with DNA methylation and demethylation. Biochem Biophys Res Commun 290:701–706 (2002).

Ohgane J, Wakayama T, Senda S, Yamazaki Y, Inoue K, Ogura A, Marh J, Tanaka S, Yanagimachi R, Shota K: The *Sall3* locus is an epigenetic hotspot of aberrant DNA methylation associated with placentomegaly of cloned mice. Genes Cells 9:253–260 (2004).

Okamoto K, Okazawa H, Okuda A, Sakai M, Muramatsu M, Hamada H: A novel octamer binding transcription factor is differentially expressed in mouse embryonic cells. Cell 60:461–472 (1990).

Okano M, Xie S, Li E: Cloning and characterization of a family of novel mammalian DNA (cytosine-5) methyltransferases. Nat Genet 19:219–220 (1998).

Okano M, Bell DW, Haber DA, Li E: DNA methyltransferases Dnmt3a and Dnmt3b are essential for de novo methylation and mammalian development. Cell 99:247–257 (1999).

Okazaki Y, Okuizumi H, Sasaki N, Ohsumi T, Kuromitsu J, Hirota N, Muramatsu M, Hayashizaki Y: An expanded system of restriction landmark genomic scanning (RLGS Ver. 1.8). Electrophoresis 16:197–202 (1995).

Okazawa H, Okamoto K, Ishino F, Ishino-Kaneko T, Takeda S, Toyoda Y, Muramatsu M, Hamada H: The *Oct3* gene, a gene for an embryonic transcription factor, is controlled by a retinoic acid repressible enhancer. EMBO J 10:2997–3005 (1991).

Olivera A, Kohama T, Edsall L, Nava V, Cuvillier O, Poulton S, Spiegel S: Sphingosine kinase expression increases intracellular sphingosine-1-phosphate and promotes cell growth and survival. J Cell Biol 147:545–558 (1999).

Pao MM, Tsutsumi M, Liang G, Uzvolgyi E, Gonzales FA, Jones PA: The endothelin receptor B (EDNRB) promoter displays heterogeneous, site specific methylation patterns in normal and tumor cells. Hum Mol Genet 10:903–910 (2001).

Postma FR, Jalink K, Hengeveld T, Moolenaar WH: Sphingosine-1-phosphate rapidly induces Rho-dependent neurite retraction: action through a specific cell surface receptor. EMBO J 15:2388–2392 (1996).

Razin A, Cedar H: DNA methylation and genomic imprinting. Cell 77:473–476 (1994).

Razin A, Webb C, Szyf M, Yisraeli J, Rosenthal A, Naveh-Many T, Sciaky-Gallili N, Cedar H: Variations in DNA methylation during mouse cell differentiation in vivo and in vitro. Proc Natl Acad Sci USA 81:2275–2279 (1984).

Resnick JL, Bixler LS, Cheng L, Donovan PJ: Long-term proliferation of mouse primordial germ cells in culture. Nature 359:550–551 (1992).

Riggs AD: X inactivation, differentiation, and DNA methylation. Cytogenet Cell Genet 14:9–25 (1975).

Robertson KD, Ait-Si-Ali S, Yokochi T, Wade PA, Jones PL, Wolffe AP: DNMT1 forms a complex with Rb, E2F1 and HDAC1 and represses transcription from E2F-responsive promoters. Nat Genet 25:338–342 (2000).

Rosner MH, Vigano MA, Ozato K, Timmons PM, Poirier F, Rigby PW, Staudt LM: A POU-domain transcription factor in early stem cells and germ cells of the mammalian embryo. Nature 345:686–692 (1990).

Rountree MR, Bachman KE, Baylin SB: DNMT1 binds HDAC2 and a new co-repressor, DMAP1, to form a complex at replication foci. Nat Genet 25:269–277 (2000).

Russo U, Martienssen R, Riggs A: Epigenetic Mechanisms of Gene Regulation. Plainview, pp 1–4 (Cold Spring Harbor Laboratory Press, Cold Spring Harbor 1996).

Schöler HR, Ruppert S, Suzuki N, Chowdhury K, Gruss P: New type of POU domain in germ line-specific protein Oct4. Nature 344:435–439 (1990).

Schulz WA, Elo JP, Florl AR, Pennanen S, Santourlidis S, Engers R, Buchardt M, Seifert HH, Visakorpi T: Genomewide DNA hypomethylation is associated with alterations on chromosome 8 in prostate carcinoma. Genes Chrom Cancer 35:58–65 (2002).

Shemer R, Birger Y, Dean WL, Reik W, Riggs AD, Razin A: Dynamic methylation adjustment and counting as part of imprinting mechanisms. Proc Natl Acad Sci USA 93:6371–6376 (1996).

Shiota K, Yanagimachi R: Epigenetics by DNA methylation for development of normal and cloned animals. Differentiation 69:162–166 (2002).

Shiota K, Kogo Y, Ohgane J, Imamura T, Urano A, Nishino K, Tanaka S, Hattori N: Epigenetic marks by DNA methylation specific to stem, germ and somatic cells in mice. Genes Cells 7:961–969 (2002).

Solter D: Mammalian cloning: advances and limitations. Nat Rev Genet 1:199–207 (2000).

Strichman-Almashanu LZ, Lee RS, Onyango PO, Perlman E, Flam F, Frieman MB, Feinberg AP: A genome-wide screen for normally methylated human CpG islands that can identify novel imprinted genes. Genome Res 12:543–554 (2002).

Suzuki Y, Tsunoda T, Sese J, Taira H, Mizushima-Sugano J, Hata H, Ota T, Isogai T, Tanaka T, Nakamura Y, Suyama A, Sakaki Y, Morishita S, Okubo K, Sugano S: Identification and characterization of the potential promoter regions of 1031 kinds of human genes. Genome Res 11:677–684 (2001).

Sylvester I, Schöler HR: Regulation of the *Oct4* gene by nuclear receptors. Nucl Acids Res 22:901–911 (1994).

Szyf M: The DNA methylation machinery as a target for anticancer therapy. Pharmacol Ther 70:1–37 (1996).

Tachibana M, Sugimoto K, Fukushima T, Shinkai Y: Set domain-containing protein, G9a, is a novel lysine-preferring mammalian histone methyltransferase with hyperactivity and specific selectivity to lysines 9 and 27 of histone H3. J Biol Chem 276:25309–25317 (2001).

Tachibana M, Sugimoto K, Nozaki M, Ueda J, Ohta T, Ohki M, Fukuda M, Takeda N, Niida H, Kato H, Shinkai Y: G9a histone methyltransferase plays a dominant role in euchromatic histone H3 lysine 9 methylation and is essential for early embryogenesis. Genes Dev 16:1779–1791 (2002).

Takizawa T, Nakashima K, Namihira M, Ochiai W, Uemura A, Yanagisawa M, Fujita N, Nakao M, Taga T: DNA methylation is a critical cell-intrinsic determinant of astrocyte differentiation in the fetal brain. Dev Cell 1:749–758 (2001).

Tamashiro KL, Wakayama T, Blanchard RJ, Blanchard DC, Yanagimachi R: Postnatal growth and behavioral development of mice cloned from adult cumulus cells. Biol Reprod 63:328–334 (2000).

Tanaka S, Kunath T, Hadjantonakis AK, Nagy A, Rossant J: Promotion of trophoblast stem cell proliferation by FGF4. Science 282:2072–2075 (1998).

Tanaka S, Oda M, Toyoshima Y, Wakayama T, Tanaka M, Yoshida N, Hattori N, Ohgane J, Yanagimachi R, Shiota K: Placentomegaly in cloned mouse concepti caused by expansion of the spongiotrophoblast layer. Biol Reprod 65:1813–1821 (2001).

Taylor SM, Jones PA: Multiple new phenotypes induced in 10T1/2 and 3T3 cells treated with 5-azacytidine. Cell 17:771–779 (1979).

Tribioli C, Tamanini F, Patrosso C, Milanesi L, Villa A, Pergolizzi R, Maestrini E, Rivella S, Bione S, Mancini M, et al: Methylation and sequence analysis around *Eag*I sites: identification of 28 new CpG islands in XQ24–XQ28. Nucl Acids Res 20:727–733 (1992).

Ushijima T, Watanabe N, Okochi E, Kaneda A, Sugimura T, Miyamoto K: Fidelity of the methylation pattern and its variation in the genome. Genome Res 13:868–874 (2003).

Van den Veyver IB: Genetic effects of methylation diets. Annu Rev Nutr 22:255–282 (2002).

Venter JC, et al: The sequence of the human genome. Science 291:1304–1351 (2001).

Wade PA: Methyl CpG-binding proteins and transcriptional repression. Bioessays 23:1131–1137 (2001).

Wainfan E, Poirier LA: Methyl groups in carcinogenesis: effects on DNA methylation and gene expression. Cancer Res 52:2071s–2077s (1992).

Wakayama T, Yanagimachi R: Cloning of male mice from adult tail-tip cells. Nat Genet 22:127–128 (1999).

Wakayama T, Perry AC, Zuccotti M, Johnson KR, Yanagimachi R: Full-term development of mice from enucleated oocytes injected with cumulus cell nuclei. Nature 394:369–374 (1998).

Walsh CP, Bestor TH: Cytosine methylation and mammalian development. Genes Dev 13:26–34 (1999).

Weiss A, Keshet I, Razin A, Cedar H: DNA demethylation in vitro: involvement of RNA. Cell 86:709–718 (1996).

Yan J, Tanaka S, Oda M, Makino T, Ohgane J, Shiota K: Retinoic acid promotes differentiation of trophoblast stem cells to a giant cell fate. Dev Biol 235:422–432 (2001).

Yen PH, Patel P, Chinault AC, Mohandas T, Shapiro LJ: Differential methylation of hypoxanthine phosphoribosyltransferase genes on active and inactive human X chromosomes. Proc Natl Acad Sci USA 81:1759–1763 (1984).

Yoshida M, Kijima M, Akita M, Beppu T: Potent and specific inhibition of mammalian histone deacetylase both in vivo and in vitro by trichostatin A. J Biol Chem 265:17174–17179 (1990).

Cytogenet Genome Res 105:335–345 (2004)
DOI: 10.1159/000078206

Identification and properties of imprinted genes and their control elements

R.J. Smith, P. Arnaud, and G. Kelsey

Developmental Genetics Programme, The Babraham Institute, Cambridge (UK)

Abstract. Imprinted genes have the unusual characteristic that the copy from one parent is destined to remain inactive. Though few in number they nonetheless constitute a functionally important part of the mammalian genome. With their memory of parental origin, imprinted genes represent an important model for the epigenetic regulation of gene function and will provide invaluable paradigms to test whether we can predict epigenetic state from DNA sequence. Since their first discovery, systematic screens and some good fortune have led to identification of over seventy imprinted genes in the mouse and human: recent microarray analysis may reveal many more. With a significant number of imprinted genes now identified and completion of key mammalian genome sequences, we are able systematically to examine the organization of imprinted loci, properties of their control elements and begin to recognize common themes in imprinted gene regulation.

Copyright © 2004 S. Karger AG, Basel

Genomic imprinting as a phenomenon recognized in mammals is barely twenty years old (Barton et al., 1984; McGrath and Solter, 1984; Surani et al., 1984; Cattanach and Kirk, 1985); the first endogenous imprinted genes in the mouse were identified a dozen or so years ago (Barlow et al., 1991; De Chiara et al., 1991). In mammals, imprinting results in the predictable silencing of one copy or allele of specific autosomal genes, according to the sex of the transmitting parent. It is a reversible epigenetic mechanism, so that whilst a memory of parental origin is retained in all somatic cells, silencing is reversed when the allele passes through the germline of the opposite parental sex (Surani, 1998; Reik and Walter, 2001a). More than seventy imprinted genes have been catalogued in the mouse (Beechey et al., 2003) and most of those investigated are also imprinted in humans (Morison et al., 2001). Imprinted genes are involved in

a range of physiological processes, notably the regulation of growth of the fetus and placenta and behavior (Reik and Walter, 2001a; Tycko and Morison, 2002). Many human disorders are now known to result from the inappropriate expression of imprinted genes (Morison et al., 2001) and there are others, such as autism, for which parent-of-origin effects have been described (Davies et al., 2001). The existence of genes constitutionally expressed from only one allele represents a sacrifice of diploid protection and argues for selective advantage in parental-allele specific silencing. A number of theories have been proposed to explain why genomic imprinting may have evolved (Hurst, 1997), and whilst no one theory can explain all of the available data, perhaps the most widely held hypothesis is that genomic imprinting represents a "tug-of-war" between the interests of maternal and paternal genomes in the fitness of offspring and the demands on maternal resources (Moore and Haig, 1991). This theory now finds support from the phenotypic effects of many mouse imprinted gene knock-outs (Tycko and Morison, 2002; Wilkins and Haig, 2003).

In this review, we shall focus on two issues: (1) How imprinted genes have been discovered and the recent opportunities for imprinted gene identification that have arisen from availability of human and mouse genome sequence and new technologies such as microarrays; (2) the nature of imprinting control elements (ICEs) and the various means by which monoallelic expression can be executed.

Supported by the BBSRC and MRC (G.K.). P.A. holds a Marie Curie Individual Fellowship from the European Community Programme in Human Potential (under contract number HPMF-CT-2001-01122).

Received 29 October 2003; manuscript accepted 10 December 2003.

Request reprints from Gavin Kelsey, The Babraham Institute
Cambridge, CB2 4AT (UK); telephone: +44-1223 496 332
fax: +44-1223 496 022; e-mail: gavin.kelsey@bbsrc.ac.uk

Table 1. Screens for imprinted gene identification

Type	Method	Source RNA/DNA	Genes identified	References
Expression	Subtractive hybridization	Pg embryos	*Peg1/Mest, Peg3/Pw1, Peg5/Nnat, Peg8/Igf2as*	Kaneko-Ishino et al., 1995 Kuroiwa et al., 1996 Kagitani et al., 1997
		Ag embryos Ag/Pg embryo fibroblasts	*Meg1/Grb10, Meg3/Gtl2* *Sgce, Zac1/Plagl1*	Miyoshi et al., 1998, 2000 Piras et al., 2000
	Microarrays	Pg embryos Ag/Pg embryos	*Peg9/Dlk1, Peg12/Frat3* *Asb4, Dcn, Slc38a4/Ata3*	Kobayashi et al., 2000, 2002 Mizuno et al., 2002 Nikaido et al., 2003
		Chr. 7/11 uniparental duplication tissues	Unnamed candidates	Choi et al., 2001
	Differential display	Ag/Pg embryos	*Peg5/Nnat*	Kikyo et al., 1997
	Allelic message display (AMD)	C57BL/6 × CAST/Ei hybrids Peromyscus hybrids	*Impact* *Peg9/Dlk1, Gatm*	Hagiwara et al., 1997 Schmidt et al., 2000 Sandell et al., 2003
Methylation	Restriction landmark genome scanning (RLGS)	C57BL/6 × DBA2 hybrids	*U2af1-rs1, Rasgrf1*	Hatada et al., 1993 Hayashizaki et al., 1994 Plass et al., 1996
		Human Pg blood, complete hydatidiform mole (Ag)	XLαS, ZAC/PLAGLI	Hayward et al., 1998 Kamiya et al., 2000
	Methylation-sensitive representational difference analysis (Me-RDA)	Chr. 2/8 uniparental duplication embryos Pg embryos	*Gnas (Gnasxl, Nesp), Nespas* *Nap1l5, Peg13, Slc38a4/Ata3, Zac1/Plagl1*	Peters et al., 1999 Kelsey et al., 1999 Smith et al., 2002, 2003
	*Eag*I library of methylated CpG islands	Wilms' tumor	TCEB3C (ElonginA3), HYMAI	Strichman-Almashanu et al., 2002

Identification of imprinted genes

One of the first imprinted genes described was discovered purely by serendipity. It was found that mice inheriting a targeted allele of *Igf2* paternally were growth deficient, whereas mice that inherited the targeted allele from mothers were indistinguishable from wild-type littermates, indicating that only the paternal allele is functional (De Chiara et al., 1991). Such fortunate discoveries have not been the rule and most imprinted genes have been identified in dedicated searches – either in regions known or suspected to contain imprinted genes, or from features imprinted genes are known to share.

A phenotypic screen for imprinting effects in the mouse has been conducted over many years by crossing translocation carriers to create offspring that inherit both copies of particular chromosomal regions from one parent – uniparental disomies or duplications (Cattanach and Beechey, 1997). These experiments laid the groundwork for our current "imprinting map" of the mouse (Beechey et al., 2003) by identifying a number of regions for which uniparental inheritance gave rise to lethality or anomalous phenotypes; the majority of imprinted genes since identified maps to these regions. In humans, uniparental disomies have also been identified which result in developmental abnormalities and many of these are homologous to mouse imprinted regions (Morison et al., 2001). For example, the Beckwith-Wiedemann syndrome can be caused by paternal uniparental duplications of chromosome 11p15.5 (Henry et al., 1989). The homologous region in mouse, distal chromosome 7, has phenotypic consequences when inherited uniparentally (Searle and Beechey, 1990) and is now known to harbor many imprinted genes (Onyango et al., 2000; Paulsen et al., 2000). As

can be seen, discovery of an imprinted locus is a powerful incentive to test for monoallelic expression of neighboring genes, and some genes long held to be isolated turn out to be within clusters on closer reexamination (Zwart et al., 2001). But a variety of screens has been developed over the past decade to identify imprinted genes independent of their genomic location and many have been highly productive (Table 1). These can be categorized into those based on imprinted expression, those that detect an epigenetic feature common to imprinted loci (differential methylation), and those that identify genes with sequence attributes shared with imprinted genes.

Screens based on imprinted expression

Most screens based on expression require the separation of maternal and paternal alleles into different mRNA sources. This can be achieved with uniparental duplications (as above), which focus on specific chromosomal regions, or uniparental embryos, generated by oocyte activation or pronuclear transfers (Dean et al., 2001). Parthenogenetic (Pg) or gynogenetic (Gg) embryos contain chromosomes entirely of maternal origin and androgenetic (Ag) embryos of paternal origin. With these separate resources, subtractive hybridization of cDNAs becomes feasible and was the basis of the first successful screens. Subtraction of Pg and normal fertilized embryo cDNAs has yielded a number of paternally expressed genes (*Peg*s; Kaneko-Ishino et al., 1995); an analogous experiment using Ag embryo cDNA isolated maternally expressed genes (*Meg*s; Miyoshi et al., 1998). An alternative to subtraction is differential display, which utilizes multiple primer pairs, at least one of which is arbitrary in sequence, to amplify from cDNA, the PCR products being resolved by denaturing gel electrophoresis. By this means Ag and Pg cDNAs have been compared, leading to identification of *Nnat* (Kikyo et al., 1997). The use of uniparental embryos allows the entire genome to be screened at once; however, the early arrest of these embryos necessarily limits such screens to genes expressed at early developmental stages and is likely to exclude those with tissue-restricted monoallelic expression. A further possible compound is that some imprinted genes may be inappropriately expressed (Sotomaru et al., 2001). In addition to the skill required in constructing uniparental embryos, a further challenge is the small amount of material. An attempt to circumvent this problem (but not limitations of cell-type specificity) has been the creation of Ag and Pg embryonic fibroblasts for subtractive hybridization, and these cells seem to retain faithful imprinting of many genes (Piras et al., 2000).

Oligonucleotide or cDNA microarrays are ideally suited to comparison of expression levels and application to uniparental material will result in isolation of imprinted genes. A limitation has been the number of genes represented on microarrays; however, the most recent of these screens was an ambitious analysis of over 27,000 transcripts comprising the FANTOM2 full-length mouse cDNA clone set (Nikaido et al., 2003). As with other expression-based techniques, expression profiling using microarrays will also identify non-imprinted transcripts whose abundance differs in uniparental mRNAs, some of which may be downstream genes (important information in its own right), and may be vulnerable to false positives arising from incomplete matching of samples. Ag and Pg embryos fail early in post-implantation development with multiple developmental anomalies, a potential source of spurious variation. It is essential, therefore, to validate candidates through independent techniques, such as analysis of expression in interspecific hybrids. From the FANTOM2 screen 2,114 candidates were found, of which five have to date been confirmed as imprinted by this means (Mizuno et al., 2002; Nikaido et al., 2003). An annotated list of candidates together with a single nucleotide polymorphism (SNP) database that will facilitate verification of imprinting status is available (http://fantom2.gsc.riken.go.jp/imprinting). Whilst a considerable proportion may be spurious, comparison of these genes with those from other screens may be a particularly powerful approach to home in on prime candidates. In addition to providing a potentially comprehensive survey of imprinted expression, the high throughput nature of microarrays may now facilitate assessment of multiple tissues to identify genes with tissue-restricted imprinted monoallelic expression (Choi et al., 2001).

As an alternative to separating the maternal and paternal genomes or chromosome regions into separate resources, polymorphic variants between mouse strains can be exploited as markers of parental origin. A derivative of differential display, allelic message display (AMD), has been developed which looks for allelic expression differences in F1 hybrid mice on the basis of polymorphisms (Hagiwara et al., 1997). AMD is geared towards the 3′ untranslated regions of transcripts, the richest source of sequence divergence. Through the use of different strains and analysis of different tissues and developmental stages, there is no theoretical limit (but obvious practical limits)

to the completeness of this screen. AMD has been applied to hybrids between the inbred strain C57BL/6J and an inbred line derived from *Mus musculus molossinus*, leading to the identification of *Impact*, the first imprinted gene on Chromosome 18 (Hagiwara et al., 1997), and to comparison of placental mRNAs from crosses of North American deer mice of the genus *Peromyscus* (Schmidt et al., 2000).

Combining the high throughput of microarrays and utility of polymorphisms, SNP arrays have multiple probes for each SNP to detect the sequence differences. A recent survey of 1063 transcribed human SNPs indicated that there is allelic expression bias for many autosomal genes (Lo et al., 2003). Such expression variation may be transmitted with Mendelian inheritance, rather than with parental origin, as has been seen previously (Yan et al., 2002). Allelic bias was shown for four out of five imprinted genes with SNPs informative in the samples tested and, although this survey of the human SNPs was not designed specifically to identify novel imprinted genes, its potential is obvious.

Screens based on epigenetic features

An important complement to expression screens have been those based on DNA methylation differences of maternal and paternal alleles (differentially methylated regions, DMRs; Constância et al., 1998). Whilst this is not the only epigenetic property distinguishing parental alleles of imprinted genes, it is the most amenable for analysis. Methylation-based screens are more likely to reveal imprinted genes rather than downstream genes and may be less vulnerable to imperfect matching of materials (DMRs tend to be all or nothing, and methylation differences owing to differences in tissue composition are relatively trivial in comparison). In addition, as methylation-based screens are directed towards DMRs, they should identify candidate control elements. On the other hand, some imprinted genes within clusters are not marked by DMRs (Yatsuki et al., 2002) and one locus appears totally to lack a DMR (Sandell et al., 2003).

Restriction landmark genome scanning (RLGS) provides a display of unmethylated sites, mostly within CpG islands, through the use of rare-cutting restriction enzymes and two-dimensional (2D) gel electrophoresis. In its first applications to imprinting, RLGS relied on restriction fragment length polymorphisms (RFLPs) between strains and interspecific hybrids to score differences in parental allele methylation (Hatada et al., 1993; Hayashizaki et al., 1994). The large amount of DNA required means that RLGS cannot feasibly be applied to Ag and Pg mouse embryos. It has since been applied to uniparental DNA of human origin, where this limitation was overcome using DNA from peripheral blood of a patient whose blood cells are entirely Pg in origin and DNA from complete hydatidiform moles, which are Ag conceptuses (Hayward et al., 1998; Kamiya et al., 2000). RLGS requires highly reproducible 2D gel technique and can only display a fraction of CpG islands. An earlier technical difficulty in cloning the specific DNA underlying spots of interest on radiolabeled 2D displays has been overcome with plasmid reference libraries (Smiraglia et al., 1999).

An alternative technique for identifying DMRs is methylation-sensitive representational difference analysis (Me-RDA) (Kelsey et al., 1999; Smith et al., 2003). In this technique a whole genome PCR is performed from DNA digested with a methylation-sensitive restriction enzyme and subtractive hybridization used to enrich differences between these PCR products, which correspond to differences in the methylation state of the starting DNAs. Me-RDA requires that at least one of the samples contains uniparental DNA (the entire genome or a particular chromosome region) and its basis in PCR means that small amounts of DNA from Ag and Pg embryos are amenable to screens. As with all subtraction-based approaches, the kinetics of reassociation may favor enrichment of some sequence differences at the expense of others. Whilst this may suggest that the technique cannot form the basis of an exhaustive screen, each DMR should be represented by multiple fragments recoverable by Me-RDA (Kelsey et al., 1999).

As DMRs are thought to be specific to imprinted genes, and DMRs are similar to CpG islands in sequence composition, it follows that a proportion of normally methylated autosomal CpG islands will be located at imprinted loci. From this premise a set of methylated human CpG islands has been isolated using a sequential restriction enzyme digestion strategy designed to clone a library of *Eag*I fragments resistant to *Hpa*II cleavage (Strichman-Almashanu et al., 2002). Amongst these were twelve candidate DMRs, based on their methylation patterns in complete hydatidiform moles and ovarian teratoma DNAs, one a known and one a novel imprinted gene. This screen was not designed specifically for imprinted genes, but comparison of the methylated CpG island set with expression data from the FANTOM2 microarray may be rewarding.

Screens based on sequence features

As more imprinted genes have been identified and genomic sequence data has become available, attempts have been made to find sequence features that are common and unique to imprinted genes, and we can expect considerable effort in the coming years in pursuing bioinformatics screens. Many imprinted genes are associated with tandem repeats (see below), and a screen has been performed by hybridizing a genomic library with tandem pentanucleotide repeats from the *H19* gene (Miura et al., 1999). However, since there does not seem to be a sequence common to repeat units at imprinted genes, a hybridization approach will be of limited use. Instead, there is obvious scope for informatics approaches based on direct repeats, but it is still unclear to what extent such repeats are over-represented at imprinted genes.

One sequence composition feature described for imprinted genes and gene clusters is that they reside in genomic sequence with a relatively low abundance of SINEs (Engemann et al., 2000; Greally, 2002). SINE Alu elements are reduced in both the upstream and downstream regions of imprinted genes (Ke et al., 2002a), particularly those of the AluJ and AluY classes. These observations have been used to try and determine whether genes whose imprinting status is unclear, or which were located in imprinted regions but not themselves known to be imprinted, fitted the expected sequence characteristics of an imprinted gene (Ke et al., 2002b), but this study did not go as far as testing expression. However, if the predictions hold true, there would be potential to derive a probability function to survey genomic sequence, whose predictions could be further refined by cross-reference with expression evidence from the FANTOM2 microarray screen.

Location and organization of imprinted genes

The majority of imprinted genes occur in clusters. Many of the regions described from the genetic tests of Cattanach and colleagues have since been found to contain multiple imprinted loci (Beechey et al., 2003). The organization of the genes within the clusters is complex and can include maternally and paternally expressed genes, protein-coding transcripts, non-coding RNAs and transcripts that are antisense to protein-coding genes (Reik and Walter, 2001a). At least in the case of the distal Chr 7 cluster, non-imprinted genes can be interspersed within the cluster (Paulsen et al., 2000). Clustering of imprinted genes may reflect mechanisms by which they are coordinately regulated: it is also possible that some genes are "innocent bystanders", imprinted because they are close to genes which are selected to become imprinted, but whose own functional haploidy presents little selective disadvantage. Alternatively, it has been suggested that organization in clusters reflects the evolution of imprinted genes – that imprinting may have arisen from mechanisms used to silence paralogous genes in duplicated genomic segments, or that imprinting originally arose on a common ancestral chromosome, from which imprinted clusters have been transposed (Walter and Paulsen, 2003). It is certainly the case that a single mouse autosome, chromosome 7, contains three extensive clusters, which account for 32 of the 72 imprinted loci. However, if a significant proportion of the FANTOM2 candidate genes turn out to be imprinted, and these seem to be more evenly distributed throughout the genome (Nikaido et al., 2003), some of our perceptions of imprinted gene organization and evolution may need to be revised.

Imprinted genes can also exist outside of clusters. The origin of some of these singletons appears to be through retrotransposition. This is an interesting finding, which attracts parallels with parent-of-origin effects that accompany some intracisternal A-particle (IAP) insertions (Duhl et al., 1994). Is the retrotransposition process prone to imprinting as part of genome defense against transposed elements (Barlow, 1993); do such mobile elements carry within them sufficient *cis*-acting sequences to confer imprinting; or does imprinting arise because they insert into a susceptible genomic region? A number of singleton imprinted genes are located within an intron of a biallelically expressed gene, for which the term "micro-imprinted" domain has been coined (Evans et al., 2001). In each case, the imprinted gene is paternally expressed and associated with a maternally methylated DMR (Nabetani et al., 1997; Evans et al., 2001; Smith et al., 2003).

It may also be the case that some imprinted genes appear to be isolated because the imprinting status of neighboring genes remains to be properly ascertained. In the case of *Igf2r*, one of the first imprinted genes identified, recent investigations have

178

Table 2. Imprinting control elements and mechanisms of monoallelic expression

Locus	Methylated allele	Deletion phenotype(s)	Genes affected	Mechanism	References
H19-DMD	Paternal	Maternal	De-repression of *Igf2*	Methylation-sensitive block to promoter:enhancer interactions; silencer of *H19* promoter	Thorvaldsen et al., 1998
		Paternal	De-repression of *H19* (tissue-restricted)		Bell and Felsenfeld, 2000
					Hark et al., 2000
					Drewell et al., 2000
Igf2r	Maternal	Maternal	none	Promoter of antisense transcript *Air*; mechanism of bi-directional action unclear	Wutz et al., 2001
		Paternal	De-repression in *cis* of *Igf2r, Slc22a2, Slc22a3*; effect bidirectional		Zwart et al., 2001
KvDMR1	Maternal	Maternal	none	Promoter of antisense transcript *Kcnq1ot1*; mechanism of bi-directional action unclear	Fitzpatrick et al., 2002
		Paternal	De-repression in *cis* of *Phlda2* (*Tssc3*), *Slc22a18* (*Scl22a1l*), *Cdkn1c, Kcnq1, Tssc4, Ascl2*; effect bidirectional		
Snrpn	Maternal	Maternal	Increased expression of *Ube3a*?	Promoter of transcript including *Ipw, Ube3a-as* and other non-coding RNAs; mechanism of effect on upstream genes unclear	Bielinska et al., 2000
		Paternal	Loss of expression in *cis* of *Zfp127, Ndn, Ipw*; effect bidirectional		Bressler et al., 2001
Gtl2 IG-DMR	Paternal	Maternal	De-repression in *cis* of *Dlk1, Rtl1, Dio3*; effect bidirectional	Possible promoter of paternal non-coding transcripts; mechanism of bi-directional action unclear	Lin et al., 2003
		Paternal	none		
Rasgrf1	Paternal	Maternal	none	Direct repeats necessary for sperm-derived methylation; expression mechanism unclear	Yoon et al., 2002
		Paternal	Loss of *Rasgrf1* expression in *cis*		

revealed that two genes *(Slc22a2, Slc22a3)* over 100 kb 3′ to *Igf2r* are also monoallelically expressed, at least in placenta (Zwart et al., 2001). The wealth of available genomic sequence information, including databases of non-coding and antisense transcripts with a potential role in regulation (Kiyosawa et al., 2003), should accelerate the process of determining the extent of imprinted gene clusters.

Methylation marks at imprinted genes and the properties of imprinting control elements

Genetic analysis from human imprinted gene syndromes such as the Prader-Willi and Angelman syndromes (PWS, AS) revealed that a single *cis*-acting element can be responsible for setting the epigenetic properties of an imprinted gene cluster (Buiting et al., 1995). In the mouse, knock-out experiments have defined imprinting control elements (ICEs) at six loci (Table 2; Thorvaldsen et al., 1998; Bielinska et al., 2000; Wutz et al., 2001; Fitzpatrick et al., 2002; Yoon et al., 2002; Lin et al., 2003). Fundamental to imprinting is the marking of the ICE specifically in one germline: such marks must be maintained in the embryo and lead to alternative states of gene activity of the maternal and paternal chromosomes. An ICE may be local and affect the activity of a single associated gene, or direct the imprinting of multiple genes distributed over hundreds of kilobases, and can act bidirectionally (Zwart et al., 2001; Fitzpatrick et al., 2002; Lin et al., 2003). A key component of the mark is DNA methylation at cytosine in CpG dinucleotides, and the six proven ICEs are DMRs with methylation present in sperm or oocyte (germline DMRs). A larger number of DMRs has been shown to have gametic methylation differences, which mark them as candidate ICEs. There is a marked asymmetry in that the majority of germline DMRs have oocyte-derived methylation (Reik and Walter, 2001b), perhaps a reflection of the

comprehensive and active demethylation of the sperm-derived chromosomes in the fertilized oocyte (Oswald et al., 2000; Santos et al., 2002). It is likely that DNA methylation is not acting in isolation and that chromatin modifications are an important factor in determining DNA methylation patterns, but in this review we confine ourselves to DNA methylation.

CpG islands and direct repeats

At the sequence level, germline DMRs coincide with CpG islands or CpG island-like elements, but methylation of CpG islands is an unusual property shared almost exclusively by imprinted genes and genes on the inactive X chromosome. In the case of germline methylation at imprinted genes, it poses the additional challenge that the mutability of methyl-cytosine would be expected to lead to erosion of CpG content. In addition to germline DMRs, there are CpG islands within imprinted gene clusters which either remain unmethylated, or acquire methylation only after fertilization (Yatsuki et al., 2002; Coombes et al., 2003), and these can exist in close juxtaposition with germline DMRs (Liu et al., 2000). What unique sequence attributes of these CpG islands direct or permit their germline-specific methylation? As intimated above, identification of unifying features of ICEs could assist in silico identification of imprinted genes. One possible distinguishing feature is the presence of direct repeats, first proposed on the basis of a small set of imprinted genes, imprinted transgenes and IAP insertions resulting in parent-of-origin effects on associated genes (Neumann et al., 1995). Subsequent analysis has shown that all ICEs and most germline DMRs have direct repeat content (Frevel et al., 1999; Mitsuya et al., 1999; Reinhart et al., 2002; Takada et al., 2002; Yoon et al., 2002), but this deserves reexamination with the availability of full genome sequence. Despite the frequent association, direct repeats do not seem to provide a common sequence motif that acts as a methylation signal (repeat sequences are unrelated, differ in length, organization and copy number), instead they may confer structural organization to the DNA which is recognized by the methylation machinery. Repeats are associated with both sperm- and oocyte-derived DMRs, therefore, maternal or paternal germline specificity may be deemed to reside elsewhere. There are some provocative associations. The imprinted mouse gene *Impact* has an intronic CpG island with direct repeat content, whereas the human homologue is not imprinted and contains an unrelated CpG island devoid of repeats (Okamura et al., 2000). A similar correlation exists amongst rodents with imprinting of the *Rasgrf1* gene (Pearsall et al., 1999). There are also exceptions: at the mouse *Grb10* locus the germline DMR is an intronic CpG island with repeat content, whilst the human GRB10 gene (which has more restricted monoallelic expression) has a homologous CpG island lacking repeats but nevertheless has maternal allele methylation (Arnaud et al., 2003; Hikichi et al., 2003).

The role of direct repeats needs to be tested experimentally. At *Rasgrf1*, gene targeting showed that the direct repeat block (which is itself unmethylated) is required for sperm-derived methylation of the DMR (Yoon et al., 2002). In contrast, a G-rich direct repeat downstream of the *H19* DMR is dispensible for imprinting (Reed et al., 2001; Thorvaldsen et al., 2002). Deletions may not necessarily distinguish between effects of repeats and the role of the sequence they occupy, therefore, transfer of repeats to ectopic locations may be a more definitive assay. This has been done in the context of the *RSVIgmyc* transgene, a synthetic construction which displays an imprinted effect on methylation irrespective of site of integration or copy number of the transgene. The transgene includes some reiterated sequences of pBR322 and RSV provenance. Direct repeats from the *Air/Igf2r* ICE are able to replace these reiterated elements and confer methylation to the transgene on maternal transmission (Reinhart et al., 2002). Paradoxically, the *Igf2r/Air* ICE when used as a transgene by itself fails to adopt imprinted methylation (Sleutels and Barlow, 2001). Thus, whether direct repeats are sufficient to direct methylation patterns remains to be determined.

ICEs are complex elements: acquiring germline-specific methylation is only the first role they need to fulfill. Therefore, there may be other reasons for a preponderance of direct repeats, such as providing multiple, methylation-regulated binding sites for transcription or insulator factors, such as CTCF at the *H19* DMR (Bell and Felsenfeld, 2000; Hark et al., 2000; Szabó et al., 2000). An alternative explanation for GC-rich direct repeats arises from the fact that methylated cytosine is susceptible to deamination leading to replacement by thymidine (the most frequent base change in the mammalian genome). Duplication of sequences within DMRs may offer one mechanism by which methyl-CpG density can be restored if methylation of the region is critical for monoallelic silencing of a gene.

Remote elements acting on ICEs

There is evidence that ICEs may not be autonomous elements, but rely on sometimes quite remote sequences for appropriate acquisition of methylation. Imprinting mutations arising from microdeletions in PWS and AS led to the concept of a bipartite imprinting center (Buiting et al., 1995; Nicholls and Knepper, 2001). The SNURF-SNRPN exon 1 DMR carries the maternal germline methylation mark (Geuns et al., 2003), but microdeletions leading to AS reveal that a small region 35 kb upstream is necessary for establishing the germline mark at the DMR. The upstream element itself is not differentially methylated (Schumacher et al., 1998). In the imprinted endocrine disorder pseudohypoparathyroidism type 1b (PHP1b), a consistent finding has been loss of methylation of the exon A/B DMR at the GNAS1 locus (Bastepe et al., 2001). A common cause of this methylation defect has now been identified as a 3-kb microdeletion 220 kb upstream of the DMR. Again, the region subject to deletion does not have differential methylation (Bastepe et al., 2003). In neither case is it yet clear how these remote elements promote methylation at the DMRs; in PHP1b the specificity of the element is also remarkable, as methylation at a more proximal DMR at XLαS is not influenced (Bastepe et al., 2003). The necessity of long-range elements may account for the frequent failure of short ICE transgenes to recapitulate imprinting, as in the case of *Igf2r/Air* (Sleutels and Barlow, 2001).

Factors marking ICEs

Existing imprints must be erased during passage through the germline and then reset in accordance with germline sex. Erasure occurs in primordial germ cells by embryonic day 11.5, concomitant with migration into the genital ridges (Szabó and Mann, 1995; Hajkova et al., 2002; Lee et al., 2002). Competence for imprinting is re-established differentially in the male and female germlines. In oocytes, it is acquired during the postnatal growth phase with different imprinted clusters becoming imprinted at different times (Obata et al., 1998; Obata and Kono, 2002). Limited analysis of the *Snrpn* DMR indicates that methylation is established in this time window (Lucifero et al., 2002). Paternal methylation imprints (as evidenced for the *H19* DMR) are laid down in mitotically arrested prospermatogonia, between 13.5 and 18.5 days post-coitum (Davis et al., 2000).

The activities responsible for marking ICEs have been defined, at least in part. A key factor is the Dnmt3l protein. Embryos from female mice homozygous for *Dnmt3l* null alleles arrest by embryonic day 10.5: oocyte-derived DMRs are not methylated, resulting in deregulated expression of associated imprinted genes, whilst methylation of the remainder of the genome is apparently unaffected (Bourc'his et al., 2001; Hata et al., 2002). A requirement of imprinted loci in the male germline for Dnmt3l is not yet clear. Although highly related to the de novo methyltransferases Dnmt3a and Dnmt3b, Dnmt3l lacks a functional methyltransferase domain. Dnmt3l may interact with one or both de novo enzymes, and germline methylation of DMRs seems to be imparted by Dnmt3a (Hata et al., 2002). Furthermore, Dnmt3l can stimulate the activity of Dnmt3a in cultured cells (Chédin et al., 2002), but how target sequence specificity is conferred remains to be shown.

A global failure to establish methylation marks in the female germline has been found in humans, in a condition known as biparental complete hydatidiform mole (BiCHM; Judson et al., 2002). Complete hydatidiform moles are normally sporadic and have an androgenetic genome, but in BiCHM affected females have recurrent molar pregnancies with maternal and paternal chromosome sets. Linkage analysis has assigned one BiCHM locus to chromosome 19q13 (Moglabey et al., 1999; Hodges et al., 2003). Molecular analysis has excluded involvement of the human DNMT3L and other known methyltransferase loci (Hayward et al., 2003), implying the existence of factors independent of or co-operating with DNMT3L in conferring maternal germline methylation.

It is currently difficult to conclude whether the germline selectivity of DMRs resides in active targeting of a de novo methylation complex in one germline or its exclusion from the DMR sequence in the other. Apart from the above, inappropriate methylation of DMRs in response to lack of or impaired binding of factors has only affected methylation patterns postfertilization (Howell et al., 2001; Schoenherr et al., 2002; Mager et al., 2003).

Mechanisms of monoallelic expression

How does an imprint mark result in monoallelic expression of the associated gene(s)? DNA methylation can be present on the inactive or active allele; a single methylated element can regulate a single gene or a whole cluster of genes. Methylation, or other epigenetic modifications for which it is a marker (Fournier et al., 2002), can therefore have a direct or indirect effect on imprinted genes. Are unifying themes needed? If silencing of one allele is all that counts, how this is arrived at in molecular terms may be immaterial, and different mechanisms that translate germline imprints into monoallelic expression in somatic cells may have evolved at different genes. Despite this implied laxity, remarkable similarities in mechanisms of monoallelic expression are becoming apparent. The simplest mechanism is the "direct" silencing of the methylated copy because the DMR coincides with the promoter-associated CpG island. In this case, methylation may directly prevent binding of transcription activating factors or lead to recruitment of a repressive chromatin organization. Examples of this direct regulation may include *Plagl1/Zac1* (Arima et al., 2001; Varrault et al., 2001) and *U2af1-rs1* (Fournier et al., 2002). This simple silencing also lies at the heart of the indirect mechanisms involving regulatory non-coding RNAs. For indirect mechanisms, two paradigms have emerged (Fig. 1, Table 2). Both involve non-coding RNA (in some capacity) and imply competition or reciprocity between an imprinting factor and protein-coding imprinted genes. This is an intriguing concept which resonates with the conflict theory of the evolution of imprinting, whereby paternally and maternally expressed imprinted genes have opposite effects on growth phenotypes. At a molecular level, growth suppressors are active only on the maternal allele because the paternal allele is silencing by non-coding transcripts, whose monoallelic expression is imposed by methylation of the maternal allele in the germline (Reik and Walter, 2001b). Below, we provide a brief overview of mechanisms; for a comprehensive recent review the reader is referred to Verona et al. (2003).

Chromatin boundary model
As the first example of linked imprinted genes (Zemel et al., 1992), the paternally expressed *Igf2* and maternally expressed non-coding RNA gene *H19* have been the object of more mechanistic investigations than any other imprinted loci (reviewed in Arney, 2003), and here we can provide little more than a cursory review. Their coordinate regulation but reciprocal imprinting has motivated a variety of competition models. The current model favors competition for *cis*-acting enhancers mediated by a chromatin boundary, but the regulation of an imprinted locus as functionally important as *Igf2* is likely to have complexity beyond this simple scheme (e.g., Constância et al., 2000; Arney, 2003). There is no compelling evidence that *H19* has a function other than acting as a regulatory decoy for *Igf2*. The ICE is a paternally methylated DMR ~ 2 kb upstream of the *H19* promoter (Tremblay et al., 1995; Thorvaldsen et al., 1998). On the unmethylated maternal allele the ICE provides multiple binding sites for the enhancer-blocking protein CTCF, which assembles a boundary element to prevent access of downstream enhancers to the *Igf2* promoters (Bell and Felsenfeld, 2000; Hark et al., 2000; Szabó et al., 2000). On the paternal allele, methylation inhibits CTCF binding, allowing the enhancers to act exclusively on the *Igf2* promoters. The DMR must also contain silencer function separable from the CTCF binding domain to cause repression in *cis* of the paternal *H19* promoter (Drewell et al., 2000). Key component of this model is the existence of enhancers capable of activating both *H19* and *Igf2*, with endoderm and mesoderm specificity, downstream of *H19* (Leighton et al., 1995; Kaffer et al., 2001; Davies et al., 2002).

No other imprinted locus has yet been shown in vivo to rely on a boundary element; however, this is an attractive mechanism for regulation at several other loci. CTCF sites have been widely sought in imprinted gene clusters, and in vitro evidence for CTCF binding and methylation-sensitive silencer or insulator function has been found (e.g., at *Grb10*, Hikichi et al., 2003; at KCNQ1OT1, Du et al., 2003). At the *Peg3* locus, reiterated methylation-sensitive binding sites for the factor YY1 coinciding with insulator elements have been found (Kim et al., 2003). YY1 binding specifically to the unmethylated paternal allele has been revealed by chromatin immunoprecipitation assays.

Regulation by antisense transcription
The concept of antisense regulation of imprinting emerged from pioneering studies of the mouse *Igf2r* locus. The ICE for *Igf2r*, an intronic CpG island methylated on the maternal allele, was identified as the start site of a large, unspliced and non-coding transcript running antisense to *Igf2r*, which has been termed *Air* (Lyle et al., 2000). Deletion of the DMR containing the *Air* promoter and, more tellingly, a mutation causing premature polyadenylation of the *Air* transcript, result in derepression of the paternal allele of *Igf2r* and failure to methylate the *Igf2r* promoter CpG island (Wutz et al., 2001; Sleutels et al., 2002). However, contrary to facile models of promoter inference, monoallelic expression of the two imprinted genes *Slc22a2* and *Slc22a3* downstream of *Igf2r* is also dependent on an intact *Air* promoter and transcript, although *Air* does not traverse these genes (Sleutels et al., 2002). Therefore, the mechanism of *Air*-dependent silencing remains obscure. If the non-

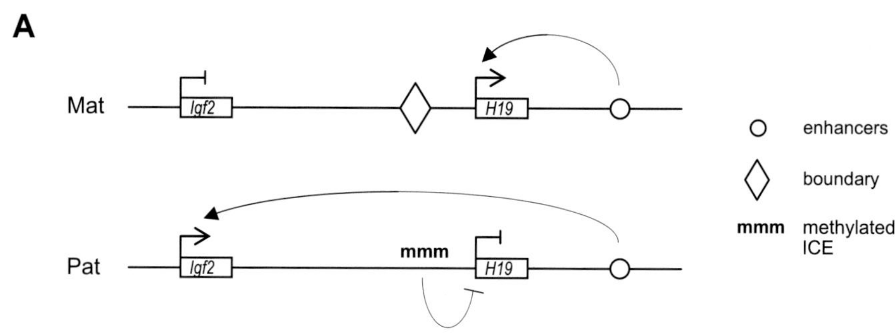

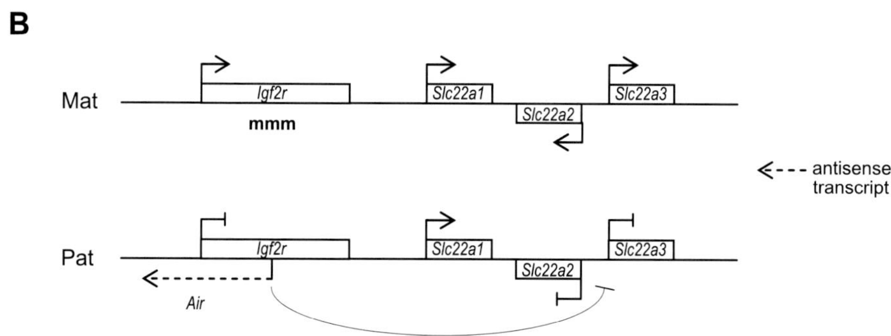

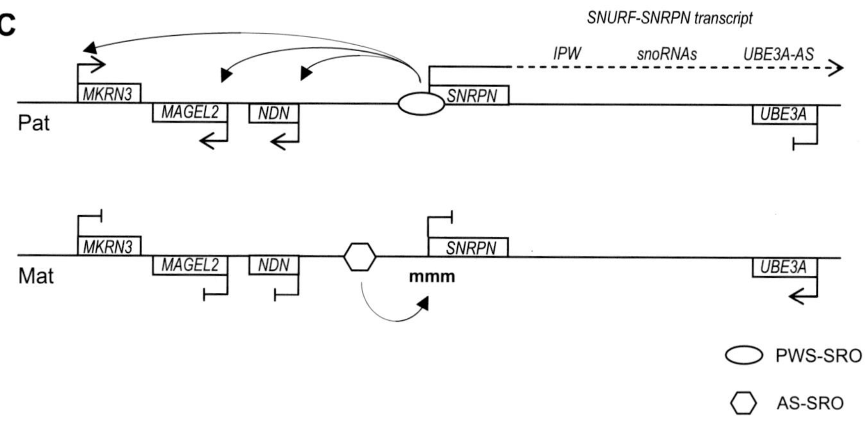

Fig. 1. Models of monoallelic expression exemplified at three imprinted gene clusters. (**A**) The *Igf2/H19* locus. The ICE comprises a methylation-sensitive chromatin boundary that prevents downstream enhancers from activating the *Igf2* promoters. Repression of the paternal *H19* promoter may also require action of a local silencer. (**B**) The *Igf2r/Air* locus. Transcription of the antisense RNA *Air* is associated with silencing of the paternal *Igf2r* promoter. *Air* is also able to silence the downstream *Slc22a2* and *Slc22a3* loci in *cis*. (**C**) The PWS/AS domain. The PWS-SRO contains the promoter of the long SNURF-SNRPN transcript, which may also encode the UBE3A-AS transcript associated with silencing of the UBE3A gene. The PWS-SRO also sets the epigenetic properties of the upstream region to ensure activity of the three upstream genes. The AS-SRO is required for maternal germline methylation of the SNURF-SNRPN exon 1 DMR.

coding RNA itself is the silencing factor, it will need to be explained why its effect is restricted to the allele in *cis*, but is not limited to promoters overlapped by *Air*, and why one gene in the cluster *(Slc22a1)* escapes silencing. It remains possible that different but interdependent mechanisms are responsible for controlling genes overlapped or not by *Air*.

Non-coding antisense transcripts have been identified at numerous other imprinted loci (Rougeulle et al., 1998; Lee et al., 1999; Smilinich et al., 1999; Wroe et al., 2000). Analogous to the case of *Air*, deletion of the *KvDMR1*, which ablates the *Kcnq1ot1/Lit1* antisense transcript, results in bidirectional loss of imprinting of six neighboring genes (Fitzpatrick et al., 2002). The collation of a database of non-coding and antisense transcripts (Kiyosawa et al., 2003) should accelerate the discovery of these potentially regulating RNAs, and may help pinpoint new imprinted gene clusters.

Long-range regulation

The most enigmatic imprinting cluster is the 2–3-Mb PWS and AS domain on human chromosome 15q11→q13. A variety of genetic anomalies give rise to these disorders (Nicholls and Knepper, 2001), which in the case of PWS result in the lack of expression of paternal gene products (PWS may be a contiguous gene syndrome), whilst AS results from deficiency of the UBE3A gene, which has brain-specific maternal allele expression. For either syndrome small subsets of patients have been described as having imprinting mutations, in that they have biparental inheritance of the 15q11→q13 interval. Many such cases are attributed to microdeletions, paternally transmitted in PWS and maternally transmitted in AS (Buiting et al., 1995): the respective shortest regions of overlap (SRO) are separable, implying the existence of a bipartite element controlling the epigenetic status of the entire domain. The PWS-SRO contains a germline DMR with maternal methylation coinciding with exon 1 of the paternally expressed SNURF-SNRPN gene (Geuns et al., 2003). SNURF-SNRPN is a long transcript with a complex splicing pattern and serves as a host transcript for snoRNAs and other non-protein-coding RNAs (Runte et al., 2001). At its 3′ end, it overlaps in the antisense orientation with the UBE3A gene, and may control monoallelic expression of UBE3A (Yamasaki et al., 2003). However, PWS-SRO deletions result not only in loss of all paternal RNAs derived from the SNURF-SNRPN transcript, but also cause loss of expression and inappropriate paternal allele methylation of the paternally expressed protein-coding genes MKRN3, MAGEL2 and NDN located far upstream of SNURF-SNRPN. The cause of this epigenotype switch is not understood and, despite the diversity of PWS causing mutations, experiments to elucidate a mechanism may need to be done in mice. A number of knock-outs have been made, which indicate that deletions need to include more than just the mouse *Snrpn* promoter for partial or complete silencing of upstream genes (Bielinska et al., 2000; Bressler et al., 2001). The AS-SRO is currently defined as a <1-kb region 35 kb upstream of SNURF-SNRPN exon 1 (Buiting et al., 1999). As noted earlier, it is required for establishing the maternal epigenetic profile of the domain, possibly by determining maternal methylation at SNURF-SNRPN exon 1, and ultimately for ensuring maternal allele expression of UBE3A.

Another example highlights the possible involvement of unconventional RNAs. The *Dlk1-Gtl2*-imprinted cluster on mouse Chromosome 12 comprises three paternally expressed protein-coding genes *(Dlk1, Rtl1* and *Dio3)* distributed over ~ 1 Mb, interspersed with maternally expressed non-protein-coding transcripts (such as *Gtl2*, snoRNAs and microRNAs). The non-coding transcripts may belong to a single transcription unit, as the RNAs are transcribed in the same orientation and individual start sites have not been determined. The controlling element, the IG-DMR, lies between *Dlk1* and *Gtl2* and has methylation on the paternal allele (Takada et al., 2002). Deletion of the IG-DMR from the maternal allele results in activation of the *Dlk1, Rtl1* and *Dio3* genes and silencing of the non-protein-coding transcripts, conferring on the maternal allele the epigenetic properties and transcriptional profile of the paternal allele (Lin et al., 2003). It is possible that this locus is also regulated by a mechanism involving silencing by maternal non-coding RNAs, operating bidirectionally akin to *Air*, whose activity depends on the unmethylated IG-DMR.

Acknowledgements

We thank Wolf Reik for valuable comments on this review.

References

Arima T, Drewell RA, Arney KL, Inoue J, Makita Y, et al: A conserved imprinting control region at the *HYMAI/ZAC* domain is implicated in transient neonatal diabetes mellitus. Hum Mol Genet 10:1475–1483 (2001).

Arnaud P, Monk D, Hitchins M, Gordon E, Dean W, et al: Conserved methylation imprints in the human and mouse GRB10 genes with divergent allelic expression suggests differential reading of the same mark. Hum Mol Genet 12:1005–1019 (2003).

Arney KL: *H19* and *Igf2* – enhancing the confusion? Trends Genet 19:17–23 (2003).

Barlow DP: Methylation and imprinting: from host defense to gene regulation? Science 260:309–310 (1993).

Barlow DP, Stöger R, Herrmann BG, Saito K, Scheifer N: The mouse insulin-like growth factor type-2 receptor is imprinted and closely linked to the *Tme* locus. Nature 349:84–87 (1991).

Barton SC, Surani MA, Norris ML: Role of paternal and maternal genomes in mouse development. Nature 311:374–376 (1984).

Bastepe M, Pincus JE, Sugimoto T, Tojo K, Kanatani M, et al: Positional dissociation between the genetic mutation responsible for pseudohypoparathyroidism type 1b and the associated methylation defect at exon A/B: evidence for a long-range regulatory element within the imprinted *GNAS1* locus. Hum Mol Genet 10:1231–1241 (2001).

Bastepe M, Frohlich LF, Hendy GN, Indridason OS, Josse RG, et al: Autosomal dominant pseudohypoparathyroidism type Ib is associated with a heterozygous microdeletion that likely disrupts a putative imprinting control element of *GNAS*. J Clin Invest 112:1255–1263 (2003).

Beechey CV, Cattanach BM, Blake A: MRC Mammalian Genetics Unit, Harwell, Oxfordshire. World Wide Web Site – Mouse Imprinting Data and References (http://www.mgu.har.mrc.ac.uk/imprinting/imprinting.html) (2003).

Bell A, Felsenfeld G: Methylation of a CTCF-dependent boundary controls imprinted expression of the *Igf2* gene. Nature 405:482–485 (2000).

Bielinska B, Blaydes SM, Buiting K, Yang T, Krajewska-Walasek M, et al: *De novo* deletions of *SNRPN* exon 1 in early human and mouse embryos result in a paternal to maternal imprint switch. Nat Genet 25:74–78 (2000).

Bourc'his D, Xu GL, Lin CS, Bollman B, Bestor TH: Dnmt3L and the establishment of maternal genomic imprints. Science 294:2536–2539 (2001).

Bressler J, Tsai TF, Wu MY, Tsai SF, Ramirez MA: The SNRPN promoter is not required for genomic imprinting of the Prader-Willi/Angelman domain in mice. Nat Genet 28:232–240 (2001).

Buiting K, Saitoh S, Gross S, Dittrich B, Schwartz S, et al: Inherited microdeletions in the Angelman and Prader-Willi syndromes define an imprinting centre on human chromosome 15. Nat Genet 9:395–400 (1995).

Buiting K, Lich C, Cottrell S, Barnicoat A, Horsthemke B: A 5-kb imprinting center deletion in a family with Angelman syndrome reduces the shortest region of deletion overlap to 880 bp. Hum Genet 105:665–666 (1999).

Cattanach BM, Beechey CV: Genomic imprinting in the mouse: possible final analysis, in Reik W, Surani A (eds): Genomic Imprinting, pp 118–145 (Oxford University Press, Oxford 1997).

Cattanach BM, Kirk M: Differential activity of maternally and paternally derived chromosome regions in mice. Nature 315:496–498 (1985).

Chédin F, Lieber MR, Hsieh CL: The DNA methyltransferase-like protein DNMT3L stimulates *de novo* methylation by Dnmt3a. Proc Natl Acad Sci USA 99:16916–16921 (2002).

Choi JD, Underkoffler LA, Collins JN, Marchegiani SM, Terry NA, et al: Microarray expression profiling of tissues from mice with uniparental duplications of Chromosomes 7 and 11 to identify imprinted genes. Mamm Genome 12:758–764 (2001).

Constância M, Dean W, Lopes S, Moore T, Kelsey G, Reik W: Deletion of a silencer element in *Igf2* results in loss of imprinting independent of *H19*. Nat Genet 26:203–206 (2000).

Constância M, Pickard B, Kelsey G, Reik W: Imprinting mechanisms. Genome Res 8:881–900 (1998).

Coombes C, Arnaud P, Gordon E, Dean W, Coar EA, et al: Epigenetic properties and identification of an imprint mark in the Nesp-Gnasxl domain of the mouse *Gnas* imprinted locus. Mol Cell Biol 23:5475–5488 (2003).

Davies K, Bowden L, Smith P, Dean W, Hill D, et al: Disruption of mesodermal enhancers for *Igf2* in the minute mutant. Development 129:1657–1668 (2002).

Davies W, Isles AR, Wilkinson LS: Imprinted genes and mental dysfunction. Ann Med 33:428–436 (2001).

Davis TL, Yang GJ, McCarrey JR, Bartolomei MS: The *H19* methylation imprint is erased and reestablished differentially on the parental alleles during male germ cell development. Hum Mol Genet 9:2885–2894 (2000).

Dean WL, Kelsey G, Reik W: Generation of monoparental embryos for investigations into genomic imprinting, in Ward A (ed): Genomic Imprinting Protocols, pp 1–19 (The Humana Press, Inc. Totowa, NJ 2001).

DeChiara TM, Robertson EJ, Efstratiadi A: Parental imprinting of the mouse insulin-like growth factor II gene. Cell 64:849–859 (1991).

Drewell RA, Brenton JD, Ainscough JF, Barton SC, Hilton KJ, et al: Deletion of a silencer element disrupts *H19* imprinting independently of a DNA methylation epigenetic switch. Development 127:3419–3428 (2000).

Du M, Beatty LG, Zhou W, Lew J, Schoenherr C, Weksberg R, Sadowski PD: Insulator and silencer sequences in the imprinted region of human chromosome 11p15.5. Hum Mol Genet 12:1927–1939 (2003).

Duhl DM, Vrieling H, Miller KA, Wolff GL, Barsh GS: Neomorphic *agouti* mutations in obese yellow mice. Nat Genet 8:59–65 (1994).

Engemann S, Strodicke M, Paulsen M, Franck O, Reinhardt R, et al: Sequence and functional comparison in the Beckwith-Wiedemann region: implications for a novel imprinting centre and extended imprinting. Hum Mol Genet 9:2691–2706 (2000).

Evans HK, Wylie AA, Murphy SK, Jirtle RL: The neuronatin gene resides in a "micro-imprinted" domain on human chromosome 20q11.2. Genomics 77:99–104 (2001).

Fitzpatrick GV, Soloway PD, Higgins MJ: Regional loss of imprinting and growth deficiency in mice with a targeted deletion of *KvDMR1*. Nat Genet 32:426–431 (2002).

Fournier C, Goto Y, Ballestar E, Delaval K, Hever AM, Esteller M, Feil R: Allele-specific histone lysine methylation marks regulatory regions at imprinted mouse genes. EMBO J 21:6560–6570 (2002).

Frevel MA, Hornberg JJ, Reeve AE: A potential imprint control element: identification of a conserved 42 bp sequence upstream of *H19*. Trends Genet 15:216–218 (1999).

Geuns E, De Rycke M, Van Steirteghem A, Liebaers I: Methylation imprints of the imprint control region of the *SNRPN*-gene in human gametes and preimplantation embryos. Hum Mol Genet 12:2873–2879 (2003).

Greally JM: Short interspersed transposable elements (SINEs) are excluded from imprinted regions in the human genome. Proc Natl Acad Sci USA 99:327–332 (2002).

Hagiwara Y, Hirai M, Nishiyama K, Kanazawa I, Ueda T, Sakaki Y, Ito T: Screening for imprinted genes by allelic message display: identification of a paternally expressed gene *Impact* on mouse chromosome 18. Proc Natl Acad Sci USA 94:9249–9254 (1997).

Hajkova P, Erhardt S, Lane N, Haaf T, El-Maarri O, et al: Epigenetic reprogramming in mouse primordial germ cells. Mech Dev 117:15–23 (2002).

Hark AT, Schoenherr CJ, Katz DJ, Ingram RS, Levorse JM, Tilghman SM: CTCF mediates methylation-sensitive enhancer-blocking activity at the *H19/Igf2* locus. Nature 405:486–489 (2000).

Hata K, Okano M, Lei H, Li E: Dnmt3L cooperates with the Dnmt3 family of de novo DNA methyltransferases to establish maternal imprints in mice. Development 129:1983–1993 (2002).

Hatada I, Sugama T, Mukai T: A new imprinted gene cloned by a methylation-sensitive genome scanning method. Nucleic Acids Res 21:5577–5582 (1993).

Hayashizaki Y, Shibata H, Hirotsune S, Sugino H, Okazaki Y, et al: Identification of an imprinted U2af binding protein related sequence on mouse chromosome 11 using the RLGS method. Nat Genet 6:33–40 (1994).

Hayward BE, Kamiya M, Strain L, Moran V, Campbell R, Hayashizaki Y, Bonthron DT: The human *GNAS1* gene is imprinted and encodes distinct paternally and biallelically expressed G proteins. Proc Natl Acad Sci USA 95:10038–10043 (1998).

Hayward BE, De Vos M, Judson H, Hodge D, Huntriss J, et al: Lack of involvement of known DNA methyltransferases in familial hydatidiform mole implies the involvement of other factors in establishment of imprinting in the human female germline. BMC Genet 4:2 (2003).

Henry I, Jeanpierre M, Couillin P, Barichard F, Serre JL, et al: Molecular definition of the 11p15.5 region involved in Beckwith-Wiedemann syndrome and probably in predisposition to adrenocortical carcinoma. Hum Genet 81:273–277 (1989).

Hikichi T, Kohda T, Kaneko-Ishino T, Ishino F: Imprinting regulation of the murine *Meg1/Grb10* and human *GRB10* genes; roles of brain-specific promoters and mouse-specific CTCF-binding sites. Nucleic Acids Res 31:1398–1406 (2003).

Hodges MD, Rees HC, Seckl MJ, Newlands ES, Fisher RA: Genetic refinement and physical mapping of a biparental complete hydatidiform mole locus on chromosome 19q13.4. J Med Genet 40:e95 (2003).

Howell CY, Bestor TH, Ding F, Latham KE, Mertineit C, Trasler JM, Chaillet JR: Genomic imprinting disrupted by a maternal effect mutation in the *Dnmt1* gene. Cell 104:829–838 (2001).

Hurst LD: Evolutionary theories of genomic imprinting, in Reik W, Surani A (eds): Genomic Imprinting, pp 211–237 (Oxford University Press, Oxford 1997).

Judson H, Hayward BE, Sheridan E, Bonthron DT: A global disorder of imprinting in the human female germ line. Nature 416:539–542 (2002).

Kaffer CR, Grinberg A, Pfeifer K: Regulatory mechanisms at the mouse Igf2/H19 locus. Mol Cell Biol 21:8189–8196 (2001).

Kagitani F, Kuroiwa Y, Wakana S, Shiroishi T, Miyoshi N, et al: *Peg5/Neuronatin* is an imprinted gene located on sub-distal chromosome 2 in the mouse. Nucleic Acids Res 25:3428–3432 (1997).

Kamiya M, Judson H, Okazaki Y, Kusakabe M, Muramatsu M, et al: The cell cycle control gene *ZAC/PLAGL1* is imprinted-a strong candidate gene for transient neonatal diabetes. Hum Mol Genet 9:453–460 (2000).

Kaneko-Ishino T, Kuroiwa Y, Miyoshi N, Kohda T, Suzuki R, et al: *Peg1/Mest* imprinted gene on chromosome 6 identified by cDNA subtraction hybridization. Nat Genet 11:52–59 (1995).

Ke X, Thomas NS, Robinson DO, Collins A: The distinguishing sequence characteristics of mouse imprinted genes. Mamm Genome 13:639–645 (2002a).

Ke X, Thomas NS, Robinson DO, Collins A: A novel approach for identifying candidate imprinted genes through sequence analysis of imprinted and control genes. Hum Genet 111:511–520 (2002b).

Kelsey G, Bodle D, Miller HJ, Beechey CV, Coombes C, Peters J, Williamson CM: Identification of imprinted loci by methylation-sensitive representational difference analysis: Application to mouse distal chromosome 2. Genomics 62:129–138 (1999).

Kikyo N, Williamson CM, John RM, Barton SC, Beechey CV, et al: Genetic and functional analysis of *neuronatin* in mice with maternal or paternal duplication of distal Chr 2. Dev Biol 190:66–77 (1997).

Kim J, Kollhoff A, Bergmann A, Stubbs L: Methylation-sensitive binding of transcription factor YY1 to an insulator sequence within the paternally expressed imprinted gene, *Peg3*. Hum Mol Genet 12:233–245 (2003).

Kiyosawa H, Yamanaka I, Osato N, Kondo S, Hayashizaki Y, et al: Antisense transcripts with FANTOM2 clone set and their implications for gene regulation. Genome Res 13:1324–1334 (2003).

Kobayashi S, Wagatsuma H, Ono R, Ichikawa H, Yamazaki M, et al: Mouse *Peg9/Dlk1* and human *PEG9/DLK1* are paternally expressed imprinted genes closely located to the maternally expressed imprinted genes: mouse *Meg3/Gtl2* and human *MEG3*. Genes Cells 5:1029–1037 (2000).

Kobayashi S, Kohda T, Ichikawa H, Ogura A, Ohki M, Kaneko-Ishino T, Ishino F: Paternal expression of a novel imprinted gene, *Peg12/Frat3*, in the mouse 7C region homologous to the Prader-Willi syndrome region. Biochem Biophys Res Commun 290:403–408 (2002).

Kuroiwa Y, Kaneko-Ishino T, Kagitani F, Kohda T, Li LL, et al: *Peg3* imprinted gene on proximal chromosome 7 encodes for a zinc finger protein. Nat Genet 12:186–190 (1996).

Lee J, Inoue K, Ono R, Ogonuki N, Kohda T, et al: Erasing genomic imprinting memory in mouse clone embryos produced from day 11.5 primordial germ cells. Development 129:1807–1817 (2002).

Lee MP, DeBraun MR, Mitsuya K, Galonek HL, Brandenburg S, Oshimura M, Feinberg AP: Loss of imprinting of a paternally expressed transcript, with antisense orientation to KVLQT1, occurs frequently in Beckwith-Wiedemann syndrome and is independent of insulin-like growth factor II imprinting. Proc Natl Acad Sci USA 96:5203–5208 (1999).

Leighton PA, Saam JR, Ingram RS, Stewart CL, Tilghman SM: An enhancer deletion affects both *H19* and *Igf2* expression. Genes Dev 9:2079–2089 (1995).

Lin SP, Youngson N, Takada S, Seitz H, Reik W, et al: Asymmetric regulation of imprinting on the maternal and paternal chromosomes at the *Dlk1-Gtl2* imprinted cluster on mouse chromosome 12. Nat Genet 35:97–102 (2003).

Liu J, Yu S, Litman D, Chen W, Weinstein LS: Identification of a methylation imprint mark within the mouse *Gnas* locus. Mol Cell Biol 20:5808–5817 (2000).

Lo HS, Wang Z, Hu Y, Yang HH, Gere S, Buetow KH, Lee MP: Allelic variation in gene expression is common in the human genome. Genome Res 13:1855–1862 (2003).

Lucifero D, Mertineit C, Clarke HJ, Bestor TH, Trasler JM: Methylation dynamics of imprinted genes in mouse germ cells. Genomics 79:530–538 (2002).

Lyle R, Watanabe D, te Vruchte D, Lerchner W, Smrzka OW, et al: The imprinted antisense RNA at the *Igf2r* locus overlaps but does not imprint *Mas1*. Nat Genet 25:19–21 (2000).

Mager J, Montgomery ND, de Villena FP, Magnuson T: Genome imprinting regulated by the mouse Polycomb group protein Eed. Nat Genet 33:502–507 (2003).

McGrath J, Solter D: Completion of mouse embryogenesis requires both the maternal and paternal genomes. Cell 37:179–183 (1984).

Mitsuya K, Meguro M, Lee MP, Katoh M, Schulz TC, et al: *LIT1*, an imprinted antisense RNA in the human *KvLQT1* locus identified by screening for differentially expressed transcripts using monochromosomal hybrids. Hum Mol Genet 8:1209–1217 (1999).

Miura K, Miyoshi O, Yun K, Inazawa J, Miyamoto T, et al: Repeat-directed isolation of a novel gene preferentially expressed from the maternal allele in human placenta. J Hum Genet 44:1–9 (1999).

Miyoshi N, Kuroiwa Y, Kohda T, Shitara H, Yonekawa H, et al: Identification of the *Meg1/Grb10* imprinted gene on mouse proximal chromosome 11, a candidate for the Silver-Russell syndrome gene. Proc Natl Acad Sci USA 95:1102–1107 (1998).

Miyoshi N, Wagatsuma H, Wakana S, Shiroishi T, Nomura M, et al: Identification of an imprinted gene, *Meg3/Gtl2* and its human homologue *MEG3*, first mapped on mouse distal chromosome 12 and human chromosome 14q. Genes Cells 5:211–220 (2000).

Mizuno Y, Sotomaru Y, Katsuzawa Y, Kono T, Meguro M, et al: *Asb4*, *Ata3*, and *Dcn* are novel imprinted genes identified by high-throughput screening using RIKEN cDNA microarray. Biochem Biophys Res Commun 290:1499–505 (2002).

Moglabey YB, Kircheisen R, Seoud M, El Mogharbel N, Van den Veyver I, Slim R: Genetic mapping of a maternal locus responsible for familial hydatidiform moles. Hum Mol Genet 8:667–671 (1999).

Moore T, Haig D: Genomic imprinting in mammalian development: a parental tug-of-war. Trends Genet 7:45–49 (1991).

Morison IM, Paton CJ, Cleverley SD: The imprinted gene and parent-of-origin effect database. Nucleic Acids Res 29:275–276 (2001).

Nabetani A, Hatada I, Morisaki H, Oshimura M, Mukai T: Mouse *U2af1-rs1* is a neomorphic imprinted gene. Mol Cell Biol 17:789–798 (1997).

Neumann B, Kubicka P, Barlow DP: Characteristics of imprinted genes. Nat Genet 9:12–13 (1995).

Nicholls RD, Knepper JL: Genome organization, function, and imprinting in Prader-Willi and Angelman syndromes. Annu Rev Genomics Hum Genet 2:153–175 (2001).

Nikaido I, Saito C, Mizuno Y, Meguro M, Bono H, et al: Discovery of imprinted transcripts in the mouse transcriptome using large-scale expression profiling. Genome Res 13:1402–1409 (2003).

Obata Y, Kono T: Maternal primary imprinting is established at a specific time for each gene throughout oocyte growth. J Biol Chem 277:5285–5289 (2002).

Obata Y, Kaneko-Ishino T, Koide T, Takai Y, Ueda T, et al: Disruption of primary imprinting during oocyte growth leads to the modified expression of imprinted genes during embryogenesis. Development 125:1553–1560 (1998).

Okamura K, Hagiwara-Takeuchi Y, Li T, Vu TH, Hirai M, et al: Comparative genome analysis of the mouse imprinted gene impact and its nonimprinted human homolog IMPACT: toward the structural basis for species-specific imprinting. Genome Res 10:1878–1889 (2000).

Onyango P, Miller W, Lehoczky J, Leung CT, Birren B, et al: Sequence and comparative analysis of the mouse 1-megabase region orthologous to the human 11p15 imprinted domain. Genome Res 10:1697–1710 (2000).

Oswald J, Engemann S, Lane N, Mayer W, Olek A, et al: Active demethylation of the paternal genome in the mouse zygote. Curr Biol 10:475–478 (2000).

Paulsen M, El-Maarri O, Engemann S, Strodicke M, Franck O, et al: Sequence conservation and variability of imprinting in the Beckwith-Wiedemann syndrome gene cluster in human and mouse. Hum Mol Genet 9:1829–1841 (2000).

Pearsall RS, Plass C, Romano MA, Garrick MD, Shibata H, Hayashizaki Y, Held WA: A direct repeat sequence at the *Rasgrf1* locus and imprinted expression. Genomics 55:194–201 (1999).

Peters J, Wroe SF, Wells CA, Miller HJ, Bodle D, et al: A cluster of novel oppositely imprinted transcripts at the *Gnas* locus in the distal imprinting region of mouse chromosome 2. Proc Natl Acad Sci USA 96:3830–3835 (1999).

Piras G, El Kharroubi A, Kozlov S, Escalante-Alcalde D, Hernandez L, et al: *Zac1 (Lot1)*, a potential tumor suppressor gene, and the gene for epsilon-sarcoglycan are maternally imprinted genes: identification by a subtractive screen of novel uniparental fibroblast lines. Mol Cell Biol 20:3308–3315 (2000).

Plass C, Shibata H, Kalcheva I, Mullins L, Kotelevtseva N, et al: Identification of *Grf1* on mouse chromosome 9 as an imprinted gene by RLGS-M. Nat Genet 14:106–109 (1996).

Reed MR, Riggs AD, Mann JR: Deletion of a direct repeat element has no effect on *Igf2* and *H19* imprinting. Mamm Genome 12:873–876 (2001).

Reik W, Walter J: Genomic imprinting: parental influence on the genome. Nat Rev Genet 2:21–32 (2001a).

Reik W, Walter J: Evolution of imprinting mechanisms: the battle of the sexes begins in the zygote. Nat Genet 27:255–256 (2001b).

Reinhart B, Eljanne M, Chaillet JR: Shared role for differentially methylated domains of imprinted genes. Mol Cell Biol 22:2089–2098 (2002).

Rougeulle C, Cardoso C, Fontes M, Colleaux L, Lalande M: An imprinted antisense RNA overlaps *UBE3A* and a second maternally expressed transcript. Nat Genet 19:15–16 (1998).

Runte M, Huttenhofer A, Gross S, Kiefmann M, Horsthemke B, Buiting K: The IC-*SNURF-SNRPN* transcript serves as a host for multiple small nucleolar RNA species and as an antisense RNA for *UBE3A*. Hum Mol Genet 10:2687–2700 (2001).

Sandell LL, Guan XJ, Ingram R, Tilghman SM: Gatm, a creatine synthesis enzyme, is imprinted in mouse placenta. Proc Natl Acad Sci USA 100:4622–4627 (2003).

Santos F, Hendrich B, Reik W, Dean W: Dynamic reprogramming of DNA methylation in the early mouse embryo. Dev Biol 241:172–82 (2002).

Schmidt JV, Matteson PG, Jones BK, Guan XJ, Tilghman SM: The *Dlk1* and *Gtl2* genes are linked and reciprocally imprinted. Genes Dev 14:1997–2002 (2000).

Schoenherr CJ, Levorse JM, Tilghman SM: CTCF maintains differential methylation at the *Igf2/H19* locus. Nat Genet 33:66–69 (2002).

Schumacher A, Buiting K, Zeschnigk M, Doerfler W, Horsthemke B: Methylation analysis of the PWS/AS region does not support an enhancer-competition model. Nat Genet 19:324–325 (1998).

Searle AG, Beechey CV: Genome imprinting phenomena on mouse chromosome 7. Genet Res 56:237–244 (1990).

Sleutels F, Barlow DP: Investigation of elements sufficient to imprint the mouse *Air* promoter. Mol Cell Biol 21:5008–5017 (2001).

Sleutels F, Zwart R, Barlow DP: The non-coding *Air* RNA is required for silencing autosomal imprinted genes. Nature 415:810–813 (2002).

Smilinich NJ, Day CD, Fitzpatrick GV, Caldwell GM, Lossie AC, et al: A maternally methylated CpG island in *KvLQT1* is associated with an antisense paternal transcript and loss of imprinting in Beckwith-Wiedemann syndrome. Proc Natl Acad Sci USA 96:8064–8069 (1999).

Smiraglia DJ, Fruhwald MC, Costello JF, McCormick SP, Dai Z, et al: A new tool for the rapid cloning of amplified and hypermethylated human DNA sequences from restriction landmark genome scanning gels. Genomics 58:254–262 (1999).

Smith RJ, Arnaud P, Konfortova G, Dean WL, Beechey CV, Kelsey G: The mouse *Zac1* locus: basis for imprinting and comparison with human *ZAC*. Gene 292:101–112 (2002).

Smith RJ, Dean W, Konfortova G, Kelsey G: Identification of novel imprinted genes in a genome-wide screen for maternal methylation. Genome Res 13:558–569 (2003).

Sotomaru Y, Kawase Y, Ueda T, Obata Y, Suzuki H, et al: Disruption of imprinted expression of U2afbp-rs/U2af1-rs1 gene in mouse parthenogenetic fetuses. J Biol Chem 276:26694–26698 (2001).

Strichman-Almashanu LZ, Lee RS, Onyango PO, Perlman E, Flam F, Frieman MB, Feinberg AP: A genome-wide screen for normally methylated human CpG islands that can identify novel imprinted genes. Genome Res 12:543–554 (2002).

Surani MA: Imprinting and the initiation of gene silencing in the germ line. Cell 93:309–312 (1998).

Surani MA, Barton SC, Norris ML: Development of reconstituted mouse eggs suggests imprinting of the genome during gametogenesis. Nature 308:548–550 (1984).

Szabó PE, Mann JR: Biallelic expression of imprinted genes in the mouse germ line: implications for erasure, establishment, and mechanisms of genomic imprinting. Genes Dev 9:1857–1868 (1995).

Szabó P, Tang SH, Rentsendorj A, Pfeifer GP, Mann JR: Maternal-specific footprints at putative CTCF sites in the *H19* imprinting control region give evidence for insulator function. Curr Biol 10:607–610 (2000).

Takada S, Paulsen M, Tevendale M, Tsai CE, Kelsey G, et al: Epigenetic analysis of the *Dlk1-Gtl2* imprinted domain on mouse chromosome 12: implications for imprinting control from comparison with *Igf2-H19*. Hum Mol Genet 11:77–86 (2002).

Thorvaldsen JL, Duran KL, Bartolomei MS: Deletion of the *H19* differentially methylated domain results in loss of imprinted expression of *H19* and *Igf2*. Genes Dev 12:3693–3702 (1998).

Thorvaldsen JL, Mann MRW, Nwoko O, Duran KL, Bartolomei MS: Analysis of sequence upstream of the endogenous *H19* gene reveals elements both essential and dispensable for imprinting. Mol Cell Biol 22:2450–2462 (2002).

Tremblay KD, Saam JR, Ingram RS, Tilghman SM, Bartolomei MS: A paternal-specific methylation imprint marks the alleles of the mouse *H19* gene. Nat Genet 9:407–413 (1995).

Tycko B, Morison IM: Physiological functions of imprinted genes. J Cell Physiol 192:245–258 (2002).

Varrault A, Bilanges B, Mackay DJ, Basyuk E, Ahr B, et al: Characterization of the methylation-sensitive promoter of the imprinted *ZAC* gene supports its role in transient neonatal diabetes mellitus. J Biol Chem 276:18653–18656 (2001).

Verona RI, Mann MRW, Bartolomei MS: Genomic imprinting: intricacies of epigenetic regulation in clusters. Annu Rev Cell Dev Biol 19:237–259 (2003).

Walter J, Paulsen M: The potential role of gene duplications in the evolution of imprinting mechanisms. Hum Mol Genet 12:R215–R220 (2003).

Wilkins JF, Haig D: What good is genomic imprinting: the function of parent-specific gene expression. Nat Rev Genet 4:359–368 (2003).

Wroe SF, Kelsey G, Skinner JA, Bodle D, Ball ST, et al: An imprinted transcript, antisense to *Nesp*, adds complexity to the cluster of imprinted genes at the mouse *Gnas* locus. Proc Natl Acad Sci USA 97:3342–3346 (2000).

Wutz A, Theussl HC, Dausman J, Jaenisch R, Barlow DP, Wagner EF: Non-imprinted *Igf2r* expression decreases growth and rescues the *Tme* mutation in mice. Development 128:1881–1887 (2001).

Yamasaki K, Joh K, Ohta T, Masuzaki H, Ishimaru T, et al: Neurons but not glial cells show reciprocal imprinting of sense and antisense transcripts of *Ube3a*. Hum Mol Genet 12:837–847 (2003).

Yan H, Yuan W, Velculescu VE, Vogelstein B, Kinzler KW: Allelic variation in human gene expression. Science 297:1143 (2002).

Yatsuki H, Joh K, Higashimoto K, Soejima H, Arai Y, et al: Domain regulation of imprinting cluster in Kip2/Lit1 subdomain on mouse chromosome 7F4/F5: large-scale DNA methylation analysis reveals that DMR-Lit1 is a putative imprinting control region. Genome Res 12:1860–1870 (2002).

Yoon BJ, Herman H, Sikora A, Smith LT, Plass C, Soloway PD: Regulation of DNA methylation of *Rasgrf1*. Nat Genet 30:92–96 (2002).

Zemel S, Bartolomei MS, Tilghman SM: Physical linkage of two mammalian imprinted genes, *H19* and insulin-like growth factor 2. Nat Genet 2:61–65 (1992).

Zwart R, Sleutels F, Wutz A, Schinkel AH, Barlow DP: Bidirectional action of the *Igf2r* imprint control element on upstream and downstream imprinted genes. Genes Dev 15:2361–2366 (2001).

Cytogenet Genome Res 105:346–350 (2004)
DOI: 10.1159/000078207

Endogenous reverse transcriptase: a mediator of cell proliferation and differentiation

C. Spadafora

Istituto Superiore di Sanità, Rome (Italy)

Abstract. Endogenous, non-telomeric Reverse Transcriptase (RT) is encoded by two classes of repeated genomic elements, retrotransposons and endogenous retroviruses, and is an essential component of the retrotransposition machinery of both types of elements. Expression of RT-coding genes is generally repressed in non-pathological, terminally differentiated cells, but is active in early embryos, germ cells, embryo and tumor tissues, all of which have a high proliferative potential. To clarify whether reverse transcription is functionally implicated in control of cell growth, differentiation and in embryogenesis, recent experiments have been undertaken to inactivate the endogenous RT activity. RT was inhibited in normal and transformed cell lines by exposure to nevirapine, a non-nucleosidic RT inhibitor. The endogenous RT was also blocked in murine embryos by microinjection of an anti-RT antibody. Both experimental approaches yielded a dramatic inhibition of proliferation. Murine embryos arrested at pre-implantation stages. Transformed cell lines underwent a significant reduction in the rate of cell growth, concomitant with the induction of differentiation. In addition, RT inhibition induced an extensive reprogramming of the gene expression profile both in cultured cell lines and in preimplantation embryos. From these studies, endogenous RT begins to emerge as a key function with a driving role in normal and pathological developmental processes.

The historical notion of reverse transcriptase (RT) activity comes from its association with replication of retroviral genomes (Baltimore, 1970; Temin and Mizutani, 1970). However, in the cells of higher eukaryotes, RT activity is associated with a wide spectrum of biological processes, both pathological and physiological. It is well known that cellular RTs are expressed at low levels in normal somatic tissues (Salganik et al., 1985; Medstrand and Blomberg, 1993; Banerjee and Thampan, 2000), whereas high levels of expression are detected in transformed cell types (Deragon et al., 1990; Friedlander and Patarca, 1999). Consistent with this, it has been noticed that the mobilization of retroelements is implicated in the genesis of many cellular abnormalities, including tumors (Miki, 1998; Ostertag and Kazazian, 2001). Increased expression of retroelements, however, is not exclusively associated with pathologies, but is a distinctive feature of embryonic tissues and organisms undergoing differentiation. In fact retrotransposons and retroviral genes are actively expressed in preimplantation embryos (Poznanski and Calarco, 1991; Packer et al., 1993) and embryonic tissues such as placenta (Mwenda, 1993). In the latter situation, endogenous retroviral genes have played a fundamental role by providing the product of the env gene, which has become a constitutive functional component of placenta (Mi et al., 2000). These observations support the view that retroviruses can play evolutionary roles by providing genes that contribute novel physiological functions. The mammalian genital tract (Kiessling, 1984; Kiessling et al., 1987, 1989; DeHaven et al., 1998), germ cells (Branciforte and Martin, 1994; Dupressoir and Heidmann, 1996) and gametes (Kattstrom et al., 1989; Nilsson et al., 1992; Miller, 2000) are other preferential sites of expression of retroviral/retrotransposon genes. Retrotransposition of cellular mRNAs is thought to generate new biologically active retrogenes in the germline (McCarrey and Thomas, 1987; Ashworth et al., 1990; Hendriksen et al., 1997; Kleene et al.,1998).

In the mouse genome, a relevant source of reverse transcriptase (RT) activity is provided by the product of the ORF2 sequence of L1 elements, which are estimated to constitute roughly some 3100 interspersed elements (Ostertag and Kazazian, 2001). At this stage, however, it is unclear whether all of them, or only a fraction, do actively contribute to the cellular RT function. The latter is indeed the case in the human genome, in which most of LINE/L1 element retrotransposition is attributable to six highly active L1 elements, while the remaining 80–100 elements play a minor role in genome plasticity (Brouha et al., 2003). The lines of evidence summarized above suggest that the retroviral/retrotransposon gene machin-

Received 11 September 2003; accepted 7 October 2003.

Request reprints from: Dr. Corrado Spadafora, Istituto Superiore di Sanità
   Viale Regina Elena 299, 00161 Rome (Italy)
   telephone: +39 06 49903117/49902972
   fax: +39 06 49387120; e-mail: cspadaf@tin.it

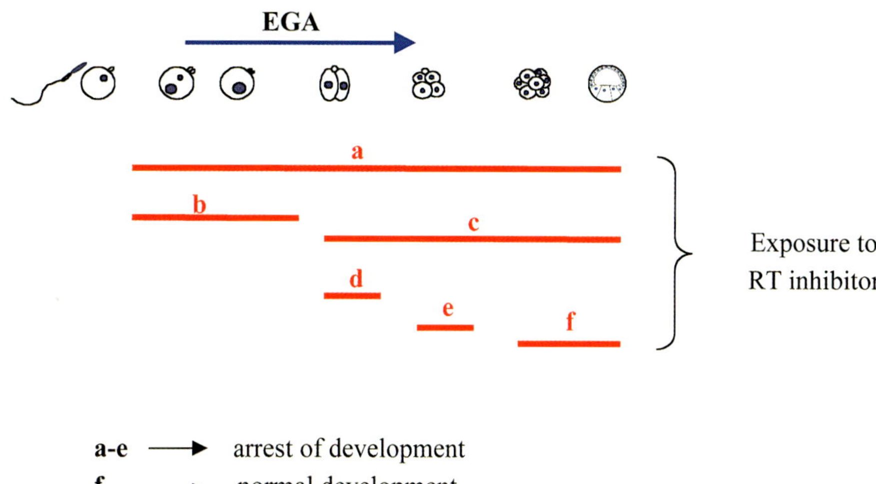

Fig. 1. Schematic summary of the effects of nevirapine treatment on mouse early development. The normal progression of pre-implantation development is schematized on the top. The arrow indicates the time of embryonic gene activation (EGA). Developmental windows during which IVF embryos were exposed to nevirapine (400 μM) are represented by the horizontal bars (designed a–f): (a) from insemination throughout development; (b) 5 h after insemination to the 2-cell stage; (c) 2-cell stage throughout development; (d) 2-cell stage only; (e) 4-cell stage only; (f) 8-cell stage throughout development. The ensuing development was followed up for 5 days (during which controls reached the blastocyst stage).

ery, and specifically the RT-encoding genes, are organized in a transcriptionally competent state in undifferentiated, highly replicating cells, both normal (embryos) and transformed (tumors), while being barely expressed or silent in terminally differentiated tissues. Available data are suggestive of a direct correlation between the expression level of RT-coding genes and the cell state, but do not indicate whether such a modulation is the cause, or the consequence, of the proliferative/differentiative state of the cell. In other words, the unanswered question is whether the retroviral/retrotransposon machinery does or does not play a role in the process of normal (embryogenesis) or pathological (tumorigenesis) cell growth. In the last few years a number of groups have tackled these issues. In our laboratory, we have taken the approach of inhibiting the endogenous RT activity in mouse preimplantation embryos and undifferentiated cell lines, and have studied the effects on proliferation and differentiation. This review summarizes the results obtained in both studies and proposes a possible role of endogenous RT in control of cell proliferation and differentiation.

Endogenous RT activity is required for mouse early embryonic development

In the last years, we have been interested to define the molecular mechanisms underlying sperm-mediated gene transfer for the generation of transgenic animals using sperm cells (see reviews by Spadafora, 1998; Chan et al., 2000; Celebi et al., 2003). In the course of our studies to characterize the mechanism of interaction between sperm cells and exogenous DNA molecules, we have found that these cells are endowed with an unexpected network of metabolic functions that are normally silent, yet are activated at fertilization (Spadafora, 1998). Among those, we have identified molecules reactive to anti-RT antibodies, which are associated with the nuclear scaffold of sperm cells as revealed by immunoelectron microscopy (Giordano et al., 2000). These molecules can retrotranscribe cDNA copies from exogenous RNA molecules incubated with spermatozoa; the retrotranscribed cDNA copies can be transferred from spermatozoa to oocytes in in vitro fertilization (IVF) assays and are further detected in a proportion of the resulting embryos at the two-cell stage (Giordano et al., 2000).

The finding that mature spermatozoa contain active RT molecules was not expected. During murine spermatogenesis, nuclear chromatin undergoes a major restructuring process, yet a small fraction retains an "active" configuration, characterized by a nucleosomal conformation and nuclease sensitivity (Pittoggi et al., 1999; Wykes and Krawetz, 2003). We have found that this fraction is highly enriched in hypomethylated *LINE* ORF2 sequences, i.e. the gene encoding RT (Pittoggi et al., 1999; Pittoggi et al., 2000). This suggests that *LINE* elements are organized in a transcriptionally competent state in this fraction of the sperm genome, and therefore are a potential source for retrotransposition in early embryos. These observations raise the question of whether an endogenous RT activity is also present in early embryos and plays a role in formation of the zygote and/or the earliest steps of development.

In recent work, an endogenous RT enzymatic activity was indeed detected in lysates from mouse zygotes, two-cell and four-cell embryos in an in vitro PCR-based assay using commercial, phage-derived RNA as the template (Pittoggi et al., 2003). In order to assess whether this endogenous RT activity is necessary for early development, we followed up the developmental progression of mouse early embryos in which RT activity was inhibited. To achieve this, two complementary strategies were followed: IVF embryos were either exposed to nevirapine, a non-nucleoside RT inhibitor widely used in the treatment of AIDS (Merluzzi et al., 1990), or microinjected with an inactivating anti-RT antibody. As can be seen from the schematic of these experiments (Fig. 1), drug-mediated inhibition of RT results in a dramatic arrest of embryo development: none of the exposed embryos reached the stage of blastocyst, in contrast to control embryos (cultured under the same conditions, but not exposed to nevirapine) which reached the blastocyst stage in a proportion of 50–60%. A comparable inhibition was obtained when the drug was present throughout development (a) or added at the one- (b), two- or four-cell stages (c–e). Interestingly, no inhibition was instead observed when embryos were exposed later to nevirapine, i.e. from the eight-cell stage onwards (f). Developmental arrest is also induced in RT inhibition experiments by microinjecting an anti-RT antibody (but not with injection of non-specific IgG) in one blastomere of two-cell

Table 1. Nevirapine-induced variations in gene expression in murine embryos and teratocarcinoma cells

Analyzed genes	IVF embryos[a]	F9 cells[b]
Gapd	unchanged	unchanged
Actb	silenced[c]	unchanged
Spin	silenced	not analysed
Aqp8	silenced	not analysed
E2f5	silenced	not analysed
Ranbp1	silenced	not analysed
Hsp70-1	unchanged	not analysed
Bcl2	silenced	up 3-fold
Trp53	silenced	unchanged
p21	not analysed	up 1.8-fold
p16	not analysed	up 6.5-fold
p27	not analysed	up 3-fold
Ccnd1	unchanged	down 8-fold
Ccnd3	not analysed	down 2-fold
Rb-1/pRb	not analysed	up 1.8-fold
Rb-2/p130	not analysed	up 3-fold

[a] Data from semi-quantitative RT-PCR (Pittoggi et al., 2003).
[b] Data from semi-quantitative RT-PCR (Mangiacasale et al., 2003).
[c] mRNA levels are undetectable and therefore nevirapine-dependent fold variations are not measurable.

embryos. Since both experimental approaches comparably arrest development, we conclude that the developmental arrest is specifically due to the inhibition of RT activity. The arrest of embryonic development is irreversible, since development is not restored after removal of the RT inhibitor (our unpublished results). Based on the finding that the sensitivity to nevirapine is developmentally regulated, being restricted to a window between 15 and 60 h after fertilization, we can conclude that developmental arrest is caused by interference with an early RT-dependent event required for the ensuing development (Pittoggi et al., 2003). Remarkably, embryos that arrest developmental progression following RT inhibition do not degenerate, nor do they undergo apoptosis (unlike arrested embryos exposed to most external sources of damage), but maintain the normal morphology of the stage at which they arrest, even after several days of culture. It is worth pointing out that nevirapine-induced developmental arrest cannot possibly reflect the inhibition of telomerase-associated reverse transcriptase (TERT): in fact, TERT is not required for embryo development, as TERT–/– mice are viable and fertile for at least six generations (Blasco et al., 1997), unlike mice exposed to nevirapine or injected with anti-RT antibody.

One intriguing observation arising from the RT inhibition experiments summarized above is that the developmental window of sensitivity to nevirapine overlaps with that of embryonic gene activation (EGA). After fertilization, the earliest transcription events occur in the late one-cell embryo, mainly in the male pronucleus and, to a lesser extent, in the female pronucleus of the zygote (Renard, 1998; Latham, 1999). After the first cell division, EGA continues in two-cell stage embryos, by which stage most embryonic genes are activated (Aoki et al., 1997). It is tempting to speculate that endogenous RT activity is functionally implicated in control of the gene expression program. In an effort to clarify the molecular mechanism(s) underlying RT-dependent modulation of gene expression, we have analyzed the expression of a panel of ten genes, both of the developmentally regulated and constitutively expressed types, in nevirapine-arrested and untreated control embryos. From the results in Table 1, it can be appreciated that seven out of ten analyzed genes are repressed in nevirapine-arrested embryos:

these results, albeit requiring confirmation through a microarray-based genome-wide analysis, currently support the conclusion that endogenous RT is indeed implicated in the mechanism(s) controlling gene expression in the earliest steps of embryogenesis.

Inhibition of RT activity reduces cell proliferation and promotes differentiation in progenitor and transformed cell types

Early work pointed out that embryos and transformed cells share a number of common biochemical and molecular features (Tsonis, 1987). In this framework, the observation that RT is required for early embryo development suggests that RT activity may also be linked to cell transformation. As briefly recalled in the Introduction, a shared feature of embryos and tumor cells is the increased expression of endogenous RT compared to differentiated tissues. Indeed, we have recently found that RT inhibition by nevirapine has a great impact on a variety of cell lines, of both murine and human origin, and can cause a significant decrease of cell growth concomitant with the stimulation of differentiation. In experiments with murine differentiating cell systems, nevirapine did: (i) significantly reduce proliferation of progenitor (myogenic C2C7) and multipotent (F9 teratocarcinoma) cell lines; (ii) facilitate the morphological differentiation of cells: C2C7 myoblasts fused to form multinucleated myotubes, and undifferentiated, round-shaped F9 cultures differentiated into lozenge-shaped cell populations, with increased adhesion properties and cytoskeletal reorganization; and (iii) induced the appearance of differentiation markers specific for these cell lines, i.e. myosin heavy chain in myogenic cells and collagen type IV (α1) chain in F9 cells, respectively (Mangiacasale et al., 2003) (also see Fig. 2).

To begin to address the possibility that tumor cells also respond to nevirapine, we used human acute myeloid leukemia (AML) cell lines and primary blasts from AML patients. These leukemia types originate from a block at different stages of myeloid differentiation, such that cells circulate when still functionally and biochemically immature. In our experiments, all tested myeloid leukemia types cultured in the presence of nevirapine did undergo phenotypic (appearance of lineage-specific

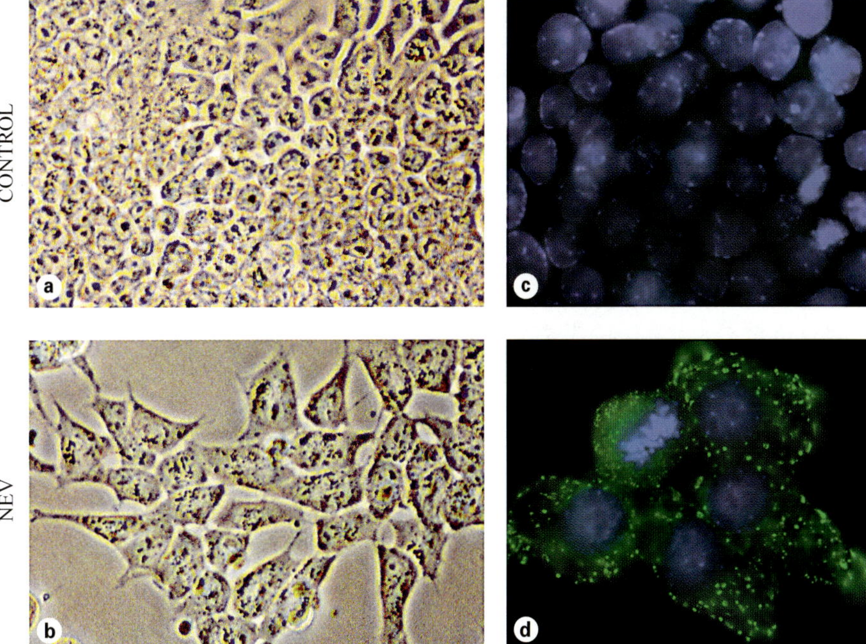

Fig. 2. Differentiation of F9 murine teratocarcinoma cells cultured in the presence of nevirapine (96 h). Control F9 cultures cultured with DMSO (**a, c**) or nevirapine (**b, d**) were examined in vivo by bright-field inverted microscopy to record the induction of morphological reorganization (panels **a** and **b**, 60× objective). Samples were then fixed and processed for immunofluorescence to collagen type IV (α1) chain (fluorescein-conjugated antibody, green) and DAPI staining of nuclei (blue). Merged pictures (**c** and **d**) are shown (100× objective).

membrane markers) and biochemical (synthesis of specific metabolic enzymes) differentiation (Mangiacasale et al., 2003). Clearly these results now open up the promising possibility that at least certain tumor types, where the loss of specialized functions is key to the transformed phenotype, can be targeted by anti-RT compounds.

In cultured teratocarcinoma cells, as seen with developing embryos, the inhibition of endogenous RT causes a significant reprogramming of the expression of cellular genes. However, while most examined genes were virtually silenced in nevirapine-arrested embryos, nevirapine exposure yields a differential modulation of specific genes in cell cultures (Table 1). Most significant changes involve the cyclin D1 gene, the overexpression of which is key to cellular transformation (Hinds et al., 1994; Hall and Peters, 1996; Lamb et al., 2003), which is heavily down-regulated in nevirapine-exposed cultures; conversely, the gene encoding p16, the major inhibitor of cyclin D-dependent kinases, is massively up-regulated upon nevirapine exposure. The expression of "housekeeping" (*Gapdh, Actin*) or apoptotic regulatory genes (p53, Bcl-2) is instead unaffected. Therefore, the modulation of nevirapine-targeted genes correlates well with the reduction of cell growth and induction of differentiation. It is important to stress that cell differentiation and gene reprogramming are both reversible effects induced in cells in which the endogenous RT activity is maintained under constant inhibition. Cells restore their original duplication and differentiation conditions within a few days upon removal of RT inhibitor (unpublished results). We have recently obtained evidence to suggest that RT inhibition by nevirapine is accompanied by a substantial reorganization of functional compartments in the nuclei of exposed cells, which likely underlies RT-dependent modulation of gene expression (manuscript in preparation).

As a whole, the picture emerging from the work with early embryos and cell lines supports the conclusion that endogenous RT plays a central role in regulating the transition between the differentiated and undifferentiated cellular states. Endogenous RT activity appears to exert such a role by modulating the expression of specific genes that, in turn, determine the fate of the cell.

Concluding remarks

Endogenous retroviruses and retrotransposons have been traditionally regarded as "junk" DNA accumulating in the genome and representing mere evolutionary genetic remnants. The "junk" DNA view has been extended to include genes that are constitutive components of the retroposon/retroviral machinery, particularly the genes encoding RT enzymes, which catalyze the retrotranscription and integration steps, the most essential functions for survival of these elements in the host genome. In sharp conflict with this view, the results summarized here suggest that endogenous RT plays an active role in cell growth and differentiation, both in normal development (embryogenesis) and in altered cell growth (tumorigenesis). Both can be viewed as mirror processes progressing in opposite directions and possibly sharing a common molecular mechanism in which RT plays a central role. Although further work is required to thoroughly define the role of endogenous RT, it is tempting to speculate that a RT-depending mechanism is implicated in control of genes that act in cell growth and differentiation during early embryogenesis and tumorigenesis.

The results summarized in this review can be interpreted in the framework of the well established pattern of differential RT activity typical of embryonic, germ and somatic cells recalled in the Introduction, and provide ground to hypothesize that a non-random retrotransposition activity is at work in early embryos and in cells endowed with a differentiating potential. This activity would provide a dynamic mechanism that can be exploited by cells to sequentially adapt to the progression of the developmental program. In this respect, endogenous RT activity can be viewed as a powerful, stage-specific retrotransposition-controlling tool (Luning Prak et al., 2003), influencing gene expression in a dual manner: genetic, through retrotransposal integration events (Ostertag and Kazazian, 2001); and epigenetic, perhaps through an expression-interference mechanism (Whitelaw and Martin, 2001). Somatic deregulation, or resumption of the RT activity at "embryonic" levels, may induce uncontrolled proliferation and loss of specialized functions in differentiated cells, leading to the onset of neoplastic growth. This dynamic functional component is likely contributed by the fraction of L1 elements that are biologically active in the mouse genome. To conclude, it is important to recall here that a crucial role of endogenous RT in the process of cell differentiation, in both normal (embryogenesis) and pathological (tumorigenesis), was predicted by Temin in a pioneering article as far back as 1971 (Temin, 1971). Available experimental data thus far fulfill that prediction.

References

Aoki F, Worrad DM, Schultz RM: Regulation of transcriptional activity during the first and second cell cycles in the pre-implantation mouse embryo. Devl Biol 181:296–307 (1997).

Ashworth A, Skene B, Swiff S, Lovell-Badge R: Zfa is an expressed retroposon derived from an alternative transcript of the *Zfx* gene. EMBO J 9:1529–1534 (1990).

Baltimore D: RNA-dependent DNA polymerase in virions of RNA tumor viruses. Nature 226:1209–1211 (1970).

Banerjee S, Thampan RV: Reverse transcriptase activity in bovine bone marrow: purification of a 66-kDa enzyme. Biochim biophys Acta 1480:1–5 (2000).

Blasco MA, Lee HW, Hande MP, Samper E, Lansdorp PM, DePinho RA, Greider CW: Telomerase shortening and tumor formation by mouse cells lacking telomerase RNA. Cell 91:25–34 (1997).

Branciforte D, Martin SL: Developmental and cell type specificity of LINE-1 expression in mouse testis: implication for transposition. Mol Cell Biol 14:2584–2592 (1994).

Brosius J, Tiedge H: Reverse transcriptase: mediator of gene plasticity. Virus Genes 11:163–179 (1996).

Brouha B, Schustak J, Badge RM, Lutz-Prigge S, Farley AH, Moran JV, Kazazian Jr HH: Hot L1s account for the bulk of retrotransposition in the human population. Proc natl Acad Sci USA 100:5280–5285 (2003).

Celebi C, Guillaudeux T, Auvray P, Vallet-Erdtmann V, Jegou B: The making of 'Transgenic Spermatozoa'. Biol Reprod 68:1477–1483 (2003).

Chan AWS, Lutjens CM, Schatten GP: Sperm-mediated gene transfer. Curr Topic Dev Biol 50:89–102 (2000).

De Haven JE, Schwartz DA, Dahm MW, Hazard III ES, Trifiletti R, Lacy ER, Norris JS: Novel retroviral sequences are expressed in the epididymis and uterus of Syrian hamsters. J Gen Virol 79:2687–2694 (1998).

Deragon JM, Sinnett D, Labuda D: Reverse transcriptase activity from human embryonal carcinoma cells Ntera2D1. EMBO J 9:3363–3368 (1990).

Dupressoir A, Heidmann T. Germ line-specific expression of intracisternal A-particle retrotransposons in transgenic mice. Mol Cell Biol 16:4495–4503 (1996).

Friedlander A, Patarca R: Endogenous proviruses. Crit Rev Oncogenesis 10:129–159 (1999).

Giordano R, Magnano AR, Zaccagnini G, Pittoggi C, Moscufo N, Lorenzini R, Spadafora C: Reverse transcriptase activity in mature spermatozoa of mouse. J Cell Biol 148:1107–1113 (2000).

Hall M, Peters G: Genetic alterations of cyclins, cyclin-dependent kinases, and cdk inhibitors in human cancer. Adv Cancer Res 68:67–108 (1996).

Hendriksen PJM, Hoogerbrugge JW, Baarends WM, de Boer P, Vreeburg JTM, Vos EA, van der Lende T, Grootegoed JA: Testis-specific expression of a functional retroposon encoding glucose-6-phosphate dehydrogenase in the mouse. Genomics 41:350–359 (1997).

Hinds PW, Dowdy SF, Eaton EN, Arnold A, Weinberg RA: Function of a human cyclin gene as an oncogene. Proc natl Acad Sci, USA 91:709–713 (1994).

Kattstrom PO, Bjermeroth G, Nilsson BO, Holmdahl R, Larsson E: A retroviral gp70 related protein is expressed at specific stages during mouse oocyte maturation and in pre-implantation embryos. Cell Diff Dev 28:47–54 (1989).

Kiessling AA: Evidence that reverse transcriptase is a component of murine epididymal fluid. Proc Soc Exp Med 176:175–182 (1984).

Kiessling AA, Crowell RC, Connell R: Sperm-associated retroviruses in the mouse epididymis. Proc natl Acad Sci, USA 84:8667–8671 (1987).

Kiessling AA, Crowell R, Fox C: Epididymis is a principal site of retrovirus expression in the mouse. Proc natl Acad Sci, USA 86:5109–5113 (1989).

Kleene KC, Mulligan E, Steiger D, Donohue K, Mastrangelo MA: The mouse gene encoding the testis-specific isoform of poly(A) binding protein (Pabp2) is an expressed retroposon: intimations that the gene expression in spermatogenic cells facilitates the creation of new genes. J Mol Evol 47:275–281 (1998).

Lamb J, Ramaswamy S, Ford HL, Contreras B, Martinez RV, Kittrell FS, Zahnow CA, Patterson N, Golub TR, Ewen ME: A mechanism of cyclin D1 action encoded in the patterns of gene expression in human cancer. Cell 114:323–334 (2003).

Latham KE: Mechanisms and control of embryonic genome activation in mammalian embryos. Int Rev Cytol 193:71–124 (1999).

Luning Prak ET, Dodson AW, Farkash EA, Kazazian Jr HH: Tracking an embryonic L1 retrotransposition event. Proc natl Acad Sci, USA 100:1831–1837 (2003).

Mangiacasale R, Pittoggi C, Sciamanna I, Careddu A, Mattei E, Lorenzini R, Travaglini L, Landriscina M, Barone C, Nervi C, Lavia P, Spadafora C: Exposure of normal and transformed cells to nevirapine, a Reverse Transcriptase inhibitor, reduces cell growth and promotes differentiation. Oncogene 22:2750–2761 (2003).

McCarrey JR, Thomas K: Human testis-specific PGK gene lacks introns and possesses characteristics of a processed gene. Nature 326:501–505 (1987).

Medstrand P, Blomberg J: Characterization of novel reverse transcriptase encoding endogenous retroviral sequences to type A and type B retroviruses: differential transcription in normal human tissues. J Virol 67:6778–6787 (1993).

Merluzzi VJ, Hargrave KD, Labadia M, Grozinger K, Skoog M, Wu JC, Shih CK, Eckner K, Hattox S, Adams J: Inhibition of HIV-1 replication by a nonnucleoside reverse transcriptase inhibitor. Science 250:1411–1413 (1990).

Mi S, Lee X, Li X, Veldman GM, Finnerty H, Racie L, LaVallie E, Tang X, Edouard P, Howes S, Keith jr JC, McCoy JM: Syncytin is a captive retroviral envelope protein involved in human placental morphogenesis. Nature 403:785–789 (2000).

Miki Y: Retrotransposal integration of mobile genetic elements in human diseases. J hum Genet 43:77–84 (1998).

Miller D: Analysis and significance of messenger RNA in human ejaculated spermatozoa. Mol Reprod Dev 56:259–264 (2000).

Mwenda JM: Biochemical characterization of a reverse transcriptase activity associated with retroviral-like particles isolated from human placental villous tissue. Cell Mol Biol 39:317–328 (1993).

Nilsson BO, Kattstrom PO, Sundstrom P, Jaquemin P, Larsson E: Human oocytes express murine retroviral equivalent. Virus Genes 6:221–227 (1992).

Ostertag EM, Kazazian HH Jr: Biology of mammalian L1 retrotransposons. Annu Rev Genet 35:501–538 (2001).

Packer AI, Manova K, Bacharova RF: A discrete LINE-1 transcript in mouse blastocysts. Devl Biol 157:281–283 (1993).

Pittoggi C, Renzi L, Zaccagnini G, Cimini D, Degrassi F, Giordano R, Magnano AR, Lorenzini R, Lavia P, Spadafora C: A fraction of mouse sperm chromatin is organized in nucleosomal hypersensitive domains enriched in retroposon DNA. J Cell Sci 112:3537–3548 (1999).

Pittoggi C, Zaccagnini G, Giordano R, Magnano AR, Lorenzini R, Spadafora C: Nucleosomal domains of mouse spermatozoa chromatin as potential sites for retroposition and foreign DNA integration. Mol Reprod Dev 56:248–251 (2000).

Pittoggi C, Sciamanna I, Mattei E, Beraldi R, Lobascio AM, Mai A, Quaglia MG, Lorenzini R, Spadafora C: Role of endogenous reverse transcriptase in murine early embryo development. Mol Reprod Dev 66:225–236 (2003).

Poznanski AA, Calarco PG: The expression of intracisternal A particle genes in the pre-implantation mouse embryo. Devl Biol 143:271–281 (1991).

Renard JP: Chromatin remodelling and nuclear reprogramming at the onset of embryonic development in mammals. Reprod Fertil Dev 10:573–580 (1998).

Salganick RI, Tomsons VP, Pyrinova GB, Korokhov NP, Kiseleva EV, Khristolyubova NB: Reverse transcriptase of rat liver associated with the endogenous retrovirus related to the mouse intracisternal A-particles. Biochem biophys Res Comm 131:492–499 (1985),

Spadafora C: Sperm cells and foreign DNA: a controversial relation. Bioessays 20:955–964 (1998).

Temin HM: The provirus hypothesis. Speculations on the significance of RNA-directed DNA synthesis from normal development and for carcinogenesis. J natl Cancer Inst 46:56–60 (1971).

Temin HM, Mizutani S: RNA-dependent DNA polymerase in virions of Rous sarcoma virus. Nature 226:1211–1213 (1970).

Tsonis PA: Embryogenesis and carcinogenesis: order and disorder. Anticancer Res 7:617–623 (1987).

Whitelaw E, Martin DIK: Retrotransposons as epigenetic mediators of phenotypic variation in mammals. Nature Genet 27:361–365 (2001).

Wykes SM, Krawetz SA: The structural organization of sperm chromatin. J biol Chem 278:29471–29477 (2003).

Cytogenet Genome Res 105:351–362 (2004)
DOI: 10.1159/000078208

Factors regulating endogenous retroviral sequences in human and mouse

D. Taruscio,[a] and A. Mantovani[b]

[a] National Centre on Rare Diseases; [b] Department of Food Safety and Veterinary Public Health, Istituto Superiore di Sanità, Rome (Italy)

Abstract. Endogenous retroviruses (ERVs) are stably integrated in the genome of vertebrates and inherited as Mendelian genes. The several human ERV (HERV) families and related elements represent up to 5–8% of the DNA of our species. ERVs may be involved in the regulation of adjacent genomic loci, especially promoting the tissue-specific expression of genes; some HERVs may have functional roles, e.g., coding for the placental fusogenic protein, syncytin. This paper reviews the growing evidence about factors that may modulate ERVs, including: cell and tissue types (with special attention to placenta and germ cells), processes related to differentiation and aging, cytokines, agents that disrupt cell functions (e.g., DNA hypomethylating agents) and steroids. Special attention is given to HERVs, due to their possible involvement in autoimmunity and reproduction, as well as altered expression in some cancer types; moreover, different HERV families may deserve specific attention, due to remarkable differences concerning, e.g., expression in tissues. A comparison with factors interacting with murine ERV-related sequences indicates that the mouse may be a useful model for studying some patterns of HERV regulation. Overall, the available evidence identifies the diverse, potential interactions with endogenous or exogenous factors as a promising field for investigating the roles of ERVs in physiology and disease.

Copyright © 2004 S. Karger AG, Basel

Endogenous retroviruses (ERVs) are stably integrated in the genome of vertebrates; their origin may derive either from ancient germ-cell infections by exogenous retroviruses or from ancestral retroelements by transposition and recombination (reviewed by Gifford and Tristem, 2003). ERV sequences in different species share a genomic organization described already in Larsson et al. (1989). The typical ERV consists of 5′- and 3′-Long Terminal Repeats (LTRs), containing regulatory sequences (e.g., promoter, enhancer regions, polyadenylation signal), and internal coding regions including *gag* (nuclear core protein), *pol* (RNA-dependent DNA polymerase or reverse transcriptase) and *env* (envelope glycoprotein). In general, the presence of *env* distinguishes ERVs from retrotransposons showing significant structural homologies with retroviral proviruses (Löwer et al., 1996; Deininger and Batzer, 2002). ERVs use reverse transcriptase for their replicative cycle and may code for retrovirus-like particles, but are generally not infectious; in fact, most sequences are highly defective and unable to code anymore, due to the accumulation of mutations, frameshifts and deletions.

The sequencing of the human genome has disclosed that a considerable fraction (estimated between 5 and 8%) is made up of thousands of sequences of human ERVs (HERVs) (Gifford and Tristem, 2003). According to sequence homologies, HERVs are grouped into several families with a highly variable number of copies, from few to more than 1000 per haploid genome. HERV families are identified by names (HERV-E, HERV-K, etc.) according to the tRNA normally used to prime reverse transcription (for nomenclature and classification, see

The present paper has been elaborated within the frame of the following Italian National Health Project: "Human Exposure to Xenobiotics with Potential Endocrine Activities: Evaluation of Reproductive and Developmental Risks", "National Centre on Rare Diseases."

Received 12 November 2003; manuscript accepted 23 December 2003.

Request reprints from Alberto Mantovani
Department of Food Safety and Veterinary Public Health
Istituto Superiore di Sanità, viale Regina Elena 299
IT–00161 Rome (Italy); telephone: ++39 06 4990 2565
fax: ++39 06 4938 7077; e-mail: alberto@iss.it
websites: http://www.iss.it/sitp/dist/, http://www.endodisru.iss.it

Tristem, 2000; Gifford and Tristem, 2003). Moreover, the human genome contains high copy numbers of LTR-retrotransposons, including solitary LTRs, which share structural similarities with HERVs but lack genomic components required for envelope and viral capsule (reviewed by Deininger and Batzer, 2002). Unlike mouse ERVs, most HERVs appear to originate from ancient insertions and to display little or no retrotransposition activity; no HERVs have yet been shown to be infectious (Gifford and Tristem, 2003).

Since the mouse is by far the most important laboratory species used in experimental research on genetics, it is obviously important to characterize ERVs in the murine genome. In fact the murine genome differs from the human one, since retroelements, though representing a fraction only slightly higher, are much more active (Dennis, 2002). Murine retroelements include important LTR elements with ERV-like characteristics, which are present also in other rodent species (e.g., the rat), such as VL30 (viruslike 30S RNA sequence) and, most noticeable, Intracisternal A-type Particles (IAPs) (in Urnovitz and Murphy, 1996). IAPs are closely related to ERVs, possess 5′- and 3′-LTRs flanking *gag-pol* open reading frames (ORFs), and encode intracellular, non-infectious particles. Several classes of IAP sequences have been identified and at least some of them can still transpose at both germ line and somatic levels. Among mouse ERVs, a remarkable, in fact unique, example is offered by the mouse mammary tumour virus (MMTV), ERV type b, responsible for mammary gland tumorigenesis in some strains of laboratory mice. MMTV is transmitted horizontally through the milk to susceptible offspring, as exogenous virus, or vertically through the germ line as endogenous provirus; exogenously acquired and some endogenous mouse mammary tumor viruses are coexpressed at high levels in lactating mammary glands (Golovkina et al., 1994; Sarkar et al., 2004).

ERVs are receiving attention due to their several potential roles, including the contribution to genomic plasticity through reverse transcription, the regulation of adjacent genes and the involvement in physiological and pathological processes, relevant in particular to autoimmunity and reproduction (Löwer et al., 1996; Taruscio and Mantovani, 1998; Obermayer-Straub and Manns, 2001); however, their actual importance in these different aspects is still a matter of debate and it may actually show significant interspecies differences (Dennis, 2002; Gifford and Tristem, 2003). A most interesting aspect, shown by a number of studies, is that ERVs and ERV-like elements are highly susceptible to regulation from stimuli from the cell/tissue environment as well as from endogenous and exogenous factors such as cytokines, steroids and some chemicals. The recent, impressive increase of knowledge about the presence and mapping of ERVs in the mammalian genome highlights the need for a better understanding. In particular, increasing data (sometimes stemming from older studies) consider ERVs from the standpoints of transcriptional activity and how it is regulated. The present review, therefore, aims at evaluating ERVs as potential targets for environmental stimuli. Particular attention will be given to HERVs, due to their possible role in human physiology and diseases, as well as to murine ERVs, due to the importance of this species as research model, both inside and outside genetics.

ERV expression in tissues

Schon et al. (2001) and Vinogradova et al. (2001) showed that HERV LTRs display specific transcription patterns according to cell types, whereas patterns were consistent among different cell lines derived from the same type of tissue. Moreover, cells associated in the same tissue may display significant differences; for instance HERV-R is highly expressed in human endothelial and vascular smooth muscle cells in vitro but not in fibroblasts (Katsumata et al., 1999).

Seifarth et al. (2003) used a pol-specific array to investigate the transcription profiles of elements from different HERV families in cell lines of diverse tissue origin. In general, cell lines from liver or kidney showed higher HERV transcription, but tissue-specific patterns were also observed. Whereas some HERVs (HERV-KC4, or HERV-E[4-1]) were active in almost all cell types examined, HERV-W and HERV-E LTRs showed different cell type preferences; moreover, HERV-W showed more consistent patterns than HERV-H, likely due to the higher homogeneity of their sequence structures. Through a systematic survey of human genome, De Parseval et al. (2003) identified sixteen retroviral genes with coding capacity for complete envelope (Env) proteins; due to their potential activities (e.g., fusogenic, immunosuppressive) Env can be highly relevant for tissue-specific expression of HERVs. Noticeably, six genes belonged to HERV-K(HML-2), a family that did generate new proviral copies in recent evolutionary times. All sixteen genes are expressed in at least some healthy tissues albeit at highly different levels. Remarkably, all genes disclosed significant expression in the testis; three of them (envR, envW, envFRD) have a very high level of expression in the placenta; also, one (envT) is specifically expressed in the thyroid; finally, a low expression of envR only was observed in heart and liver.

Different studies consistently support the regulatory role of tissue microenvironment on ERV expression. Under this regard, due attention should be paid to specific genomic integration patterns, i.e., non-random integration of ERVs and their relationships with adjacent sequences. Therefore, in the following paragraphs we will give special attention to the influences of the genomic context and to ERV expression in gonads, placenta as well as in differentiating tissues.

Influence of the genomic context
Although ERVs are widespread throughout the mammalian genome, available data do not support a random integration. For instance, the distribution of HERV-K (HML-2) LTR along human chromosome 21 is roughly correlated with gene density, even though LTRs have apparently been inserted randomly into the chromosome relative to each other (Kurdyukov et al., 2001). A high level of polymorphism was reported by Lavrentieva et al. (1999) in genomic sequences flanking insertion sites of HERV LTRs. Sequences located downstream of the LTR-U3 region may modulate the functional activity of LTRs as observed by Baust et al. (2001) concerning HERV-K-T47D-related LTRs. Mapping of several HERVs on, or close to, fragile sites, chromosomal breakpoints or radiation hot spots has been consistently reported, including ERV-9 (Svensson et al., 2001), HERV-E(4-1) (Taruscio and Manuelidis, 1991; Tarus-

cio and Mantovani, 1996), HERVK-10 (Meese et al., 1996) as well as HRES1, an endogenous sequence seemingly unrelated to other HERVs (Tristem, 2000), which maps on the fragile site 1q42 (Perl et al., 1991). As regards the mouse genome, retroinsertion hotspots have also been identified by, e.g., Peterfy et al. (1998). Therefore, it is worthwhile to increase knowledge of host genes that control retroviral insertion; an example in mouse is the nuclear export factor *Nxf1*, providing a link between pre-mRNA processing and mRNA export receptor (Floyd et al., 2003).

The possible regulatory role of ERVs toward the expression of an adjacent gene is a major topic in order to understand any potential consequences of factors regulating ERVs. In the mouse insertion of ERVs or ERV-like transposable elements has been reported concerning, e.g., *Hebc1*, a gene involved in the homeostasis of iron (Ilyin et al., 2003) as well as a new mutant allele of the Brachyury locus (Goldin and Papaioannou, 2003).

In humans, two recognized examples are HERV-E elements which, respectively, contribute a tissue-specific enhancer to the salivary amylase gene (Ting et al., 1992) and insert into the growth factor pleiotrophin (PTN) gene, thus generating functional HERV-E-PTN fusion transcripts expressed in both choriocarcinomas and gestational trophoblastic tissue (Schulte et al., 2000). More recent data provide increasing instances of HERV sequences involved in the tissue-specific expression of genes. Examples include: HERV-E LTRs providing alternative promoters to the apolipoprotein C1 and endothelin B receptor genes in liver and in a placental cell line (Medstrand et al., 2001); a HERV-K regulating the placenta-specific expression of a gene of the insulin family, INSL4 (Bieche et al., 2003); a HERV LTR acting as dominant promoter for human beta1,3-galactosyltransferase in colon and mammary gland (Dunn et al., 2003). The association of HERV elements with genomic loci which are associated with congenital defects might be noteworthy, as reported by Taruscio and Mantovani (1996) for several integration sites of the HERV-E(4-1) sequence, and by Landry et al. (2002) for the fusion transcript of a HERV-E LTR with *Mid1*, the gene of the Opitz syndrome.

Finally, the contribution of some HERVs as regulatory elements of genes involved in the response to environmental stimuli is remarkable, these HERVs include HERV-R (ERV3) and the 5′-LTR of HERV-Fb for Kruppel-related zinc finger genes (Abrink et al., 1998; Kjellman et al., 1999); HERV-K LTRs for the steroid 17 alpha-hydroxylase gene CYP17 (Maghsoudlou et al., 1995) and the human leptin receptor (Kapitonov and Jurka, 1999); an element related to ERV-9 LTR regulating the tissue-specific expression of alcohol dehydrogenase 1C, an enzyme metabolizing ethanol, in fetal kidney, epididymis, and adult stomach mucosa (Chen et al., 2002). Further hints can be provided by analysis of gene mapping in the same band as HERV integration sites, as performed, e.g., by Taruscio et al. (2002) as regards sequences belonging to HERV-E family.

Germ cells and gonads

The expression of all HERV *env* in the testis, as detected by De Parseval et al. (2003), may not be unexpected. In fact, germline expression is a common feature of diverse transposable ele-

ments in other species, including IAPs in mice (Branciforte and Martin, 1994; Dupressoir and Heidmann, 1996). One possible evolutionary explanation could be that retroviral elements expressed in the germ cells have a high chance of insertion in the germline. ERV expression in the male reproductive tract might be relevant to mechanisms enhancing genomic plasticity. In several lines of transgenic mice the expression of IAPs (Dupressoir and Heidmann, 1996) as well as of an isolated HERV-K/HTDV LTR (Casau et al., 1999) was consistently restricted to premeiotic, less differentiated stages of spermatogenesis. Testis was also the only adult tissue where IAP LTRs were hypomethylated (Dupressoir and Heidmann, 1996).

Moreover, high levels of retroviral particles have been reported by Kiessling et al. (1989) in the mouse epididymal epithelium; Rothenfluh (1995) speculated about an involvement of ERVs in a soma-to-germline feedback loop with memory lymphocytes occurring in the epididymis, but more recent experimental data on ERVs in the epididymis are apparently unavailable. However, further investigation might be prompted by the detection of reverse transcriptase in murine epididymal spermatozoa (Giordano et al., 2000) as well as the characterization of a nucleosomal subfraction of mouse sperm chromatin that is highly enriched in unmethylated retroposon DNA from a variety of families (Pittoggi et al., 2000). Foreign DNA is spontaneously taken up by mouse epididymal sperm cells and is further internalized. Integration of exogenous DNA molecules, and the endogenous retroposition activity appear to occur in the same chromatin sites (Pittoggi et al., 2000); this may be a critical stage to investigate either the actual role, if any, of ERVs in shaping genomic plasticity and the endogenous or exogenous factors that may modulate such a role.

The testis is a preferential site of expression for other HERVs, including the widely distributed ERV-9 (Svensson et al., 2001) and the retroviral element provisionally named HERV-K-MEL (Schiavetti et al., 2002). The association of HERV-K with germ cell tumours and germ cell tumour cell lines is also noticeable and will be discussed below.

Placenta

Placenta is recognized as a preferential site for the expression of specific HERVs (reviewed by Taruscio and Mantovani, 1998); nevertheless, as confirmed by De Parseval et al. (2003), preferential expression in placenta is not a common feature to all HERV families. The single-copy HERV-R (ERV3), mapping on 7q11.2, shows both high expression levels throughout pregnancy and chorion-specific primary transcripts; such activity is specifically restricted to cells undergoing fusion and peaks together with synthesis of beta-human chorionic gonadotropin, a marker of trophoblastic differentiation (Boyd et al., 1993; Venables et al., 1995). Also HERV-E elements, such as the sequence (4-1) (Taruscio et al., 2002), are preferentially expressed in placenta, especially in syncytiotrophoblasts and vascular endothelia (Kitamura et al., 1994). The HERV-E promoter of *Mid1* strongly and specifically contributes to the expression of the gene in the placenta (Landry et al., 2002). The full-length HERV-F (XA34) is also preferentially expressed in placenta (Kjellman et al., 1999).

The recently discovered HERV-W encodes an Env protein, syncytin, that is specifically expressed in syncytiotrophoblasts (Blond et al., 2000). Syncytin appears as a fusigenic factor involved in human placental morphogenesis (Mi et al., 2000; Frendo et al., 2003); reduced syncytin expression has been correlated with inadequate function of syncytiotrophoblast in pre-eclampsia and similar conditions (Lee et al., 2001; Knerr et al., 2002). The expression of HERV-W in placenta is regulated by the pregnancy course; a 3.1-kb mRNA transcript remains unchanged, whereas an 8-kb transcript increases in term placenta. A marked reduction of syncytin synthesis occurs in late pregnancy; therefore, the relative levels of the two transcripts in trophoblasts could regulate syncytin synthesis in late pregnancy, possibly by competition of the two RNA species for translational apparatus (Smallwood et al., 2003). The HERV-K element promoting the placenta-specific expression of INSL4 is also upregulated during the differentiation of cytotrophoblast into syncytiotrophoblast (Bieche et al., 2003). Thus, the morphogenesis of syncytiotrophoblast appears as a critical phase of expression for some HERV sequences and/or adjacent genes. On the other hand, the HERV-E-PTN fusion transcripts seem associated with the proliferating and invading properties of the trophoblast (Schulte et al., 2000); a similar pattern of expression in the placenta has been reported by Luton et al., (1997) for c-proto-oncogenes.

Mouse retroviral elements do not show a potential regulatory role in placenta comparable to that of HERVs. An exception is the mouse IAP-promoted placental gene, *Ipp*, transcriptionally regulated by a solo IAP LTR in some, but not all, mouse strains (Chang-Yeh et al., 1993). Moreover, previous observations in feral mice suggested that the expression of ERV genes in placental tissues might have a protective role against infection by exogenous retroviruses through receptor interference by ERV Env (Gardner et al., 1991).

Differentiating cells and tissues

Earlier studies (reviewed by Taruscio and Mantovani, 1998) already showed a significant expression of ERV-related particles, including IAPs, during distinct stages of the mouse preimplantation development, with the two-cell stage as a critical phase. According to Dupressoir and Heidmann (1996) the expression of IAPs in early mouse embryos is dependent upon the mouse strain and thus is likely to result from activation by nearby genes rather than from involvement in basic embryogenetic process. However, reverse transcriptase may be a critical factor in early mouse embryogenesis. Pittoggi et al. (2003) have shown that the impairment of reverse transcriptase (using either the drug nevirapine or an antibody) between the late one-cell and the four-cell stage induces a developmental arrest before the blastocyst stage. In contrast, development was not affected when embryos were exposed to nevirapine after the eight-cell stage. Moreover, in nevirapine-arrested two-cell embryos the authors detected an extensive reprogramming of expression of both developmentally regulated and constitutively expressed genes. Thus, any factor modulating reverse transcriptase might affect the very developmental stages in mouse embryos; however, further investigation is needed to clarify whether this may occur through interference with a role (if any) of ERV-related elements in early murine embryogenesis.

Interestingly, IAP expression in the early embryo may be modulated also by patterns of sperm-oocyte interaction. Hayashi et al. (2003) observed that the expression of retrovirus-1 mobile element IAP from two-cell through to blastocyst stage was markedly elevated in mouse embryos derived by round spermatid injection as compared with intracytoplasmic sperm injection.

Following implantation of the embryo, IAP expression is regulated throughout the dynamic processes of development, growth and aging. In conceptuses from transgenic mice lines, IAP LTR expression was increasingly evident in early to term fetuses and it was consistently restricted to primitive gonocytes of the immature seminiferous tubules; no expression was found in female germ cells at developmentally analogous stages (Dupressoir and Heidmann, 1996). This finding may be consistent with those in adult mice, showing a presence of ERV-related transcripts restricted to premeiotic cells of the seminiferous tubules (Dupressoir and Heidmann, 1996; Casau et al., 1999). Dragani et al. (1987) and Gaubatz et al. (1991) observed significant increases of IAP-related transcripts in the early postnatal life, possibly associated with the mitotic rate of tissues. Age-related demethylation and desilencing of an IAP sequence by repetitive induction has been reported in aging mice. Although the functional significance remains to be determined, the progressive demethylation and turning on of the IAP sequence may result, as an epigenetic and stochastic process, from the transient, daily activation of its promoter through a timescale of months to years (Barbot et al., 2002).

The data on HERV expression in human embryos and fetuses are obviously more limited than in laboratory rodents. However, HERV-R(ERV3) is expressed in an organ-specific way in human fetuses, with highest levels in adrenal cortex, and also in renal tubules, tongue, heart, liver, and central nervous system; a less evident pattern was observed for HERV-K rec (former cORF) transcripts, whereas HERV-K pol/int was not possible to quantitate (Andersson et al., 2002). Another interesting observation by Landry et al. (2002) reports that, besides in the placenta, the HERV-E LTR regulating-Mid1, is a strong and specific enhancer of the gene also in embryonic kidney.

In vitro studies provide information on HERV modulation in differentiating tissues following treatment with retinoic acid or other factors. La Mantia et al. (1991) observed that ERV-9 is downregulated during retinoic acid-induced differentiation; moreover, the solitary LTRs of ERV-9 have a strong enhancer activity in cells from placenta, embryonic liver and kidney and lowest in adult non-haematopoietic cells (Ling et al., 2002). On the contrary, HERV-R (ERV3) is upregulated upon retinoic acid differentiation of U-937 cells (Larsson et al., 1997), consistent with its expression in the placenta, restricted to the differentiated syncytiotrophoblast (Venables et al., 1995). HERV-K, which is highly expressed in germ cell tumours, is significantly upregulated during differentiation of NT2D1 human embryonic carcinoma cells treated with bone morphogenetic proteins or retinoic acid; nevertheless, the upregulation is transient, with transcripts becoming undetectable in fully differentiated cells. Moreover, other factors (activin, transforming growth factor-beta) are unable to modulate HERV-K in NT2D1 cells (Caricasole et al., 2000). Interestingly, Beyer et al. (2002) indicated apoptosis as a possible regulatory mechanism in differentiating tissues; in teratocarcinoma cell line Tera-1 HERV-K10 Gag proteins are cleaved during the execution phase of apoptosis in a way similar to nuclear autoantigens.

HERV and tumours

In the mouse MMTV represents a well recognized instance of ERV associated with neoplasia (see, e.g., Hartig et al., 1993). MMTV is highly steroid-responsive and able to act as heritable somatic mutagen whose target range is limited. In fact, the tumorigenic capacity of MMTV is restricted to mammary gland, hinting to a specific relationship with the micro-environment of mammary epithelium, and it is related to LTRs enhancing the expression of adjacent genes, such as proto-oncogenes. Expression of endogenous MMTV proviruses can lead to the release of infectious viruses and reinfection of the same tissue, leading to the rare occurrence of neoplastic transformation (Callahan and Smith, 2000). Due to its marked steroid regulation, MMTV will be discussed below, in the section of endocrine modulation of ERVs.

The HERV-K family is closely associated with germ cell tumour, both in vitro and in vivo. While HERV-K sequences are preferentially expressed in cell lines derived from teratocarcinomas, HERV-K10 Gag proteins are consistently expressed in different trophoblastic and germ cell tumours, especially seminomas as well as gonadoblastomas. On the contrary, no, or very low, positivity was observed in previous seminoma patients after therapy, as well as in patients with different conditions (neoplasms, autoimmune diseases, immunosuppression) and healthy controls (Sauter et al., 1995; Herbst et al., 1996, 1999). In seminomas expression of the Gag protein was restricted to tumour cells, while it was undetectable in the surrounding testicular tissue (Sauter et al., 1995). High positivity was found also in so called in situ carcinomas of the testis, considered to be precursors of testicular germ cell tumours. On the contrary, teratoma samples showed a consistent absence of expression, opposite to cell lines in vitro (Herbst et al., 1996). The authors suggested that downregulation of HERV-K in teratomas was related to the differentiation in three germ layers, which distinguishes teratomas from other germ-cell tumours. Besides Gag proteins, elective differential display of RNAs containing interspersed repeats showed that unexpected large numbers of HERV-K LTRs are also transcribed in testicular germ cell tumours, with patterns distinctly different from normal tissues (Vinogradova et al., 2002).

Although preferentially associated with germ cell tumours, HERV-K may be activated in several other neoplastic tissues; in fact, HERV-K LTRs are actively expressed in various cancer cell lines (Kim et al., 2001). Florl et al. (1999) observed the hypomethylation of HERV-K, as well as of LINE-1 retroposons, in bladder carcinomas, whereas no activation was observed in renal carcinomas. Transcripts of an HERV-K gene (Np9 with homology to HERV-K 101 env, GenBank accession no. AF164609) were detected in various tumour tissues (breast carcinomas, seminomas, lymphocytes from leukaemia patients) but not in normal, nontransformed cells; since the protein, on the contrary, is difficult to detect, Armbruester et al. (2002) suggested a post-transcriptional regulatory mechanism. A recent paper by Wang-Johanning et al. (2003) indicates that HERV-E(4-1) env transcripts are upregulated in prostate adenocarcinoma, while they are undetectable in normal control specimens.

Some studies indicate a possible relationship between ERV expression and cancer cells or lymphoid or myeloid origin in either mice and humans. Novel endogenous type D retroviral particles were revealed at high level in the murine thymoma cell line ThyE1M6, whereas only low levels were expressed in normal mice (Ristevski et al., 1999). An intact HERV, HERV-H/F, located on 6q13, is specifically expressed in cancer cell-lines of B-lymphoid and myeloid origin, whereas no expression was observed in normal tissues (Patzke et al., 2002). A relative overexpression of HERV-K has been detected in blood cells of leukaemia patients (Depil et al., 2002); moreover, a fusion transcript between HERV-K and fibroblast receptor 1 kinase has been described in the 8p12 stem-cell myeloproliferative disorder (Guasch et al., 2003).

Finally, Schiavetti et al. (2002) identified a HERV element, showing homologies to HERV-K, which provides a further example of association with specific tumours. The element, provisionally named HERV-K-MEL, encodes an antigen recognized by cytolytic T lymphocytes on ocular and cutaneous melanoma, naevi and a minority of carcinomas and sarcomas; on the contrary HERV-K-MEL is not expressed in normal tissues, with the exception of testis and some skin samples.

ERVs and immunity

HERVs may be implied in the development of autoimmunity through, e.g., the coding of superantigens (Obermayer-Straub and Manns, 2001). In fact, HERVs are upregulated in some autoimmune conditions (Obermayer-Straub and Manns, 2001). HERV LTR elements are associated with the polymorphic 5′ flanking region of major histocompatibility complex HLA-DQB1 (6p21.3) (Donner et al., 2000; Pascual et al., 2001). Pani et al. (2002a) observed that an LTR13 insertion within the HLA region was significantly more frequent in subjects with Addison's disease, whereas no association between type 1 diabetes and HERV-K (C4) insertion was found (Pani et al., 2002b). Antibody levels toward HERV-K Env were not associated with type 1 diabetes or autoimmune thyroiditis (Badenhoop et al., 1999), but increased significantly in systemic lupus erythematosus (Hervé et al., 2002). HERV-E(4-1) gag-transcripts were also significantly upregulated in subjects with systemic lupus erythematosus. Tamura et al. (1997) detected the HERV-E(4-1) Env in alveolar macrophages of a small group of patients affected by interstitial lung diseases (idiopathic pulmonary fibrosis, sarcoidosis), but the actual significance of this finding has yet to be assessed. HERV upregulation in some autoimmune conditions might just represent an epiphenomenon induced by, e.g., inflammatory processes. Thus, to assess the actual role of specific HERVs, it is important to investigate in detail the effects of factors involved in the immune response network.

King et al. (1990) reported that MMTV transcription is upregulated during B-cell differentiation, mostly because of the specifically enhanced expression of a single proviral gene, *Mtv9*. Krieg et al. (1992) investigated the effects of lymphocyte mitogens on the transcription of HERV-R (ERV3), HERV-H (also called RTVL-H), HRES1, several HERV-K sequences (NMWV 1–9) and the retrotransposon EHS-1 in cultured peripheral blood mononuclear cells. Different responses were obtained: pokeweed mitogen (active on both B and T cells) was a general upregulator except for HRES1; concanavalin A (T cell mitogen) was effective as well, but for HERV-R (ERV3) and NMWV-4. Phorbol myristate acetate enhanced only NMWV-7 and, to a lesser extent, HERV-H. Corticosteroid treatment effectively impaired the upregulation of HERV-E(4-1) gag-transcripts observed by Ogasawara et al. (2001) in subjects with systemic lupus erythematosus.

ERV modulation by cytokines may provide important information. In the presence of high glucose concentration, interleukin (IL)-1beta markedly increased the expression of IAPs and type C ERVs in the pancreatic islets from mice of the diabetes susceptible NOD strain (Tsumara et al., 1994). In IAP-mediated transformation of FDC-P1 cells, IL-3 induces a significant and specific increase of a major 7.4-kb transcript (Blumenstein et al., 1998).

IL-1beta was a specific enhancer of HERV-R (ERV3) in proximal tubular kidney cells but not in synovial cells from rheumatoid arthritis patients (Takeuchi et al., 1995). IL-1 alpha, as well as IL-1 beta and tumour-necrosis factor-alpha, upregulated HERV-R (ERV3) also in vascular endothelial cells (Katsumata et al., 1999). IL-1 is highly present in placenta and it is a mediator of both inflammation and parturition (Baergen et al., 1994), which might be noteworthy since HERV-R (ERV3) is specifically expressed in syncytiotrophoblast. HERV-R (ERV3) was upregulated by interferon-gamma or phorbol esters in the monocytic U-937 cell line (Larsson et al., 1996), whereas in vascular endothelial cells, interferon-gamma acted as a down-regulator (Katsumata et al., 1999). Overall, the available data on modulation of HERV-R (ERV3) indicate an interplay between specific cytokines and tissue environments. HERV-K (C4) transcripts are significantly downregulated in cells expressing the complement component C4; downregulation is further modulated by interferon-gamma stimulation of C4 expression (Schneider et al., 2001).

Johnston et al. (2001) suggested that the increased expression of an HERV-W detected in the brains of subjects with multiple sclerosis can be a secondary effect triggered by macrophage activation, in particular, by tumour necrosis factor alpha. Actually, interferon gamma and tumour necrosis factor-alpha (as well as, to a lesser extent, IL-4 and IL-6) upregulated the expression of a HERV-W associated with multiple sclerosis in peripheral blood mononuclear cells in vitro. On the other hand, interferon beta, used in the therapy of multiple sclerosis, acted as down-regulator (Serra et al., 2003). Firouzi et al. (2003) also observed that upregulation of HERV-W was associated with tumor necrosis factor-alpha and interferon gamma in immunodeficient mice grafted with human lymphocytes. More

information is also desirable about the interactions of HERVs with coinfecting viruses. Sutkowski et al. (2001) report that the Epstein-Barr virus transcriptionally activates the env gene of HERV-K18, possessing superantigen activity; the env might, in turn, play a role in the pathogenesis of Epstein-Barr virus infection.

Agents that disrupt cell functions

Cytotoxic and/or genotoxic chemicals enhanced ERV RNA expression in murine cell lines, as reported by a number of earlier studies (reviewed by Taruscio and Mantovani, 1998). DNA hypomethylating agents (e.g., 5-azacytidine) and halogenated pyrimidines (iododeoxyuridine) were particularly effective, but ERV enhancers included other carcinogens (e.g., aflatoxin B1) and cytotoxic agents (e.g., selenomethionine). An in vitro study by Hsieh and Weinstein (1990) showed that ERV upregulation in rat embryo fibroblasts was dependent upon toxicological mechanisms. In particular the decreasing gradient of effect was DNA hypomethylation (5-azacytidine) > inhibition of protein synthesis (cycloheximide) > DNA damage (benzopyrene diol epoxide), whereas no upregulation was induced by activation of protein kinase C (12-O-tetradecanoylphorbol-13-acetate).

Murine ERV elements may show different responses. Type C ERV particles were induced by treatment of K BALB mouse cells with 5-iododeoxyuridine either alone or, to a greater extent, in combination with 5-azacytidine; 5-azacytidine alone enhanced activation of type A particles (Khan et al., 2001).

As regards HERVs, some in vitro studies indicated the influence of the cell microenvironment. The chromatin-modifying agent n-butyrate or 5-azacytidine increased HERV-K Gag protein levels and induced specific undermethylation of the gag gene and adjacent 5′-LTR in one (Tera 1) out of two teratocarcinoma cell lines (Gotzinger et al., 1996). Moreover, cycloheximide (reported as an ERV enhancer by Hsieh and Weinstein, 1990) inhibited the transcription of HERV-H in normal human T lymphocytes, similarly to the immunosuppressant cyclosporin (Kelleher et al., 1996). Regulation associated with cytotoxicity may not be confined to chemicals. The transcription patterns of HERV-K LTRS in cell lines may show significant qualitative changes following cell stress by heat shock (Vinogradova et al., 2001).

Unfortunately, there are not many studies on ERV regulation in vivo. Dragani et al. (1987) observed that treatment of adult mouse with single doses of two promoters of hepatocarcinogenesis (carbon tetrachloride and 1,4-bis[2-(3,5-dichloropyridyloxy)] benzene) induced a differential upregulation of VL30 and IAP transcription in liver; the effect was roughly related both to the increased mitotic rate to repair tissue damage, and to the infiltration by mononuclear cells, which was induced by carbon tetrachloride only. The increased transcription of ERV-related elements preceded the increase of mitotic activity. This latter finding might have been consistent with a concurrent experiment with cycloheximide; this chemical did not cause apparent histological lesions, but induced a marked, even though transient, increase of ERV-related transcripts. Accordingly, Dragani et al. (1987) hypothesized that cycloheximide could affect labile proteins involved in the regulation of ERV transcription; this early hypothesis might be consistent with the already quoted indications by Armbruester et al. (2002) on post-transcriptional regulatory mechanisms for HERV-K.

As regards DNA hypomethylation in vivo, Saavedra et al. (1996) reported that in p-BOR-IL-3 transgenic mice 5-azacytidine interacted with the transcription of IL-3 driven by an ERV LTR, thus increasing the incidence of thymic lymphomas.

The above data suggest that, in addition to the regulation of ERVs in neoplastic tissues, it could also be worthwhile investigating their response to treatments with chemotherapeutic agents as well as any potential involvement in secondary adverse effects of anticancer drugs. Interestingly, Okada et al. (2002) observed that 5-azacytidine increased the transcripts of HERV-E(4.1) in healthy subjects, but not in patients with systemic lupus erythematosus; the authors hypothesize that DNA demethylation of HERV-E(4.1) DNA might have already occurred in patients, possibly related to the pathogenesis of the disease.

Most in vitro and in vivo data concern markedly toxic agents. However, there is increasing evidence that murine ERVs may also be influenced by more diluted stressors. Feeding pregnant mouse dams with a diet supplemented with methyl donors caused a shift in coat colour in the offspring; although the mechanism is still unclear, the shift was correlated to an increased IAP methylation (Morgan et al., 1999; Cooney et al., 2002). No data are yet available on a possible regulation by environmental xenobiotics. However, Lu and Ramos (2003) have recently provided some interesting hints concerning the mouse retrotransposon L1Md-A2. Environmental hydrocarbons which interact with the aryl-hydrocarbon (AhR) receptor (benzo-a-pyrene and 2,3,7,8-tetrachlorodibenzo-p-dioxin) enhanced the expression of the retrotransposon in vascular smooth muscle cells; a redox-sensitive mechanism was activated which, in fact, was antagonized by antioxidants. On the other hand, a more direct oxidant agent such as hydrogen peroxide failed to transactivate the L1Md-A2 promoter. Thus, the modulation of redox homeostasis through AhR signalling (a well-known mechanism in toxicology) appears as a requisite for the activation of the retrotransposon. It may be well worth assaying whether similar mechanisms can occur during ERV regulation.

Endocrine modulation of ERVs

ERVs can be responsive to steroid regulation, MMTV representing a special model. During pregnancy and lactation MMTV is highly expressed in mammary glands of mice of several strains, and the offspring becomes reinfected through the milk, resulting in the reintegration of MMTV in the host genome. Progesterone and glucocorticoids, which modulate proliferation and differentiation of mammary epithelium, regulate MMTV expression as well; in fact, MMTV has glucocorticoid response elements in both the left and right LTRs (Hartig et al., 1993). Corticosteroids enhanced MMTV expression also in macrophage-like cells (Fiegl et al., 1995). Archer et al. (1994)

observed that the native structure of the MMTV proximal promoter region permits activation by glucocorticoid, not by progesterone. In fact, progestins antagonize glucocorticoid activation of MMTV (Deroo and Archer, 2002). Moreover, the transiently expressed glucocorticoid receptor is an effective activator of the MMTV promoter also when it acquires an ordered chromatin structure as an endogenous, replicating gene (Smith et al., 1997). The transiently expressed progesterone receptor is unable to activate the MMTV as ordered chromatin structure; however, the receptor can do so when expressed continuously (Smith et al., 2000). Mulholland et al. (2003) report that trichostatin A, a histone acetylase inhibitor, downregulates the transcription of MMTV, either in the presence or absence of glucocorticoids; rather than an effect on the chromatin organization of the promoter, the authors suggest a mechanism involving acetylation of nonhistone proteins required for basal transcription. Moreover, Qin et al. (1999) showed that the expression of MMTV in the mammary gland can be regulated also by a prolactin-inducible transcription factor in the LTR.

The presence of a glucocorticoid response element may be important also for murine retroviruses other than ERVs. In Friend murine leukaemia virus the presence of the element, that is responsive to both glucocorticoids and glucocorticoid antagonist, is a major determinant for sex differences in susceptibility to infection in vivo (Bruland et al., 2003). Dexamethasone and, to a lesser extent, estradiol, progesterone and the synthetic estrogen diethylstilbestrol, upregulated VL30 in NIH3T3 mouse cells, possibly resulting in the activation of transcription and retrotransposition of these retroelements. In the same model, dexamethasone only was able to upregulate retroviral reverse transcriptase; nevertheless, 5-azacytidine had a significantly stronger upregulating effect on both parameters than steroids (Tzavaras et al., 2003). Upregulation of mouse retrotransposons by an AhR agonist with endocrine toxicity such as 2,3,7,8-tetrachlorodibenzo-p-dioxin may also be noteworthy (Lu and Ramos, 2003).

Preferential expression in endocrine tissues is often a feature of HERVs. In a broad assay on transcription profiles of HERV-H and HERV-W in different cell lines, the steroid-responsive breast cancer cell line T47D showed a high rate of HERV transcription; in particular, it was the only cell line with high expression of the HERV-F-RD *pol* gene (Schon et al., 2001). Studies investigating HERVs in different tissue samples showed that adrenal cortex is a preferential site for ERV-9, which is expressed also in the medulla, HERV-R (ERV3) and HERV-E(4-1) (Katsumata et al., 1998; Andersson et al., 2002; Svensson et al., 2001); an HERV-E element mapping to 17q11 is preferentially expressed in thyroid and pancreas (Shiroma et al., 2001). Most noticeably, the retroviral envelope protein Env T in thyroid is a unique instance of specific expression in a healthy organ, and it may be associated with the hormone-producing status of the tissue (De Parseval et al., 2003). Moreover, the preferential expression in testis and placenta of many HERVs, also detected by De Parseval et al. (2003) may hint at steroid modulation. Nevertheless, no HERV has yet been investigated as thoroughly as MMTV concerning steroid responsiveness. Ono et al. (1987) and Vogetseder et al. (1995) reported HERV-K activation in the human breast carcinoma

Table 1. Summary of endogenous and exogenous agents influencing the expression of human endogenous retroviruses (HERVs)

Agent	Herv(s)	System	Effect	Reference(s)
Morphogenetic factors				
Retinoic acid	ERV-9	Embryonic carcinoma cells	Downregulation	La Mantia et al. (1991)
	HERV-K	Embryonic carcinoma cells	Upregulation	Caricasole et al. (2000)
	HERV-R(ERV3)	Monocytic U-937 cells	Upregulation	Larsson et al. (1997)
Bone morphogenetic proteins	HERV-K	Embryonic carcinoma cells	Upregulation	Caricasole et al. (2000)
Lymphocyte mitogens				
Concanavalin A	HERV-H, HERV-K (NMWV -1-3 and 5-9), HRES-1	Peripheral blood monocytes	Upregulation	Krieg et al. (1992)
Phorbol myristate acetate	HERV-H, HERV-K (NMWV-7)	Peripheral blood monocytes	Upregulation	Krieg et al. (1992)
Pokeweed mitogen	HERV-R(ERV3), HERV-H, HERV-K	Peripheral blood monocytes	Upregulation	Krieg et al. (1992)
Cytokines				
Complement component C4	HERV-K(C4)	White blood cells expressing C4	Downregulation	Schneider et al. (2001)
IL-1alpha and beta	HERV-R(ERV3)	Proximal tubular kidney cells, Vascular endothelial cell	Upregulation Upregulation	Takeuchi et al. (1995), Katsumata et al. (1999)
IL-4, IL-6	HERV-W (associated with multiple sclerosis)	Peripheral blood monocytes	Moderate upregulation	Serra et al. (2003)
Interferon-beta	HERV-W (associated with multiple sclerosis)	Peripheral blood monocytes	Downregulation	Serra et al. (2003)
Interferon-gamma	HERV-R(ERV3)	Monocytic U-937 cells Vascular endothelial cell	Upregulation Downregulation	Larsson et al. (1996) Katsumata et al. (1999)
	HERV-K(C4)	White blood cells expressing C4	Downregulation	Schneider et al. (2001)
	HERV-W (associated to multiple sclerosis)	Peripheral blood monocytes	Upregulation	Serra et al. (2003)
Tumour-necrosis factor-alpha	HERV-R(ERV3)	Vascular endothelial cell	Upregulation	Katsumata et al. (1999)
	HERV-W (associated with multiple sclerosis)	Peripheral blood monocytes	Upregulation	Serra et al. (2003)
Cytotoxic and genotoxic agents				
5-azacytidine	HERV-E(4.1)	Healthy subjects and patients with systemic lupus erythematosus	Upregulation in healthy subjects, not in patients	Okada et al. (2002)
	HERV-K	Tera 1 teratocarcinoma cell line	Increased Gag protein, undermethylation of the gag gene	Gotzinger et al. (1996)
N-butyrate	HERV-K	Tera 1 teratocarcinoma cell line	Increased Gag protein, undermethylation of the gag gene	Gotzinger et al. (1996)
Cycloheximide	HERV-H	T lymphocytes	Downregulation	Kelleher et al. (1996)
Cyclosporin	HERV-H	T lymphocytes	Downregulation	Kelleher et al. (1996)
Steroids				
Corticosteroid	HERV-E (4-1)	Subjects with systemic lupus erythematosus	Down-regulation of gag-transcripts	Ogasawara et al. (2001)
Estradiol priming followed by progesterone	HERV-K	Breast carcinoma cell line T47D	Upregulation	Ono et al. (1987) Vogetseder et al. (1995)
Progesterone	HERV-W	Endometrium of Rhesus monkey, in vivo	Temporal regulation of the Env protein, syncytin	Okulicz and Ace (2003)

cell line T47D with estradiol followed by progesterone; the authors hypothesized that priming with estradiol may facilitate the binding of progesterone receptor complexes on certain HERV-K LTRs. The interplay between steroids and ERVs in some reproductive events is supported by a recent paper, showing the role of progesterone in the temporal regulation of the HERV-W Env protein, syncytin, in the endometrium of rhesus monkey (Okulicz and Ace, 2003). Further hints that might also deserve attention include the possible relationship with autoimmune disorders of endocrine tissues, such as adrenals (Pani et al., 2002a) as well as the regulatory role of an HERV LTR for human beta1,3-galactosyltransferase gene in two tissues with abundant estrogen receptors such as the colon and the mammary gland; Dunn et al. (2003) suggested the interaction with the binding site of hepatocyte nuclear factor 1 as mechanism for the tissue-specific activation of this LTR promoter.

Conclusions

ERVs should not be considered as an undifferentiated whole. In humans the different HERV families, as well as specific sequences, show different patterns concerning, e.g., expression in cells and tissues as well as possible biological roles (Gifford and Tristem, 2003; De Parseval et al., 2003). Major determinants of such differences are obviously related to the genomic insertion patterns, such as non-random targeting of distinctive chromosomal regions (Bushman, 2003), including fragile sites (Perl et al., 1991; Taruscio and Manuelidis, 1991; Meese et al., 1996; Taruscio and Mantovani, 1996; Svensson et al., 2001); adjacent sequences that may be regulated by ERV elements; insertional polymorphisms as observed by Turner et al. (2001) for the most recent HERV family, HERV-K. The human transcriptome map reveals significant clusters of highly expressed or weakly expressed genes, with both types of domains showing tissue-specific modulation (Versteeg et al., 2003). Therefore, the influence of genomic context can be highly relevant to ERV regulation. In particular, concerning the relationships with adjacent genomic sequences, it may be noted that the main regulatory role of HERV elements relates to tissue specific expression (Chen et al., 2002; Dunn et al., 2003); moreover, HERVs regulate a number of genes involved in the response to environmental stimuli (Abrink et al., 1998; Kapitonov and Jurka, 1999; Kjellman et al., 1999; Chen et al., 2002). Thus, it is important to improve the analysis of genes adjacent to HERV integration sites (Taruscio et al., 2002).

Moreover, different ERVs also show differential responsiveness to environmental stimuli, which include endogenous factors from the cell/tissue environment (e.g., cytokines, retinoic acid) but also agents (e.g., drugs) which may become available from the external environment. Table 1 summarizes the available evidence for factors regulating HERVs.

Tissue-specific expression is a major feature of HERVs (Schon et al., 2001; Vinogradova et al., 2001; De Parseval et al., 2003); some HERV products may have physiological roles, such as the putative fusogenic factor syncytin in placenta (Knerr et al., 2002; Frendo et al., 2003). The factors underlying tissue specificity still need to be clarified. Nevertheless, cell type-specific activation of adjacent genes may be an important determinant, but direct modulation by, e.g., steroids, may also be considered. A most noticeable feature of HERVs is preferential expression in critical differentiating tissues, such as the syncytiotrophoblast and the earlier stages of spermatogenic cells (De Parseval et al., 2003). Exposure to retinoic acid in vitro provides information on HERV behaviour in differentiating cells, e.g., ERV-9, is downregulated (Ling et al., 2002), HERV-R (ERV3) is upregulated (Larsson et al., 1997), whereas HERV-K may be upregulated by retinoic acid but not by other morphogenetic factors (Caricasole et al., 2000). Finally, Beyer et al. (2002) pointed out apoptosis as a possible regulatory mechanism of HERV-K in differentiating tissues, a topic which may well deserve further consideration.

The preferential expression of HER-K and HERV-E(4.1) in cancers of the reproductive system (Sauter et al., 1995; Vinogradova et al., 2002; Wang-Johanning et al., 2003) as well as of HERV-K and HERV-H/F in cancers of lymphoid or myeloid origin (Depil et al., 2002; Patzke et al., 2002) is also noteworthy. HERV-K is actually upregulated in several, but not all, cancer types, e.g., in bladder but not in renal carcinomas (Florl et al., 1999). The available evidence does not allow a conclusion about a possible involvement in the process of carcinogenesis. It may be possible that HERV-K is simply triggered within certain cancer cells; in fact, post-transcriptional regulatory mechanisms may exist (Armbruester et al., 2002). Nevertheless, the use of HERVs as disease markers might be explored, as suggested, e.g., by Wang-Johanning et al. (2003).

The data available on the regulation of HERVs in different cells/tissues by cytokines (Larsson et al., 1996; Katsumata et al., 1999; Schneider et al., 2001; Serra et al., 2003), DNA hypomethylators and other cytotoxic agents (Gotzinger et al., 1996; Kelleher et al., 1996) and steroids (Ono et al., 1987; Vogetseder et al., 1995; Schon et al., 2001) may indicate a most promising field for research. Deeper and more systematic investigations

about the interaction of HERVs with endogenous or exogenous stimuli might provide significant clarifications about the possible roles of HERVs in autoimmunity and reproduction, including their relationships with exogenous viruses. Moreover, such interactions should be viewed in the context of the cell (or tissue) type and function. Several papers on HERV regulation by cytokines (Krieg et al., 1992; Takeuchi et al., 1995) or cytotoxic agents (Gotzinger et al., 1996; Kelleher et al., 1996) already indicated the interplay between HERV family, tissue and agent. HERV responsiveness to steroids is interesting, although it may not be surprising; steroid receptors are transcription factors inducing rapid and reversible changes in chromatin structure, which are accompanied by transcriptional activation; noticeably, agents interfering with acetylation of histones and other transcriptional proteins are able to downregulate MMTV (Mulholland et al., 2003). Also, the recent, scattered data showing upregulation of murine retroelements by xenobiotics acting on nuclear receptors (Lu and Ramos, 2003; Tzavaras et al., 2003) may prompt further investigation on the modulation of ERV expression by environmental or dietary factors with endocrine activities, such as the so-called endocrine disrupting chemicals (for a brief review, see Mantovani, 2002).

Murine ERVs, as already mentioned, are different from human HERVs; for instance, they are significantly more active and some of them are infectious; moreover, there are important unique elements like MMTV and the abundant, ERV-related IAPs (Dennis, 2002; Gifford and Tristem, 2003). Nevertheless, as regards potential modulation by endogenous and exogenous factors, one may say that differences are reduced. In fact, insertion in genes relevant to metabolism has been reported (Ilyin et al., 2003); testis may be a preferential expression site (Dupressoir and Heidmann, 1996); there is an evident regulation during the processes of differentiation and aging (Dupressoir and Heidmann, 1996; Casau et al., 1999; Barbot et al., 2002), although with unknown functional consequences, if any; several ERV-related sequences are upregulated in cancers of lymphoid origin (Ristevski et al., 1999); there is good evidence for regulation from cytokines (Tsumara et al., 1994; Blumenstein et al., 1998) and hypomethylating agents (Saavedra et al., 1996; Morgan et al., 1999; Khan et al., 2001; Cooney et al., 2002); finally, MMTV is an obvious model for steroid regulation of ERVs (Callahan and Smith, 2000). Nevertheless, a remarkable difference is provided by expression in placenta. Mouse ERV-related elements are not preferentially expressed in placenta, nor are there hints toward major functional roles, although there are instances of tissue-specific regulation of adjacent genes (Chang-Yeh et al., 1993); also a protective role against exogenous retroviruses has been suggested (Gardner et al., 1991). Human and murine placenta have different structure and duration, even though many genes appear conserved (Cross et al., 2003); thus differences between human and murine ERVs in this critical tissue might be viewed within the overall context of evolutionary divergence between rodents and primates. Thus studies on primates may provide useful information on the regulation of evolutionarily-conserved ERVs in human placenta, as shown by Okulicz and Ace (2003). Overall, with a caveat for placenta, murine models apparently can provide information on ERV regulation that may be useful to understand HERV behaviour in instances (e.g., embryonic and foetal development) where human samples cannot be used.

There is an increasing need to investigate on a broader basis the role of HERVs in the genome and tissues through transcriptomics and proteomics approaches. The extensive genomic survey performed by De Parseval et al. (2003) identified only 16 HERV elements with complete and coding env. Thus, studies on the possible biological roles of HERV products may focus on a limited number of elements. Moreover, a DNA chip-based assay has been recently described as a tool to investigate HERV activities and replication strategies; the assay combines multiplex polymerase chain reaction and chip hybridization (Seifarth et al., 2003).

ERVs are an integral (even though somewhat peculiar) part of the mammalian genome and more broadly of mammalian physiology. The organism works to maintain its homeostasis in a bath of environmental stimuli, reacting through its interconnected endocrine, immune and neural networks. Sometimes, things go wrong. For instance, it cannot be excluded that improper messages from the organism and/or from the external environment might turn some putative beneficial activities of ERVs into detriments, such as increased genomic plasticity or tissue-specific regulatory actions. Overall, the available evidence identifies the diverse, potential interactions with endogenous or exogenous factors as an interesting field for investigating the roles of ERVs in physiology and disease.

References

Abrink M, Larsson E, Hellman L: Demethylation of ERV3, an endogenous retrovirus regulating the Kruppel-related zinc finger gene H-plk, in several human cell lines arrested during early monocyte development. DNA Cell Biol 17:27–37 (1998).

Andersson AC, Venables PJ, Tonjes RR, Scherer J, Eriksson L, Larsson E: Developmental expression of HERV.R (ERV3) and HERV.K in human tissue. Virology 297:220–225 (2002).

Archer TK, Lee HL, Cordingley MG, Mymryk JS, Fragoso G, Berard DS, Hager GL: Differential steroid hormone induction of transcription from the mouse mammary tumor virus promoter. Mol Endocrinol 8:568–576 (1994).

Armbruester V, Sauter M, Krautkraemer E, Meese E, Kleiman A, Best B, Roemer K, Mueller-Lantzsch N: A novel gene from the human endogenous retrovirus K expressed in transformed cells. Clin Cancer Res 8:1800–1807 (2002).

Badenhoop K, Donner H, Neumann J, Herwig J, Kurth R, Usadel KH, Tonjes RR: IDDM patients neither show humoral reactivities against endogenous retroviral envelope protein nor do they differ in retroviral mRNA expression from healthy relatives or normal individuals. Diabetes 48:215–218 (1999).

Baergen B, Benirschke K, Ulich TR: Cytokine expression in the placenta. The role of interleukin 1 and interleukin 1 receptor antagonist expression in chorioamnionitis and parturition. Arch Pathol Lab Med 118:52–55 (1994).

Barbot W, Dupressoir A, Lazar V, Heidmann T: Epigenetic regulation of an IAP retrotransposon in the aging mouse: progressive demethylation and desilencing of the element by its repetitive induction. Nucl Acids Res 30:2365–2373 (2002).

Baust C, Seifarth W, Schon U, Hehlmann R, Leib-Mosch C: Functional activity of HERV.K-T47D-related long terminal repeats. Virology 283:262–272 (2001).

Beyer TD, Kolowos W, Dumitriu IE, Voll RE, Heyder P, Gaipl US, Kalden JR, Herrmann M: Apoptosis of the teratocarcinoma cell line Tera-1 leads to the cleavage of HERV.K10gag proteins by caspases and/or granzyme B. Scand J Immunol 56:303–309 (2002).

Bieche I, Laurent A, Laurendeau I, Duret L, Giovangrandi Y, Frendo JL, Olivi M, Fausser JL, Evain-Brion D, Vidaud M: Placenta-specific INSL4 expression is mediated by a human endogenous retrovirus element. Biol Reprod 68:1422–1429 (2003).

Blond JL, Lavillette D, Cheynet V, Bouton O, Oriol G, Chapel-Fernandes S, Mandrand B, Mallet F, Cosset FL: An envelope glycoprotein of the human endogenous retrovirus HERV.W is expressed in the human placenta and fuses cells expressing the type D mammalian retrovirus receptor. J Virol 74:3321–3329 (2000).

Blumenstein M, Tessmer U, Hossfeld DK, Duhrsen U: Intracisternal A-particle (IAP)-mediated leukemogenesis: levels and stability of IAP mRNA in FDC-P1 cells exposed to the conditions of an irradiated environment. Cell Biol Int 22:563–574 (1998).

Boyd MT, Bax CM, Bax BE, Bloxam DL, Weiss RA: The human endogenous retrovirus ERV-3 is upregulated in differentiating placental trophoblast cells. Virology 196:905–909 (1993).

Branciforte D, Martin SL: Developmental and cell type specificity of LINE-1 expression in mouse testis: implications for transposition. Mol Cell Biol 14:2584–2592 (1994).

Bruland T, Lavik LA, Dai HY, Dalen A: A glucocorticoid response element in the LTR U3 region of Friend murine leukaemia virus variant FIS-2 enhances virus production in vitro and is a major determinant for sex differences in susceptibility to FIS-2 infection in vivo. J Gen Virol 84:907–916 (2003).

Bushman FD: Targeting survival: integration site selection by retroviruses and LTR-retrotransposons. Cell 115:135–138 (2003).

Callahan R, Smith GH: MMTV-induced mammary tumorigenesis: gene discovery, progression to malignancy and cellular pathways. Oncogene 19:992–1001 (2000).

Caricasole A, Ward-van Oostwaard D, Mummery C, van den Eijnden-van Raaij A: Bone morphogenetic proteins and retinoic acid induce human endogenous retrovirus HERV.K expression in NT2D1 human embryonal carcinoma cells. Dev Growth Differ 42:407–411 (2000).

Casau AE, Vaughan JE, Lozano G, Levine AJ: Germ cell expression of an isolated human endogenous retroviral long terminal repeat of the HERV.K/HTDV family in transgenic mice. J Virol 73:9976–9983 (1999).

Chang-Yeh A, Mold DE, Brilliant MH, Huang RC: The mouse intracisternal A particle-promoted placental gene retrotransposition is mouse-strain-specific. Proc Natl Acad Sci, USA 90:292–296 (1993).

Chen HJ, Carr K, Jerome RE, Edenberg HJ: A retroviral repetitive element confers tissue-specificity to the human alcohol dehydrogenase 1C (ADH1C) gene. DNA Cell Biol 21:793–801 (2002).

Cooney CA, Dave AA, Wolff GL: Maternal methyl supplements in mice affect epigenetic variation and DNA methylation of offspring. J Nutr 132:2393S–2400S (2002).

Cross JC, Baczyk D, Dobric N, Hemberger M, Hughes M, Simmons DG, Yamamoto H, Kingdom JC: Genes, development and evolution of the placenta. Placenta 24:123–130 (2003).

Deininger PL, Batzer MA: Mammalian retroelements. Genome Res 12:1455–1465 (2002).

Dennis C: Mouse genome: A forage in the junkyard. Nature 420:458–459 (2002).

De Parseval N, Lazar V, Casella JF, Benit L, Heidmann T: Survey of human genes of retroviral origin: identification and transcriptome of the genes with coding capacity for complete envelope proteins. J Virol 77:10414–10422 (2003).

Depil S, Roche C, Dussart P, Prin L: Expression of a human endogenous retrovirus, HERV.K, in the blood cells of leukemia patients. Leukemia 16:254–259 (2002).

Deroo BJ, Archer TK: Differential activation of the IkappaBalpha and mouse mammary tumor virus promoters by progesterone and glucocorticoid receptors. J Steroid Biochem Mol Biol 81:309–317 (2002).

Donner H, Tonjes RR, Bontrop RE, Kurth R, Usadel KH, Badenhoop K: MHC diversity in Caucasians, investigated using highly heterogeneous noncoding sequence motifs at the DQB1 locus including a retroviral long terminal repeat element, and its comparison to nonhuman primate homologues. Immunogenetics 51:898–904 (2000).

Dragani TA, Manenti G, Della Porta G, Weinstein IB: Factors influencing the expression of endogenous retrovirus-related sequences in the liver of B6C3 mice. Cancer Res 47:795–798 (1987).

Dunn CA, Medstrand P, Mager DL: An endogenous retroviral long terminal repeat is the dominant promoter for human (beta)1,3-galactosyltransferase 5 in the colon. Proc Natl Acad Sci USA 100:12841–12846 (2003).

Dupressoir A, Heidmann T: Germ line-specific expression of intracisternal A-particle retrotransposons in transgenic mice. Mol Cell Biol 16:4495–4503 (1996).

Fiegl M, Strasser-Wozak E, Geley S, Gsur A, Drach J, Kofler R: Glucocorticoid-mediated immunomodulation: hydrocortisone enhances immunosuppressive endogenous retroviral protein (p15E) expression in mouse immune cells. Clin Exp Immunol 101:259–264 (1995).

Firouzi R, Rolland A, Michel M, Jouvin-Marche E, Hauw JJ, Malcus-Vocanson C, Lazarini F, Gebuhrer L, Seigneurin JM, Touraine JL, Sanhadji K, Marche PN, Perron H: Multiple sclerosis-associated retrovirus particles cause T lymphocyte-dependent death with brain hemorrhage in humanized SCID mice model. J Neurovirol 9:79–93 (2003).

Florl AR, Lower R, Schmitz-Drager BJ, Schulz WA: DNA methylation and expression of LINE-1 and HERV.K provirus sequences in urothelial and renal cell carcinomas. Br J Cancer 80:1312–1321 (1999).

Floyd JA, Gold DA, Concepcion D, Poon TH, Wang X, Keithley E, Chen D, Ward EJ, Chinn SB, Friedman RA, Yu HT, Moriwaki K, Shiroishi T, Hamilton BA: A natural allele of *Nxf1* suppresses retrovirus insertional mutations. Nat Genet 35:221–228 (2003).

Frendo JL, Olivier D, Cheynet V, Blond JL, Bouton O, Vidaud M, Rabreau M, Evain-Brion D, Mallet F: Direct involvement of HERV-W Env glycoprotein in human trophoblast cell fusion and differentiation. Mol Cell Biol 23:3566–3574 (2003).

Gardner MB, Kozak CA, O'Brien SJ: The Lake Casitas wild mouse: evolving genetic resistance to retroviral disease. Trends Genet 7:22–27 (1991).

Gaubatz JW, Arcement B, Cutler RG: Gene expression of an endogenous retrovirus-like element during murine development and aging. Mech Ageing Dev 57:71–85 (1991).

Gifford R, Tristem M: The evolution, distribution and diversity of endogenous retroviruses. Virus Genes 26:291–315 (2003).

Giordano R, Magnano AR, Zaccagnini G, Pittoggi C, Moscufo N, Lorenzini R, Spadafora C: Reverse transcriptase activity in mature spermatozoa of mouse. J Cell Biol 148:1107–1113 (2000).

Goldin SN, Papaioannou VE: Unusual misregulation of RNA splicing caused by insertion of a transposable element into the T (Brachyury) locus. BMC Genomics 4:14 (2003).

Golovkina TV, Jaffe AB, Ross SR: Coexpression of exogenous and endogenous mouse mammary tumor virus RNA in vivo results in viral recombination and broadens the virus host range. J Virol 68:5019–5026 (1994).

Gotzinger N, Sauter M, Roemer K, Mueller-Lantzsch N: Regulation of human endogenous retrovirus-K Gag expression in teratocarcinoma cell lines and human tumors. J Gen Virol 77:2893–2990 (1996).

Guasch G, Popovici C, Mugneret F, Chaffanet M, Pontarotti P, Birnbaum D, Pebusque MJ: Endogenous retroviral sequence is fused to FGFR1 kinase in the 8p12 stem-cell myeloproliferative disorder with t(8;19)(p12;q13.3). Blood 101:286–288 (2003).

Hartig E, Nierlich B, Mink S, Nebl G, Cato ACB: Regulation of expression of mouse mammary tumor virus through sequences located in the hormone response element: involvement of cell-cell contact and a negative regulatory factor. J Virol 67:813–821 (1993).

Hayashi S, Yang J, Christenson L, Yanagimachi R, Hecht NB: Mouse preimplantation embryos developed from oocytes injected with round spermatids or spermatozoa have similar but distinct patterns of early messenger RNA expression. Biol Reprod 69:1170–1176 (2003).

Herbst H, Sauter M, Mueller-Lantzsch N: Expression of human endogenous retrovirus K elements in germ cells and trophoblastic tumors. Am J Pathol 149:1727–1735 (1996).

Herbst H, Kuhler-Obbarius C, Lauke H, Sauter M, Mueller-Lantzsch N, Harms D, Loning T: Human endogenous retrovirus (HERV)-K transcripts in gonadoblastomas and gonadoblastoma-derived germ cell tumours. Virchows Arch 434:11–15 (1999).

Herve CA, Lugli EB, Brand A, Griffiths DJ, Venables PJ: Autoantibodies to human endogenous retrovirus-K are frequently detected in health and disease and react with multiple epitopes. Clin Exp Immunol 128:75–82 (2002).

Hsieh L-L, Weinstein IB: Factors influencing the expression of endogenous retrovirus-like sequences in Rat 6 cells. Mol Carcinogen 3:344–349 (1990).

Ilyin G, Courselaud B, Troadec MB, Pigeon C, Alizadeh M, Leroyer P, Brissot P, Loreal O: Comparative analysis of mouse hepcidin 1 and 2 genes: evidence for different patterns of expression and co-inducibility during iron overload. FEBS Lett 542:22–26 (2003).

Johnston JB, Silva C, Holden J, Warren KG, Clark AW, Power C: Monocyte activation and differentiation augment human endogenous retrovirus expression: implications for inflammatory brain diseases. Ann Neurol 50:434–442 (2001).

Kapitonov VV, Jurka J: The long terminal repeat of an endogenous retrovirus induces alternative splicing and encodes an additional carboxy-terminal sequence in the human leptin receptor. J Mol Evol 48:248–251 (1999).

Katsumata K, Ikeda H, Sato M, Harada H, Wakisaka A, Shibata M, Yoshiki T: Tissue-specific high-level expression of human endogenous retrovirus-R in the human adrenal cortex. Pathobiology 66:209–215 (1998).

Katsumata K, Ikeda H, Sato M, Ishizu A, Kawarada Y, Kato H, Wakisaka A, Koike T, Yoshiki T: Cytokine regulation of env gene expression of human endogenous retrovirus-R in human vascular endothelial cells. Clin Immunol 93:75–80 (1999).

Kelleher CA, Wilkinson DA, Freeman JD, Mager DI, Gelfand EW: Expression of novel transposon containing mRNAs in human T cells. J Gen Virol 77:1101–1110 (1996).

Khan AS, Muller J, Sears JF: Early detection of endogenous retroviruses in chemically induced mouse cells. Virus Res 79:39–45 (2001).

Kiessling AA, Crowell R, Fox C: Epididymis is a principal site of retrovirus expression in the mouse. Proc Natl Acad Sci USA 86:5109–5113 (1989).

Kim HS, Yi JM, Jeon SH: Isolation and phylogenetic analysis of HERV.K long terminal repeat cDNA in cancer cells. AIDS Res Hum Retroviruses 17:987–990 (2001).

King LB, Lund FE, White DA, Sharma S, Corley RB: Molecular events in B lymphocyte differentiation. Inducible expression of the endogenous mouse mammary tumor proviral gene, Mtv-9. J Immunol 144:3218–3227 (1990).

Kitamura M, Mruyama N, Shirasawa T, Nagasawa R, Watanabe K, Tateno M, Yoshiki T: Expression of a endogenous retroviral gene product in human placenta. Int J Cancer 58:836–840 (1994).

Kjellman C, Sjogren HO, Salford LG, Widegren B: HERV.F (XA34) is a full-length human endogenous retrovirus expressed in placental and fetal tissues. Gene 239:99–107 (1999).

Knerr I, Beinder E, Rascher W: Syncytin, a novel human endogenous retroviral gene in human placenta: evidence for its dysregulation in preeclampsia and HELLP syndrome. Am J Obstet Gynecol 186:210–213 (2002).

Krieg AM, Gourley MF, Klinman DM, Perl A, Steinberg AD: Heterogenous expression and coordinate regulation of endogenous retroviral sequences in human peripheral blood mononuclear cells. AIDS Res Hum Retroviruses 8:1991–1998 (1992).

Kurdyukov SG, Lebedev YB, Artamonova II, Gorodentseva TN, Batrak AV, Mamedov IZ, Azhikina TL, Legchilina SP, Efimenko IG, Gardiner K, Sverdlov ED: Full-sized HERV.K (HML-2) human endogenous retroviral LTR sequences on human chromosome 21: map locations and evolutionary history. Gene 273:51–61 (2001).

La Mantia G, Maglione D, Pengue G, Di Cristofano A, Simeone A, Lanfrancone L, Lania L: Identification and characterization of novel human endogenous retroviral sequences preferentially expressed in undifferentiated embryonal carcinoma cells. Nucl Acids Res 19:1513–1520 (1991).

Landry JR, Rouhi A, Medstrand P, Mager DL: The Opitz syndrome gene *Mid1* is transcribed from a human endogenous retroviral promoter. Mol Biol Evol 19:1934–1942 (2002).

Larsson E, Kato N, Cohen M: Human endogenous proviruses. Curr Top Microbiol Immunol 148:115–132 (1989).

Larsson E, Venables PJ, Andersson AC, Fan W, Rigby S, Botling J, Oberg F, Cohen M, Nilsson K: Expression of the endogenous retrovirus ERV3 (HERV-R) during induced monocytic differentiation in the U-937 cell line. Int J Cancer 67:451–456 (1996).

Larsson E, Venables P, Andersson AC, Fan W, Rigby S, Botling J, Oberg F, Cohen M, Nilsson K: Tissue and differentiation specific expression on the endogenous retrovirus ERV3 (HERV.R) in normal human tissues and during induced monocytic differentiation in the U-937 cell line. Leukemia 11:142–144 (1997).

Lavrentieva I, Broude NE, Lebedev Y, Gottesman II, Lukyanov SA, Smith CL, Sverdlov ED: High polymorphism level of genomic sequences flanking insertion sites of human endogenous retroviral long terminal repeats. FEBS Lett 443:341–347 (1999).

Lee X, Keith JC Jr, Stumm N, Moutsatsos I, McCoy JM, Crum CP, Genest D, Chin D, Ehrenfels C, Pijnenborg R, van Assche FA, Mi S: Downregulation of placental syncytin expression and abnormal protein localization in pre-eclampsia. Placenta 22:808–812 (2001).

Ling J, Pi W, Bollag R, Zeng S, Keskintepe M, Saliman H, Krantz S, Whitney B, Tuan D: The solitary long terminal repeats of ERV-9 endogenous retrovirus are conserved during primate evolution and possess enhancer activities in embryonic and hematopoietic cells. J Virol 76:2410–2423 (2002).

Löwer R, Löwer J, Kurth R: The viruses in all of us: characteristics and biological significance of human endogenous retrovirus sequences. Proc Natl Acad Sci USA 93:5177–5184 (1996).

Lu KP, Ramos KS: Redox regulation of a novel L1Md-A2 retrotransposon in vascular smooth muscle cells. J Biol Chem 278:28201–28209 (2003).

Luton D, Sibony O, Oury JF, Blot P, Dieterlen-Lievre F, Pardanaud L: The c-ets1 protooncogene is expressed in human trophoblast during the first trimester of pregnancy. Early Hum Dev 47:147–156 (1997).

Maghsoudlou SS, Hughes TR, Hornsby PJ: Analysis of the distal 5′ region of the human CYP17 gene. Genome 38:845–849 (1995).

Mantovani A: Hazard identification and risk assessment of endocrine disrupting chemicals with regard to developmental effects. Toxicology 181–182:367–370 (2002).

Medstrand P, Landry JR, Mager DL: Long terminal repeats are used as alternative promoters for the endothelin B receptor and apolipoprotein C-I genes in humans. J Biol Chem 276:1896–1903 (2001).

Meese E, Gottert E, Zang KD, Sauter M, Schommer S, Mueller-Lantzsch N: Human endogenous retroviral element k10 (HERV.k10): chromosomal localization by somatic hybrid mapping and fluorescence in situ hybridization. Cytogenet Cell Genet 72:40–42 (1996).

Mi S, Lee X, Li X, Veldman GM, Finnerty H, Racie L, LaVallie E, Tang XY, Edouard P, Howes S, Keith JC Jr, McCoy JM: Syncytin is a captive retroviral envelope protein involved in human placental morphogenesis. Nature 403:785–789 (2000).

Morgan HD, Sutherland HG, Martin DI, Whitelaw E: Epigenetic inheritance at the agouti locus in the mouse. Nat Genet 23:314–318 (1999).

Mulholland NM, Soeth E, Smith CL: Inhibition of MMTV transcription by HDAC inhibitors occurs independent of changes in chromatin remodeling and increased histone acetylation. Oncogene 22:4807–4818 (2003).

Obermayer-Straub P, Manns MP: Hepatitis C and D, retroviruses and autoimmune manifestations. J Autoimmun 16:275–285 (2001).

Ogasawara H, Naito T, Kaneko H, Hishikawa T, Sekigawa I, Hashimoto H, Kaneko Y, Yamamoto N, Maruyama N, Yamamoto N: Quantitative analyses of messenger RNA of human endogenous retrovirus in patients with systemic lupus erythematosus. J Rheumatol 28:533–538 (2001).

Okada M, Ogasawara H, Kaneko H, Hishikawa T, Sekigawa I, Hashimoto H, Maruyama N, Kaneko Y, Yamamoto N: Role of DNA methylation in transcription of human endogenous retrovirus in the pathogenesis of systemic lupus erythematosus. J Rheumatol 29:1678–1682 (2002).

Okulicz WC, Ace CI: Temporal regulation of gene expression during the expected window of receptivity in the rhesus monkey endometrium. Biol Reprod 69:1593–1599 (2003).

Ono M, Kawakami M, Ushikuto H: Stimulation of expression of the human endogenous retrovirus genome by female steroid hormones in human breast cancer cell line T47D. J Virol 61:2059–2062 (1987).

Pani MA, Seidl C, Bieda K, Seissler J, Krause M, Seifried E, Usadel KH, Badenhoop K: Preliminary evidence that an endogenous retroviral long-terminal repeat (LTR13) at the HLA-DQB1 gene locus confers susceptibility to Addison's disease. Clin Endocrinol (Oxf) 56:773–777 (2002a).

Pani MA, Wood JP, Bieda K, Toenjes RR, Usadel KH, Badenhoop K: The variable endogenous retroviral insertion in the human complement C4 gene: a transmission study in type I diabetes mellitus. Hum Immunol 63:481–484 (2002b).

Pascual M, Martin J, Nieto A, Giphart MJ, van der Slik AR, de Vries RR, Zanelli E: Distribution of HERV.LTR elements in the 5′-flanking region of HLA-DQB1 and association with autoimmunity. Immunogenetics 53:114–118 (2001).

Patzke S, Lindeskog M, Munthe E, Aasheim HC: Characterization of a novel human endogenous retrovirus, HERV.H/F, expressed in human leukemia cell lines. Virology 303:164–173 (2002).

Perl A, Isaacs CM, Eddy RL, Byers MG, Sait SNJ, Shows TB: The human T-cell leukemia virus-related endogenous sequence (HRES1) is located on chromosome 1 at q42. Genomics 11:1172–1173 (1991).

Peterfy M, Gyuris T, Antonio L, Takacs L: Characterization and chromosomal mapping of two pseudogenes of the mouse *Pafaha/Lis1* gene: retrointegration hotspots in the mouse genome. Gene 216:225–231 (1998).

Pittoggi C, Zaccagnini G, Giordano R, Magnano AR, Baccetti B, Lorenzini R, Spadafora C: Nucleosomal domains of mouse spermatozoa chromatin as potential sites for retroposition and foreign DNA integration. Mol Reprod Dev 56:248–251 (2000).

Pittoggi C, Sciamanna I, Mattei E, Beraldi R, Lobascio AM, Mai A, Quaglia MG, Lorenzini R, Spadafora C: Role of endogenous reverse transcriptase in murine early embryo development. Mol Reprod Dev 66:225–236 (2003).

Qin W, Golovkina TV, Peng T, Nepomnaschy I, Buggiano V, Piazzon I, Ross SR: Mammary gland expression of mouse mammary tumor virus is regulated by a novel element in the long terminal repeat. J Virol 73:368–376 (1999).

Ristevski S, Purcell DF, Marshall J, Campagna D, Nouri S, Fenton SP, McPhee DA, Kannourakis G: Novel endogenous type D retroviral particles expressed at high levels in a SCID mouse thymic lymphoma. J Virol 73:4662–4669 (1999).

Rothenfluh HS: Hypothesis: a memory lymphocyte-specific soma-to-germline genetic feedback loop. Immunol Cell Biol 73:174–180 (1995).

Saavedra HI, Wang TH, Hoyt PR, Popp D, Yang WK, Stambrook PJ: Interleukin-3 increases the incidence of 5-azacytidine-induced thymic lymphomas in pBOR-Il-3 mice. Cell Immunol 173:116–123 (1996).

Sarkar NH, Golovkina T, Uz-Zaman T: RIII/Sa mice with a high incidence of mammary tumors express two exogenous strains and one potential endogenous strain of Mouse Mammary Tumor Virus. J Virol 78:1055–1062 (2004).

Sauter M, Schommer S, Kremmer E, Remberger K, Dolken G, Lemm I, Buck M, Best B, Neumann-Haefelin D, Mueller-Lantzsch N: Human endogenous retrovirus K10: expression of gag protein and detection of antibodies in patients with seminomas. J Virol 69:414–421 (1995).

Schiavetti F, Thonnard J, Colau D, Boon T, Coulie PG: A human endogenous retroviral sequence encoding an antigen recognized on melanoma by cytolytic T lymphocytes. Cancer Res 62:5510–5516 (2002).

Schneider PM, Witzel-Schlomp K, Rittner C, Zhang L: The endogenous retroviral insertion in the human complement C4 gene modulates the expression of homologous genes by antisense inhibition. Immunogenetics 53:1–9 (2001).

Schon U, Seifarth W, Baust C, Hohenadl C, Erfle V, Leib-Mosch C: Cell type-specific expression and promoter activity of human endogenous retroviral long terminal repeats. Virology 279:280–291 (2001).

Schulte AM, Malerczyk C, Cabal-Manzano R, Gajarsa JJ, List HJ, Riegel AT, Wellstein A: Influence of the human endogenous retrovirus-like element HERV.E.PTN on the expression of growth factor pleiotrophin: a critical role of a retroviral Sp1-binding site. Oncogene 19:3988–3998 (2000).

Seifarth W, Spiess B, Zeilfelder U, Speth C, Hehlmann R, Leib-Mosch C: Assessment of retroviral activity using a universal retrovirus chip. J Virol Methods 112:79–91 (2003).

Serra C, Mameli G, Arru G, Sotgiu S, Rosati G, Dolei A: In vitro modulation of the multiple sclerosis (MS)-associated retrovirus by cytokines: implications for MS pathogenesis. J Neurovirol 9:637–643 (2003).

Shiroma T, Sugimoto J, Oda T, Jinno Y, Kanaya F: Search for active endogenous retroviruses: identification and characterization of a HERV.E gene that is expressed in the pancreas and thyroid. J Hum Genet 46:619–625 (2001).

Smallwood A, Papageorghiou A, Nicolaides K, Alley MK, Jim A, Nargund G, Ojha K, Campbell S, Banerjee S: Temporal regulation of the expression of syncytin (HERV.W), maternally imprinted PEG10, and SGCE in human placenta. Biol Reprod 69:286–293 (2003).

Smith CL, Htun H, Wolford RG, Hager GL: Differential activity of progesterone and glucocorticoid receptors on mouse mammary tumor virus templates differing in chromatin structure. J Biol Chem 272:14227–14235 (1997).

Smith CL, Wolford RG, O'Neill TB, Hager GL: Characterization of transiently and constitutively expressed progesterone receptors: evidence for two functional states. Mol Endocrinol 14:956–971 (2000).

Sutkowski N, Conrad B, Thorley-Lawson DA, Huber BT: Epstein-Barr virus transactivates the human endogenous retrovirus HERV.K18 that encodes a superantigen. Immunity 15:579–589 (2001).

Svensson AC, Raudsepp T, Larsson C, Di Cristofano A, Chowdhary B, La Mantia G, Rask L, Andersson G: Chromosomal distribution, localization and expression of the human endogenous retrovirus ERV-9. Cytogenet Cell Genet 92:89–96 (2001).

Takeuchi K, Katsumata K, Ikeda H, Wakisaka A, Yoshiki T: Expression of endogenous retroviruses, ERV-3 and 4-1, in synovial tissues from patients with rheumatoid arthritis. Clin Exp Immunol 99:338–344 (1995).

Tamura N, Iwase A, Suzuki K, Maruyama N, Kira S: Alveolar macrophages produce the Env protein of a human endogenous retrovirus, HERV.E 4-1, in a subgroup of interstitial lung diseases. Am J Respir Cell Mol Biol 16:429–437 (1997).

Taruscio D, Mantovani A: Eleven chromosomal integration sites of a human endogenous retrovirus (HERV 4-1) map close to known loci of thirteen hereditary malformation syndromes. Teratology 54:108–110 (1996).

Taruscio D, Mantovani A: Human endogenous retroviral sequences: possible roles in reproductive physiopathology. Biol Reprod 59:713–724 (1998).

Taruscio D, Manuelidis L: Integration site preferences of endogenous retroviruses. Chromosoma 101: 141–156 (1991).

Taruscio D, Floridia G, Zoraqi GK, Mantovani A, Falbo V: Organization and integration sites in the human genome of endogenous retroviral sequences belonging to HERV.E family. Mammal Genome 13:216–222 (2002).

Ting C-N, Rosenberg MP, Snow CM, Samuelson LC, Meisier MH: Endogenous retroviral sequences are required for tissue-specific expression of a human salivary amylase gene. Genes Develop 6:1457–1465 (1992).

Tristem M: Identification and characterization of novel human endogenous retrovirus families by phylogenetic screening of the human genome mapping project database. J Virol 74:3715–3730 (2000).

Tsumara H, Wang JZ, Ogawa S, Ohota H, Komada H, Ito Y, Shimura K: IL-1 induces intracisternal type A virus and retrovirus type C in pancreatic beta-cells of NOD mice. J Exp Anim Sci 36:141–150 (1994).

Turner G, Barbulescu M, Su M, Jensen-Seaman MI, Kidd KK, Lenz J: Insertional polymorphisms of full-length endogenous retroviruses in humans. Curr Biol 11:1531–1535 (2001).

Tzavaras T, Eftaxia S, Tavoulari S, Hatzi P, Angelidis C: Factors influencing the expression of endogenous reverse transcriptases and viral-like 30 elements in mouse NIH3T3 cells. Int J Oncol 23:1237–1243 (2003).

Urnovitz HB, Murphy WH: Human endogenos retroviruses: nature, occurrence, and clinical implications in human disease. Clin Microbiol Rev 9:72–99 (1996).

Venables PJ, Brookes SM, Griffiths D, Weiss RA, Boyd MT: Abundance of an endogenous retroviral envelope protein in placental trophoblasts suggests a biological function. Virology 211:589–592 (1995).

Versteeg R, van Schaik BD, van Batenburg MF, Roos M, Monajemi R, Caron H, Bussemaker HJ, van Kampen AH: The human transcriptome map reveals extremes in gene density, intron length, GC content, and repeat pattern for domains of highly and weakly expressed genes. Genome Res 13:1998–2004 (2003).

Vinogradova TV, Leppik LP, Nikolaev LG, Akopov SB, Kleiman AM, Senyuta NB, Sverdlov ED: Solitary human endogenous retroviruses-K LTRs retain transcriptional activity in vivo, the mode of which is different in different cell types. Virology 290:83–90 (2001).

Vinogradova T, Leppik L, Kalinina E, Zhulidov P, Grzeschik KH, Sverdlov E: Selective differential display of RNAs containing interspersed repeats: analysis of changes in the transcription of HERV.K LTRs in germ cell tumors. Mol Genet Genomics 266:796–805 (2002).

Vogetseder W, Feng J, Haibach C, Mayerl W, Dierich MP: Detection of a 67-kD glycoprotein in human tumor cell lines by a monoclonal antibody established against a recombinant human endogenous retrovirus-K envelope-gene-encoded protein. Exp Clin Immunogenet 12:96–102 (1995).

Wang-Johanning F, Frost AR, Jian B, Azerou R, Lu DW, Chen DT, Johanning GL: Detecting the expression of human endogenous retrovirus E envelope transcripts in human prostate adenocarcinoma. Cancer 98:187–197 (2003).

Cytogenet Genome Res 105:363–374 (2004)
DOI: 10.1159/000078209

An integrative genomics approach to the reconstruction of gene networks in segregating populations

J. Zhu,[a] P.Y. Lum,[a] J. Lamb,[a] D. GuhaThakurta,[a] S.W. Edwards,[a] R. Thieringer,[b] J.P. Berger,[c] M.S. Wu,[d] J. Thompson,[e] A.B. Sachs,[a] and E.E. Schadt[a]

[a] Rosetta Inpharmatics, Seattle, WA; [b] Department of Atherosclerosis and Endocrinology,
[c] Department of Metabolic Disorders, [d] Department of Pharmacology,
[e] Department of Molecular Profiling, Merck & Co, Rahway, NJ (USA)

Abstract. The reconstruction of genetic networks in mammalian systems is one of the primary goals in biological research, especially as such reconstructions relate to elucidating not only common, polygenic human diseases, but living systems more generally. Here we propose a novel gene network reconstruction algorithm, derived from classic Bayesian network methods, that utilizes naturally occurring genetic variations as a source of perturbations to elucidate the network. This algorithm incorporates relative transcript abundance and genotypic data from segregating populations by employing a generalized scoring function of maximum likelihood commonly used in Bayesian network reconstruction problems. The utility of this novel algorithm is demonstrated via application to liver gene expression data from a segregating mouse population. We demonstrate that the network derived from these data using our novel network reconstruction algorithm is able to capture causal associations between genes that result in increased predictive power, compared to more classically reconstructed networks derived from the same data.

Copyright © 2004 S. Karger AG, Basel

Complex disease research has benefited significantly from the completion of the sequencing of several genomes and from high-throughput functional genomics technologies like microarrays for molecular profiling. The complete human and mouse genomic sequences have allowed researchers to more rapidly identify genes underlying susceptibility loci for common human diseases like schizophrenia (Stefansson et al., 2002) and autoimmune disorders (Ueda et al., 2003). Further, gene expression microarrays and other high-throughput molecular profiling technologies have been used to identify complex disease subtypes (van 't Veer et al., 2002; Schadt et al., 2003), to directly identify genes underlying susceptibility loci for common human diseases like asthma (Karp et al., 2000) and cyto-

chrome c oxidase deficiency (Mootha et al., 2003), and to identify biomarkers for clinical trials (van de Vijver et al., 2002). More recently, Schadt et al. (2003) have combined gene expression and genetic data in segregating populations to elucidate complex diseases by treating gene expression as a quantitative trait and mapping quantitative trait loci (QTL) for those traits in mouse models for common human diseases. By looking at patterns of co-localization between disease trait QTL and gene expression QTL, Schadt et al. (2003) demonstrated how candidate genes for complex diseases can be identified in a completely objective fashion.

The integration of genotype, transcription, and clinical trait data to elucidate pathways associated with complex disease traits can be more generally modeled using graphical structures constructed from experimental data. Graphical models have the potential to efficiently identify and represent the key gene-gene interactions driving the complex disease traits. The causal inferences that can be derived from QTL data, where causality follows from the central dogma of biology (i.e., DNA variations lead to changes in transcription regulation/protein function,

Received 3 November 2003; manuscript accepted 9 December 2003.

Request reprints from Eric E. Schadt, Rosetta Inpharmatics
    401 Terry Ave N, Seattle, WA 98109 (USA); telephone: 206-802-7368
    fax: 206-802-6477; e-mail: eric_schadt@merck.com

which in turn cause variations in disease phenotypes), provide a novel source of information that complement gene expression data and that can be incorporated into methods that seek to identify graphical models (gene networks) of gene interactions. Several approaches exist for the systematic study of biological systems that ultimately result in the construction of these graphical models. A number of methods utilize protein-protein binding information to construct gene networks (Marcotte et al., 2001; Xenarios et al., 2002). These networks, termed association networks, establish gene-gene interactions by examining binding domains shared between protein pairs. While these approaches have been effective in associating genes involved in common pathways, they are not able to determine genes that are causative for other genes in a given pathway, nor are they able to predict outcomes of perturbations to a given system, thus limiting their utility. Other methods used to systematically characterize interaction data include differential equations for dynamic systems (Davidson et al., 2002) and multiple linear equations for near steady-state systems (Gardner et al., 2003). One drawback of these approaches is that they require extensive data and other quantitative information in addition to the gene expression data, making them suitable for only small, focused networks/pathways.

More recently, significant research interest has shifted to the use of Bayesian networks to study causal interaction networks of biological systems based on gene expression data from time series and gene knockout experiments, protein-protein interaction data derived from predicted genomics features, and on other direct experimental interaction data (Pe'er et al., 2001; Jansen et al., 2003). Bayesian networks represent directed acyclic graphs, and so are capable of not only depicting important interactions among genes, but they can also represent causal associations between genes since the graphs are directed. In the biological systems context the nodes of these graphs represent genes, and the edges are weighted and directed based on an associated set of conditional probabilities that represent the extent and direction of the association between nodes connected by an edge. The conditional probabilities can be represented by a discrete or continuous probability distribution. To estimate the conditional probabilities used to construct a Bayesian network, perturbations that cover all possible conditions are needed.

Typically, the multiple conditions needed to estimate the conditional probabilities are generated by "artificial" genetic perturbations, such as gene knockouts, transgenics, siRNA, and mutagenesis. Environmental perturbations such as changes in nutrition and temperature (Ideker et al., 2001) can also be used to perturb a network. In addition to genetic and environmental perturbations, it is reasonable to assume a temporal dimension for any given experimental condition. Therefore, sampling a series of time points for a given experimental condition may represent multiple conditions that can be used to estimate conditional probabilities for network reconstruction. Although others have demonstrated that when gene expression data are used to estimate these conditional probabilities over different time points, the causal relationships inferred from time series data may be less reliable than those derived from the competing methods just discussed, given absolute

mRNA levels are confounded by variations in degradation ratesamong the different mRNA (Gordon et al., 1988; Yang et al., 2003).

To systematically study interaction networks in experimental systems, genes can be systematically knocked out, inhibited by drug compounds that target specific genes, or inhibited/activated using chemical or siRNA technologies for every gene in the system under study. Some of these techniques are time-consuming and lack the multifactorial context needed to achieve many complex phenotypes of interest. Chemical and siRNA inhibition can be accomplished efficiently, but these techniques frequently give rise to off-target effects that cannot be resolved without additional experimentation (Jackson et al., 2003). Naturally occurring genetic variations, however, are one source of perturbations that address many of the shortcomings just noted. This type of perturbation has yet to be exploited on a genome-wide scale using gene expression or other molecular profiling data, despite it being one of the more comprehensive forms of multifactorial perturbations available in mammalian systems. Others have already demonstrated that these genetic perturbations are easily detected in segregating populations, and that they suggest causal links between gene expression and disease traits (Brem et al., 2002; Schadt et al., 2003). This form of perturbation potentially provides for other advantages in that these perturbations are naturally occurring, and so, have more naturally evolved in the system under study. Therefore, the expression levels represent steady states of the system rather than transient states provided by most of the artifical perturbation techniques.

Bayesian networks, or graphical models more generally, can be applied to gene expression data to reconstruct interaction networks. However, because of the limited expression data typically available for any particular system in a given state, network reconstruction processes typically result in the identification of multiple networks that explain the data equally well. In fact, in most cases causal relationships can not be reliably inferred from gene expression data alone, since for any particular network changing the direction of the edge between any two genes has little effect on the model fit. To reliably infer causal relationships, additional information is required. Patterns of overlapping gene expression quantitative trait loci (eQTL) can help distinguish between gene expression changes representing downstream (i.e., reactive) effects and those representing upstream (i.e., causal) effects (Schadt et al., 2003). We use large-scale liver microarray and genotypic data from the segregating mouse population described by Schadt et al. (2003) to establish criteria for discriminating among alternative interactions, and then use these results to guide reconstruction of an interaction network for the mouse liver system. We demonstrate the utility of the resulting network in this system by examining the gene expression behavior of 11-beta hydroxysteroid dehydrogenase 1 *(Hsd11b1)* in the reconstructed network. The predictive capabilities of the network are assessed by comparing the set of genes predicted by the network to respond to perturbations in the expression of *Hsd11b1*, with the set of genes observed to change in response to activating the peroxisome proliferators activated receptor alpha (Ppara) using a Ppara agonist in an independent experiment, where the *Hsd11b1* expression is down-regulated in response to the Ppara agonist (Hermanow-ski-Vosatka et al., 2000). Further, by refining the predicted network based on these combined experimental results, we demonstrate how these data can be used to identify the key steps involved in *Hsd11b1* regulating downstream responders.

Materials and methods

Experimental design and data collection

The F2 mouse population and gene expression data used in this study have been previously described by Drake et al. (2001) and Schadt et al. (2003). Briefly, an F2 population consisting of 111 mice was constructed from two inbred strains of mice, C57BL/6J and DBA/2J. Only female mice were maintained in this population. All mice were housed under conditions meeting the guidelines of the Association for Accreditation of Laboratory Animal Care. Mice were on a rodent chow diet up to 12 months of age, and then switched to an atherogenic high-fat, high-cholesterol diet for another 4 months. At 16 months of age the mice were phenotyped and their livers extracted for gene expression profiling. The mice were genotyped at 139 microsatellite markers uniformly distributed over the mouse genome to allow for the genetic mapping of the gene expression and disease traits.

At the time of sacrifice, livers were removed from the F2 animals and RNA isolated for gene expression profiling, as described by Schadt et al. (2003). Prepared RNA from each F2 animal was hybridized against a pool of RNA, constructed from equal aliquots of RNA from each F2 animal, using a comprehensive 23,574 gene microarray manufactured by Agilent Technologies. In all, 18,131 genes gave rise to intensity measures that were significantly above the background noise ($P < 0.05$). However, only 1088 of these genes were detected as significantly differentially expressed based on criteria that expression levels change at least 1.5-fold with an associated P value less than 0.01 for at least 25 % of mice. These 1088 genes were used in the network reconstruction study and are referred to as the F2 data set.

To assess the accuracy of the predicted network and further refine it, we performed a perturbation experiment on *Ppara*. *Ppara* is one of the 1088 genes in the network constructed from the F2 data set, and perturbations in *Ppara* activity can be seen to down regulate *Hsd11b1* (Hermanowski-Vosatka et al., 2000). Nineteen male C57BL/6J mice between the ages of 10 and 13 weeks were used in the experiment. The following three treatments were administered orally for 7 days: 1) 6 mice treated with 200 mg/kg of Fenofibrate, 2) 6 mice treated with 30 mg/kg of WY-14643, and 3) 7 mice treated as vehicle controls. After the 7 day treatment regimen, which allowed the mice to reach a new steady state with respect to *Ppara* activity, all mice were sacrificed and RNA was extracted from the livers of each animal for profiling on gene expression microarrays. RNA from randomly formed pairs of animals in the treatment groups was pooled, resulting in three replicate RNA samples for each treatment group. These RNA samples were hybridized against a pool of RNA constructed from equal aliquots of RNA from each control mouse using microarrays manufactured by Agilent Technologies.

Network reconstruction using Bayesian networks

Forming candidate relationships among genes was carried out using an extension of standard Bayesian network reconstruction methods (Friedman et al., 2000). In our extension of these methods we incorporated QTL information for the transcript abundances of each gene considered in the network. It is well known that searching for the best possible network linking a moderately sized set of genes is an NP-hard problem. Exhaustively searching for the optimal network with hundreds of genes is presently a computationally intractable problem. Therefore, various simplifications are typically applied to reduce the size of the search space and to reduce the number of parameters that need to be estimated from the data. Here we employed two simplifying assumptions to achieve such reductions. First, we assumed that while any gene in a biological system can control many other genes, a given gene was not allowed to be controlled by more than three genes. Second, we allowed only a subset of candidate genes to be considered as possible causal drivers (parent nodes) for a given gene, instead of allowing for the possibility of any gene in the complete gene set to be so considered.

To select potential parents for each gene, we employed a strategy to assess the extent of genetic overlap between any two gene expression traits. If RNA levels for two genes are tightly associated, or if such levels are genetically controlled by a similar set of loci, then their eQTL should overlap. We used the genome-wide eQTL computed by Schadt et al. (2003) to determine whether any two genes in our 1088 gene set had overlapping eQTL. The extent of QTL overlap was measured by computing the correlation coefficient between vectors of LOD scores associated with eQTL identified over entire chromosomes for each gene. If we assume no epistasis between eQTL for a given expression trait, then the eQTL can be considered independent between chromosomes, and a measure of genetic relatedness over all chromosomes for any two traits can be subsequently computed as a weighted average of correlations for each individual chromosome:

$$r = \sum_c w(c) * r(c), \tag{1}$$

where $r(c)$ is the correlation coefficient for chromosome c, and $w(c)$ is the chromosome-specific weight. The chromosome-specific correlation coefficient is given by

$$r(c) = \frac{\sum_l x_c(l) * y_c(l) * I_c(l)}{\sum_l x_c(l) * x_c(l) + \sum_l y_c(l) * y_c(l)}. \tag{2}$$

In this equation, $x_c(l)$ and $y_c(l)$ are LOD scores at locus l on chromosome c, and $I_c(I)$ is an indicator function defined as:

$$I_c(l) = \left\{ \begin{array}{l} 1, \text{ if } x_c(l) > 1.5, y_c(l) > 1.5 \\ 0, \text{ otherwise} \end{array} \right\}, \tag{3}$$

which was incorporated to eliminate low LOD scores that would have likely only contributed noise to the correlation measures. The chromosome-specific weight terms are given by

$$w(c) = \max_l(\min(x_c(l),10) * \min(y_c(l),10)) \tag{4}$$

In effect, this equation provides for those chromosomes with high LOD scores to more significantly influence the overall correlation measure. This heuristic weight is intuitively appealing since gene expression traits with common significant eQTL have a larger percentage of their overall variation explained by these common genetic effects. For each gene, the correlation coefficients described above were computed for all genes in the set of 1088 genes of interest, and the 80th percentile of the rank-ordered list of correlations was arbitrarily chosen as the cutoff for genes to be considered as parental nodes in the network for a given gene of interest.

To differentiate between co-localization of QTL for a pair of traits due to common genetic effects (pleiotropy) versus multiple closely linked QTL, we assessed the extent of phenotypic (RNA levels) association as measured by the mutual information measure

$$mi(A, B) = \sum_{i,j} p(a_i, b_j) \log \frac{p(a_i, b_j)}{p(a_i)p(b_j)}, \tag{5}$$

where $p(x)$ is the probability density function for the expression of gene X in the system of interest. As was done for the correlation calculations, the mutual information measure was computed for all gene pairs in the F2 data set, and again this list was rank ordered and those genes in the 80th percentile were chosen as candidate parental nodes for a given gene of interest.

The selection of the genes in the 80th percentile of the rank-ordered lists generated from the correlation and mutual information measures provides the prior evidence that two genes may be causally related. The QTL data provide the causal anchors that allow this type of inference to be made. That is, by definition, a QTL controlling for the expression of two gene expression traits implies that DNA variations in the QTL lead to variations in the expression of the associated gene traits. Therefore, it must be the case that any gene expression trait pair controlled by a common QTL is either 1) independently driven by the same QTL, or 2) causally associated in that one of the two traits is driven by the QTL, while the other trait responds to the trait driven by the QTL. Below we shall see how the conditional mutual information measure helps resolve which of these cases holds.

In addition to utilizing the QTL information as prior information to restrict the types of relationships that can be established among genes, the QTL information can be more intimately integrated into the network reconstruction process. Because correlation measures are symmetric, they can

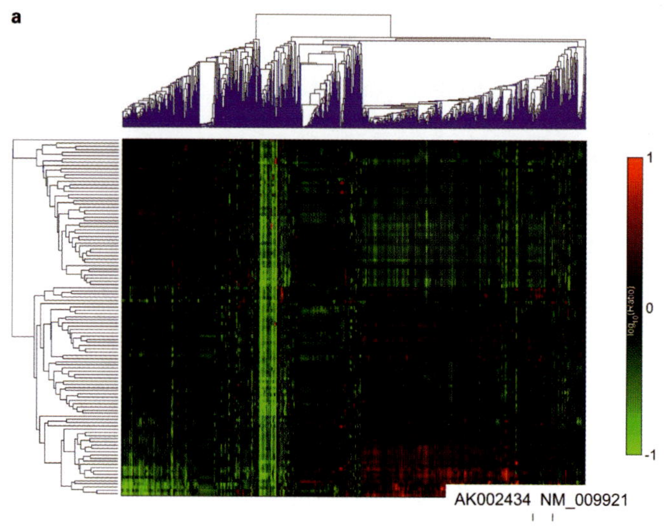

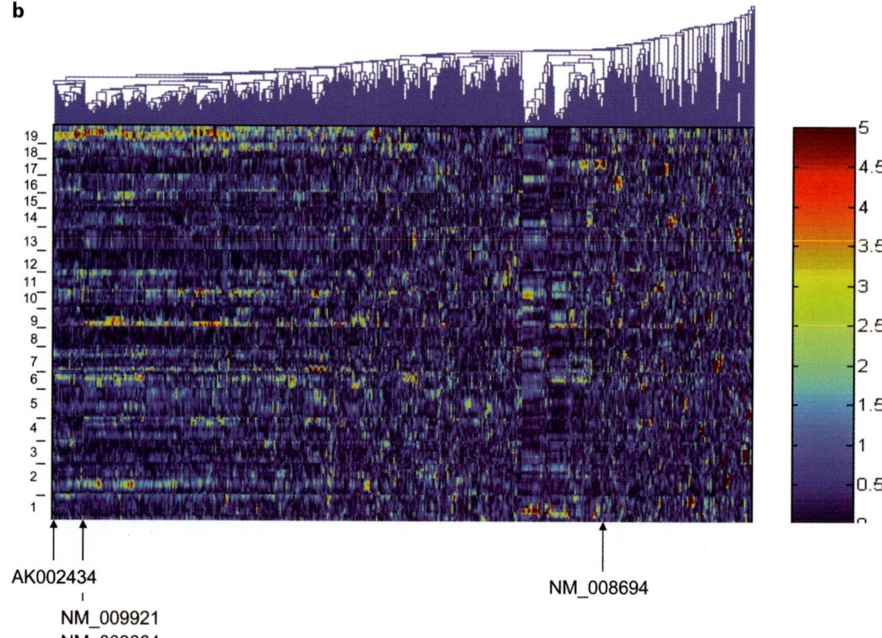

Fig. 1. Hierarchical clustering of the F2 data set in the gene expression and eQTL dimensions. (**a**) Color matrix display of cluster results in the gene expression and experiment dimensions. The horizontal and vertical axes represent the gene expression and experiment dimensions, respectively. The color matrix elements are color coded according to the mean log ratio of the expression values. Red indicates a gene is up regulated relative to the reference pool, green indicates a gene is down-regulated relative to the reference pool, black indicates a gene is not differentially expressed relative to the reference pool, and grey indicates the expression ratio could not be computed for the mouse/gene pair. Four genes that cluster tightly together are highlighted. (**b**) Color matrix display of cluster results in the QTL dimension, given along the horizontal axis. The vertical axis is ordered according to chromosomal location. Each element of the color matrix represents a LOD score for a given gene expression trait at a given chromosomal location. The LOD scores for each element are color coded to highlight regions with significant QTL activity. The four genes highlighted in **a** are also highlighted in this plot. While these four genes grouped tightly together in the expression cluster, here we see that NM_009864 does not group well with the other three genes, indicating the genetic component of NM_009864 does not contribute significantly to the overall correlation structure between this gene and the other three genes.

indicate association but not causality. However, as indicated above, QTL mapping information for the gene expression traits can be used to help sort out causal relationships as well. For example, suppose gene expression trait X has two high confidence eQTL at loci L_1 and L_2, while gene expression trait Y has a single eQTL at L_1 that is more significant than the eQTL for X at L_1. In this instance, it is reasonable to infer that Y may control X (or is "causal" for), since if X were controlling Y, we would expect Y to have an eQTL at L_2 in addition to the eQTL at L_1, given X has an eQTL at L_2. Further, the asymmetry in the significance of the eQTL at L_1 for X and Y also favors Y as being causal for X. Thus, it is possible to use the eQTL overlap information to infer causality by defining the prior for a candidate relationship as

$$p(X \rightarrow Y) = r(X, Y) \frac{N(Y)}{N(X) + N(Y)}, \qquad (6)$$

where N is a gene expression trait's complexity as measured by the number of significant eQTL mapped for the given gene expression trait. If X and Y have coincident QTL, but the overall complexity of Y is greater than the overall

complexity of X, then we infer that X causes Y $(X \rightarrow Y)$, as opposed to $Y \rightarrow X$. Implicit in this weighting scheme is the assumption that traits driven by common QTL are causally related (i.e., one trait drives the other), even though it is possible that the traits could be independently driven by the same set of QTL. However, in cases where multiple traits are independently driven by a common set of QTL, the correlation between the traits will be smaller than when the traits are causally related, so the prior will carry less weight. Further, the conditional mutual information measure discussed below will also serve to prevent, in at least some cases, causal links from being made between genes that are independently driven by a common set of QTL.

The causality relationships between gene expression traits can be further assessed by considering whether the QTL for a particular gene expression trait is *cis* acting. If a gene expression trait X has an eQTL at locus L that is coincident with the gene's physical location, we say the gene has a *cis*-acting eQTL at L. This relationship indicates it is likely that DNA variations in the gene itself at least partially explain variations in the gene's observed RNA levels. Therefore, we can infer that a gene with a *cis*-acting eQTL is at least partially under the control of the gene itself. In this particular situation, where another gene expression trait Y has an eQTL mapping to the physical location of gene X, we can infer that X may be causal for Y, and that Y is likely not causal for X. As a result, when this condition holds we set $p(Y \rightarrow X)$ equal to 0.

With the various constraints and measures defined above, our goal is to find a graphical model M (a gene network) that best represents the relationships between genes, given a gene expression data set, D, of interest. That is, given data D, we seek to find the model with the highest posterior probability $P(D|M)$, where

$$P(M|D) \propto P(D|M)P(M). \tag{7}$$

In this relationship, $P(D|M)$ is the likelihood of D given M, and $P(M)$ is the prior probability of model M

$$P(M) = \prod_{X \rightarrow Y} p(X \rightarrow Y), \tag{8}$$

taken over all paths in the network (M) under consideration. The algorithm we employed to search through all possible models to find the network that best fits the data is similar to the local maximum search algorithm implemented by Friedman et al. (2000).

Results

Exploratory clustering in the gene expression and QTL dimensions

Similarities for the transcript abundances of the 1088 genes in the F2 data set were broadly assessed using hierarchical clustering in the gene expression and eQTL dimensions. Fig. 1a depicts hierarchical clustering results for the gene expression and experiment dimensions, where the Pearson correlation measure on relative transcript abundances was used as the similarity metric, as described by Schadt et al. (2003). Fig. 1b depicts the hierarchical clustering results in the eQTL dimension, where in this case the similarity metric used is the correlation measure defined in equation (1).

From Fig. 1a we note the relatively tight associations among the genes that break the mice roughly into two general groups. From the eQTL clustering in Fig. 1b, it is apparent there is no preferred grouping of the genes, but instead genes driven by the same set of QTL, which can be seen as the bright bands running along the chromosome-location axis of the color display matrix, appear to group most tightly together. That is, while genes that are highly correlated group together in the standard gene expression cluster shown in Fig. 1a, for eQTL clustering those genes driven by common genetic loci are more closely associated, indicating that the genetic component of the overall correlation structure between all gene expression trait pairs may give different preferred orderings, thereby highlighting different relationships among genes that are not apparent from the gene expression data alone. For example, NM_009921 *(Camp)* and NM_008694 *(Ngp)* are genes involved in defense response that are seen to cluster together in Figs. 1a and 1b, indicating at least part of the correlation structure between these two genes is due to shared genetic components. However, AK002434 *(Krt27)* and NM_009864 *(Cdh1)* cluster together in Fig. 1a, but are spatially separated in Fig. 1b, indicating their correlation structure was largely a consequence of non-genetic components. We sought to exploit these genetic relationships among genes, in addition to the raw gene expression relationships, to reconstruct gene networks.

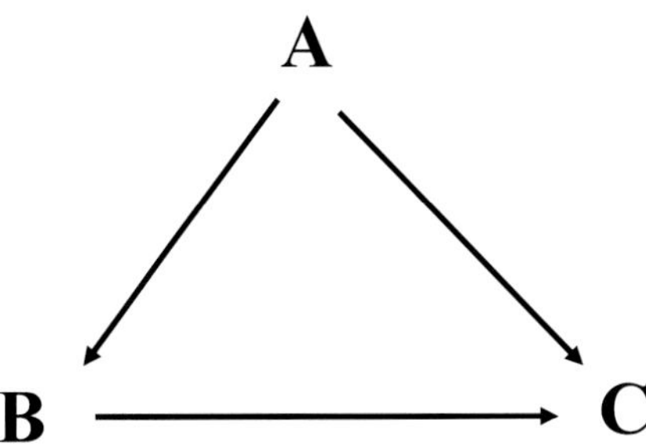

Fig. 2. Example of a sub-network that potentially over fits the data. Node A is seen as causal for nodes B and C, and node B is seen as causal for C. To determine whether the causal association between B and C is a consequence of A, we compute the conditional mutual information between B and C given A. If this value does not differ significantly from 0 (as described in the text), then the link between B and C is removed.

Consensus networks

The 1088 genes in the F2 data set were input into a Bayesian network program written in the C++ programming language and based on the reconstruction algorithm described above. We constructed 1000 networks based on this set of genes and derived a consensus network from this set of networks by identifying links between gene pairs that existed in more than 40 % of the networks. Each link was assigned a confidence value corresponding to the number of times it appeared in the 1000 networks considered. Cycles in the resulting consensus network were broken by removing links in the cycle associated with the lowest confidence values.

Starting with the consensus network, we sought to identify the most parsimonious model given the data, using the maximum likelihood method described above. Although we introduced priors for network structure to penalize more complex topologies, we recognized that optimal networks derived from this process may still over fit the data. To lessen the likelihood that the resulting networks over fit the data, we removed links in the network to simplify relationships among genes using a conditional information measure. For instance, if the type of sub-network shown in Fig. 2 was found to exist in the larger network, we computed the conditional mutual information between B and C to determine if B and C were still found to be dependent given information on node A. The conditional mutual information in this instance is given by

$$MI(B,C|A) = \sum_{i,j,k} p(b_i, c_j, a_k) \log \left(\frac{p(b_i, c_j|a_k)}{p(b_i|a_k)p(c_j|a_k)} \right). \tag{9}$$

If $MI(B,C|A)$ is not significantly different from 0, then the determination is made that the link $B \rightarrow C$ can be safely removed. Pearl (1988) provides more details on this type of procedure.

After removing links causing cycles and links that lead to over fitting of the data, the resulting consensus network contained links for 909 genes. Interestingly, there were 179 genes that did not have any links in the network, suggesting these 179 genes are not informative for this network reconstruction problem. Clustering the expression values for these 179 genes (data

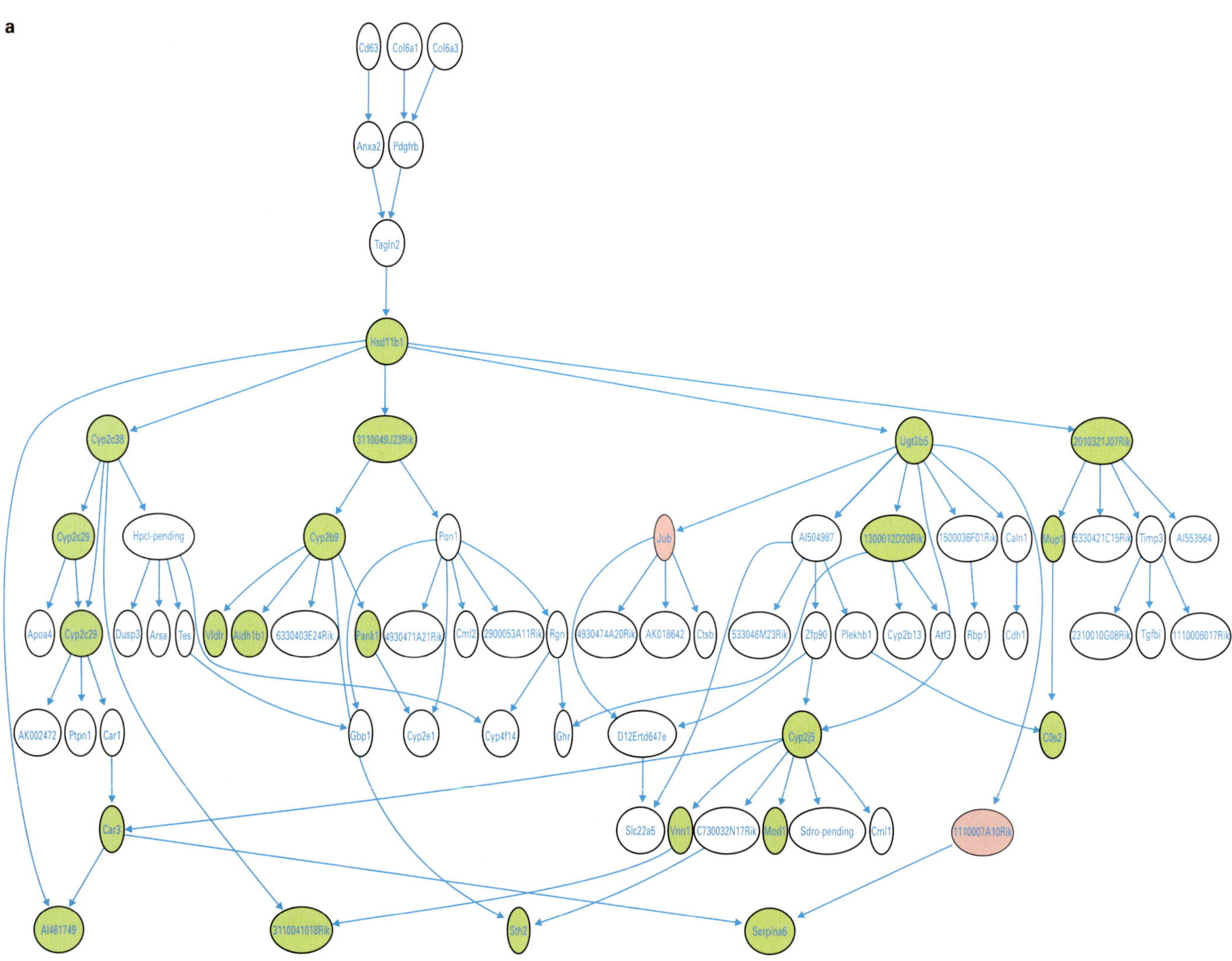

Fig. 3. M2-type sub-networks associated with *Hsd11b1*. (**a**) The predicted sub-network centered at the *Hsd11b1* node. The nodes (genes) making up this sub-network were identified by requiring that they have a path to *Hsd11b1* no longer than three links. The gene expression state of each node is colored according to the predicted state when *Hsd11b1* is in the down-regulated state. Red indicates a gene is up-regulated relative to the reference pool, green indicates a gene is down-regulated relative to the reference pool, and no fill indicates a gene is not differentially expressed relative to the reference pool. (**b**) An *Hsd11b1* sub-network related to the sub-network given in **a**. This sub-network is defined by those nodes in the complete network whose gene expression is predicted to change given the down-regulated state of *Hsd11b1*. Pictured here are the predicted expression states of 33 genes, given *Hsd11b1* is in the down-regulated state. The stars indicate the 20 genes represented in the Ppara signature.

Table 1. Goodness of fit statistics for the two model types described in the text

Network\Z-score	F^a Pair 1	F^a Pair 2	F^a Pair 3	W^b Pair 1	W^b Pair 2	W^b Pair 3
Network w/o genetics (*M1*)	8.50	9.95	9.07	9.88	10.17	8.75
Network w/ genetics (*M2*)	7.67	8.71	8.19	8.78	9.85	8.75

[a] Fenofibrate treated.
[b] WY-14,643 treated.

not shown) indicated there were no clear patterns of expression that distinguished the mice in any meaningful way. One explanation for the lack of strong association between these 179 genes and the other 909 genes in the F2 data set could be that these 179 genes are an artifact of the microarray experiments.

Assessing the significance of predicted networks

It is of interest to determine whether our network reconstruction method involving expression and QTL data in a segregating population leads to optimal networks that possess greater predictive power than similar networks derived from the expression data alone. A series of comparisons were per-

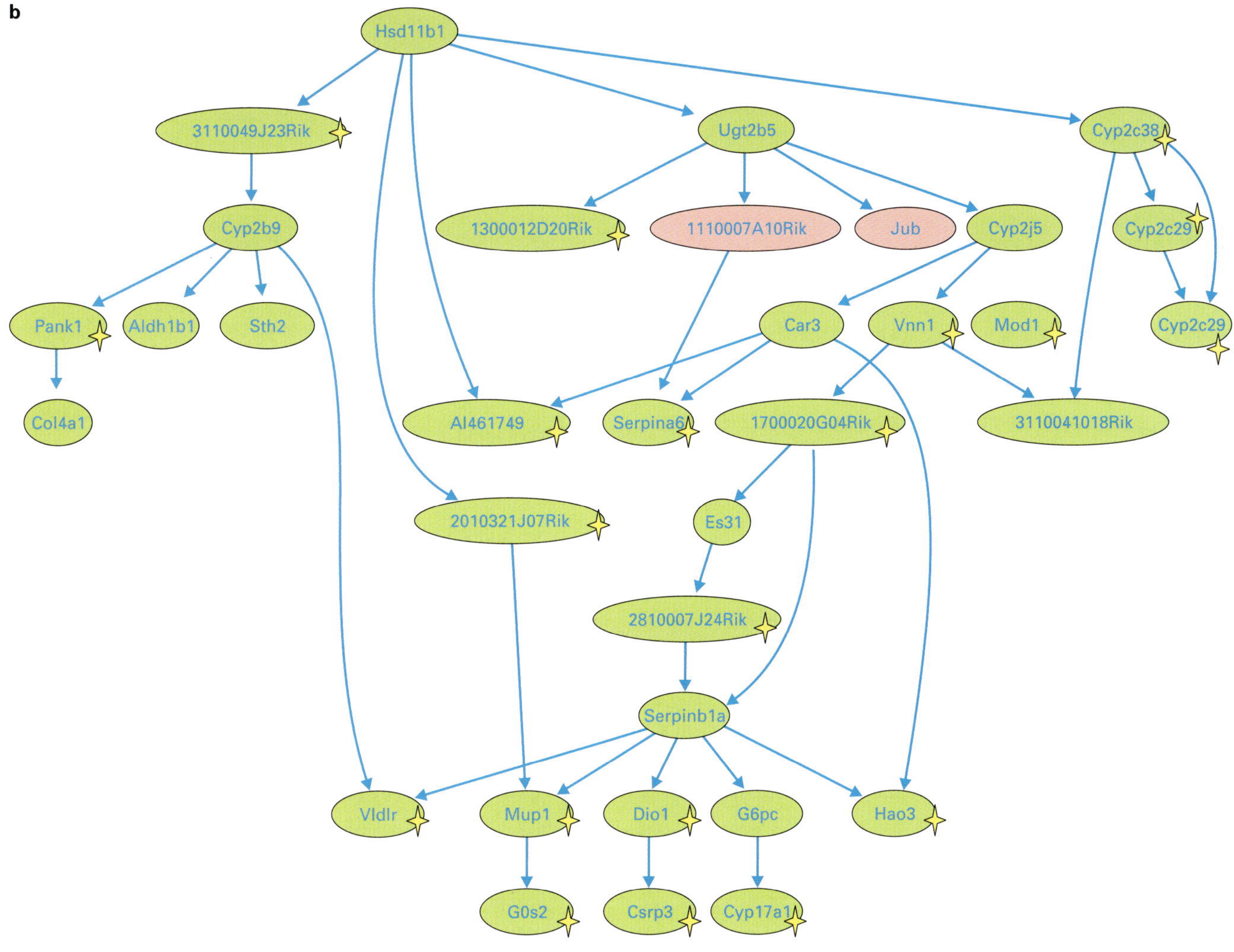

formed to assess the utility of the networks derived from our network reconstruction procedure. First, after fitting the model to the F2 data, we used goodness of fit tests to assess whether this model fit the independent Ppara agonist data set well. Second, we tested whether this model was able to predict the expression response for a given gene perturbed in the independent Ppara agonist data set. Finally, we compared the predictive power of our novel model based on the combined gene expression/QTL data, with a more basic model based on the expression data alone.

We can assess the goodness of fit of a model on a particular set of data using the likelihood $P(D|M)$, where D represents the data and M the model (i.e., those models fitting the data better will yield a higher likelihood score). Here we consider two model types: 1) *M1* based on the expression data alone, and 2) *M2* based on the genotypic data as well as the expression data. Given relative transcript abundances for the 1088 genes in the F2 data set as measured in the six Ppara agonist treated mouse pairs, we compared the likelihoods of the optimal networks identified for each model type with the likelihoods generated for each model type based on 1000 random permutations of the gene states. The resulting z-scores for these tests for each of the Ppara agonist treated mouse pairs were computed and are given in Table 1. Both model types are seen to fit the experimental data well, with a P value less than 8.7e(–15) reported for the best fits.

This goodness of fit test demonstrates that the expression levels of different genes are related, as we already observed in Fig. 1, and that Bayesian networks (or general graphic models) can well capture these inter-gene relationships. However, the more important question is whether the pathways inferred from these reconstructions are able to capture causal information reliably. If a network truly captures the state of a biological system, it should be able to answer one important question: what genes will change if a gene is perturbed in the system? One of the genes represented in the optimal networks identified for each model type was *Hsd11b1*. Figure 3a represents the sub-network derived from the *M2* network by restricting attention to nodes that were within a path of length 3 to the *Hsd11b1* node (i.e., an *Hsd11b1*-centered sub-network showing nodes that are "close" to the *Hsd11b1* node). The *Hsd11b1* sub-network indicates that *Hsd11b1* plays a key regulatory role in the network, given it is controlled by relatively few genes and ultimately controls a large number of genes. This sub-network is of further interest because *Hsd11b1* is significantly correlated with fat pad masses, insulin levels, leptin levels, cholesterol levels and other related phenotypes collected on the F2 mice making up the F2 data set (Drake et al., 2001; Colinayo et al., 2003; Schadt et al., 2003).

The ideal experiment to assess the predictive power of the *Hsd11b1* sub-networks for each model type would be to perturb the activity of *Hsd11b1* directly in an independent mouse model to identify the set of genes that change in response to inhibiting *Hsd11b1* activity. This gene set could then be compared to the predictions made from the networks of each type (*M1* and *M2*). Since no genome-wide liver expression profiles exist in the public domain from *Hsd11b1* knockout or *Hsd11b1* inhibitor-treated animals, we were not able to perform this comparison directly. However, *Hsd11b1* was seen to be down-regulated in the Ppara agonist experiments described above, a result that is consistent with those reported by others (Yamazaki et al., 2002). In addition to being regulated by fatty acids, Ppara's expression is also regulated by glucocorticoid (Kroetz et al., 1998). Because *Hsd11b1* converts cortisone to cortisol, which in turn binds to glucocorticoid receptors, Ppara can also fall under the control of *Hsd11b1*, as we have detected in the network we constructed from the 1088 gene set (data not shown). Therefore, we would expect gene expression signatures induced by perturbations to *Hsd11b1* and Ppara to overlap given this connection between them. *Pck1* and *G6pc* were reported to change in *Hsd11b1* knockout mice compared to wild-type (Kotelevtsev et al., 1997) as well as in *Hsd11b1* inhibitor-treated mice, compared to untreated mice (Alberts et al., 2002, 2003). In the Ppara agonist experiments described here, *G6pc* and *Pck1* were down-regulated, providing direct confirmation that the *Hsd11b1* and Ppara perturbation signatures overlap.

There were 1206 genes identified in the *Ppara* gene expression signature, and 322 of these overlapped the 1088 making up the F2 data set. This overlap is very statistically significant, given the P value that this overlap could have occurred by chance is effectively equal to 0 using the Fisher Exact Test. This significant overlap indicates that Ppara-targeted genes in the F2 set explain a significant proportion of the transcriptional variation in the F2 mice. Next we identified those genes predicted by the network to change in response to down-regulation of *Hsd11b1* expression, since *Hsd11b1* is down-regulated in the Ppara signature. For model type *M2* (the expression/QTL-based model) there were 33 genes predicted by the network to change in response to changes in *Hsd11b1*, as shown in Fig. 3b. Of the 33 genes predicted to change, 20 overlapped the set of 322 genes in the Ppara agonist signature, as shown in Fig. 3c. The statistical significance of this overlap can be assessed using the Fisher Exact Test. The P value for the null hypothesis that the overlap between the genes predicted to change by the network and the genes observed to change in response to down-regulation of *Hsd11b1* by the Ppara agonist, was approximately 1.7e(–4). It is of note that one of the two expression markers for *Hsd11b1* inhibition identified by Alberts et al. (2002), *G6pc*, was similarly predicted by the network in Fig. 3b to be down-regulated in response to *Hsd11b1* inhibition. The other marker, *Pck1*, did

Table 2. Summary of the top 12 genes with the most children (nodes in the graph that descend from other nodes) in the full M2 network described in the text

Name	No. of children	Symbol	Description
AJ251685	623	*Gpnmb*	glycoprotein (transmembrane) nmb
AK018527	583	9030425E11Rik	adipocyte-specific protein 5
NM_026158	477	0610042E07Rik	RIKEN Gene
AF345951	413	*Dusp16*	dual specificity phosphatase 16
AI662255	291	AI662255	CYTOCHROME P450 2c40
NM_018881	138	*Fmo2*	flavin containing monooxygenase 2
NM_008394	56	*Isgf3g*	interferon dependent positive acting transcription factor 3 gamma
AK017566	41	*Zdhhc2*	zinc finger, DHHC domain containing 2
NM_007799	21	*Ctse*	cathepsin E
NM_019393	20	*Pmscl1*	polymyositis/scleroderma autoantigen 11
AK002319	19	*Tcea3*	transcription elongation factor A (SII), 3
NM_008620	12	*Mpa2*	macrophage activation 2

not have a valid probe represented on the microarray used to generate the F2 data set, so the gene was not in the set of 1088 genes used to construct the network. On the other hand, only five genes were predicted to change by the best network of type *M1* (the model based on the expression data alone) identified from the reconstruction process, and three of these overlapped the set of 322 Ppara-influenced genes (P value = 0.16). These results clearly indicate that model type *M2* has superior prediction capabilities compared to model *M1*, which does no better than we would expect by chance.

These tests demonstrate that network models based on gene expression and QTL data are better able to capture causal relationships. While both model types capture interaction data well, only the genetics-based model was able to capture causal relationships in a statistically significant way. It is of further note that the genes most causally associated with *Hsd11b1* are not necessarily from the set of genes whose RNA levels are most correlated with *Hsd11b1* RNA levels. For example, the top 33 genes in the network most correlated with *Hsd11b1* yield a minimum coefficient of determination equal to 0.836. Thirteen of these top 33 most correlated and anti-correlated genes actually overlap the Ppara agonist signature (P value = 0.15). This again demonstrates that association based on expression data alone does not yield the causal information required to order pathways.

Sub-networks and cis-acting QTL

The QTL data not only allow for stronger causal inferences to be made in the type of network reconstruction problem described above, these data also provide gene-specific perturbation information that can be utilized to identify genes affected by other genes, as highlighted for *Hsd11b1*. While genes with *cis*-acting eQTL likely have polymorphisms in the gene itself that result in variations in RNA levels, these causative polymorphisms will also lead to changes in the RNA levels of other genes, much in the same way that knock outs, transgenics and siRNA experiments targeting a particular gene lead to changes in the expression of other genes responding to the perturbation. The DNA variations provide definitive causal information that is useful in reconstructing pathways.

Table 3. Genes controlled by MPA2 in the full M2 network. Genes with an asterisk are predicted to change when MPA2 is in the up-regulated state

Name	Symbol	Description
NM_019440*	*Gtpi*	interferon-g induced GTPase
AK018544*	*Stat1*	signal transducer and activator of transcription 1
AK009386	2310016F22Rik	RIKEN cDNA 2310016F22 gene
NM_018738*	*Igtp*	interferon gamma induced GTPase
NM_021792	*Iigp*	interferon-inducible GTPase
BC004064	*Sox9*	SRY-box containing gene 9
NM_018734*	*Gbp3*	guanylate nucleotide binding protein 3
NM_011579*	*Tgtp*	T-cell specific GTPase
NM_008331	*Ifit1*	interferon-induced protein with tetratricopeptide repeats 1
AK018585*	9130002C22Rik	RIKEN cDNA 9130002C22 gene
AW909491		hypothetical protein MGC41320
NM_011316	*Saa4*	serum amyloid A 4

Of the 1088 genes in the F2 data set, 108 had *cis*-acting eQTL (eQTL located within 15 cM of the physical location of the gene and having an associated LOD score greater than 10). Of these 108 genes, 44 had children in the network, while 13 had more than 10 children, as shown in Table 2. Two of the genes represented in this table, *Tcea3* and *Isgf3g*, are transcription-factor related genes. This functional role potentially explains why these two genes were identified as controlling for the expression of many other genes in the predicted network.

As discussed above, it is known that one of the functional roles of *Hsd11b1* is to convert cortisone to cortisol. Cortisol acts as an immunosuppressor by reducing T cell proliferation, reducing complement synthesis, and increasing the rate of B cell death (Weyts et al., 1998). Therefore, while we would expect genes associated with immune response to vary in response to perturbations in *Hsd11b1*, we would expect these immune response genes to be under the control of many other biochemical and physiological processes in a genetically more diverse setting, like that which exists in an F2 population. We would then expect that genes downstream of the immune response processes would be harder to link to *Hsd11b1* expression in the network reconstruction process.

As an example, consider the sub-network associated with guanylate-binding protein Mpa2, which is induced by interferon-gamma during macrophage induction (Nguyen et al., 2002). However, there are many biochemical and physiological steps sitting between down-regulating *Hsd11b1*, the subsequent decrease of cortisol, and the ultimate increase in T cell proliferation, which in turn results in the induction of Mpa2. Because the F2 mice were on an atherogenic diet, many of the transcriptionally active genes in the liver were associated with immune response. The sub-network associated with *Mpa2* is given in Fig. 4. Table 3 lists the 12 genes that are given as nodes in Fig. 4. Six of the genes are predicted to respond when *Mpa2* is in the up-regulated state, and these genes can be seen to be involved in immune response functions. Therefore, from the *Hsd11b1* and *Mpa2* sub-networks, we see that the complexity of the immune response in the F2 data set results in our failing to link *Hsd11b1* as a driver of *Mpa2* expression, most likely because of a diversity of immune response pathways that are active in this data set and that are unaffected by *Hsd11b1* activity. This example helps explain why a larger fraction of genes that may be under the control of *Hsd11b1* were not identified as

being under the control of *Hsd11b1* in the sub-network predicted from the F2 data set.

Linkage disequilibrium (LD) as a confounding factor in the reconstruction of gene networks

The network reconstruction procedures described here are strongly dependent on the correlation structures observed between gene expression traits and the associated pattern of overlapping eQTL to establish causal relationships. Any factors influencing these correlation structures, but that are independent of the actual functional relationships of interest between any two genes, could be expected to impact the structure of a predicted network. Perhaps the most striking example of such a confounding effect in an F2 population is linkage disequilibrium. Linkage disequilibrium (LD), or gametic phase disequilibrium, is a concept that describes the association of alleles across two or more loci. For example, consider two loci A and B with alleles (A1/A2 and B1/B2) that are polymorphic between two parental strains used to construct an F2 cross. Further suppose these two loci are physically proximal on the same chromosome. Then, we would expect to see haplotypes A1B1 (or A1B2) and A2B2 (or A2B1) more often than haplotypes A1B2 and A2B1, since there would be few recombinations observed between these two genes, in an F2 population, given their close proximity. This type of effect is very pronounced in an F2 intercross, given all animals obtain from a single F1 founder, with only two meiotic events separating any two mice in the F2 population. If DNA variations at both loci cause variations in the RNA levels for the two genes, then the correlation structure between these RNA levels will be at least partially explained by gametic phase disequilibrium, as opposed to being explained by meaningful functional relationship between the genes.

While this issue has been well studied, the practical consequences of it have never before been observed in the context of tens of thousands of gene expression traits. As an example of how the correlation structure between two genes can be almost fully explained by linkage disequilibirum, we consider all genes on chromosome 1 with *cis*-acting eQTL with LOD scores over 10.0, as given by Schadt et al. (2003). Figure 5 depicts a plot of all pairwise correlations between the transcript abundances of these genes as a function of LD. From this figure it is quite apparent that the correlations between these genes are largely a consequence of 1) being under strong genetic *cis*-acting control,

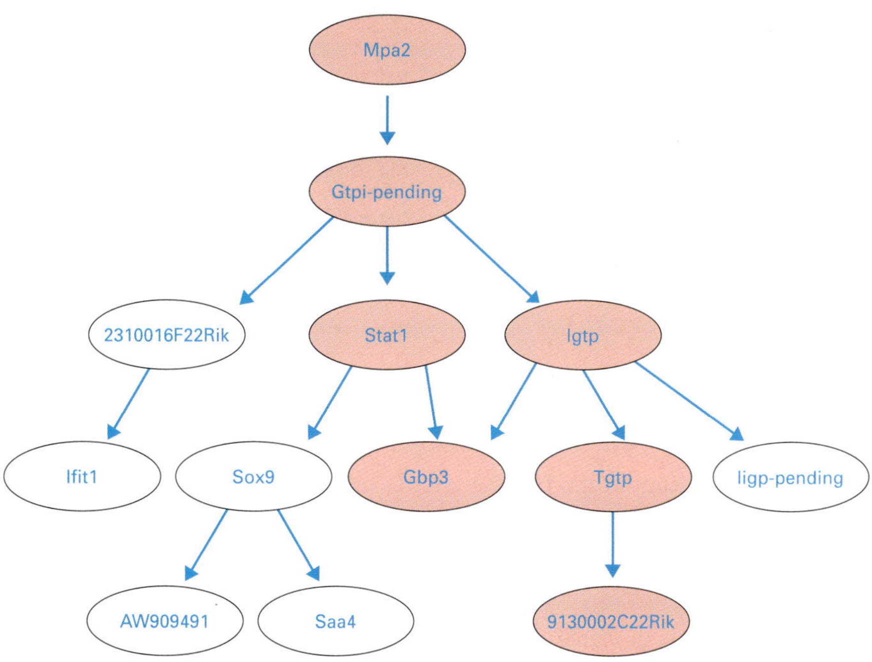

Fig. 4. Sub-network of *Mpa2* (NM_008620). The predicted states for genes responding to *Mpa2* when *Mpa2* is up-regulated. Red indicates the gene is up-regulated and no fill indicates no expression change relative to the reference pool.

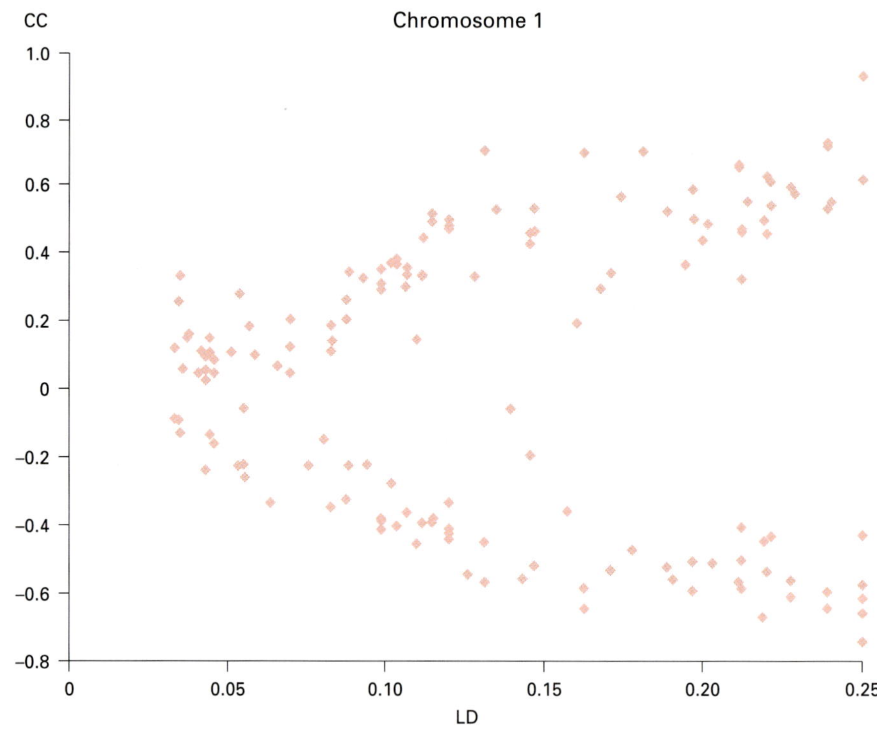

Fig. 5. Pairwise correlations of mRNA levels as a function of gametic phase disequilibrium between the associated genes. Twenty genes physically residing on chromosome 1 were identified with strong *cis*-acting eQTL (corresponding LOD score > 10.0) from the eQTL analysis provided by Schadt et al. (2003) on the F2 data set described in the main text. Pearson correlation coefficients were computed for the mean log expression ratios between each of the 190 pairs of genes. Each of the correlations is plotted here against the linkage disequilibrium values corresponding to each gene pair. The pattern in this plot indicates that the magnitude of correlation is directly proportional to the linkage disequilibrium values, a relationship we would expect if the correlation structures were largely attributed to linkage disequilibrium.

and 2) being in LD with other genes that are under strong *cis*-acting genetic control.

The practical consequence of this confounding effect in the context of network reconstruction can be seen as follows. Suppose that variations in the expression of gene g1 cause changes in the expression of gene g3, that g1 is physically close to gene g2, and that both g1 and g2 have strong *cis*-acting eQTL. Given this scenario, depending on the strength of the cis-acting eQTL and the genetic distance between them, it could be difficult to identify whether g1 or g2 was causal for g3, since g1 and g2 could be potentially significantly correlated. As a result, we may recover the relationship g2→g3 as well as the true relationship g1→g3 due to the effect of LD.

Refinement based on additional experiments

Because there are only 111 experiments in the F2 data set from which the 1088 nodes (genes) in the network were derived, there is the possibility that the network over fits the data. Even though we added priors to constrain the network topology based on QTL information, there are still many networks that fit the data equally well when the topologies between these different directed networks are similar. In addition, some links established in the reconstruction process may be due to the type of LD confounding effects just described. To refine the structure of a given network we can either integrate more data into the reconstruction process or design additional experiments to refine the causal relationships.

By incorporating additional experimental evidence, we can update a given network to reflect the information the additional experiments provide, which may then result in better discriminating power to identify the best network out of the most likely network structures identified as part of the original reconstruction process. The Ppara agonist data can be used for these purposes. If a gene is in the Ppara agonist signature set, then we can weight the link between *Hsd11b1* and this gene with a higher prior probability to indicate *Hsd11b1* activity may be causative for variations in the RNA levels of the gene. Applying the Ppara agonist data in this way, we obtained a refinement to the network given in Fig. 3, fitted from the 1088 genes in the F2 data set. The refined network predicts 41 genes responding to *Hsd11b1* in the down-regulated state, and 26 of these overlap with the Ppara signature, a statistically significant result with a *P* value equal to 5.0e(−6). This overlap is even more significant than the results discussed for Fig. 3. These results further demonstrate the utility and flexibility of the Bayesian network approach in incorporating different data to elucidate pathway structures.

Discussion

We have presented the first ever use of combined gene expression and genotypic data in a segregating population to reconstruct genetic networks. Our method considers eQTL data in addition to gene expression data to constrain the possible types of relationships between any two genes in a system of interest, and then more formally integrates these data into classic Bayesian network reconstruction methods. We have demonstrated that our novel network reconstruction algorithm is able to represent causal relationships among genes by validating predictions made from networks derived from the F2 data set, using the independent Ppara agonist data set. We further demonstrated that these same causal relationships could not be reliably inferred using existing expression based network reconstruction methods.

Our methods based on Bayesian networks are better suited to study large global networks, compared to the dynamic modeling methods used to study small sub-networks (Tegner et al.,

2003). The structure of Bayesian networks can be automatically estimated from the data without any manual intervention. Also, predictions based on the fitted networks can be made in the context of the models we have developed here. However, there are also limitations to our approach as well. First, Bayesian networks are acyclic, directed graphs, and so, they can not represent feedback loops. In a stable biological system, many processes are under negative feedback control, which presents difficulties in establishing causality since such loop structures cannot be represented using our methods. Further, if a gene's RNA levels are self-regulated, then these levels may vary in a narrow range, potentially resulting in only a marginally significant LOD score for the associated *cis*-acting eQTL. In this situation, without explicit information on the self-regulatory role of the gene, the potential exists for such a gene to be identified as a downstream responder of those genes that appear to be influencing its expression. For example, Ppara is selected as a downstream responder in the general network constructed from the 1088 genes in the F2 data set. However, Ppara knockout and agonist data suggest that Ppara may actually be causative for many of the genes falling upstream of it in our network, given these genes are in the Ppara knockout and agonist gene expression signatures (data not shown). Therefore, while we have demonstrated that eQTL information aids in sorting out causative relationships, ultimately additional data or more specific experimentation may be required to refine these relationships.

Despite these shortcomings and our working in the context of a limited data set, the significance of the predictions surrounding *Hsd11b1* is encouraging, especially given the animals analyzed in the F2 and Ppara agonist experiments were genetically distinct and raised under dramatically different environmental conditions. Animals in the F2 set were more than one year old, all female, and they had been on a high-fat, atherogenic diet for 16 weeks. Genetically, the chromosomes of the F2 mice are comprised of randomly assorted segments from the B6 and DBA inbred strains. Therefore, in the F2 data set, the observed variations in *Hsd11b1* stem from a complicated mixture of extreme environmental conditions (given the atherogenic component of the diet) and different genetic backgrounds (each F2 animal represents a different genetic background), and with only 111 animals, the power to detect causal signals was likely minimal. On the other hand, the Ppara agonist experiment involved young, male mice (less than 4 months old) from a single genetic background under more natural environmental conditions, where just a single gene was strongly perturbed. It is remarkable then that we were able to achieve the extent of overlap we did between the genetic network and the Ppara signature set. The 26 genes in the overlap likely represent the key, most primary responders to *Hsd11b1*. Many of the other genes identified in the Ppara agonist experiment that did not overlap the network predictions may be the result of off-target effects and responders to these off-target effects, or they could represent more distant responders to the primary *Hsd11b1* responders that are easily seen in a single gene perturbation experiment, but that get lost in the noise in a more complex, multifactorial perturbation experiment such as that achieved in an F2 cross. Additional experiments that involve directly perturbing

Hsd11b1 are needed to better understand the *Hsd11b1* mechanism of action and to better access the prediction accuracy of the Bayesian network approach we have presented.

Integrating genetic and genomic data using the Bayesian network algorithm developed here offers a promising approach to understanding the complex network of gene changes that are associated with complex traits, and that more generally underlie the complexity of living systems. The results presented in this study indicate for the first time that combining eQTL and gene expression information may allow for the possibility of ordering pathways associated with complex traits.

Acknowledgements

We thank Peggy E. McCann, Thomas W. Doebber, and David E. Moller for their support in generating and interpreting the Ppara data.

References

Alberts P, Engblom L, et al: Selective inhibition of 11beta-hydroxysteroid dehydrogenase type 1 decreases blood glucose concentrations in hyperglycaemic mice. Diabetologia 45:1528–1532 (2002).

Alberts P, Nilsson C, et al: Selective inhibition of 11 beta-hydroxysteroid dehydrogenase type 1 improves hepatic insulin sensitivity in hyperglycemic mice strains. Endocrinology 144:4755–4762 (2003).

Brem RB, Yvert G, et al: Genetic dissection of transcriptional regulation in budding yeast. Science 296:752–755 (2002).

Colinayo VV, Qiao JH, et al: Genetic loci for diet-induced atherosclerotic lesions and plasma lipids in mice. Mammal Genome 14:464–471 (2003).

Davidson EH, Rast JP, et al: A genomic regulatory network for development. Science 295:1669–1678 (2002).

Drake TA, Schadt E, et al: Genetic loci determining bone density in mice with diet-induced atherosclerosis. Physiol Genomics 5:205–215 (2001).

Friedman N, Linial M, et al: Using Bayesian networks to analyze expression data. J Comput Biol 7:601–620 (2000).

Gardner TS, di Bernardo D, et al: Inferring genetic networks and identifying compound mode of action via expression profiling. Science 301:102–105 (2003).

Gordon DA, Shelness GS, et al: Estrogen-induced destabilization of yolk precursor protein mRNAs in avian liver. J Biol Chem 263:2625–2631 (1988).

Hermanowski-Vosatka A, Gerhold A, et al: PPARalpha agonists reduce 11beta-hydroxysteroid dehydrogenase type I in the liver. Biochem Biophys Res Commun 279:330–336 (2000).

Ideker T, Galitski T, et al: A new approach to decoding life: systems biology. Annu Rev Genomics Hum Genet 2:343–372 (2001).

Jackson AL, Bartz SR, et al: Expression profiling reveals off-target gene regulation by RNAi. Nat Biotechnol 21:635–637 (2003).

Jansen R, Yu H, et al: A Bayesian networks approach for predicting protein-protein interactions from genomic data. Science 302:449–453 (2003).

Karp CL, Grupe A, et al: Identification of complement factor 5 as a susceptibility locus for experimental allergic asthma. Nat Immunol 1:221–226 (2000).

Kotelevtsev Y, Holmes MC, et al: 11beta-hydroxysteroid dehydrogenase type 1 knockout mice show attenuated glucocorticoid-inducible responses and resist hyperglycemia on obesity or stress. Proc Natl Acad Sci USA 94:14924–14929 (1997).

Kroetz DL, Yook P, et al: Peroxisome proliferator-activated receptor alpha controls the hepatic CYP4A induction adaptive response to starvation and diabetes. J Biol Chem 273:31581–31589 (1998).

Marcotte EM, Xenarios I, et al: Mining literature for protein-protein interactions. Bioinformatics 17:359–363 (2001).

Mootha VK, Lepage P, et al: Identification of a gene causing human cytochrome c oxidase deficiency by integrative genomics. Proc Natl Acad Sci USA 100:605–610 (2003).

Nguyen TT, Hu Y, et al: Murine GBP-5 a new member of the murine guanylate-binding protein family is coordinately regulated with other GBPs in vivo and in vitro. J Interferon Cytokine Res 22:899–909 (2002).

Pearl J: Probabilistic Reasoning in Intelligent Systems: Net-works of Plausible Inference (Morgan Kaufmann Publishers, San Mateo 1988).

Pe'er D, Regev A, et al: Inferring subnetworks from perturbed expression profiles. Bioinformatics 17(suppl 1):S215–S224 (2001).

Schadt EE, Monks SA, et al: Genetics of gene expression surveyed in maize, mouse and man. Nature 422:297–302 (2003).

Stefansson H, Sigurdsson E, et al: Neuregulin 1 and susceptibility to schizophrenia. Am J Hum Genet 71:877–892 (2002).

Tegner J, Yeung MK, et al: Reverse engineering gene networks: integrating genetic perturbations with dynamical modeling. Proc Natl Acad Sci USA 100:5944–5949 (2003).

Ueda H, Howson JM, et al: Association of the T-cell regulatory gene CTLA4 with susceptibility to autoimmune disease. Nature 423:506–511 (2003).

van de Vijver MJ, He YD, et al: A gene-expression signature as a predictor of survival in breast cancer. New Engl J Med 347:1999–2009 (2002).

van 't Veer LJ, Dai H, et al: Gene expression profiling predicts clinical outcome of breast cancer. Nature 415:530–536 (2002).

Weyts FA, Flik G, et al: Cortisol induces apoptosis in activated B cells not in other lymphoid cells of the common carp *Cyprinus carpio* L. Dev Comp Immunol 22:551–562 (1998).

Xenarios I, Salwinski L, et al: DIP the Database of Interacting Proteins: a research tool for studying cellular networks of protein interactions. Nucl Acids Res 30:303–305 (2002).

Yamazaki K, Kuromitsu J, et al: Microarray analysis of gene expression changes in mouse liver induced by peroxisome proliferator-activated receptor alpha agonists. Biochem Biophys Res Commun 290:1114–1122 (2002).

Yang E, van Nimwegen E, et al: Decay rates of human mRNAs: correlation with functional characteristics and sequence attributes. Genome Res 13:1863–1872 (2003).

Cytogenet Genome Res 105:375–384 (2004)
DOI: 10.1159/000078210

Chromosomes and speciation in *Mus musculus domesticus*

E. Capanna and R. Castiglia

Dipartimento di Biologia animale e dell'Uomo, Roma (Italy)

Abstract. Thirty years after its identification, the model of chromosomal speciation in *Mus musculus domesticus* is reevaluated using the methods of population biology, molecular cytogenetics and functional genomics. Three main points are considered: (1) the structural predisposition of *M. m. domesticus* chromosomes to Robertsonian fusion; (2) the impediment of structural heterozygosity to gene flow between populations of mice with karyotypes rearranged by Robertsonian fusion and between them and populations with the standard all-acrocentric 40-chromosome karyotype; (3) the selective advantage of chromosomal novelty, essential for the attainment of homozygosis and the rapid fixation of the new karyotype in the population.

Copyright © 2004 S. Karger AG, Basel

"For a biologist the alternative to not thinking in evolutionary terms is not to think at all" – P. and J. Medawar: "The imperfections of man" in: P. Medawar "The uniqueness of the individual," 1957.

In 1969, Alfred Gropp (Gropp et al., 1969, 1970) described the "Tobacco mouse" karyotype, characterized by 26 chromosomes produced by seven Robertsonian fusions, in a population of house mice from the Swiss canton of Grisons. This led to the discovery of the dramatic karyotypic Robertsonian variability of *Mus musculus domesticus* (Gropp et al., 1972) in several geographically isolated population systems in Italy (Capanna et al., 1973, 1976; Capanna and Corti, 1982), northern Scotland (Nash et al., 1983), western Europe (Adolph and Klein, 1981), northern Africa (Said et al., 1986), the Balkans (Tichy and Vucak, 1987; Winking et al., 1988) and the Middle East (Gündüz et al., 2000), that is, in the whole distribution area of this semispecies (Thaler et al., 1981; Auffray et al., 1990). In addition, a surprising complex of populations was recently discovered in Madeira Island (Britton-Davidian et al., 2000).

The mouse model of chromosomal variability has been widely used for decades as an excellent experimental model for chromosome-derived neonatal pathologies (Gropp, 1978, 1982; Gropp et al., 1974, 1975; Johannisson et al., 1983; Nielsen et al., 1985) and chromosome-derived human infertility. Indeed it will likely be used even more after the publication of the fully sequenced mouse and human genomes. In this model, the ability to cross carriers of selected karyotypes allowed the design of peculiar meiotic configurations (single or multiple independent trivalents, chain or ring configurations) variably subject to malsegregation events, thus leading to chromosome-derived subfertility or sterility and to the production of trisomic (and monosomic) animals. This opportunity opened the field of dynamic embryological studies in reproductive and developmental biology (Harris et al., 1986; Mittwoch et al., 1984; Redi et al., 1984, 1985, 1988, 1991, 1993; Redi and Capanna, 1988; Redi and Garagna, 1992; Garagna et al., 1990, 2002; Wallace et al., 1991b).

I believe it would be superfluous to provide an analytical description of all these systems of karyotypic variability and all the populations in the semispecies *M. m. domesticus*, since they are certainly well known and already the subject of thorough reviews (Sage et al., 1992; Nachman and Searle, 1995; Searle, 1998).

Supported by the Italian MIUR-COFIN 2001-2002.

Received 3 October 2003; accepted 27 October 2003.

Request reprints from: Dr. E. Capanna
   Dipartimento di Biologia animale e dell'Uomo
   Università di Roma "La Sapienza", via A. Borelli 50
   IT–00161 Roma (Italy); telephone: +39-(0)6-49918008
   fax +39-(0)6-4457516; e-mail:ernesto.capanna@uniroma1.it.

The thanks to – in spite of paradox

The chromosomal variability in the house mouse provided the scientific community with a "paradigmatic" model of microevolution based on post-mating reproductive isolation related to hypofertility of hybrids with structural heterozygosity. In the last few decades, it has been well accepted by evolutionary cytogeneticists that chromosomal changes can lead to speciation in natural populations, at least those characterized by the *R* reproductive strategy and low vagility (White, 1968, 1978a, b, 1982; Capanna, 1982, 1985; Backer and Bickham, 1986). The mechanism was considered to be very simple: the appearance of a structural rearrangement provides a group of individuals with the opportunity to build up a post-mating reproductive barrier. However, this simplistic mode of speciation was severely contrasted (Mayr, 1963, 1982; Carson, 1982). In fact, what causes difficulties in understanding "chromosomal speciation" is that we have to accept that a factor, i.e. structural rearrangements, acts negatively on a physiological process, i.e. gametogenesis, but that it also overcomes the detrimental effect and is established in the population. We have called this the *"thanks to – in spite of"* paradox (Capanna and Redi, 1994): *thanks to* the chromosomally derived subfertility of the structural hybrids and *in spite of* it, chromosomal rearrangements can become fixed in a population and speciation events can occur.

Consequently, it was necessary to carefully reflect on the role of different demographic and geographical parameters involved in this speciation mode (Wilson et al., 1975; Lande, 1979, 1981; Capanna et al., 1984), as well as to revise the old model based on schemes of White's (1968) "stasipatric" speciation.

All models of speciation based on post-mating reproductive isolation are obliged to include selective advantage of the carriers of this genome configuration among the factors essential for fixation of the new homokaryotype (Hedrick, 1981; Walsh, 1982; Capanna et al., 1985; Chesser and Baker, 1986). This is undoubtedly the most difficult problem of "chromosomal speciation", since it involves aspects of functional genomics that are not easily tackled experimentally. In other words, we must try to understand the cytological and developmental bases of the selective advantage of carriers of the rearranged karyotype that could explain the possibility, and rapidity, of fixation of the chromosomal novelty in the population and its diffusion despite the negative heterosis of hybrids.

Other questions about the paradigmatic model of chromosomal speciation have been postulated but not fully resolved: (1) what is the true efficacy of the post-mating reproductive barrier?, and (2) what are the molecular bases of Robertsonian fusion that can explain the proneness of the karyotype of *M. m. domesticus*, and only of this semispecies, to molecular rearrangements?

The molecular machinery of fusion

We begin with the second question, about the proneness of the *M. m. domesticus* karyotype to centric fusion.

As early as 1986, Redi (Redi et al., 1986, 1989, 1990) proposed a molecular model of Robertsonian fusion based on two main stochastic events: (1) a random favourable spatial relationship of DNA sequences in the telomeric regions of two chromosomes in the *S* phase of the cell cycle of a primitive germ cell, and (2) chemical-physical properties of DNA related to inherent genomic traits of *M. domesticus*, i.e. the identity of satellite DNA sequences in the centromere of all acrocentric autosomes. Wrong recombination between homologous sequences, such as satDNA on non-homologous chromosomes, has been suggested as a possible mode of formation of Robertsonian translocation in mice and in humans (Therman et al., 1989; Vissel and Choo, 1989; Page et al., 1996).

The pericentromeric region of the telocentric chromosomes in *M. musculus* is constituted by a large block of major satellite DNA (between 6 and 17 megabases) (Pardue and Gall, 1970; Wong et al., 1990; Kipling et al., 1991), a smaller block of minor satellite DNA (between 250 and 500 kilobases) (Wong and Rattner, 1988) and, at the physical end of the chromosome, a small block of telomeric sequences (from 10 to 60 kb). In the semispecies *domesticus*, and only in it (Garagna et al., 1993), the AT-rich major satellite DNA is amplified, constituting around 10% of the genome; it is located, with identical sequences, on all the chromosomes except the Y. Another characteristic of the organization of the satDNA in the semispecies *domesticus* is the absence of interspersion with other non-satellite sequences (Garagna et al., 2002). This organization, which we will call "long-range", has been demonstrated by application of the fibre-FISH technique. Using the same technique, Garagna et al. (2002) also clearly demonstrated the physical continuity between major and minor satellite DNA. The two families of sequences are organized in blocks without interspersion between them or with other DNA; in fact, the hybridisation signal appears continuous, without any interruption. The same type of organization was shown by analysis of the chromatin fibres after hybridisation with minor satDNA and telomeric sequences. A further aspect must be underlined: most of the centromeres tend to cluster together with the nucleus in the interphase (Rattner and Lin, 1985). Consequently, mouse centromeres can cluster during interkinesis due to recognition of identical sequences. If chromosomal breakages occur in adjacent centromeres, two DNA strands could rejoin because of the proximity and sequence identity of their satellite DNA. In this way, unpaired single strands of two different chromosomes could occasionally base pair during DNA synthesis. A 5′ parental strand base pairs with a homologous sequence on a 3′ parental strand of another chromosome to give a double helix with the B DNA structure reconstituted. The heteroduplex DNA so formed is cut and rejoined by nicking-closing enzymes. Telomeric sequences and part of the minor satDNA are lost in the newly formed Rb metacentric chromosome.

This proposed mechanism of Robertsonian chromosome formation has received wide consensus. Nonetheless, as correctly stated by King (1993), the *"model remains untested, although it appears to agree with our understanding of the structure of Mus chromosomes."*

As a consequence, our work in the last ten years has focused on testing the proposed molecular mechanism in *M. m. domesticus*. Direct observation of the involvement of the centromeric region in the process of centric fusion has been possible with FISH (Garagna et al., 1995; Nanda et al., 1995) and fibre-FISH (Garagna et al., 2002) preparations with centromeric satellite DNA probes. During the process of fusion, the telo-

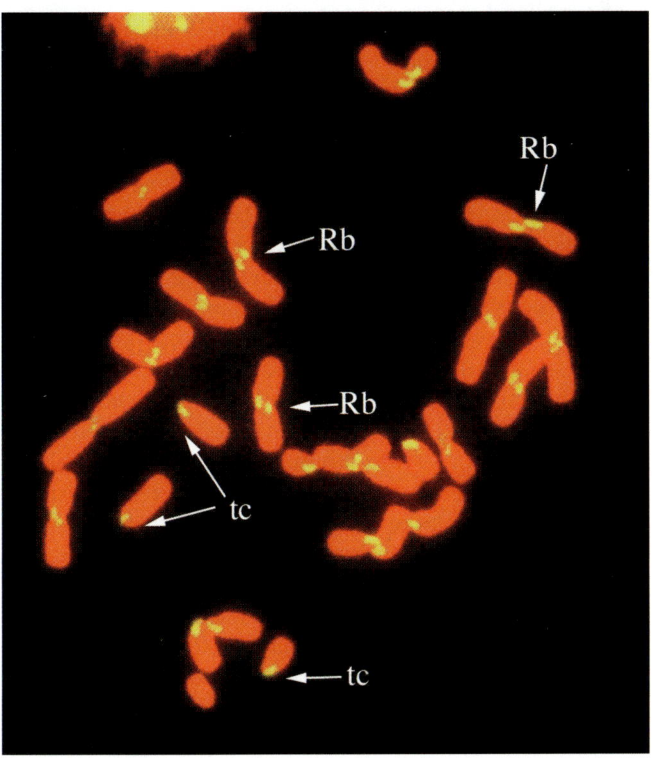

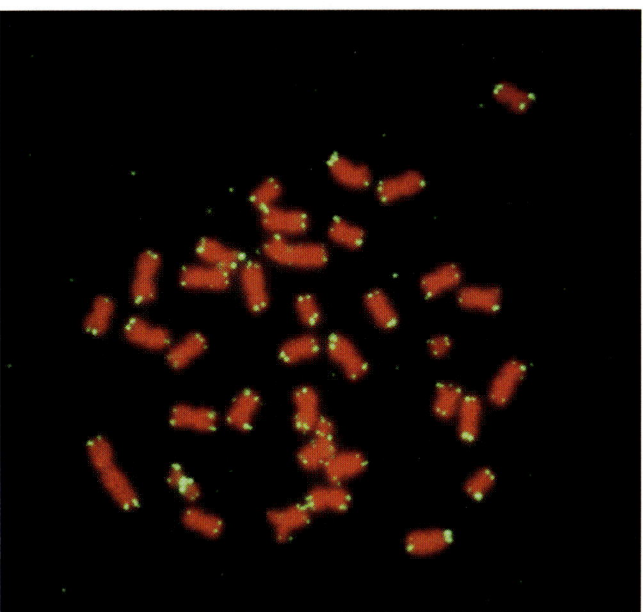

Fig. 1. CO-FISH staining of a chromosome spread from a Poschiavo mouse (2n = 26) using a major satellite oligonucleotide. A contralateral signal was obtained for Rb translocations (examples: Rb arrows) and a single lateral signal on telocentrics (examples: tc arrows), indicating that the satDNA repeated sequences are oriented head-to-tail towards the centromere.

Fig. 2. In situ hybridisation with the (TTAGGG)$_n$ probe on metaphase chromosomes of *Acomys dimidiatus*. Fluorescence signals are present at the telomeric regions of every chromosome. No signals are present in centromeric regions of Rb metacentric chromosomes.

meric sequences are completely lost, while a hybridisation signal is still present when a probe is used to show the minor satellite DNA. This demonstrates that the telomeric sequences are lost in the process of fusion and that the fusion occurs within the context of the repeated sequences of the minor satellite DNA.

A direct prediction of the proposed model was that mouse Robertsonian chromosomes must show contralateral asymmetry between the thymidine-rich strands (since acrocentrics exhibit lateral asymmetry), with the proper centric orientation of the satellite DNA. They do, in fact, show this, as Garagna et al. (2001a) recently proved using the CO-FISH/PRINS/FISH method. This method combines the FISH (Goodwin and Meyne, 1993) and PRINS techniques (Kock et al., 1989) and allows the precise localisation of the breakpoints in telocentrics involved in Robertsonian translocation. It was also demonstrated that both the major and minor satDNA tandem repeated sequences are oriented head-to-tail in telocentrics and in metacentrics, and their polarity is always the same relative to the centromere (Fig. 1). Garagna et al. (2001a) also showed that in Robertsonian chromosomes the two telocentric chromosomes contribute symmetrically, with 10–30 kilobases of minor satDNA each, to the newly formed centromeric region.

Nonetheless, we have to admit that the above process is undoubtedly valid for *M. m. domesticus* but not for all mammals. Volobouev (1998) demonstrated that in *Sorex araneus*, another mammal in which the dramatic rearrangement of the

karyotype involves Robertsonian fusions, the metacentrics retain the telomeric sequences. A similar conservative pattern was observed by Sharma and Sharma (1998) in three "chromosomal" species of the *Mus terricolor* complex. We (Castiglia et al., 2002) tested the Rb metacentrics in *Nannomys minutoides*, a species pertaining to the famous *Evantail Robertsonien* (Robertsonian Fan) of *Leggada* described by Robert Matthey in 1966 (Matthey, 1966, 1970). In this species, most Rb metacentrics preserve telomeric sequences, but not a pair of large metacentrics. Garagna (unpublished data, Fig. 2) observed that in *Acomys dimidiatus* (2n = 38), whose karyotype evolution mainly involved Rb translocations (Volobouev et al., 1991, 1996), the Rb metacentric chromosomes do not show telomeric sequences in the centromeric regions.

This does not decrease the validity of the proposed model for *M. m. domesticus*, in which the exceptional size of the satDNA explains the predisposition of the pericentromeric area to centric fusion. We must ask why the "erroneous base pairing" does not occur in the major satDNA, which is much more extensive and whose repeat identical sequences are much more numerous than in the minor satDNA. We cannot exclude a priori that this may be observed in other mammals. However, a possible reason is that if this occurred the entire fraction of minor satDNA would be lost and the absence of this non-coding DNA could lead to defective operation of the whole centromeric compartment. In fact, the heterochromatin is considered to be important in the correct functioning of the genome, as convincingly discussed by Redi et al. (2001).

Rb chromosomes and earthquakes

If this is the intrinsic molecular machinery causing chromosome rearrangement, we wondered whether extrinsic factors could trigger it.

Vorontsov and Liapunova (1983) stressed the astonishing circumstance that the major systems of mammalian chromosomal variation, i.e. *Ellobius, Spalax, Thomomys* and also *M. domesticus*, are located in areas of intense seismic activity. Hence the variation could be a consequence of mutagenic effects accompanying earthquakes, radiation, radon and so on.

In 1997 a strong earthquake shocked the Umbrian Apennines, with the epicentre located along the tectonic depression of Colfiorito. On that occasion, a seismologist, Dr. Tigran Sadoyan, asked us to collaborate in searching for new Robertsonian metacentrics, which according to the Vorontsov-Liapunova hypothesis might have been produced by the recent seismic activity. Three mouse populations living along the active Leonessa-Colfiorito fault were found to have Rb translocations in their karyotypes (Sadoyan et al., 2003). In contrast, no Rb metacentrics were detected in surrounding populations living a few kilometres from the active fault.

Do these findings validate Vorontsov's theory? Can mutagenic radiation produced during seismic activity really trigger the molecular machinery? Although we cannot exclude this possibility, alternative hypotheses can be proposed. The Colfiorito population could be at an extreme fringe of the hybrid area of the 24-chromosome ACR population located nearly 30 km to the south; the Robertsonian 10.12 is, in fact, present in the 24-chromosome ACR karyotype. Nonetheless, Rb metacentrics 15.17 and 9.14 are absent in all the known Apennine populations, but the Colfiorito population could pertain to a hybrid area of a new nondescribed population of the more northerly Apennine system.

The problem undoubtedly remains unresolved, especially as we have no data on the presence of Rb populations in those districts before the seismic events of 1997. Nevertheless, these observations permit us to make some epistemological considerations about the use of correlations between natural phenomena. From the formal point of view, the correlation between seismic activity and systems of chromosomal polymorphism, as suggested by Vorontsov and Liapunova (1983) and supported by our data, is correct; however, it is likely that there are also positive correlations with other environmental phenomena, both physical and biological.

Whole-arm reciprocal translocations

Capanna (1982, 1985) proposed two mechanisms to explain the progressive accumulation of Rb metacentrics and the characteristic sharing of homologous metacentrics in geographically adjoining populations in complex systems, such as those of the Raethian Alps (Capanna and Riscassi, 1978; Gropp et al.,

1982; Capanna and Corti, 1982; Capanna et al., 1985). These two mechanisms involved successive mutations in very small isolated demes and subsequently, hybridisation of adjoining populations, whose karyotypes, however, lacked metacentrics with monobrachial homology. This hypothesis was tested by both parsimony and compatibility analysis (Corti et al., 1986) and it was found to be consistent with the geographical distribution of the nine populations of the Raethian Alps system. However, Hauffe and Piàleck (1997) re-examined the question on the hypothesis that the complicated sharing of Robertsonian metacentrics was based on whole-arm reciprocal translocations (WARTs). On this basis, they recalculated the parameters of the trees, which showed higher parsimony values. Moreover, since the molecular model of Robertsonian fusion was first presented (Redi et al., 1989), we have stressed that the model entails the possibility of chromosomal arm exchanges between pre-existing Robertsonian chromosomes, as in Hauffe's hypothesis. Both predictions have now been tested: mice carriers of WARTs have been found at Seveso in northern Italy (Garagna et al., 1997), within the complex of "Robertsonian" races in Lombardy, and in the "Apennine" system (Capanna and Redi, 1995; Castiglia and Capanna, 1999a). In all cases, the WARTs were found in a hybrid contact area between a Robertsonian population and the all-acrocentric 40-chromosome one surrounding it (Capanna and Redi, 1995) or at a site of parapatric contact between two mouse populations with Robertsonian rearranged karyotypes (Castiglia and Capanna, 1999b) (Fig. 3).

The new karyotype of the Seveso mice described by Garagna in 1997 is likely derived from the "Milan 2" population – in the parapatric contact area with "Milan 1," by means of 3 WARTs. But what is exceptional here is the opportunity to assign a time scale to such karyotype restructuring: no more than 20 years. In fact, complete desertification of the Seveso area, following the ecological disaster of the escape of dioxins from the ICMESA chemical plant, occurred in 1976 (Garagna et al., 2001c).

So, now we know that chromosomal speciation can occur in a very short time, a "flash" in terms of evolutionary time. This finding reconciles all the contradictory data we have on mouse speciation and strengthens the hypothesis that incipient speciation events can be established suddenly thanks to genomic properties, thanks to genome functioning.

Effectiveness of the post mating barrier

The success of molecular biology and functional genomics (Fields et al., 1999) in the analysis of speciation events in *M. m. domesticus* cannot hide the fact that microevolutionary processes must be analysed in terms of population biology and genetics. Accordingly, let us consider the second problem we proposed: the possibility of gene flow through natural hybridisation zones.

Most research on the hypofertility of structural heterozygotes has consisted of laboratory studies conducted since the first discovery of the tobacco mouse karyotype (Redi and Capanna, 1988 and references herein). Field and laboratory studies on structural heterozygotes derived from wild popula-

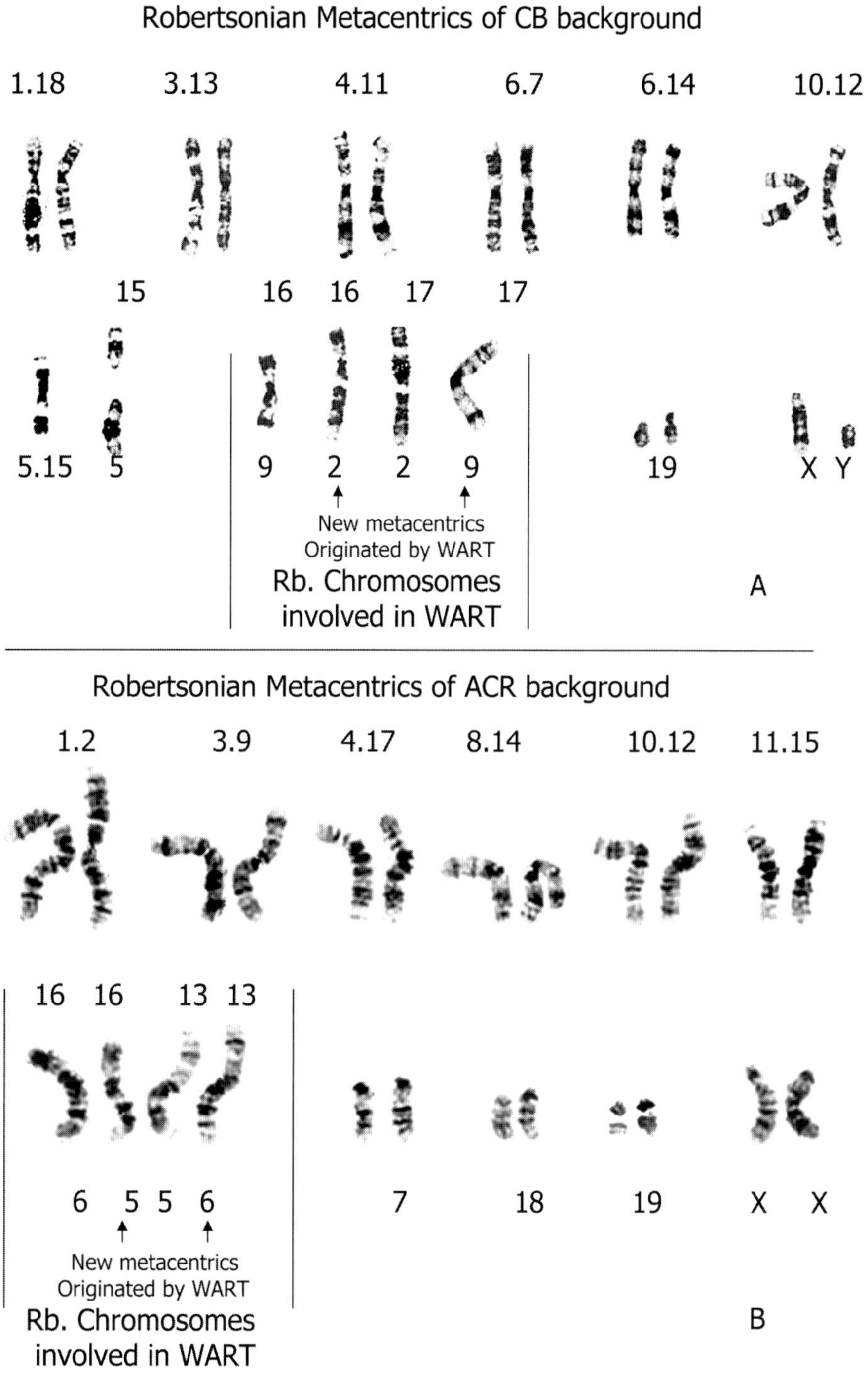

Fig. 3. Whole-arm reciprocal translocations (WART) in contact areas: (**A**) between a CB-22-chromosome population and an all-acrocentric one (Ascoli Satriano, Foggia); (**B**) between two Rb populations, i.e. ACR-24-chromosome and CD-22-chromosome (Pizzoli, L'Aquila).

tions have shown that the presence of one, two or three trivalents at meiosis may have little effect on fertility (Winking et al., 1988; Britton-Davidian et al., 1990; Scriven, 1992; Viroux and Bauchau, 1992; Wallace et al., 1992). In contrast, many trivalents, in multiple single heterozygotes or long chains or rings in complex heterozygotes, may reduce fertility to the point of sterility. However, it is crucial to directly explore these aspects in natural hybridisation zones where chromosomally differentiated populations come into contact. Only Hauffe and

Searle (1998) have carried out a complete analysis of the fertility of hybrid mice belonging to the same mottled hybrid zone in the Raethian Alp system. Another paper by these authors is included in the present issue.

We have recently studied two areas in the central Apennines that reflect two different aspects of the problem. The first is an area of natural hybridisation between the Cittaducale (CD) 2n = 22-chromosome population and the population with the standard 40-chromosome karyotype (Castiglia and Capanna,

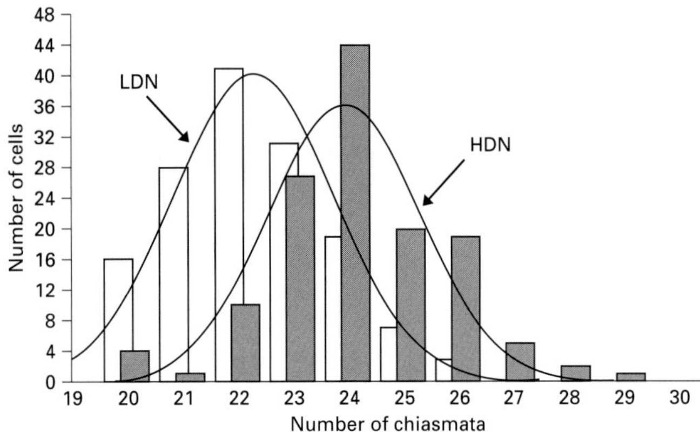

Fig. 4. Number of chiasmata in High Diploid Number (HDN) and Low Diploid Number (LDN) individuals in the hybrid zone between a CD-22-chromosome population and an all-acrocentric area.

1999a, 2000, 2002). This transect is the same one studied 25 years ago by Spirito and co-workers (Spirito et al., 1980). The second is a zone of parapatric contact between two populations characterized by Robertsonian rearranged karyotypes: the CD population with 22 chromosomes and the Ancarano (ACR) with 24 chromosomes (Castiglia and Capanna, 1999, 2002).

The hybrid zone

The hybrid zone between the CD race and the standard race is characterized by clinal variation for all the metacentrics. The passage from homozygous CD population to homozygous standard population occurs within 10 km. Structural heterozygosity reaches its highest level in the centre of the zone. The analysis of fertility was performed on mice from this zone, considering both the litter size of hybrid pairs and the anaphase first non-disjunction (NDJ) of male hybrids. Our observations show that these levels of structural heterozygosity have a detrimental effect on fertility, anaphase I NDJ increases with the number of Rb metacentrics, but generally less than previously believed on the basis of laboratory hybrids. The presence of such structural hybrids should have the effect of slowing down, but not impeding, the passage of genetic information.

In the same zone, we performed an analysis of chiasma number and distribution in heterozygotes and homozygotes (Castiglia and Capanna, 2002), which can provide indirect information about the amount of gene flow. It was found that chiasma repatterning occurs along the transect, with a reduction of the mean number of chiasmata from the standard mice to the CD ones (Fig. 4).

Moreover, the metacentric configurations (trivalent and metacentric bivalent) have a high number of chiasmata in mice with the CD chromosomal background, while the reverse occurs for telocentric bivalents. The reduction of chiasmata from the standard side to the CD side of the zone causes a decrease of the recombination rate, and this asymmetry corresponds to a higher value of the barrier to gene flow towards the CD populations. This indicates that genes could pass from one race to another in an asymmetrical way: from metacentric races to the standard population. Gene flow in the opposite direction could be slowed by metacentric bivalents with a low number of chiasmata.

The parapatric contact area

The contact zone between the Cittaducale 22-chromosome (CD) and Ancarano 24-chromosome (ACR) metacentric races has a quite different pattern of introgression (Castiglia and Capanna, 2002).

We have studied 17 sites in villages located along 25 km of the Aterno River valley and we have only found homozygous populations. The 24-chromosome ACR race (11 sites) is distributed in the northern part of the study area, while the 22-chromosome CD race is found in the southern part (6 sites). Both chromosomal races are found in the village of Pizzoli, extending along the main road for about 3 km. In this village, no mixed population was found at the same trapping site. The nearest sites (no. 11 ACR-24 and no. 12 CD-22) are about 700 metres apart. The contact zone does not correspond to any geographical or ecological barrier; to the contrary, it is located in a zone with a high density of human buildings, a habitat favourable to mice.

At site 11, we found a female with a karyotype not conforming to those of the two parental races. This mouse, whose karyotype was previously discussed concerning WARTs (Fig. 2), is characterized by a complex heterozygosity, originating a ring of four metacentrics in meiotic diakinesis, in an ACR chromosomal background. The differences consist in two metacentrics that are not present in the ACR complement: one derives from the CD karyotype, i.e. the Robertsonian 6.13, and one is a new metacentric deriving by whole-arm reciprocal translocation, the Robertsonian 5.16.

The absence of known genetic markers demonstrable by electrophoresis, which could differentiate between the chromosomal races, compelled us to carry out an analysis of the gene flow in the parapatric contact zone using the sequence of the hypervariable region of the D-loop of mitochondrial DNA. The extent of hybridisation was assessed by cytogenetic and sequence analyses of a segment (468 base pairs) obtained from 41 mice, representing a subset of the mice subjected to chromosomal analysis.

Gene introgression is rare and when present is limited to the area between sites 10–11 and site 12. Thus gene flow between the chromosomal races is drastically reduced. Sporadic hybridisation events can occur, as suggested by the presence of the previously mentioned individual of the ACR race carrying a metacentric belonging to the CD race and an ACR individual carrying a CD-like mitochondrial haplotype. The origin of these two introgressed mice is not easily identified because presumptive F1 males carrying a large ring of 14 paired chromosomes and a multivalent chain of 5 chromosomes are considered sterile on the basis of laboratory tests. However, it has been reported (Redi and Capanna, 1988) that females show less impairment of gametogenesis than males and are occasionally fertile. Hence we cannot exclude the sporadic presence of F1 females with a certain level of fertility in our parapatric contact zone.

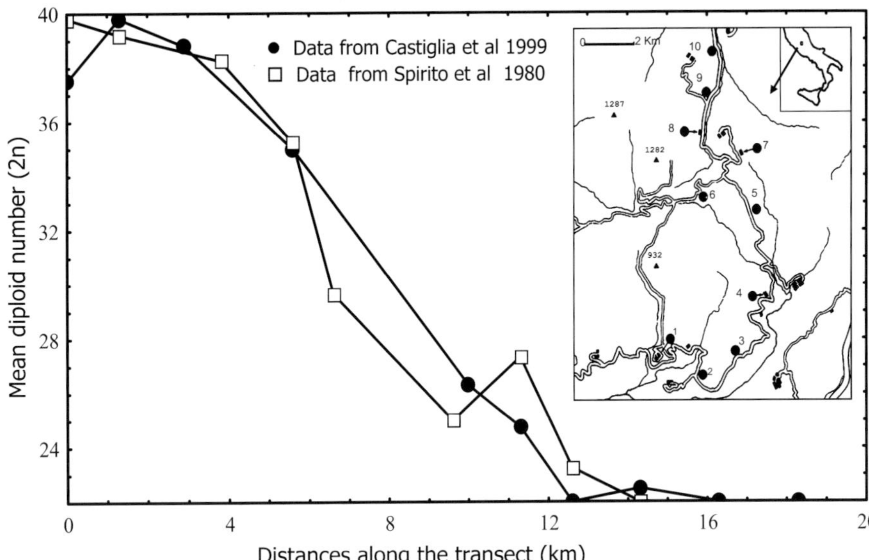

Fig. 5. Comparison of mean diploid numbers of hybrids along the same transect of a hybridisation zone: data from Spirito et al. (1980) and Castiglia and Capanna (1999b).

The selective advantage of the chromosomal novelty

The selective value of the metacentric is a very interesting topic, and several mathematical models (Hedrick, 1981; Walsh, 1982; Capanna et al., 1985) show that this evolutionary condition is necessary for fixation of the new homokaryotype in a population. This is really the focal point of our *"thanks to-in spite of"* paradox.

Some indications in this regard come from the temporal analysis of the hybrid zones. The idea that tension zones are unstable is based on the assumption that they are ephemeral (Harrison, 1990). When hybrid individuals are better equipped, the hybrid zone can be considered stable. But when one of the two cytotypes contributing to the tension zone is better adapted to local ecological conditions, the hybrid zone is considered both spatially and temporally dynamic (Moore and Bauchanan, 1985). Mathematical models describing the dynamics of such tension zones provide the possibility to evaluate the changes in their position and/or width (Barton, 1979; Barton and Hevitt, 1985).

The best way to test these hypotheses is to perform direct long-term observations of hybrid zones with a long history of genetic studies. We had this chance (Castiglia and Capanna, 1999b). We analysed 75 animals along the 18-km transect crossing the hybrid zone between the CD 22-chromosome and standard all-acrocentric karyotypes which Spirito et al. (1980) analysed 25 years ago (data were collected in 1976); this time period corresponds to about one hundred generations in mice. The results show that this long hybridisation time has not had a significant effect on the position of the hybrid zone nor on the location of its extreme margin (Fig. 5). The absence of a significant shift of the tension zone excludes a possible imbalance between the two cytotypes below some hypothesized value; in fact, a fitness superiority of one homozygote greater than 0.1 would have produced a shift of the zone of more than two thousand metres, according to Barton's model (1979). The stability of the hybrid zone could be a consequence of heterosis of the hybrid, which overcomes per se the detrimental effect of the structural heterozygosity. An alternative view is that the parental cytotypes are in some way adapted to different ecological conditions. This hybrid zone is located in an area of transition between a mountain zone and the alluvial plain of the Tiber River. Thus ecological constraints could help to maintain the separation of these chromosomal races.

We can consider the previously discussed reduction of crossing-over in the metacentric configurations of homozygotes from this perspective. In fact, reduction of crossing-over can have an adaptive value in conserving groups of co-adapted genes and lowering the possibility of genetic variability. Another way by which the metacentric can be selectively favoured is through meiotic drive. Nonetheless, meiotic drive has not been directly demonstrated in wild house mice. Pardo-Manuel de Villena and Sapienza (2001) recently reviewed 1,170 karyotypes in several mammalian orders and provided evidence that karyotypic evolution may be driven by non-random segregation during meiosis in females. If demonstrated in the house mouse, meiotic drive would represent a strong force for fixation of the metacentric in the population.

In addition to this point of view, which we consider "meiotic-dependent", the selective advantage of the new homokaryotype can be considered an aspect of functional genomics. Indeed, it is important to add to these factors the consequence of the complex functioning of a linear genome when it is integrated into a spatial map of the chromosomes in the nucleus.

The idea that chromosomes occupy exclusive territories in the interkinetic nucleus can be related to the conjectures of Theodor Boveri in the late 1800s (Boveri, 1877). Yet a precise doctrine of a close correlation between nuclear architecture, gene expression and cell function has only recently been formulated (Manuelidis, 1990, 1997; Cremer et al., 1993; Cremer and Cremer, 2001; Boyle et al., 2001) and supported by experimental results (Wakayama et al., 1998).

Within the nucleus, the chromatin is divided into morphological and functional compartments and its spatial organization plays a crucial role in the regular functioning of the cellular machinery. Robertsonian fusion causes substantial changes in the nuclear territories, since the loss of telomeric sequences sequesters the chromosomes from the original sub-regions.

We tested this theory by means of whole chromosome painting techniques (Garagna et al., 2001). We studied the spatial organization of chromosomes 5, 11, 13, 15, 16, 17, X and Y in germ cells taken from male telocentric homozygotes, metacentric homozygotes and hybrids obtained from crosses between the two types of homozygotes. We found that the chromosomes occupy truly specific territories inside the nucleus at each moment of the cell cycle. Moreover, the heterozygosities cause significant changes in the nuclear territories, particularly an alteration of the state of condensation of the X and Y chromosomes in Sertoli cells, a somatic element of the gonad that plays a fundamental role in spermiohistogenesis.

In both types of homozygous animals (telocentrics and metacentrics), the hybridisation signals of the X and Y chromosomes in the Sertoli cells are diffuse, with a decondensed state in their chromatin. This suggests active gene expression. In the structural heterozygotes, the X and Y signals appear condensed, which is related to transcriptional inactivation.

It is known that condensation of the X and Y chromosomes in the Sertoli cells of mice (Guttenbach et al., 1996) and humans (Speed et al., 1993) is dynamic: the X and Y chromosomes are condensed during the pre-pubertal period and decondensed in fertile adults, while they have always been found to be condensed in sterile individuals. What occurs in the somatic cells of the gonad suggests a hypothesis for other functional and differentiational activities of somatic cells regulated, and deregulated, by reorganization of the nuclear topography due to the centric fusion per se.

We have also demonstrated that structural heterozygosities in hybrids between different mouse cytotypes have a negative effect on the cytodifferentiation of both the male and female germ cells, not only in the post-meiotic phase (Redi et al., 1985; Redi and Capanna, 1988; Garagna et al., 1990). In fact, the deregulation of spermatogenesis induced by heterozygosities in different animal models occurs at pre-meiotic stages, and it has been ascribed mainly to two factors: interference with inactivation of the X chromosome and non-saturation of the pairing sites.

The problem of a selective advantage of the new homokaryotype is still unresolved; in models of speciation, it remains a presupposition demonstrated ab absurdo. The idea of using functional genomics to resolve this problem is plausible (Capanna and Redi, 1994), but for now it has only been able to certainly identify detrimental effects related to changes in the nuclear territories caused by Robertsonian fusions.

Nevertheless, the road to take has been identified. Our studies, based on the role of pericentromeric satellite DNA in the process of chromosomal rearrangement, seek to link the biology of satellite DNA with organization of the genome and with processes of development, such as differentiation of the germ cells. They also seek to explain, within a coherent demographic and biogeographical scenario, how germ cell differentiation influences evolutionary processes.

However, evolution, and the process of speciation that underlies it, is a complex phenomenon, and it is illusive to explore it solely with a reductionist approach. Cytogenetics and molecular biology alone will not be able to clarify the problem if they do not tackle the epiphenomenal aspects, looking for correlations between organization of the genomic processes of development and population dynamics. One cannot pretend to reduce speciation and evolution only to the ultimate mechanisms of nucleic acid dynamics, ignoring the impact these messages have at the chromosomal, cellular, organismal and population levels.

References

Adolph S, Klein J: Robertsonian variation in *Mus musculus* from Central Europe, Spain and Scotland. J Hered 72:219–221 (1981).

Auffray J, Marshal JT, Thaler L, Bonhomme F: Focus on the nomenclature of European species of *Mus*. Mouse Genet 88:7–8 (1990).

Backer RJ, Bickham JW: Speciation by monobrachial centric fusions. Proc Natl Acad Sci USA 83:2845–2848 (1986).

Barton NH: The dynamics of hybrid zones. Heredity 43:341–359 (1979).

Barton NH, Hevitt GM: Analysis of hybrid zones. Ann Rev Ecol Syst 16:113–148 (1985).

Boveri T: Ueber Differenzierung der Zellkerne während der Furchung des Eies von *Ascaris megalocephala*. Anat Anz 2:288–293 (1877).

Boyle S, Gilchrist S, Bridger JM, Perry P, Mahy NL, Ellis JA, Bickmore WA: The spatial organization of human chromosomes with the nuclei of normal and emerin-mutant cells. Human Mol Gen 10:211–219 (2001).

Britton Davidian J, Sonjaya H, Catalan J, Cattaneo Berberi G: Robertsonian heterozygosity in wild mice; fertility and transmission rates in Rb (16.17) translocation heterozygotes. Genetica 80:171–174 (1990).

Britton-Davidian J, Catalan J, da Graca Ramalhinho M, Ganem G, Auffray JC, Capela R, Biscoito M, Searle JB, da Luz Mathias M: Rapid chromosomal evolution in island mice. Nature 403:158 (2000).

Capanna E: Robertsonian numerical variation in animal speciation. *Mus musculus*, an emblematic model, in Barigozzi C (ed): Mechanism of speciation, pp 155–177 (Alan Liss, New York 1982).

Capanna E: Karyotype variability and chromosomal transilience in rodents: the case of the genus *Mus*, in: Luckett P, Hartenberger JL (eds): Evolutionary relationships among rodents: a multidisciplinary analysis, pp 643–669 (Plenum Press, New York 1985).

Capanna E, Corti M: Reproductive isolation between two chromosomal races of *Mus musculus* in the Rhaetian Alps (Northern Italy). Mammalia 46: 107–109 (1982).

Capanna E, Redi CA: Chromosomes and microevolutionary processes. Boll Zool 61:285–294 (1994).

Capanna E, Redi CA: Whole arm reciprocal translocation (WART) between Robertsonian chromosomes: finding of Robertsonian heterozygous mouse with karyotype derived through WARTs. Chrom Res 3:135–137 (1995).

Capanna E, Riscassi E: Robertsonian karyotype variability in *Mus musculus* populations in the Lombardy area of the Po valley. Boll Zool 45:63–71 (1978).

Capanna E, Civitelli MV, Cristaldi M: Una popolazione Appenninica di *Mus musculus* caratterizzata da un cariotipo a 22 cromosomi. Rend Fis Acc Lincei, S VIII 54:981–984 (1973).

Capanna E, Gropp A, Winking H, Noack G, Civitelli MV: Robertsonian metacentric in the mouse. Chromosoma 58:341–353 (1976).

Capanna E, Corti M, Mainardi D, Parmigiani S, Brain PF: Karyotype and intermale aggression in wild house mice: Ecology and speciation. Behaviour Genetics 14:195–208 (1984).

Capanna E, Corti M, Nascetti G: Role of contact areas in chromosomal speciation of the European long-tailed house mouse *(Mus musculus domesticus)*. Boll Zool 52:97–119 (1985).

Carson HL: Speciation as a major reorganisation of polygenic balance, in Barigozzi C (ed): Mechanism of Speciation, pp 411–433 (Alan Liss, New York 1982).

Castiglia R, Capanna E: Whole arm reciprocal translocation (WART) in a feral population of mice. Chromosome Res 7:493–495 (1999a).

Castiglia R, Capanna E: Contact zone between chromosomal races of *Mus musculus domesticus*. 1. Temporal analysis of an hybrid zone between the CD chromosomal race and population with standard karyotype. Heredity 83:319–326 (1999b).

Castiglia R, Capanna E: Contact zone between chromosomal races of *Mus musculus domesticus*. 2. Fertility and segregation in laboratory-reared and wild mice heterozygous for multiple Robertsonian rearrangements. Heredity 85:147–156 (2000).

Castiglia R, Capanna E: Chiasma repatterning across a chromosomal hybrid zone between chromosomal races of *Mus musculus domesticus*. Genetica 114:35–40 (2002).

Castiglia R, Gornung E, Corti M: Cytogenetic analysis of chromosomal rearrangements in *Mus minutoides/musculoides* from North-West Zambia through mapping the telomeric sequences (TTACCC)n and banding techniques. Chromosome Res 10:399–406 (2002).

Chesser RK, Baker RJ: On factors affecting the fixation of chromosomal mutations and neutral genes: computer simulations. Evolution 40:625–632 (1986).

Corti M, Capanna E, Estabrook GF: Micro evolutionary sequences in house mouse chromosomal speciation. Syst Zool 35:163–175 (1986).

Cremer T, Cremer C: Chromosomes territories, nuclear territories and gene regulation in mammalian cells. Nat Rev Genet 2:292–301 (2001).

Cremer T, Kurz A, Zirbel R, Dietzel S, Rinke B, Schrock E, Speicher MR, Mathieu U, Jauch A, Emmerich P, Schertan H, Ried T, Cremer C, Lichter P: The role of chromosome territories in the functional compartmentalization of the cell nucleus. Cold Spring Harb Symp Quant Biol 58:777–792 (1993).

Fields S, Kohara Y, Lockhart DJ: Functional genomics. Proc Natl Acad Sci USA 96:8825–8826 (1999).

Garagna S, Redi CA, Zuccotti M, Britton-Davidian J, Winking H: Kinetics of oogenesis in mice heterozygous for Robertsonian translocations. Differentiation 42:167–171 (1990).

Garagna S, Redi CA, Capanna E, Andayani N, Alfano RM, Doi P, Viale G: Genome distribution, chromosomal allocation, and organization of the major and minor satellite DNA in 11 species and subspecies of the genus *Mus*. Cytogenet Cell Genet 64:247–255 (1993).

Garagna S, Broccoli D, Redi CA, Searle JB, Cooke HJ, Capanna E: Robertsonian metacentrics of the house mouse lose telomeric sequences but retain some minor satellite DNA in the pericentromeric area. Chromosoma 103:685–692 (1995).

Garagna S, Zuccotti M, Redi CA, Capanna E: Trapping speciation. Nature 390:241–242 (1997).

Garagna S, Marziliano N, Zuccotti M, Searle JB, Capanna E, Redi CA: Pericentromeric organization at the fusionpoint of mouse Robertsonian translocation chromosomes. Proc Natl Acad Sci USA 98:171–175 (2001a).

Garagna S, Zuccotti M, Thornhill A, Fernandez-Donoso R, Berrios S, Capanna E, Redi CA: Alteration of nuclear architecture in male germ cells of chromosomally derived subfertile mice. J Cell Sci 114:4429–4434 (2001b).

Garagna S, Zuccotti M, Vecchi ML, Rubini PG, Capanna E, Redi CA: Human dominant ecosistem and restoration ecology: Seveso today. Chemosphere 43:577–585 (2001c).

Garagna S, Zuccotti M, Capanna E, Redi CA: High-resolution organization of mouse telomeric and pericentromeric DNA. Cytogenet Genome Res 96:125–129 (2002).

Goodwin E, Meyne J: Strand-specific FISH reveals orientation of chromosome 18 alphoid DNA. Cytogenet Cell Genet 63:126–127 (1993).

Gropp A: Relevance of phases of development for expression of abnormality. Perspectives drawn from experimentally induced chromosome aberrations, in F Naftolin (ed): Abnormal Fetal Growth; Biological Basis and Consequences, pp 85–100 (Dahlem Konferenzen, Berlin 1978).

Gropp A: Value of an animal model for trisomy. Virchows Arch Pathol Anat 395:117–131 (1982).

Gropp A, Tettenborn U, von Lehemann E: Chromosomenuntersuchungen bei der Tabakmaus *(M. poschiavinus)* und bei den Hybriden mit der Laboratoriummaus. Experientia 25:875–876 (1969).

Gropp A, Tettennborn U, von Lehmann E: Chromosomenvariation vom Robertsonschen Typus bei der Tabakmaus, *Mus poschiavinus*, und ihren Hybriden mit Laboratoriummäusen. Cytogenetics 9:9–23 (1970).

Gropp A, Winking H, Zech L, Muller H: Robertsonian chromosomal variation and identification of metacentric chromosomes in feral mice. Chromosoma 39:265–288 (1972).

Gropp A, Giers D, Kolbus U: Trisomy in the fetal backcross progeny of male and female metacentric heterozygotes of the mouse. I Cytogenet Cell Genet 13:511–535 (1974).

Gropp A, Kolbus U, Giers D: Systematic approach to the study of trisomy in the mouse. II. Cytogenet Cell Genet 14:42–62 (1975).

Gropp A, Winking H, Redi CA, Capanna E, Britton E, Davidian J, Noack G: Robertsonian karyotype variation in wild house nice in Rhaeto-Lombardia. Cytogenet Cell Genet 34:67–77 (1982).

Gündüz I, Coskun T, Searle JB: House mice with metacentric chromosomes in the Middle East. Hereditas 133:175–177 (2000).

Guttenbach M, Martinez EMJ, Engel W, Schmid M: Interphase chromosome arangement in Sertoli cells of adult mice. Biol Reprod 54:980–986 (1996).

Harris MJ, Wallace ME, Evans EP: Aneuploidy in the embrionic progeny of females heterozygous for the Robertsonian chromosome (9.12) in genetical wild Peru-Coppock mice *(Mus musculus)*. J Reprod Fertl 76:193–203 (1986).

Harrison RG: Hybrid zones: Windows on the evolutionary process. Oxford Surv Evol Biol 7:69–128 (1990).

Hauffe HC, Piàleck J: Evolution of chromosomal races of *Mus musculus domesticus* in the Rhaetian Alps: the role of whole-arm reciprocal translocation and zonal raciacion. Biol J Linnean Soc 62:255–278 (1997).

Hauffe HC, Searle JB: Chromosomal heterozygosity and fertility in house mouse *(Mus musculus domesticus)* in Northern Italy. Genetics 150:1143–1154 (1998).

Hedrick PW: The establishment of a chromosomal variant. Evolution 33:322–332 (1981).

Johannisson R, Gropp A, Winking H, Coerdt W, Rehder H, Schwinger E: Down's syndrome in the male. Reproductive pathology and meiotic studies. Hum Genet 63:132–138 (1983).

King M: Species evolution; the role of chromosome changes (Cambridge Univiversity Press, Cambridge 1993).

Kipling D, Ackford HE, Taylor BA, Cooke HJ: Mouse minor satellite DNA genetically maps to the centromere and is physically linked to the proximal telomere. Genomics 11:235–241 (1991).

Kock JE, Kolvraa S, Petersen KB, Gregersen N, Bolund L: Oligonucleotide-priming methods for the chromosome-specific labelling of alpha satellite DNA in situ. Chromosoma 98:259–265 (1989).

Lande RL: Effective deme size during long term evolution estimates from rates of chromosomal rearrangements. Evolution 34:234–251 (1979).

Lande RL: Models of speciation by sexual selection on polygenic traits. Proc Natl Acad Sci USA 78:3721–3725 (1981).

Leger-Silvestre I, Trumtel S, Noaillac-Depeyre J, Gas N: Functional compartmentalization of the nucleus in the budding yeast *Saccharomyces cerevisiae*. Chromosoma 108:103–113 (1999).

Manuelidis L: A view of interphase chromosomes. Science 250:1533–1540 (1990).

Manuelidis L: Interphase chromosome positions and structure during silencing, transcription and replication, in van Driel R, Otte AP (eds): Nuclear Organization Chromatin Structure, and Gene Expression, pp 145–168 (Oxford Univiversity Press, Oxford 1997).

Matthey R: Le polymorphisme chromosomique des *Mus* Africains du genre *Leggada*. Révision générale portant sur l'analyse de 213 individus. Rev Suisse Zool 73:586–607 (1966).

Matthey R: L'évental Robertsonienne chez le *Mus (Leggada)* groupe *minutoides/musculoides*. Rev Suisse Zool 77:625–629 (1970).

Mayr E: Animal Species and Evolution (Cambridge University Press, Cambridge 1963).

Mayr E: Processes of speciation in animals, in Barigozzi C (ed): Mechanism of Speciation, pp 1–20 (Alan Liss, New York 1982).

Mittwoch U, Mahadevaiah S, Setterfield LA: Chromosomal anomalies that cause male sterility in the mouse also reduce ovary size. Genet Res 44:219–224 (1984).

Moore WS, Bauchanan DB: Stability of the Northern flicker hybrid zone in historical times: Implications for adaptive speciation theory. Evolution 39:135–151 (1985).

Nachman MW, Searle JB: Why is the house mouse karyotype so variable? Trends Ecol Evol 10:397–402 (1995).

Nanda I, Schneider-Rasp S, Winking H, Schmid M: Loss of telomeric sites in the chromosomes of *Mus musculus domesticus* during Robertsonian rearrangements. Chromosome Res 3:399–409 (1995).

Nash HR, Brooker PC, Davis SJM: The Robertsonian translocation house mouse populations of North East Scotland. A study on their origin and evolution. Heredity 50:303–310 (1983).

Nielsen K, Marcus M, Gropp A: In vitro kinetics of mouse trisomies 12 and 19. Hereditas 102:77–84 (1985).

Page SL, Shin JC, Han JY, Choo KHA, Shaffer LG: Breakpoint diversity illustrates distinct mechanisms for Robertsonian translocation formation. Hum Mol Gen 5:1279–1288 (1996).

Pardo-Manuel de Villena F, Sapienza C: Female meiosis drives karyotypic evolution in mammals. Genetics 159:1179–1189 (2001).

Pardue ML, Gall JG: Chromosomal localization of mouse satellite DNA. Science 168:1356–1358 (1970).

Redi CA, Capanna E: Robertsonian heterozygotes in the house mouse and the fate of their germ cells, in Daniel A (ed): The Cytogenetics of Mammalian Autosomal Rearrangements, pp 315–359 (Alan Liss, New York 1988).

Redi CA, Garagna S: Chromosome variability and germ cell development in the house mouse. Andrologia 24:11–16 (1992).

Redi CA, Garagna S, Pellicciari C, Manfredi Romanini MG, Capanna E, Winking H, Gropp A: Spermatozoa of chromosomally heterozygous mice and their fate in male and female genital tracts. Gamete Res 9:273–286 (1984).

Redi CA, Garagna S, Hilscher B, Winking H: The effect of some Robertsonian chromosome combinations in the seminiferous epithelium of the mouse. J Embryol Exp Morphol 85:1–19 (1985).

Redi CA, Garagna S, Mazzini G, Winking H: Pericentromeric heterochromatin and A-T contents during Robertsonian fusion in the house mouse. Chromosoma 94:31–35 (1986).

Redi CA, Garagna S, Winking H: Kinetics of spermatogenesis of some T-mutant mice. J Fetal Med 8:59–60 (1988).

Redi CA, Garagna S, Capanna E: Satellite DNA and chromosome translocation: a hypothesis regarding Robertsonian chromosome formation. Rend Fis Acc Lincei 83:319–326 (1989).

Redi CA, Garagna S, Zuccotti M: Robertsonian chromosome formation and fixation: The genomic scenario, in Berry RJ, Corti M (eds): Inherited Variation and Evolution in the House Mouse, pp 235–255 (Academic Press, New York-London 1990).

Redi CA, Garagna S, Zuccotti M: Chromosomal aspects of spermatogenesis, in Baccetti B (ed): Comparative Spermatology 20 Years after Serono Symposia on Reproduction, pp 683–687 (Raven Press, New York 1991).

Redi CA, Garagna S, Winking H: Further examination of the kinetics of gonadal devolopment in XY female mice. Cell Mol Biol 39:509–514 (1993).

Redi CA, Garagna S, Zacharias H, Zuccotti M, Capanna E: The other chromatin. Chromosoma 110:136–147 (2001).

Sadoyan T, Castiglia R, Capanna E, Serva L: Robertsonian polymorphism in house mouse *Mus musculus domesticus* from area of intense seismic activity. Acta Theriol 48:189–195 (2003).

Sage RD, Atchley WR, Capanna E: House mouse as a model in systematic biology. Syst Biol 42:523–561 (1992).

Said K, Jaquart T, Montgelartd C, Sojaya H, Helal AV, Britton-Davidian J: Robertsonian house mouse population in Tunisia: A karyological and biochemical study. Genetica 68:151–156 (1986).

Said K, Saad A, Auffray J-Ch, Britton Davidian J: Fertility estimates in the Tunisinian all-acrocentrics and Robertsonian populations of the house mouse and their chromosomal hybrids. Heredity 71:532–538 (1993).

Scriven PN: Robertsonian translocation introduced into an island population of house mice. J Zool 227:493–502 (1992).

Searle JB: Speciation, chromosomes and genomes. Chromosome Res 8:1–3 (1998).

Sharma GG, Sharma T: Unusual chromosomal organization of telomeric sequences and expeditious karyotypic differentiation in the recently evolved *Mus terricolor* complex. Cytogenet Cell Genet 80:204–208 (1998).

Speed RM, Vogt P, Kohler MR, Hargreave TB, Chandley AC: Chromatin condensation behaviour of the Y chromosome in the human testis. Chromosoma 102:421–427 (1993).

Spirito F, Modesti A, Perticone P, Cristaldi, Federici R, Rizzoni M: Mechanism of fixation and accumulation of centric fusions in natural population of *Mus musculus*. 1. Karyological analysis of an hybrid zone between two populations in Central Apennines. Evolution 34:453–466 (1980).

Thaler L, Bonhomme F, Britton Davidian J: Processes of speciation and semi-speciation in the house mouse. Symp Zool Soc London 47:27–31 (1981).

Therman E, Susmann B, Dennison C: The nonrandom participation of human acrocentric chromosomes in Robertsonian translocations. Ann Hum Gen 53:49–65 (1989).

Tichy H, Vucak C: Chromosomal polymorphism in the house mouse *(Mus domesticus)* of Greece and Yugoslavia. Chromosoma 95:31–36 (1987).

Viroux MC, Bauchau V: Segregation and fertility of *Mus musculus domesticus* (wild mice) heterozygous for the Rb (4.12) translocation. Heredity 68:131–134 (1992).

Vissel B, Choo KH: Mouse major (γ) satellite DNA is highly conserved and organized into extremely long tandem arrays: Implications for recombinations beetween non-homologous chromosomes. Genomics 5: 407–414 (1989).

Volobouev V: Parallel observation of C-heterochromatin and telomeric sequences during Rb.process in the *Sorex araneus* species group (Insectivora, Soricidae). Euro-American Mammal Congress, Santiago de Compostela, Spain, July 19–24, pp 57–58 (1998).

Volobouev V, Tranier M, Dutrillaux B: Chromosome evolution in the genus *Acomys*: chromosome banding analysis of *Acomys cf. dimidiatus* (Rodentia, Muridae). Bonn Zoo Beitr 42:253–260 (1991).

Volobouev V, Gautum JC, Sicard B, Tranier M: The chromosome complement of *Acomys* spp (Rodentia, Muridae) from Oursi, Burkina Faso – the ancestral karyotype of the *cahirinus-dimidiatus* group? Chromosome Res 4:526–530 (1996).

Vorontsov NN, Liapunova EA: Explosive chromosomal speciation in seismic active regions. Chromosome Today 8:279–294 (1983).

Wakayama T, Perry ACE, Zuccotti M, Johnson KR, Yanagimachi R: Full term development of mice from enucleated oocytes injected with cumulus cell nuclei. Nature 394:369–374 (1998).

Wallace BM , Searle JB, Everett CA: Male meiosis and gametogenesis in wild house mouse *(Mus musculus domesticus)* from a chromosomal hybrid zone. A comparison between "simple" Robertsonian heterozygotes and homozygotes. Cytogenet Cell Genet 6:211–220 (1992).

Walsh JB: Rate of accumulation of reproductive isolation by chromosome arrangements. Am Naturalist 120:331–352 (1982).

White MJD: Models of speciation. Science 159:1065–1070 (1968).

White MJD: Modes of Speciation (W.H. Freeman and Co., San Francisco 1978a).

White MJD: Chain processes in chromosomal speciation. Syst Zool 27:285–295 (1978b).

White MJD: Rectangularity, speciation, and chromosome architecture, in Barigozzi C (ed): Mechanims of Speciation, pp 57–74 (Alan Liss, New York 1982).

Wilson AL, Bush GL, Case SM, King MG: Social structuring of mammalian populations and rates of chromosomal evolution. Proc Natl Acad Sci USA 72:5061–5065 (1975).

Winking H, Dulic B, Bulfield G: Robertsonian karyotype variation in European house mouse, *Mus musculus*: survey of present knowledge and new observations. Z Säugetierk 53:148–161 (1988).

Wong AKC, Biddle FG, Rattner JE: The chromosomal distribution of the major and minor satellite is not conserved in the genus *Mus*. Chromosoma 99:190–195 (1990).

Wong AKC, Rattner JE: Sequence organization and cytological localization of the minor satellite of the mouse. Nucl Acids Res 16:11645–11661 (1988).

Cytogenet Genome Res 105:385–394 (2004)
DOI: 10.1159/000078211

Mouse biodiversity in the genomic era

N. Galtier, F. Bonhomme, C. Moulia, K. Belkhir, P. Caminade, E. Desmarais, J.J. Duquesne, A. Orth, B. Dod, and P. Boursot

UMR 5171 – "Génome, Populations, Interactions, Adaptation", Université Montpellier 2, Montpellier (France)

Abstract. Comparative genomics has developed by comparison of distantly related genomes, for which the link between the reported evolutionary changes and species development/physiology/ecology is not obvious. It is argued that the mouse (genus *Mus*) is an optimal model for microevolutionary genomics in vertebrates. This is because the mouse genome sequence, physical and genetic map have been completed, because mouse genetics, morpho-anatomy, pathology, behavior and ecology are well-studied, and because the *Mus* genus is a diverse, well-documented taxon, allowing comparative studies at the level of individual, population, subspecies, and species. The potential of the interaction between mouse genome and mouse biodiversity is illustrated by recent studies of speciation in the house mouse *Mus musculus*, and studies about the evolution of isochores, the peculiar pattern of GC-content variation across mammalian genomes.

Copyright © 2004 S. Karger AG, Basel

The field of molecular evolution has greatly benefited from the achievement of full-genome sequencing projects. Our knowledge about bacterial evolution, for example, has progressed at an impressive rate during the last ten years with the massive release of full genomes in public databases. The study of vertebrate evolution is also undergoing a genomic revolution. The genomes of human, mouse, and fugu are completed, and others (tetraodon, zebrafish, *Xenopus*, chicken, rat, chimpanzee) should follow soon. These data have helped addressing up to now unattainable help in understanding the dynamics of isochores (the peculiar variation of local GC content among regions of the genome; Eyre-Walker and Hurst, 2001), of repeated elements (IHGSC, 2001), of intron/exon structure (Kondrashov and Koonin, 2003), of codon usage (Duret, 2002), and of many specific genes or gene families of interest (e.g. Escriva Garcia et al., 2003). They also have revived the debate about whole genome duplications, and their possible link to vertebrate diversification and body plan evolution (Amores et al., 1998).

Such studies are highly valuable in describing patterns of genome variability. Inferences about the underlying evolutionary processes, however, can be limited by the specificities of current whole-genome comparisons: they involve few, distantly related species, whose biology and ecology are often poorly known. The evolutionary forces applying to genomes are best understood at a recent time scale. Comparing the patterns of within-species polymorphism and divergence between closely related species potentially allows to distinguish between mutational and selective effects, while patterns of variation between distantly related species reflect the combination of the two, integrated over a long period of time. This view was applied with great success in *Drosophila*, in which population genomics has yielded important results concerning the processes of molecular evolution (e.g. Begun and Aquadro, 1992; McVean and Charlesworth, 2000; Gibson and Mackay, 2002). In humans, key arguments about the origin and evolution of isochores resulted from studies of intra-specific variation (Smith and Eyre-Walker, 2001; Lercher et al., 2002; present report). Usage of population data (or data from closely related species) for molecular evolutionary prospects, however, is still infrequent in vertebrates. Many evolutionary hypotheses formulated from whole-genome comparisons lack confirmation on a recent time scale.

There are several reasons why the mouse *(Mus musculus)* is probably the best candidate among vertebrates for such microevolutionary genomics. First, the mouse genome sequence is already assembled and publicly available (Gregory et al., 2002; MGSC, 2002; Clamp et al., 2003). Secondly, mouse develop-

Received 8 October 2003; manuscript accepted 8 December 2003.

Request reprints from N. Galtier, UMR 5171 –
"Génome, Populations, Interactions, Adaptation"
Université Montpellier 2 – CC 63, Place E. Bataillon
34095 Montpellier (France); telephone: (+33) 467 14 46 85
fax: (+33) 467 14 45 54; e-mail: galtier@univ-montp2.fr

ment, morpho-anatomy, pathology, behavior and ecology are well studied, both as a laboratory animal and in the field, so that links can be made between genomic variation patterns and phenotypic evolution. Besides the well-known and long-standing laboratory strains, wild-derived mouse strains are now kept in the laboratory (Guénet and Bonhomme, 2003), providing a unique (at least in vertebrate) renewable sample of the naturally-occurring genetic variation. Thirdly, and most importantly, the highly studied house mouse *M. musculus* belongs to a diversified, well-characterized taxonomic group, allowing comparative analyses at the level of the individual, subspecies, and closely-related species, rat (*Rattus norvegicus*, genome sequencing in progress, diverged from mouse around 12 million years ago) serving as a useful outgroup. Of particular interest is the occurrence of a hybrid zone between two recently diverged subspecies *(M. musculus musculus* and *M. musculus domesticus)* in Europe.

In this paper, we examine various perspectives offered by the combination of these favorable factors, that is, how genomic data help in understanding mouse evolution and biodiversity, and how mouse biodiversity makes it a useful model to approach the processes of genome evolution in a mammal.

I. Mouse biodiversity

The genus *Mus* diverged from its closest relatives *(Rattus, Apodemus, Mastomys)* around 10–15 million years ago (Mya), and diversified as four subgenera (see Fig. 1 in She et al., 1990b). Comparative studies centered on the house mouse *M. musculus* can therefore make use of a panel of outgroups, involving variable levels of divergence, and distinct ecological features. *M. musculus* has the peculiarity of being the most, if not the only commensal mouse species. It is originally a species inhabiting steppic grassland habitats, but now occurs outside its original range in all sorts of human dwellings, and shows a worldwide geographic distribution, most likely as a consequence of human migrations. Its closest relatives are the North African short-tailed *M. spretus*, with which it occasionally hybridizes (Orth et al., 2002), and two Near-Eastern sister species, *M. macedonicus* and the mound building *M. spicilegus*. The divergence date between these palearctic mice was estimated as 1–3 Mya. More divergent species are found in East Asia: the *M. famulus/M. fragilicauda* group diverged from the palearctic lineage about 5 Mya and the *M. cooki/M. caroli/ M. cervicolor* group somewhat earlier. Three other subgenera, *Pyromis, Coelomys* and *Nannomys*, having diverged from the subgenus *Mus* sensu stricto about 8–12 Mya, encompass several species that are most distantly related to *M. musculus*. These results are supported by molecular phylogenies based on mitochondrial and nuclear genes, and by DNA-DNA hybridization experiments (She et al., 1990b; Boursot et al., 1993; Lundrigan et al., 2002).

It should be noted that mouse taxonomy is a dynamic field. Murinae are one of the mammalian families in which new species are still recurrently discovered. The oriental *M. famulus* (Chevret et al., 2003) and *M. fragilicauda* (Auffray et al., 2003) were recently added to the mouse tree. They are the closest known outgroups to the four palearctic species, and might help in resolving the *M. musculus/spretus/macedonicus/spicilegus* radiation.

The history of the *M. musculus* species complex is characterized by recent expansions of various lineages, all probably originating from the Indian subcontinent (Boursot et al., 1993). These expansions generated peripheral subspecies, the best-known being *M. musculus domesticus* (Africa, Western Europe and Near-East), *M. m. musculus* (Eastern Europe, Russia, northern China, Japan), and *M. m. castaneus* (South East Asia). Note that a fourth mitochondrial lineage (*M. m. gentilulus*) was recently reported in Yemen (Prager et al., 1998) and Madagascar (Duplantier et al., 2002). These subspecies undergo genetic exchanges at the boundary of their geographic ranges. The best documented cases are those occurring between *M. m. musculus* and *M. m. castaneus* in Japan, where the hybridization was high enough to give rise to a unique population often referred to as *M. m. molossinus* (Yonekawa et al., 1988), and between *M. m. musculus* and *M. m. domesticus* in Europe, along a narrow hybrid zone.

Remarkably, standard mouse laboratory strains are actually recombinant strains derived (in unequal percentages) from the three major *M. musculus* subspecies. The published mouse genome sequence is therefore a mosaic, as demonstrated by a sliding-window analysis of the similarity between the mouse assembly and various wild strains (Wade et al., 2002). This mosaicism, and the elevated homozygosity of laboratory strains, has important functional consequences, as discussed by Guénet and Bonhomme (2003): the genotype/phenotype relationship of laboratory strains is probably distinct from that of wild mice. From an evolutionary point of view, the mosaicism of the mouse genome sequence highlights the importance of the population sampling in mouse evolutionary genomics. Single nucleotide polymorphism (SNP) data based on the mouse genome project, for example, should be treated with caution; their evolutionary relevance (that is, to what extent do they reflect the sequence polymorphism present in natural populations) strongly depends on the genomic region surveyed, and on the sample within which variation was screened for.

As has started to prove most invaluable in *Drosophila*, mouse genomics would presumably gain much by having at hand a second complete genome sequence collinear to the one already available. *Mus spretus* is certainly a good candidate for this, since it is the only other species for which several laboratory strains, including completely inbred ones, exist, and is cross-fertile with *M. musculus* through the female lineage. It is a clear outgroup to the *M. musculus* complex, and yet very close (mean nuclear sequence divergence in the order of 2%), so that the mutational pathways since the common ancestor of the two species can easily be reconstructed.

II. Mouse evolutionary genetics: selection, speciation

The mouse offers a good model for comparative evolutionary genetics analyses: it is abundant in nature, it is taxonomically very diversified, with various levels of divergence (see above), and it can easily be bred in the laboratory. Despite these advantages, however, it has not much been used as a model to address important evolutionary questions. It is a well-developed model for molecular and cellular genetics, but *Drosophila* and humans remain by far the most important evolutionary models in population and evolutionary genetics. We will here review some questions that are well addressed in either or both of these organisms, and for which some data are available on the mouse, and we will attempt to illustrate why the mouse is a good model.

Recombination and selection across the genome

A correlation between local recombination rate and degree of polymorphism in the genome was first described in *Drosophila* (Begun and Aquadro, 1992), and later in several other eukaryotes (for review see Nachman, 2002). This could either result from a correlation between mutation and recombination, or from the interaction between recombination and selection. Because in *Drosophila* recombination did not seem to correlate with interspecific divergence, the latter hypothesis was favored, implying that selection was a major determinant of polymorphism, and much of the debate has since centered on the relative importance of hitchhiking by favorable mutations as compared to background selection against deleterious mutations. However, recent analyses at a genomic scale have revealed a yet unnoticed correlation between recombination rates in humans and divergence with the chimpanzee and baboon, suggesting that in humans mutation rather than selection explained the correlation between recombination and polymorphism (Hellmann et al., 2003). It could be that human populations are too small for weak selective effects to have a significant impact. Because mouse population sizes are presumably intermediate between those of human and *Drosophila*, such studies on the mouse will be of great interest. There is evidence that recombination rates vary across the mouse genome (Nachman, 2002), and limited evidence for a correlation between recombination and levels of polymorphism (Nachman, 1997). This approach is complicated by the recent suggestions that recombination in humans is concentrated to hotspots, which would create blocks of linkage disequilibrium (reviewed in Wall and Pritchard, 2003). Mouse recombination hotspots have been described (Shiroishi et al., 1982; Yauk et al., 2003; and references therein), but the data needed to test the block structure of recombination from natural polymorphisms are essentially lacking in the mouse. Claims of a block structure of polymorphism in the mouse (Wiltshire et al., 2003) for the moment are based on the study of a limited number of genomes, from laboratory strains that share common ancestries. Thus the block structure could result from this history rather than from the distribution of recombination (Wade et al., 2002).

Evolution of reproductive isolation

The expression "genetics of speciation" is often used by evolutionary biologists and covers many different aspects, central to which are the questions of the genetic architecture of incompatibilities between diverged genomes, and the nature of the evolutionary forces leading to this divergence. In this context, hybrid sterility and inviability have been the focus of many studies, most detailed in *Drosophila*. In all studied cases, the genetic architecture of these incompatibilities turned out to involve several factors with strong epistatic interactions. The number of factors involved varies from a few for closely related taxa (e.g. Orr and Irving, 2001) to many when more distant species are confronted (e.g. Noor et al., 2001; Tao et al., 2003). Thus the question of the nature of the genes involved in the acquisition of reproductive isolation remains largely unsolved, as well as the associated question of the role of selection in their divergence between species. However, in two cases where a gene contributing to hybrid sterility (Ting et al., 1998) or hybrid inviability (Presgraves et al., 2003) could be identified, it was shown to evolve fast, with a pattern evoking positive Darwinian selection.

Several male hybrid sterility factors have been identified in the mouse. One of them was found in crosses between laboratory mouse and wild-derived *M. m. musculus* (*Hst1*, review in Forejt, 1996). It is located on chromosome 17 and seems to be involved in epistatic interaction with several other loci to produce the sterility phenotype. However, the sterility allele does not appear to be fixed in natural populations. Thus the importance of this gene for reproductive isolation between the two subspecies of mice remains to be assessed. The other sterility factors reported concerned the interaction between the genomes of the laboratory mouse and its close relative, *M. spretus*. One sterility effect was located at the tip of the X chromosome near the pseudoautosomal boundary *(Hst3)*, but it is not clear whether this phenotype is linked to allelic variation at a locus or to the structural differences that exist in this region of the chromosome between the two species (Forejt, 1996). Additional sterility factors between these species were discovered incidentally when using interspecific crosses to identify the genes responsible for male sterility of homozygotes for the *t*-haplotype. Four such factors (Pilder et al., 1991, 1993, 1997; review by Schimenti, 2000) were identified in the *t*-haplotype region. They produce hybrid sterility phenotypes analogous to those associated with sperm inactivation causing transmission distortion of the *t*-haplotype. This raises the interesting possibility that these hybrid sterility genes coincide with the *t*-haplotype distorter genes, and evolved as by-products of the selection exerted by the selfish *t*-haplotype distorter in the *M. musculus* lineage (Hurst, 1993). All strong candidates for these distortion/sterility genes encode dynein chains involved in sperm motility, and the role of at least one of them seems to differ between *M. spretus* and the laboratory mouse (Fossella et al., 2000). Little is yet known of the pattern of evolution of these genes in the mouse lineages. Interestingly, Nurminsky et al. (1998) report the discovery of a newly evolved sperm-specific gene encoding an axonemal dynein in *Drosophila melanogaster*, and provide evidence that, like many male-specific genes, it appears to have evolved under strong positive Darwinian selection.

The interesting question of whether hybrid male sterility could evolve as a by-product of genomic conflicts (existing between the sexes or between selfish genes such as transmission distorters) thus remains valid. The control of embryonic development by the placenta in mammals is a particular circumstance in which theory also predicts potential genetic conflicts, between the genes of paternal and maternal origins (Haig,

1996). It has been hypothesized that genomic imprinting could be used, or could even have evolved, to restrain such conflicts (Haig, 1997; Hurst and McVean, 1997), and many genes expressed in the placenta are imprinted (Georgiades et al., 2001). The placenta is among the most variable organs between different mammals (Carter, 2001), and such conflicts could have triggered rapid evolution of its developmental control, thus promoting rapid divergence between species. The mouse has been a particularly good model to address this question. Hybrids between the laboratory mouse and either of its close relatives, *M. macedonicus* or *M. spretus*, show defects in placental growth control, resulting in hyper- or hypoplasia. A major determinant of this phenotype is located on the X chromosome (Zechner et al., 1996), in addition to several autosomal factors. The Y chromosome was also shown to influence placental size in these hybrids (Hemberger et al., 2001b). However, a large region of the X chromosome is implied, so that up to now it turned out to be impossible to narrow down the locus and identify candidate genes (Hemberger et al., 1999). Interestingly, a very similar syndrome was observed in hybrids between two species of the American rodent genus *Peromyscus*. As in the mouse, a major effect of the X chromosome was detected. In addition, imprinting was shown to be disrupted in hybrids (Vrana et al., 1998), and loss of imprinting at the *Peg3* locus to be involved in placental hybrid dysplasia, a remarkable realization of the theoretical prediction (Vrana et al., 2000). In the mouse, no evidence for the implication of this gene or of a general loss of imprinting in hybrids could be found (Zechner et al., 2004). However, as hyperplasic mouse interspecies hybrids and cloned placentas exhibit very similar phenotypes (Zechner et al., 1996; Tanaka et al., 2001) it can be assumed that disruption of epigenetic states, such as loss of imprinting, is involved. It will be interesting to compare the different evolutionary routes along which these very similar hybrid breakdown syndromes have been acquired in these two lineages of rodents. Increased knowledge of the mouse genome and new technologies will certainly help decipher the complex metabolic pathways controlling placental development (Hemberger et al., 2001a), and allow to test hypotheses about the forces driving the evolution of genes controlling placental development.

Hybridization between natural populations represents a powerful sieve of the regions of the genome involved in hybrid incompatibility, integrating many generations of recombination and natural selection, which is not the case in laboratory experiments. The house mouse subspecies appear to be a good model because (i) they are very closely related, (ii) they show incomplete reproductive isolation, and (iii) they form natural hybrid zones in various places in the world. The European hybrid zone between *M. m. domesticus* and *M. m. musculus* has been studied in most detail. Studies in several regions of this hybrid zone, from Denmark to the Black Sea, have shown a clear and geographically limited genetic transition between the two subspecies, using diagnostic allozyme markers (Hunt and Selander, 1973; Boursot et al., 1984; Sage et al., 1986b; Vanlerberghe et al., 1986, 1988a), indicating that free gene flow was impeded by the counter selection of hybrid genotypes. The most compelling evidence of counter selection in this hybrid zone concerns the sex chromosomes which in several regions studied show extremely limited introgression, and are thus presumably strongly involved in hybrid unfitness (Vanlerberghe et al., 1986, see Fig. 2; Tucker et al., 1992; Dod et al., 1993; Prager et al., 1997). Indirect evidence was also gathered that some chromosomal incompatibilities might be involved in hybrid incompatibilities. The first such evidence came from the observation that Robertsonian translocations segregating in Danish *domesticus* populations never reached the center of the hybrid zone or the *musculus* territory (Fel-Clair et al., 1996), and that a centromeric marker did not show any introgression either (Fel-Clair et al., 1998). Further studies are needed to confirm these chromosomal effects, the problem being to assess the diagnostic status of the markers studied. Even using a more representative sample of the studied species, extending beyond the hybridizing populations, distinguishing between ancestral shared polymorphisms and secondary introgression can be difficult, especially so for recently divergent taxa (e.g. Hudson and Coyne, 2002). This requires both a detailed phylogeography of the marker and a fine study of the introgression pattern. For instance, while the latter approach reported the fixation of mitochondrial DNA of *domesticus* origin into Scandinavian *musculus* (Ferris et al., 1983), the former approach could attribute this observation to a historical colonization accident rather than extensive gene flow across the hybrid zone (Vanlerberghe et al., 1988b; Prager et al., 1993, 1997).

How many regions of the genome are involved in strong incompatibilities between these genomes is a major question. The knowledge of the mouse genome and its variations offer new perspectives for this kind of study, and it becomes possible to scan the genome for diagnostic markers. Strategies to identify numerous and easy to score markers must be set up. Munclinger et al. (2003) have shown that short interspersed repetitive elements could provide such tools. Furthermore, the mouse genome projects are starting to produce SNP data that could be used. Although most of the effort is put on laboratory mice rather than natural populations, some of the polymorphisms between laboratory strains could turn out to be useful diagnostic markers between the subspecies. The genetic evidence that laboratory mice are hybrids between *M. m. domesticus* and other oriental subspecies is clear (e.g. Bishop et al., 1985), and has consequences on the overall diversity among these strains (Fitch and Atchley, 1985; Bonhomme et al., 1987; Atchley and Fitch, 1991). This fact is now well established and recent high-resolution analyses show that the genome of old inbred laboratory strains is a mosaic of segments derived from mice of *domesticus* and Asian origin (presumably *molossinus*, which is close to *musculus*; Wade et al., 2002), although this is not always accounted for in comparative genome-wide studies (Wiltshire et al., 2003). Such comparative studies define thousands of markers that vary between the genomes of the different house mouse subspecies, and presumably many of them could be used to characterize differential patterns of introgression in the hybrid zone. However, it is not yet clear what proportion of these markers differentiate between the subspecies rather than represent shared ancestral polymorphisms. It will be interesting to quantify these proportions and to determine what fractions of the genome of the different subspecies have reached reciprocal monophyly.

The power of such an approach to ultimately pinpoint the genes responsible for hybrid incompatibility, and the functions that are affected in the hybrid genotypes, remains to be assessed. It is thus also important to address the question from a more integrated point of view, by trying to identify the phenotypic traits subject to hybrid breakdown. Hybrid male sterility has been reported to occur in some experimental crosses between the laboratory mouse and wild-derived *musculus* (Forejt, 1996). Crosses between wild-derived *domesticus* and *musculus* mice revealed decreased fertility or apparent sterility of some F1 males but also females (Alibert et al., 1997). Although these experiments suggest that some sterility genes are segregating in natural populations, many more data are needed to assess their importance and ubiquity.

The higher parasite load in hybrids than in parental natural populations of the hybrid zone (Sage et al., 1986a; Moulia et al., 1991) is another indication of hybrid unfitness. The higher susceptibility of hybrid genotypes was confirmed in experimental infections with the mouse nematode *Aspicularis teraptera* of artificial hybrids between these two subspecies, thus confirming the genetic nature of the hybrid susceptibility (Moulia et al., 1993), and suggesting a breakdown of the coadaptation of immunity genes (Moulia et al., 1996). Interestingly, hybrids proved to be highly susceptible to some, but not all kinds of parasites (Derothe et al., 1999a, 2001) which could reflect diverse host/parasite coevolutionary histories, or the involvement of different immune pathways in the reaction to various infections (Porcherie, unpublished results). However, other characters might be suggestive of hybrid vigor rather than hybrid depression between these subspecies. In natural populations from the hybrid zone and in experimental crosses, hybrid genotypes display less fluctuating asymmetry in tooth morphology than the parental genotypes (Alibert et al., 1994, 1997). If fluctuating asymmetry is taken as an indicator of the quality of the development process, this could be taken as evidence of heterosis in hybrids, rather than hybrid breakdown.

Although prezygotic isolation is obviously not an absolute barrier between these subspecies, it could play an important role in impeding gene flow, and the question is posed of how it could evolve in conjunction with post-zygotic isolation once it has evolved. Mechanisms of inter-individual recognition and mate choice are extremely complex in mice, but there is increasing evidence that important signals are delivered in the urine, that allow mice to perform subspecies discrimination (Smadja and Ganem, 2002). Major urinary proteins (MUPs) appear to be good candidates for the making up of that signal (Hurst et al., 2001; Beynon and Hurst, 2003), but belong to a complex multigene family, thus making knowledge of the genome extremely useful.

Genetic introgression, selection and evolution
Studying patterns of polymorphism and coalescence of genes between populations, subspecies and species is another way to infer the forces driving the evolution of the genome during speciation processes. Here again the mouse appears an excellent model because of the numerous closely related species and subspecies. Because they are of recent origin and have retained the capacity to hybridize, the subspecies of the house mouse are expected to show mosaic and reticulate patterns of gene coalescence. Detailed genealogic information is still available only for a handful of well-studied genes or regions of the genome, but show interesting patterns. For instance the distribution of the major lineages of Y chromosome and mtDNA among the house mouse subspecies appear to be extremely contrasting (Boissinot and Boursot, 1997) and suggest the recent spread of new Y chromosome lineages. This is an interesting observation considering that the Y chromosome appears strongly involved in hybrid incompatibilities in the hybrid zone (Vanlerberghe et al., 1986; Tucker et al., 1992; Dod et al., 1993; Prager et al., 1997). Another interesting case is the evolution of the androgen binding protein (ABP), a protein excreted in large amounts in mouse saliva. The evolution and polymorphism of the gene for the alpha subunit of this protein was studied in detail. Contrary to many other protein genes tested, it shows a unique pattern of fixation of a single allele in each of the peripheral subspecies of the house mouse. The absence of intronic polymorphism in the subspecies, and the high inter-subspecies and inter-species non-synonymous substitution rates in the exon suggest that it evolves under positive Darwinian selection (Karn and Nachman, 1999; Karn et al., 2002). However, intron phylogeny provides evidence that interspecific transfers have occurred in the past evolution of this gene among the Palearctic mouse species (*M. musculus, M. spretus, M. macedonicus* and *M. spicilegus*; Karn et al., 2002).

Evidence of interspecific secondary exchanges between the Palearctic mouse species also came from the reconstruction of the evolutionary history of L1 LINEs (long interspersed elements). The many copies of these retrotransposons in the genome can be classified into families of related sequences, thought to result from successive events of expansion of an active copy in the genome. Most LINEs are truncated pseudogenes and the age of a given family can be estimated from the neutral mutations that accumulate after retroposition. LINEs are good markers of secondary introgressions when different (sub)species share families younger than species divergence. This is the case for certain recently discovered L1 families in the mouse, suggesting that introgression has occurred between the house mouse and *M. spretus*, and possibly *M. spicilegus* (Rikke et al., 1995; Zhao et al., 1998; Hardies et al., 2000). Evidence for introgression of unique *M. spretus* sequences into *M. musculus* has also been reported (Greene-Till et al., 2000). Because *M. spretus* and *M. m. domesticus* are presently sympatric over the whole range of the former, it is possible to test whether and how frequently hybridization occurs. A large-scale survey of populations of these two species with mtDNA and nuclear markers revealed sporadic introgressions in both directions, and clear evidence of recent hybridization in at least one Moroccan population (Orth et al., 2002).

Although such interspecific exchanges remain exceptional, and probably only affect the evolution of the genomes of these species marginally, they could have important evolutionary consequences if they result in the spread of strongly selected genes (adaptation genes or selfish elements). Such a hypothesis has been put forward to explain the reticulate pattern of evolution of the *Fv1* gene, conferring resistance to the Murine Leukemia Virus (Qi et al., 1998).

The mammalian genome is characterized by a strong spatial heterogeneity in base composition (Bernardi, 1993). The GC content of long (>100 kb) fragments of the human genome typically ranges between 35 and 60%, a range much wider than in, say, *Xenopus* or fish genomes. This was called the isochore structure (Bernardi et al., 1985). Isochores correlate with many features of mammalian genome organization, potentially relevant to its function. GC-rich isochores in human show a higher coding sequence density (shorter introns, shorter intergenic regions; Duret et al., 2002), and a lower density of repeated elements (with the noticeable exception of *Alu* repeats; IHGSC, 2001). They correspond to early-replicating DNA (Watanabe et al., 2002), to specific chromosomal bands (Saccone et al., 1993), and to highly recombining regions (Fullerton et al., 2001; Kong et al., 2002).

Probably because of these many correlations, the origin and evolutionary significance of isochores have long been debated (Eyre-Walker and Hurst, 2001). Three major models compete. Bernardi has long argued that isochores have been selected as an adaptation to homeothermy (Bernardi, 1993). This hypothesis, however, has received no empirical support, and was essentially dismissed by the discovery of isochores in cold-blooded vertebrates (Hughes et al., 1999). The existence of a variable mutation bias across genomic regions is the most natural neutralist alternative to selectionist scenarios: the ratio of (neutral) AT→GC vs. GC→AT mutations could vary spatially, resulting in variable base composition. Analysis of polymorphism data in human led to rejection of this hypothesis (Eyre-Walker, 1999; Duret et al., 2002; Lercher et al., 2002): AT→GC and GC→AT mutations undergo distinct fixation processes, as measured by the distribution of allele frequencies across loci. GC-rich regions are so because AT→GC mutations in these regions fix preferably, not because they are more numerous. This (somewhat unexpected) result revived a third hypothesis potentially explaining the existence of isochores: biased gene conversion (Eyre-Walker, 1993).

Gene conversion is a mechanism by which two similar but distinct DNA fragments of a genome become identical: one fragment (the donor) is copied/pasted onto the other one (the recipient). Gene conversion can occur between paralogous fragments of a haploid genome (ectopic gene conversion) or between the two alleles of a given locus in a diploid genome (allelic gene conversion). Allelic gene conversion occurs at meiosis, as a part of the mechanism of homologous recombination. The biased gene conversion (BGC) hypothesis invokes a recombination-associated repair bias toward GC, providing an evolutionary "advantage" to GC alleles, just like natural selection would (Nagylaki, 1983). This would result in a GC enrichment of highly recombining regions of the genome. A number of observations are compatible with, but do not demonstrate, the effectiveness of BGC in shaping GC-content variation in mammals (Galtier et al., 2001; Birdsell, 2002; Marais, 2003, and references therein). We now review two recent studies that provide a strong support for the BGC hypothesis. The first one makes use of full-genome data, the second one of mouse genetics and biodiversity.

One prediction of the BGC hypothesis is that sequences undergoing frequent ectopic gene conversion should be GC-enriched. To check this prediction, Galtier (2003) analyzed histone gene sequences in human and mouse. There are four distinct replication-dependent core histones in mammals (H2a, H2b, H3, H4), each encoded in human and mouse by 10 to 30 paralogous genes. Histone genes of a given family encode the very same protein, but can differ at synonymous positions. The pattern of sequence variation between histone paralogues, indeed, reflects the dynamics of ectopic gene conversion. Some histone genes belong to sub-families of nearly identical copies; these genes are probably undergoing recurrent gene conversion, homogenizing their nucleotide sequences (the so-called "concerted evolution" process). Other paralogues are more or less unique, i.e. distantly related from any distinct gene copy; such genes must have escaped gene conversion. In agreement with the BGC model, it was found that the GC content at third codon positions (GC3) of histone gene copies involved in sub-families of nearly identical genes was significantly higher than the GC3 of "unique" histone gene copies (Galtier, 2003). This result demonstrates the efficiency of BGC in significantly increasing genomic GC content.

This effect (GC3-enriched sub-families of nearly identical histone paralogues) was found stronger in mouse than in human. This is probably because mouse histone genes are more clustered than human ones. Physically linked gene copies tended to share high sequence similarity, and a high GC content, in agreement with the notion that gene conversion occurs more frequently between close than between distant DNA sequences. It is worth noting that histones are an exception in that the GC3 of GC-rich genes is generally lower in mouse than in human (Mouchiroud and Gautier, 1988). We now know that GC-rich isochores are gradually being eroded from mammalian genomes (Duret et al., 2002) and for some reason this is occurring faster in rodents than in primates.

The histone studies provide strong support for the BGC model of GC content evolution in mammals. They do not explain, however, why GC content correlates with many features of mammalian genome organization, including gene density. This was addressed by investigating the molecular evolution of *Fxy (Mid1)*, a gene recently translocated into the highly recombining mouse pseudoautosomal region (Montoya-Burgos et al., 2003).

The pseudoautosomal region (PAR) is a short region of homology between the X and Y chromosomes that behaves like an autosome at meiosis. The PAR undergoes one obligatory crossing-over event per generation, resulting in a high recombination rate per nucleotide (Soriano et al., 1987). The *Fxy (Mid1)* gene has recently moved in *M. musculus* from an X-specific location to a new location where it overlaps the pseudoautosomal boundary: the 3′ part of the gene, including exons 4 to 10, now lies in the PAR (Perry and Ashworth, 1999). This translocation resulted in a dramatic increase of GC content: the average GC3 of exons 4–10 is 50% in the X-linked copy of short-tailed mouse *Mus spretus*, but 73% in the pseudoautosomal copy of house mouse *M. musculus*. Recombination-associated BGC is most probably the reason of this GC increase. After having clarified the history of this gene by sequencing three exons in nine Murinae species, Montoya-Burgos et al. (2003) compared intronic and exonic sequences from the X-specific (from *M. spretus*, henceforth called *Fxy*_X) vs. pseudoautosomal (from *M. musculus*, henceforth called *Fxy*_PAR) copy. This was done using publicly available (ENSEMBL) and original genomic data. Intron length, intron vs. third codon position GC content, and minisatellite density were analyzed.

It was found that the introns of the 3′ part of *Fxy* have been dramatically shortened after being translocated to the highly recombining PAR. The average intron length is much higher in the X-linked 5′ part (~ 20 kb) than in the pseudoautosomal 3′ part (~2 kb) of *M. musculus Fxy*_PAR. This discrepancy is actually a consequence of the translocation since the 3′ part of *Fxy*_X in *M. spretus* has long introns. This is reminiscent of the much lower intron size reported in GC-rich than in GC-poor regions of the mammalian genome. The GC content in introns (GCi) did not increase as fast as GC3 did, resulting in a considerable difference between the base compositions of introns vs. synonymous sites in *Fxy*_PAR (GC3-GCi is about 30% in *M. musculus* vs. 14% in *M. spretus*). A high GC3-GCi is a well-known characteristic of GC-rich isochores (Duret and Hurst, 2001) that was enhanced in *Fxy* after it was translocated to the PAR.

A notable feature of the translocated *Fxy* gene is the high number of minisatellites: seven minisatellite loci (i.e., tandem repeats of 7- to 121-base-long motifs) were found in introns of *M. musculus* 3′ *Fxy*_PAR, but none in *M. spretus* 3′ *Fxy*_X. This led to the following prediction concerning the genomic distribution of minisatellites in mammalian genomes: if *Fxy* evolution actually reflected the dynamics of isochores, one would expect a higher number of minisatellites in GC-rich than in GC-poor regions of the genome. This was checked by scanning human DNA fragments of variable GC content. The number of detected minisatellites was strongly positively correlated to contig GC content ($P < 0.001$), a yet unreported result consistent with the above prediction.

In summary, the mouse 3′ *Fxy*, a typical GC-poor piece of DNA, quickly came to resemble a GC-rich isochore shortly after it was translocated into the highly-recombining PAR in *M. musculus*: increased GC content, shortened introns, higher GC3-GCi, newly born minisatellites – a remarkably high number of typical features of GC-rich isochores that co-evolved in less than 3 million years. These results strongly suggest that variation of local recombination rate is the primary determinant of the isochore structure of mammalian genomes. It was proposed that the increase of gene density occurred through the deletion of large GC-enriched non-coding genomic fragments (Montoya-Burgos et al., 2003).

IV. Mouse biodiversity and immune response

The most promising perspective of the mouse genome/biodiversity interaction, however, might belong to the medical field. The mouse is one of the important animal models in medicine and pharmacology. Laboratory strains are frequently tested for their ability to overcome infections due to virus, bacteria, metazoan parasites (Skamene et al., 1984; Mock et al., 1990; Wassom and Kelly, 1990). Such studies are fundamental for understanding the processes of infection in mammals. However, if we are to understand infectious processes at genetic and population level as well as coevolutionary outcome of mammal/pathogen interactions, investigations have to be extended to wild mice. Laboratory strains are of mosaic origin and represent only a small proportion of the genetic diversity found in the wild (Guénet and Bonhomme, 2003).

This discrepancy is illustrated by the general tendency of wild-derived strains to be more resistant to experimental infections than laboratory ones (Guénet and Bohomme, 2003). However, laboratory stocks were found to be more resistant to the common oxyuroid *Aspiculuris tetraptera* than wild-derived strains, probably because of the differential impact of laboratory breeding on the mouse parasite interactions (Derothe et al., 1997, 1999b).

One of the main steps for the establishment of the immune response against pathogens is the recognition of the foreign antigen by the T cell, which initiates and coordinates the subsequent response (Janeway, 1993). This depends on the T cell receptor (TCR) and the class II major histocompatibility complex (MHC). As in primates, analyses of the allelic diversity in wild mice showed the maintenance of ancestral polymorphism within various lineages and through speciation events (Figueroa et al., 1988). Ancestral sequences are mixed by intraexonic recombination which generates numerous combinations in natural populations (McConnell et al., 1988; She et al., 1990a, 1991). As predicted, the polymorphism is mainly located within the antigen binding site (She et al., 1990a; Hughes and Yeager, 1998).

Susceptibility to parasites has been linked to some MHC haplotypes during experimental infection of laboratory mice (Else et al., 1990; Behnke and Wahid, 1991). Diversity of pathogens is viewed as one of the main sources of selective pressure on these loci (Red Queen hypothesis). Recently, multiple infections of MHC congenic mice revealed a higher resistance in heterozygous offspring than in homozygous parents (Penn et al., 2002; McClelland et al., 2003). These works performed on laboratory stocks need to be completed by analyses in wild populations, with the idea of linking past and present parasite/bacterial/viral richness to level of polymorphism at MHC loci in the field.

A few studies of natural polymorphism at other immune loci were performed: minor histocompatibility antigens (Rammensee and Klein, 1983; Roopenian et al., 1993), TCR (Roger et al., 1993), and constant region of immunoglobulin genes (Jouvin-Marche et al., 1989; Morgado et al., 1993). They showed limited polymorphism when compared to MHC. In a remarkable study of the Lake Casitas mainly *domesticus* population (California), Gardner et al. (1991) showed that the resistance to current epizooty was conferred by the relic of an ancestral virus of, presumably, *castaneus* origin. This study illustrates the interest of epidemiological surveys in wild mouse populations for the understanding of the evolutionary dynamics of the immune system.

V. Conclusion

These examples illustrate, we believe, the high potential of mouse as a model for evolutionary genetics/genomics. Mouse is the "mammalian *Drosophila*". It provides a good opportunity to understand the way the mammalian genome evolves and adapts. Several general questions of evolutionary genomics deserve to be addressed using the mouse model, including the evolutionary fate of gene duplications, the dynamics of repeated elements, chromosomal evolution, sexual evolution and genomic conflicts, detection of adaptive evolution using hitchhiking mapping (Schlotterer, 2003), among others. It should be noted that a decisive argument in the 25-year-old issue of isochore evolution arose from the comparative analysis of a relatively small locus sampled in the appropriate mouse species, while the analysis of the two complete mammalian genomes available had not solved the problem.

Conversely, a number of questions specific to mouse evolution are still pending, and will probably benefit from the recently published full mouse genome. An interesting peculiarity of mouse is karyotype plasticity: many karyotypic races characterized by various acrocentric (Robertsonian) fusions between chromosomes have been reported (Britton-Davidian et al., 1989; Boursot et al., 1993). Surprisingly, all these variants belong to the *M. m. domesticus* lineage – a pattern raising the question of the forces underlying chromosome evolution. Mouse ecology and behavior are also potential targets for genomic approaches. How did commensalism evolve? This major ecological shift occurred recently (i.e., posterior to the emergence of agriculture), and must have left traces in the mouse genome. Finally, we do not yet know how many genes, and which ones, are involved in hybrid dysgenesis between the subspecies *M. m. musculus* and *M. m. domesticus*. Genomic mapping of introgression patterns across the hybrid zone, combined to a candidate-gene approach, should help in solving this puzzling question.

References

Alibert P, Renaud S, Dod B, Bonhomme F, Auffray J-C: Fluctuating asymmetry in the *Mus musculus* hybrid zone: a heterotic effect in disrupted co-adapted genomes. Proc R Soc Lond B Biol Sci 258:53–59 (1994).

Alibert P, Fel-Clair F, Manolakou K, Britton-Davidian J, Auffray J-C: Developmental stability, fitness, and trait size in laboratory hybrids between European subspecies of the house mouse. Evolution 51:1284–1295 (1997).

Amores A, Force A, Yan YL, Joly L, Amemiya C, Fritz A, Ho RK, Langeland J, Prince V, Wang YL, et al: Zebrafish hox clusters and vertebrate genome evolution. Science 282:1711–1714 (1998).

Atchley WR, Fitch WM: Gene trees and the origins of inbred strains of mice. Science 254:554–558 (1991).

Auffray JC, Orth A, Catalan J, Gonzalez J-P, Desmarais E, Bonhomme F: Phylogenetic position and description of a new species of subgenus *Mus* (Rodentia, Mammalia) from Thailand. Zoologica Scripta 32:119–127 (2003).

Begun DJ, Aquadro CF: Levels of naturally occurring DNA polymorphism correlate with recombination rates in *D. melanogaster*. Nature 356:519–520 (1992).

Behnke JM, Wahid FN: Immunological relationships during primary infection with *Heligmosomoides polygyrus (Nematospiroides dubius)*: H-2 linked genes determine worm survival. Parasitology 103:157–163 (1991).

Bernardi G: The isochore organization of the human genome and its evolutionary history – a review. Gene 135:57–66 (1993).

Bernardi G, Olofsson B, Filipski J, Zerial M, Salinas J, Cuny G, Meunier-Rotival M, Rodier F: The mosaic genome of warm-blooded vertebrates. Science 228:953–958 (1985).

Beynon RJ, Hurst JL: Multiple roles of major urinary proteins in the house mouse, *Mus domesticus*. Biochem Soc Trans 31:142–146 (2003).

Birdsell JA: Integrating genomics, bioinformatics, and classical genetics to study the effects of recombination on genome evolution. Mol Biol Evol 19:1181–1197 (2002).

Bishop CE, Boursot P, Bonhomme F, Hatat D: Most classical *Mus musculus domesticus* laboratory mouse strains carry a *Mus musculus musculus* Y chromosome. Nature 315:70–72 (1985).

Boissinot S, Boursot P: Discordant phylogeographic patterns between the Y chromosome and mitochondrial DNA in the house mouse: selection on the Y chromosome? Genetics 146:1019–1033 (1997).

Bonhomme F, Guénet J-L, Dod B, Moriwaki K, Bulfield G: The polyphyletic origin of laboratory inbred mice and their rate of evolution. Biol J Linn Soc 30:51–58 (1987).

Boursot P, Bonhomme F, Britton-Davidian J, Catalan J, Yonekawa H, Orsini P, Guerasimov S, Thaler L: Introgression différentielle des génomes nucléaires et mitochondriaux chez deux semi-espèces de souris. Comptes Rendus de l'Académie des Sciences, Paris 299:365–370 (1984).

Boursot P, Auffray JC, Britton-Davidian J, Bonhomme F: The evolution of house mouse. Annu Rev Ecol Syst 24:119–152 (1993).

Britton-Davidian J, Nadeau JH, Croset H, Thaler L: Genic differentiation and origin of Robertsonian populations of the house mouse (*Mus musculus domesticus* Rutty). Genet Res 53:29–44 (1989).

Carter AM: Evolution of the placenta and fetal membranes seen in the light of molecular phylogenetics. Placenta 22:800–807 (2001).

Chevret P, Jenkins P, Catzeflis F: Evolutionary systematics of the Indian mouse *Mus famulus* Bonhote, 1898: molecular (DNA/DNA hybridization and 12S rRNA sequences) and morphological evidence. Zool J Linn Soc 137:385–401 (2003).

Clamp M, Andrews D, Barker D, Bevan P, Cameron G, Chen Y, Clark L, Cox T, Cuff J, Curwen V, et al: Ensembl 2002: accommodating comparative genomics. Nucl Acids Res 31:38–42 (2003).

Derothe JM, Loubes C, Orth A, Renaud F, Moulia C: Comparison between patterns of pinworm infection *(Aspiculuris tetraptera)* in wild and laboratory strains of mice, *Mus musculus*. Int J Parasitol 27:645–651 (1997).

Derothe JM, Loubès C, Perriat-Sanguinet M, Orth A, Moulia C: Experimental trypanosomiasis of natural hybrids between house mouse subspecies. Int J Parasitol 29:1011–1016 (1999a).

Derothe JM, Le Brun N, Loubes C, Perriat-Sanguinet M, Moulia C: *Trypanosoma musculi*: compared levels of parasitosis in wild and laboratory strains of *Mus musculus* mice. Exp Parasitol 91:196–198 (1999b).

Derothe JM, Le Brun N, Loubes C, Perriat-Sanguinet M, Moulia C: Susceptibility of natural hybrids between house mouse subspecies to *Sarcocystis muris*. Int J Parasitol 31:15–19 (2001).

Dod B, Jermiin LS, Boursot P, Chapman VM, Nielsen JT, Bonhomme F: Counterselection on sex chromosomes in the *Mus musculus* European hybrid zone. J Evol Biol 6:529–546 (1993).

Duplantier JM, Orth A, Catalan J, Bonhomme F: Evidence for a mitochondrial lineage originating from the Arabian peninsula in the Madagascar house mouse *(Mus musculus)*. Heredity 89:154–158 (2002).

Duret L: Evolution of synonymous codon usage in metazoans. Curr Opin Genet Dev 12:640–649 (2002).

Duret L, Hurst LD: The elevated G and C content at exonic third sites is not evidence against neutralist models of isochore evolution. Mol Biol Evol 18:757–762 (2001).

Duret L, Semon M, Piganeau G, Mouchiroud D, Galtier N: Vanishing GC-rich isochores in mammalian genomes. Genetics 162:1837–1847 (2002).

Else KJ, Wakelin D, Wassom DL, Hauda KM: The influence of genes mapping within the major histocompatibility complex on resistance to *Trichuris muris* infections in mice. Parasitology 101:61–67 (1990).

Escriva Garcia H, Laudet V, Robinson-Rechavi M: Nuclear receptors are markers of animal genome evolution. J Struct Funct Genomics 3:177–184 (2003).

Eyre-Walker A: Recombination and mammalian genome evolution. Proc R Soc Lond B Biol Sci 252:237–243 (1993).

Eyre-Walker A: Evidence of selection on silent site base composition in mammals: potential implications for the evolution of isochores and junk DNA. Genetics 152:675–683 (1999).

Eyre-Walker A, Hurst LD: The evolution of isochores. Nat Rev Genet 2:549–555 (2001).

Fel-Clair F, Lenormand T, Catalan J, Grobert J, Orth A, Boursot P, Viroux M-C, Britton-Davidian J: Genomic incompatibilities in the hybrid zone between house mice in Denmark: evidence from steep and non-coincident chromosomal clines for Robertsonian fusions. Genet Res 67:123–134 (1996).

Fel-Clair F, Catalan J, Lenormand T, Britton-Davidian J: Centromeric incompatibilities in the hybrid zone between house mouse subspecies from Denmark: evidence from patterns of NOR activity. Evolution 52:592–603 (1998).

Ferris SD, Sage RD, Huang C-M, Nielsen JT, Ritte U, Wilson AC: Flow of mitochondrial DNA across a species boundary. Proc Natl Acad Sci USA 80:2290–2294 (1983).

Figueroa F, Gunther E, Klein J: MHC polymorphism pre-dating speciation. Nature 335:265–267 (1988).

Fitch WM, Atchley WR: Evolution in inbred strains of mice appears rapid. Science 228:1169–1175 (1985).

Forejt J: Hybrid sterility in the mouse. Trends Genet 12:412–417 (1996).

Fossella J, Samant SA, Silver LM, King SM, Vaughan KT, Olds-Clarke P, Johnson KA, Mikami A, Vallee RB, Pilder SH: An axonemal dynein at the hybrid sterility 6 locus: implications for t haplotype-specific male sterility and the evolution of species barriers. Mamm Genome 11:8–15 (2000).

Fullerton SM, Bernardo Carvalho A, Clark AG: Local rates of recombination are positively correlated with GC content in the human genome. Mol Biol Evol 18:1139–1142 (2001).

Galtier N: Gene conversion drives GC-content evolution in mammalian histones. Trends Genet 19:65–68 (2003).

Galtier N, Piganeau G, Mouchiroud D, Duret L: GC-content evolution in mammalian genomes: the biased gene conversion hypothesis. Genetics 159:907–911 (2001).

Gardner MB, Kozak CA, O'Brien SJ: The Lake Casitas wild mouse: evolving genetic resistance to retroviral disease. Trends Genet 7:22–27 (1991).

Georgiades P, Watkins M, Burton GJ, Ferguson-Smith AC: Roles for genomic imprinting and the zygotic genome in placental development. Proc Natl Acad Sci USA 98:4522–4527 (2001).

Gibson G, Mackay TF: Enabling population and quantitative genomics. Genet Res 80:1–6 (2002).

Greene-Till R, Zhao Y, Hardies SC: Gene flow of unique sequences between *Mus musculus domesticus* and *Mus spretus*. Mamm Genome 11:225–230 (2000).

Gregory SG, Sekhon M, Schein J, Zhao S, Osoegawa K, Scott CE, Evans RS, Burridge PW, Cox TV, Fox CA, et al: A physical map of the mouse genome. Nature 418:743–750 (2002).

Guénet JL, Bonhomme F: Wild mice: an ever-increasing contribution to a popular mammalian model. Trends Genet 19:24–31 (2003).

Haig D: Altercation of generations – genetic conflicts of pregnancy. Am J Reprod Immunol 35:226–232 (1996).

Haig D: Parental antagonism, relatedness asymmetries, and genomic imprinting. Proc R Soc Lond B Biol Sci 264:1657–1662 (1997).

Hardies SC, Wang LP, Zhou LX, Zhao YP, Casavant NC, Huang SJ: LINE-1 (L1) lineages in the mouse. Mol Biol Evol 17:616–628 (2000).

Hellmann I, Ebersberger I, Ptak SE, Paabo S, Przeworski M: A neutral explanation for the correlation of diversity with recombination rates in humans. Am J Hum Genet 72:1527–1535 (2003).

Hemberger MC, Pearsall RS, Zechner U, Orth A, Otto S, Ruschendorf F, Fundele R, Elliott R: Genetic dissection of X-linked interspecific hybrid placental dysplasia in congenic mouse strains. Genetics 153:383–390 (1999).

Hemberger M, Cross JC, Ropers HH, Lehrach H, Fundele R, Himmelbauer H: UniGene cDNA array-based monitoring of transcriptome changes during mouse placental development. Proc Natl Acad Sci USA 98:13126–13131 (2001a).

Hemberger M, Kurz H, Orth A, Otto S, Luttges A, Elliott R, Nagy A, Tan SS, Tam P, Zechner U, Fundele RH: Genetic and developmental analysis of X-inactivation in interspecific hybrid mice suggests a role for the Y chromosome in placental dysplasia. Genetics 157:341–348 (2001b).

Hudson RR, Coyne JA: Mathematical consequences of the genealogical species concept. Evolution 56:1557–1565 (2002).

Hughes AL, Yeager M: Natural selection at major histocompatibility complex loci of vertebrates. Annu Rev Genet 32:415–435 (1998).

Hughes S, Zelus D, Mouchiroud D: Warm-blooded isochore structure in Nile crocodile and turtle. Mol Biol Evol 16:1521–1527 (1999).

Hunt WG, Selander RK: Biochemical genetics of hybridisation in European house mice. Heredity 31:11–33 (1973).

Hurst LD: A Model for the mechanism of transmission ratio distortion and for t-associated hybrid sterility. Proc R Soc Lond B Biol Sci 253:83–91 (1993).

Hurst LD, McVean GT: Growth effects of uniparental disomies and the conflict theory of genomic imprinting. Trends Genet 13:436–443 (1997).

Hurst JL, Payne CE, Nevison CM, Marie AD, Humphries RE, Robertson DHL, Cavaggioni A, Beynon RJ: Individual recognition in mice mediated by major urinary proteins. Nature 414:631–634 (2001).

IHGSC, International Human Genome Sequencing Consortium: Initial sequencing and analysis of the human genome. Nature 409:860–921 (2001).

Janeway CA Jr: How the immune system recognizes invaders. Sci Am 269:72–79 (1993).

Jouvin-Marche E, Morgado MG, Leguern C, Voegtle D, Bonhomme F, Cazenave PA: The mouse Igh-1a and Igh-1b H chain constant regions are derived from two distinct isotypic genes. Immunogenetics 29:92–97 (1989).

Karn RC, Nachman MW: Reduced nucleotide variability at an androgen-binding protein locus (Abpa) in house mice: evidence for positive natural selection. Mol Biol Evol 16:1192–1197 (1999).

Karn RC, Orth A, Bonhomme F, Boursot P: The complex history of a gene proposed to participate in a sexual isolation mechanism in house mice. Mol Biol Evol 19:462–471 (2002).

Kondrashov FA, Koonin EV: Evolution of alternative splicing: deletions, insertions and origin of functional parts of proteins from intron sequences. Trends Genet 19:115–119 (2003).

Kong A, Gudbjartsson DF, Sainz J, Jonsdottir GM, Gudjonsson SA, Richardsson B, Sigurdardottir S, Barnard J, Hallbeck B, Masson G, Shlien A, Palsson ST, Frigge ML, Thorgeirsson TE, Gulcher JR, Stefansson K: A high-resolution recombination map of the human genome. Nat Genet 31:241–247 (2002).

Lercher MJ, Smith NG, Eyre-Walker A, Hurst LD: The evolution of isochores: evidence from SNP frequency distributions. Genetics 162:1805–1810 (2002).

Lundrigan BL, Jansa SA, Tucker PK: Phylogenetic relationships in the genus *Mus*, based on paternally, maternally, and biparentally inherited characters. Syst Biol 51:410–431 (2002).

Marais G: Biased gene conversion: implications for genome and sex evolution. Trends Genet 19:330–338 (2003).

McClelland EE, Penn DJ, Potts WK: Major histocompatibility complex heterozygote superiority during coinfection. Infect Immun 71:2079–2086 (2003).

McConnell TJ, Talbot WS, McIndoe RA, Wakeland EK: The origin of MHC class II gene polymorphism within the genus *Mus*. Nature 332:651–654 (1988).

McVean GA, Charlesworth B: The effects of Hill-Robertson interference between weakly selected mutations on patterns of molecular evolution and variation. Genetics 155:929–944 (2000).

MGSC, Mouse Genome Sequencing Consortium: Initial sequencing and comparative analysis of the mouse genome. Nature 420:520–562 (2002).

Mock B, Krall M, Blackwell J, O'Brien A, Schurr E, Gros P, Skamene E, Potter M: A genetic map of mouse chromosome 1 near the Lsh-Ity-Bcg disease resistance locus. Genomics 7:57–64 (1990).

Montoya-Burgos JI, Boursot P, Galtier N: Recombination explains isochores in mammalian genomes. Trends Genet 19:128–130 (2003).

Morgado MG, Jouvin-Marche E, Gris-Liebe C, Bonhomme F, Anand R, Talwar GP, Cazenave PA: Restriction fragment length polymorphism and evolution of the mouse immunoglobulin constant region gamma loci. Immunogenetics 38:184–192 (1993).

Mouchiroud D, Gautier C: High codon-usage changes in mammalian genes. Mol Biol Evol 5:192–194 (1988).

Moulia C, Aussel JP, Bonhomme F, Boursot P, Nielsen JT, Renaud F: Wormy mice in a hybrid zone a genetic control of susceptibility to parasite infection. J Evol Biol 4:679–688 (1991).

Moulia C, Le Brun N, Dallas J, Orth A, Renaud F: Experimental evidence of genetic determinism in high susceptibility to intestinal pinworm infection in mice: A hybrid zone model. Parasitology 106:387–393 (1993).

Moulia C, Le Brun N, Renaud F: Mouse-parasite interactions: from gene to population. Adv Parasitol 38:120–167 (1996).

Munclinger P, Boursot P, Dod B: B1 insertions as easy markers for mouse population studies. Mamm Genome 14:359–366 (2003).

Nachman MW: Patterns of DNA variability at X-linked loci in *Mus domesticus*. Genetics 147:1303–1316 (1997).

Nachman MW: Variation in recombination rate across the genome: evidence and implications. Curr Opin Genet Dev 12:657–663 (2002).

Nagylaki T: Evolution of a finite population under gene conversion. Proc Natl Acad Sci USA 80:6278–6281 (1983).

Noor MAF, Grams KL, Bertucci LA, Almendarez Y, Reiland J, Smith KR: The genetics of reproductive isolation and the potential for gene exchange between *Drosophila pseudoobscura* and *D. persimilis* via backcross hybrid males. Evolution 55:512–521 (2001).

Nurminsky DI, Nurminskaya MV, De Aguiar D, Hartl DL: Selective sweep of a newly evolved sperm-specific gene in Drosophila. Nature 396:572–575 (1998).

Orr HA, Irving S: Complex epistasis and the genetic basis of hybrid sterility in the *Drosophila pseudoobscura* Bogota-USA hybridization. Genetics 158:1089–1100 (2001).

Orth A, Belkhir K, Britton-Davidian J, Boursot P, Benazzou T, Bonhomme F: Natural hybridization between two sympatric species of mice, *Mus musculus domesticus* L. and *Mus spretus* Lataste. Comptes Rendus de l'Académie des Sciences, Biologie 325:89–97 (2002).

Penn DJ, Damjanovich K, Potts WK: MHC heterozygosity confers a selective advantage against multiple-strain infections. Proc Natl Acad Sci USA 99:11260–11264 (2002).

Perry J, Ashworth A: Evolutionary rate of a gene affected by chromosomal position. Curr Biol 9:987–989 (1999).

Pilder SH, Hammer MF, Silver LM: A novel chromosome *17* hybrid sterility locus: implications for the origin of *t* haplotypes. Genetics 129:237–246 (1991).

Pilder SH, Olds-Clarke P, Phillips DM, Silver LM: Hybrid sterility-6: A mouse t complex locus controlling sperm flagellar assembly and movement. Dev Biol 159:631–642 (1993).

Pilder SH, Olds-Clarke P, Orth JM, Jester WF, Dugan L: Hst7: a male sterility mutation perturbing sperm motility, flagellar assembly, and mitochondrial sheath differentiation. J Androl 18:663–671 (1997).

Prager EM, Sage RD, Gyllensten U, Thomas WK, Hübner R, Jones CS, Noble L, Searle JB, Wilson AC: Mitochondrial DNA sequence diversity and the colonization of Scandinavia by house mice from East Holstein. Biol J Linn Soc 50:85–122 (1993).

Prager EM, Boursot P, Sage RD: New assays for Y chromosome and p53 pseudogene clines among East Holstein house mice. Mamm Genome 8:279–281 (1997).

Prager EM, Orrego C, Sage RD: Genetic variation and phylogeography of central Asian and other house mice, including a major new mitochondrial lineage in Yemen. Genetics 150:835–861 (1998).

Presgraves DC, Balagopalan L, Abmayr SM, Orr HA: Adaptive evolution drives divergence of a hybrid inviability gene between two species of Drosophila. Nature 423:715–719 (2003).

Qi CF, Bonhomme F, Buckler-White A, Buckler C, Orth A, Lander MR, Chattopadhyay SK, Morse HC 3rd: Molecular phylogeny of *Fv1*. Mamm Genome 9:1049–1055 (1998).

Rammensee HG, Klein J: Polymorphism of minor histocompatibility genes in wild mice. Immunogenetics 17:637–647 (1983).

Rikke BA, Zhao Y, Daggett LP, Reyes R, Hardies SC: *Mus spretus* LINE-1 sequences in the *Mus musculus* inbred strain C57BL/6J using LINE-1 DNA probes. Genetics 139:901–906 (1995).

Roger T, Pepin LF, Jouvin-Marche E, Cazenave PA, Seman M: New T-cell receptor gamma haplotypes in wild mice and evidence for limited Tcrg-V gene polymorphism. Immunogenetics 37:161–169 (1993).

Roopenian DC, Christianson GJ, Davis AP, Zuberi AR, Mobraaten LE: The genetic origin of minor histocompatibility antigens. Immunogenetics 38:131–140 (1993).

Saccone S, De Sario A, Wiegant J, Raap AK, Della Valle G, Bernardi G: Correlations between isochores and chromosomal bands in the human genome. Proc Natl Acad Sci USA 90:11929–11933 (1993).

Sage RD, Heyneman D, Lim KC, Wilson AC: Wormy mice in a hybrid zone. Nature 324:60–63 (1986a).

Sage RD, Whitney JB III, Wilson AC: Genetic analysis of a hybrid zone between Domesticus and Musculus mice (*Mus musculus* complex): hemoglobin polymorphisms, in Potter M, Nadeau JH, Cancro MP (eds): The Wild Mouse in Immunology, pp 75–85 (Springer-Verlag, Berlin 1986b).

Schimenti J: Segregation distortion of mouse *t* haplotypes – the molecular basis emerges. Trends Genet 16:240–243 (2000).

Schlotterer C: Hitchhiking mapping – functional genomics from the population genetics perspective. Trends Genet 19:32–38 (2003).

She JX, Boehme SA, Wang TW, Bonhomme F, Wakeland EK: The generation of MHC class II gene polymorphism in the genus *Mus*. Biol J Linn Soc 41:141–161 (1990a).

She JX, Bonhomme F, Boursot P, Thaler L, Catzeflis F: Molecular phylogenies in the genus *Mus*: Comparative analysis of electrophoretic, scnDNA hybridization, and mtDNA RFLP data. Biol J Linn Soc 41:83–103 (1990b).

She JX, Boehme SA, Wang TW, Bonhomme F, Wakeland EK: Amplification of major histocompatibility complex class II gene diversity by intraexonic recombination. Proc Natl Acad Sci USA 88:453–457 (1991).

Shiroishi T, Sagai T, Moriwaki K: A new wild-derived H-2 haplotype enhancing K-Ia recombination. Nature 300:370–372 (1982).

Skamene E, Gros P, Forget A, Patel PJ, Mesbitt MN: Regulation of resistance to leprosy by chromosome 1 locus in mouse. Immunogenetics 19:117–124 (1984).

Smadja C, Ganem G: Subspecies recognition in the house mouse: a study of two populations from the border of a hybrid zone. Behav Ecol 13:312–320 (2002).

Smith NGC, Eyre-Walker A: Synonymous codon bias is not caused by mutation bias in human. Mol Biol Evol 18:982–986 (2001).

Soriano P, Keitges EA, Schorderet DF, Harbers K, Gartler SM, Jaenisch R: High-rate of recombination and double crossovers in the mouse pseudoautosomal region during male meiosis. Proc Natl Acad Sci USA 84:7218–7220 (1987).

Tanaka S, Oda M, Toyoshima Y, Wakayama T, Tanaka M, Yoshida N, Hattori N, Ohgane J, Yanagimachi R, Shiota K: Placentomegaly in cloned mouse concepti caused by expansion of the spongiotrophoblast layer. Biol Reprod 65:1813–1821 (2001).

Tao Y, Zeng ZB, Li J, Hartl DL, Laurie CC: Genetic dissection of hybrid incompatibilities between *Drosophila simulans* and *D. mauritiana*. II. Mapping hybrid male sterility loci on the third chromosome. Genetics 164:1399–1418 (2003).

Ting CT, Tsaur SC, Wu ML, Wu CI: A rapidly evolving homeobox at the site of a hybrid sterility gene. Science 282:1501–1504 (1998).

Tucker PK, Sage RD, Warner J, Wilson AC, Eicher EM: Abrupt cline for sex chromosomes in a hybrid zone between two species of mice. Evolution 46:1146–1163 (1992).

Vanlerberghe F, Dod B, Boursot P, Bellis M, Bonhomme F: Absence of Y-chromosome introgression across the hybrid zone between *Mus musculus domesticus* and *Mus musculus musculus*. Genet Res 48:191–197 (1986).

Vanlerberghe F, Boursot P, Catalan J, Gerasimov S, Bonhomme F, Botev BA, Thaler L: Analyse génétique de la zone d'hybridation entre les deux sous-espèces de souris *Mus musculus domesticus* et *Mus musculus musculus* en Bulgarie. Genome 30:427–437 (1988a).

Vanlerberghe F, Boursot P, Nielsen JT, Bonhomme F: A steep cline for mitochondrial DNA in Danish mice. Genet Res 52:185–193 (1988b).

Vrana PB, Guan XJ, Ingram RS, Tilghman SM: Genomic imprinting is disrupted in interspecific *Peromyscus* hybrids. Nat Genet 20:362–365 (1998).

Vrana PB, Fossella JA, Matteson P, del Rio T, O'Neill MJ, Tilghman SM: Genetic and epigenetic incompatibilities underlie hybrid dysgenesis in *Peromyscus*. Nat Genet 25:120–124 (2000).

Wade CM, Kulbokas III EJ, Kirby AW, Zody MC, Mullikin JC, Lander ES, Lindblad-Toh K, Daly MJ: The mosaic structure of variation in the laboratory mouse genome. Nature 420:574–578 (2002).

Wall JD, Pritchard JK: Haplotype blocks and linkage disequilibrium in the human genome. Nat Rev Genet 4:587–589 (2003).

Wassom DL, Kelly EA: The role of the major histocompatibility complex in resistance to parasite infections. Crit Rev Immunol 10:31–52 (1990).

Watanabe Y, Fujiyama A, Ichiba Y, Hattori M, Yada T, Sakaki Y, Ikemura T: Chromosome-wide assessment of replication timing for human chromosomes 11q and 21q: disease-related genes in timing-switch regions. Hum Mol Genet 11:13–21 (2002).

Wiltshire T, Pletcher MT, Batalov S, Barnes SW, Tarantino LM, Cooke MP, Wu H, Smylie K, Santrosyan A, Copeland NG, et al: Genome-wide single-nucleotide polymorphism analysis defines haplotype patterns in mouse. Proc Natl Acad Sci USA 100:3380–3385 (2003).

Yauk CL, Bois PR, Jeffreys AJ: High-resolution sperm typing of meiotic recombination in the mouse MHC Ebeta gene. EMBO J 22:1389–1397 (2003).

Yonekawa H, Moriwaki K, Gotoh O, Miyashita N, Matsushima Y, Shi LM, Cho WS, Zhen XL, Tagashira Y: Hybrid origin of Japanese mice "*Mus musculus molossinus*": evidence from restriction analysis of mitochondrial DNA. Mol Biol Evol 5:63–78 (1988).

Zechner U, Reule M, Orth A, Bonhomme F, Strack B, Guénet J-L, Hameister H, Fundele R: An X-chromosome linked locus contributes to abnormal placental development in mouse interspecific hybrids. Nat Genet 12:398–403 (1996).

Zechner U, Shi W, Hemberger M, Himmelbauer H, Otto S, Orth A, Kalscheuer V, Fischer U, Elango R, Reis A, Vogel W, Ropers H, Rüschendorf F, Fundele R: Divergent genetic and epigenetic postzygotic isolation mechanisms in *Mus* and *Peromyscus*. J Evol Biol 17:453–460 (2004).

Zhao Y, Greene-Till R, Hardies SC: *Mus spretus* LINE-1s in C57BL/6J map to at least two different chromosomes. Mamm Genome 9:679–680 (1998).

Cytogenet Genome Res 105:395–405 (2004)
DOI: 10.1159/000078212

The tobacco mouse and its relatives: a "tail" of coat colors, chromosomes, hybridization and speciation

H.C. Hauffe,[a,b] T. Panithanarak,[a] J.F. Dallas,[c] J. Piálek,[d] İ. Gündüz,[e] and J.B. Searle[a]

[a] Department of Biology, University of York, York (UK); [b] Centre for Alpine Ecology, Viote del Monte Bondone (Italy);
[c] School of Biological Sciences, University of Aberdeen, Aberdeen (UK); [d] Institute of Vertebrate Biology,
Studenec (Czech Republic); [e] Department of Biology, University of Balıkesir, Balıkesir (Turkey)

Abstract. The article reviews over 30 years' study of the chromosomal variation of the western house mice *(Mus musculus domesticus)* from the neighboring valleys of Poschiavo and Valtellina on the Swiss-Italian border. This is done in the context of the social and political history of this area, on the grounds that mice, as commensals, are influenced by human history. The chromosomal study of mice in this area was initiated because their unusual black coat color led a 19th century naturalist to describe the "tobacco mice" from Val Poschiavo as a separate species *(Mus poschiavinus)*. The special coloration of the Val Poschiavo mice is matched by their chromosomes: they have 26 chromosomes instead of the usual 40. The Val Poschiavo mice are not a separate species according to the Biological Species Concept; instead they constitute a chromosome race (the "Poschiavo", POS) that is related to other races with reduced chromosome numbers that occur in N Italy (of which only those races in Val Poschiavo and Upper Valtellina have black coats). A phylogenetic analysis of mitochondrial DNA sequences suggests that the lineage of chromosome races found in N Italy was not formed during an extreme population bottleneck, although such bottlenecks have apparently occurred during the origin of individual races and certainly have influenced single populations. In one small, isolated population in Valtellina (Migiondo), two chromosome races (the POS and the "Upper Valtellina", UV, 2n = 24) became reproductively isolated from each other. In another small population (Sernio) bottlenecking led to fixation of a hybrid form with the UV karyotype and coat color, but with allozyme and microsatellite alleles characteristic of mice with the standard 40-chromosome karyotype. Two of the chromosome races in Valtellina (the UV and the "Mid Valtellina", MV, 2n = 24) also appear to be the product of hybridization. The dynamic history and patchy distribution of the house mouse chromosome races in Val Poschiavo and Valtellina in part reflects extinction-recolonization events; the formation of the UV and MV races and the introduction of the pale brown Standard race mice are believed to reflect such events. Dynamism in the chromosomal constitution of single populations is also evident from 25 years of data on the population in Migiondo. Due to change in agricultural practices, house mice in Valtellina and Val Poschiavo are becoming rarer, which is likely to have further impacts on the distribution and characteristics of the chromosome races in this area.

Copyright © 2004 S. Karger AG, Basel

Coat colors, chromosomes and valleys full of mice

This article is about the house mice in a small region of the Rhaetian Alps on the border between SE Switzerland and N Italy (Fig. 1). It was Fatio in 1869 who first highlighted the interest of the mice in this area. He described the mice that thrived in the houses and barns of the Poschiavo valley of Switzerland as a new species: the "tobacco mouse", *Mus poschiavinus.* He did this on the basis of their unusual coloration (black with a browny tint, a coloration due to homozygosity of the E^{tob} allele at the *extension (e)* locus; Lyon et al., 1996). Unusual coloration has been described elsewhere in the house mouse (indeed, black mice are known from other parts of Switzerland; Hübner, 1992) and it is no longer common practice to describe a species of mammal on the basis of such features. However, the designation *M. poschiavinus* was resurrected a hundred years after Fatio when Gropp et al. (1969) showed that the mice

Supported by the Natural Environment Research Council and the Royal Thai Government.

Received 4 December 2003; revision accepted 5 December 2003.

Request reprints from Jeremy B. Searle, Department of Biology (area 2)
University of York, PO Box 373, York YO10 5YW (Great Britain)
telephone: +44-1904-328615; fax: +44-1904-328505; e-mail: jbs3@york.ac.uk

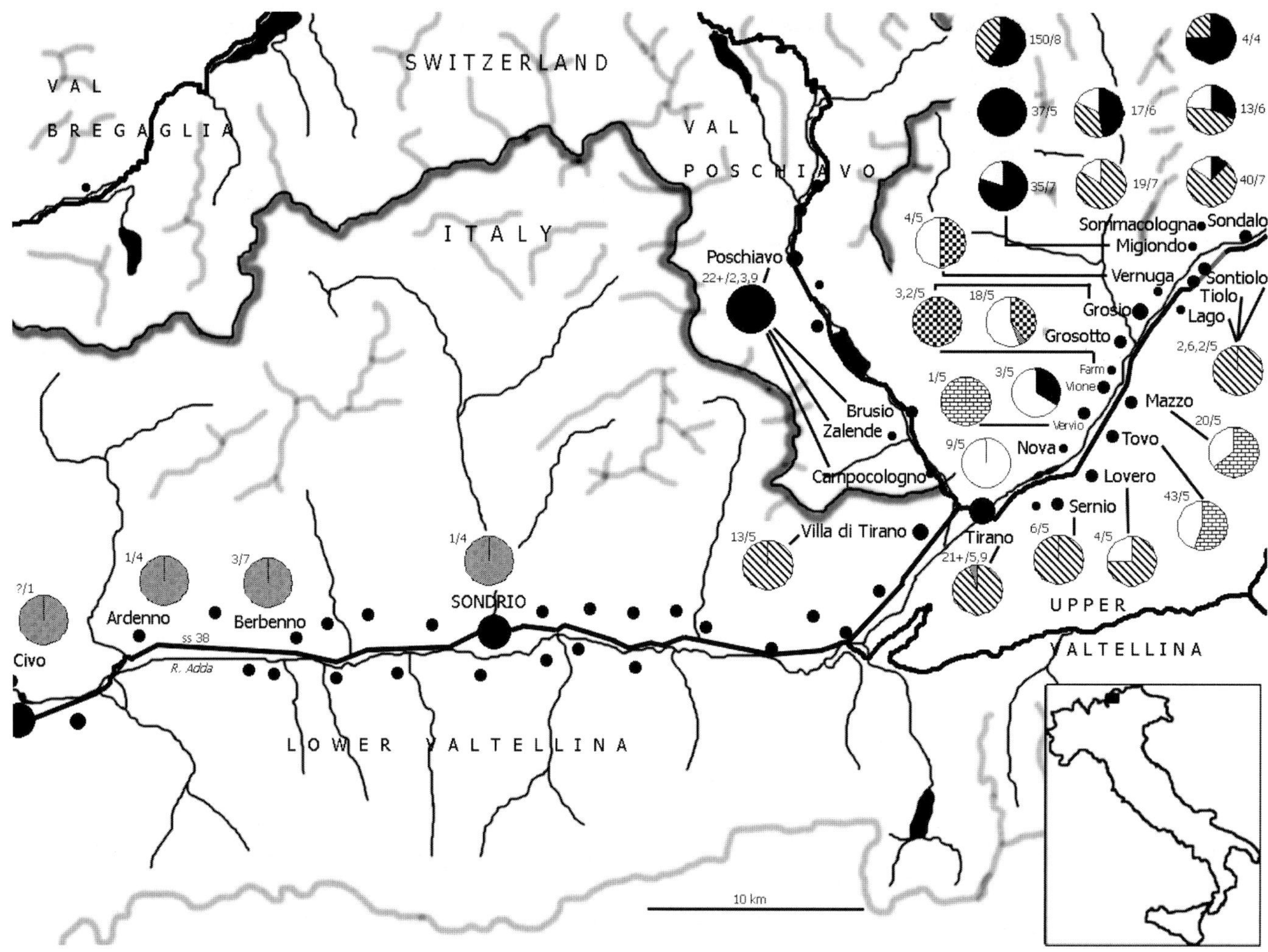

Fig. 1. Map of Valtellina and Val Poschiavo showing distribution of chromosome races of house mice and interracial hybrids (all published and unpublished data from 1969 to present). Main rivers are shown with fine black lines; paved roads with thick black line; mountains over 2000 m high with grey lines; border between Italy and Switzerland with grey/black line. Legend for pie diagrams: bricks: AA; black: POS; diagonal stripes: UV; checkerboard: MV; fine dots: LV; white: hybrids. Samples from Sondalo, Migiondo and Sommacologna are divided by sampling occasion to show changes in race composition. Near each pie diagram numbers of individuals karyotyped/reference(s) are indicated. References are as follows: 1, Capanna and Valle, 1977; 2, Gropp et al., 1970; 3, Gropp et al., 1972; 4, Gropp et al., 1982; 5, Hauffe and Searle, 1993; 6, Hauffe and Searle, all data up to 1991; 7, Hauffe et al., unpubl. data 2000–2001; 8, Mainardi et al., 1986; 9, The Jackson Laboratory, personal communication. Villages other than those sampled are shown as unnamed black dots. The samples from Val Poschiavo came from various locations, but information is incomplete and a total of 20 mice were collected from Poschiavo, Brusio and Campocologno (refs. 2 and 3), and at least two from Tirano around 1982 (ref. 9).

from Val Poschiavo had a diploid number of 26 chromosomes (seven pairs of metacentrics and six pairs of acrocentrics) whilst all other wild house mice studied hitherto had had 40 chromosomes (20 pairs of acrocentrics). At that time this was an extraordinary discovery, and it was assumed that mice that differed so drastically from the norm in chromosome number should be considered a separate species, particularly given the impact that heterozygous chromosomes may have on fertility (White, 1968). However, even in this case, the specific name has since been shown to be inappropriate. The Val Poschiavo population is only one of many in N Italy and E Switzerland to have unusual chromosomal characteristics (see Table 1 for data focusing on N Italy). Furthermore, house mice in various parts of the Mediterranean basin and W Europe have also been shown to have a non standard chromosome complement (Nachman and Searle, 1995; Piálek et al., submitted). In fact, these chromosome forms are currently classified as metacentric "chromosome races" within the W European subspecies of the house mouse *Mus musculus domesticus* (see Hausser et al., 1994 for a definition of "chromosome race"). According to the Biological Species Concept of Mayr (1963) this nomenclature is justified because chromosomally differentiated populations of *M. m. domesticus* are able to interbreed with *M. m. domesticus* of the standard chromosome complement, and *M. m. domesticus* are able to interbreed with other subspecies of *M. musculus* including the E European subspecies *M. m. musculus* (Boursot et al., 1993; Searle, 1993, 1998).

Table 1. The metacentric races of house mouse in N Italy (references for first descriptions in Piálek et al., submitted)

Race	2n	Karyotype[a]										
Poschiavo[b]	26	1.3			4.6	5.15			8.12	9.14		11.13	16.17
Upper Valtellina	24	1.3	2.8		4.6	5.15				9.14	10.12	11.13	16.17
Mid Valtellina	24	1.3			4.6	5.15		7.18	8.12	9.14		11.13	16.17
Lower Valtellina	22	1.3	2.8		4.6	5.15		7.18		9.14	10.12	11.13	16.17
Cremona	22	1.6	2.8	3.4		5.15		7.18		9.14	10.12	11.13	16.17
Gallarate	24		2.4	3.6		5.15		7.8		9.14	10.12	11.13	16.17
Binasco	24		2.8	3.4		5.15	6.7			9.14	10.12	11.13	16.17
Seveso	24		2.12	3.4		5.15	6.7		8.11	9.14	10.13		16.17
Luino	24		2.4	3.8		5.13	6.7			9.14	10.12	11.18	16.17

[a] The metacentrics that characterize each race are listed in format x.y, where x and y are individual autosomes of the standard house mouse karyotype that are found comined together in a metacentric.

[b] The Poschiavo race was discovered in Val Poschiavo in Switzerland that borders N Italy.

All this might seem to diminish the biological significance of the mice of Val Poschiavo, but in fact, as the first metacentric race of mice discovered, and thus the initial source of valuable variant chromosomes, the Poschiavo race mice have become well known in biomedical and developmental genetics research. The individual Poschiavo metacentric chromosomes were introduced into a laboratory mouse background and have been involved, for example, in the analysis of the developmental arrest of aneuploids (Epstein, 1986) and the discovery of chromosomal imprinting (Cattanach and Kirk, 1985).

Before returning to the Poschiavo metacentric race in its natural setting, we need to set the scene in terms of geography, human history and mouse ecology. In particular for this review, we have researched the social and political history of the region of study because, as emphasized by Capanna (1982), an understanding of human history is essential to appreciate the chromosomal history of the house mouse. In Europe, house mice are generally commensal; their population size in any particular area depends on the habitat and food made available by humans, and their long distance movements almost entirely reflect accidental transport of stowaways by humans (Pocock et al., submitted). Therefore, the history of human use of Val Poschiavo and its vicinity is vital for an understanding of the Poschiavo and neighboring chromosome races.

Although it is currently within the Swiss Canton of Graubünden, Val Poschiavo is geographically and historically more closely allied to the neighboring area of Italy known as Valtellina, which is the valley of the River Adda above Lake Como (Bagiotti, 1958; Tognina, 1975; Benetti and Giudetti, 1990). The Poschiavino River drains into the Adda at Tirano (Fig. 1), so that Val Poschiavo is a side-valley of Valtellina. Val Poschiavo is south of the long and difficult Bernina Pass (2338 m), which was never a major trade route and only became commercially viable in the 1400s. Although Valtellina has been inhabited for over 5000 years, it was also relatively isolated from the rest of Italy until the early 1800s. Consequently, the people of Valtellina have, until recently, been virtually self-sufficient, and their main trading partner has always been Val Poschiavo. In addition, before 500 AD, Valtellina and Val Poschiavo were probably both ruled by the Romans, and up to 1408 from either Milan and/or Chur; then from 1512 to 1797, the valleys were governed by the Graubünden administration. In fact, the Italian dialect spoken in Val Poschiavo is closely related to that of Valtellina, and both valleys used the same currency until the 1850s. Hay and livestock were regularly exported from Val Poschiavo to Valtellina up to the Second World War (with wine and grain going in the opposite direction). Therefore, there are many grounds to consider Val Poschiavo and Valtellina as a single geographical and historical unit, even though at the current time they belong to different countries and their people generally practice different forms of Christianity (Val Poschiavo: Protestant, Valtellina: Roman Catholic).

However, while at a coarse level Val Poschiavo and Valtellina have long been integrated through politics and trade, at a finer scale the degree of human interaction between villages even a few kilometers apart has often been very low, such that villages or groups of villages have tended to be isolated from each other. This particularly applies up until the mid nineteenth century, when the valley bottoms would have been boggy and subject to frequent flooding, such that contact between many closely located villages would have been difficult. It was not until 1841 that the river started to be constrained, and the first major roads and the railway line were constructed in 1848 and 1902, respectively (Benetti and Giudetti, 1990). Even today, the distinct villages in Val Poschiavo and Valtellina should be considered as discontinuous patches of mouse habitat.

In this paper, we will not consider the other major side-valley of Valtellina, Valchiavenna, because the chromosome complements of the mice there consist mainly of hybrid karyotypes and metacentrics which appear to have been introduced from central Switzerland, Val Mesolcina and around Lake Como. This probably reflects the long and prosperous history of Valchiavenna as an international trade route between present-day Italy and the rest of Europe (the Splügen Pass was heavily used for commercial purposes from 200 AD to the late 19th century); culturally, Valtellina and Valchiavenna probably began to diverge as early as 2500 BC (Benetti and Giudetti, 1990).

In recent centuries house mice would certainly have thrived in Val Poschiavo and Valtellina. The people in this area are traditionally livestock farmers, maintaining domesticated animals (cattle, goats, sheep, pigs and horses) in villages during the winter, usually under or adjoining human habitations where

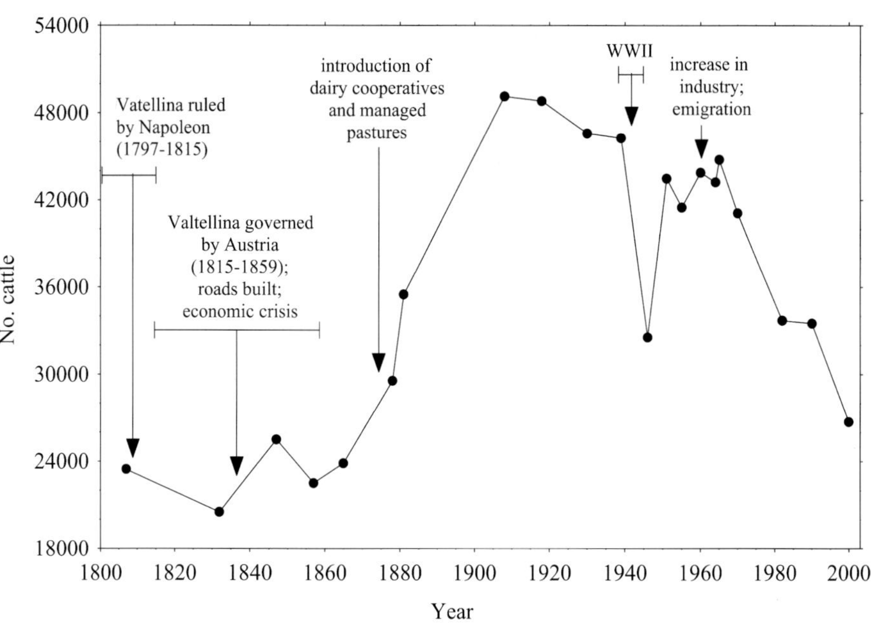

Fig. 2. Fluctuation in total number of cattle in the province of Sondrio (Valtellina and Valchiavenna) from 1800 to the present day, and major historical landmarks recorded as possible causes of these changes (Benetti and Giudetti, 1990; Torricelli, 1990). Where there are cattle, there are almost invariably hay and grain and, consequently, rodent pests; thus, fluctuations in number of cattle probably reflect changes in house mouse populations in this area (since cattle were and still are the most numerous and valuable domesticated animals in Valtellina, these estimates are more accurate than those recorded for horses, goats, sheep and pigs). Superimposed on these long-term fluctuations were seasonal changes in population sizes (see text). Unfortunately, these data cannot reveal the ratio of house mice to rats, the latter of which often exclude mice from certain parts of farms and would have been particularly common before the 1960s (when sewage systems were built and regular garbage collection commenced: Denotte, 1963).

they act as a supplementary heating system. Both livestock feed and human foodstuffs can be consumed by mice. Through a capture-mark-recapture study in a traditional village setting in Valtellina in 1993–1995, we estimated the population size of house mice to be about 100–150 per village of 500 people and 50 cattle (Hauffe et al., 2000). The numbers of cattle and mice would almost certainly have been higher earlier in the twentieth century. Fig. 2 illustrates the fluctuation in numbers of cattle over the last two centuries, which is also likely to be a good index for the numbers of mice.

As we have already emphasized, long-distance movements of house mice in N Italy (between villages in Val Poschiavo and Valtellina, or more distant movements than that), can be attributed to stowaways transported by humans (Pocock et al., submitted). Voluntary movements by commensal house mice are much more restricted (Pocock et al., submitted). Our field studies in a Valtellina village showed that mice rarely move away from buildings, and even then they will generally move less than 30 m (Hauffe et al., 2000). One female did disperse 320 m across agricultural land between one set of buildings and another, but this was an unusual example of a voluntary long-distance movement.

While Val Poschiavo and Valtellina have been exceptional house mouse habitat for hundreds of years, this situation is changing rapidly. In 1989 two of us (Hauffe and Searle) visited Val Poschiavo, but we found little suitable habitat, and detected no animals. In 1989, Valtellina was still a rich source of small farms teeming with mice, but in the last 15 years such farms have been abandoned, mechanized or refurbished as residences, and mouse populations in almost all villages have been reduced substantially.

More chromosomes and more coat colors

The particular metacentrics that characterize the Poschiavo (POS) race and differentiate it from the standard all-acrocentric (AA) race are shown in Table 1. The occurrence of metacentrics reflects Robertsonian (Rb) fusion events, whereby acrocentric autosomes fuse at their centromeres (Garagna et al., 1995). However, the particular arm combinations may also reflect whole-arm reciprocal translocations (WARTs) that cause the swapping of arms between metacentrics or between metacentrics and acrocentrics (Hauffe and Piálek, 1997).

Although POS is the only chromosome race that has been found in Val Poschiavo, there is more variety in Valtellina (Fig. 1). Gropp et al. (1982) reported that the POS race also occurred there and described two other races that are now known as the Upper and Lower Valtellina races (UV and LV, 2n = 24 and 22, respectively). These share most of their metacentrics with the POS race, as does a further race, the Mid Valtellina (MV, 2n = 24; Table 1). This latter race and the AA race were discovered in previously untrapped areas of Valtellina during our studies (Hauffe and Searle, 1993).

The chromosomes shared between the races in Valtellina and Val Poschiavo almost certainly reflect recent common ancestry (Corti et al., 1986; Bauchau et al., 1990; Hauffe and Piálek, 1997; Piálek et al., submitted). In fact, considering the metacentric races of N Italy in general (Table 1), they all form a single lineage in a chromosomal phylogeny of *M. m. domesticus* (Piálek et al., submitted). All the races share metacentrics 9.14 and 16.17, some share other metacentrics, and WARTs most reasonably link some of the remaining race-specific metacentrics (Piálek et al., submitted). However, in an analysis involving a wide range of *M. m. domesticus* mtDNA haplotypes, we found that representatives of the N Italian metacentric chromosome lineage are dispersed over the resulting mtDNA phylogenetic tree (Fig. 3), confirming previous results obtained by

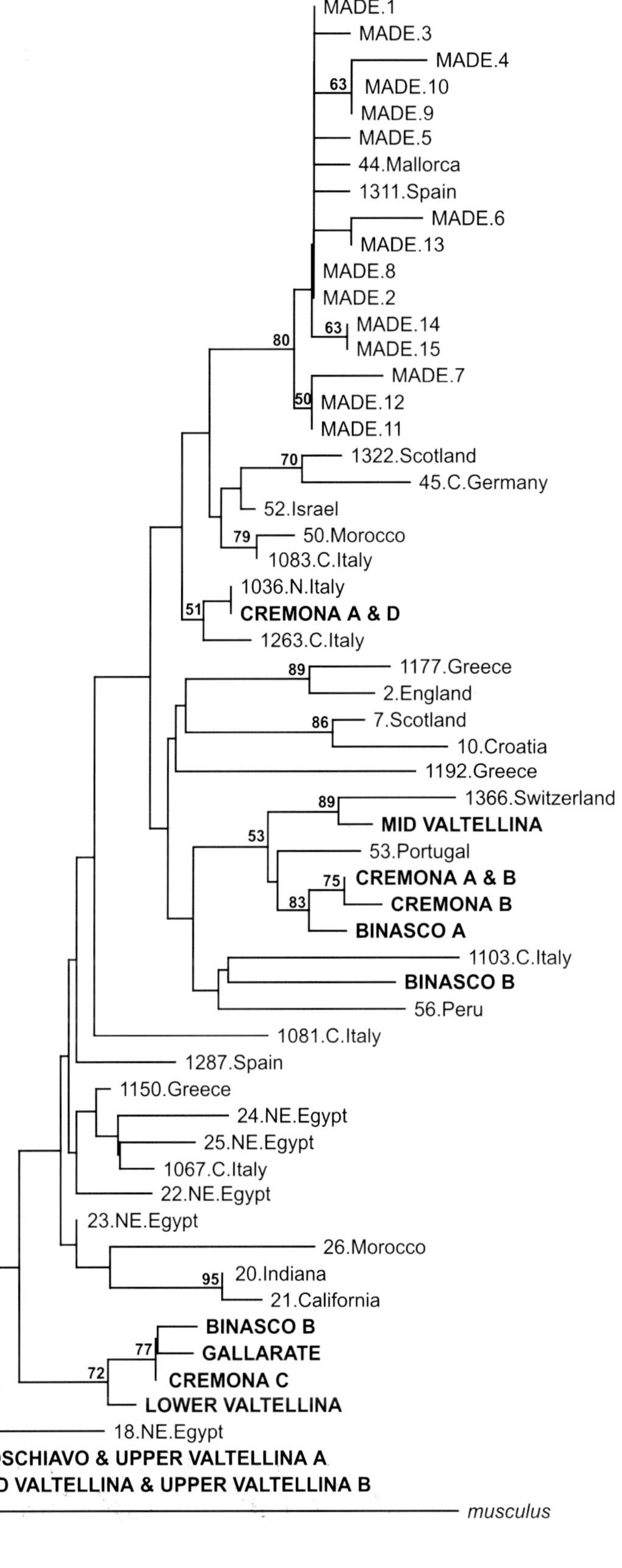

Fig. 3. Phylogenetic tree showing the relationship of mtDNA D-loop haplotypes (933 bp) from seven out of the nine N Italian metacentric races in comparison with haplotypes from other parts of the distribution of *M. m. domesticus*. The N Italian haplotypes are shown in bold and labeled according to chromosome race (see Table 1); the places of capture (labeled A, B, C, D where more than one) and number of mice/haplotypes are as follows (the data from Nachman et al., 1994 are included where appropriate): Poschiavo (Migiondo, n = 2/1), Upper Valtellina A (Tiolo, n = 2/1), Upper Valtellina B (Sernio, n = 4/1 including 2 specimens from Nachman et al., 1994), Lower Valtellina (Berbenno, n = 2/1), Mid Valtellina (Grossotto, n = 2/2), Cremona A (Fengo, n = 2/2), Cremona B (Seriate, n = 2/2), Cremona C (Pedrengo, n = 2/1), Cremona D (Cremona, n = 2/1 from Nachman et al., 1994), Gallarate (Tradate, n = 2/1), Binasco A (Casarile, n = 2/1), Binasco B (Binasco, n = 2/2 from Nachman et al., 1994). (The Poschiavo and Upper Valtellina A haplotypes were the same; the Upper Valtellina B haplotypes were identical to one of the Mid Valtellina haplotypes, one of the Cremona A haplotypes was the same as one of the Cremona B haplotypes and the other Cremona A haplotype was identical to the Cremona D haplotype.) The tree was constructed by the neighbor joining method (Saitou and Nei, 1987) using the program PAUP (Swofford, 1998). The *M. m. domesticus* haplotypes used for comparative purposes are the same set as used by Gündüz et al. (submitted) together with the Madeiran haplotypes (MADE.1, etc.) from Gündüz et al. (2001). A *M. m. musculus* sequence (Nachman et al., 1994) was used as outgroup. The most appropriate DNA substitution models for our data set were established using MODELTEST (Posada and Crandall, 1998). For the data including the outgroup, the hierarchical likelihood ratio tests and the Akaike information criterion estimates (Akaike, 1974) both supported the HKY model (Hasegawa et al., 1985) with the frequency of invariant sites set at 0.7744 and gamma correction of 0.8356. For the data without the outgroup, the same model (HKY) but with the frequency of invariable sites set at 0.8001 and gamma correction of 0.9095 was selected by both the hierarchical likelihood ratio tests and the Akaike information criterion estimates. The trees constructed using these parameters gave similar results; we display here the tree produced with the HKY model with the frequency of invariable sites set at 0.7744 and gamma correction of 0.8356. Ten thousand bootstrap replicates were performed using the same setting but with 10 random addition replicates. All bootstrap values above 50 % are shown on the tree. Our new D-loop sequences were obtained by methods described in Gündüz et al. (submitted).

Table 2. Nucleotide diversity (π) of mtDNA sequences for house mice of the N Italian lineage of metacentric chromosome races in comparison with other groupings of *M. m. domesticus*

Grouping	N[a]	mtDNA sequence	π	Reference
N Italy	20	D-loop, flanking tRNAs (1013 bp)	0.0091	this study
Madeira	44	D-loop, flanking tRNAs (1013 bp)	0.0014	Gündüz et al., 2001
W Europe	56	D-loop, ND3 (1449 bp)	0.0084	Nachman et al., 1994
Worldwide	152	D-loop, flanking tRNAs (1008 bp)	0.0044	Prager et al., 1996

[a] Number of individuals sequenced.

Table 3. Unbiased estimates of average heterozygosity (H; Nei, 1987) for samples of house mice from villages within Upper Valtellina for 21 allozyme loci (Hauffe et al., 2002) and 13 microsatellite loci (Panithanarak et al., 2004). The villages are listed in geographic sequence and characterized according to presence of chromosome races and hybrids between those races. Those samples including the AA race or hybrids involving the AA race are highlighted in bold

Village	Year of sampling	Chromosomal characteristics[a]	Allozymes[b] N	Allozymes[b] H	Microsatellites[b] N	Microsatellites[b] H
Sondalo	1989–1991	UV, POS, met × met	11	0.012	10	0.42
	2000–2001	UV, POS, met × met	n.a.	n.a.	35	0.41
Sommacologna	1989–1991	UV, POS, met × met	11	0.004	10	0.40
	2000–2001	UV, met × met	n.a.	n.a.	9	0.38
Migiondo	1989–1991	POS	22	0.046	23	0.33
	2000–2001	POS, met × met	n.a.	n.a.	26	0.31
Sontiolo	1989–1991	UV	4	0	3	0.21
Tiolo	1989–1991	UV	7	0	6	0.36
Lago	1989–1991	UV	2	0	2	0.14
Grosio	1989–1991	MV, met × met	9	0.005	9	0.48
Grosotto	1989–1991	LV, MV, met × met	16	0.014	16	0.42
Farm Via Prada	1989–1991	MV	5	0	5	0.24
Vione	**1989–1991**	**POS, AA × met**	**3**	**0.054**	**2**	**0.36**
Mazzo	**1989–1991**	**AA, AA × met**	**12**	**0.076**	**5**	**0.44**
Vervio	**1989–1991**	**AA**	**1**	**0.048**	n.a.	n.a.
Tovo	**1989–1991**	**AA, AA × met**	**13**	**0.059**	**5**	**0.57**
Nova	**1989–1991**	**AA × met**	**2**	**0.056**	n.a.	n.a.
Lovero	**1989–1991**	**UV, AA × met**	**4**	**0**	**4**	**0.50**
Sernio	1989–1991	UV	5	0	5	0.12
Biolo	1989–1991	UV	2	0	2	0.37
Villa di Tirano	1989–1991	UV	n.a.	n.a.	2	0.45

[a] met × met: hybrids between metacentric races; AA × met: hybrids between the all-acrocentric race and one or more metacentric races.
[b] n.a.: not analyzed.

Nachman et al. (1994) with fewer N Italian haplotypes. This contrasts, for example, with the mice of Madeira, which comprise six related metacentric chromosome races (2n = 22–28; Britton-Davidian et al., 2000) and which form a single mtDNA clade in the same analysis (Fig. 3). The mtDNA phylogenetic result and the substantial mtDNA nucleotide diversity in N Italian metacentric races (Table 2) argue against an extreme population bottleneck during the formation of the N Italian metacentric chromosome lineage. According to the classical view, chromosomal variants that make heterozygotes unfit can only become fixed in small populations (Lande, 1979). In fact, it has been shown more recently that house mice heterozygous for a single Rb fusion can be only marginally unfit (Britton-Davidian et al., 1990; Wallace et al., 1992; Hauffe and Searle, 1998), so a restricted population size would not be necessary if the chromosome races within the N Italian lineage were formed by accumulation of one or a few Rb fusions at a time (Capanna, 1982). Judging by data from other small mammals (Banaszek et al., 2000), a single WART may also not necessarily cause substantial unfitness. Additionally, meiotic drive may enable fixation of Rb fusions and WARTs without a reduction in population size (Pardo-Manuel de Villena and Sapienza, 2001; but see Piálek et al., submitted).

The metacentric races in Valtellina and Val Poschiavo have lower allozyme diversity than house mice elsewhere in Europe, so their formation might have involved population bottlenecks (Table 3; Hauffe et al., 2002). In this case, population bottlenecks would undoubtedly *enhance* fixation of chromosomal rearrangements, even though they are not *essential*, and they are certainly likely during the colonization of small alpine valleys.

One of the striking features of the chromosomal variation in Valtellina and Val Poschiavo is that the races are distributed in patches (Fig. 1) so that different races predominate in different groups of villages. For example, two patches of the UV race (villages northeast and including Lago, and villages west of and including Sernio, respectively) are separated from each other by a patch of the AA race (Tovo, Mazzo). The POS race patch in Val Poschiavo is even further separated from populations with a high frequency of POS race individuals in Valtellina (Sondalo, Migiondo).

This patchy pattern demonstrates that there is relatively little movement on a generation-by-generation basis between the villages, since high rates of dispersal would tend to homogenize chromosomal characteristics over the whole area. Nonetheless, there is a need to explain how the patchiness in Valtellina and Val Poschiavo arose in the first place.

We believe that extinction-recolonization events have been particularly important. Taking the recolonizations first, these are easily explained in house mice because habitat empty of mice will readily become colonized by long-distance stowaways, as long as there is enough food and living space in the empty habitat and enough movement of agricultural products into the habitat to ensure that some mice will be inadvertently brought in (Pocock et al., submitted). If a village empty of mice is colonized by a race that is distinct from those found in neighboring villages, i.e. from a source beyond those villages, a new patch is generated. Regarding the mouse extinction events in Val Poschiavo and Valtellina, these most likely reflect temporary absence of humans and their livestock from a village. Traditionally there is a substantial migration of humans and livestock from villages to alpine meadows in the summer, which could regularly have caused mouse extinctions, or at the very least, bottlenecking of their populations. Events that affect people such as epidemics of fatal diseases could have caused temporary abandonment of villages, and consequent loss of their mouse populations (bubonic plague was especially common between 1400 and 1650; Benetti and Giudetti, 1990). A single important event in the history of Valtellina, that we have highlighted in the past (Hauffe and Searle, 1993), was a landslide in 1807 that blocked the river Adda below Sernio and caused catastrophic flooding in Lovero and Tovo for over a year, that would most likely have exterminated house mouse populations in these villages (Fig. 1).

One racial patch in the area of Val Poschiavo and Valtellina illustrates the extinction-recolonization process particularly well, and that is the AA patch centered on the villages of Tovo and Mazzo (see Fig. 1). We believe that the AA race colonized after the area had been occupied by metacentric mice and, indeed, probably after the origin of *all* the metacentric races found in the area. We have made morphological and genetic comparisons between the AA race of Upper Valtellina with the metacentric races that predominate in that section of river valley: POS, MV and UV. The AA race is distinctly different from the metacentric races in every characteristic we have examined: mandible shape, body measurements, allozyme alleles, microsatellite alleles, D-loop haplotypes (Hauffe et al., 2002; Panithanarak, unpublished data). Both for allozymes and microsatellites, heterozygosity levels are particularly high in Tovo and Mazzo, which are the villages where AA is most dominant (Table 3). Also, the coat color of the AA mice is different from the metacentric mice in Upper Valtellina. While all the metacentric races in Upper Valtellina are characterized by the black-brown tobacco mouse coloration first described in Val Poschiavo (see above), the AA mice have the pale brown coloration that is more typical for house mice. (Interestingly, the metacentric mice of the LV race in Lower Valtellina are also pale brown.)

Their distinctiveness suggests that AA mice are not closely related to any of the metacentric races in Upper Valtellina, and that they are relatively recent arrivals in the valley. The 1807 landslip would have exterminated house mice from those villages where AA mice are currently found, so is a candidate extinction event that preceded the AA colonization event. As we have pointed out before (Hauffe and Searle, 1993), following a disaster of this magnitude, supplies from far afield may have been brought into the stricken villages, thereby increasing the probability that mice were brought in from outside of Val Poschiavo and Valtellina, and thereby potentially from an area dominated by AA mice.

The origin of the metacentric races and the particular importance of hybridization

While the colonization of the AA race into the area of Val Poschiavo and Valtellina is relatively easily explained, the diversification of metacentric races that apparently preceded it was a more complex process. Starting with Capanna (1980), various models have been developed to explain the evolution and distribution of the metacentric races in Val Poschiavo and Valtellina. Our most recent phylogenetic analysis of the chromosome races of the house mouse (Piálek et al., submitted) strongly suggest that the POS and LV races were ancestral among the metacentric races in the area. So, we take this as the starting point in this latest reconsideration of the origin of chromosome races in Val Poschiavo and Valtellina.

The POS and LV races evolved from a common ancestor characterized by metacentrics 1.3, 4.6, 5.15, 9.14, 11.13 and 16.17 (Table 1). The specific chromosomes that distinguish the POS and LV races are 2.8, 7.18 and 10.12 in LV and 8.12 in POS. These differences arose either by Rb fusions or by a mixture of Rb fusions and WARTs (Hauffe and Piálek, 1997; Piálek et al., submitted). However, it is most likely that the POS and LV races evolved their chromosomal differences during a substantial period in allopatry. Diagnostic allelic differences in microsatellite loci at the centromeres of chromosomes 10 and 12 have been demonstrated between the POS version of these chromosomes (10 as an acrocentric, 12 as part of metacentric 8.12) and the LV version (metacentric 10.12; see Panithanarak et al., 2004).

In discussing the social history of Val Poschiavo and Valtellina, we have emphasized the discontinuity of settlements. Going back in time, the more severe this discontinuity would have been. It is not difficult to envisage, therefore, in the distant past that there was enough of a gap and insufficient migration events, to allow formation of two chromosome races in allopatry, with LV evolving somewhere in Lower Valtellina and POS somewhere in Val Poschiavo and/or Upper Valtellina. However, the details of precisely when these races would have evolved as distinctive forms and precisely where are unknown.

We believe that further raciation in Val Poschiavo and Valtellina followed a breakdown of the allopatry that allowed formation of the LV and POS races. Both the MV and UV races can be derived from the POS and LV races by hybridization, and it is most parsimonious to consider that they evolved in this way (Hauffe and Piálek, 1997; Piálek et al., submitted). Again, it is difficult to date when the MV and UV races were first formed (indeed they may have been produced on several occasions), but we envisage that their formation followed extinction-recolonization events involving house mice in Upper Valtellina (Hauffe and Searle, 1993; Piálek et al., 2001). If under these circumstances both the POS and LV races managed to colonize villages that had been emptied of mice (presumably by passive dispersal from different sources), on one or more occasions the MV race became fixed, and likewise for the UV race. Under realistic assumptions, the fixation of the hybrid homozygous forms can occur with a moderately high probability (Piálek et al., 2001). Once formed, further extinction-recolonization events would have allowed the spread of these races to other villages. Currently, the UV race is the most widespread race of Upper Valtellina (Fig. 1).

In summary, different aspects of the social history of Val Poschiavo and Valtellina can help explain the origin and spread of the chromosome races of house mice that are found there, reflecting the fact that humans create the habitat that mice can occupy and that they accidentally move mice between one locality and another. Firstly, the colonization of the area by a single ancestral chromosome race is consistent with the historical integrity of Val Poschiavo and Valtellina, and historical domination of the area by people from the south (what is currently Italy). The chromosome race that colonized Val Poschiavo and Valtellina had a close relationship with other chromosome races in N Italy, and was undoubtedly of Italian origin. Secondly, the discontinuity of settlements in the area was at some stage sufficient to allow the formation of the LV and POS races in allopatry, and the concomitant extinction of the ancestral race in the area. However, the spread of the LV and POS races from their sites of formation would have required movements of mice between villages. Trade and other interactions between people of different villages would have promoted this process. Thirdly, there is evidence for past events in villages in Val Poschiavo and Valtellina that led to the temporary removal of people, their livestock and their foodstuffs, and associated extinction of house mice. When people reoccupied these villages, they would have had to bring in agricultural products from other villages, and house mice would have arrived as stowaways. This could have led to multiple colonizations of single villages by house mice. In this way we believe the LV and POS races came into contact in Upper Valtellina, and on hybridization generated the MV and UV races (in different villages). We also believe that an extinction-recolonization event led to the arrival of the standard AA race in Upper Valtellina, which apparently came as stowaways from beyond Val Poschiavo and Valtellina.

This process of race formation and colonization has resulted in multiple races of house mouse in Val Poschiavo and Valtellina, with a particularly high density of interacting forms in Upper Valtellina. An area where races meet and hybridize is termed a "hybrid zone". In Upper Valtellina, we propose that there was initially a hybrid zone between two races (POS and LV), but that two other races have originated within the hybrid zone. The process of formation of new races within a hybrid zone we term "zonal raciation" (Hauffe and Searle, 1993; Searle, 1993), and so we propose that the MV and UV races have originated in this way. There is evidence for another instance of a new hybrid homozygous type, associated with one particular village, Sernio (Fig. 1). Although only a small sample of individuals was studied (five mice), the results are striking. All individuals were homozygous for the UV race and had the expected tobacco mouse coloration for that race. However, they were also all homozygous for an allozyme allele *(Amy1^u)* and a microsatellite allele (172 at locus *D10Mit180*) associated with the AA race (see Hauffe et al., 2002; Panithanarak et al., 2004). Presumably this population was derived by hybridization between the UV and AA races, and then suffered a bottleneck (or multiple bottlenecks) leading to fixation of a combination of UV and AA race characteristics. The hete-

rozygosity of the Sernio sample for all genetic markers is very low (Table 3).

As we have described above, the patchy distribution of races in Upper Valtellina indicate movement of house mice within the area or into the area, in association with extinction-recolonization events. However, as well as mass movements of house mice, sometimes over a long distance, associated with recolonization events, the patterns of genetic variation in Upper Valtellina also indicate the occurrence of more limited movements between neighboring villages (0.5 –1.0 km). These movements may reflect house mice moving on their own between villages or, more likely, short-range passive dispersal with humans.

Thus, although villages tend to be dominated by one chromosome race, the occurrence of hybrids in many villages indicates immigration of a different race and successful interbreeding of the immigrant(s) and resident(s). These hybrids are not only detected chromosomally (Fig. 1), they can be demonstrated with other markers. Because of the allozyme and microsatellite differences between the AA and metacentric races in Upper Valtellina, we have been able to demonstrate gene flow between the two racial categories (Hauffe et al., 2002; TP, unpublished data). Furthermore, there are microsatellite differences at chromosomes 10 and 12 between the POS and MV races on one hand and LV and UV races on the other, and again we have demonstrated gene flow between the two (Panithanarak et al., 2004).

The frequency of hybrids in mouse populations in Upper Valtellina will not only reflect the degree of movement between populations, it also depends upon hybrid fertility. The fertility of hybrids between the UV and POS races and between both these races and the AA race has been studied in detail through laboratory crosses and examination of wild-caught individuals (Hauffe and Searle, 1998). The fertility of hybrids in all these cases is reduced relative to pure race individuals; however, none of the hybrid categories are routinely sterile. Therefore, it is not surprising that hybrids are observed in nature: they will be generated by interracial crosses and higher generation hybridization events. The higher generation hybrids between metacentric races and the AA race (heterozygous for 1–3 Rb fusions) show greater fertility than the F$_1$ hybrids (heterozygous for 7–9 Rb fusions; Hauffe and Searle, 1998), and it is these higher generation hybrids that are present at Vione, Mazzo, Tovo, Nova and Lovero (Fig. 1). These are the villages that we propose were recolonized by AA mice following the 1807 flood, and their neighbors.

We have described the multiple hybrid zone in Upper Valtellina as a patchy or "mottled" hybrid zone (Hauffe and Searle, 1993), and there is a need to consider what is the impact of the patchy structure on gene flow. This is interesting in relation to the centromeric regions of chromosomes 2, 8, 10 and 12, comparing a situation where races carrying metacentric 8.12 and acrocentrics 2 and 10 (POS or MV) are in direct contact with a race carrying 2.8 and 10.12 (LV or UV) and a situation where the races occur in separated villages because of the mottled structure. There is an expectation that recombination in chromosomal heterozygotes will be reduced in the vicinity of the rearrangement breakpoint (Searle, 1993), i.e. at the centromere for Rb fusions and WARTs. Therefore, on this basis, the degree of gene flow between the POS and MV races on the one hand and the LV and UV races on the other should be at its minimum in the centromeric regions of chromosomes 2, 8, 10 and 12. The centromeres of chromosomes 2, 8, 10 and 12 also represent the "unfitness loci" of POS × LV, MV × LV, POS × UV and MV × UV hybrids, which might also be expected to reduce gene flow in the centromeric regions of these chromosomes. The various hybrids produce a chain-of-five configuration at meiosis I involving chromosomes 2, 8, 10 and 12; it is this configuration that causes the reduced fertility, and hence unfitness, of the hybrids (Hauffe and Searle, 1993, 1998).

In Upper Valtellina, we managed to obtain data on the interaction of the POS and MV races with the UV race for centromeric regions of chromosomes 10 and 12 (Panithanarak et al., 2004). While there was evidence of a barrier to gene flow for the centromeric markers of chromosomes 10 and 12 when the POS and MV races are in different villages from the UV race (reducing the degree of interaction between the races), this same barrier is not evident for loci located elsewhere in the genome or for the centromeric loci when the POS and UV races occur in the same village (Sondalo, Sommacologna). Therefore, it is the combination of unfitness, reduced recombination and the mottled zone structure that is sufficient to create a barrier to gene flow for certain regions of the genome in the Upper Valtellina hybrid zone.

Speciation

We began this article denouncing the designation of *M. poschiavinus*, but for a short time, the POS race apparently did achieve reproductive isolation, at least in one village. Capanna and Corti (1982) clearly demonstrated that the POS and UV races occurred together in the village of Migiondo, but found no hybrids in a sample of 150 individuals caught between 1978 and 1983 (see also Fig. 1; Mainardi et al., 1986). However, since the completion of Capanna and Corti's (1982) study, one of the races (UV) has become extinct in Migiondo, so it is no longer possible to study the interaction of the reproductively isolated forms (Hauffe and Searle, 1992). The POS and UV races make contact elsewhere in Valtellina (Sondalo, Sommacologna) and hybrids are produced. It is reasonable to presume, therefore, that the two races initially were able to interbreed when they came into contact, and that reproductive isolation has arisen in situ. The unfitness of POS × UV hybrids (Hauffe and Searle, 1998) would have created a selection pressure for assortative mating that could have led to the observed absence of hybrids. It is unlikely that assortative mating was the result of ecological differentiation, since both races occur in the same habitats. Therefore, we have argued that the local speciation event that occurred in Migiondo was the result of reinforcement (Hauffe and Searle, 1992, 1993). (By "local speciation", we mean that the races could not breed with each other within one village; however, they may have been able to breed with other populations, raising the possibility of gene flow via intermediates. The POS and UV races in Migiondo were therefore probably not strictly species under the Biological Species Concept; Searle, 1998.)

There have been very few well-documented examples of reinforcement (Marshall et al., 2002), so the speciation event in Migiondo is of considerable interest. Why reinforcement should have occurred in Migiondo but not the other villages where the POS and UV races occur together, is uncertain. However, Migiondo is a small and isolated village and is, for example, characterized by *Mod1* alleles that are not found elsewhere in Upper Valtellina (Fraguedakis-Tsolis et al., 1997). Small populations may evolve quickly and in unusual directions through genetic drift and the isolation means that any evolutionary change in Migiondo would not readily be disrupted by immigration of alternative alleles into the population (Butlin, 1987). One condition of reinforcement is that specific assortative mating alleles should be associated with each race, and so the loci concerned need to be protected from recombination between the races (Butlin, 1987). As we have seen, loci located close to the centromeres of chromosomes 2, 8, 10 or 12 could be protected in this way, as long as there is rather little interbreeding between the races. Therefore, we suggest that reinforcement could have occurred in Migiondo if the UV and POS populations that colonized the village already showed a degree of assortative mating (as has been demonstrated in mice coming from different sources; Cox, 1984), thereby minimizing the initial interaction between the races. If there were assortative mating loci located close to the centromeres of chromosomes 2, 8, 10 or 12, this assortment may have strengthened, leading ultimately to local speciation. We are presently in the process of investigating possible behavioral differences between the UV and POS races that could have led to assortative mating and a reduction in the production of unfit hybrids.

Migiondo is notable among house mouse populations in that reasonably large samples of individuals have been karyotyped at intervals over 25 years. It is clear that the population is volatile in terms of chromosome variation (Fig. 1). Between 1978 and 1983 there were high proportions of both the POS and UV races (0.6 and 0.4, respectively; Mainardi et al., 1986); by 1989, the UV race had gone extinct and very recently (within the last 5 years) the MV race has colonized the village, probably from Vernuga (Fig. 1). This volatility may reflect boom-bust dynamics in Valtellina relating, for example, to the movement of people and livestock out of villages in the spring (creating a population "bust") and return in the autumn (creating a population "boom"). It also emphasizes again the dispersal (both active and passive) and consequent colonizing capabilities of the house mouse. Researchers should be aware that similar volatility in chromosomal characteristics of mouse populations may occur elsewhere, with implications, for example, in the modeling of chromosome evolution.

The "tail" does not end here

Although chromosome variation in wild house mouse was first described in Val Poschiavo and was subsequently shown to extend into Valtellina, other interesting patterns of chromosome variation have since been described in various parts of western Europe and the Mediterranean basin (Nachman and Searle, 1995; Piálek et al., submitted). Together these systems form an exciting model for understanding chromosomal evolution and the role of chromosomes in speciation (Sage et al., 1993). We hope that we have demonstrated that the founding Val Poschiavo-Valtellina system is still one of the most interesting, and has offered particular insight into hybrid zone structure and its impact on gene flow, selection against hybrids between chromosomal races and the possibility of reinforcement, zonal raciation and the dynamism of chromosomal variation within sites. The continuing reduction in numbers of house mice within Val Poschiavo and Valtellina associated with recent changes in agricultural practices may promote future changes in the pattern of chromosome variation in the area, with exciting consequences which may further increase our insight into these fascinating systems.

Acknowledgements

We are very grateful to Silvia Garagna, both for the invitation to write this article and for her kind help in aspects of this work. Both Carlo Redi and Silvia Garagna were very supportive in setting up the "Valtellina project" and we offer our heartfelt thanks.

We also thank Anita Glover and Richard Ward for help with the microsatellite studies.

References

Akaike H: A new look at the statistical model identification. IEEE Trans Automatic Control 19:716–723 (1974).

Bagiotti T: Storia Economica della Valtellina e Valchiavenna. Banca Popolare di Sondrio (1958).

Benetti D, Giudetti M: Storia di Valtellina e Valchiavenna. Una introduzione (Editoriale Jaca Book SpA, Milano 1990).

Banaszek A, Fedyk S, Szałaj KA, Chêtnicki W: Reproductive performance in two hybrid zones between chromosome races of the common shrew Sorex araneus in Poland. Acta Theriol (suppl 1) 45:69–78 (2000).

Bauchau V: Phylogenetic analysis of the distribution of chromosomal races of Mus musculus domesticus Rutty in Europe. Biol J Linn Soc 41:171–192 (1990).

Boursot P, Auffray J-C, Britton-Davidian J, Bonhomme F: The evolution of house mice. Ann Rev Ecol Syst 24:119–152 (1993).

Britton-Davidian J, Sonjaya H, Catalan J, Cattaneo-Berrebi G: Robertsonian heterozygosity in wild mice: fertility and transmission rates in Rb (16.17) translocation heterozygotes. Genetica 80:171–174 (1990).

Britton-Davidian J, Catalan J, Ramalhinho MG, Ganem G, Auffray J-C, Capela R, Biscoito M, Searle JB, Mathias ML: Rapid chromosomal evolution in island mice. Nature 403:158 (2000).

Butlin R: Speciation by reinforcement. Trends Ecol Evol 2:8–13 (1987).

Capanna E: Chromosomal rearrangement and speciation in progress in Mus musculus. Folia Zool 29:43–57 (1980).

Capanna E: Robertsonian numerical variation in animal speciation: Mus musculus, an emblematic model, in Barigozzi C (ed): Mechanisms of Speciation, pp 155–177 (Liss, New York 1982).

Capanna E, Corti M: Reproductive isolation between two chromosomal races of Mus musculus in the Rhaetian Alps (northern Italy). Mammalia 46:107–109 (1982).

Capanna E, Valle M: A Robertsonian population of Mus musculus L. in the Orobian Alps. Atti della Accad Nazionale Dei Lincei, Rendiconti della Classe di Scienze Fisiche, Matematiche e Naturali, Serie VIII 62:680–684 (1977).

Cattanach BM, Kirk M: Differential activity of maternally and paternally derived chromosome regions in mice. Nature 315:496–498 (1985).

Corti M, Capanna E, Estabrook GF: Microevolutionary sequences in house mouse chromosomal speciation. Syst Zool 35:163–175 (1986).

Cox TP: Ethological isolation between local populations of house mice (Mus musculus) based on olfaction. Anim Behav 32:1068–1077 (1984).

Denotte G: L'agricoltura in Valtellina nell'ultimo secolo. Tesi di laurea. University of the Sacred Heart, Milan. Faculty of Economics and Commerce (1963).

Epstein CJ: The Consequences of Chromosome Imbalance: Principles, Mechanisms, and Models (Cambridge University Press, Cambridge 1986).

Fatio V: Faune des Vertébrés de la Suisse, Vol. 1 (H Georg, Libraire-Editeur, Geneva/Basel 1869).

Fraguedakis-Tsolis S, Hauffe HC, Searle JB: Genetic distinctiveness of a village population of house mice: relevance to speciation and chromosomal evolution. Proc R Soc Lond B Biol Sci 264:355–360 (1997).

Garagna S, Broccoli D, Redi CA, Searle JB, Cooke HJ, Capanna, E: Robertsonian metacentrics of the house mouse lose telomeric sequences but retain some minor satellite DNA in the pericentromeric area. Chromosoma 103:685–692 (1995).

Gropp A, Tettenborn U, von Lehmann E: Chromosomenuntersuchungen bei der Tabakmaus (M. poschiavinus) und bei den Hybriden mit der Laboratoriumsmaus. Experientia 25:875–876 (1969).

Gropp A, Tettenborn U, von Lehmann E: Chromosomenvariation vom Robertson'schen Typus bei der Tabakmaus, M. poschiavinus, und ihren Hybriden mit der Laboratoriumsmaus. Cytogenetics 9:9–23 (1970).

Gropp A, Winking H, Zech L, Muller H: Robertsonian chromosomal variation and identification of metacentric chromosomes in feral mice. Chromosoma 39:265–288 (1972).

Gropp A, Winking H, Redi C, Capanna E, Britton-Davidian J, Noack G: Robertsonian karyotype variation in wild house mice from Rhaeto-Lombardia. Cytogenet Cell Genet 34:67–77 (1982).

Gündüz İ, Auffray J-C, Britton-Davidian J, Catalan J, Ganem G, Ramalhinho MG, Mathias ML, Searle JB: Molecular studies on the colonization of the Madeiran archipelago by house mice. Mol Ecol 10:2023–2029 (2001).

Hasegawa M, Kishino K, Yano T: Dating the human-ape splitting by a molecular clock of mitochondrial DNA. J Mol Evol 22:160–174 (1985).

Hauffe HC, Piálek J: Evolution of the chromosomal races of Mus musculus domesticus in the Rhaetian Alps: the roles of whole-arm reciprocal translocation and zonal raciation. Biol J Linn Soc 62:255–278 (1997).

Hauffe HC, Searle JB: A disappearing speciation event? Nature 357:26 (1992).

Hauffe HC, Searle JB: Extreme karyotypic variation in a Mus musculus domesticus hybrid zone: the tobacco mouse story revisited. Evolution 47:1374–1395 (1993).

Hauffe HC, Searle JB: Chromosomal heterozygosity and fertility in house mice (Mus musculus domesticus) from northern Italy. Genetics 150:1143–1154 (1998).

Hauffe HC, Piálek J, Searle JB: The house mouse chromosomal hybrid zone in Valtellina (SO): a summary of past and present research. Hystrix 11:17–25 (2000).

Hauffe HC, Fraguedakis-Tsolis S, Mirol PM, Searle JB: Studies of mitochondrial DNA, allozyme and morphometric variation in a house mouse hybrid zone. Genet Res 80:117–129 (2002).

Hausser J, Fedyk S, Fredga K, Searle JB, Volobouev V, Wójcik JM, Zima J: Definition and nomenclature of the chromosome races of Sorex araneus. Folia Zool 43(suppl 1):1–9 (1994).

Hübner R: Chromosomal and biochemical variation in wild mice from Switzerland: relevance for models of chromosomal evolution in European house mice. (DPhil thesis, Oxford 1992).

Lande R: Effective deme sizes during long-term evolution estimated from rates of chromosomal rearrangement. Evolution 33:234–251 (1979).

Lyon MF, Rastan S, Brown SDM: Genetic Variants and Strains of the Laboratory Mouse, 3rd ed. (Oxford University Press, Oxford 1996).

Mainardi D, Parmigiani S, Jones SE, Brain PF, Capanna E, Corti M: Social conflict and chromosomal races of feral house mice: an assessment of conflicting laboratory and field investigations. Proceedings of the International Meetings on variability and behavioural evolution. Accademia Nazionale dei Lincei, Rome, 1983 (23–26 Nov.) Quaderno N. 259 (1986).

Marshall JL, Arnold ML, Howard DJ: Reinforcement: the road not taken. Trends Ecol Evol 17:558–563 (2002).

Mayr E: Animal Species and Evolution (Harvard University Press, Cambridge, Massachusetts 1963).

Nachman MW, Searle JB: Why is the house mouse karyotype so variable? Trends Ecol Evol 10:397–402 (1995).

Nachman MW, Boyer SN, Searle JB, Aquadro CF: Mitochondrial DNA variation and the evolution of Robertsonian chromosomal races of house mice, Mus domesticus. Genetics 136:1105–1120 (1994).

Nei M: Molecular Evolutionary Genetics (Colombia University Press, New York 1987).

Panithanarak T, Hauffe HC, Dallas JF, Glover A, Ward RG, Searle JB: Linkage-dependent gene flow in a house mouse chromosomal hybrid zone. Evolution (in press).

Pardo-Manuel de Villena F, Sapienza C: Female meiosis drives karyotypic evolution in mammals. Genetics 159:1179–1189 (2001).

Piálek J, Hauffe HC, Rodríguez-Clark KM, Searle JB: Raciation and speciation in house mice from the Alps: the role of chromosomes. Mol Ecol 10:613–625 (2001).

Piálek J, Hauffe HC, Searle JB: Chromosomal variation in the house mouse: a review. Biol J Linn Soc (submitted).

Posada D, Crandall KA: Modeltest: testing the model of DNA substitution. Bioinfomatics 14:817–818 (1998).

Prager EM, Tichy H, Sage RD: Mitochondrial DNA sequence variation in the eastern house mouse, *Mus musculus*: comparison with other house mice and report of a 75-bp repeat. Genetics 143:427–446 (1996).

Sage RD, Atchley WR, Capanna E: House mice as models in systematic biology. Syst Biol 42:523–561 (1993).

Saitou N, Nei M: The neighbor-joining method: a new method for reconstructing phylogenetic trees. Mol Biol Evol 4:406–425 (1987).

Searle JB: Chromosomal hybrid zones in eutherian mammals. In Harrison RG (ed.): Hybrid Zones and the Evolutionary Process, pp 309–353 (Oxford University Press, New York 1993).

Searle JB: Speciation, chromosomes, and genomes. Genome Res 8:1–3 (1998).

Swofford DL: PAUP*: Phylogenetic Analysis Using Parsimony and Other Methods (Sinauer, Sunderland, Massachusetts 1998).

Tognina R: Il Comun Grande di Poschiavo e Brusio. Tipografia Meneghini, Poschiavo (1975).

Torricelli GP: Territoire et Agricolture en Valteline. Géographie et groupes de relations. PhD Thesis. Faculty of Social and Economic Science of the University of Geneva. Le concept moderne/Editions. Geneva (1990).

Wallace BMN, Searle JB, Everett CA: Male meiosis and gametogenesis in wild house mice *(Mus musculus domesticus)* from a chromosomal hybrid zone; a comparison between "simple" Robertsonian heterozygotes and homozygotes. Cytogenet Cell Genet 61:211–220 (1992).

White MJD: Models of speciation. Science 159:1065–1070 (1968).

Cytogenet Genome Res 105:406–411 (2004)
DOI: 10.1159/000078213

Chromosome painting in the long-tailed field mouse provides insights into the ancestral murid karyotype

R. Stanyon,[a] F. Yang,[b,c] A.M. Morescalchi,[d] and L. Galleni[e]

[a] Comparative Molecular Cytogenetics, National Cancer Institute, Frederick, MD (USA);
[b] Centre for Veterinary Science, University of Cambridge, Cambridge (UK);
[c] Key Laboratory of Cellular and Molecular Evolution, The Chinese Academy of Sciences, Kunming Institute of Zoology, Kunming, Yunnan (PR China);
[d] Dipartimento di Scienze della Vita, Seconda Università degli Studi di Napoli, Caserta;
[e] Dipartimento di Chimica e Biotecnologie agrarie, Università di Pisa, Pisa (Italy)

Abstract. We report on the hybridization of mouse chromosomal paints to *Apodemus sylvaticus*, the long-tailed field mouse. The mouse paints detected 38 conserved segments in the *Apodemus* karyotype. Together with the species reported here there are now six species of rodents mapped with *Mus musculus* painting probes. A parsimony analysis indicated that the syntenies of nine *M. musculus* chromosomes were most likely already formed in the muroid ancestor: 3, 4, 7, 9, 14, 18, 19, X and Y. The widespread occurrence of syntenic segment associations of mouse chromosomes 1/17, 2/13, 7/19, 10/17, 11/16, 12/17 and 13/15 suggests that these associations were ancestral syntenies for muroid rodents. The muroid ancestral karyotype probably had a diploid number of about 2n = 54. It would be desirable to have a richer phylogenetic array of species before any final conclusions are drawn about the Muridae ancestral karyotype. The ancestral karyotype presented here should be considered as a working hypothesis.

Cross-species chromosome painting has provided essential information for mapping chromosomal homology and reconstructing ancestral genomes in many mammalian orders (Frönicke et al., 1996; Goureau et al., 1996; Wienberg et al., 1997; Ferguson-Smith et al., 1998; O'Brien et al., 1999; Yang et al., 1999, 2000a, 2003; Chowdhary and Raudsepp, 2001; Murphy et al., 2001). Most of these reports used human chromosome paints in unidirectional painting in order to relate animal genomes to the human karyotype. In a few cases reciprocal painting using whole chromosome probes from two or more species provided more detailed information on subchromosomal homologies and translocation breakpoints (Wienberg et al., 1997; Nash et al., 1998; Korstanje et al., 1999).

In contrast to some other mammalian orders, reports of chromosome painting in rodents are few and far between. It is quite amazing that in spite of intense interest in the genomics of the mouse, and the availability of the complete mouse sequence, there has never been a genome-wide cytogenetic FISH map directly linking the mouse and human genomes. Indeed, it was just recently that the chromosomes of any rodent species were mapped using human chromosome paints (Richard et al., 2003; Stanyon et al., 2003).

When a complete set of mouse paints first became available, it was thought that they would be of use for evolutionary comparison, similar to the great utility that human probes had provided for mapping chromosomal homology between species and even between mammalian orders. However, mouse probes proved to be of limited value when painting phylogenetically distant species and in particular for inter-ordinal FISH.

The highly fragmented and rearranged genome of the mouse may be one reason why chromosome painting between mouse and humans has only limited success (Scherthan et al., 1994; Ferguson-Smith et al., 1998). Gene mapping and sequencing data amply illustrate that the most extensively analyzed rodent genomes, the mouse and rat genomes, are highly rearranged compared to humans and other mammals (Nadeau and Taylor, 1984; Nilsson et al., 2001; Kent et al., 2003). The evolutionary

This article is a US Government work and, as such, is in the public domain in the United States of America.

Received 16 October 2003; revision accepted 20 November 2003.

Request reprints from Roscoe Stanyon, Comparative Molecular Cytogenetics Core Genetics Branch, National Cancer Institute, Building 560, Room 11-74A Frederick, MD 21702-1201 (USA); telephone +1 301 846 5440 fax: +1 301-846-6315; e-mail: stanyonr@ncifcrf.gov

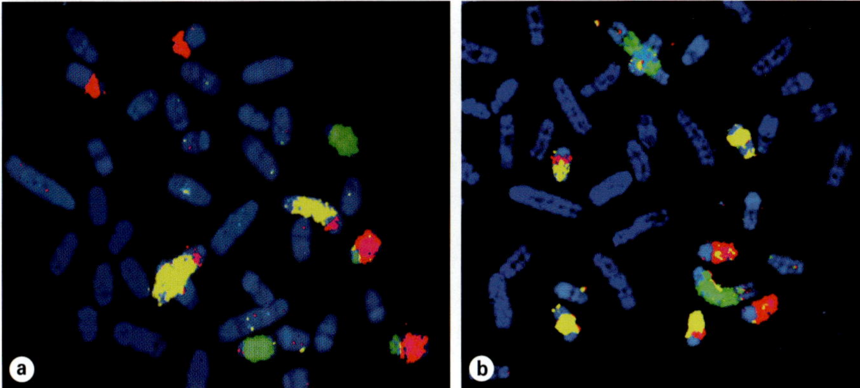

Fig. 1. Examples of triple-color hybridizations of mouse chromosome-specific paints on *Apodemus sylvaticus:* (**a**) 11 = green, 5 = red, 6 = yellow and (**b**) 7 = green, 15 = red and 13 = yellow.

rate of chromosome translocations between these two rodents appears to be up to ten times greater than that found between humans and cats, or between humans and chimpanzees where over the last 5–6 million years just one translocation (a tandem fusion) has occurred (Stanyon et al., 1999).

The rodents in general display the highest degree of karyotype variability so far found in mammals. Chromosome numbers range from nine or ten in a South American *Akodon* species (Silva and Yonenaga-Yassuda, 1998) to the highest chromosome number known in mammals (2n = 102) in the red viscacha rat *Tympanoctomys barrerae* (Contreras et al., 1990; Gallardo et al., 1999). Some taxa were extremely variable. For example, Fagundes et al. (1997) described 24 different karyotypes (2n = 14–16; FN = 18–26) for the South American *Akodon cursor* species, and Volobouev et al. (1995, 2002) found 20 different karyotypes for *Gerbillus nigeriae* (2n = 62–73).

The wide range of karyotypic variation may be partially explained by the fact that the rodents constitute the largest mammalian order (about 40% of mammals are rodents). However, other unique features also contribute to karyotype variability in rodents. Isochore studies indicate that the Muridae genomes may have undergone a very special type of genome evolution (Bernardi et al., 1985) and chromosome evolution in rodents may have involved the unique tolerance of both chromosome polymorphisms and tetraploidization (Gallardo et al., 1999).

Up to now only five rodent species have been painted with mouse chromosome-specific paint probes: *Rattus norvegicus* (Grutzner et al., 1999; Guilly et al., 1999; Stanyon et al., 1999), *Cricetulus griseus* (Yang et al., 2000b), *Rattus rattus* (Cavagna et al., 2002), *Rhabdomys pumilio* (Rambau and Robinson, 2003) and *Mus platythrix* (Matsubara et al., 2003). Here we report on the hybridization of mouse chromosomal paints to an additional rodent, *Apodemus sylvaticus*, the long-tailed field mouse.

A. sylvaticus belongs to the subfamily Murinae and is found throughout Europe. The long-tailed field mouse usually inhabits woodlands and forests as well as grassy fields, cultivated areas, and it is often considered an agricultural pest. *A. sylvaticus* is one of the principal reservoirs of Lyme disease in Europe and it is also known to harbor hantaviruses (Humair et al, 1999; Heyman et al., 2002).

Materials and methods

Cell culture and chromosome preparation

Primary fibroblast cell cultures were obtained from ear punches of four *A. sylvaticus*, the long-tailed field mouse, captured in the Bosco Negri, a residual of an antique forest in the Po valley near Pavia, Italy.

Preparation of metaphase chromosomes of *A. sylvaticus* followed standard procedures. To facilitate chromosome identification, most chromosome preparations were G-banded prior to in situ hybridization (Klever et al., 1991). Further, the identification of chromosomes was also made using DAPI-banding concurrently with in situ hybridization. Chromosomes were arranged according to length.

In situ hybridization

Mouse chromosome-specific probes were made by degenerate oligonucleotide-primed PCR (DOP-PCR) from flow-sorted chromosomes using PCR primers, amplification and labeling conditions as previously described (Telenius et al., 1992; Rabbitts et al., 1995; Ferguson-Smith et al., 1998). Chromosome sorting was performed using a dual-laser cell sorter (FACStar Plus; Becton Dickinson Immuno-Cytometry Systems, Cambridge) or a Facs-DiVa (Frederick). Both systems allowed a bivariate analysis of the chromosomes by size and base pair composition. In situ hybridization and detection of painting probes to *Apodemus* was as previously described (Stanyon et al., 1999). Digital images were obtained using a Photometrics CCD camera and the software SmartCaptureVP (DigitalScientific).

Results

The *A. sylvaticus* karyotype was composed of 23 autosomal pairs of chromosomes and the sex chromosomes for a diploid number of 2n = 48. All chromosomes are acrocentric and therefore the fundamental number (48) is identical to the diploid number. The sex chromosomes are unusual; the X chromosome is the second largest chromosome in the karyotype and the Y chromosome is about 60% as large as the X. There are abundant blocks of heterochromatin in the karyotype. Both sex chromosomes have large blocks of heterochromatin and the Y chromosome is mostly composed of heterochromatin. *A. sylvaticus* chromosomes 10, 16, 21, 22 and 23 all have notable heterochromatic blocks, which remained unhybridized with any mouse probe.

Paints specific for each mouse chromosome, except the Y, were used to delimit the homologous chromosomal segments on *A. sylvaticus* metaphases (Fig. 1). The 20 mouse paints detected 38 conserved segments on the long-tailed field mouse

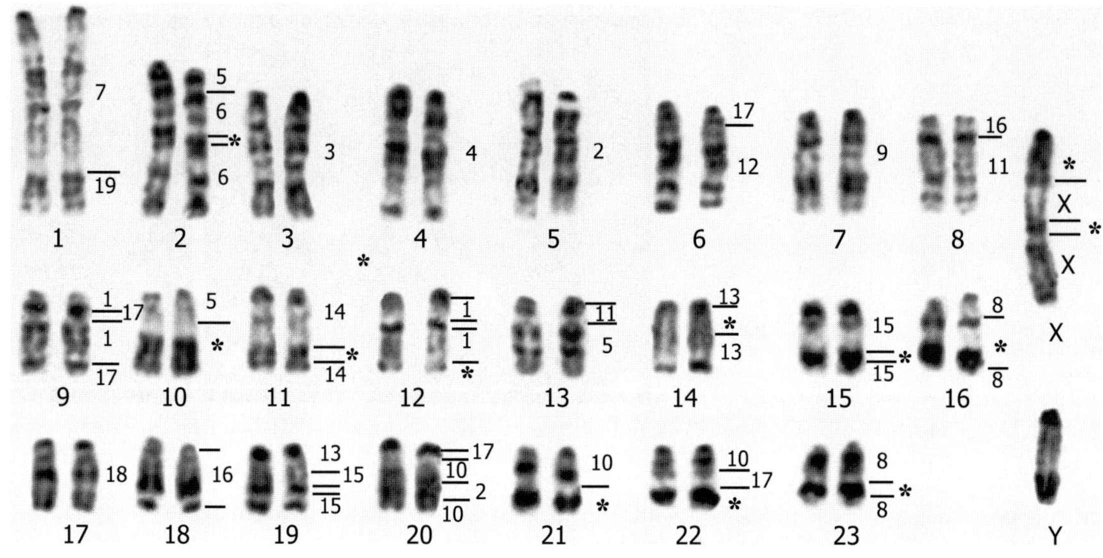

Fig. 2. The G-banded karyotype of *Apodemus sylvaticus*, with the *A. sylvaticus* chromosomes numbered below, and the mouse hybridization signals of mouse probes assigned to the right. Asterisks denote heterochromatic areas which remained unhybridized by mouse paints.

Table 1. The number of different chromosomes of the target species that are hybridized by each of the 21 mouse chromosomal paints. In parentheses is the total number of signals

Mouse paints	MPL[a]	ASY[a]	RNO[a]	RPU[a]	CGR[a]
1	1	2 (3)	2	3	3
2	1	2	2	2	2
3	1	1	1	1	1
4	1	1	1	1	1
5	1	3	3	2	2
6	2	1	2	2	2
7	1	1	1	1	1
8	1	2	2	2	2
9	1	1	1	2	1
10	1	3 (4)	3	4	2 (3)
11	1	2	2	2	2 (3)
12	1	1	1	2	2
13	1	2	3	2	2
14	4	1	2	1	1
15	1	2	2	2	2
16	1	2	2	2	2
17	2	4 (5)	4 (5)	6 (7)	4 (9)
18	1	1	1	1	1
19	1	1	1	1	1
X	1	1	1	1	1
Y	1	1	1	1	1

[a] MPL = *Mus platythrix* (data from Matsubara et al., 2003), ASY = *Apodemus sylvaticus* (data from present report), RNO = *Rattus norvegicus* (data from Cavagna et al., 2002), RPU = *Rhabdomys pumilio* (data from Rambau and Robinson, 2003 and our unpublished data), CGR = *Cricetulus griseus* (data from Yang et al., 2000b).

karyotype (Fig. 2). Fifteen *Apodemus* chromosomes were each hybridized by a single mouse paint, eight were each hybridized by two paints, one was hybridized by three paints. *Apodemus* chromosomes painted by multiple mouse paints produced nine associations of segments homologous to mouse (MMU) chromosomes or MMU chromosome segments: MMU 1/17(twice), 2/10 (twice) 5/6, 5/11, 7/19, 10/17 (twice), 11/16, 12/17 and 13/15. Table 1 shows that some mouse chromosome paints produced up to five (MMU 17) signals. Signals on the same chromosome separated only by unhybridized segments of heterochromatin were counted as single signals.

The in situ hybridization pattern of mouse probes on *A. sylvaticus* reveals a notable number of differences between these two genomes. The syntenies of only 9 out of 19 mouse autosomes have been conserved (MMU 3, 4, 6, 7, 9, 12, 14, 18, and 19). Four of these autosomes (MMU 6, 7, 12, and 19) are found associated with other chromosomes or chromosome segments. The most likely explanation for the alternating pattern of mouse paints on *A. sylvaticus* chromosomes 9 and 20 is that inversions may have occurred on these chromosomes.

Discussion

The in situ hybridization of mouse paints onto chromosomes of the long-tailed field mouse allowed us to map the chromosomal homology between these two species. We can anticipate that in cases of conserved synteny we can transfer gene assignments with greater than 94% accuracy (Stanyon et al., 1999). That is, from the well-known *Mus musculus* genome to a species in which no gene mapping has been done, *A. sylvaticus*.

Table 2. Syntenic associations of mouse chromosomes found in various rodents

Associations	MPL[a] 2n = 26	ASY[a] 2n = 48	RNO[a] 2n = 42	RPU[a] 2n = 46,48	CGR[a] 2n = 22
7/19	X	X	X	X	X
13/15	X	X	X	X	X
10/17		X	X	X	X
12/17		X	X	X	X
1/17		X	X	X	X
11/16		X	X	X	
2/13			X	X	X
5/6		X	X		
11/17				X	X
5/11		X	X		X
Additional associations	11	none	5	7	16

[a] MPL = *Mus platythrix* (data from Matsubara et al., 2003), ASY =*Apodemus sylvaticus* (data from present report), RNO = *Rattus norvegicus* (data from Cavagna et al., 2002), RPU = *Rhabdomys pumilio* (data from Rambau and Robinson, 2003 and our unpublished data), CGR = *Cricetulus griseus* (data from Yang et al., 2000b).

Together with this report there are now six species of rodents mapped with *Mus musculus* painting probes. Recently published molecular phylogenies allow us to order these species in an increasing phylogenetic distance from *M. musculus* (Michaux et al., 1996, 2002; Dubois et al., 1999): *Mus musculus – Mus platythrix – Apodemus sylvaticus – Rattus norvegicus/R. rattus – Rhabdomys pumilio – Cricetulus griseus.*

Using parsimony as the guiding principle and the hamster as an outgroup, a working hypothesis can be developed about the ancestral muroid karyotype. Parsimony indicates that it is unlikely that the same syntenic groups can be brought together independently in different lineages, but that syntenies may often be disrupted independently. Therefore, when chromosomal synteny is found intact between a wide range of species, this condition is probably ancestral.

We can compare the hybridization patterns in *A. sylvaticus* with the other rodent species using two sets of data. The first set of data is the number of hybridization signals provided by each mouse paint on the haploid karyotype of the target species (Table 1). Parsimony indicates that nine *M. musculus* chromosomes were most likely already syntenic in the muroid ancestor: MMU 3, 4, 7, 9, 14, 18, 19, X and Y.

The second set of data is the associations of conserved segments revealed by mouse chromosomes in each species (Table 2). For example, nine such associations were found in *A. sylvaticus*: MMU 1/17 (twice), 2/10 (twice) 5/6, 5/11, 7/19, 10/17 (twice), 11/16, 12/17 and 13/15. Only one association MMU 2/10 is not found in at least one other species. Six of these associations are found in at least four species.

The widespread associations (MMU 1/17, 7/19, 10/17, 11/16, 12/17 and 13/15) and an association MMU 2/13 not found in *Apodemus* are probably ancestral syntenies for muroid rodents. Association MMU 5/11 may also be ancestral especially if its presence was missed in RPU. All other associations found in each species are likely apomorphic and not present in the muroid ancestor.

Given the phylogenetically widespread occurrence of the two associations of MMU 10/17, there may have also been two chromosomes in the muroid ancestor containing these associations. The association MMU 1/17 is also found twice in all species except that in rat where it is on a single chromosome. The double occurrence of the MMU 1/17 association on this chromosome is likely the result of subsequent inversions. Reciprocal painting with rat probes on the mouse supports this conclusion, because the two MMU 17 signals paint a single contiguous area on mouse 17 (Stanyon et al., 1999).

In summary, the muroid ancestral karyotype probably had a diploid number of 2n = 54 and contained the following chromosomes that are homologous to *M. musculus*:
1a, 1b/17a, 2a, 2b/13a, 3, 4, 5a, 5b/11, 6a, 6b, 7/19, 8a, 8b, 9, 10a, 10b/17b, 10c/17c, 11a, 11b/16a, 12a, 12/17d, 13b/15, 14, 15b, 16b, 18, X and Y.

It would be desirable to have a richer phylogenetic array of species before any firm conclusions are drawn on the Muridae ancestral karyotype. The ancestral karyotype presented above should be considered as a working hypothesis. For instance, it is not clear if mouse chromosome 1 is present as two or three segments in the ancestral karyotype.

However, if the ancestral muroid karyotype presented here is taken as a departure point, we can make additional conjectures about the type of inter-chromosomal rearrangements that formed the mouse genome. A derivation hypothesis can be proposed for each mouse chromosome. The most striking examples are MMU 10 and MMU 17. MMU 10 probe paints up to 4 chromosome segments in the various species, while MMU 17 paints from 2 to 9 signals in the five species. The sequence of events in the formation of MMU 17 and MMU 10 are clear and fairly straightforward given the ancestral muroid karyotype proposed above. A reciprocal translocation between MMU 10b/17b and MMU 10c/17c would form a part of mouse chromosomes 10 (10b/10c) and the first half of MMU 17 (17b/17c). Although the exact order of all events is unknown, the presence of two 17 homologs in MPL indicates that the formation of MMU 10 by a fusion of 10a with 10b/10c was completed first. There were fissions of chromosomes MMU 1b/17a and MMU 12/17d. The lack of a MMU 12/17 association in MPL indicates that first MMU 17a/17d was formed and then tandemly fused with MMU 17b/17c to form MMU 17.

At this point we have attempted no speculation on the presence and importance of intra-chromosomal rearrangements such as inversions, duplications and deletions. These rearrangements are difficult to assess with chromosome painting. FISH with subregional probes such as BACS and YACs or fiber FISH can demonstrate inversions and changes in gene order. Further we have made no attempt to determine the morphology of the ancestral chromosome (i.e. metacentric, acrocentric, etc.).

The ancestral rodent genome

Contrary to expectations derived from gene mapping data in mouse and rat, extensive conservation of chromosomal syntenies has recently been reported between humans and squirrels (Richard et al., 2003; Stanyon et al., 2003). Both reports show that the genomes of squirrels are not far removed from that of the presumed ancestral eutherian karyotype. On the basis of only gene mapping data, a few derived associations found in squirrel are also found in rat or mouse. Squirrel, rat and mouse have the associations of HSA 1/10 and 8/12 (Nilsson et al., 2001). These associations could have been present in the last common ancestor of squirrels and murids. However, the high number of rearrangements in murid genomes makes convergence more likely and weakens this hypothesis. We can conclude that few derived translocations characterized the evolutionary origin of the rodents, but others may be found when other conserved rodent species are studied. Although comparative chromosome painting is possible between distantly related placental orders, the higher quality of FISH signals between intra-ordinal species certainly allows for an easier and more precise comparative analysis. The availability of a complete set of rodent chromosome paint probes, derived from the conserved squirrel karyotype, and eventually from other rodents as

well, may allow us in the near future to connect by multidirectional chromosome painting, in a stepping-stone manner, the muroid genome to the human genome. At the very least, chromosome painting allows a survey of animal genomes which are unlikely to be sequenced and would therefore provide invaluable data, especially in the species-rich Rodentia, on the relationship between chromosome rearrangements, speciation and a host of other questions.

Acknowledgements

The authors would like to thank Dr. Luca Canova (University of Pavia, Italy) for the ear punches of *Apodemus sylvaticus* from Bosco Negri. We thank Patricia C.M. O'Brien (Cambridge) and Gary Stone (Frederick) for the flow sorting of mouse chromosomes. R.S. thanks Francesca Bigoni and Marta Svartman for comments and suggestions.

References

Bernardi G, Olofsson B, Filipski J, Zerial M, Salinas J, Cuny G, Meunier-Rotival M, Rodier F: The mosaic genome of warm-blooded vertebrates. Science 228:953–958 (1985).

Cavagna P, Stone G, Stanyon R: Black rat *(Rattus rattus)* genomic variability characterized by chromosome painting. Mamm Genome 13:157–163 (2002).

Chowdhary BP, Raudsepp T: Chromosome painting in farm, pet and wild animal species. Methods in Cell Science 23:37–55 (2001).

Contreras LC, Torres-Mura JC, Spotorno AE: The largest known chromosome number for a mammal, in a South American desert rodent. Experientia 46:506–508 (1990).

Dubois JY, Catzeflis FM, Beintema JJ: The phylogenetic position of "Acomyinae" (Rodentia, Mammalia) as sister group of a Murinae + Gerbillinae clade: evidence from the nuclear ribonuclease gene. Mol Phylogenet Evol 13:181–192 (1999).

Fagundes V, Scalzi-Martin JM, Sims K, Hozier J, Yonenaga-Yassuda Y: ZOO-FISH of a microdissection DNA library and G-banding patterns reveal the homeology between the Brazilian rodents *Akodon cursor* and *A. montensis*. Cytogenet Cell Genet 78:224–228 (1997).

Ferguson-Smith MA, Yang F, O'Brien PC: Comparative Mapping Using Chromosome Sorting and Painting. ILAR J 39:68–76 (1998).

Frönicke L, Chowdhary BP, Scherthan H, Gustavsson I: A comparative map of the porcine and human genomes demonstrates ZOO-FISH and gene mapping-based chromosomal homologies. Mamm Genome 7:285–290 (1996).

Gallardo MH, Bickham JW, Honeycutt RL, Ojeda RA, Kohler N: Discovery of tetraploidy in a mammal. Nature 401:341 (1999).

Goureau A, Yerle M, Schmitz A, Riquet J, Milan D, Pinton P, Frelat G, Gellin J: Human and porcine correspondence of chromosome segments using bidirectional chromosome painting. Genomics 36:252–262 (1996).

Grutzner F, Himmelbauer H, Paulsen M, Ropers HH, Haaf T: Comparative mapping of mouse and rat chromosomes by fluorescence in situ hybridization. Genomics 55:306–313 (1999).

Guilly MN, Fouchet P, de Chamisso P, Schmitz A, Dutrillaux B: Comparative karyotype of rat and mouse using bidirectional chromosome painting. Chromosome Res 7:213–221 (1999).

Heyman P, Van Mele R, De Jaegere F, Klingstrom J, Vandenvelde C, Lundkvist A, Rozenfeld F, Zizi M: Distribution of hantavirus foci in Belgium. Acta Trop 84:183–188 (2002).

Humair PF, Rais O, Gern L: Transmission of *Borrelia afzelii* from *Apodemus* mice and *Clethrionomys* voles to *Ixodes ricinus* ticks: differential transmission pattern and overwintering maintenance. Parasitology 118(Pt 1):33–42 (1999).

Kent WJ, Baertsch R, Hinrichs A, Miller W, Haussler D: Evolution's cauldron: duplication, deletion, and rearrangement in the mouse and human genomes. Proc Natl Acad Sci USA 100:11484–11489 (2003).

Klever M, Grond-Ginsbach C, Scherthan H, Schroeder-Kurth TM: Chromosomal in situ suppression hybridization after Giemsa banding. Hum Genet 86:484–486 (1991).

Korstanje R, O'Brien PC, Yang F, Rens W, Bosma AA, van Lith HA, van Zutphen LF, Ferguson-Smith MA: Complete homology maps of the rabbit *(Oryctolagus cuniculus)* and human by reciprocal chromosome painting. Cytogenet Cell Genet 86:317–322 (1999).

Matsubara K, Nishida-Umehara C, Kuroiwa A, Tsuchiya K, Matsuda Y: Identification of chromosome rearrangements between the laboratory mouse *(Mus musculus)* and the Indian spiny mouse *(Mus platythrix)* by comparative FISH analysis. Chromosome Res 11:57–64 (2003).

Michaux JR, Filippucci MG, Libois RM, Fons R, Matagne RF: Biogeography and taxonomy of *Apodemus sylvaticus* (the woodmouse) in the Tyrrhenian region: enzymatic variations and mitochondrial DNA restriction pattern analysis. Heredity 76(Pt 3):267–277 (1996).

Michaux JR, Chevret P, Filippucci MG, Macholan M: Phylogeny of the genus *Apodemus* with a special emphasis on the subgenus *Sylvaemus* using the nuclear IRBP gene and two mitochondrial markers: cytochrome b and 12S rRNA. Mol Phylogenet Evol 23:123–136 (2002).

Murphy WJ, Stanyon R, O'Brien SJ: Evolution of mammalian genome organization inferred from comparative gene mapping. Genome Biol 2: REVIEWS0005 (2001).

Nadeau JH, Taylor BA: Lengths of chromosomal segments conserved since divergence of man and mouse. Proc Natl Acad Sci USA 81:814–818 (1984).

Nash WG, Wienberg J, Ferguson-Smith MA, Menninger JC, O'Brien SJ: Comparative genomics: tracking chromosome evolution in the family Ursidae using reciprocal chromosome painting. Cytogenet Cell Genet 83:182–192 (1998).

Nilsson S, Helou K, Walentinsson A, Szpirer C, Nerman O, Stahl F: Rat-mouse and rat-human comparative maps based on gene homology and high-resolution zoo-FISH. Genomics 74:287–298 (2001).

O'Brien SJ, Eisenberg JF, Miyamoto M, Hedges SB, Kumar S, Wilson DE, Menotti-Raymond M, Murphy WJ, Nash WG, Lyons LA, Menninger JC, Stanyon R, Wienberg J, Copeland NG, Jenkins NA, Gellin J, Yerle M, Andersson L, Womack J, Broad T, Postlethwait J, Serov O, Bailey E, James MR, Marshall Graves JA, et al.: Genome maps 10. Comparative genomics. Mammalian radiations. Wall chart. Science 286:463–478 (1999).

Rabbitts P, Impey H, Heppell-Parton A, Langford C, Tease C, Lowe N, Bailey D, Ferguson-Smith M, Carter N: Chromosome specific paints from a high resolution flow karyotype of the mouse. Nat Genet 9:369–375 (1995).

Rambau RV, Robinson TJ: Chromosome painting in the African four-striped mouse *Rhabdomys pumilio*: detection of possible murid specific contiguous segment combinations. Chromosome Res 11:91–98 (2003).

Richard F, Messaoudi C, Bonnet-Garnier A, Lombard M, Dutrillaux B: Highly conserved chromosomes in an Asian squirrel *(Menetes berdmorei*, Rodentia: Sciuridae) as demonstrated by ZOO-FISH with human probes. Chromosome Res 11:597–603 (2003).

Scherthan H, Cremer T, Arnason U, Weier HU, Lima-de-Faria A, Fronicke L: Comparative chromosome painting discloses homologous segments in distantly related mammals. Nat Genet 6:342–347 (1994).

Silva MJ, Yonenaga-Yassuda Y: Karyotype and chromosomal polymorphism of an undescribed *Akodon* from Central Brazil, a species with the lowest known diploid chromosome number in rodents. Cytogenet Cell Genet 81:46–50 (1998).

Stanyon R, Yang F, Cavagna P, O'Brien PC, Bagga M, Ferguson-Smith MA, Wienberg J: Reciprocal chromosome painting shows that genomic rearrangement between rat and mouse proceeds ten times faster than between humans and cats. Cytogenet Cell Genet 84:150–155 (1999).

Stanyon R, Stone G, Garcia M, Froenicke L: Reciprocal chromosome painting shows that squirrels, unlike murid rodents, have a highly conserved genome organization. Genomics 82:245–249 (2003).

Telenius H, Carter NP, Bebb CE, Nordenskjold M, Ponder BAJ, Tunnacliffe A: Degenerate Oligonucleotide-Primed PCR: General Amplification of Target DNA by a Single Degenerate Primer. Genomics 13:718–725 (1992).

Volobouev V, Vogt N, Viegas-Pequignot E, Malfoy B, Dutrillaux B: Characterization and chromosomal location of two repeated DNAs in three *Gerbillus* species. Chromosoma 104:252–259 (1995).

Volobouev VT, Aniskin VM, Lecompte E, Ducroz JF: Patterns of karyotype evolution in complexes of sibling species within three genera of African murid rodents inferred from the comparison of cytogenetic and molecular data. Cytogenet Genome Res 96:261–275 (2002).

Wienberg J, Stanyon R, Nash WG, O'Brien PC, Yang F, O'Brien SJ, Ferguson-Smith MA: Conservation of human vs. feline genome organization revealed by reciprocal chromosome painting. Cytogenet Cell Genet 77:211–217 (1997).

Yang F, O'Brien PC, Milne BS, Graphodatsky AS, Solanky N, Trifonov V, Rens W, Sargan D, Ferguson-Smith MA: A complete comparative chromosome map for the dog, red fox, and human and its integration with canine genetic maps. Genomics 62:189–202 (1999).

Yang F, Graphodatsky AS, O'Brien PC, Colabella A, Solanky N, Squire M, Sargan DR, Ferguson-Smith MA: Reciprocal chromosome painting illuminates the history of genome evolution of the domestic cat, dog and human. Chromosome Res 8:393–404 (2000a).

Yang F, O'Brien PC, Ferguson-Smith MA: Comparative chromosome map of the laboratory mouse and Chinese hamster defined by reciprocal chromosome painting. Chromosome Res 8:219–227 (2000b).

Yang F, Alkalaeva EZ, Perelman PL, Pardini AT, Harrison WR, O'Brien PC, Fu B, Graphodatsky AS, Ferguson-Smith MA, Robinson TJ: Reciprocal chromosome painting among human, aardvark, and elephant (superorder Afrotheria) reveals the likely eutherian ancestral karyotype. Proc Natl Acad Sci USA 100:1062–1066 (2003).

Cytogenet Genome Res 105:412–421 (2004)
DOI: 10.1159/000078214

RNAi knock-down mice: an emerging technology for post-genomic functional genetics

D. Prawitt,[a] L. Brixel,[a] C. Spangenberg,[a] L. Eshkind,[b] R. Heck,[b] F. Oesch,[b] B. Zabel,[a] and E. Bockamp[b]

[a] Children's Hospital, Johannes Gutenberg-Universität Mainz;
[b] Laboratory of Molecular Mouse Genetics, Institute of Toxicology, Johannes Gutenberg-Universität Mainz, Mainz (Germany)

Abstract. RNA interference (RNAi) has been extensively used for sequence-specific silencing of gene function in mammalian cells. The latest major breakthrough in the application of RNAi technology came from experiments demonstrating RNAi-mediated gene repression in mice and rats. After more than two decades of functional mouse research aimed at developing and continuously improving transgenic and knock-out technology, the advent of RNAi knock-down mice represents a valuable new alternative for studying gene function in vivo. In this review we provide some basic insight as to how RNAi can induce gene silencing to then focus on recent findings concerning the applicability of RNAi for regulating gene function in the mouse. Reviewed topics will include delivery methods for RNAi-mediating molecules, a comparison between traditional knock-out and innovative transgenic RNAi technology and the generation of graded RNAi knock-down phenotypes. Apart from the exciting possibilities RNAi provides for studying gene function in mice, we discuss several caveats and limitations to be considered. Finally, we present prospective strategies as to how RNAi technology might be applied for generating conditional and tissue-restricted knock-down mice.

Copyright © 2004 S. Karger AG, Basel

"By having the sequence in hand, we're at the end of the beginning". Only just confronted with the enormous wealth of information arising from the human genome sequencing program, this statement by the director of the National Human Genome Research Institute, Francis Collins (given during an interview by the BBC on the 5th of January 2001), perfectly portrays the present state of affairs in post-genomic biomedical research. Indeed, after the completion of the human and mouse genome sequences it became immediately apparent that annotated sequences will often not be sufficient for providing clues about gene function. One very elegant way of translating nucleotide sequences into biological function is to use tailor-made mouse models harbouring specific genetic modifications. Since the first report of transgenic mice produced by injecting DNA into the pronucleus of one-cell mouse embryos (Gordon et al., 1980), the generation of transgenic and knock-out mouse models has been constantly improved, providing the scientific community with a large number of invaluable animal models (for reviews on this topic Lewandoski, 2001; Bockamp et al., 2002; van der Weyden et al., 2002). The first publication of transgenic knock-down mice generated by virtue of RNAi (for RNA interference)-mediated gene silencing has been the latest exciting addition to the already exquisite toolbox of functional mouse genomics (Hasuwa et al., 2002).

This review will focus on the most recent findings concerning the applicability of RNAi-mediated gene silencing for studying gene function in the mouse. In addition, we will provide a few basic insights as to how RNAi induces gene silencing. However, as it is not intended to give a comprehensive

Research in the Bockamp and Zabel laboratories is supported by the Deutsche Forschungsgemeinschaft, the European Union, the Stiftung Rheinland-Pfalz for Innovation, the MAIFOR Program of the Johannes Gutenberg-Universität Mainz and the Dr. Mildred Scheel Stiftung for Cancer Research.

Received 1 October 2003; manuscript accepted 20 November 2003.

Request reprints from Dirk Prawitt, Molecular Genetics Laboratory
Children's Hospital, Johannes Gutenberg-Universität Mainz
Obere Zahlbacher Strasse 63, D–55131 Mainz (Germany)
telephone: +49 6131-393-3335; fax: +49 6131-393-0227
e-mail: prawitt@molgen.medizin.uni-mainz.de

D.P. and L.B. contributed equally to this work.

view of the rapidly expanding field of RNAi, the reader is encouraged to explore other recent reviews on this topic (Denli and Hannon, 2003; Dykxhoorn et al., 2003; Shi, 2003). It is also of note that the article by Svoboda in this issue will provide an excellent overview about dsRNA-mediated gene repression in mouse oocytes and early embryos (Svoboda, 2004).

The RNAi principle: short RNAs that repress gene expression

The term RNAi was coined five years ago, when the groups of Fire and Mello discovered that double stranded RNAs (dsRNAs) induced gene-specific silencing in the nematode *Caenorhabditis elegans*. They called this phenomenon RNA interference or RNAi (Fire et al., 1998). RNAi has subsequently been demonstrated to function as a powerful cellular surveillance mechanism in such diverse organisms as protozoa, insects, plants, fungi and vertebrates. Careful analysis of the basic principles governing RNAi-mediated gene repression resulted in a clearer picture of how short siRNAs bring about gene silencing.

The RNAi silencing cascade is initiated by dsRNAs which are homologous to the subsequently repressed RNA. It is generally assumed that RNAi originally evolved as a defense mechanism of the cellular machinery against invading viruses or active transposable elements (Matzke et al., 2001; Vance and Vaucheret, 2001). Indeed, many foreign RNAs differ from endogenous transcripts in that they are not single-stranded but consist of two anti-parallel RNA strands. In case dsRNAs appear in the cell, a particular ribonuclease, named Dicer, is in charge of specifically recognizing these dsRNAs, cutting them into RNA duplexes of about 21 nucleotides which contain characteristic 2-nucleotide 3′-overhangs: siRNAs (Zamore et al., 2000). The next stage of cellular RNAi-mediated degradation involves a multi-protein complex called RISC (for RNA-induced silencing complex (Hammond et al., 2000)). Functional RISC is believed to contain at least four different subunits including a helicase, an endonuclease, an exonuclease and a "homology searching" component. Upon binding of siRNA to RISC the helicase unwinds the double-stranded siRNA molecule resulting in two short single strands (Nykanen et al., 2001). In doing so, activated RISC is poised for "homology searching". One single strand of the siRNA is used as bait for searching and binding additional complementary RNA strands (Nykanen et al., 2001; Martinez et al., 2002). Finally, RISC-bound target RNA is degraded through action of the exo- and endonuclease subunits. A general summary as to how RNAi mediated gene silencing is assumed to work is depicted in Fig. 1.

Plants, insects and worms have developed an additional defense mechanism to even more efficiently fight off double-stranded RNAs. As viral attacks often result in attempts to "highjack" the cell by reprogramming the transcriptional machinery to produce large quantities of viral transcripts, these organisms fight back by a strategy termed transitive or secondary RNAi. Using RNA-dependent RNA polymerases (RdRp) the initial siRNA is assumed to provide sequence-specific primers using homologous RNAs as a template. As a consequence the total number of RNAi molecules will be amplified thus increasing the chances for complete clearance of target RNA molecules altogether (Dalmay et al., 2000; Lipardi et al., 2001; Nishikura, 2001; Sijen et al., 2001). By contrast, this amplification loop does not exist in mammalian oocytes and possibly also not in mammals (Stein et al., 2003a). In an attempt to enhance the overall efficiency of a 317-bp dsRNA transcript to disrupt γ-globin function in transgenic mice, the tomato RdRp transgene was stably introduced into these animals. Bi-transgenic mice expressing both the 317-bp dsRNA and the RdRp gene did not display any measurable knock-down effect (de Wit et al., 2002). However, it remains to be clarified if expression of RdRp alone or in concert with other components of the transitive or secondary RNAi amplification loop can contribute to enhance RNAi mediated gene repression in mice.

Initially, the use of RNAi for studying gene function was restricted to plants, *Drosophila* and *C. elegans*. However, when Tuschl and coworkers showed that gene function was selectively repressed in mammalian cells by virtue of small in vitro synthesized dsRNAs, the tremendous impact RNAi might have on biomedical research became clear (Elbashir et al., 2001). The fact that long dsRNAs are inducing an interferon response in most mammalian cells but that 21–23-bp-long siRNA did appear to not trigger this pathway made the use of RNAi as a tool for reverse genetics and for therapeutic applications even more attractive (Caplen et al., 2001; Elbashir et al., 2001). All that is required for efficiently knocking-down gene function is the sequence of the target gene, an optimised siRNA and an efficient method ensuring the proper delivery of the siRNA to the target. For that reason RNAi technology already represents a powerful alternative to other genetic inactivating tools including long dsRNAs, antisense oligonucleotides, ribozymes or the laborious genetic manipulation of murine embryonic stem cells (ES cells) for generating loss-of-function mice.

RNAi delivery: how to reach the target ?

For cell culture systems effective delivery of siRNAs is relatively simple as introduction of siRNA into the cell is easily achieved by physical or chemical transfection (Elbashir et al., 2002). When studying more complex model organisms such as worms and flies appropriate delivery of the siRNA molecules can for example be facilitated by either soaking the whole animal in a solution containing synthesized siRNAs or by adding bacteria that express complementary dsRNAs to the diet. The effectiveness of such strategies for large scale functional genomic studies has recently been demonstrated for *C. elegans* resulting in a knock-down coverage of expressed genes of about 86 % (Kamath et al., 2003).

In mammals, such simple uptake or soaking strategies can not be applied. Instead the first method for implementing RNAi repression in mice simply made use of repeated injections of large amounts of siRNA molecules into the animal (Lewis et al., 2002; McCaffrey et al., 2002; Sorensen et al., 2003). These pioneering experiments showed that RNAi can suppress the function of endogenous and viral RNAs in mice.

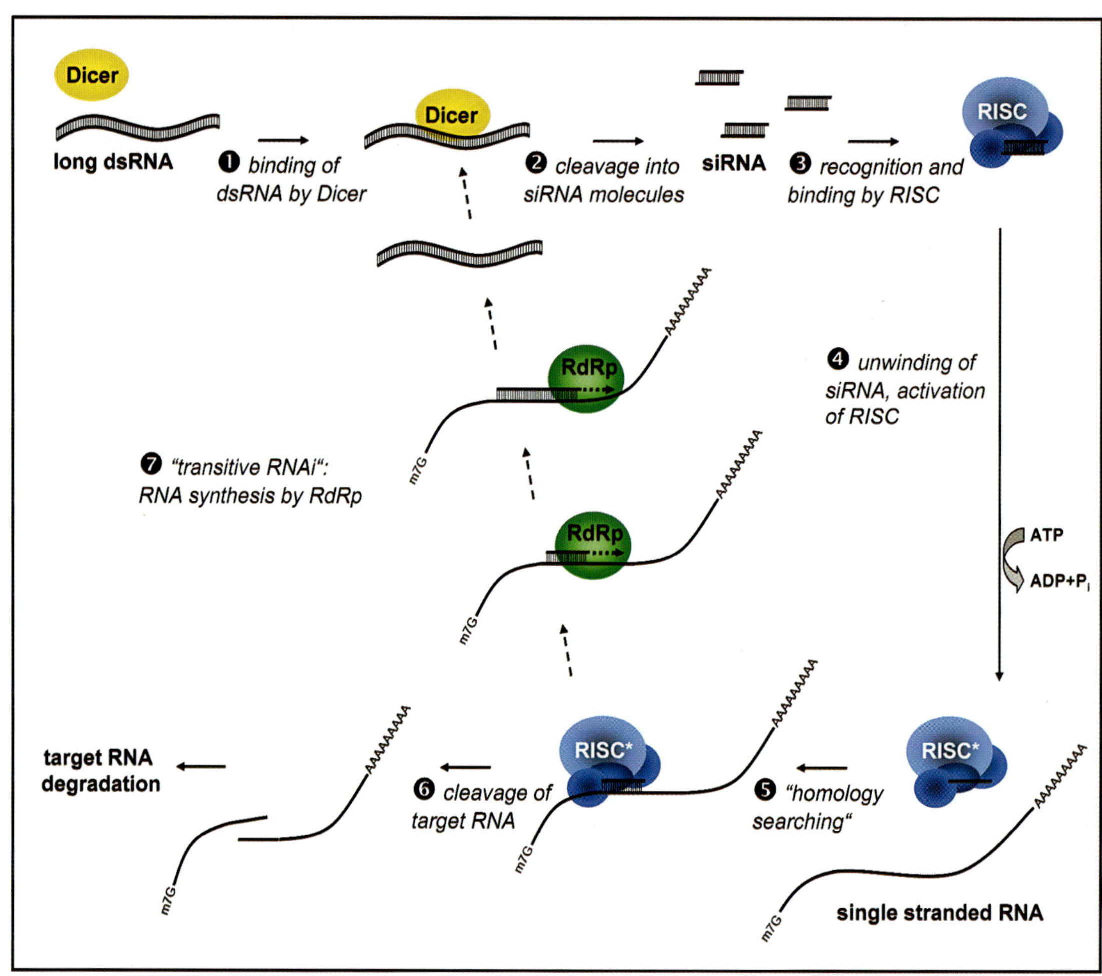

Fig. 1. Proposed RNAi pathway: Long dsRNA molecules are recognized by the cellular RNase Dicer (**1**). Subsequently the dsRNA is cleaved resulting in 21-nt RNA duplexes, the siRNAs (**2**). These siRNA molecules are then incorporated into the RISC (RNA-induced silencing complex) multiprotein complex (**3**) where they are unwound by an ATP-dependent process, transforming the complex into an active state (**4**). Activated RISC (RISC*) uses one strand of the RNA as a bait to bind homologous RNA molecules ("homology searching" **5**). Next the target RNA is cleaved and degraded (**6**). For *C. elegans* and plants it has been shown that new siRNA molecules are synthesized by RNA-dependent RNA polymerases (RdRp) in a process called transitive or secondary RNAi amplification loop (**7**).

However, to obtain a long-term knock-down effect in the animal, the inoculations of siRNA have to be systemic or repeated at regular frequencies. To overcome these limitations, vector systems were developed that used RNA polymerase I or III promoters for intracellular expression of siRNAs. These specifically designed vectors were capable of expressing short RNA molecules lacking a polyadenylation signal and the 5′ cap (Brummelkamp et al., 2002b; Lee et al., 2002; McManus et al., 2002; Miyagishi and Taira, 2002a, b; Paddison et al., 2002; Paul et al., 2002; Sui et al., 2002; Yu et al., 2002; Kawasaki and Taira, 2003). In order to efficiently trigger the RNAi degradation cascade, these short RNAs either fold into siRNA-like hairpin structures, so called short-hairpin RNAs (shRNAs), or form siRNA duplexes consisting of two complementary short RNA strands. Comparing the efficiency of hairpin shaped shRNAs to duplex siRNAs, shRNAs were reported to repress gene func-

tion more efficiently than duplex siRNAs in tissue culture (Yu et al., 2002).

To induce long-term knock-down phenotypes in mice H1 or U6 RNA polymerase III transcription elements driving expression of shRNAs were stably integrated into the genome of mice by microinjecting DNA into one cell mouse embryos or transfecting ES cells (Hasuwa et al., 2002; Kunath et al., 2003). These proof of principle experiments demonstrated that the desired knock-down phenotype was apparent in the founder generation also being efficiently transmitted to the germline. Alternatively, viral transduction methods using retro-, lenti- or adenovirus-associated delivery systems were applied for stably introducing siRNA expression elements into the genome of transformed cell lines, primary cells or mice (Brummelkamp et al., 2002a; Rubinson et al., 2003; Tiscornia et al., 2003). The use of lentiviral vectors for introducing shRNA expression ele-

Table 1. Examples of published RNAi-mediated knock-down experiments in mouse and rat

Reference	Target gene[a]	Knock-down inducer	Delivery method	Effects / Consequences
Lewis et al., 2002	EGFP (t)	siRNA molecules	tail vein injection ("hyrodynamic transfection method")	injection of high dosages of siRNA molecules has to be repeated regularly to obtain the desired knock-down effect
Wianny et al., 2002	*c-Mos* (e) in the oocyte; E-cadherin (e) and EGFP (t) in the zygote	dsRNA molecules (>500bp)	injection into mouse oocytes and zygotes, embryo implantation in pseudopregnant animals	phenotypes of the c-*Mos* and E-cadherin RNAi mice are comparable to those of null mutants; the RNAi effect persisted for 6 cell divisions after injection; RNAi per se does not interfere with the normal embryonal development
McCaffrey et al., 2002	luciferase (t); HCV NS5B fused with luciferase (t)	shRNA expression construct; siRNA molecules	tail vein injection	siRNA molecules and plasmid encoded shRNAs induce potent and specific RNAi responses against human pathogenic (HCV) RNA in adult mice
McCaffrey et al., 2003	replicative intermediates of HBV (t)	shRNA expression construct with U6 promoter	tail vein injection	RNAi inhibits HBV replication in mice
Hasuwa et al., 2002	EGFP (t)	shRNA expression construct with H1 promoter	pronucleus injection into fertilized eggs, embryo implantation in pseudopregnant animals	ubiquitous knock-down; successful also in rat, where ES-cell lines remain unestablished to date
Stein et al., 2003b	*Mos* (e)	long hairpin dsRNA (~500bp) expression construct with oocyte-specific Zp3 promoter	pronucleus injection into fertilized eggs, embryo implantation in pseudopregnant animals	RNAi animals show *Mos* knock-out phenotype long hairpin dsRNA-mediated RNAi is suitable to study oocyte-synthesized genes during oocyte development and early embryogenesis, in contrast to somatic cells that typically undergo apoptosis when exposed to dsRNA >30 bp
Carmell et al., 2003	*Neil1* (e)	shRNA expression vector with H1 promoter	electroporation of ES-cells, injection into mouse embryos, implantation in pseudopregnant animals	RNAi can be used to create germline transgenic RNAi mice
Kunath et al., 2003	*Rasa1* (e)	shRNA expression construct with H1 promoter	electroporation of ES-cells; embryos derived completely from ES-cells by tetraploid aggregation method, implantation in pseudopregnant animals	RNAi animals show *Rasa1* null mutation phenotype
Hemann et al., 2003	p53 (e) in hematopoietic stem cells	retroviral shRNA expression vector with U6 promoter	retroviral infection of hematopoietic stem cells and injection into lethally irradiated recipient mice	stable knock-down of genes in stem cells and reconstituted organs generation of different phenotypes, ranging from mild to severe ("epi-allelic series") by using RNAi molecules of different knock-down efficiency
Song et al., 2003	*Fas* (e)	siRNA molecules	tail vein injection	RNAi is successfully used to protect mice from fulminant hepatitis
Rubinson et al., 2003	p53, Bim, CD8α and CD25 (e)	lentiviral shRNA expression constructs with U6 promoter	lentiviral infection of cells and embryos, implantation of embryos in pseudopregnant animals	stable knock-down in ES-cell derived mice and in various, otherwise unefficiently transfected cell types: cycling and non-cycling mammalian cells, stem cells, zygotes and their differentiated progeny
Tiscornia et al., 2003	GFP (t)	lentiviral shRNA expression constructs with H1 promoter	lentiviral infection of eggs from GFP-positive transgenic mice	lentivirus vector system capable of expressing siRNA can efficiently transduce preimplantation mouse embryos. The resulting progeny expressed siRNA and showed reduced expression of target gene.
Sorensen et al., 2003	GFP (t) TNF-alpha (e)	siRNA molecules	cationic liposome-based intravenous injection in mice of plasmid encoding the green fluorescent protein (GFP) with its cognate siRNA, intraperitoneal injection of anti-TNF-alpha siRNA	stable inhibition of GFP expression in various organs, intraperitoneal injection of anti-TNF-alpha siRNA inhibited specifically lipopolysaccharide-induced TNF-alpha expression. Use of RNAi as pharmacological inducers (development of sepsis in mice following a lethal dose of lipopolysaccharide injection was significantly inhibited by pre-treatment of the animals with anti-TNF-alpha siRNAs)

[a]   (e) = endogenous, (t) = transgene.

ments into the genome of mice might turn out to be very useful as lentiviruses have been demonstrated to efficiently transduce fertilized eggs at different preimplantation stages (Lois et al., 2002; Tiscornia et al., 2003).

In conclusion, as expression and delivery tools necessary for genetic RNAi targeting have now been developed, it is possible to effectively implement RNAi-induced degradation in mice for studying gene function.

Knock-down versus knock-out

With continuously expanding sequence databases complemented by the ever growing data provided by in silico genomic and proteomic profiling techniques, it is important to further develop rapid, efficient and cost-saving methods for analysing gene function in vivo. For this reason the genetically modified mouse has gained increasing interest as a tool for facilitating data on gene function. Until recently, researchers were confined to the use of naturally occurring or experimentally engineered knock-out mouse models. The generation of sequence-specific mutant mice did rely on either large scale mouse mutagenesis programs or on homologous recombination in ES cells. Both methods are laborious, time-consuming and very costly. Following the advent of RNAi technology, it was possible to target repression of individual genes in transgenic knock-down mice without having to rely on large scale mutagenesis programs or homologous recombination in ES cells. An overview of recently published RNAi mouse models illustrating the versatility and flexibility of RNAi for studying gene function is shown in Table 1. It is also to be expected that very soon high throughput functional genomic studies will greatly benefit from RNAi thus circumventing the costly and time-consuming gene mapping analysis and nucleotide sequencing efforts required to identify the involved target genes in traditional mouse or ES cell mutant libraries.

An additional important issue to consider before embarking on a knock-out or knock-down mouse project is the genetic background. Commonly used germline transmitting ES cell lines are all derived from 129 mouse backgrounds (usually 129Sv, 129/SVEv or 129/Ola), a strain known to have very bad breeding qualities. To overcome this problem, knock-out mice are usually crossed back to strains with better reproductive performance. As a consequence, the defined genetic background is completely lost in the offspring therefore making it in many cases difficult to properly interpret the observed effects. Indeed, careful comparison of the consequences brought about by identical genetic changes in different mouse strains revealed the huge and often forgotten importance of genetic background effects for the function or loss-of-function of a particular gene (Bowers et al., 1999; Phillips et al., 1999; Sanford et al., 2001). This matter is further complicated by the fact that in general 2–3 years of backcross to a defined strain are required before the resulting genetic backgrounds are statistically >99 % homogeneous (Sigmund, 2000). By contrast, RNAi knock-down targeting is not restricted to specific cell types, strains or species thus allowing us to use clearly defined genetic backgrounds. Most importantly, RNAi targeting has been successfully applied for gene silencing in rats (Hasuwa et al., 2002). This finding represents a major breakthrough because before RNAi it was not possible to generate knock-out rats due to the lack of germline-transmitting ES cells. It is to be expected that the use of RNAi knock-down technology will be extended to other laboratory animals especially those for which ES cells are not available. A schematic overview comparing the generation of transgenic RNAi knock-down mice and knock-out mice is shown in Fig. 2.

Epi-allelic series: RNAi as a tool for generating graded loss-of-function

Initially, RNAi knock-down efficiencies were assumed to not entirely extinguish gene function in mammalian cells suggesting that RNAi-mediated gene inactivation might not be suitable to produce true loss-of-function phenotypes (Shi, 2003). However, reports from different laboratories showed that RNAi-mediated gene repression could lead to undetectable expression levels of the targeted gene product (Hasuwa et al., 2002; Kunath et al., 2003). Moreover, a recent report demonstrated that knock-down embryos generated by genetic RNAi targeting of the GTPase-activating protein produced an identical phenotype as the previously reported null mutant (Kunath et al., 2003). This result clearly established that RNAi-mediated gene silencing can be sufficient for generating loss-of-function phenotypes in mice.

So far all reports published on RNAi knock-down mice showed a marked phenotypic inter-individual variation ranging from partial to complete knock-down of the targeted gene product. The observation that RNAi knock-down efficiencies vary has several explanations. First, as the insertion of the RNAi construct occurs at random, it is to be expected that positional integration effects will directly determine the absolute levels of siRNA expression, an effect known as positional variegation (Wilson et al., 1990). Indeed, dissimilar knock-down efficiencies were observed in different founder lines which had been produced by pronucleus injection using the same RNAi construct (Hasuwa et al., 2002 and Brixel et al., unpublished data). A representative epi-allelic series illustrating different knock-down efficiencies in RNAi transgenic mice generated with an identical Trpm5 RNAi construct from our laboratory is shown in Fig. 3A. In addition, semi-quantitative RT-PCR analysis of Trpm5 knock-down efficiencies in different tissues of the same founder revealed different levels of target gene suppression (Fig. 3B, compare repression levels between tongue, liver and testis). The view that positional variegation effects, namely the integration of the microinjected RNAi expression construct into transcriptionally active or more quiescent chromosomal locations, will directly determine the knock-down efficiency, was further supported by ES cell derived knock-down mice or mice generated by infection of blastocysts with shRNA-expressing lentiviruses. In these animals marked inter-individual knock-down silencing efficiencies were observed although the same RNAi construct or recombinant lentivirus was used (Carmell et al., 2003; Kunath et al., 2003; Tiscornia et al., 2003). These results indicate that graded knock-down efficiencies are to be expected if the knock-down animals are generated by stochastic integration of the RNAi expression vector. Moreover, it is also very likely that the number of stably inte-

Fig. 2. Generation of knock-out versus transgenic knock-down mice: (**A**) Generation of knock-out mice using homologous recombination in embryonic stem (ES) cells. (1) Generation of the knock-out targeting construct (a): The wild-type genomic locus containing five coding exons (I–V) has to recombine with the targeting construct, thus introducing a marker for positive selection ((+) selection cassette) and negative selection ((–) selection cassette). Recombination of the knock-out construct (b) within the genomic locus of the targeted gene deletes exon II and III and is promoted by two flanking homology arms containing exact matching genomic DNA sequences (5′ and 3′ homology arms). The negative selection cassette offers the possibility to counter select against clones which did not recombine within the desired genomic locus. The recombined knock-out locus lacking exon II and III is depicted below (c). (2) Ex vivo manipulation of ES cells: The targeting construct is introduced into an appropriate ES cell line. Subsequently correctly recombined clones have to be isolated, characterized and amplified.

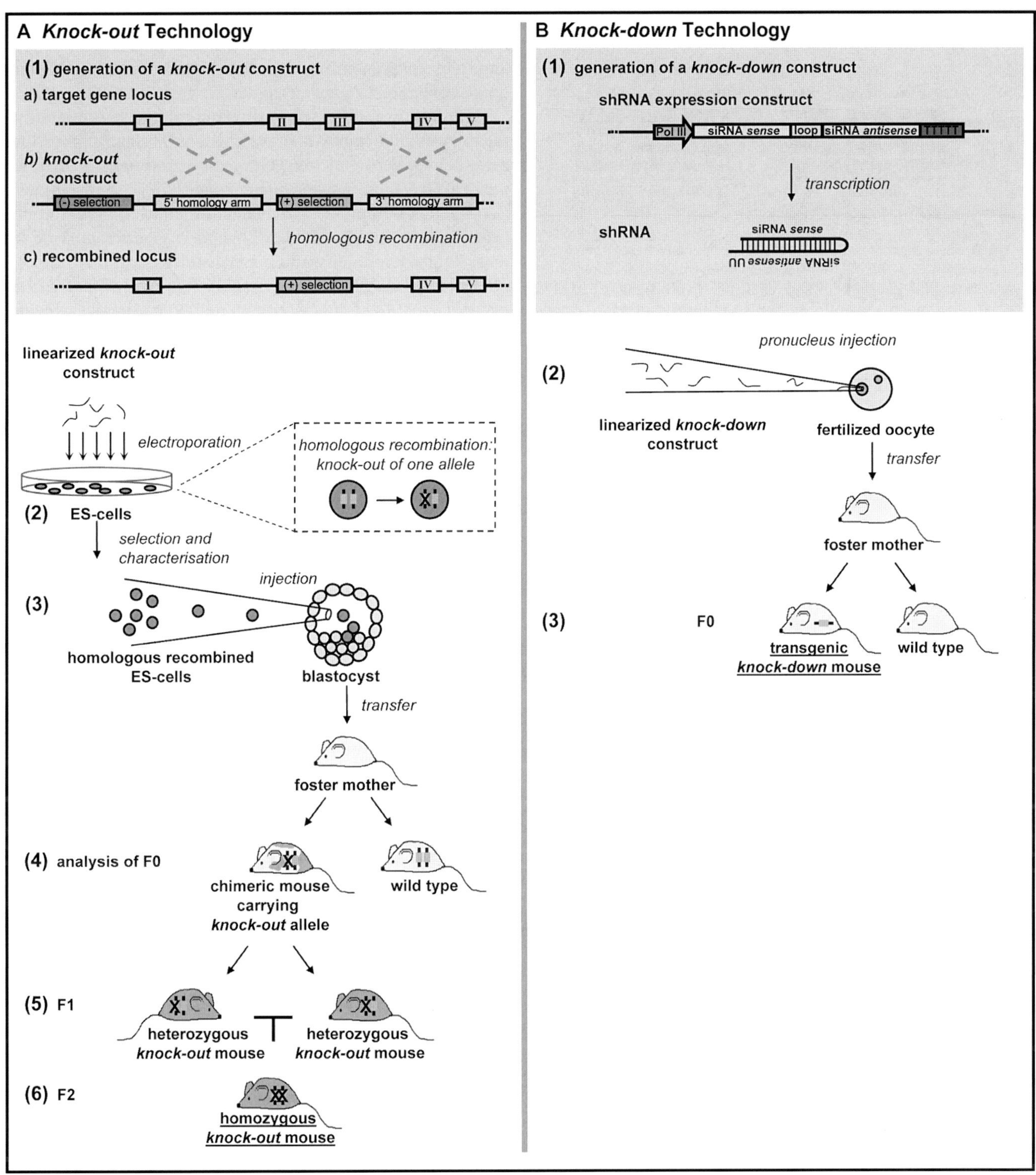

(3) Injection and transfer: Homologous recombined ES cells are injected into blastocysts and transferred into a pseudopregnant recipient mother. (4) Chimera: As the injected ES cells contribute to all tissues, chimeric mice can be generated. (5) Germline transmission: The previously introduced recombination is transmitted to the germline resulting in hemizygous knock-out mice which can be intercrossed. (6) Homozygous knock-out mouse: Knock-out null mutant needed for establishing the colony. (**B**) Generation of knock-down transgenic mice by injection of an shRNA expression construct into the pronucleus of fertilized one-cell mouse embryos. (1) Generation of the transgenic shRNA expression construct: A synthetic duplex oligonucleotide, containing sense and antisense DNA sequences homologous to the target mRNA is cloned into an appropriate shRNA expression vector. Transcription by RNA polymerase I or III will generate shRNA molecules needed for degradation of the target mRNA (below). (2) Pronucleus injection: Using microinjection the shRNA expression vector is introduced into the pronucleus of fertilized oocytes which are subsequently transferred into a foster mother. (3) Knock-down founder: Transgenic founders are identified and bred to wild-type mice to establish a knock-down colony. Pol III, RNA polymerase III promoter; F0, founder, F1, first daughter generation; F2 second daughter generation.

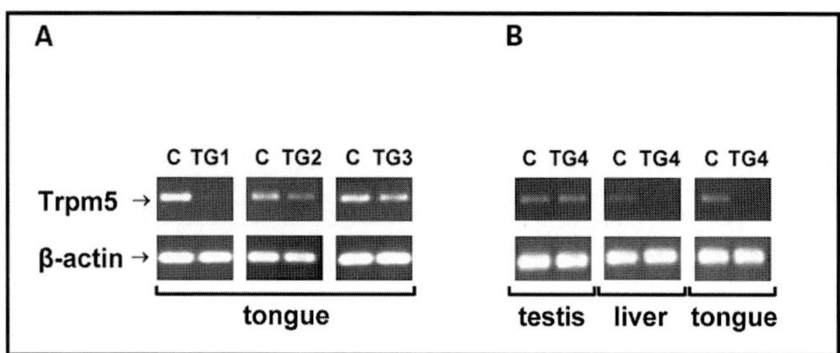

Fig. 3. ShRNA suppression in transgenic Trpm5 knock-down mice varied between tissues and different founder lines. (**A**) Epi-allelic series for Trpm5: Three independent transgenic mouse lines (TG1-TG3) generated by microinjection with the same Trpm5 shRNA expression construct showed marked knock-down differences in the same target tissue. RT-PCR analysis of mRNA extracted from the tongue revealed almost complete repression of the Trpm5 mRNA in the transgenic TG1 mouse line, mild reduction in TG2 and only marginal silencing in line TG3. Also compare the unchanged RT-PCR levels for β-actin from the same RT-PCR reactions. (**B**) Expression of the shRNA transgene resulted in variable knock-down levels in different organs. RT-PCR analysis of mRNA from liver and tongue showed significant knock-down efficiencies of the endogenous Trpm5 gene in these tissues. By contrast, when testis derived mRNA levels of the same founder were analysed, no reduction of Trpm5 message was observed. C, control non-transgenic mouse from the same litter; TG1–4, individual transgenic mouse lines expressing Trpm5 shRNA.

in stem cells and reconstituted organs. Most importantly, these experiments established that RNAi-targeting provides an excellent tool for generating epi-allelic knock-down series with graded phenotypes ranging from mild to severe.

The ability of RNAi to induce different levels of gene silencing in individual knock-down mice has important implications for studying gene function. Graded phenotypes produced by virtue of RNAi knock-down technology will be extremely useful for studying gene dosage-related questions. In addition, the generation of mild knock-down phenotypes might provide the opportunity to obtain viable animals for studying partial loss-of-gene function thus overcoming the frequently reported embryonic lethality of knock-out null mutants. Moreover, the use of RNAi allows us to specifically target single exons thus providing an experimental strategy for studying gene function of different splice forms.

The important up-front decision to use knock-down epi-allelic series or complete knock-down/knock-in strategies for studying gene function will obviously depend on the scientific questions to be asked. However, a major advantage of RNAi-mediated transgenesis is that the generation of RNAi transgenic founders is more flexible and less cost- and labour-intensive than the production of knock-out mouse models, which has to involve the manipulation of murine ES cells (see also Fig. 2).

grated RNAi expression constructs into the genome might affect the degree of the resulting knock-down (Brixel et al., unpublished data).

A second possibility to produce graded knock-down phenotypes in mice is to use shRNA expression vectors directed against different sequences within the same target gene. This strategy was convincingly applied for studying the role of the tumour suppressor gene p53 in an Eμ-myc leukaemia mouse model. Using haematopoietic stem cells from Eμ-Myc transgenic mice that had been engineered to express different p53 shRNAs, the groups of Hannon and Lowe reconstituted lethally irradiated mice and compared the induced blood cell pathology in the resulting recipients (Hemann et al., 2003). Prior to generating the different mouse recipients, the ability of each individual RNAi construct for suppressing p53 was tested in tissue culture cells. These experiments showed that depending on the construct low, medium, or high levels of p53 knock-down efficiencies were induced. In all subsequently generated p53 knock-down mouse lines the severity and type of disease correlated exactly with the ability of each individual RNAi knock-down construct to down-regulate p53 expression as measured previously in tissue culture. The following patho-phenotypic analysis revealed that the RNAi construct, triggering an extensive knock-down resulted in aggressive lymphomas capable of invasively infiltrating lung, liver and spleen tissues. Intermediate p53 knock-down levels induced by the second construct caused less rapidly developing tumours without infiltration and considerable levels of apoptosis. Finally, transgenic mice derived from the low knock-down construct did not develop lymphomas faster than Eμ-Myc control mice. This pioneering study showed that RNAi (as an alternative to a conventional gene knock-out) can be used to stably suppress gene expression

Caveats and limitations: To be or not to be on target?

This is a critical question for any RNAi-mediated experiment as siRNA targeting completely relies on high selectivity and has to be exclusively limited to the target gene. Cross-hybridisation of the antisense siRNA strand with additional transcripts containing full or partial identity with the targeted mRNA will ultimately lead to off-target repression. In this situation, the observed phenotype will represent a mixture of multiple gene silencing events rather than produce a clear picture of a gene specific knock-down. To comprehensively evaluate if and to what extend RNAi-mediated gene silencing is specific for the target mRNA, several independent studies have applied microarray-assisted gene profiling techniques (Chi et al., 2003; Jackson et al., 2003; Semizarov et al., 2003; van de Wetering et al., 2003). In these studies synthetic siRNAs targeting a specific mRNA as well as different mRNAs were transfected into mammalian cells followed by a detailed microarray analysis of the siRNA-induced expression profiles. The results of these experiments were controversial. In one analysis, the RNAi-induced transcript profile changes revealed considerable off-target gene regulation which seemed to be siRNA specific rather than target specific. For this reason the authors infered that cross-hybridisation of single stranded siRNA molecules to transcripts of similar sequence were responsible for the observed off-target effect (Jackson et al., 2003). By contrast, two additional very similar conducted studies did not see any off-target effects on their microarrays leading to the conclusion that RNAi-induced knock-down signatures were unique and highly specific for each targeted gene (Chi et al., 2003; Semizarov et al., 2003). In a similar study, shRNA was stably expressed in cell lines and RNAi-induced transcriptional changes monitored by microarray expression profiling. The results of these experiments demonstrated a high specificity for the β-catenin/TCF target mRNA, therefore suggesting an exquisite selectivity and high specificity of the knock-down effect (van de Wetering et al., 2003). At present it is difficult to draw general conclusions about the specificity of RNAi silencing as each study targeted different genes and the experimental set-up varied. However, stable expression of shRNAs is probably less harsh than transient transformation of synthetic siRNAs. For this reason it is tempting to speculate that stable siRNA/shRNA expression might have induced less non-specific effects as transient transfection.

General non-specific effects induced through RNAi-mediated gene silencing might also include the possibility of inducing the interferon system. Interferons are acting at the front line of the cell's defense machinery to counteract viral attacks (Stark et al., 1998; Grandvaux et al., 2002). Normally, the induction of the interferon defense pathway is set off by long dsRNAs. Although siRNAs were generally assumed to be too short for triggering an interferon response in mammalian cells (Caplen et al., 2001; Elbashir et al., 2001), two recent reports demonstrated that expression of shRNAs in mammalian cells did induce target genes of the interferon pathway (Bridge et al., 2003; Sledz et al., 2003). These two recent reports are a note of caution to RNAi technology in general and might be of importance for interpreting RNAi effects both in tissue culture as well as in mouse experiments.

The future: inducible and tissue-restricted knock-down mouse models

All published transgenic RNAi knock-down mouse models have so far made use of constitutive RNAi expression elements (see Table 1 for a comprehensive list). However, for many future applications it will be crucial to develop inducible as well as tissue-restricted RNAi knock-down mouse models. Why is that so important? In a worst case scenario constitutive RNAi expression in transgenic knock-down mice might provoke an embryonic lethal phenotype, precluding any further functional analysis during adulthood. Furthermore, pleiotropic effects compensating the constitutive knock-down phenotype might also obscure or prevent a clear-cut analysis. Finally, it might be important to exclusively silence gene expression at a specific developmental time point or in a particular tissue. This is compounded by the fact that many genes have a wide expression pattern and constitutive knock-down silencing might thus induce a highly complicated accumulative phenotype involving multiple tissues. Consequently, for many experimental applications it will be of importance to develop knock-down mouse models allowing for inducible and tissue-restricted gene silencing.

Several systems for exogenous inducible and also tissue-specific RNAi-mediated gene silencing are to be envisaged some of which have already been used in mammalian tissue culture cells or transgenic plants. The first strategy makes use of tetracycline-inducible shRNA expression systems (Miyagishi and Taira, 2002a; Chen et al., 2003; Matsukura et al., 2003; Sarkar

and Das, 2003; van de Wetering et al., 2003; Wiznerowicz and Trono, 2003). In these conditional knock-down RNAi vectors the tetracycline operon sequence (tetO) was integrated into H1 or U6 RNA polymerase III promoter sequences. In the absence of tetracycline (or the commercially available low-cost alternative doxycycline, DOX) the tetracycline repressor TetR (Hillen and Berens, 1994) is bound to the tetO sequence resulting in complete extinction of siRNA expression. By contrast, addition of DOX to this system induces a conformational change in the TetR repressor, preventing DNA binding and leading to expression of RNAi. As tetracycline-mediated gene-regulatory systems have been extremely useful for generating conditional transgenic and knock-out mice variants (for a recent review see Bockamp et al., 2002), it is also to be expected that the "tet on/off" system will be very suitable for generating conditional RNAi-mediated knock-down mice. A second and probably not less valuable method for generating conditional RNAi knock-down mice exploits site-specific recombination systems (Lakso et al., 1992; Orban et al., 1992; Rodriguez et al., 2000). Based on the enzymatic activity of the Cre (for "causes recombination") recombinase, it was possible to induce expression of shRNA in mammalian cells and also in plants (Guo et al., 2003; Kasim et al., 2003). These "Cre-on" siRNA-expression systems hold great promise to be also widely used for generating conditional knock-down mice. However, "tet on/off" based knock-down systems have significant advantages over the one-way "Cre-on" conditional shRNA expression in that they allow for completely switchable and reversible RNAi expression. A further advantage of tetracycline-controlled shRNA expression strategies is that the level of induced knock-down in a clonal cell population was shown to depend on the concentration of the inducer DOX (Matsukura et al., 2003). Translated to transgenic knock-down mice, it may well be possible to generate dissimilar phenotypes using a single transgenic founder line by simply adding different concentrations of DOX to the drinking water. However, it remains to be experimentally proven if applying different amounts of DOX can be a fast and easy method for studying graded knock-down phenotypes in the mouse.

In conclusion, both site-specific recombinase-based as well as tetracycline-mediated RNAi strategies have been established for conditionally knocking-down gene expression in mammalian cells. It will be only a matter of time when these systems can also be effectively applied for studying gene function in the mouse.

Perspective

Gene-function studies require a carefully selected and representative set-up to ensure that the experimentally determined findings really reflect the role a gene plays in the organism. In this respect Aristotle already realized that "The whole is more than the sum of its parts" (Aristotle, Metaphysica, 1953). Keeping this in mind it seems very reasonable to establish adequate animal models for deciphering gene function. At present the preferred model organism for studying mammalian gene function is the laboratory mouse. Linking genomic sequencing data with mouse functional genomics holds the promise for efficiently translating nucleotide sequence into functional data. To satisfy the increasing need for in vivo models for studying functional genomics, it is necessary to simplify the generation of suitable murine models. The advent of RNAi-mediated knock-down technology as a tool for gene silencing in mice and rats will therefore substantially improve the experimental possibilities to answer post-genomic questions. However, there are still many caveats and unresolved issues concerning RNAi technology itself.

First of all, it should be of prime importance to resolve the controversial subject about possible RNAi side effects including the induction of off-target knock-down silencing and the triggering of the interferon defense system. As RNAi also seems to have a key role in epigenetic gene regulation (Grewal and Moazed, 2003; Matzke and Matzke, 2003), a clear dissection of the underlying principles and mechanism involved is crucial. This will eventually lead to the establishment of general rules ensuring the selectivity and specificity of RNAi targeting. A second and no less important issue will be the development of inducible and tissue-specific knock-down mice. For example, several tetracycline-based as well as recombinase-mediated RNAi vectors have recently been used for exogenously inducing gene silencing in mammalian cells. It needs to be seen if these or alternative systems will also be amenable for conditionally expressing shRNAs in mice. Finally, it is also to be postulated that micro RNAs (miRNAs) will very soon emerge as possible additional tools for generating knock-down phenotypes in mice (Hutvagner and Zamore, 2002).

At present it is to be expected that functional post-genomic research will greatly benefit from RNAi technology in mice even if there are still some limitations to be overcome and inconsistencies to be resolved which may represent a rate-limiting step. However, unlike with the completion of the human and murine sequencing program, RNAi technology is only at the start of the beginning.

Acknowledgements

We thank members of the Bockamp and Zabel laboratories, including Cecilia Antunes, Stephan Fees, Stuart Frazer, Louise Griffin, Dorothe Hameyer, Ekkehart Lausch, Swetlana Ohngemach, Dirk Reutzel, Steffen Schmitt and Tatjana Trost for helpful comments in the preparation of this review. We apologize if any significant contribution to this field was unintentionally omitted from this review.

References

Aristotle: Metaphysica. Revised text with introduction and commentary translated by WD Ross, Aristotle's Metaphysica (10f-1045a) (Oxford, 1953).

Bockamp E, Maringer M, Spangenberg C, Fees S, Fraser S, Eshkind L, Oesch F, Zabel B: Of mice and models: improved animal models for biomedical research. Physiol Genomics 11:115–132 (2002).

Bowers BJ, Owen EH, Collins AC, Abeliovich A, Tonegawa S, Wehner JM: Decreased ethanol sensitivity and tolerance development in gamma-protein kinase C null mutant mice is dependent on genetic background. Alcohol Clin Exp Res 23:387–397 (1999).

Bridge AJ, Pebernard S, Ducraux A, Nicoulaz AL, Iggo R: Induction of an interferon response by RNAi vectors in mammalian cells. Nature Genet 34:263–264 (2003).

Brummelkamp TR, Bernards R, Agami R: Stable suppression of tumorigenicity by virus-mediated RNA interference. Cancer Cell 2:243–247 (2002a).

Brummelkamp TR, Bernards R, Agami R: A system for stable expression of short interfering RNAs in mammalian cells. Science 296:550–553 (2002b).

Caplen NJ, Parrish S, Imani F, Fire A, Morgan RA: Specific inhibition of gene expression by small double-stranded RNAs in invertebrate and vertebrate systems. Proc Natl Acad Sci USA 98:9742–9747 (2001).

Carmell MA, Zhang L, Conklin DS, Hannon GJ, Rosenquist TA: Germline transmission of RNAi in mice. Nature Struct Biol 10:91–92 (2003).

Chen Y, Stamatoyannopoulos G, Song CZ: Down-regulation of CXCR4 by inducible small interfering RNA inhibits breast cancer cell invasion in vitro. Cancer Res 63:4801–484 (2003).

Chi JT, Chang HY, Wang NN, Chang DS, Dunphy N, Brown PO: Genomewide view of gene silencing by small interfering RNAs. Proc Natl Acad Sci USA 100:6343–6346 (2003).

Dalmay T, Hamilton A, Rudd S, Angell S, Baulcombe DC: An RNA-dependent RNA polymerase gene in *Arabidopsis* is required for posttranscriptional gene silencing mediated by a transgene but not by a virus. Cell 101:543–553 (2000).

Denli AM, Hannon GJ: RNAi: an ever-growing puzzle. Trends Biochem Sci 28:196–201 (2003).

de Wit T, Grosveld F, Drabek D: The tomato RNA-directed RNA polymerase has no effect on gene silencing by RNA interference in transgenic mice. Transgenic Res 11:305–310 (2002).

Dykxhoorn DM, Novina CD, Sharp PA: Killing the messenger: short RNAs that silence gene expression. Nature Rev Mol Cell Biol 4:457–467 (2003).

Elbashir SM, Harborth J, Lendeckel W, Yalcin A, Weber K, Tuschl T: Duplexes of 21-nucleotide RNAs mediate RNA interference in cultured mammalian cells. Nature 411:494–498 (2001).

Elbashir SM, Harborth J, Weber K, Tuschl T: Analysis of gene function in somatic mammalian cells using small interfering RNAs. Methods 26:199–213 (2002).

Fire A, Xu S, Montgomery MK, Kostas SA, Driver SE, Mello CC: Potent and specific genetic interference by double-stranded RNA in *Caenorhabditis elegans.* Nature 391:806–811 (1998).

Gordon JW, Scangos GA, Plotkin DJ, Barbosa JA, Ruddle FH: Genetic transformation of mouse embryos by microinjection of purified DNA. Proc Natl Acad Sci USA 77:7380–7384 (1980).

Grandvaux N, tenOever BR, Servant MJ, Hiscott J: The interferon antiviral response: from viral invasion to evasion. Curr Opin Infect Dis 15:259–267 (2002).

Grewal SI, Moazed D: Heterochromatin and epigenetic control of gene expression. Science 301:798–802 (2003).

Guo HS, Fei JF, Xie Q, Chua NH: A chemical-regulated inducible RNAi system in plants. Plant J 34:383–392 (2003).

Hammond SM, Bernstein E, Beach D, Hannon GJ: An RNA-directed nuclease mediates post-transcriptional gene silencing in *Drosophila* cells. Nature 404:293–296 (2000).

Hasuwa H, Kaseda K, Einarsdottir T, Okabe M: Small interfering RNA and gene silencing in transgenic mice and rats. FEBS Lett 532:227–230 (2002).

Hemann MT, Fridman JS, Zilfou JT, Hernando E, Paddison PJ, Cordon-Cardo C, Hannon GJ, Lowe SW: An epi-allelic series of p53 hypomorphs created by stable RNAi produces distinct tumor phenotypes in vivo. Nature Genet 33:396–400 (2003).

Hillen W, Berens C: Mechanisms underlying expression of Tn10 encoded tetracycline resistance. A Rev Microbiol 48:345–369 (1994).

Hutvagner G, Zamore PD: A microRNA in a multiple-turnover RNAi enzyme complex. Science 297:2056–2060 (2002).

Jackson AL, Bartz SR, Schelter J, Kobayashi SV, Burchard J, Mao M, Li B, Cavet G, Linsley PS: Expression profiling reveals off-target gene regulation by RNAi. Nature Biotechnol 21:635–637 (2003).

Kamath RS, Fraser AG, Dong Y, Poulin G, Durbin R, Gotta M, Kanapin A, Le Bot N, Moreno S, Sohrmann M, Welchman DP, Zipperlen P, Ahringer J: Systematic functional analysis of the *Caenorhabditis elegans* genome using RNAi. Nature 421:231–237 (2003).

Kasim V, Miyagishi M, Taira K: Control of siRNA expression utilizing Cre-loxP recombination system. Nucl Acids Res Suppl 255–256 (2003).

Kawasaki H, Taira K: Short hairpin type of dsRNAs that are controlled by tRNA(Val) promoter significantly induce RNAi-mediated gene silencing in the cytoplasm of human cells. Nucl Acids Res 31:700–707 (2003).

Kunath T, Gish G, Lickert H, Jones N, Pawson T, Rossant J: Transgenic RNA interference in ES cell-derived embryos recapitulates a genetic null phenotype. Nature Biotechnol 21:559–561 (2003).

Lakso M, Sauer B, Mosinger B Jr, Lee EJ, Manning RW, Yu SH, Mulder KL, Westphal H: Targeted oncogene activation by site-specific recombination in transgenic mice. Proc Natl Acad Sci USA 89:6232–6236 (1992).

Lee NS, Dohjima T, Bauer G, Li H, Li MJ, Ehsani A, Salvaterra P, Rossi J: Expression of small interfering RNAs targeted against HIV-1 rev transcripts in human cells. Nature Biotechnol 20:500–505 (2002).

Lewandoski M: Conditional control of gene expression in the mouse. Nature Rev Genet 2:743–755 (2001).

Lewis DL, Hagstrom JE, Loomis AG, Wolff JA, Herweijer H: Efficient delivery of siRNA for inhibition of gene expression in postnatal mice. Nature Genet 32:107–108 (2002).

Lipardi C, Wei Q, Paterson BM: RNAi as random degradative PCR: siRNA primers convert mRNA into dsRNAs that are degraded to generate new siRNAs. Cell 107:297–307 (2001).

Lois C, Hong EJ, Pease S, Brown EJ, Baltimore D: Germline transmission and tissue-specific expression of transgenes delivered by lentiviral vectors Science 295:868–872 (2002).

Martinez J, Patkaniowska A, Urlaub H, Luhrmann R, Tuschl T: Single-stranded antisense siRNAs guide target RNA cleavage in RNAi. Cell 110:563–574 (2002).

Matsukura S, Jones PA, Takai D: Establishment of conditional vectors for hairpin siRNA knockdowns. Nucl Acids Res 31:e77 (2003).

Matzke M, Matzke AJ: RNAi extends its reach. Science 301:1060–1061 (2003).

Matzke M, Matzke AJ, Kooter JM: RNA: guiding gene silencing. Science 293:1080–1083 (2001).

McCaffrey AP, Meuse L, Pham TT, Conklin DS, Hannon GJ, Kay MA: RNA interference in adult mice. Nature 418:38–39 (2002).

McCaffrey AP, Nakai H, Pandey K, Huang Z, Salazar FH, Xu H, Wieland SF, Marion PL, Kay MA: Inhibition of hepatitis B virus in mice by RNA interference. Nature Biotechnol 21:639–644 (2003).

McManus MT, Haines BB, Dillon CP, Whitehurst CE, van Parijs L, Chen J, Sharp PA: Small interfering RNA-mediated gene silencing in T lymphocytes. J Immunol 169:5754–5760 (2002).

Miyagishi M, Taira K: Development and application of siRNA expression vector. Nucl Acids Res Suppl: 113–114 (2002a).

Miyagishi M, Taira K: U6 promoter-driven siRNAs with four uridine 3′ overhangs efficiently suppress targeted gene expression in mammalian cells. Nature Biotechnol 20:497–500 (2002b).

Nishikura K: A short primer on RNAi: RNA-directed RNA polymerase acts as a key catalyst. Cell 107:415–418 (2001).

Nykanen A, Haley B, Zamore PD: ATP requirements and small interfering RNA structure in the RNA interference pathway. Cell 107:309–321 (2001).

Orban PC, Chui D, Marth JD: Tissue- and site-specific DNA recombination in transgenic mice. Proc Natl Acad Sci USA 89:6861–6865 (1992).

Paddison PJ, Caudy AA, Bernstein E, Hannon GJ, Conklin DS: Short hairpin RNAs (shRNAs) induce sequence-specific silencing in mammalian cells Genes Dev 16:948–958 (2002).

Paul CP, Good PD, Winer I, Engelke DR: Effective expression of small interfering RNA in human cells. Nature Biotechnol 20:505–508 (2002).

Phillips TJ, Hen R, Crabbe JC: Complications associated with genetic background effects in research using knockout mice. Psychopharmacology (Berl) 147:5–7 (1999).

Rodriguez CI, Buchholz F, Galloway J, Sequerra R, Kasper J, Ayala R, Stewart AF, Dymecki SM: High-efficiency deleter mice show that FLPe is an alternative to Cre-loxP. Nature Genet 25:139–140 (2000).

Rubinson DA, Dillon CP, Kwiatkowski AV, Sievers C, Yang L, Kopinja J, Rooney DL, Ihrig MM, McManus MT, Gertler FB, Scott ML, Van Parijs L: A lentivirus-based system to functionally silence genes in primary mammalian cells stem cells and transgenic mice by RNA interference. Nature Genet 33:401–406 (2003).

Sanford LP, Kallapur S, Ormsby I, Doetschman T: Influence of genetic background on knockout mouse phenotypes. Methods Mol Biol 158:217–225 (2001).

Sarkar SN, Das HK: Regulatory roles of presenilin-1 and nicastrin in neuronal differentiation during in vitro neurogenesis. J Neurochem 87:333–343 (2003).

Semizarov D, Frost L, Sarthy A, Kroeger P, Halbert DN, Fesik SW: Specificity of short interfering RNA determined through gene expression signatures. Proc Natl Acad Sci USA 100:6347–6352 (2003).

Shi Y: Mammalian RNAi for the masses. Trends Genet 19:9–12 (2003).

Sigmund CD: Viewpoint: are studies in genetically altered mice out of control? Arterioscler Thromb Vasc Biol 20:1425–1429 (2000).

Sijen T, Fleenor J, Simmer F, Thijssen KL, Parrish S, Timmons L, Plasterk RH, Fire A: On the role of RNA amplification in dsRNA-triggered gene silencing. Cell 107:465–476 (2001).

Sledz CA, Holko M, de Veer MJ, Silverman RH, Williams BR: Activation of the interferon system by short-interfering RNAs. Nature Cell Biol 5:834–839 (2003).

Song E, Lee SK, Wang J, Ince N, Ouyang N, Min J, Chen J, Shankar P, Lieberman J: RNA interference targeting Fas protects mice from fulminant hepatitis. Nature Med 9:347–351 (2003).

Sorensen DR, Leirdal M, Sioud M: Gene silencing by systemic delivery of synthetic siRNAs in adult mice. J Molec Biol 327:761–766 (2003).

Stark GR, Kerr IM, Williams BR, Silverman RH, Schreiber RD: How cells respond to interferons. A Rev Biochem 67:227–264 (1998).

Stein P, Svoboda P, Anger M, Schultz RM: RNAi: mammalian oocytes do it without RNA-dependent RNA polymerase. RNA 9:187–192 (2003a).

Stein P, Svoboda P, Schultz RM: Transgenic RNAi in mouse oocytes: a simple and fast approach to study gene function. Dev Biol 256:187–193 (2003b).

Sui G, Soohoo C, Affar el B, Gay F, Shi Y, Forrester WC: A DNA vector-based RNAi technology to suppress gene expression in mammalian cells. Proc Natl Acad Sci USA 99:5515–5520 (2002).

Svoboda P: Long dsRNA and silent genes strike back: RNAi in mouse oocytes and early embryos. Cytogenet Genome Res 105:422–434 (2004).

Tiscornia G, Singer O, Ikawa M, Verma IM: A general method for gene knockdown in mice by using lentiviral vectors expressing small interfering RNA. Proc Natl Acad Sci USA 100:1844–1848 (2003).

van de Wetering M, Oving I, Muncan V, Pon Fong MT, Brantjes H, van Leenen D, Holstege FC, Brummelkamp TR, Agami R, Clevers H: Specific inhibition of gene expression using a stably integrated inducible small-interfering-RNA vector. EMBO Rep 4:609–615 (2003).

van der Weyden L, Adams DJ, Bradley A: Tools for targeted manipulation of the mouse genome. Physiol Genomics 11:133–164 (2002).

Wianny F, Zernicka-Goetz M: Specific interference with gene function by double-stranded RNA in early mouse development. Nature Cell Biol 2:70–75 (2000).

Vance V, Vaucheret H: RNA silencing in plants – defense and counterdefense. Science 292:2277–2280 (2001).

Wilson C, Bellen HJ, Gehring WJ: Position effects on eukaryotic gene expression. A Rev Cell Biol 6:679–714 (1990).

Wiznerowicz M, Trono D: Conditional suppression of cellular genes: lentivirus vector-mediated drug-inducible RNA interference. J Virol 77:8957–8961 (2003).

Xia H, Mao Q, Paulson HL, Davidson BL: siRNA-mediated gene silencing in vitro and in vivo. Nature Biotechnol 20:1006–1010 (2002).

Yu JY, DeRuiter SL, Turner DL: RNA interference by expression of short-interfering RNAs and hairpin RNAs in mammalian cells. Proc Natl Acad Sci USA 99:6047–6052 (2002).

Zamore PD, Tuschl T, Sharp PA, Bartel DP: RNAi: double-stranded RNA directs the ATP-dependent cleavage of mRNA at 21 to 23 nucleotide intervals. Cell 101:25–33 (2000).

Cytogenet Genome Res 105:422–434 (2004)
DOI: 10.1159/000078215

Long dsRNA and silent genes strike back:RNAi in mouse oocytes and early embryos

P. Svoboda

Department of Biology, University of Pennsylvania, Philadelphia, PA (USA)

Abstract. RNA interference (RNAi) refers to the selective degradation of mRNA induced by double-stranded RNA (dsRNA), first discovered in *Caenorhabditis elegans*. Homology-dependent silencing phenomena related to RNAi have been observed in many species from all eukaryotic kingdoms. RNAi and related mechanisms share several conserved components. The hallmark of these phenomena is the presence of short dsRNA molecules (21–25 bp long), termed short interfering RNA (siRNA), which are generated from dsRNA by the activity of Dicer, a specific type III RNAse. These molecules serve as a template for the recognition and cleavage of the cognate mRNA. As it is beyond the scope of a single review to cover all aspects of RNAi, this review will focus on certain steps of the pathway relevant to mammals and on the use of long dsRNA to specifically silence genes in mammalian cells permissive to this technique, such as oocytes and early embryos.

Two different pathways that respond to dsRNA in mammals

Mammals exhibit two different pathways that respond to dsRNA (Fig. 1). The RNAi pathway is the more ancient, dating back to the origin of eukaryotic kingdoms in the Precambrium. Another mechanism that recognizes dsRNA and mounts an orchestrated response to it, the PKR/interferon pathway, evolved relatively recently and is specific to mammals. The principal difference between these two pathways is that RNAi mounts a sequence-specific response while the other pathway is sequence independent.

The PKR/interferon pathway

The PKR/interferon response to dsRNA in mammalian cells (Fig. 1) was discovered more than a quarter of a century ago. Hunter et al. (1975) described that exposing mammalian cells to dsRNA, regardless of its sequence, triggers a global repression of protein synthesis, and eventually leads to apoptosis. In most mammalian somatic cells, exposure to dsRNA activates protein kinase R (PKR), which catalyzes phosphorylation of translation initiation factor eIF2α, which in turn inhibits translation. PKR is also involved in regulating NF-κB, which plays a key role in interferon induction. Interferon and dsRNA also activate 2′,5′-oligoadenylate synthetase (2′,5′-OAS), which produces 2′,5′ oligoadenylates with 5′-terminal triphosphate residues that subsequently induce activation of RNAse L, which is responsible for general RNA degradation (reviewed in Barber, 2001). PKR and 2′,5′-OAS mutant mice demonstrate that these two components are essential for the apoptotic response to dsRNA (Der et al., 1997; Zhou et al., 1997).

This sequence-nonspecific response to dsRNA led to strong initial skepticism about RNAi in mammals. However, long dsRNA induced RNAi but not the PKR/interferon response in initial experiments in mammalian oocytes and embryos (Svo-

This work was supported by a grant from the National Institutes of Health (HD22681) and an EMBO Long Term Fellowship (ALTF 2003-199).

Received 6 October 2003; accepted 14 November 2003.

Request reprints from: Dr. Petr Svoboda, Friedrich Miescher Institute
    Maulbeerstrasse 66, CH-4058, Basel (Switzerland)
    telephone: +41 61 697-4128; fax: +41 61 697-3976; e-mail: svoboda@fmi.ch

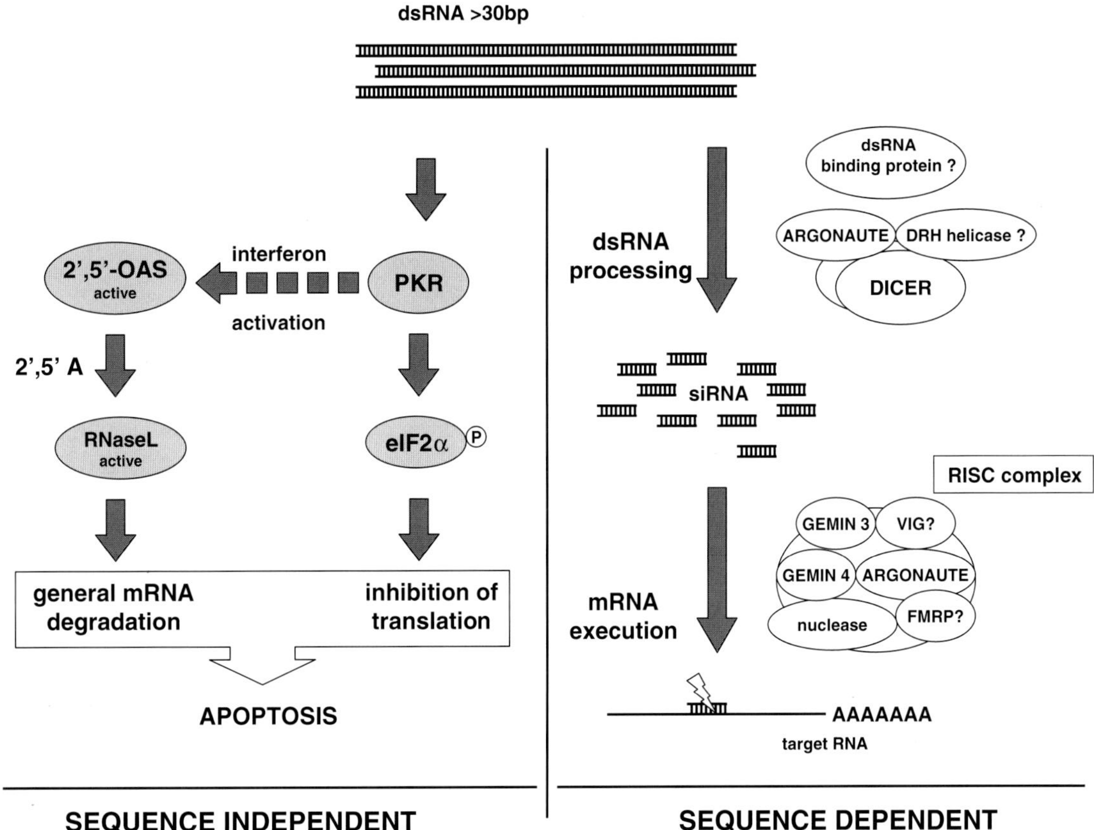

Fig 1. Pathways responding to dsRNA in mammalian cells. In the sequence-independent response (left) dsRNA activates protein kinase R (PKR), which catalyzes phosphorylation of translation initiation factor eIF2α, which in turn inhibits translation. PKR is also involved in interferon induction (through NF-κB). Interferon and dsRNA also activate 2′,5′-oligoadenylate synthetase (2′,5′-OAS) which produces 2′,5′ oligoadenylates (2′,5′ A) with 5′-terminal triphosphate residues. Oligoadenylates subsequently induce activation of RNAse L, which is responsible for general RNA degradation. Both PKR and 2′,5′-OAS are essential for the apoptotic response to dsRNA, as has been demonstrated in null mutant mice (Der et al., 1997; Zhou et al., 1997). The sequence-specific response (right), also known as RNAi, starts with the processing of dsRNA into small interfering RNA (siRNA) by the Dicer complex. siRNAs serve as guide sequences for RISC (RNA-induced silencing complex), which recognizes and cleaves the cognate mRNA.

boda et al., 2000; Wianny and Zernicka-Goetz, 2000; Billy et al., 2001; Yang et al., 2001). A better understanding of the RNAi mechanism also allowed the elimination of a nonspecific response to dsRNA in somatic cells because although siRNAs can induce RNAi, they are believed to be too short to trigger the PKR/2′,5′-OAS pathways (Zamore et al., 2000; Elbashir et al., 2001a).

The RNAi pathway

Interference with gene function in mammals was first shown in oocytes and early embryos in which dsRNA microinjection resulted in inhibition of expression of targeted genes (Wianny and Zernicka-Goetz, 2000). Subsequently, it was demonstrated that this effect is directed by sequence-specific mRNA degradation of cognate mRNA sequences (Svoboda et al., 2000). Functional homology to RNAi was later confirmed by detecting Dicer activity in mammalian cell lysates (Billy et al., 2001), and by the ability of siRNA to induce sequence specific mRNA degradation (Elbashir et al., 2001a) in mammalian cells.

Substantial insights into the mechanism of RNAi came from studies of RNAi-resistant *Caenorhabditis elegans* mutants (Ketting et al., 1999; Tabara et al., 1999, 2002) and from the development of an in vitro RNAi system in *Drosophila* (Tuschl et al., 1999; Zamore et al., 2000). To date, numerous genes have been associated with RNAi and related silencing mechanisms in different organisms (reviewed in Denli and Hannon, 2003). Not all genes associated with RNAi, however, are conserved in mammals. For instance, mammalian RNAi lacks an RNA-dependent RNA polymerase step, as well as a dsRNA uptake and transport mechanism, which greatly simplifies dsRNA delivery into cells in *C. elegans*.

In this section, three mechanistic features of RNAi relevant to the mammalian pathway (Fig. 1) will be discussed: (1) processing of dsRNA into siRNA; (2) recognition and cleavage of the cognate mRNA; and (3) absence of an RNA-dependent RNA polymerase.

Processing of dsRNA into siRNA by Dicer

RNAi is initiated by the processing of dsRNA into siRNA by Dicer, an RNAse III (Bernstein et al., 2001), first discovered as an ATP-dependent dsRNA processing activity in *Drosophila* (Zamore et al., 2000). A hypothetical model proposes that Dicer functions as an antiparallel dimer to produce ~ 21–23 nt siRNAs and that only two active sites are involved in the cleavage (Hannon, 2002). Interestingly, studies using cell lysates and recombinant protein suggest that mammalian Dicer preferentially cleaves dsRNA at the termini in an ATP-independent manner (Billy et al., 2001; Provost et al., 2002; Zhang et al., 2002).

Dicer is also implicated in producing microRNAs (miRNA, also known as small temporal RNA, stRNA), regulatory RNA molecules implicated in developmental timing in *C. elegans* through their ability to inhibit translation (Grishok et al., 2001; Hutvagner et al., 2001; Knight and Bass, 2001). Homologs of miRNAs were found among animals with bilateral symmetry, suggesting that regulation of development by miRNA is a conserved mechanism (reviewed in Rougvie, 2001).

Two other genes were identified in RNAi deficient (Rde) mutants *rde-1* and *rde-4* in *C. elegans* (Tabara et al., 1999; Parrish and Fire, 2001). RDE-1 is a member of the Argonaute protein family. Members of this family (reviewed in Carmell et al., 2002) are highly similar to the translation initiation factor eIF4C, and are involved in posttranscriptional silencing (Fagard et al., 2000) and in miRNA-mediated translational repression (Grishok et al., 2001). Recent work in *C. elegans* showed that RDE-4 is a dsRNA-binding protein that interacts with the trigger dsRNA and does not show any similarity to mammalian sequences (Tabara et al., 2002). RDE-4 also interacts in vivo with DCR-1 (Dicer), RDE-1, and a conserved DExH box helicase DRH-1. These findings suggest a model in which RDE-4 and RDE-1 cooperate to detect and retain foreign dsRNA and to present this dsRNA to DCR-1 for processing (Tabara et al., 2002).

Recognition and cleavage of the cognate mRNA

The siRNAs produced by Dicer serve as guide sequences for the RNA-induced silencing complex (RISC), a multicomponent nuclease that cleaves the target mRNA. In the final RNAi step, the RISC complex recognizes and cleaves the cognate mRNA one or more times at sites corresponding to the middle of the guiding nucleotide sequences (this nuclease activity is sometimes called Slicer), and the cleaved mRNA is ultimately degraded (Elbashir et al., 2001b). The RISC complex can efficiently target sense as well as antisense RNA strands, but not dsRNA homologous to the "triggering" sequence (Elbashir et al., 2001c). Target recognition is highly specific – mismatches in the center of the siRNA duplex prevent cognate RNA cleavage (Elbashir et al., 2001c). The exact mechanisms of recognition of the targeted mRNA and its cleavage are still unknown. Remarkably, although mRNA degradation mediated by siRNAs and translational repression by miRNA are different processes, miRNA can mediate mRNA degradation if it has perfect homology with the cognate mRNA (Hutvagner and Zamore, 2002).

Purified RISC is a large ribonucleoprotein complex biochemically separable from Dicer (Hammond et al., 2000). To date, six components of the RISC complex have been identified in *Drosophila* and mammals. The first identified constituent of the RISC complex was AGO2 in *Drosophila* (Hammond et al., 2001). AGO2 is a homolog of RDE-1 and is one of the five Argonaute proteins in *Drosophila*, four of which were implicated in RNAi or in potentially related silencing phenomena (reviewed in Carmell et al., 2002). Because some Argonaute mutations also exhibited phenotypes other than silencing defects, RNAi may be associated with other functions. However, such connections remain to be established.

Another component of the RISC complex is dFXR, the *Drosophila* homolog of the human fragile X mental retardation protein (FMRP) (Caudy et al., 2002; Ishizuka et al., 2002). DFXR is the single FMRP homolog in *Drosophila* and has similar biochemical properties to mammalian FMRP (Wan et al., 2000). The human FMRP is an RNA-binding protein that contains several structural motifs implicated in RNA interactions. FMRP associates with translating polyribosomes as part of a large ribonucleoprotein complex (Feng et al., 1997), suggesting a role in regulating translation. The exact role of dFXR in RNAi and the possible role of RNAi in the fragile X syndrome are not known. It has been hypothesized that rather than being a constitutive component of the transcript-cleaving RISC complex, dFMR1 might be an accessory factor that associates with RISC but uses another mechanism to regulate its targets (Carthew, 2002).

Another protein found in the *Drosophila* RISC complex is VIG, which is a conserved protein without identifiable protein domains except for an RGG box (an RNA binding motif). The precise role of VIG in the RISC complex is also unknown.

The most recent RISC component discovered in *Drosophila*, *C. elegans* and mammals is Tudor-SN (staphylococcal nuclease), the first nuclease associated with the RISC (Caudy et al., 2003). However, it remains unclear whether Tudor-SN is the Slicer because there are several inconsistencies between the biochemical properties of Tudor-SN and the activity of Slicer (Caudy et al., 2003).

In addition, biochemical analysis of the RISC-like complex in mammals identified an Argonaute protein (hAgo2) and two novel proteins, Gemin 3 and Gemin 4, which were not found in other model systems (Mourelatos et al., 2002; Hutvagner and Zamore, 2002). Gemin 3 is a DEAD box helicase associated with the survival of motor neurons (SMN) complex while Gemin 4 does not contain any known motifs. The exact role of both Gemin 3 and 4 also remains to be established.

The mammalian RNAi pathway does not utilize RNA dependent RNA polymerase

It is conceivable that during the time since the branching of mammalian ancestors from other phyla, mammalian RNAi may have evolved specific features. One such feature is the lack of an RNA dependent RNA polymerase (RdRp) component, which is present in the homologous pathways in *C. elegans*, *Arabidopsis* and *Neurospora* but not in *Drosophila* (Schwarz et al., 2002; Stein et al., 2003a).

The absence of an RdRp in mammalian RNAi has an important implication for its practical use. It has been suggested that RdRp serves as an amplifier of the RNAi response by extending the siRNA on the cognate mRNA template, thus generating additional dsRNA (and thus extending the cognate region upstream). The lack of the RdRp component in mammals suggests that only sequences complementary to the experimentally introduced dsRNA will be targeted, as there is no additional dsRNA produced from the cognate mRNA template. However, the lack of the RdRp cannot solely explain why RNAi in mammalian cells appears less efficient than in *C. elegans* or *Drosophila* (Svoboda et al., 2000; Ui-Tei et al., 2000) because *Drosophila* also very likely lacks the RdRp component (Schwarz et al., 2002; Zamore et al., 2000).

The role of RNAi in mammalian cells

The RNAi pathway is present and functional in most mammalian cell types, yet exposure of most of these cells to dsRNA triggers the PKR/interferon pathway resulting in a general translational arrest, nonspecific mRNA degradation, and, eventually, apoptosis, making any RNAi effect meaningless. Only a few cell types exhibit preference for the RNAi response to some form of long dsRNA (microinjected, transfected, expressed). These include oocytes and preimplantation embryos (Wianny and Zernicka-Goetz, 2000), undifferentiated ES cells (Yang et al., 2001), embryonic carcinoma cells (Billy et al., 2001), neuroblastoma cells (Gan et al., 2002), differentiated myotubes (Yi et al., 2003), and the NIH 3T3 and HEK 293 cell lines (Wang et al., 2003).

As mentioned above, the induction of the PKR/interferon pathway can be avoided for experimental purposes by using siRNA molecules, which are too short to trigger RNAi. But why would mammals maintain a functional RNAi pathway in the presence of the PKR/interferon pathway? There are several interesting observations related to this problem, which I believe are worth mentioning.

First, some components of the RNAi pathway are involved in other essential pathways (e.g. Dicer in production of miRNAs, Argonaute proteins in the initiation of translation, development and the miRNA pathway). But would this be sufficient selective pressure to maintain an intact RNAi pathway? RNAi is a multistep process, and as its components were involved in other pathways there would be selective pressure to maintain their roles in those pathways. But if RNAi itself is dispensable, it is unlikely that selective pressure would exist to maintain its machinery in toto (unless one of the other pathways, such as the miRNA pathway, requires the complete RNAi machinery). But mammals have intact machinery for siRNA generation, mRNA recognition, and its destruction in most cells. Interestingly, miRNA can induce mRNA degradation if it has perfect homology with the cognate mRNA (Hutvagner and Zamore, 2002). This raises an interesting question: is it possible that small hairpins could act as substrates for RNAi and that some of their cognate mRNAs could be degraded instead of blocked for translation? Recent work identified two maternally expressed miRNA genes complementary to a paternally expressed retrotransposon-like gene (Rtl1) (Seitz et al., 2003). Since these miRNAs show perfect homology to Rtl1, the authors raise the question of whether these miRNAs could regulate Rtl1 mRNA levels when coexpressed.

Second, the PKR pathway does not necessarily interfere with RNAi. There is a difference in PKR sensitivity to exogenous and endogenous (expressed in the nucleus) dsRNA. Experiments used to provide evidence that dsRNA induces a PKR-mediated apoptotic response are typically based on exposure of cells or lysates to exogenous dsRNA. Reported nonspecific effects of expressed dsRNA include non-specific mRNA reduction (Yang et al., 2001) and activation of interferon response genes by some (but not all) plasmids expressing short hairpin RNA (shRNA) (Bridge et al., 2003). However, the evidence for PKR activation (and apoptosis) by dsRNA expression is inconclusive as dsRNA has been detected in somatic cells in several instances (Kim and Wold, 1985; Kramerov et al., 1985; Schmitt et al., 1986; Okano et al., 1991). Interestingly, sense and antisense RNA coexpression in NIH 3T3 cells induced a specific gene knockdown (Wang et al., 2003) while it is known that exposure of NIH 3T3 cells to exogenous dsRNA leads to the PKR/interferon response and apoptosis (McMillan et al., 1995; Srivastava et al., 1998). Presumably in mammalian somatic cells, RNAi is not involved in responding to dsRNA entering the cell – this response is controlled by the PKR/interferon pathway. However, cells could express levels of dsRNA, which would not be sufficient to trigger the PKR/interferon pathway but would enter the RNAi pathway instead.

This leads to another unanswered question: does RNAi naturally regulate gene expression by recognizing dsRNA and eliminating homologous transcripts? As discussed above, this role would be independent of the PKR/interferon response, which typically recognizes exogenous dsRNA. It is known that the mammalian genome produces a significant number of overlapping transcripts and, in some cases, these transcripts appear to be functionally related (Shendure and Church, 2002; Yelin et al., 2003). However, there is no direct experimental evidence supporting this speculation. On the contrary, a rigorous attempt to detect siRNAs derived from two overlapping transcripts failed (Houbaviy and Sharp, 2002), and so far, the isolation of siRNAs/miRNAs from mammalian cells has not yielded good evidence either. Finally, while affecting such regulation of gene expression would likely generate a distinct phenotype, it is known that the absence of RNAi in *C. elegans* does not necessarily produce one (Tabara et al., 1999).

RNAi does not appear to be essential for the survival of individual organisms. RNAi mutants in *C. elegans* appear otherwise healthy (Tabara et al., 1999), and a germline RNAi-resistant population of *C. elegans* has been found in nature (Tijsterman et al., 2002). Nevertheless, the fact that the RNAi pathway is found in so many species suggests that it confers some benefits leading to its conservation throughout eukaryotic evolution.

RNAi (and related mechanisms) in other species is viewed as a protective mechanism against parasitic sequences such as viruses and mobile elements. RNAi is associated with the suppression of transposable elements in several species. For example, transposable elements are mobilized in some RNAi-resistant mutants of *C. elegans* (Ketting et al., 1999), and transposon "taming" in *Drosophila* is homology-dependent and mediated by RNA (Jensen et al., 1999). Furthermore, siRNAs derived from retrotransposons were found in *Trypanosoma* (Djikeng et al., 2001) and *Drosophila* (Caudy and Hannon, 2002). The nexus between retrotransposon silencing and dsRNA-induced RNAi, however, is incomplete. The absence of mobilization of transposable elements in some RNAi mutants such as *rde-1* and *rde-4* suggests that a part of the RNAi pathway is not involved in transposon silencing (Tabara et al., 1999). In other words, transposable elements in *C. elegans* may be recognized and directed to the RNAi pathway differently than dsRNA fed to or injected into animals and processed by RDE-4/RDE-1 (Tabara et al., 2002). Nevertheless, the presence of siRNAs from transposable elements in *Trypanosoma* and *Drosophila* implies that a conserved RNAi-related mechanism generating siRNAs operates on transposable elements in eukaryotes. But experimental evidence that RNAi would play a similar role in mammalian cells is lacking. While our findings from early embryos suggest a correlation between RNAi inhibition and increased mRNA levels of two retrotransposons (Svoboda et al., 2004), no other observations supporting the role of RNAi in repression of mobile elements in mammals have been made so far.

Although links between RNAi and other mechanisms controlling gene expression have been found in other species, they remain unclear in mammals. The RNAi pathway or its components were recently linked to (1) heterochromatin formation in fission yeast (Hall et al., 2002; Volpe et al., 2002; Schramke and Allshire, 2003), (2) transcriptional silencing and sequence-specific methylation in plants (Mette et al., 2000; Morel et al., 2000; Zilberman et al., 2003), (3) transcriptional silencing in *Drosophila* (Pal-Bhadra et al., 2002), and (4) repression mediated by several polycomb group (PcG) proteins in *C. elegans* (Dudley et al., 2002).

As mentioned above, one of the observed effects of dsRNA expression in plants is methylation of homologous DNA sequences (Mette et al., 2000; Sijen et al., 2001). But analysis of oocytes expressing dsRNA from a transgene for four weeks revealed no change in methylation of the cognate DNA sequence (Svoboda et al., unpublished). However, the existence of this mechanism in mammals cannot be ruled out based on these results. One reason is that oocytes do not replicate DNA, and the methylation mechanism may require DNA replication.

Experimental gene silencing with long dsRNA in mammalian cells

This section reviews the experimental use of long dsRNA (> 500 bp) in mammals. This approach is typically omitted in mammalian RNAi reviews because it is applicable to only a small number of cell types. The short dsRNA (siRNA or shRNA) methods, generally applicable to all cell types with functional RNAi pathway, have been extensively reviewed elsewhere (McManus and Sharp, 2002), including an article in this issue (Prawitt et al., 2004). Long dsRNA induces efficient

RNAi but does not trigger apoptosis in oocytes and early embryos (Svoboda et al., 2000; Wianny and Zernicka-Goetz, 2000). Compared to siRNA, long dsRNA is inexpensive to generate in large amounts. Because long dsRNA is processed into many different siRNAs, it assures more consistent and predictable results than an individual siRNA (nevertheless, efficient siRNA design is rapidly improving). In addition, expression of long dsRNA, which can also efficiently induce RNAi, can be targeted to a specific cell type by using a tissue-specific Pol II promoter (Stein et al., 2003). To date, such specific RNAi targeting in a single cell type in a living animal has not been achieved with a short RNA system. Therefore, long dsRNA is quite suited to studies in oocytes and early embryos and it can compete here with widespread short RNA approaches. Although oocytes and early embryos (including embryonic stem cells) may appear as study subjects infrequently, they are essential model systems for studies of fertilization, meiosis, totipotency and genome reprogramming.

RNAi has been used as an experimental tool ever since its discovery and it is currently a favorite knockdown approach. RNAi is often chosen over other knockdown approaches due to its successful use on numerous occasions and because of many modifications allowing an extremely wide range of applications. RNAi has several benefits and drawbacks when compared to morpholinos and antisense oligonucleotides, two alternative knockdown approaches (reviewed in Heasman, 2002 for morpholinos; Lavery and King, 2003, for antisense oligos). All three approaches inhibit gene expression posttranscriptionally. Antisense oligos and dsRNA induce mRNA degradation while morpholinos inhibit translation. Therefore in an ideal case, these methods would produce a comparable effect, which would mostly depend on the stability of the expressed protein. Morpholinos and long dsRNA were both used to inhibit the *Miss* gene in mouse oocytes. They produced a phenotype, although RNAi appeared to be a more robust approach in this case (Lefebvre et al., 2002). A slightly better RNAi effect compared to morpholinos has also been observed in early postimplantation mouse embryos (Mellitzer et al., 2002) but this higher efficiency of RNAi is not always the case, as observed for example in HeLa cells (Munshi et al., 2002).

Therefore, differences in the mode of action, efficiency of delivery, toxicity, and cost may lead to different choices in different situations. For example, antisense oligos may be chosen over siRNA to knock down genes in cells, which are difficult to transfect. The morpholino approach requires knowledge of the translation start, so it may be difficult to use it to target genes that are not well characterized. On the other hand, the inhibitory effect of morpholinos may be faster than that of antisense oligos and RNAi, which depends on recognition and degradation of mRNA. Toxicity and possible nonspecific effects of these three agents may be additional considerations in specific cases.

A major limitation of RNAi stems from the knockdown principle of the approach. Therefore, critical factors for a successful RNAi experiment are (1) the minimal protein level to achieve a phenotype; (2) the stability of a protein from a targeted gene; and (3) the amount of mRNA being translated before it is destroyed by RNAi. The limitations of RNAi (and

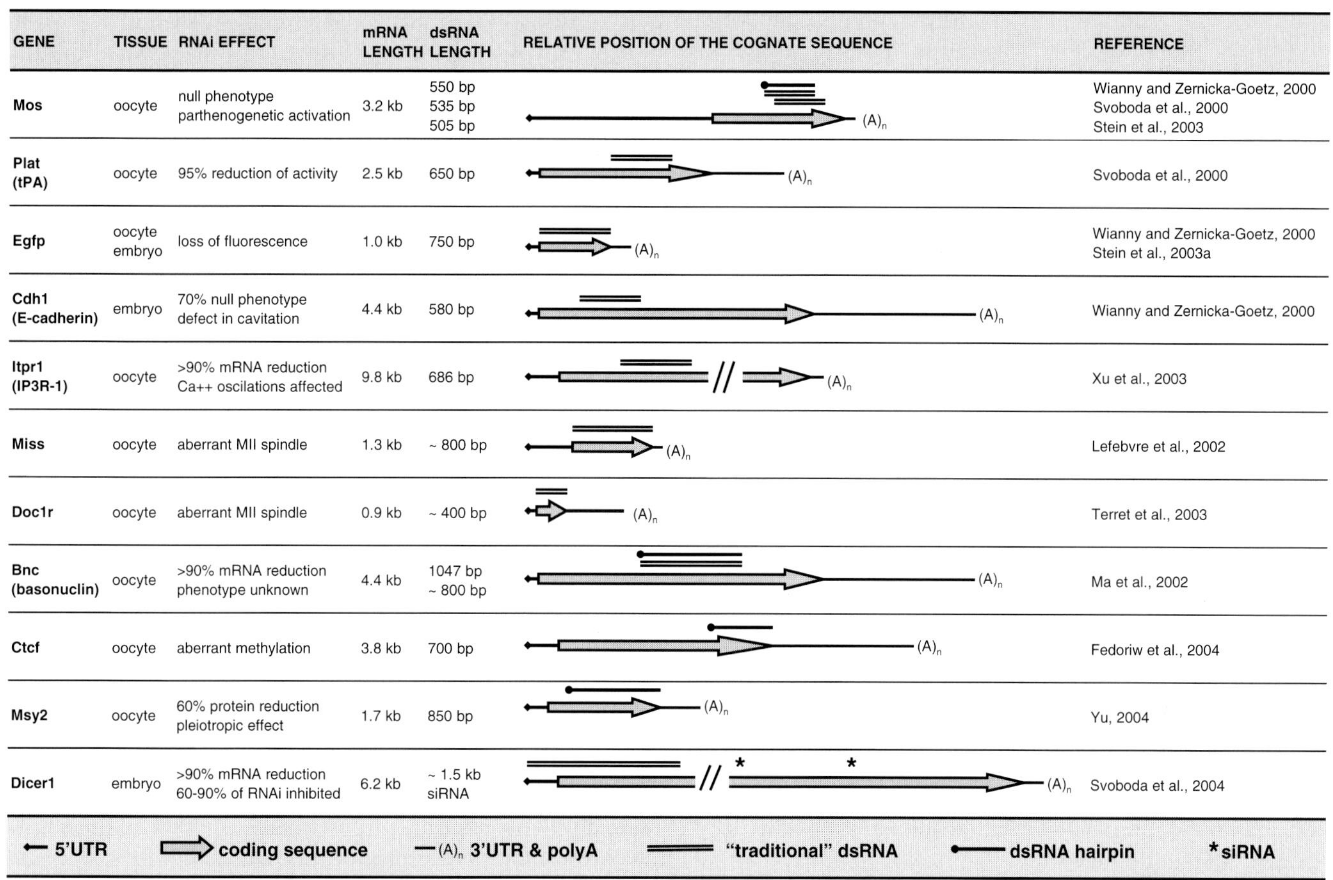

GENE	TISSUE	RNAi EFFECT	mRNA LENGTH	dsRNA LENGTH	RELATIVE POSITION OF THE COGNATE SEQUENCE	REFERENCE
Mos	oocyte	null phenotype parthenogenetic activation	3.2 kb	550 bp 535 bp 505 bp		Wianny and Zernicka-Goetz, 2000 Svoboda et al., 2000 Stein et al., 2003
Plat (tPA)	oocyte	95% reduction of activity	2.5 kb	650 bp		Svoboda et al., 2000
Egfp	oocyte embryo	loss of fluorescence	1.0 kb	750 bp		Wianny and Zernicka-Goetz, 2000 Stein et al., 2003a
Cdh1 (E-cadherin)	embryo	70% null phenotype defect in cavitation	4.4 kb	580 bp		Wianny and Zernicka-Goetz, 2000
Itpr1 (IP3R-1)	oocyte	>90% mRNA reduction Ca++ oscilations affected	9.8 kb	686 bp		Xu et al., 2003
Miss	oocyte	aberrant MII spindle	1.3 kb	~ 800 bp		Lefebvre et al., 2002
Doc1r	oocyte	aberrant MII spindle	0.9 kb	~ 400 bp		Terret et al., 2003
Bnc (basonuclin)	oocyte	>90% mRNA reduction phenotype unknown	4.4 kb	1047 bp ~ 800 bp		Ma et al., 2002
Ctcf	oocyte	aberrant methylation	3.8 kb	700 bp		Fedoriw et al., 2004
Msy2	oocyte	60% protein reduction pleiotropic effect	1.7 kb	850 bp		Yu, 2004
Dicer1	embryo	>90% mRNA reduction 60-90% of RNAi inhibited	6.2 kb	~ 1.5 kb siRNA		Svoboda et al., 2004

Fig. 2. Position and length of cognate sequences in various transcripts targeted by long dsRNA. Oocytes and early embryos were microinjected with 5–10 pl of 0.2–2 µg/µl dsRNA. Different types of dsRNA molecules (siRNA, "traditional" dsRNA (annealed sense and antisense RNA), and dsRNA hairpins) used on different occasions are depicted. mRNA lengths and transcript schematics are based on data from GenBank.

knockdown approaches in general) may be more prominent in studies in which the experimental timeframe cannot be extended to achieve a more pronounced RNAi effect. This can be problematic, especially for studies in preimplantation embryos that develop in a limited time period. Furthermore, it has been observed that genes coding for more stable proteins (such as cytoplasmic actin) are less amenable to the application of RNAi than proteins with relatively short lives (~ 10 hours), which are efficiently silenced (Wei et al., 2000). Interestingly, Harborth and colleagues showed in cultured mammalian cells that targeting of cytoplasmic β and γ actin can lead to a dramatic (yet incomplete) elimination of protein and causes an observable phenotype (Harborth et al., 2001).

Mammalian genes efficiently targeted by RNAi code for proteins with many different functions, including signaling proteins, nuclear and cytoplasmic structural proteins, secreted enzymes, and ion channels. Based on the literature, over one hundred mammalian genes have been efficiently knocked down by RNAi, most of them in tissue culture (some of them are reviewed in Dykxhoorn et al., 2003; McManus and Sharp, 2002). However the number of mammalian genes efficiently knocked down by RNAi will be much higher, considering ongoing genome scale RNAi projects (Hannon, 2002) and routine use of RNAi in drug target validation (Lavery and King, 2003; Deveraux et al., 2003). The number of genes efficiently knocked down in oocytes and early embryos is considerably smaller. Eleven genes knocked down in oocytes or early embryos by RNAi (mostly with microinjection of long dsRNA) have been reported to date (summarized in Fig. 2). Typically, dsRNA used in these experiments was 500–1500 bp long and targeted sequence within the coding region.

Considering published data as well as unpublished results from several sources, it appears that efficient mRNA reduction can be achieved with almost every long dsRNA (in contrast to siRNA in tissue culture). I am aware of one unpublished example from mammalian oocytes in which a particular long dsRNA (600 bp in length) did not induce efficient mRNA degradation. A second dsRNA (also 600 bp in length), which was homologous to a sequence upstream of the first one, induced >80% mRNA reduction. Both cognate regions were located adjoining in the 3′ UTR of the mRNA.

Obtaining a phenotype, however, is a separate issue. For example, targeting of the maternal *Itpr1* gene in mouse oocytes resulted in a dramatic reduction of mRNA and a specific phenotype but the ITPR1 protein level was only mildly decreased (Xu et al., 2003). Similarly, *Msy2* targeting by transgenic RNAi

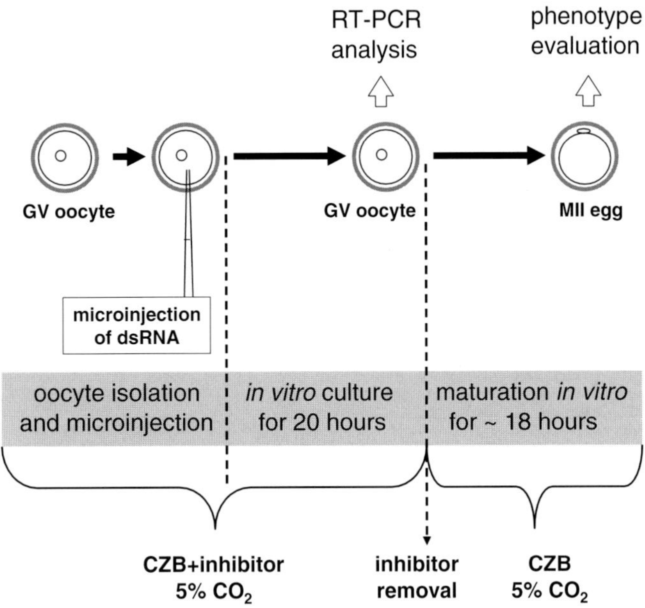

Fig. 3. Schematic of an RNAi experiment in mouse oocytes. Oocytes are collected, microinjected and cultured in a medium containing an inhibitor of meiotic resumption (e.g. isobutyl-1-methyl-1-xanthine or milrinone). Using such inhibitors allows for extension of the exposure of oocytes to dsRNA. Oocytes are typically cultured in an inhibitor-containing medium for 20 h. If needed, this culture can be extended up to 40 h. However, it should be noted that longer culture can affect the developing potential of cultured oocytes. All cultures are performed in CZB medium in an atmosphere of 5 % CO_2 in air.

resulted in up to 70 % reduction of the extremely abundant MSY2 protein and a pleiotropic phenotype including reduced fertility (Yu et al., 2004). Similar inhibition of some genes would not produce a phenotype. It is likely that *Msy2* and *Itpr1* inhibition could be improved by using, for example, transgenic RNAi to study *Itpr1* or by screening more transgenic RNAi lines with the *Msy2* construct. In any case, investigators attempting to use RNAi should be aware that this approach may require troubleshooting to achieve the desired goal and that the targeted gene may be "resistant" to an efficient protein knockdown in some experimental designs.

Finally, there is a growing concern about "off target" regulation and activation of the interferon system in mammalian cells (Bridge et al., 2003; Jackson et al., 2003; Sledz et al., 2003). Experiments documenting these phenomena were performed in somatic cells using short dsRNA – either siRNA transfection or shRNA expression. Unfortunately, similar information from mouse oocytes or early embryos is lacking because such experiments require significant amounts of material. Therefore, we must rely on indirect evidence suggesting that dsRNA in oocytes and early embryos does not cause significant non-specific effects. First, oocytes microinjected with dsRNA undergo normal meiotic maturation (which requires translation), and early embryos microinjected with dsRNA develop to the blastocyst stage and beyond (Wianny and Zernicka-Goetz, 2000, Svoboda et al., unpublished). Second, the phenotypes obtained with different dsRNAs in oocytes and early embryos are differ-

ent, and some of these phenotypes represent phenocopies of the null phenotype (Fig. 2). This suggests that the possible nonspecific response to dsRNA in oocytes and early embryos is not an issue that would hamper analysis of gene function using this approach. However, if there is any concern about the specificity of the observed phenotype, experimental controls should contain not only some non-homologous dsRNA but also some dsRNA targeting a different part of the cognate RNA as well.

dsRNA microinjection of oocytes and early embryos

In 2000, two groups independently published the first evidence that RNAi exists in mammals (Svoboda et al., 2000; Wianny and Zernicka-Goetz, 2000). It has been demonstrated that long dsRNA induces specific interference with gene function in oocytes and early embryos and that the effect lasts for the entire duration of preimplantation development in the mouse (Wianny and Zernicka-Goetz, 2000). However, unless pre-processed into siRNAs (Yang et al., 2002; Myers et al., 2003), long dsRNA is useless for RNAi experiments in most somatic cells due to the PKR/interferon response (discussed above). Thus, the use of long dsRNA became confined to experiments in oocytes and early embryos, which apparently do not exhibit the PKR/interferon response.

RNAi induced by dsRNA (long dsRNA or siRNA) microinjection is an excellent tool for studying the role of maternal transcripts recruited either during oocyte maturation or embryo development. These transcripts accumulate in the oocyte but are not translated; therefore, the stability of the coded protein does not affect the efficiency of RNAi. When needed, inhibition of oocyte maturation with compounds such as 3-isobutyl-l-methyl-xanthine (IBMX) can extend the exposure to dsRNA, so transcripts that would be recruited during oocyte maturation can be more efficiently degraded (Svoboda et al., 2000). A general experimental design for RNAi experiments in oocytes is shown in Fig. 3. Detailed protocols for this application can be found in Chapter 15 of the RNAi manual published recently (Stein and Svoboda, 2003). It should be noted that extending the exposure to dsRNA by using inhibitors of meiotic maturation is not a strict rule because several genes have been effectively targeted without it (Wianny and Zernicka-Goetz, 2000; Lefebvre et al., 2002; Terret et al., 2003).

The RNAi approach in early embryos is generally similar to that in oocytes except the exposure to dsRNA cannot be extended if needed. In addition, dsRNA delivery by microinjection is mostly restricted to 1-cell, eventually, 2-cell stages and the use of dsRNA expression is hampered by the lack of sufficient transcription between the GV-oocyte and early 2-cell stages. Thus the RNAi approach in early embryos may be less successful than in oocytes. Nevertheless, successful targeting of three genes in early embryos demonstrates that early embryos are accessible by RNAi (Fig. 2).

There are currently two alternatives to dsRNA microinjection: electroporation (Grabarek et al., 2002) and transgenic RNAi (described in a later section). Notably, experiments with dsRNA transfection, an approach routinely used for somatic cells, failed, mainly due to the toxicity of transfection reagents (Stein et al., unpublished). Electroporation is an attractive alternative to dsRNA microinjection because it allows the induction of the RNAi effect in a higher number of oocytes or early embryos and it also allows for the simultaneous delivery of dsRNA into individual blastomeres during preimplantation development (Grabarek et al., 2002). Importantly, Grabarek and coworkers developed protocols for electroporation of zona-enclosed embryos allowing studies involving embryo transfer of treated samples. Electroporation of pre-processed siRNA has been also successfully used in early postimplantation embryos (Mellitzer et al., 2002)

Finally, it should be noted that experiments in the oocyte showed decreased RNAi efficiency when two genes were targeted simultaneously, suggesting that the RNAi pathway can be readily saturated so that simultaneous targeting of several genes may be difficult to achieve at this stage (Stein and Svoboda, 2003).

Expression of dsRNA in mammalian cells

Expression of dsRNA to specifically inhibit gene function is another modification of the RNAi approach. Although it has been successfully used on many occasions, it should be noted that RNAi induced by dsRNA expressed in transgenic metazoan organisms (discussed below) seems to produce weaker phenotypes than RNAi induced by microinjected dsRNA (Martinek and Young, 2000; Piccin et al., 2001; Tavernarakis et al., 2000).

I will only briefly introduce short hairpin RNA-expressing systems, which are reviewed elsewhere (McManus and Sharp, 2002; Dykxhoorn et al., 2003), including this issue (Prawitt et al., 2004). These systems were designed to avoid possible problems with the nonspecific PKR-mediated response to long dsRNA in somatic cells. Interestingly, the finding by Bridge and colleagues that some of the shRNA plasmids indeed induce the interferon response (Bridge et al., 2003) suggests caution when these plasmids are employed. Despite their successful use in tissue culture, shRNA systems have two basic disadvantages. First, there is an unpredictable variability in knockdown efficiency of different individual siRNAs or shRNAs. Second, the Pol III-based expression systems that are used to express shRNA cannot be used in a tissue-specific manner because there are no Pol III tissue-specific promoters available. This is a significant disadvantage because, for example, genes with a lethal phenotype cannot be studied in oocytes or early embryos using this approach. A possible tissue-specific shRNA approach could be developed from a method using miRNA precursors included in Pol II transcripts (Zeng et al., 2002) or by using a loxP system producing a functional Pol III promoter in a tissue-specific manner, but such approaches are not yet available.

Long dsRNA can theoretically be generated by using one of three different methods: (1) an inverted repeat (IR) can be transcribed from a single promoter; (2) one DNA fragment can be transcribed in both directions using two opposing promoters; or (3) a bidirectional promoter module can be used. It should be kept in mind that expression of an IR may in some cases fail to induce RNAi because regulation of nuclear expression of such constructs is likely very complex. Antisense expression may result in activation of cryptic splice or polyadenylation sites. Although such cases were not well documented, some observa-tions suggest that expression of dsRNA from an inverted repeat may be altered and the RNAi effect may be lost. For example, experiments with the Mos IR demonstrated that the RNAi effect was sensitive to the position of an intron in the expressed sequence (Svoboda et al., 2001).

Several reports describe the successful use of long dsRNA expression in different cell types (Svoboda et al., 2001; Yang et al., 2001; Paddison et al., 2002; Shinagawa and Ishii, 2003; Wang et al., 2003; Yi et al., 2003), and a large number of reports describing various systems expressing short hairpins or siRNAs is available. Both long and short expressed hairpins were also successfully used to obtain an RNAi effect in transgenic mice (Yang et al., 2001; Hasuwa et al., 2002; Carmell et al., 2003; Kunath et al., 2003; Rubinson et al., 2003; Shinagawa and Ishii, 2003; Stein et al., 2003b; Wang et al., 2003;).

Numerous different targeting constructs have been reported for the expression of long dsRNA in mammals (Fig. 4). More systematic comparisons of the various types of plasmids have been done in plants (Smith et al., 2000; Wesley et al., 2001). In general, although an IR is sometimes difficult to clone, it better assures formation of dsRNA. It has been suggested that includ-ing a spacer between repeats facilitates cloning (Piccin et al., 2001), and in mammals, constructs with spacers as long as 700 bp have succesfully been used (Yi et al., 2003). Reports using plants and *Drosophila* suggest that the targeting efficiency of expressed dsRNA can be increased by employing constructs in which an inverted repeat is interrupted by an intron. The intron acts as a spacer to increase cloning efficiency, but because it is spliced following transcription, the product is a loopless hairpin dsRNA (Smith et al., 2000; Wesley et al., 2001; Kalidas and Smith, 2002).

An alternative to IR plasmids is a dual promoter system. However, the use of dual promoter plasmids may not be the best choice for mammals even though the cloning of the target sequence is simple. Dual promoter plasmids express sense and antisense strands separately; therefore, their efficiency depends on the annealing of single-stranded RNAs in vivo. Considering the observed lower efficiency of RNAi in mammals as com-pared to *Drosophila* and *C. elegans* and the lower efficiency of expressed dsRNA in metazoan organisms, a dual promoter sys-tem may be a risky choice. On the other hand, a dual promoter system (using a bidirectional module, see also Fig. 4) worked well in NIH 3T3 and HEK 293 cells (Wang et al., 2003), and there are no studies that make a good comparison of the efficiency of dual promoter versus IR systems in mammals.

Transgenic RNAi

The development of transgenic RNAi logically followed the development of dsRNA expression systems. Historically, the first transgenic RNAi was done in *C. elegans* (Tavernarakis et al., 2000). Successful transgenic experiments employing shRNA or long hairpin were subsequently done in mammals (Hasuwa et al., 2002; Carmell et al., 2003; Kunath et al., 2003; Rubinson et al., 2003; Shinagawa and Ishii, 2003; Stein et al., 2003b). To date, there are two published transgenic experiments demonstrating efficient use of long dsRNA expressed by

SCHEMATIC DESIGN   PLASMID   TRANSGENE   REFERENCE

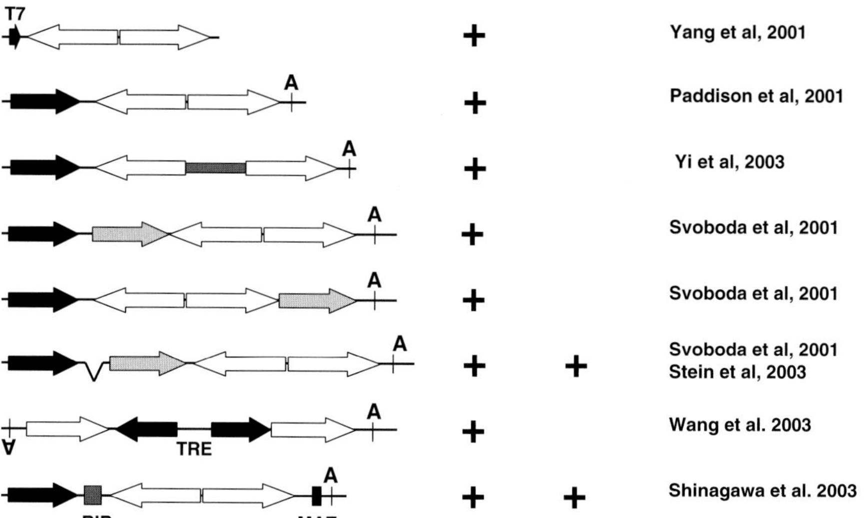

Fig. 4. Different constructs used to express long dsRNA in mammalian cells. This figure summarizes the design of constructs used to express long dsRNA in mammalian cells. White arrow, cognate sequence; gray arrow, EGFP coding sequence; long black arrow, eukaryotic promoter; short black arrow denoted T7, T7 polymerase promoter; gray rectangle, a DNA spacer; A, polyadenylation site; gray square denoted RIB, cis-acting ribozyme; black rectangle denoted MAZ, MAZ binding site (see text for details); TRE, tetracyclin responsive element used for induction of the dual promoter system.

a Pol II (Shinagawa and Ishii, 2003; Stein et al., 2003b; Fedoriw et al., 2004; Yu et al., 2004).

The first three reports describe efficient RNAi in transgenic mice expressing capped and polyadenylated long RNA hairpin in oocytes. The other report by Shigawa and Ishii (2003) represents a clever modification of the long dsRNA approach for somatic cells. The authors were concerned about the possibility that expressed dsRNA would be transported to the cytoplasm and trigger the PKR/interferon response. To block proper mRNA processing (thus blocking the export into the cytoplasm) they designed an intron-less transgene with a cis-acting hammerhead ribozyme inserted upstream of the IR and a MAZ sequence (binding site of the zinc finger protein MAZ) located downstream of the IR. The ribozyme cleaves off the 5′ CAP structure and the MAZ sequence mediates Pol II transcriptional pausing. Using the transcriptional co-repressor Ski as a target, they demonstrated that a CMV promoter-driven construct produced a phenotype similar to the one observed in Ski deficient embryos. Notably, this paper not only provides a secure approach for using Pol II-driven expression of long dsRNA in mammalian cells, but also raises questions about localization of dsRNA processing.

A transgenic construct targeting genes in oocytes (the development of this design is described in detail in Svoboda et al., 2001) contains a shortened version of the ZP3 promoter followed by a short first non-coding exon, an intron, and a second exon containing the EGFP coding sequence (CDS), the inverted repeat, and a poly A signal. Insertion of the EGFP CDS upstream of the IR serves two purposes. First, it functions as a spacer between the intron and the IR because our initial experiments showed that close positioning of an intron interfered with efficiency of the RNAi targeting (Svoboda et al., 2001). Second, it provides a "handle" for genotyping and detection of transgene expression. It should be noted that EGFP fluorescence in incompetent oocytes (either microinjected with this plasmid or from transgenic animals) is very low (for some constructs barely detectable); thus it may be difficult to use it for genotyping (Svoboda et al., unpublished).

To simplify the generation of transgenic constructs, we have developed a transgenic cassette for inserting an inverted repeat (Fig. 5). To generate a transgenic RNAi construct, one first chooses the target sequence. We recommend using 0.5–1.0 kb of cDNA sequence free of repeats and strong homologies to other expressed sequences to minimize nonspecific targeting. This suggestion for the minimal and maximal length is based on five different effective transgenic constructs. The length can vary but it was noted that Dicer exhibits lower affinity to shorter dsRNA substrates (Bernstein et al., 2001) and inverted repeats that are too long may not be stable in living organism. Once the targeting sequence is chosen, one of several strategies for cloning an IR can be used. The cloning process can be facilitated by using inverted repeats with spacers (20–200 bp), which are easier to clone and maintain in bacterial cells. Three of the successful general strategies are described in Fig. 6. After the IR is cloned, it is inserted into the *Xba*I restriction site downstream of the EGFP CDS in the transgenic cassette containing the rest of the construct (Fig. 5).

Transgenic RNAi in mouse oocytes (and in other tissues) has several attractive features:

Simplicity: Generating the construct for transgenic RNAi is simple. At least 0.5 kb of cDNA sequence is needed as starting material. First, an inverted repeat is produced from this sequence (Fig. 6). After an inverted repeat is cloned, it is inserted into the transgenic cassette (Fig. 5). Transgenic founder animals are produced by pronuclear microinjection of one-cell embryos with a purified transgene, followed by culture and subsequent embryo transfer. The F1 generation can be immediately analyzed with no need for selection of homozygotes. The RNAi effect occurs only in tissues where the dsRNA is expressed. This is extremely useful for studies of oocytes

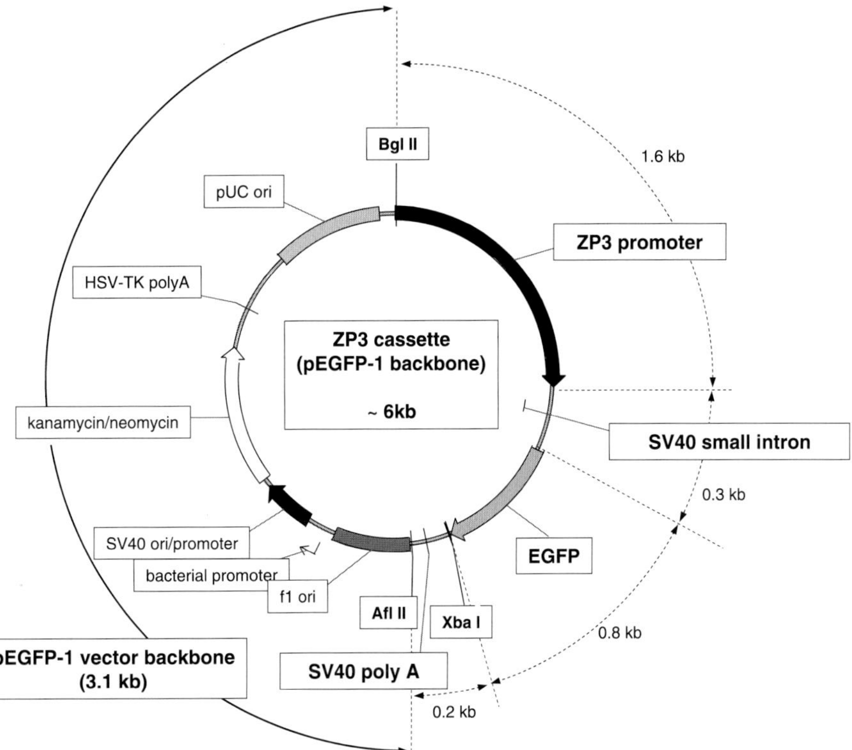

Fig. 5. Plasmid cassette for transgenic RNAi in mouse oocytes. This plasmid is derived from the pEGFP-1 plasmid (Clontech). The ZP3 promoter is inserted in the *Bgl*II and *Hind*III sites and the SV40 small intron (obtained by PCR in a 300-bp fragment from the pGL2 plasmid (Promega)) is inserted in the *Sma*I site in pEGFP-1's multiple cloning site. Shown are restriction sites for cloning into the vector *(Xba*I) and sites for removal of the expression cassette from the vector (*Bgl*II and *Afl*II). Note that the *Xba*I site is blocked by overlapping dam/dcm methylation in common *E. coli* strains. This plasmid produces bright fluorescence in incompetent oocytes 24 h after microinjection (Svoboda et al., 2001).

because specific expression of dsRNA in oocytes allows the study of a gene in living animals even if it is essential for other cells and its classic knockout would not be viable.

Speed: Targeting a gene by transgenic RNAi is limited mostly by the mouse reproductive period. In an ideal case, it is possible to obtain a phenotype within half a year from the decision to analyze a particular gene. In the "best case" scenario, a transgene can be prepared within 2 weeks (3–4 if the transgene is tested by microinjection in incompetent oocytes before it is submitted to the transgenic facility); 2–3 months are needed for producing, genotyping, analyzing and mating founder animals. If enough founders are obtained, the F1 generation can be produced and analyzed for a phenotype within another 2–3 months. Especially attractive is the case in which a transgenic male is repeatedly mated to different females, thus rapidly obtaining enough F1 animals for analysis.

When producing the first transgenic RNAi mice, it took about 9 months to test different transgene designs and produce the final transgenic construct (Svoboda et al., 2001). Transgenic mice were obtained from quarantine 3 months after the transgene was submitted to the facility. Genotyping, breeding and F1 phenotype analysis took altogether another 4 months.

Efficiency: It was discussed above that the RNAi approach is a knockdown approach. Although a null phenotype can be clearly achieved with transgenic RNAi, it should be pointed out that the null phenotype is not achieved in every transgenic line and most likely not for every gene studied. To date, three transgenic RNAi experiments demonstrated that some transgenic lines exhibit a stronger phenotype than others (Stein et al., 2003b; Fedoriw et al., 2004; Yu et al., 2004). A phenotype was observed in all three experiments; however, as discussed above,

the MSY2 protein level in one of these studies (Yu et al., 2004) decreased only by 70% (the best transgenic line established so far) but MSY2 is extremely abundant in oocytes (2–4% of the total protein). The other two experiments showed a much stronger RNAi effect. Inhibition of *Mos* resulted in ~90% reduction of MAP kinase activity (*Mapk* is downstream of *Mos* in the MAP kinase pathway) (Stein et al., 2003b) and the CTCF study yielded >95% reduction of the protein level (estimated by Western blot, Fedoriw et al., 2004). Therefore, the transgenic RNAi approach appears to be a very efficient knockdown method to study genes in developing oocytes.

Throughput: Transgenic RNAi can yield a lot of material for subsequent analysis. Hundreds of transgenic oocytes can be obtained from females by increasing the number of matings with transgenic males. When compared to alternatives such as microinjection (antisense RNA, dsRNA, morpholino, deficient protein or blocking antibody) or tissue-specific lox-P knockouts, the advantage of transgenic RNAi is clear. Also, generating several different transgenic lines with different constructs (e.g. for studies of pathways) can be achieved in a time not much longer than it takes to produce a single transgenic line.

Phenotype variability: Transgenic RNAi may produce different phenotypes because dsRNA expression and RNAi effect may differ in different transgenic lines. This may, in fact, be an advantage in many cases. Different levels of gene repression may aid in the understanding of that gene's function. In an ideal case, the strongest RNAi phenotype would resemble the null phenotype, while other phenotypes would provide insights into threshold levels for gene function(s). Such a situation was observed for the *Mos* transgenic RNAi mice, where one transgenic line exhibited low MAP kinase activity and parthenogen-

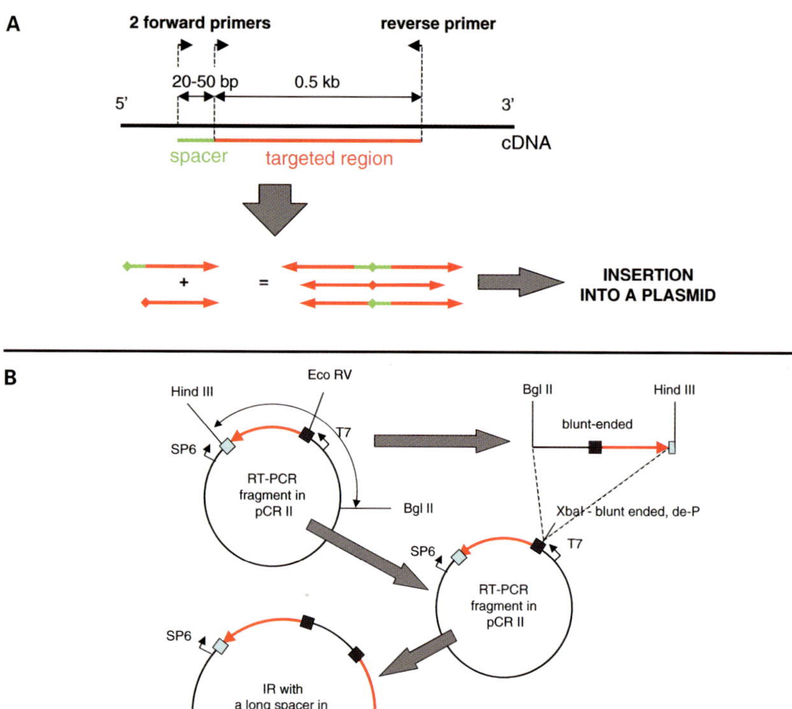

Fig. 6. Different strategies to clone an inverted repeat. (**A**) Cloning by ligation of two PCR products of two different lengths. Primers are designed such that one of the PCR products is slightly longer at the 5′ end with respect to the cDNA orientation. The inverted repeat is made by ligation of the two PCR products. Ligation is facilitated by using restriction sites added at the 5′ end of the forward primers. After ligating the two PCR products of different sizes at their 5′ ends, three possible outcomes are produced: short-short, long-long (both perfect inverted repeats), and long-short. The last product contains a spacer, which is defined by the excess of the sequence from the longer PCR product. The inverted repeat is then digested with a restriction enzyme, which cuts at a site engineered into the reverse primer, and the IR is inserted into a plasmid. (**B**) Inverted repeat assembly with a long spacer "in and out". The main advantage of this strategy is that it eliminates most of the empty, single-arm and head to tail background clones, which appear when an in vitro ligated inverted repeat is inserted into a plasmid. An inverted repeat with a huge spacer (1–2 kb) is generated, and the spacer is removed to obtain a perfect inverted repeat. Alternatively, one can replace the long spacer with a set of small spacers to identify the smallest spacer size, which is well-tolerated by bacteria. The scheme shows an RT-PCR product cloned into the pCR II plasmid as starting material and the creation of an inverted repeat with a long spacer in a single ligation step. The long spacer can then be removed (or replaced) using restriction sites from the "gray" portion of MCS. Alternatively, by cutting close to the 5′ end of the insert, the whole IR could be removed using enzymes cutting the "left" part of MCS or close to the 3′ end of the insert. (**C**) Inverted repeat assembly around a spacer in two steps. An inverted repeat is assembled in a plasmid carrying a spacer. In this scheme, the spacer (e.g. a short EGFP PCR fragment in or an intron) is cloned first. In the next step, the first arm of the IR is blunt-end cloned into the EcoRV site, and clones with the appropriate orientation are selected. In the second step, the second arm is cloned into suitable restriction sites on the other side of the spacer. All three strategies produced in screens on average 10–30 % desired clones. Symbols: Red arrow, RT-PCR poduct (arrow indicates its orientation as the orientation of 5′-3′ cDNA); light blue box, "left" part of pCR II MCS (closer to the SP6 promoter); black box, "right" part of MCS (closer to the T7 promoter); green line, spacer.

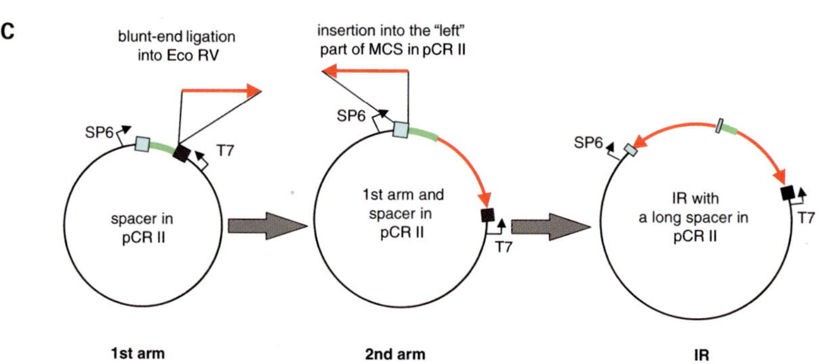

etic activation, while another showed low MAP kinase activity without parthenogentic activation. The difference in MAPK activities in metaphase II (M II) eggs between these two lines was approximately 10 % (Stein et al., 2003b). The CTCF transgenic experiment also produced an array of different phenotypes depending on the level of RNAi effect (Fedoriw et al., 2004). In addition, shRNA transgenic RNAi produced similar results as well (Hemann et al., 2003).

Concluding remarks

Five years ago, RNAi was "that mysterious phenomenon" in *C. elegans*. Today, RNAi is a widespread approach employed in almost every experimental model system. The dramatic expansion of this field speaks for itself. Our understanding of RNAi is rapidly improving and yet, we still know little about underlying molecular mechanisms or about biological functions of RNAi and related pathways. However, the incomplete understanding of RNAi is not necessarily a bad thing. It means that the exciting RNAi story is still far from over.

Acknowledgements

The author thanks Dr. Wesley Rose, Dr. Paula Stein and Dr. Richard Schultz for help with preparation of the manuscript.

References

Barber GN: Host defense, viruses and apoptosis. Cell Death Differ 8:113–126 (2001).

Bernstein E, Caudy AA, Hammond SM, Hannon GJ: Role for a bidentate ribonuclease in the initiation step of RNA interference. Nature 409:363–366 (2001).

Billy E, Brondani V, Zhang H, Muller U, Filipowicz W: Specific interference with gene expression induced by long, double-stranded RNA in mouse embryonal teratocarcinoma cell lines. Proc Natl Acad Sci USA 98:14428–14433 (2001).

Bridge AJ, Pebernard S, Ducraux A, Nicoulaz AL, Iggo R: Induction of an interferon response by RNAi vectors in mammalian cells. Nat Genet 34:263–264 (2003).

Carmell MA, Xuan Z, Zhang MQ, Hannon GJ: The Argonaute family: tentacles that reach into RNAi, developmental control, stem cell maintenance, and tumorigenesis. Genes Dev 16:2733–2742 (2002).

Carmell MA, Zhang L, Conklin DS, Hannon GJ, Rosenquist TA: Germline transmission of RNAi in mice. Nat Struct Biol 10:91–92 (2003).

Carthew RW: RNA interference: the fragile X syndrome connection. Curr Biol 12:R852–854 (2002).

Caudy AA, Hannon GJ: Endogenous Targets of RNA Interference in Cultured Cells, in: Keystone Symposia 2002, RNA Interference, Cosuppression and Related Phenomena (Taos, New Mexico, 2002).

Caudy AA, Myers M, Hannon GJ, Hammond SM: Fragile X-related protein and VIG associate with the RNA interference machinery. Genes Dev 16:2491–2496 (2002).

Caudy AA, Ketting RF, Hammond SM, Denli AM, Bathoorn AMP, Tops BBJ, Silva JM, Myers MM, Hannon GJ, Plasterk RHA: A microccocal nuclease homologue in RNAi effector complexes. Nature 425:411–414 (2003).

Denli AM, Hannon GJ: RNAi: an ever-growing puzzle. Trends Biochem Sci 28:196–201 (2003).

Der SD, Yang YL, Weissmann C, Williams BR: A double-stranded RNA-activated protein kinase-dependent pathway mediating stress-induced apoptosis. Proc Natl Acad Sci USA 94:3279–3283 (1997).

Deveraux QL, Aza-Blanc P, Wagner KW, Bauerschlag D, Cooke MP, Hampton GM: Exposing oncogenic dependencies for cancer drug target discovery and validation using RNAi. Semin Cancer Biol 4:293–300 (2003).

Djikeng A, Shi H, Tschudi C, Ullu E: RNA interference in *Trypanosoma brucei:* cloning of small interfering RNAs provides evidence for retroposon-derived 24–26-nucleotide RNAs. RNA 7:1522–1530 (2001).

Dudley NR, Labbe JC, Goldstein B: Using RNA interference to identify genes required for RNA interference. Proc Natl Acad Sci USA 99:4191–4196 (2002).

Dykxhoorn DM, Novina CD, Sharp PA: Killing the messenger: short RNAs that silence gene expression. Nat Rev Mol Cell Biol 4:457–467 (2003).

Elbashir SM, Harborth J, Lendeckel W, Yalcin A, Weber K, Tuschl T: Duplexes of 21-nucleotide RNAs mediate RNA interference in cultured mammalian cells. Nature 411:494–498 (2001a).

Elbashir SM, Lendeckel W, Tuschl T: RNA interference is mediated by 21- and 22-nucleotide RNAs. Genes Dev 15:188–200 (2001b).

Elbashir SM, Martinez J, Patkaniowska A, Lendeckel W, Tuschl T: Functional anatomy of siRNAs for mediating efficient RNAi in *Drosophila melanogaster* embryo lysate. EMBO J 20:6877–6888 (2001c).

Fagard M, Boutet S, Morel JB, Bellini C, Vaucheret H: AGO1, QDE-2, and RDE-1 are related proteins required for post-transcriptional gene silencing in plants, quelling in fungi, and RNA interference in animals. Proc Natl Acad Sci USA 97:11650–11654 (2000).

Fedoriw AM, Stein P, Svoboda P, Schultz RM, Bartolomei MS: Transgenic RNAi reveals essential function for CTCF in *H19* gene imprinting. Science 303:238–240 (2004)

Feng Y, Absher D, Eberhart DE, Brown V, Malter HE, Warren ST: FMRP associates with polyribosomes as an mRNP, and the I304N mutation of severe fragile X syndrome abolishes this association. Mol Cell 1:109–118 (1997).

Fire A, Xu S, Montgomery MK, Kostas SA, Driver SE, Mello CC: Potent and specific genetic interference by double-stranded RNA in *Caenorhabditis elegans.* Nature 391:806–811 (1998).

Gan L, Anton KE, Masterson BA, Vincent VA, Ye S, Gonzalez-Zulueta M: Specific interference with gene expression and gene function mediated by long dsRNA in neural cells. J Neurosci Methods 121:151–157 (2002).

Grabarek JB, Plusa B, Glover DM, Zernicka-Goetz M: Efficient delivery of dsRNA into zona-enclosed mouse oocytes and preimplantation embryos by electroporation. Genesis 32:269–276 (2002).

Grishok A, Pasquinelli AE, Conte D, Li N, Parrish S, Ha I, Baillie DL, Fire A, Ruvkun G, Mello CC: Genes and mechanisms related to RNA interference regulate expression of the small temporal RNAs that control *C. elegans* developmental timing. Cell 106:23–34 (2001).

Hall IM, Shankaranarayana GD, Noma K, Ayoub N, Cohen A, Grewal SI: Establishment and maintenance of a heterochromatin domain. Science 297:2232–2237 (2002).

Hammond SM, Bernstein E, Beach D, Hannon GJ: An RNA-directed nuclease mediates post-transcriptional gene silencing in *Drosophila* cells. Nature 404:293–296 (2000).

Hammond SM, Boettcher S, Caudy AA, Kobayashi R, Hannon GJ: Argonaute2, a link between genetic and biochemical analyses of RNAi. Science 293:1146–1150 (2001).

Hannon GJ: RNA interference. Nature 418:244–251 (2002).

Harborth J, Elbashir SM, Bechert K, Tuschl T, Weber K: Identification of essential genes in cultured mammalian cells using small interfering RNAs. J Cell Sci 114:4557–4565 (2001).

Hasuwa H, Kaseda K, Einarsdottir T, Okabe M: Small interfering RNA and gene silencing in transgenic mice and rats. FEBS Lett 532:227–230 (2002).

Heasman J: Morpholino oligos: making sense of antisense? Dev Biol 243:209–214 (2002).

Hemann MT, Fridman JS, Zilfou JT, Hernando E, Paddison PJ, Cordon-Cardo C, Hannon GJ, Lowe SW: An epi-allelic series of p53 hypomorphs created by stable RNAi produces distinct tumor phenotypes in vivo. Nat Genet 33:396–400 (2003).

Houbaviy HB, Sharp PA: Small RNA Northerns suggest that RNA interference is not involved in the downregulation of *Xist* by *Tsix*, in: Keystone Symposia 2002, RNA Interference, Cosuppression and Related Phenomena (Taos, New Mexico 2002).

Hunter T, Hunt T, Jackson RJ, Robertson HD: The characteristics of inhibition of protein synthesis by double-stranded ribonucleic acid in reticulocyte lysates. J Biol Chem 250:409–417 (1975).

Hutvagner G, Zamore PD: A microRNA in a multiple-turnover RNAi enzyme complex. Science 297:2056–2060 (2002).

Hutvagner G, McLachlan J, Pasquinelli AE, Balint E, Tuschl T, Zamore PD: A cellular function for the RNA-interference enzyme Dicer in the maturation of the let-7 small temporal RNA. Science 293:834–838 (2001).

Ishizuka A, Siomi MC, Siomi H: A *Drosophila* fragile X protein interacts with components of RNAi and ribosomal proteins. Genes Dev 16:2497–2508 (2002).

Jackson AL, Bartz SR, Schelter J, Kobayashi SV, Burchard J, Mao M, Li B, Cavet G, Linsley PS: Expression profiling reveals off-target gene regulation by RNAi. Nat Biotechnol 21:635–637 (2003).

Jensen S, Gassama MP, Heidmann T: Taming of transposable elements by homology-dependent gene silencing. Nat Genet 21:209–212 (1999).

Kalidas S, Smith DP: Novel genomic cDNA hybrids produce effective RNA interference in adult *Drosophila.* Neuron 33:177–184 (2002).

Ketting RF, Haverkamp TH, van Luenen HG, Plasterk RH: Mut-7 of *C. elegans,* required for transposon silencing and RNA interference, is a homolog of Werner syndrome helicase and RNaseD. Cell 99:133–141 (1999).

Kim SK, Wold BJ: Stable reduction of thymidine kinase activity in cells expressing high levels of anti-sense RNA. Cell 42:129–138 (1985).

Knight SW, Bass BL: A role for the RNase III enzyme DCR-1 in RNA interference and germ line development in *Caenorhabditis elegans.* Science 293:2269–2271 (2001).

Kramerov DA, Bukrinsky MI, Ryskov AP: DNA sequences homologous to long double-stranded RNA. Transcription of intracisternal A-particle genes and major long repeat of the mouse genome. Biochim Biophys Acta 826:20–29 (1985).

Kunath T, Gish G, Lickert H, Jones N, Pawson T, Rossant J: Transgenic RNA interference in ES cell-derived embryos recapitulates a genetic null phenotype. Nat Biotechnol 21:559–561 (2003).

Lavery KS, King TH: Antisense and RNAi: powerful tools in drug target discovery and validation. Curr Opin Drug Discov Devel 6:561–569 (2003).

Lefebvre C, Terret ME, Djiane A, Rassinier P, Maro B, Verlhac MH: Meiotic spindle stability depends on MAPK-interacting and spindle-stabilizing protein (MISS), a new MAPK substrate. J Cell Biol 157:603–613 (2002).

Ma J, Zhou H, Su L, Ji W: Effects of exogenous double-stranded RNA on the basonuclin gene expression in mouse oocytes. Science in China Series C 45:593–603 (2002).

Martinek S, Young MW: Specific genetic interference with behavioral rhythms in *Drosophila* by expression of inverted repeats. Genetics 156:1717–1725 (2000).

McManus MT, Sharp PA: Gene silencing in mammals by small interfering RNAs. Nat Rev Genet 3:737–747 (2002).

McMillan NA, Carpick BW, Hollis B, Toone WM, Zamanian-Daryoush M, Williams BR: Mutational analysis of the double-stranded RNA (dsRNA) binding domain of the dsRNA-activated protein kinase, PKR. J Biol Chem 270:2601–2606 (1995).

Mellitzer G, Hallonet M, Chen L, Ang SL: Spatial and temporal "knock down" of gene expression by electroporation of double-stranded RNA and morpholinos into early postimplantation mouse embryos. Mech Dev 118:57–63 (2002).

Mette MF, Aufsatz W, van der Winden J, Matzke MA, Matzke AJ: Transcriptional silencing and promoter methylation triggered by double-stranded RNA. EMBO J 19:5194–5201 (2000).

Morel JB, Mourrain P, Beclin C, Vaucheret H: DNA methylation and chromatin structure affect transcriptional and post-transcriptional transgene silencing in *Arabidopsis.* Curr Biol 10:1591–1594 (2000).

Mourelatos Z, Dostie J, Paushkin S, Sharma A, Charroux B, Abel L, Rappsilber J, Mann M, Dreyfuss G: miRNPs: a novel class of ribonucleoproteins containing numerous microRNAs. Genes Dev 16:720–728 (2002).

Munshi CB, Graeff R, Lee HC: Evidence for a causal role of CD38 expression in granulocytic differentiation of human HL-60 cells. J Biol Chem 277:49453–49458 (2002).

Myers JW, Jones JT, Meyer T, Ferrell JE: Recombinant Dicer efficiently converts large dsRNAs into siRNAs suitable for gene silencing. Nat Biotechnol 21:324–328 (2003).

Okano H, Aruga J, Nakagawa T, Shiota C, Mikoshiba K: Myelin basic protein gene and the function of antisense RNA in its repression in myelin-deficient mutant mouse. J Neurochem 56:560–567 (1991).

Paddison PJ, Caudy AA, Hannon GJ: Stable suppression of gene expression by RNAi in mammalian cells. Proc Natl Acad Sci USA 99:1443–1448 (2002).

Pal-Bhadra M, Bhadra U, Birchler JA: RNAi related mechanisms affect both transcriptional and post-transcriptional transgene silencing in Drosophila. Mol Cell 9:315–327 (2002).

Parrish S, Fire, A: Distinct roles for RDE-1 and RDE-4 during RNA interference in Caenorhabditis elegans. RNA 7:1397–1402 (2001).

Piccin A, Salameh A, Benna C, Sandrelli F, Mazzotta G, Zordan M, Rosato E, Kyriacou CP, Costa R: Efficient and heritable functional knock-out of an adult phenotype in Drosophila using a GAL4-driven hairpin RNA incorporating a heterologous spacer. Nucleic Acids Res 29:E55–65 (2001).

Prawitt D, Brixel D, Spangenberg C, Eshkind L, Heck R, Oesch F, Zabel B, Bockamp E: RNAi knockdown mice: an emerging technology for postgenomic functional genetics. Cytogenet Genome Res 105:412–421 (2004).

Provost P, Dishart D, Doucet J, Frendewey D, Samuelsson B, Radmark O: Ribonuclease activity and RNA binding of recombinant human Dicer. EMBO J 21:5864–5874 (2002).

Rougvie AE: Control of developmental timing in animals. Nat Rev Genet 2:690–701 (2001).

Rubinson DA, Dillon CP, Kwiatkowski AV, Sievers C, Yang L, Kopinja J, Rooney DL, Ihrig MM, McManus MT, Gertler FB, et al: A lentivirus-based system to functionally silence genes in primary mammalian cells, stem cells and transgenic mice by RNA interference. Nat Genet 33:401–406 (2003).

Schmitt HP, Kuhn B, Alonso A: Characterization of cloned sequences complementary to F9 cell double-stranded RNA and their expression during differentiation. Differentiation 30:205–210 (1986).

Schramke V, Allshire R: Hairpin RNAs and retrotransposon LTRs effect RNAi and chromatin-based gene silencing. Science 301:1069–1074 (2003).

Schwarz DS, Hutvagner G, Haley B, Zamore PD: Evidence that siRNAs function as guides, not primers, in the Drosophila and human RNAi pathways. Mol Cell 10:537–548 (2002).

Seitz H, Youngson N, Lin SP, Dalbert S, Paulsen M, Bachellerie JP, Ferguson-Smith AC, Cavaille J: Imprinted microRNA genes transcribed antisense to a reciprocally imprinted retrotransposon-like gene. Nat Genet 34:261–262 (2003).

Shendure J, Church GM: Computational discovery of sense-antisense transcription in the human and mouse genomes. Genome Biol 3:RESEARCH0044 (2002).

Shinagawa T, Ishii S: Generation of Ski-knockdown mice by expressing a long double-strand RNA from an RNA polymerase II promoter. Genes Dev 17:1340–1345 (2003).

Sijen T, Vijn I, Rebocho A, van Blokland R, Roelofs D, Mol JN, Kooter JM: Transcriptional and posttranscriptional gene silencing are mechanistically related. Curr Biol 11:436–440 (2001).

Sledz CA, Holko M, De Veer MJ, Silverman RH, Williams BR: Activation of the interferon system by short-interfering RNAs. Nat Cell Biol 5:834–839 (2003).

Smith NA, Singh SP, Wang MB, Stoutjesdijk PA, Green AG, Waterhouse PM: Total silencing by intron-spliced hairpin RNAs. Nature 407:319–320 (2000).

Srivastava SP, Kumar KU, Kaufman RJ: Phosphorylation of eukaryotic translation initiation factor 2 mediates apoptosis in response to activation of the double-stranded RNA-dependent protein kinase. J Biol Chem 273:2416–2423 (1998).

Stein P, Svoboda, P: Guide to RNAi in mouse oocytes and preimplantation embryos, in Hannon GJ (ed): RNAi: A Guide to Gene Silencing, chapter 15 (Cold Spring Harbor Press, Cold Spring Harbor 2003).

Stein P, Svoboda P, Anger M, Schultz RM: RNAi: mammalian oocytes do it without RNA-dependent RNA polymerase. RNA 9:187–192 (2003a).

Stein P, Svoboda P, Schultz RM: Transgenic RNAi in mouse oocytes: a simple and fast approach to study gene function. Dev Biol 256:187–193 (2003b).

Svoboda P, Stein P, Hayashi H, Schultz RM: Selective reduction of dormant maternal mRNAs in mouse oocytes by RNA interference. Development 127:4147–456 (2000).

Svoboda P, Stein P, Schultz RM: RNAi in mouse oocytes and preimplantation embryos: Effectiveness of hairpin dsRNA. Biochem Biophys Res Commun 287:1099–1104 (2001).

Svoboda P, Stein P, Anger M, Bernstein E, Hannon GJ, Schultz RM: RNAi and expression of retrotransposons MuERV-L and IAP in preimplantation mouse embryos. Dev Biol 269:276–285 (2004).

Tabara H, Sarkissian M, Kelly WG, Fleenor J, Grishok A, Timmons L, Fire A, Mello CC: The rde-1 gene, RNA interference, and transposon silencing in C. elegans. Cell 99:123–132 (1999).

Tabara H, Yigit E, Siomi H, Mello CC: The dsRNA binding protein RDE-4 interacts with RDE-1, DCR-1, and a DExH-box helicase to direct RNAi in C. elegans. Cell 109:861–871 (2002).

Tavernarakis N, Wang SL, Dorovkov M, Ryazanov A, Driscoll M: Heritable and inducible genetic interference by double-stranded RNA encoded by transgenes. Nat Genet 24:180–183 (2000).

Terret ME, Lefebvre C, Djiane A, Rassinier P, Moreau J, Maro B, Verlhac MH: DOC1R: a MAP kinase substrate that control microtubule organization of metaphase II mouse oocytes. Development 130:5169–5177 (2003).

Tijsterman M, Okihara K, Thijssen K, Plasterk R: PPW-1, a PAZ/PIWI protein required for efficient germline RNAi, is defective in a natural isolate of C. elegans. Curr Biol 12:1535 (2002).

Tuschl T, Zamore PD, Lehmann R, Bartel DP, Sharp PA: Targeted mRNA degradation by double-stranded RNA in vitro. Genes Dev 13:3191–3197 (1999).

Ui-Tei K, Zenno S, Miyata Y, Saigo K: Sensitive assay of RNA interference in Drosophila and Chinese hamster cultured cells using firefly luciferase gene as target. FEBS Lett 479:79–82 (2000).

Volpe TA, Kidner C, Hall IM, Teng G, Grewal SI, Martienssen RA: Regulation of heterochromatic silencing and histone H3 lysine-9 methylation by RNAi. Science 297:1833–1877 (2002).

Wan L, Dockendorff TC, Jongens TA, Dreyfuss G: Characterization of dFMR1, a Drosophila melanogaster homolog of the fragile X mental retardation protein. Mol Cell Biol 20:8536–8547 (2000).

Wang J, Tekle E, Oubrahim H, Mieyal JJ, Stadtman ER, Chock PB: Stable and controllable RNA interference: Investigating the physiological function of glutathionylated actin. Proc Natl Acad Sci USA 100:5103–5106 (2003).

Wei Q, Marchler G, Edington K, Karsch-Mizrachi I, Paterson BM: RNA interference demonstrates a role for nautilus in the myogenic conversion of Schneider cells by daughterless. Dev Biol 228:239–255 (2000).

Wesley SV, Helliwell CA, Smith NA, Wang MB, Rouse DT, Liu Q, Gooding PS, Singh SP, Abbott D, Stoutjesdijk PA, et al: Construct design for efficient, effective and high-throughput gene silencing in plants. Plant J 27:581–590 (2001).

Wianny F, Zernicka-Goetz M: Specific interference with gene function by double-stranded RNA in early mouse development. Nat Cell Biol 2:70–275 (2000).

Xu Z, Williams CJ, Kopf GS, Schultz RM: Maturation-associated increase in IP3 receptor type 1: role in conferring increased IP3 sensitivity and Ca^{2+} oscillatory behavior in mouse eggs. Dev Biol 254:163–171 (2003).

Yang D, Buchholz F, Huang Z, Goga A, Chen CY, Brodsky FM, Bishop JM: Short RNA duplexes produced by hydrolysis with Escherichia coli RNase III mediate effective RNA interference in mammalian cells. Proc Natl Acad Sci USA 99:9942–9947 (2002).

Yang S, Tutton S, Pierce E, Yoon K: Specific double-stranded RNA interference in undifferentiated mouse embryonic stem cells. Mol Cell Biol 21:7807–7816 (2001).

Yelin R, Dahary D, Sorek R, Levanon EY, Goldstein O, Shoshan A, Diber A, Biton S, Tamir Y, Khosravi R, et al: Widespread occurrence of antisense transcription in the human genome. Nat Biotechnol 21:379–386 (2003).

Yi CE, Bekker JM, Miller G, Hill KL, Crosbie RH: Specific and potent RNA interference in terminally differentiated myotubes. J Biol Chem 278:934–939 (2003).

Yu J, Deng M, Medvedev S, Yang J, Hecht NB, Schultz RM: Transgenic RNAi-mediated reduction of MSY2 in mouse oocytes results in reduced fertility. Dev Biol 268:195–206 (2004).

Zamore PD, Tuschl T, Sharp PA, Bartel DP: RNAi: double-stranded RNA directs the ATP-dependent cleavage of mRNA at 21 to 23 nucleotide intervals. Cell 101:25–33 (2000).

Zeng Y, Wagner EJ, Cullen BR: Both natural and designed micro RNAs can inhibit the expression of cognate mRNAs when expressed in human cells. Mol Cell 9:1327–1333 (2002).

Zhang H, Kolb FA, Brondani V, Billy E, Filipowicz W: Human Dicer preferentially cleaves dsRNAs at their termini without a requirement for ATP. EMBO J 21:5875–5885 (2002).

Zhou A, Paranjape J, Brown TL, Nie H, Naik S, Dong B, Chang A, Trapp B, Fairchild R, Colmenares C, et al: Interferon action and apoptosis are defective in mice devoid of 2′,5′-oligoadenylate-dependent RNase L. EMBO J 16:6355–6363 (1997).

Zilberman D, Cao X, Jacobsen SE: ARGONAUTE4 control of locus-specific siRNA accumulation and DNA and histone methylation. Science 299:716–719 (2003).

Cytogenet Genome Res 105:435–441 (2004)
DOI: 10.1159/000078216

Sequence-specific modification of mouse genomic DNA mediated by gene targeting techniques

F. Sangiuolo[a] and G. Novelli[a,b]

[a] Department of Biopathology and Diagnostic Imaging, Tor Vergata University, Rome (Italy);
[b] Division of Cardiovascular Medicine, Department of Medicine, University of Arkansas for the Medical Sciences, Little Rock, AR (USA)

Abstract. The major impact of the human genome sequence is the understanding of disease etiology with deduced therapy. The completion of this project has shifted the interest from the sequencing and identification of genes to the exploration of gene function, signalling the beginning of the post-genomic era. Contrasting with the spectacular progress in the identification of many morbid genes, today therapeutic progress is still lagging behind. The goal of all gene therapy protocols is to repair the precise genetic defect without additional modification of the genome. The main strategy has traditionally been focused on the introduction of an expression system designed to express a specific protein, defective in the transfected cell. But the numerous deficiencies associated with gene augmentation have resulted in the development of alternative approaches to treat inherited and acquired genetic disorders. Among these one is represented by gene repair based on homologous recombination (HR). Simply stated, the process involves targeting the mutation in situ for gene correction and for restoration of a normal gene function. Homologous recombination is an efficient means for genomic manipulation of prokaryotes, yeast and some lower eukaryotes. By contrast, in higher eukaryotes it is less efficient than in the prokaryotic system, with non-homologous recombination being 10–50 fold higher. However, recent advances in gene targeting and novel strategies have led to the suggestion that gene correction based on HR might be used as clinical therapy for genetic disease. This site-specific gene repair approach could represent an alternative gene therapy strategy in respect to those involving the use of retroviral or lentiviral vectors to introduce therapeutic genes and linked regulatory sequences into random sites within the target cell genome. In fact, gene therapy approaches involving addition of a gene by viral or nonviral vectors often give a short duration of gene expression and are difficult to target to specific populations of cells. The purpose of this paper is to review oligonucleotide-based gene targeting technologies and their applications on modifying the mouse genome.

Augmentation is the standard approach of gene therapy and involves the delivery of a transgene to replace an existing non-functional gene. Despite being the most common gene therapy approach, it has significant drawbacks. For example, prolonged expression of the transgene requires integration into the genome of the host cell. However, random integration also increases the possibility of gene disruption, including disruption of genes involved in cell cycle or tumor suppression (Remus et al., 1999; Muller et al., 2001). For this purpose, the targeted gene repair approach has received increasing attention because of the safety and of efficacy issues encountered with more traditional gene therapy strategies, where additional copies of therapeutic genes are delivered to and are expressed in transduced cells. The potential of this type of strategy has obvious implications for maintaining genomic integrity and cell-specific expression. In fact, the direct conversion of mutant genomic sequences to a wild-type genotype, restoring the normal phenotype, has clear advantages over therapeutic cDNA. By preserving the integrity of the targeted gene, the relationship between the coding sequences and regulatory elements remains intact, and therefore the targeted gene would be expressed at

This work was supported by grants from Associazione Studio Atrofia Muscolare Spinale Infantile (ASAMSI), Fondazione Federica, Associazione Girotondo and the Italian Ministry of Health.

Received 12 October 2003; manuscript accepted 21 October 2003.

Request reprints from: Dr. Giuseppe Novelli
Department of Biopathology and Diagnostic Imaging
Tor Vergata University, Via Montpellier 1, IT–00133 Rome (Italy)
telephone: +39-(0)6-20900665; fax: +39-(0)6-20900669
e-mail: novelli@med.uniroma2.it

physiological levels in the appropriate cell type. Consequently, cell-specific expression is not altered. The recent observation that retroviral gene transfer induced T cell leukaemia in two children of ten patients treated for typical X-linked severe combined immunodeficiency (SCID-X1) has raised significant safety concerns for traditional gene strategies (Hacein-Bey-Abina et al., 2003a, b; Williams and Baum, 2003). Accordingly, the American Food and Drug Administration (FDA) placed on "clinical hold" all active gene therapy trials using retroviral vectors and suspended the enrolment of new patients in clinical trials that involve the use of retroviruses. The main risk of retrovirus-mediated gene transfer is insertional mutagenesis resulting from random retroviral integration. Hacein-Bey-Abina and colleagues (Hacein-Bey-Abina et al., 2003b) showed retrovirus vector integration in proximity to the *LMO2* proto-oncogene promoter, leading to aberrant transcription and expression of *LMO2,* suggesting that retrovirus vector insertion can trigger deregulated premalignant cell proliferation with unexpected frequency, most likely driven by retrovirus enhancer activity on the *LMO2* gene promoter. It is therefore mandatory to develop vectors with an improved safety profile and/or transfect cells in ex vivo, followed by checking them carefully by genetic analysis before infusing back into the patient.

Therefore, once adequately developed, gene targeting strategies will result in less random mutagenesis of the genome and lead to fewer mutagenic side effects than do methods that randomly insert genes into the genome (Sullenger, 2003).

Chimeric RNA/DNA oligonucleotides

Chimeraplasts or chimeric RNA/DNA oligonucleotides are designed to target a homologous genomic sequence and induce a site-specific base change (Kmiec, 1999; Igoucheva and Yoon, 2000; Wu et al., 2001). The chimera is a single-stranded molecule, usually 70–80 bases in length, with sequence complementation so that it folds into a double-hairpin configuration (Fig. 1). This configuration avoids nuclease digestion and concatenation of double-stranded molecules, two adverse molecular events that are common when these types of DNA structures are transfected into eukaryotic cells. The base composition includes both RNA and DNA nucleotides with three and four thymidine residues in the "cap turns." The RNA bases stabilize the intermediate joint molecule on one strand, while the DNA portion of the molecule starts the single-base exchange event. Finally, a unique phosphodiester bond is left unligated, so as to facilitate topological intertwining with the target site in the helix. Once one mismatch is corrected, the chimera dissociates, leaving behind a single mismatched base pair in the original targeted helix. The normal repair pathway is then activated and generates an intact DNA helix. The successful repair of the chimeraplast may depend entirely on mismatch affinity (Brown et al., 2001) and protein recognition, and so the secondary structure, the quality of synthesis and subsequent purification, the surrounding genomic sequence, and the transfection efficiency all influence the conversion rates. In an elegant study, melanocytes from albino mice were isolated and transfected with chimeraplasts to repair the genetic defect, a point mutation in the tyrosinase gene that is an essential enzyme in melanin synthesis (Alexeev and Yoon, 1998). After transfection with chimeraplasts, single clones of pigmented cells were identified and isolated. Genomic correction was established by restriction fragment length polymorphism (RFLP), immunoblotting analysis, and enzymatic activity and, moreover, genotypic and phenotypic changes resulted which were stable over numerous passages of the clones. Another paper reported in vivo studies performed to induce a point mutation in the factor IX gene, by direct injection into the tail vein of rats (Kren et al., 1998). A dose-related genomic conversion rate of 15–40% was observed together with a corresponding reduction in factor IX activity. Importantly, the altered clotting profiles and genomic change remained stable for nearly two years. The same approach and delivery system was used to promote liver-specific delivery and to repair a single base deletion in the Gunn-rat model of the human disease Crigler-Najjar syndrome type 1 (Kren et al., 1999). The mutation is a single point deletion in the UDP-glucuronosyltransferase gene (UGT1A1), producing a premature stop codon. The chimeric oligonucleotide was either complexed with polyethylenimine or encapsulated in anionic liposomes, administered i.v., and targeted to the hepatocyte via the asialoglycoprotein receptor. Following DNA repair (the efficiency reported was up to 23%) the result of the transfection was specific and stable throughout the 6-month observation period, and was associated with a reduction of serum bilirubin levels. Chimeric RNA/DNA oligonucleotides were also used to alter splicing mutations in the dystrophin gene and produce in-frame transcripts. Modification was evaluated at protein level by Western blot and immunohistochemical analysis in differentiated cells. In this way chimeraplasts can modify splice site sequences at the genomic level, transforming a severe DMD phenotype into a much milder BMD phenotype (Bertoni et al., 2003).

Successful chimeraplasty has been demonstrated in a variety of tissues, but curiously, little data are available for stem cells. A major criticism regarding this technique is that the high conversion frequencies observed in some experiments, are a consequence of polymerase chain reaction (PCR) artifacts and/or cell to cell contamination (Thomas and Capecchi, 1997; Zhang et al., 1998). Many scientists believed that the results were most likely artifactual, predicting that no one would be able to use the technique. The question of reproducibility has plagued this technique, especially when the issue of the high frequencies of gene repair was discussed. In fact the main problem is centred on the variability among the data of different research groups, varying from 0.5 to 20% even within the same laboratory (Albuquerque-Silva et al., 2001; van der Steege et al., 2001; Kmiec, 2003). However, it may be of general concern that a broad application of this technique awaits, despite the number and the extent of positive reports.

Small Fragment Homologous Replacement (SFHR)

SFHR is a gene repair strategy that involves the introduction of small DNA fragments (SDFs) (up to 1 kb) into cells. These SDFs effect homologous exchange between their se-

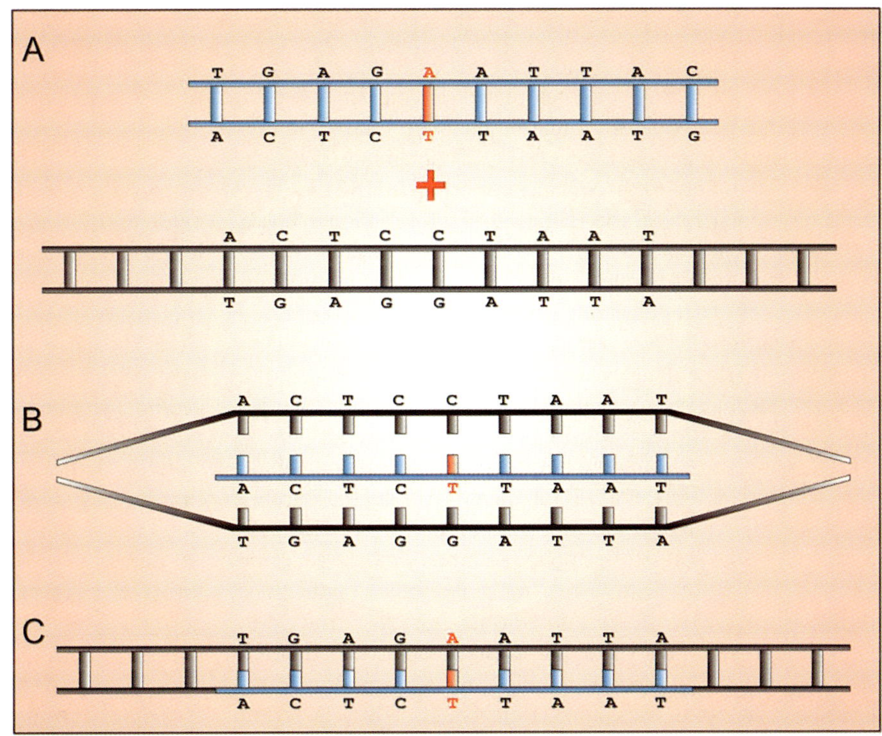

Fig. 1. Targeted gene repair directed by chimeric RNA/DNA oligonucleotides. (**A**) The RNA/DNA chimera (inset) is an oligonucleotide capped by four unpaired thymidines (green) at both ends, consisting of RNA (blue) flanking DNA residues (grey). The complementary strand is all DNA residues (black). A 3'-5' GC clamp (purple) is present at one end of the double-stranded molecule. (**B**) The molecule acts by annealing at the site in the target DNA, with the RNA section hybridizing with perfect complementation to one strand and the DNA stretch hybridizing to the other strand. The DNA repair systems of the cell are activated by the mismatch formed. (**C**) The genomic TA is replaced by CG, the complementary sequence to the chimeraplast.

Fig. 2. Small Fragments Homologous Recombination (SFHR). (**A**) SDFs PCR-amplified (blue) containing the mismatch to be introduced into homologous sequence (grey). (**B**) After pairing, single-strand SDF affects homologous exchange between their sequence and the target one. (**C**) The introduction of the desired alteration (nucleotide in red).

quences and the endogenous (genomic or episomal) ones, resulting in phenotypic changes (Fig. 2). The technique has been used to modify endogenous genomic DNA in both human and mouse cells. After entering the cells, the fragment pairs with its genomic homologue and replaces the endogenous sequence with the exogenous fragment through an, as yet, undefined mechanism (Gruenert et al., 1998; Yanez and Porter, 1998; Gruenert, 1999), that probably involves similar pathways to those of homologous recombination and/or uncharacterized pathways of DNA repair (Goncz and Gruenet 2000). It is likely that SFHR functions by targeting and replacement, involving the necessary proofreading and annealing to homologous target regions, with subsequent strand invasion, and exchange of genetic material. Genotype and phenotype analyses have shown specific modification of disease-causing genetic loci and suggested that it has potential as a therapeutic modality for the treatment of inherited disease, if not with a single treatment, with repeated applications. Different kinds of genomic mutations have been altered by using this approach, suggesting a broad range of utility in terms of target genes and cell types able to support SFHR. Moreover, the SFHR technique appears to be effective both in vivo and in vitro. In fact, genetic modification of several kinds of cells was reported, including mouse embryonic stem cells (Gruenert et al., 1998; Goncz et al., 2001, 2002; Kapsa et al., 2002). The recent application of this strategy to stem cell and demonstration that these cells can be efficiently targeted would mean that a single treatment could correct a genetic defect within a given organ for the lifetime of the patient, using an ex vivo strategy (Hatada et al., 2000; Goncz et al., 2001; Gruenert et al., 2003). By targeting gene repair to stem cell populations, it is possible that long-term correction might be achieved through clonal expansion.

Different efficiency values of SFHR-mediated modification were reported. This variability can depend on transfection efficiency, specific for each cell line. In this way any improvement in gene delivery will be fundamental to the success of gene repair, especially for somatic gene repair in vivo where transfection efficiencies are considerably lower than in vitro. The efficiency of transfection could be increased using a different way of delivery of SDF into cell nuclei. Recent studies have reported the successful use of microinjection techniques in progenitor stem hematopoietic cells (HPCs), human lung sarcoma-HT1080, and immortalized lung epithelial-16HBE14o- and primary human fibroblasts (Davis et al., 2000; Goncz et al., 2001; Schindelhauer and Laner, 2002). This technique, together with an innovative electroporation protocol, might overcome the inefficient nuclear delivery of DNA that was observed using chemical delivery vehicles. Another factor surely influencing SFHR efficiency is represented by the absence of a selection mechanism to accurately define the genetic changes induced by this strategy. For this purpose a subcloning of the transfected cells by using the limited dilution technique could be attempted (Bruscia et al., 2002), demonstrating the stability of the modification over multiple generations. The SFHR technique has been successfully applied to modify mouse genomic loci, as the *Cftr* (cystic fibrosis transmembrane conductance regulator) gene, and the *dystrophin* gene, both in vitro and in vivo. Phenotypic and genotypic changes have been demonstrated for the *Cftr* gene, while only genotypic modifications have been shown in *mdx* mice (Kapsa et al., 2002). A recent study by Goncz et al. (2001) has demonstrated the importance of the delivery during the application of SFHR-mediated protocols. The authors used the SFHR technique to introduce in the genome of a wild-type mouse a 3-bp deletion within the *Cftr* gene. SDFs were transferred in normal mouse lung via intratracheal instillation, after complexation with four different transfecting agents. Detectable levels of sequence alteration were observed, but the data suggest a gradient of these DNA vehicles

in terms of reproducibility and specifically, the artificial viral envelope (AVE) gave better results than LipofectAMINE, dimethyl-dioctadecyl-ammonium-bromide (DDAB) and Super-Fect. This study underscored the difficulties in extrapolating results from the in vitro to in vivo settings. A similar approach was used in our laboratory to introduce the same modification in mouse embryonic stem cells (D3) using cationic liposome as the SDFs transfection vehicle (manuscript submitted). Confocal microscopy was carried out to track the entry of Cy-5 labelled SDFs (red) into the nuclei (green) (Fig. 3). The 3-bp deletion (TTT) was detected at the DNA and mRNA level, after amplicon cloning, to avoid PCR artifacts. Modification efficiency of *Cftr* locus was quantified at the mRNA level by using real-time PCR, resulting in more than 5 % of transfected cells. In addition, we demonstrated that a mutated gene product was obtained using this technique. This high frequency of selective change in ES cells using SFHR is encouraging for the future development of therapeutic protocols based on cell therapy.

The Mdx mouse model of Duchenne muscular dystrophy (DMD) (Kapsa et al., 2001, 2002) was also repaired by using SFHR. A nonsense mutation in the dystrophin locus was targeted in both primary myoblast cultures and by direct injection of affected muscle (tibilias anterior). In vitro and in vivo application of a wild-type SDF (603 bp) was used to mediate a T to C conversion in exon 23 of the dystrophin gene. Different conditions in the lipofectamine complex enhanced the efficiency of SFHR-mediated modification in vitro. Conversion was observed at both the DNA and RNA levels. The conversion of mdx to wild-type sequence in vitro was about 15 % by PCR analysis, although there was no detection of normal dystrophin protein. In vivo the correction efficiency was up to 0.1 % in the tibialis anterior of a male *mdx* mice, but again there was no evidence of gene expression at either the transcript or protein level. It was suggested that the disparity between the genomic repair and protein expression was possibly due to toxicity of the transfected agent on myoblasts, or a delay in protein expression. The correction in myoblasts from mdx mice persisted at least 28 days in culture and up to 3 weeks in vivo. These genomic conversion frequencies were lower than those reported for chimeraplasty, which did result in protein expression (Bartlett et al., 2000; Rando et al., 2000), after direct injection into muscles of *mdx* mice. Two weeks after single injections into tibialis anterior muscles, the maximum number of dystrophin-positive fibers in any muscle represented 1–2 % of the total number of fibers in that muscle. The expression appears to be stable until ten weeks after single injections.

Triplex-forming oligonucleotides

Triplex-forming oligonucleotides (TFOs) bind specifically to duplex DNA and provide a strategy for site-directed modification of genomic DNA by forming stable and specific triple helical structures with homopurine-rich areas of the genome (Fig. 4A). These regions (15–30 bp) occur about every kilobase of genome, and so it is possible that the majority of genes will contain TFO targets (Vasquez et al., 2001a; Casey and Glazer,

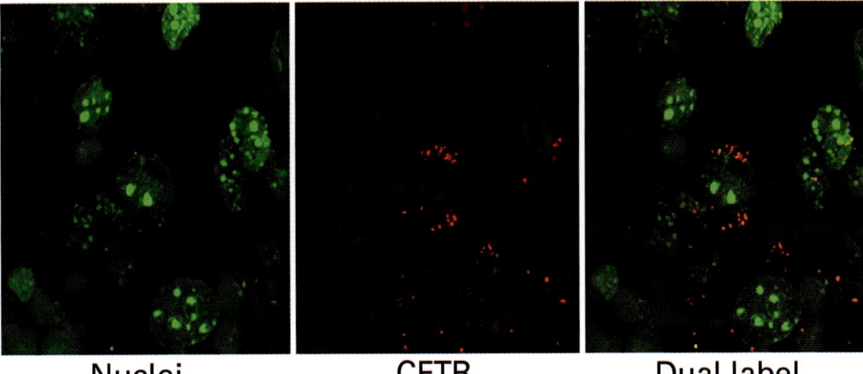

Fig. 3. Confocal microscopy analysis of D3 cells 48 h after transfection with SDFs-Cy5 labelled (red). Fragments are located inside nuclei, as shown in the right part of the figure (dual label). Nuclei are labelled with Syber-Green (green).

2001). To overcome this potential limitation, phosphodiester bonds were replaced with positively charged N,N′-diethylethylenediamine linkages, in order to enhance the intracellular activity of the TFOs and allow targeting of relatively short polypurine sites, thereby substantially expanding the number of potential triplex target sites in the genome. This study performed in mouse cells demonstrated that chemically modified TFOs were able to target a shorter (10 bp) site in a chromosomal locus inducing site-specific mutations (Vasquez et al., 2001b). TFOs containing mutagenic adjuncts, such as psoralen, have already been used to target and induce site specific chromosomal breaks within the genome (Raha et al., 1998) (Fig. 4B). The resultant DNA damage stimulates the genomic repair system and thereby increases the frequency of homologous recombination. The mechanism for correction appears to involve at least the nucleotide excision repair (NER) and transcription-coupled repair pathways (Faruqi et al., 2000). Triplex-forming oligonucleotides have successfully produced specific mutations in a variety of somatic cells (Majumdar et al., 1998; Vasquez et al., 1999). Psoralen-modified TFOs directed at a site in the *Hprt* gene were able to create stable *Hprt*-deficient clones of CHO cells. Analysis of 282 clones indicated that 85 % contained mutations in the triplex target region (deletions and insertions), within the triplex binding site, indicating a TFO-specific induction of mutagenesis. This finding demonstrates the ability of triplex-forming oligonucleotides to influence mutation frequencies at a specific site in a mammalian chromosome. Interestingly, TFOs without DNA damaging adducts have also been shown to facilitate homologous recombination, suggesting that TFO binding itself may be mutagenic and is recognized as abnormal by repair systems. The study was performed in a mouse LTK(–) cell line carrying two mutant copies of the herpes simplex virus thymidine kinase (TK) gene as direct repeats in a single chromosomal locus. Recombination between these repeats can produce a functional TK gene and occurs at a spontaneous frequency of 4×10^{-6} under standard culture conditions. When cells were microinjected with TFOs designed to bind to a 30-bp polypurine site situated between

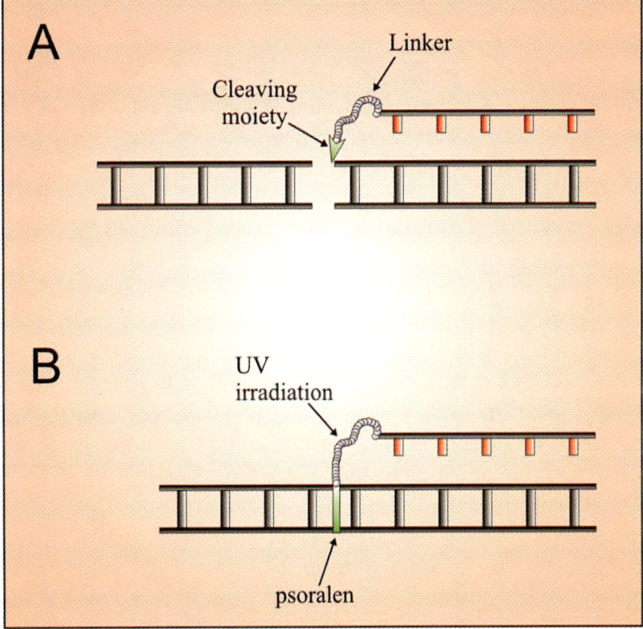

Fig. 4. TFO model. (**A**) Creation of the triple-helix region with a complementary purine-rich genomic sequence (grey). A DNA cleaving agent (green arrow) linked to TFO once activated causes strand cleavage of genomic DNA. (**B**) Psoralen (green), delivered by TFO, intercalates into genomic DNA adjacent to TFO domain and once activated by UV irradiation, cross-links the three DNA strands. The damage is then repaired via excision repair, resulting in base conversion.

the two TK genes, recombination was observed at frequencies in the range of 1 %, which is 2,500-fold above the background. Recombination was induced efficiently by injection of both psoralen-conjugated TFOs and unconjugated TFOs alone. TFOs transfected with cationic lipids also induced recombinants in a highly sequence-specific manner but were less effective, reporting recombination frequencies of 6–7 fold over background (Luo et al., 2000). Another study demonstrated the feasibility to induce mutations at specific genomic sites in somatic cells of adult mice by TFOs. The mutation frequency in the supF gene was about 5-fold greater than mice treated with a scrambled sequence control oligomer, demonstrating the site-specific genome modification by TFOs in the mouse (Vasquez et al., 2000). Recent studies reported the chemical effect of morpholino (MOR) backbone modification of TFOs by evaluating the formation of the pyrimidine motif triplex at neutral pH, which occurred 60 times more frequently than that observed with unmodified TFOs. The chemical modification consists of the replacement of ribose or deoxyribose rings characteristic of RNA and DNA with a morpholine ring. Kinetic data demonstrated that the observed increase in the binding constant by MOR resulted mainly from the considerable increase in the association rate constant. MOR backbone modification of TFO could be a key modification and may eventually lead to progress in therapeutic applications of the antigene strategy in vivo (Torigoe et al., 2003).

Conclusions and perspectives

It is imperative that gene repair be an efficient and reproducible strategy before applying it to clinical medicine. Many of the variables need to be defined and evaluated, including pathways and essential factors involved in gene targeting and repair. In fact the evaluation of which factors can be manipulated in order to improve the success of this approach should be of great interest. Improving limiting processes as delivery, targeting, and manipulation, could increase the spectrum of homologous recombination applications. It is well recognized that different cells exhibit different rates of homologous recombination and DNA repair, depending also on their functional repair pathways or cell cycle phases. Moreover, the precise mismatch and location within the genome appear to impact the efficiency of recognition and alteration of the DNA by endogenous DNA repair pathways (Modrich, 1997; Bar-Ziv and Libchaber, 2001). To improve the efficiency of gene targeting, modified nucleotides, such as peptide or locked nucleic acids, may enhance resistance of molecules to nuclease degradation. Concerning the potential to induce random mutations within the genome, there is little data directly implicating the gene repair technologies in this process (Yanez and Porter, 1998). Recently, a comparative study has been performed in human embryonic kidney epithelial cells (HEK293) and mouse embryonic stem (ES) cells, for evaluating gene repair efficiency of small DNA fragments (0.52–1.9 kb long), single-stranded modified DNA oligonucleotides (20, 35, or 80 bases) and a 68-base RNA/DNA chimeric oligonucleotide (Nickerson and Colledge, 2003). SDF gave the highest frequency of episomal gene repair, while single-stranded DNA oligonucleotides gave the highest frequency of chromosomal repair. Antisense DNA oligonucleotides gave 5-fold higher frequencies of gene repair than their sense counterparts and the RNA/DNA chimeric oligonucleotide gave little or no gene repair on either a chromosomal or episomal target. The hypothesis is that small oligonucleotides may be more efficient in chromosomal gene repair than longer SDFs fragments because of their small size increasing mobility in the cytoplasm and therefore access to the nucleus. Another limiting factor could be the relative abundance of DNA repair proteins in each cell line and the transcriptional activity of each gene. Moreover, some genes may present more difficult targets for gene repair, as they may be situated in less accessible areas of the chromosome and also delivery systems will most likely be designed for individual cell types.

A great potential for gene alterations will undoubtedly be in stem cells. Stem cells could be harvested and transduced ex vivo, and the corrected cells reintroduced into the host. The growth advantage of the corrected stem cells could easily translate into a significant improvement and possible cure for many diseases (Zhang et al., 2001; Zwaka and Thomson, 2003). The in vitro differentiation capacity of ES cells provides unique opportunities for experimental analysis of gene regulation and function during cell commitment and differentiation in early embryogenesis. Finally, targeting genes for mutation within embryonic stem cells by homologous recombination has been a powerful research tool in developing animal models of human disease (Zwaka and Thomson, 2003).

References

Albuquerque-Silva J, Vassert G, Lavinha J: Chimeraplasty validation. Nat Biotechnol 19:1011 (2001).

Alexcev V and Yoon K: Stable and inheritable changes in genotype and phenotype of albino melanocytes induced by an RNA-DNA oligonucleotide. Nat Biotechnol 16:1343–1346 (1998).

Bartlett RJ, Stockinger S, Denis MM, Bartlett WT, Inverardi L, Le TT, thi Man N, Morris GE, Bogan DJ, Metcalf-Bogan J, Kornegay JN: In vivo targeted repair of a point mutation in the canine dystrophin gene by a chimeric RNA/DNA oligonucleotide. Nat Biotechnol 18:615–622 (2000).

Bar-Ziv R, Libchaber A: Effects of DNA sequence and structure on binding of RecA to single-stranded DNA. Proc Natl Acad Sci USA 98:9068–9073 (2001).

Bertoni C, Lau C, Rando TA. Restoration of dystrophin expression in mdx muscle cells by chimeraplast-mediated exon skipping. Hum Mol Genet 12:1087–1099 (2003).

Brown J, Brown T, Fox KR: Affinity of mismatch-binding protein MutS for heteroduplexes containing different mismatches. Biochem J 354(Pt 3):627–633 (2001).

Bruscia E, Sangiuolo F, Sinibaldi, Goncz KK, Novelli G, Gruenert DC: Isolation of CF cell lines corrected at DeltaF508-CFTR locus by SFHR-mediated targeting. Gene Ther 9:683–685 (2002).

Casey BP, Glazer PM: Gene targeting via triple-helix formation. Prog Nucleic Acid Res Mol Biol 67:163–192 (2001).

Davis BR, Brown DB, Prokopishyn NL, Yannariello-Brown J: Micro-injection-mediated hematopoietic stem cell gene therapy. Curr Opin Mol Ther 2:412–419 (2000).

Faruqi AF, Datta HJ, Carroll D, Seidman MM, Glazer PM: Triple helix formation induces recombination in mammalian cells via a nucleotide excision repair-dependent pathway. Mol Cell Biol 20:990–1000 (2000).

Goncz KK, Gruenert DC: Site-directed alteration of genomic DNA by small-fragment homologous replacement. Methods Mol Biol 133:85–99 (2000).

Goncz KK, Kunzelmann K, Xu Z, Gruenert DC: Targeted replacement of normal and mutant CFTR sequences in human airway epithelial cells using DNA fragments. Hum Mol Genet 7:1913–1919 (1998).

Goncz KK, Colosimo A, Dallapiccola B, Gagne L, Hong K, Novelli G, Papahadjopoulos D, Sawa T, Schreier H, Wiener-Kronish J, Xu S, Gruenert DC: Expression of DeltaF508 CFTR in normal mouse lung after site-specific modification of CFTR sequences by SFHR. Gene Ther 8:961–965 (2001).

Goncz KK, Prokopishyn NL, Chow BL, Davis BR, Gruenert DC: Application of SFHR to gene therapy of monogenic disorders. Gene Ther 9:691–694 (2002).

Gruenert DC: Gene correction with small DNA fragments. Curr Res Molec Ther 1:607–613 (1998).

Gruenert DC: Opportunities and challenges in targeting genes for therapy. Gene Ther 6:1347–1348 (1999).

Gruenert DC, Xu Z, Wiener-Kronish J, Colledge WH, Radcliff R, Mc Vinish L, Goncz KK: Gene targeting in somatic cells: prospects for the future of gene therapy. 6th Max Delbruck Symposium on Gene Therapy, May 6–8, 1998 (Berlin, Germany, 1998).

Gruenert DC, Bruscia E, Novelli G, Colosimo A, Dallapiccola B, Sangiuolo F, Goncz KK: Sequence-specific modification of genomic DNA by small DNA fragments. J Clin Invest 112:637–641 (2003).

Hacein-Bey-Abina S, von Kalle C, Schmidt M, Le Deist F, Wulffraat N, McIntyre E, Radford I, Villeval JL, Fraser CC, Cavazzana-Calvo M, Fischer A: A serious adverse event after successful gene therapy for X-linked severe combined immunodeficiency. N Engl J Med 348:193–194 (2003a).

Hacein-Bey-Abina S, Von Kalle C, Schmidt M, McCormack MP, Wulffraat N, Lebouch P, Lim A, Osborne CS, Pawliuk R, Morillon E, Sorensen R, Forster A, Fraser P, Cohen JI, De Saint Basile G, Alexander I, Wintergerst U, Frebourg T, Aurias A, Stoppa-Lyonnet D, Romana S, Radford-Weiss I, Gross F, Valensi F, Delabesse E, Macintyre E, Sigaux F, Soulier J, Leiva LE, Wissler M, Prinz C, Rabbitts TH, Le Deist F, Fischer A, Cavazzana-Calvo M: LMO2-associated clonal T cell proliferation in two patients after gene therapy for SCID-X1. Science 302:415–419 (2003b).

Hatada S, Nikkuni K, Bentley SA, Kirby S, Smithies O: Gene correction in hematopoietic progenitor cells by homologous recombination. Proc Natl Acad Sci USA 97:13807–13811 (2000).

Kapsa R, Quigley A, Lynch GS, et al: In vivo and in vitro correction of the mdx dystrophin gene nonsense mutation by short-fragment homologous replacement. Hum Gene Ther 12:629–642 (2001).

Kapsa RM, Quigley AF, Vadolas J, Steeper K, Ioannou PA, Byrne E, Kornberg AJ: Targeted gene correction in the mdx mouse using short DNA fragments: towards application with bone marrow-derived cells for autologous remodelling of dystrophic muscle. Gene Ther 9:695–699 (2002).

Kmiec EB: Targeted gene repair. Gene Ther 6:1–3 (1999).

Kmiec EB: Targeted gene repair – in the arena. J Clin Invest 112:632–636 (2003).

Kren BT, Bandyopadhyay P, Steer CJ: In vivo site-directed mutagenesis of the factor IX gene by chimeric RNA/DNA oligonucleotides. Nat Med 4:285–290 (1998).

Kren BT, Parashar B, Bandyopadhyay P, Chowdhury NR, Chowdhury JR, Steer CJ: Correction of the UDP-glucuronosyltransferase gene defect in the gunn rat model of Crigler-Najjar syndrome type I with a chimeric oligonucleotide. Proc Natl Acad Sci USA 96:10349–10354 (1999).

Igoucheva O, Yoon K: Targeted single-base correction by RNA-DNA oligonucleotides. Hum Gene Ther 11:2307–2312 (2000).

Luo Z, Macris MA, Faruqi AF, Glazer PM: High-frequency intrachromosomal gene conversion induced by triplex-forming oligonucleotides microinjected into mouse cells. Proc Natl Acad Sci USA 97:9003–9008 (2000).

Majumdar A, Khorlin A, Dyatkina N, Lin FL, Powell J, Liu J, Fei Z, Khripine Y, Watanabe KA, George J, Glazer PM, Seidman MM: Targeted gene knockout mediated by triple helix forming oligonucleotides. Nat Genet 20:212–214 (1998).

Modrich P: Strand-specific mismatch repair in mammalian cells. J Biol Chem 272:24727–24730 (1997).

Muller K, Heller H, Doerfler W: Foreign DNA integration. Genome-wide perturbations of methylation and transcription in the recipient genomes. J Biol Chem 276:14271–14278 (2001).

Nickerson HD, Colledge WH: A comparison of gene repair strategies in cell culture using a lacZ reporter system. Gene Ther 10: 1584–1591 (2003).

Raha M, Lacroix L, Glazer PM: Mutagenesis mediated by triple helix-forming oligonucleotides conjugated to psoralen: effects of linker arm length and sequence context. Photochem Photobiol 67:289–294 (1998).

Rando TA, Disatnik M-H, Zhou LZ-H: Rescue of dystrophin expression in *mdx* mouse muscle by RNA/DNA oligonucleotides. Proc Natl Acad Sci USA 97:5363–5368 (2000).

Remus R, Kammer C, Heller H, Schmitz B, Schell G, Doerfler W: Insertion of foreign DNA into an established mammalian genome can alter the methylation of cellular DNA sequences. J Virol 73:1010–1022 (1999).

Schindelhauer D, Laner A: Visible transient expression of EGFP requires intranuclear injection of large copy numbers. Gene Ther 9:727–730 (2002).

Sullenger BA: Targeted genetic repair: an emerging approach to genetic therapy. J Clin Invest 112:310–311 (2003).

Thomas KR, Capecchi MR: Recombinant DNA technique and sickle cell anemia research. Science 275:1404–1405 (1997).

Torigoe H, Kawahashi K, Tamura Y: Promotion of triplex formation by morpholino modification: thermodynamic and kinetic studies. Nucleic Acids Res 3:157–158 (2003).

van der Steege G, Schuilenga-Hut PHL, Buys CHMC, Scheffer H, Pas HH, Jonkman MF: Persistent failures in gene repair. Nat Biotechnol 19:305–306 (2001).

Vasquez KM, Wang G, Havre PA, Glazer PM: Chromosomal mutations induced by triplex-forming oligonucleotides in mammalian cells. Nucleic Acids Res 27:1176–1181 (1999)

Vasquez KM, Narayanan L, Glazer PM: Specific mutations induced by triplex-forming oligonucleotides in mice. Science 290:530–533 (2000).

Vasquez KM, Marburger K, Intody Z, Wilson JH: Manipulating the mammalian genome by homologous recombination. Proc Natl Acad Sci USA 98:8403–8410 (2001a).

Vasquez KM, Dagle JM, Weeks DL, Glazer PM: Chromosome targeting at short polypurine sites by cationic triplex-forming oligonucleotides. J Biol Chem 276:38536–38541 (2001b).

Williams DA, Baum C: Gene therapy-new challenges ahead. Science 302:400–401 (2003).

Wu XS, Liu DP, Liang CC: Prospects of chimeric RNA-DNA oligonucleotides in gene therapy. J Biomed Sci 8:439–445 (2001).

Yanez J, Porter AC: Therapeutic gene targeting. Gene Ther 5:149–159 (1998).

Zhang Z, Eriksson M, Falk G, Graff C, Presnell SC, Read MS, Nichols TC, Blomback M, Anvret M: Failure to achieve gene conversion with chimeric circular oligonucleotides: potentially misleading PCR artifacts observed. Antisense Nucleic Acid Drug Dev 8:531–536 (1998).

Zhang SC, Wernig M, Duncan ID, Brustle O, Thomson JA: In vitro differentiation of transplantable neural precursors from human embryonic stem cells. Nat Biotechnol 19:1129–1133 (2001).

Zwaka TP, Thomson J: Homologous recombination in human embryonic stem cells. Nat Biotechnol 21:319–321 (2003).

Cytogenet Genome Res 105:442–447 (2004)
DOI: 10.1159/000078217

Sox genes and cancer

C. Dong, D. Wilhelm and P. Koopman

Institute for Molecular Bioscience, The University of Queensland, Brisbane, QLD (Australia)

Abstract. Sox genes encode transcription factors belonging to the HMG (High Mobility Group) superfamily. They are conserved across species and involved in a number of developmental processes. In vitro studies have shown at least one Sox gene to be capable of inducing oncogenic transformation of fibroblast cells. In addition, overexpression and/or amplification of Sox genes are associated with a large number of tumour types in vivo. We review here the available evidence linking Sox gene expression and cancer, and show that this link is supported by extensive EST database analysis. This work provides a basis for further studies aimed at investigating the possible role of Sox genes in the oncogenic process.

Improvements in life expectancy over the last century have led to the emergence of cancer as a major cause of death in developed countries. In recent decades the molecular and cellular processes leading to oncogenesis have become much better understood, suggesting improved methods of diagnosis and management of cancer.

Many proteins which are involved in developmental processes have been shown to play a role in the formation of malignancies, and many genes initially identified as proto-oncogenes have turned out to be important for developmental processes. These include genes encoding secreted proteins such as the platelet-derived growth factors, transmembrane proteins such as RET and NTRK1 (Alberti et al., 2003), components of signal transduction pathways such as the SHH, WNT and NOTCH pathways (Wicking et al., 1999; Classon and Harlow, 2002; Lustig and Behrens, 2003), and transcription factors such as RUNX (Lund and van Lohuizen, 2002) and AP-1 (Jochum et al., 2001), and members of the PAX (Barr, 1997) and HOX families (Lu et al., 1995; Chang et al., 1997a).

The present review focuses on a family of genes encoding developmental transcription factors known as SOX factors, with emphasis on emerging evidence that they may play a role in some cancers.

HMG proteins

Before turning our attention to SOX factors, it is pertinent to discuss the available evidence that related proteins, the high-mobility-group (HMG) proteins, can play a role in cancer. The HMG protein family is a class of architectural non-histone proteins thought to be involved in gene regulation and maintenance of chromatin structure (Grosschedl et al., 1994). One of the common characteristics of HMG proteins is the presence of a DNA-binding domain called the HMG domain, consisting of approximately 80 residues that form three alpha helices in a twisted L-shape structure (Bewley et al., 1998). The concave surface of the L-shape binds the minor groove of the DNA. Two groups of HMG domains have been classified on the basis of their DNA-binding properties: HMG1 and related proteins bind to distorted DNA structures in a sequence unspecific

C.D. was funded by the Australian-American Fulbright Association, and P.K. by an Australian Professorial Research Fellowship from the Australian Research Council.

Received 15 October 2003; manuscript accepted 19 November 2003.

Request reprints from P. Koopman, Institute for Molecular Bioscience
The University of Queensland, Brisbane, QLD 4072 (Australia)
telephone: +61 7 3346 2059; fax: +61 7 3346 2101
e-mail; p.koopman@imb.uq.edu.au

Fax + 41 61 306 12 34
E-mail karger@karger.ch
www.karger.com

Accessible online at:
www.karger.com/cgr

manner, whereas transcription factors such as SOX proteins recognize and bind to specific DNA sequences (Bewley et al., 1998).

Like HOX and PAX proteins, misregulation of members of the HMG protein family have been linked to the formation and growth of solid tumours (Hesketh, 1995). Overexpression of HMG-like proteins was found in virally transformed rat epithelial thyroid cells as well as cells transformed by murine sarcoma retrovirus (Giancotti et al., 1985, 1987). Compared to the corresponding normal tissues the HMGI proteins, HMGI(Y) and HMGI-C, are overexpressed in metastatic prostate cancer cells, thyroid carcinomas and colorectal carcinomas (Bussemakers et al., 1991; Manfioletti et al., 1991; Tamimi et al., 1993; Chiappetta et al., 1995; Fedele et al., 1996). Furthermore, levels of HMG protein expression correlate with levels of tumour progression, with malignant cells showing high protein expression levels and benign tumour cells and normal adult tissues exhibiting little or no expression (Wunderlich and Bottger, 1997).

SOX factors

Members of the Sox (Sry-like HMG box) gene family were first identified through homology of the HMG domain to the testis-determining factor, SRY (Gubbay et al., 1990). Located on the Y chromosome of mouse and human, Sry is thought to trigger differentiation of genital ridge somatic cells into Sertoli cells and is necessary and sufficient for proper development of the testes (Koopman et al., 1991). Twenty orthologous pairs of Sox genes have been identified in mouse and human and over a dozen have been found to date in invertebrates (Schepers et al., 2002). Like the HMG class of proteins discussed above, SOX factors contain a 79-amino-acid, DNA-binding HMG domain and regulate gene transcription (Wegner, 1999), recognizing the consensus motif, 5′-(A/T)(A/T)CAA(A/T)G-3′ (van de Wetering et al., 1991; Denny et al., 1992; Harley et al., 1992). When bound to DNA the structure of SOX proteins is not significantly altered, however, the DNA strand exhibits a 70–85° bend and a widening of the minor groove (Werner et al., 1995). It is likely that the role of SOX proteins is at least partly architectural, allowing other transcription factors to bind the major groove, and/or bringing together regulatory elements and thereby facilitating the formation of protein complexes (Giese et al., 1992; Koopman, 2001). In addition, it has been shown that sequences outside of the HMG box may facilitate interactions and influence the specificity of SOX proteins (Kamachi et al., 1999; Wilson and Koopman, 2002).

Sox genes are conserved across species, and show tissue-specific expression patterns. Knockout experiments demonstrated that these genes play key roles in embryonic development. In addition to sex differentiation, Sox genes regulate a number of processes including germ layer formation and nervous system development (Wegner, 1999).

At the cellular level, mounting evidence points to a role for many of the SOX factors in cell-type specification and cellular differentiation. For example, SOX9 is implicated in chondrocyte differentiation (Bi et al., 1999), SOX10 in neural crest specification (Kuhlbrodt et al., 1998), SOX17 in endoderm specification (Hudson et al., 1997), and SOX18 in endothelial cell differentiation (Pennisi et al., 2000).

Like many other developmental regulatory factors, the improper functioning of SOX genes has been linked to a number of severe clinical disorders. Mutations in SRY have resulted in XY gonadal dysgenesis (Li et al., 2001). SOX2 mutations have been shown to cause anophthalmia, a rare and severe form of structural malformation of the eye (Fantes et al., 2003). SOX3 has been implicated in X-linked mental retardation (Laumonnier et al., 2002). SOX9 is involved in campomelic dysplasia, a skeletal dysmorphology associated in most XY cases with sex reversal (Foster et al., 1994; Wagner et al., 1994), and SOX10 is linked to Waardenburg-Hirschsprung disease, congenital hypomyelinating neuropathy (CHN), and Yemenite deaf-blind hypopigmentation (Pingault et al., 1998; Bondurand et al., 1999; Inoue et al., 2002). A recent study of hereditary lymphedema identified mutations in SOX18 as a cause of both recessive and dominant forms of the disorder (Irrthum et al., 2003).

SOX factors thus represent an important, diverse and well-conserved group of developmental transcription factors that are being intensively studied in a variety of developmental contexts. We turn now to discuss the emerging body of data suggesting a link between SOX transcription factors and oncogenesis.

SOX genes and oncogenic transformation: causal links

As is often the case in studies aimed at finding the causative agents in cancers, much of the evidence linking SOX factors and cancer is correlative. However, at least one direct study assaying the oncogenic potential of a SOX gene has been carried out. In that study, ectopic *SOX3* expression was found to induce oncogenic transformation of chicken embryonic fibroblasts, and this effect was shown to depend upon both the HMG and transactivation domains of the gene (Xia et al., 2000). In addition, *Sox3* was identified, amongst several known proto-oncogenes in a genome-based analysis of retroviral insertion sites in mouse T-cell lymphomas, further suggesting that this Sox gene may play a role in tumorigenesis (Kim et al., 2003).

SOX genes and tumorigenesis: associative links

Overexpression studies
In order to determine whether other Sox genes might also be involved in cancer development, we carried out an extensive EST analysis to determine which SOX genes are expressed in tumours, cross-referencing the outcomes with published literature where possible. The results are summarised in Table 1.

Four SOX genes – SOX1, SOX2, SOX3 and SOX21 – were found in a cDNA expression library screen to be expressed in sera from small-cell lung cancer patients. EST results indicate that these genes are up-regulated in lung carcinoids and lung squamous cell carcinomas. All four genes, although expressed in early development, are significantly down-regulated in adult tissues (Güre et al., 2000), which supports the hypothesis that

Table 1. EST analysis of SOX gene expression in cancers and cancer cell lines. SOX gene expression was assessed by the presence of ESTs in available EST databases (column 2). This analysis was in many cases supported by direct studies (column 3)

SOX gene	Sources of ESTs	Related studies
SOX1	oligodendroglioma, lung carcinoid	SOX1, SOX2, SOX3 and SOX21 cDNAs have been isolated from sera of small cell lung cancer cell lines (Güre et al., 2000).
SOX2	medulloblastoma, glioblastoma, anaplastic oligodendroglioma, lung carcinoma, primary lung cystic fibrosis epithelial cells	Amplification of SOX2 was detected in human prostate cancer through comparative PCR and Southern blot analyses (Sattler et al., 2000).
SOX3	anaplastic oligodendroglioma	
SOX4	bronchioalveolar carcinoma, invasive prostate tumour, colonic mucosa from ulcerated colitis patients, invasive ductal carcinoma, prostatic intraepithelial neoplasia	SOX4 transcripts are strongly expressed in most classical medulloblastomas, but weakly expressed in desmoplastic medulloblastomas (Lee et al., 2002). Oligonucleotide microarray analysis revealed that SOX4 is highly expressed in adenoid cystic carcinomas (Frierson et al., 2002). It is also up-regulated in breast cancer cell lines (Graham et al., 1999) and has been identified as a potential regulator of HER2/neu, a human breast cancer oncogene (Chang et al., 1997b).
SOX5	testis	SOX5 maps to a region of amplification in human testicular seminomas (Zafarana et al., 2002).
SOX6	multiple sclerosis lesions, melanotic melanoma	
SOX7	cervical carcinoma cell line	SOX7 was shown to be up-regulated in pancreatic cancer cell lines and primary gastric cancer cases, but down-regulated in primary colorectal tumours and breast cancer (Katoh, 2002).
SOX8	anaplastic oligodendroglioma, glioblastoma	
SOX9	chondrosarcoma cell line, osteosarcoma cell line, glioblastoma	SOX9 expression was observed in mesenchymal chondrosarcomas and not in other small cell malignancies (Wehrli et al., 2003).
SOX10	melanotic melanoma cell line, malignant melanoma, metastic to lymph node, anaplastic oligodendroglioma	SOX10 was isolated as a tumour-associated antigen using lymphocytes obtained from a patient with clinical response to immunotherapy (Khong and Rosenberg, 2002).
SOX11	anaplastic oligodendroglioma	
SOX12	medulloblastoma, anaplastic oligodendroglioma, lung carcinoma, ovarian tumour	
SOX13	endometrium adenocarcinoma cell line, melanotic melanoma cell line	
SOX15	papillary serous carcinoma, squamous cell carcinoma	
SOX17	ovarian tumour, well-differentiated endometrial adenocarcinoma, colon tumour, squamous cell carcinoma, serous papillary carcinoma	
SOX18	ovarian tumour, well-differentiated endometrial carcinoma, colon tumour	A number of studies have shown that gain of chromosome region 20q13, where SOX18 is located, is linked to breast, ovarian and colon cancers (Collins et al., 1998; Korn et al., 1999; Diebold et al., 2000).
SOX21	medulloblastoma, oligodendroglioma, glioblastoma, squamous cell carcinoma	SOX21 is associated with small cell lung cancer (Güre et al., 2000), and in the mouse it shows expression in highly metastatic, but not in low metastatic melanoma cells (Tani et al., 1997).

cancer progression often reflects a reversal of normal developmental processes.

SOX4 and SOX11 are of interest in respect to neurological cancers. Both are strongly expressed in most medulloblastomas (Lee et al., 2002), and our EST analysis extends this to anaplastic oligodendroglioma and glioblastoma. In addition to neurological cancers, SOX4 has been implicated in several different tumour types. Expression is increased in malignancies of the pancreas and ovary (Lee et al., 2002). EST analysis confirms SOX4 expression can be found in ovarian cancers as well as invasive ductal carcinoma and prostate cancer. Oligonucleotide microarray analysis of adenoid cystic carcinoma, a type of salivary gland cancer, identified SOX4 among a diverse group of genes that may be involved in oncogenic transformation (Frierson et al., 2002). Furthermore, SOX4 is also highly expressed in subcutaneous human hepatocarcinoma Hep3B mouse xenografts. Expression of SOX4 mRNA is up-regulated in breast cancer cell lines, indicating that SOX4 may also be involved in oncogenic transformation in the breast (Graham et al., 1999). In line with these studies, SOX4 has also been identified as a potential regulator of the human breast cancer oncogene, HER2/neu (c-ErbB2) (Chang et al., 1997a).

SOX8 and SOX9 are both potential genetic markers for cancer. SOX9 is considered a master regulator gene for cartilage differentiation. A study looking at expression of this gene in paediatric small blue round-cell tumours indicates that SOX9 is expressed in mesenchymal chondrosarcomas, but not in other small cell malignancies, lymphomas and leukemias (Wehrli et al., 2003). Small blue round-cell tumours are difficult to distinguish morphologically and finding genes with differential expression would provide a useful means of improving characterisation of these cancers. *SOX8* is expressed in the immature glia of developing chick cerebellum as well as cells scattered throughout medulloblastomas, providing a useful marker for these tumours (Cheng et al., 2001).

One of the tumour antigens identified in melanomas, SOX10, was found in a patient with robust clinical response to tumour immunotherapy (Khong and Rosenberg, 2002). SOX10 is overexpressed in melanoma tumours as compared to normal tissue (Khong and Rosenberg, 2002), and is critical for regulation of melanocyte development as a transactivator of MITF (micropthalmia-associated transcription factor), a master gene for development and postnatal survival of melanocytes (Bondurand et al., 2000). Our EST analysis supports these results, revealing expression of SOX10 in melanomas as well as melanoma cell lines. Expression of this gene has been correlated with metastatic potential of K-1735 melanoma cells, showing significant levels in high metastatic cells with little or no expression in low metastatic cells (Tani et al., 1997). In addition to melanomas, our analysis revealed expression of SOX10 in anaplastic oligodendroglioma and mammary adenocarcinoma cell lines. However, further studies of SOX10 activity must be done in order to provide better insight into the development of carcinomas.

Northern blot, RNA dot blot and cDNA-PCR experiments indicate higher expression levels of SOX18 mRNA in gastric cancer cell lines, pancreatic cell lines and embryonic tumour cell lines compared to other tissues (Saitoh and Katoh, 2002). EST analysis also showed that SOX18 mRNA is expressed in a number of cancers including melanotic melanomas, neuroblastoma cells, pancreatic carcinomas and a number of ovarian and uterine cancers. It is not yet known whether SOX18 marks tumour cells themselves, or whether its expression is linked to tumour angiogenesis in these cases, since SOX18 is expressed in developing vasculature and during neovascularization (Pennisi et al., 2000; Darby et al., 2001).

Gene amplification studies

One of the characteristics of immortalised cells is the gene amplification that stems from chromosomal instability. Amplification sites on the genome indicate regions where putative oncogenes involved in tumorigenesis may be located (Savelieva et al., 1997), and studies looking at the sites where Sox genes are located may provide insight into their potential role in tumorigenesis.

SOX18 is located on chromosome arm 20q13.3, a genomic region implicated in a number of human carcinomas. Breast (Collins et al., 1998), colon (Schlegal et al., 1995) and ovarian cancers (Iwabuchi et al., 1995) all show frequent low level 20q gain and less frequent high level 20q13.2 amplification. Whether SOX18 is the causative gene in these cases is not known.

Other SOX genes, such as SOX2, SOX4 and SOX5, have also demonstrated increased copy number in tumour cells. Comparative PCR and Southern blot analysis to detect DNA copy number gains showed amplification of SOX2 in human prostate cancer (Sattler et al., 2000), and SOX4 in urinary bladder carcinoma cell lines (Bruch et al., 2000). SOX5 is one of three known genes that map to a region of amplification on the short arm of chromosome 12 in testicular seminomas and non-seminomas (Zafarana et al., 2002). However, EST analysis indicates SOX5 may also be expressed in normal testis, so further studies will be needed to determine if up-regulation occurs in malignant tissue.

Concluding remarks

It is clear from the foregoing discussion that a number of links have been found between SOX transcription factors and human cancers. These links range from direct establishment of a causal relationship between SOX3 overexpression and cellular transformation, to many examples of correlative studies linking SOX gene expression with a diverse range of tumours.

However, no examples have arisen to date where specific mutations in SOX genes have been associated with human tumorigenesis or modelled in mice to demonstrate a causal oncogenic role. Similarly, transgenic Sox overexpression experiments in mice have to date not revealed a direct causal link between Sox gene expression and cancer. In other words, while Sox gene expression is undoubtedly correlated with the progression of many tumours, the question of cause versus consequence remains to be resolved.

Nonetheless, it is clear that sufficient evidence exists to warrant further investigations to test the possibility that members of this protein family may play a direct or ancillary role in oncogenic transformation and/or transition between states of cellular differentiation in tumours. In addition to searching for possible cancer-causing mutations and performing transgenic overexpression experiments in mice, it will be instructive to test for clustering between Sox genes and genes known to be involved in oncogenesis in tumour microarray databases as these become more comprehensive.

Thorough investigation of the expression of a suite of SOX proteins in a wide variety of tumours would seem a potentially fruitful avenue of future research, with a view to establishing a correlation between levels of expression and the pathology and differentiation status of the cells in the tumour sample. This in turn may augment the battery of markers available for tumour diagnosis, classification and/or prognosis. Identifying and clarifying specific binding partners and activation signals of SOX proteins may provide additional insights and potential applications relating to human cancers. Further research along these lines could lead to the development of novel therapeutic strategies based on modulation of SOX protein function.

Acknowledgements

We thank Carol Wicking and John Nambu for comments on the manuscript, and Mark Crowe for his suggestions regarding database searches and table organization.

References

Alberti L, Carniti C, Miranda C, Roccato E, Pierotti MA: RET and NTRK1 proto-oncogenes in human diseases. J Cell Physiol 195:168–186 (2003).

Barr FG: Chromosomal translocations involving paired box transcription factors in human cancer. Int J biochem Cell Biol 29:1449–1461 (1997).

Bewley CA, Gronenborn AM, Clore GM: Minor groove-binding architectural proteins: structure, function, and DNA recognition. A Rev Biophys Biomol Struct 27:105–131 (1998).

Bi W, Deng JM, Zhang Z, Behringer RR, de Crombrugghe B: *Sox9* is required for cartilage formation. Nature Genet 22:85–89 (1999).

Bondurand N, Kuhlbrodt K, Pingault V, Enderich J, Sajus M, Tommerup N, Warburg M, Hennekam R, Read A, Wegner M, Goossens M: A molecular analysis of the Yemenite deaf-blind hypopigmentation syndrome: SOX10 dysfunction causes different neurocristopathies. Hum molec Genet 8:1785–1789 (1999).

Bondurand N, Pingault V, Goerich D, Lemort N, Sock E, Le Caignec C, Wegner M, Goossens M: Interaction among SOX10, PAX3 and MITF, three genes altered in Waardenburg syndrome. Hum molec Genet 9:1907–1917 (2000).

Bruch J, Schulz-Wolfgang A, Häussler J, Melzner I, Brüderlein S, Möller P, Kemmerling R, Vogel W, Hameister H: Delineation of the 6p22 amplification unit in urinary bladder carcinoma cell lines. Cancer Res 60:4526–4530 (2000).

Bussemakers M, van de Ven W, Debruyne F, Schalken J: Identification of high mobility group protein I(Y) as potential progression marker for prostate cancer by differential hybridization analysis. Cancer Res 51:606–611 (1991).

Chang C, de Vivo I, Cleary M: The Hox cooperativity motif of the chimeric oncoprotein E2a-Pbx1 is necessary and sufficient for oncogenesis. Mol Cell Biol 17:81–88 (1997a).

Chang CH, Scott GK, Kuo WL, Xiong X, Suzdaltseva Y, Park JW, Sayre P, Erny K, Collins C, Gray JW, Benz CC: ESX: A structurally unique Ets overexpressed early during human breast tumorigenesis. Oncogene 14:1617–1622 (1997b).

Cheng Y, Lee C, Badge R, Orme A, Scotting P: *Sox8* gene expression identifies immature glial cells in developing cerebellum and cerebellar tumours. Mol Brain Res 92:193–200 (2001).

Chiappetta G, Bandiera A, Berlingieri M, Visconti R, Manfioletti G, Battista S, Martinez-Tello F, Santoro M, Giancotti V, Fusco A: The expression of the high mobility group HMGI(Y) proteins correlates with the malignant phenotype of human thyroid neoplasias. Oncogene 10:1307–1314 (1995).

Classon M, Harlow E: The retinoblastoma tumour suppressor in development and cancer. Nature Rev Cancer 2:910–917 (2002).

Collins C, Rommens J, Kowbel D, Godfrey T, Tanner M, Hwang S, Polikoff D, Nonet G, Cochran J, Myambo K, Jay K, et al: Positional cloning of *ZNF217* and *NABC1*: Genes amplified at 20q13.2 and overexpressed in breast carcinoma. Proc natl Acad Sci, USA 95:8703–8708 (1998).

Darby I, Bisucci T, Raghoenath S, Olsson J, Muscat G, Koopman P: *Sox18* is transiently expressed during angiogenesis in granulation tissue of skin wounds with an identical expression pattern to *Flk-1* mRNA. Lab Invest 81:937–943 (2001).

Denny P, Swift S, Brand N, Dabhade N, Barton P, Ashworth A: A conserved family of genes related to the testis determining gene, SRY. Nucl Acids Res 20:2887 (1992).

Diebold J, Mosinger K, Peiro G, Pannekamp U, Kaltz C, Baretton GB, Meier W, Lohrs U: 20q13 and cyclin D1 in ovarian carcinomas. Analysis by fluorescence in situ hybridization. J Pathol 190:564–571 (2000).

Fantes J, Ragge N, Lynch S, McGill N, Richard J, Collin O, Howard-Peebles P, Hayward C, Vivian A, Williamson K, van Heyningen V, FitzPatrick D: Mutations in SOX2 cause anophthalmia. Nature Genet 33:1–2 (2003).

Fedele M, Bandiera A, Chiappetta G, Battista S, Viglietto G, Manfioletti G, Casamassimi A, Santoro M, Giancotti B, Fusco A: Human colorectal carcinomas express high levels of high mobility group HMGI(Y) proteins. Cancer Res 56:1896–1901 (1996).

Foster JW, Dominguez-Steglich MA, Guioli S, Kwok C, Weller PA, Weissenbach J, Mansour S, Young ID, Goodfellow PN, Brook JD, Schafer AJ: Campomelic dysplasia and autosomal sex reversal caused by mutations in an SRY-related gene. Nature 372:525–530 (1994).

Frierson HJ, El-Naggar A, Welsh J, Sapinoso L, Su A, Cheng J, Saku T, Moskaluk C, Hampton G: Large scale molecular analysis identifies genes with altered expression in salivary adenoid cystic carcinoma. Am J Path 161:1315–1323 (2002).

Giancotti V, Berlingieri M, Di Fiore P, Fusco A, Vecchio G, Crane-Robinson C: Changes in nuclear proteins on transformation of rat epithelial thyroid cells by a murine sarcoma retrovirus. Cancer Res 45:6051–6057 (1985).

Giancotti V, Pani B, D'Andrea P, Berlingieri M, Di Fiore P, Fusco A, Vecchio GP, R., Crane-Robinson C, Nicolas R, Wright C, Goodwin G: Elevated levels of a specific class of nuclear phosphoproteins in cells transformed with *v-ras* and *v-mos* oncogenes and by co-transfection with *c-myc* and polyoma middle T genes. EMBO J 6:1981–1987 (1987).

Giese K, Cox J, Grosschedl R: The HMG domain of lymphoid enhancer factor 1 bends DNA and facilitates assembly of functional nucleoprotein structures. Cell 69:185–195 (1992).

Graham JD, Hunt SMN, Tran N, Clarke CL: Regulation of the expression and activity by progestins of a member of the SOX gene family of transcriptional modulators. J Mol Endocrinol 22:295–304 (1999).

Grosschedl R, Giese J, Pagel J: HMG domain proteins: architectural elements in the assembly of nucleoprotein structures. Trends Genet 10:94–100 (1994).

Gubbay J, Collignon J, Koopman P, Capel B, Economou A, Münsterberg A, Vivian N, Goodfellow P, Lovell-Badge R: A gene mapping to the sex-determining region of the mouse Y chromosome is a member of a novel family of embryonically expressed genes. Nature 346:245–250 (1990).

Güre A, Stockert E, Scanlan M, Keresztes R, Jäger D, Altorki N, Old L, Chen Y: Serological identification of embryonic neural proteins as highly immunogenic tumor antigens in small cell lung cancer. Proc natl Acad Sci, USA 97:4198–4203 (2000).

Harley VR, Jackson DI, Hextall PJ, Hawkins JR, Berkovitz GD, Sockanathan S, Lovell-Badge R, Goodfellow PN: DNA binding activity of recombinant SRY from normal males and XY females. Science 255:453–456 (1992).

Hesketh R: The Oncogene Facts Book (Academic Press, London 1995).

Hudson C, Clements D, Friday RV, Stott D, Woodland HR: Xsox17a and -b mediate endoderm formation in *Xenopus*. Cell 91:397–405 (1997).

Inoue K, Shilo K, Boerkoel C, Crowe C, Sawady J, Lupski J, Agamanolis D: Congenital hypomyelinating neuropathy, central dysmyelination, and Waardenburg-Hirschsprung disease: phenotypes linked by SOX10 mutation. Annal Neurology 52:836–842 (2002).

Irrthum A, Devriendt K, Chitaya D, Matthijs G, Glade C, Steijlen P, Fryns J, Van Steensel M, Vikkula M: Mutations in the transcription factor gene SOX18 underlie recessive and dominant forms of hypotrichosis-lymphedema-telangiectasia. Am J hum Genet 72:1470–1478 (2003).

Iwabuchi H, Sakamoto M, Sakunaga H, Yen-Ming M, Carcanyin M, Pinkel D, Yang-Feng T, Gray J: Genetic analysis of benign, low-grade, and high-grade ovarian tumors. Cancer Res 55:6172–6180 (1995).

Jochum W, Passegue E, Wagner EF: AP-1 in mouse development and tumorigenesis. Oncogene 20:2401–2412 (2001).

Kamachi Y, Cheah KS, Kondoh H: Mechanism of regulatory target selection by the SOX high-mobility-group domain proteins as revealed by comparison of SOX1/2/3 and SOX9. Mol Cell Biol 19:107–120 (1999).

Katoh M: Expression of human SOX7 in normal tissues and tumors. Int J molec Med 9:363–368 (2002).

Khong H, Rosenberg S: The Waardenburg Syndrome type 4 gene, SOX10, is a novel tumor-associated antigen identified in a patient with a dramatic response to immunotherapy. Cancer Res 62:3020–3023 (2002).

Kim R, Trubetskoy A, Suzuki T, Jenkins N, Copeland N, Lenz J: Genome-based identification of cancer genes by proviral tagging in mouse retrovirus-induced T-cell lymphomas. J Virol 77:2056–2062 (2003).

Koopman P: SRY and DNA-bending proteins, Encyclopedia of Life Sciences, pp 1–6 (2001).

Koopman P, Gubbay J, Vivian N, Goodfellow P, Lovell-Badge R: Male development of chromosomally female mice transgenic for *Sry*. Nature 351:117–121 (1991).

Korn WM, Yasutake T, Kuo WL, Warren RS, Collins C, Tomita M, Gray J, Waldman FM: Chromosome arm 20q gains and other genomic alterations in colorectal cancer metastatic to liver, as analyzed by comparative genomic hybridization and fluorescence in situ hybridization. Genes Chrom Cancer 25:82–90 (1999).

Kuhlbrodt K, Herbarth B, Sock E, Enderich J, Hermans-Borgmeyer I, Wegner M: Cooperative function of POU proteins and SOX proteins in glial cells. J biol Chem 273:16050–16057 (1998).

Laumonnier F, Ronce N, Hamel B, Thomas P, Lespinasse J, Raynaud M, Paringaux C, Van Bokhoven H, Kalscheuer V, Fryns J, Chelly J, Moraine C, Briault S: Transcription factor SOX3 is involved in X-linked mental retardation with growth hormone deficiency. Am J hum Genet 71:1450–1455 (2002).

Lee C, Appleby V, Orme A, Chan W, Scotting P: Differential expression of SOX4 and SOX11 in medulloblastoma. J Neuro-Oncol 57:201–214 (2002).

Li B, Zhang W, Chan G, Jancso-Radek A, Liu S, Weiss M: Human sex reversal due to impaired nuclear localization of SRY – A clinical correlation. J biol Chem 276:46480–46484 (2001).

Lu Q, Knœpfler P, Scheele J, Wright D, Kamps M: Both Pbx1 and E2A-Pbx1 bind the DNA motif ATCAATCAA cooperatively with the products of multiple murine *Hox* genes, some of which are themselves oncogenes. Mol Cell Biol 15:3786–3795 (1995).

Lund AH, van Lohuizen M: RUNX: a trilogy of cancer genes. Cancer Cell 1:213–215 (2002).

Lustig B, Behrens J: The Wnt signaling pathway and its role in tumor development. J Cancer Res Clin Oncol 129:199–221 (2003).

Manfioletti G, Giancotti V, Bandiera A, Buratti E, Sautiere P, Cary P, Crane-Robinson C, Coles B, Goodwin G: cDNA cloning of the HMGI-C phosphoprotein, a nuclear protein associated with neoplastic and undifferentiated phenotypes. Nucl Acids Res 19:6793–6797 (1991).

Pennisi D, Gardner J, Chambers D, Hosking B, Peters J, Muscat G: Mutations in Sox18 underlie cardiovascular and hair follicle defects. Nature Genet 24:434–437 (2000).

Pingault V, Bondurand N, Kuhlbrodt K, Goerich DE, Préhu M-O, Puliti A, Herbarth B, Hermans-Borgmeyer I, Legius E, Matthijs G, Amiel J, Lyonnet S, Ceccherini I, Romeo G, Clayton Smith J, Read AP, Wegner M, Goossens M: SOX10 mutations in patients with Waardenburg-Hirschsprung disease. Nature Genet 18:171–173 (1998).

Saitoh T, Katoh M: Expression of human SOX18 in normal tissues and tumours. Int J molec Med 10:339–344 (2002).

Sattler H, Lensch R, Rohde V, Zimmer E, Meese E, Bonkhoff H, Retz M, Zwergel T, Bex A, Stoekle M, Wullich B: Novel amplification unit at chromosome 3q25 → q27 in human prostate cancer. Prostate 45:207–215 (2000).

Savelieva E, Belair C, Newton M, DeVries S, Gray J, Waldman F, Reznikoff C: 20q gain associates with immortalization: 20q13.2 amplification correlates with genome instability in human papillomavirus 16 E7 transformed human uroepithelial cells. Oncogene 14:551–560 (1997).

Schepers G, Teasdale R, Koopman K: Twenty pairs of Sox: Extent homology, and nomenclature of the mouse and human Sox transcription factor gene families. Dev Cell 3:167–170 (2002).

Schlegal J, Stumm G, Scherthan H, Bocker T, Zirngibl H, Ruschoff J, Hostadter F: Comparative genomic in situ hybridization of colon carcinomas with replication error. Cancer Res 55:6002–6005 (1995).

Tamimi Y, van der Poel H, Denyn M, Umbas R, Karthaus H, Debruyne F, Shalken J: Increased expression of high mobility group protein I(Y) in high grade prostate cancer determined by in situ hybridization. Cancer Res 53:5512–5516 (1993).

Tani M, Shindo-Okada N, Hashimoto Y, Shirioshi T, Takenoshita S, Nagamachi Y, Yokota J: Isolation of a novel Sry-related gene that is expressed in high-metastatic K-1735 murine melanoma cells. Genomics 39:30–37 (1997).

Wagner T, Wirth J, Meyer J, Zabel B, Held M, Zimmer J, Pasantes J, Bricarelli FD, Keutel J, Hustert E, Wolf U, Tommerup N, Schempp W, Scherer G: Autosomal sex reversal and campomelic dysplasia are caused by mutations in and around the SRY-related gene SOX9. Cell 79:1111–1120 (1994).

Wegner M: From head to toes: the multiple facets of Sox proteins. Nucl Acids Res 27:1409–1420 (1999).

Wehrli B, Huang W, De Crombrugghe B, Ayala A, Czerniak B: Sox9, a master regulator of chondrogenesis, distinguishes mesenchymal chondrosarcoma from other small blue round cell tumours. Human Pathol 34:263–269 (2003).

Werner MH, Huth JR, Gronenborn AM, Clore GM: Molecular basis of human 46X,Y sex reversal revealed from the three-dimensional solution structure of the human SRY-DNA complex. Cell 81:705–714 (1995).

van de Wetering M, Oosterwegel M, Dooijes D, Clevers H: Identification and cloning of TCF-1, a T lymphocyte-specific transcription factor containing a sequence-specific HMG box. EMBO J 10:123–132 (1991).

Wicking C, Smyth I, Bale A: The hedgehog signalling pathway in tumorigenesis and development. Oncogene 18:7844–7851 (1999).

Wilson M, Koopman P: Matching SOX: partner proteins and co-factors of the SOX family of transcriptional regulators. Curr Opin Genet Dev 12:441–446 (2002).

Wunderlich V, Bottger M: High-mobility-group proteins and cancer – an emerging link. J Cancer Res Clin Oncol 123:133–140 (1997).

Xia Y, Papalopulu N, Vogt P, Li J: The oncogenic potential of the high mobility group box protein Sox3. Cancer Res 60:6303–6306 (2000).

Zafarana G, Gillis A, van Gurp R, Olsson P, Elstrodt F, Stoop H, Millan J, Oosterhuis J, Looijenga L: Coamplification of DAD-R, SOX5, and EKI1 in human testicular seminomas, with specific overexpression of DAD-R, correlates with reduced levels of apoptosis and earlier clinical manifestation. Cancer Res 62:1822–1831 (2002).

Cytogenet Genome Res 105:448–454 (2004)
DOI: 10.1159/000078218

Genetic modifiers in mice: the example of the fragile X mouse model

V. Errijgers and R.F. Kooy

Department of Medical Genetics, University of Antwerp, Antwerp (Belgium)

Abstract. Modifiers play an important role in most, if not all human diseases, and mouse models. For some disease models, such as the cystic fibrosis knockout mouse model, the effect of genetic factors other than the causative mutation has been well established and a modifier gene has been mapped. For other mouse models, including those of the fragile X syndrome, a common form of inherited mental retardation, controversies between test results obtained in different laboratories have been well recognized. Yet, the possibility that modifiers could at least explain part of the discrepancies is only scarcely mentioned. In this review we compare the test results obtained in different laboratories and provide evidence that modifiers may affect disease severity in the fragile X knockout mouse.

Copyright © 2004 S. Karger AG, Basel

Modifier genes

It is becoming increasingly evident that the phenotypic expression of traits that are seemingly inherited in a simple Mendelian fashion can differ in subtle or profound ways (Nadeau, 2001). It is well known that allelic variance and environmental factors may contribute to the variation in disease severity. However, a factor commonly overlooked is the effect of modifier genes, e.g., when genes apart from the disease gene modulate the severity of the disease. Modifier genes have been mapped and cloned to a limited extent in humans and mice (Nadeau, 2003). Frequently, studies in animal models guide the search for modifier loci in humans. For instance, the discovery of a locus in mice that modifies the disease severity in cystic fibrosis *(Cftr)* knockout mice (Rozmahel et al., 1996) led to the successful description of a corresponding locus in humans (Zielenski et al., 1999). Identification of modifier genes is a challenging task, but some recent successes show the feasibility. For example, SCNM1, a putative RNA splicing factor was identified as a modifier of disease severity in mice (Buchner et al., 2003). Mutations in both copies of the *Scn8a* gene cause motor endplate disease (Burgess et al., 1995), a neurological disorder with symptoms ranging from tremor or ataxia to juvenile lethality, dependent on genetic background. One of the causative mutations, *Scn8a^medJ*, affects the splice donor site of exon 3, resulting in skipping of exon 2 and 3 and a truncated, nonfunctional protein. In *Scn8a^medJ* mutants, a mixture of incorrect and correct splice variants is apparent and the ratio correlates with disease severity. Severely affected mutant mice of the C57BL/6J strain produce less that 5 % of correctly spliced transcript, whereas mildly affected C3H mice affected produce 10 % correctly spliced transcript. Subsequently, a nonsense mutation in the SCNM1 gene was identified in mice of the C57BL/6J strain, but not in C3H or other mildly affected strains. It was hypothesized that the mutation in the modifier gene SCNM1 reduces the percentage of correctly spliced transcripts in the C57BL/6J strain below a critical threshold.

Fragile X syndrome research in Antwerp is supported by the University of Antwerp Special Research Fund (ASPEO-VIS), the Belgian National Fund for Scientific Research – Flanders (FWO), an Interuniversity Attraction Poles Program (IUAP-V), and the Conquer Fragile X Foundation.

Received 4 December 2003; manuscript accepted 23 December 2003.

Request reprints from: Dr. R. Frank Kooy, Department of Medical Genetics
University of Antwerp, Universiteitsplein 1, 2610 Antwerp (Belgium)
telephone: +32 (0)3 820 2630; fax: +32 (0)3 820 2566
e-mail: Frank.Kooy@ua.ac.be.

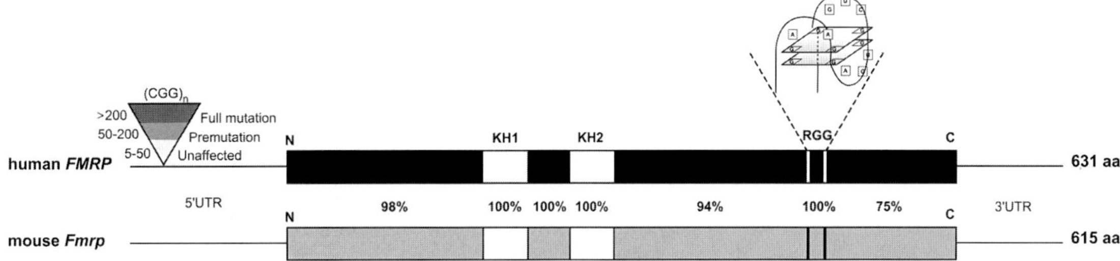

Fig. 1. Comparison of the human and murine fragile X mental retardation protein (FMRP). The similarity between the two proteins is indicated by percentages. KH, heterogeneous nuclear ribonucleoprotein K homology domain; RGG, a repeated arginine-glycine-glycine motif; aa, amino acids.

Identification of modifier genes is of primary importance. Besides providing insights in the biological pathways the disease genes are involved in, the identification of modifier genes might lead us to novel targets for therapeutic interventions in human disease.

The fragile X syndrome

The fragile X syndrome in man is characterized by mental retardation, macroorchidism (enlarged testes), and aberrant behavior (Kooy et al., 2000; Jin and Warren, 2003). The causative gene, FMR1, was identified because of its association with a fragile site at the tip of the long arm of the X chromosome (Verkerk et al., 1991). The gene has a CGG repeat in its 5′ untranslated region that is expanded in patients (Fig. 1). Expansion over a threshold length induces methylation of the promotor region of the gene, leading to a transcriptional stop and consequently, of fragile X protein (FMRP) production.

The function of the FMRP protein has long remained elusive. It is an RNA binding protein that shuttles through the cell as part of a messenger ribonucleoprotein particle (mRNP) guided by a nuclear localization (NLS) and a nuclear export signal (NES). In the cytoplasm, it is mostly associated with the (poly)ribosomes, although FMRP is not believed to be an intrinsic part of the protein making machinery. Bound RNAs include, but are not restricted to, RNAs containing a so-called G-quartet, a hydrogen-bonded structure formed by four guanosine residues in a square-plane array (Schaeffer et al., 2001; Brown et al., 2001). Many G-quartet containing genes play a role in translation and it is believed that, in neurons, the fragile X protein transports specific RNAs to the synapses, thus fine tuning the translation of a selected set of proteins.

Indications that modifier genes play a role in the fragile X syndrome

Despite the fact that >99 % of the patients share the same fragile X mutation (repeat expansion) and do not synthesize any FMRP, even within fragile X families the severity of the cognitive handicap varies enormously from patient to patient. At one end of the spectrum, some patients are able to finalize high school (albeit with intensive aid and care), while patients at the other end of the spectrum are not able to attend any school at all. Moreover, there is no evidence that the cognitive performance of the patients is related to the cognitive performance of their non-affected relatives (de Vries et al., 1996). As in the later study females were investigated, it could be argued that non-random X inactivation explains part of the differences in intellectual capacities between patients. However, even in so-called mosaic males, still synthesizing FMRP in up to 40 % of their cells, the amount of FMRP only explains part of the cognitive variability amongst patients (Bailey et al., 2001). It can therefore be concluded that genetic factors in addition to the lack of FMRP play a role in determining the cognitive capacities of fragile X patients.

The fragile X knockout mouse model

A knockout mouse model for the fragile X syndrome was generated a decade ago (Bakker et al., 1994). Inactivation of the murine fragile X gene, that shares 97 % of its amino acids with the human counterpart, was accomplished by interrupting exon 5 with a neomycin cassette. Thus, although the mutational mechanism differs, like human patients the fragile X mouse is no longer able to generate FMRP. The mouse model has been extensively studied. It appeared to be a genuine model for the human condition, showing learning deficits, macroorchidism, aberrant synaptic connections, hyperactivity and increased sensitivity to epileptic seizures (Kooy, 2003; Bakker and Oostra, 2003).

The genetic background influences the test results obtained with fragile X knockout mice

However, despite the fact that the phenotype of the fragile X mouse model is well defined, a closer comparison of the data obtained in different laboratories reveals inconsistencies among test results. While these inconsistencies might be due to the different experimental environment (Crabbe et al., 1999), little attention is given to the different genetic backgrounds that have been used to test the knockout mouse. Just as random human individuals differ by 0.1 % of their DNA sequence, inbred mouse strains (an inbred strain consists of genetically identical animals obtained by at least 20 generations of brother × sister mating) differ amongst each other by an average of 1 in

Fig. 2. Illustration of the effect of small differences in the genetic composition on the phenotype. The progeny of two brownish 129P2/OlaHsd × C57BL/6J F1 hybrids shows a wide variation in coat color.

1000 base pairs (Wade et al., 2002). Such a seemingly small difference in genetic composition may have a dramatic effect on the phenotype. Genetic diversity is illustrated by the multitude of coat colors that occur in the progeny of two brownish C57BL/6J × 129P2/OlaHsd F1 hybrids (Fig. 2).

The fragile X knockout mouse has been generated by homologous recombination in E14TG2a stem cells, derived from the 129P2/OlaHsd strain (Bakker et al., 1994). Starting from the knockout generated in the 129P2/OlaHsd background, two separate mutant lines were bred: one in a C57BL/6J background and one in a FVB/N background. Thus, mice bred in different genetic backgrounds were studied. Moreover, analysis of both lines began before the mice were fully inbred, e.g. before the completion of 10 generations of backcrossing and thus a significant proportion of 129P2/OlaHsd alleles remained present. This review aims to highlight the differences in test results obtained in different laboratories, and hypothe-

sizes that differences in genetic background might, in part, be responsible.

Morris water maze

The water maze task is used extensively to study spatial learning in small rodents. Animals are trained to escape from a large circular pool filled with opaque water by locating and climbing onto a platform hidden beneath the surface (Morris et al., 1982). A multitude of test protocols has been described (D'Hooge and De Deyn, 2001). A typical experiment with fragile X mice consisted of 12 acquisition trials followed by 4 reversal trials with the platform placed in the opposite quadrant.

Initially, highly significant differences in latency between knockout and control mice were reported, especially during the reversal trials (Bakker et al., 1994) (Fig. 3a). In subsequent experiments under identical conditions, the differences between knockouts and controls were reduced, albeit still signifi-

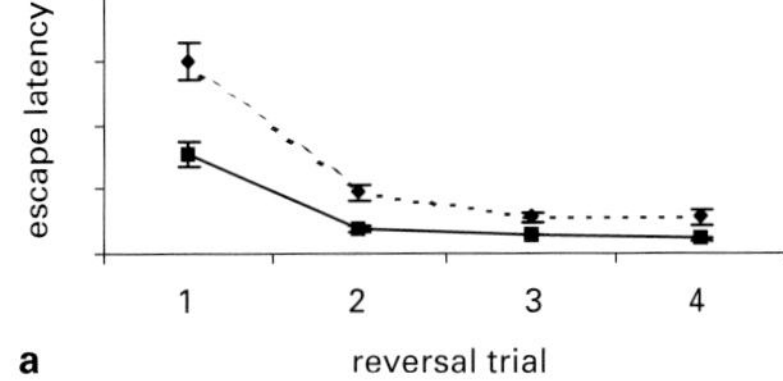

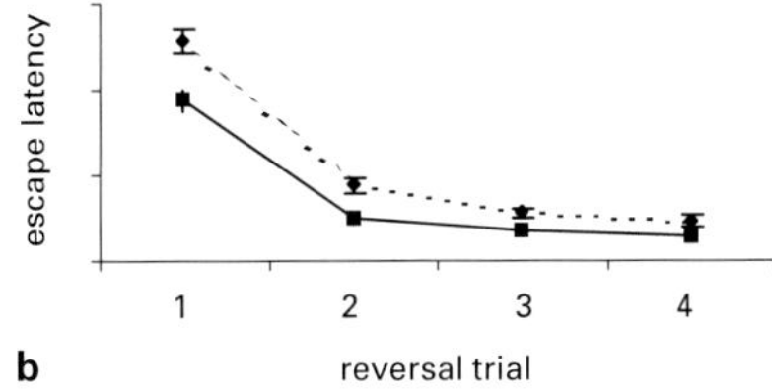

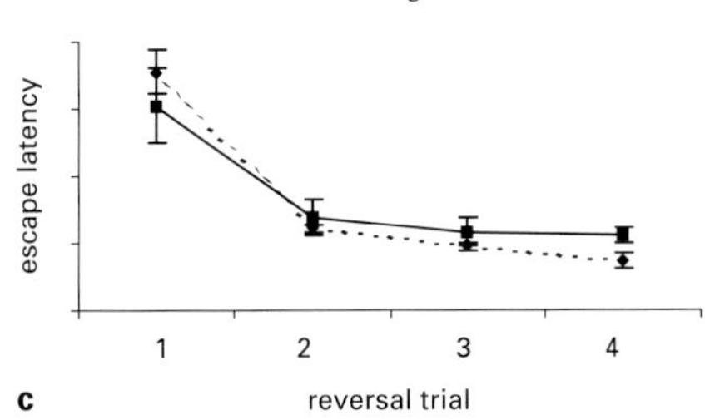

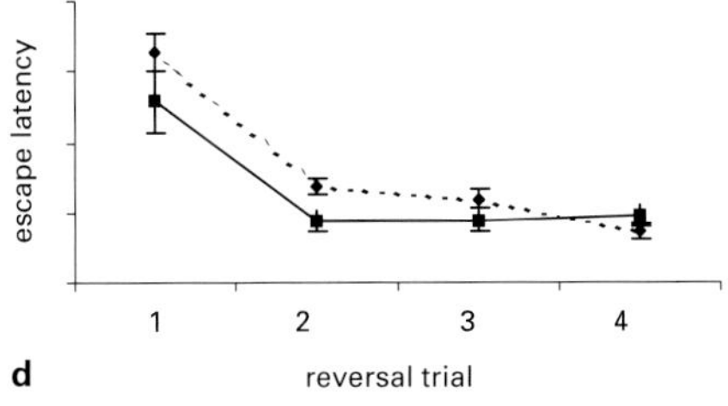

Fig. 3. Escape latency during four reversal trials in a training protocol of hidden-platform water maze learning of mutant mice (dashed line) and their normal littermates (straight line). Taken from: (**a**) Bakker et al., 1994; (**b**) D'Hooge et al., 1997; (**c**) and (**d**) Paradee et al., 1999.

cant (Kooy et al., 1996; D'Hooge et al., 1997) (Fig. 3b). Both studies were carried out in our laboratory during subsequent stages of the inbreeding process of the knockout mouse in a C57BL/6J background. While using fully congenic mutants, Paradee and colleagues (1999) did not observe a difference in Morris water maze performance between knockouts and controls using exactly the same experimental conditions, suggesting that the differences in performance reported initially were due to the presence of remaining 129P2/OlaHsd alleles in the genetic background (Fig. 3c). To test this hypothesis, fragile X mice inbred in the C57BL/6J background were crossed with 129P1Re/J (a close relative of 129P2/OlaHsd) mice and the resulting F1 progeny was tested in the Morris water maze. In accordance with predictions, significant differences in performance between knockouts and controls were observed (Paradee et al., 1999) (Fig. 3d).

Plus-shaped water maze
The plus-shaped water maze is conceptually similar to the Morris water maze task. It is a visuospatial place-learning task in which mice use distal visual cues to find a submerged escape platform in one of the four arms of a cross-shaped basin. When the performance of fragile X mice, partially inbred in a C57BL/6J background was compared with their control littermates, differences in performance could be measured, most notable when the position of the platform was altered after a series of training sessions (Van Dam et al., 2000). When Dobkin and colleagues (2000) measured the performance of nearly congenic (>97% pure) fragile X mice in a C57BL/6J background in the plus-shaped water maze, no differences between knockouts and controls were measured in the acquisition trials. A reversal trial was not performed. Surprisingly, knockout mice in a mixed FVB/N and 129P2/OlaHsd background were seemingly not able to learn to find the hidden platform over 6 consecutive acquisition trials in contrast to their wild type littermates that learned the task well (Dobkin et al., 2000).

Neuroanatomy
Opposite results were obtained when the size of the intra- and infra-pyramidal mossy fiber terminal fields (IIPMF) in the hippocampus were measured in two different strains. The studies were undertaken because the differences in spatial learning correlated positively with the size of the mossy fiber distributions in the hippocampus of inbred mouse lines (Schwegler and Crusio, 1995).

Brain sections through the hippocampus were prepared and stained according to Timm's method, a procedure specifically staining the zinc-rich mossy fibers. The size of the IIPMF fields was measured and expressed as a percentage of the total size of the terminal field. A marginally but significantly reduced size of the IIPMF fields in fragile X knockout mice bred in a C57BL/6J background when compared to control littermates was reported, while the size of the CA4 and suprapyramidal mossy fiber terminal fields remained unchanged (Mineur et al., 2002). This observation is in line with the mildly reduced performance of the knockout mice in tests of spatial learning. However, in a simultaneous study, an increased size of the

C57BL/6J

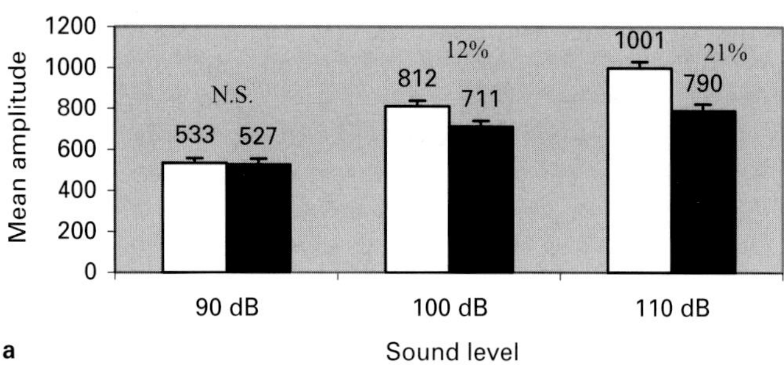

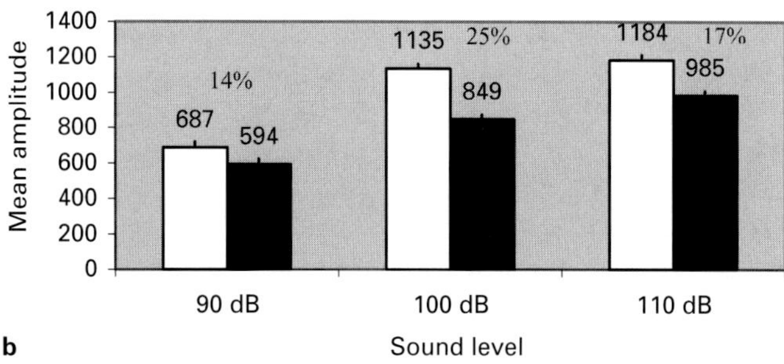

Fig. 4. Startle response of fragile X and control mice to auditory stimuli of variable intensity. Knockouts (black) were compared with control (transparent) littermates in a C57BL/6 (**a**) and in a C57BL/6J × 129P2/OlaHsd F1 background (**b**).

same hippocampal field was reported in knockout animals bred in an FVB/N background (Ivanco and Greenough, 2002).

Acoustic startle response (ASR)

The ASR is used as a behavioral tool to assess the neuronal basis of behavior. An ASR is a fast twitch of facial and body muscles evoked by a sudden, loud acoustic stimulus, mediated by a relatively simple neuronal circuit located in the lower brainstem (Koch, 1999). The magnitude of the reflex is dependent on the sound intensity. The ASR may be inhibited by low intensity acoustic stimuli presented before the high intensity startle-inducing stimulus, a phenomenon termed prepulse inhibition (PPI). PPI is mediated by a more complex forebrain circuitry.

Fragile X mice are less responsive to high-intensity stimuli than their wild type littermates in the ASR. The difference appears strain dependent. The ASR of knockout mice bred in a FVB/N background in response to a sudden noise of 115 dB was 14 % lower than the ASR of control littermates (Chen and Toth, 2001). Under different experimental conditions, knockouts bred in a C57BL6/J background for 5 generations showed significant differences in ASR magnitude when compared with control littermates at a stimulus intensity of 120 dB, but not at 100 dB and 110 dB. However, the ASR of knockouts bred in an C57BL/6J × FVB/N F1 background differed from the control

littermates at each sound level of 100, 110 as well as 120 dB (Nielsen et al., 2002). We too found evidence for strain differences in the ASR at sound intensities of 90, 100 and 110 dB. For example, a 12 % difference in ASR between fragile X and control mice was measured in a congenic C57BL/6 inbred background, while a larger, a 25 % difference was observed with mice bred in a C57BL/6J × 129P2/OlaHsd F1 background at a stimulus intensity of 100 dB (Fig. 4, Errijgers and Kooy, unpublished data). Peier and colleagues (2000) found no significant difference in ASR when knockouts and controls were measured in a pure C57BL/6J background at a stimulus intensity of 120 dB.

In addition, knockout mice bred in different genetic backgrounds show various responses to the same stimuli after prepulse inhibition. Knockouts bred in a C57BL/6J (after 5 generations of inbreeding) background (Nielsen et al., 2002) or an FVB/N background (Chen and Toth, 2001) showed evidence for enhanced prepulse inhibition when compared with wild type littermates, while prepulse inhibition did not differ between knockouts and wild types bred in an C57BL/6J × FVB/N F1 background (Nielsen et al., 2002).

Conclusion and prospects

The variation in test results is compatible with an effect of modifier genes on the fragile X phenotype (Table 1). The presence of the remaining 129 alleles in the C57BL6/J background seems to trigger the difference between knockouts and controls in tests that are dependent on visuospatial orientation, as most elegantly demonstrated by Paradee and colleagues (1999). In addition, knockouts in an FVB/N background seem not to posses any visuospatial learning strategy in the plus-shaped water maze (Dobkin et al., 2000).

Striking is the difference in size of the IIPMF field that was reduced in size in the knockout in the C57BL/6J background (Mineur et al., 2002) but enlarged in the knockout in an FVB/N background (Ivanco and Greenough, 2002). Though differences in disease are known to vary with mouse background; this is the first time to our knowledge that opposite effects in different genetic backgrounds have been reported. The reason for the discrepancy is not known, but might be related to the fact that wild type C57BL/6J mice have relatively large IIPMF fields, whereas wild type FVB/N mice have relatively small IIPMF fields. Although in each study the knockout mice exhibited a diminished ASR, differences in ASR magnitude between mice with different genetic backgrounds were observed, with or without preceding prepulse.

Thus, there is overwhelming evidence, even using a mixture of only three genetic backgrounds of mice that the phenotype of the fragile X knockout mouse is influenced by the genetic background, which suggests that modifier genes in mice may influence disease severity. The next challenge will be to map and eventually clone the responsible modifier gene(s). This will require a test that reproducibly and robustly shows substantial differences between knockouts and controls in one genetic background and not or hardly in other genetic backgrounds. Of the tests described here, in our experience only the ASR poten-

Table 1. Stain differences of the fragile X knockout mouse

Test	Strain dependent deficit	References
Morris water maze	Learning deficit dependent of the presence of remaining 129 alleles in the genetic background: no learning deficit a congenic C57BL/6J background.	Bakker et al., 1994; Kooy et al., 1996; D'Hooge et al., 1997; Paradee et al., 1999
Plus-shaped water maze	Knockouts in FVB/N (mixed with 129P2/OlaHsd) background are not able to learn the task, while knockouts C57BL/6J background are.	Van Dam et al., 2002; Dobkin et al., 2000
Neuroanatomy	Knockouts display a decreased size of the hippocampal IIPMF field in a C57BL/6J background and an enlarged size of the same hippocampal field in an FVB/N background.	Mineur et al., 2002; Ivanco and Greenough, 2002
ASR	Knockouts are less responsive to high-intensity stimuli than controls. The magnitude of the response is dependent on the genetic background. Enhanced PPI in a C57BL/6J and FVB/N background, but not in a C57BL/6J x FVB/N F1 background.	Chen and Toth, 2001; Peier et al., 2000; Nielsen et al., 2002; Errijgers and Kooy, unpublished data

tially fulfills these criteria, but successful identification of modifier genes may require the development of novel tests or of fragile X mouse models that display a more severe phenotype. For example, generation of a mouse model with a mutation corresponding to the Ile304Asn point mutation that in a fragile X patient causes an extreme phenotype with profound mental retardation and excessive macroorchidism might be helpful (De Boulle et al., 1993). Alternatively, the knockout mutation might display a more severe phenotype when bred to other genetic backgrounds than the three laboratory strains used so far. But even with a more pronounced phenotype of the fragile X mouse model, mapping of a modifier will require a well designed set up of experimental crosses and the testing of a large number of animals (Darvasi, 1998). Subsequent fine mapping is possible by setting up additional crosses. Various experimental setups have been applied (Nadeau, 2001, 2003), including crosses using so-called HS strains, outbred mice consisting of a fine grained mosaic background stemming from eight different progenitor strains. The use of HS strains enabled fine mapping of a psychological trait to a subcentimorgan range (Talbot et al., 1999). But even within a small chromosomal region, the number of genes and single nucleotide polymorphisms (SNP) that distinguish both strains may be high (Wade et al., 2002). Therefore, it is important to study the function of the fragile X protein to anticipate the genes that may affect disease severity. As the protein is part of an mRNP particle, other proteins present in the same complex, including the paralogous proteins FXR1 and FXR2, and nucleolin are potential candidates (Ceman et al., 1999). Altered structures of one of these proteins could perhaps restore some of the function of the mRNP complex caused by the loss of FMRP. An entirely different class of candidate modifiers are components of the protein translation machinery, as FMRP appears to be an inhibitor of translation (Laggerbauer et al., 2001; Li et al., 2001). SNPs in other proteins involved in this process that normally result in a decreased efficiency of protein translation might reduce the amount of extra mRNA synthesized in absence of fragile X protein.

Identification of modifier genes would not only increase our insight in the function of the fragile X protein, it might also identify novel targets for therapeutic intervention. In contrast to current efforts, targeting modifier genes aims to improve, rather than cure, the condition of fragile X patients. Given that mental retardation is the predominant hallmark of fragile X syndrome, such a strategy should not a priori be discarded.

References

Bailey DB, Hatton DD, Tassone F, Skinner M, Taylor AK: Variability in FMRP and early development in males with fragile X syndrome. Am J Ment Retard 106:16–27 (2001).

Bakker CE, Oostra BA: Understanding fragile X syndrome: insights from animal models. Cytogenet Genome Res 100:111–123 (2003).

Bakker CE, Verheij C, Willemsen R, van der Helm R, Oerlemans F, Vermey M, Bygrave A, Hoogeveen AT, Oostra BA, Reyniers E, De Boulle K, D'Hooge R, Cras P, van Velzen D, Nagels G, Martin J-J, De Deyn PP, Darby JK, Willems PJ: *Fmr1* knockout mice: a model to study fragile X mental retardation. Cell 78:23–33 (1994).

Brown V, Jin P, Ceman S, Darnell J, O'Donnell WT, Tenenbaum SA, Jin X, Feng Y, Wilkinson KD, Keene JD, Darnell RB, Warren ST: Microarray identification of FMRP-associated brain mRNAs and altered mRNA translational profiles in fragile X syndrome. Cell 107:477–487 (2001).

Buchner DA, Trudeau M, Meisler MH: SCNM1, a putative RNA splicing factor that modifies disease severity in mice. Science 301:967–969 (2003).

Burgess DL, Kohrman DC, Galt J, Plummer NW, Jones JM, Spear B, Meisler MH: Mutation of a new sodium channel gene, *Scn8a,* in the mouse mutant "motor endplate disease". Nat Genet 10:461–465 (1995).

Ceman S, Brown V, Warren ST: Isolation of an FMRP-associated messenger ribonucleoprotein particle and identification of nucleolin and the fragile X-related proteins as components of the complex. Mol Cell Biol 19:7925–7932 (1999).

Chen L, Toth M: Fragile X mice develop sensory hyper-reactivity to auditory stimuli. Neuroscience 103:1043–1050 (2001).

Crabbe JC, Wahlsten D, Dudek BC: Genetics of mouse behavior: interactions with laboratory environment. Science 284:1670–1672 (1999).

D'Hooge R, De Deyn PP: Applications of the Morris water maze in the study of learning and memory. Brain Res Rev 36:60–90 (2001).

D'Hooge R, Nagels G, Franck F, Bakker CE, Reyniers E, Storm K, Kooy RF, Oostra BA, Willems PJ, De Deyn PP: Mildly impaired water maze performance in male *Fmr1* knockout mice. Neuroscience 76:367–376 (1997).

Darvasi A: Experimental strategies for the genetic dissection of complex traits in animal models. Nat Genet 18:19–24 (1998).

De Boulle K, Verkerk AJMH, Reyniers E, Vits L, Hendrickx J, van Roy B, van den Bos F, de Graaff E, Oostra BA, Willems PJ: A point mutation in the FMR1 gene associated with fragile X mental retardation. Nat Genet 3:31–35 (1993).

de Vries BBA, Wiegers AM, Smits APT, Mohkamsing S, Duivenvoorden HJ, Fryns J-P, Curfs LMG, Halley DJJ, Oostra BA, van den Ouweland AMW, Niermijer MF: Mental status of females with an FMR1 gene full mutation. Am J Hum Genet 58:1025–1032 (1996).

Dobkin C, Rabe A, Dumas R, El Idrissi A, Haubenstock H, Brown WT: *Fmr1* knockout mouse has a distinctive strain-specific learning impairment. Neuroscience 100:423–429 (2000).

Ivanco TL, Greenough WT: Altered mossy fiber distributions in adult *Fmr1* (FVB) knockout mice. Hippocampus 12:47–54 (2002).

Jin P, Warren ST: New insights into fragile X syndrome: from molecules to neurobehaviors. Trends Biochem Sci 28:152–158 (2003).

Koch M: The neurobiology of startle. Prog Neurobiol 59:107–128 (1999).

Kooy RF, Willemsen R, Oostra BA: Fragile X syndrome at the turn of the century. Mol Med Today 6:194–199 (2000).

Kooy RF: Of mice and the fragile X syndrome. Trends Genet 19:148–154 (2003).

Kooy RF, D'Hooge R, Reyniers E, Bakker CE, Nagels G, De Boulle K, Storm K, Clincke G, De Deyn PP, Oostra BA, Willems PJ: Transgenic mouse model for the fragile X syndrome. Am J Med Genet 64:241–245 (1996).

Laggerbauer B, Ostareck D, Keidel EM, Ostareck-Lederer A, Fischer U: Evidence that fragile X mental retardation protein is a negative regulator of translation. Hum Mol Genet 10:329–338 (2001).

Li Z, Zhang Y, Ku L, Wilkinson KD, Warren ST, Feng Y: The fragile X mental retardation protein inhibits translation via interacting with mRNA. Nucleic Acids Res 29:2276–2283 (2001).

Mineur YS, Sluyter F, de Wit S, Oostra BA, Crusio WE: Behavioral and neuroanatomical characterization of the *Fmr1* knockout mouse. Hippocampus 12:39–46 (2002).

Morris RGM, Garrud P, Rawlins JNP, O'Keefe J: Place navigation impaired in rats with hippocampal lesions. Nature 297:681–683 (1982).

Nadeau JH: Modifier genes in mice and humans. Nat Rev Genet 2:165–174 (2001).

Nadeau JH: Modifier genes and protective alleles in humans and mice. Curr Opin Genet Dev 13:290–295 (2003).

Nielsen DM, Derber WJ, McClellan DA, Crnic LS: Alterations in the auditory startle response in *Fmr1* targeted mutant mouse models of fragile X syndrome. Brain Res 927:8–17 (2002).

Paradee W, Melikian HE, Rasmussen DE, Kenneson A, Conn PJ, Warren ST: Fragile X mouse: strain effects of knockout phenotype and evidence suggesting deficient amygdala function. Neuroscience 94:185–192 (1999).

Peier AM, McIlwain KL, Kenneson A, Warren ST, Paylor R, Nelson DL: (Over)correction of FMR1 deficiency with YAC transgenics: behavioral and physical features. Hum Mol Genet 9:1145–1159 (2000).

Rozmahel R, Wilschanski M, Matin A, Plyte S, Oliver M, Auerbach W, Moore A, Forstner J, Durie P, Nadeau J, Bear C, Tsui L-C: Modulation of disease severity in cystic fibrosis transmembrane conductance regulator deficient mice by a secondary genetic factor. Nat Genet 12:280–287 (1996).

Schaeffer C, Bardoni B, Mandel JL, Ehresmann B, Ehresmann C, Moine H: The fragile X mental retardation protein binds specifically to its mRNA via a purine quartet motif. EMBO J 20:4803–4813 (2001).

Schwegler H, Crusio WE: Correlations between radial-maze learning and structural variations of septum and hippocampus in rodents. Behav Brain Res 67:29–41 (1995).

Talbot CJ, Nicod A, Cherny SS, Fulker DW, Collins AC, Flint J: High-resolution mapping of quantitative trait loci in outbred mice. Nat Genet 21:305–308 (1999).

Van Dam D, D'Hooge R, Hauben U, Reyniers E, Gantois I, Bakker CE, Oostra BA, Kooy RF, De Deyn PP: Spatial learning, contextual fear conditioning and conditioned emotional response in *Fmr1* knockout mice. Behav Brain Res 117:127–136 (2000).

Verkerk AJMH, Pieretti M, Sutcliffe JS, Fu Y-H, Kuhl DPA, Pizzuti A, Reiner O, Richards S, Victoria MF, Zhang F, Eussen BE, van Ommen G-JB, Blonden LAJ, Riggins GJ, Chastain JL, Kunst CB, Galjaard H, Caskey CT, Nelson DL, Oostra BA, Warren ST: Identification of a gene (FMR-1) containing a CGG repeat coincident with a breakpoint cluster region exhibiting length variation in fragile X syndrome. Cell 65:905–914 (1991).

Wade CM, Kulbokas III EJ, Kirby AW, Zody MC, Mullikin JC, Lander ES, Lindblad-Toh K, Daly MJ: The mosaic structure of variation in the laboratory mouse genome. Nature 420:574–578 (2002).

Zielenski J, Corey M, Rozmahel R, Markiewicz D, Aznarez I, Casals T, Larriba S, Mercier B, Cutting GR, Krebsova A, Macek Jr M, Langfelder-Schwind E, Marshall BC, DeCelie-Germana J, Claustres M, Palacio A, Bal J, Nowakowska A, Ferec C, Estivill X, Durie P, Tsui L-C: Detection of a cystic fibrosis modifier locus for meconium ileus on human chromosome 19q13. Nat Genet 22:128–129 (1999).

Cytogenet Genome Res 105:455–463 (2004)
DOI: 10.1159/000078219

The role of replicative senescence in cancer and human ageing: utility (or otherwise) of murine models

S.K. Smith and D. Kipling

Department of Pathology, University of Wales College of Medicine, Cardiff (UK)

Abstract. Replicative senescence has the potential both to act as an anti-tumour mechanism, and to contribute to age-related changes in tissue function. Studies on human cells have revealed much, both about the nature of cell division counters, some of which utilize the gradual erosion of chromosomal telomeres, and the downstream signalling pathways that initiate and maintain growth arrest in senescence. A powerful test of the hypothesis that senescence is linked to either ageing or tumour prevention now requires a suitable animal model system. Here we overview the current understanding of replicative senescence in human cells, and address to what extent the senescence of murine cells in culture mirrors this phenomenon. We also discuss whether examples of telomere-independent senescence, such as those seen in mouse embryonic fibroblasts (MEFs) and several human cells types, should be viewed not as a consequence of "inadequate growth conditions", but rather as a powerful potential model system to dissect the selective pressures that occur in the early stages of tumour development, ones that we speculate lead to the observed high frequency of abrogation of p16^{INK4a} function in human cancer.

Copyright © 2004 S. Karger AG, Basel

What is replicative senescence?

Normal mammalian cells cannot divide indefinitely in culture. They have a limited proliferative lifespan, after which they enter a state of permanent growth arrest, known as cellular or replicative senescence. This process is believed to exist as an inherent barrier to excessive cell division in all normal somatic cells (with the possible exception of some stem cells), and may be an in vitro manifestation of intrinsic anti-tumour mechanisms. Cellular proliferative capacity was first examined in detail in human diploid fibroblasts by Leonard Hayflick (Hayflick and Moorhead, 1961; Hayflick, 1965), who defined the replicative lifespan of a primary culture as the number of cumulative population doublings (PD) it could sustain. The number of PD a culture will undergo depends on the cell type, but the point at which any particular cell within the population will stop dividing is stochastically determined. Cell cultures are thus comprised of populations of actively proliferating cells and post-mitotic "senescent" cells. At early passage numbers replicating cells make up the greater proportion, but this fraction decreases steadily with time as clones reach the end of their lifespan, until the whole culture is made up of senescent cells and growth stops (Faragher, 2000). In normal human fibroblast cultures, this point (the "Hayflick limit") is typically reached after 60–80 PD.

From studies of human diploid fibroblasts emerged the classical model of replicative senescence in which the telomeres – the nucleoprotein structures that cap the ends of linear chromosomes – act as a "cell-division counter." With each round of mitosis, typically 50–100 bp of DNA are lost from the chromosome ends, due to the end-replication problem, oxidative damage and C-strand processing (see below). Shortening of one or more telomeres to a certain critical length is believed to be the trigger that activates the cell's senescence machinery and imposes growth arrest (Harley et al., 1990; Allsopp and Harley, 1995; Reddel, 1998).

SKS is supported by the BBSRC's *Exploiting Genomics* Initiative.

Received 3 November 2004; manuscript accepted 4 December 2003.

Request reprints from: Dr. David Kipling, Department of Pathology
    University of Wales College of Medicine, Heath Park, Cardiff CF14 4XN (UK)
    telephone: +44-(0)-29-2074-2700; fax: +44-(0)-29-2074-4276
    e-mail: KiplingD@cardiff.ac.uk.

Senescence as an anti-cancer mechanism

Replicative senescence is thought to have evolved as a tumour suppression mechanism that allows sufficient cell division for the growth, development and daily function of the organism, whilst prohibiting the excessive clonal expansion necessary for cancer development. Tumorigenesis requires multiple successive mutations of genes in growth regulatory pathways and, normally, the senescence programme will cause a mutation-bearing clone to cease dividing before it has formed a pool of cells large enough to acquire a further oncogenic mutation by chance (Kipling, 2001). During tumour development there is strong selective pressure for cells that have lost the function of a gene in a senescence-inducing pathway, thereby acquiring a greater replicative potential (Parkinson et al., 1997; Wynford-Thomas, 1999). Indeed, the most common loss-of-function mutations in human cancers occur in genes encoding p53 and members of the pRB pathway, effectors of the cell cycle arrest at replicative senescence (Sherr, 2000; Campisi, 2001; Mathon and Lloyd, 2001). In addition, telomere stabilization appears to be a key event in acquisition of the immortal phenotype, since the majority of human tumours lose the repression of telomerase activity, or maintain the length of their telomeres by an alternative, recombination-based mechanism (Kim et al., 1994; Bryan et al., 1997). Human fibroblasts in which p53 and pRb have been inactivated bypass the senescence checkpoint and replicate for an additional 30 PD, until the ensuing telomere shortening initiates a second proliferative lifespan barrier known as "crisis", characterized by chromosomal aberrations, continued mitosis and concurrent apoptosis (Shay et al., 1991a; Bond et al., 1999). This mechanism is the key determinant of lifespan in cells like human fibroblasts, but is only possible because humans have relatively short telomeres (5–15 kb) and, in most cells, do not express significant amounts of the telomere-lengthening reverse transcriptase, telomerase. Forced expression of telomerase, which counteracts replication-associated telomere attrition, is sufficient to bypass both senescence and crisis and to immortalize human diploid fibroblasts (Bodnar et al., 1998; Counter et al., 1998; Vaziri and Benchimol, 1998). In mice the story appears to be somewhat different, as will be detailed below.

Senescence may contribute to human ageing

Cell division occurs in many tissues during life, either as a result of the day-to-day function of that tissue (e.g. skin, lining of the gut, immune system), or in response to wounding or infection. If cells have a finite proliferative capacity, there is a possibility that this division potential will be exhausted during life, which in turn may contribute to some of the changes that we see in aged tissue. These changes may reflect both a reduction in division potential (where the ability to divide is a key feature of the functioning of that tissue), and the changed gene expression profiles seen in senescent cells (Faragher and Kipling, 1998). Senescent cells in culture remain viable for long periods of time, cannot be stimulated to divide by physiological mitogens, and are resistant to triggers of apoptosis (Stein et al., 1991; Wang, 1995). At senescence, growth arrest is accompanied by dramatic changes in cellular phenotype. Senescent fibroblasts, for example, up-regulate inflammatory cytokines and various enzymes associated with extracellular matrix degradation (e.g. matrix metalloproteinases, stromelysin, and PAI-1 and 2), a gene expression pattern similar to that of activated fibroblasts during wound healing (Campisi, 1997; Shelton et al., 1999). They also express pH-6.0 β-galactosidase (SA-βgal) activity, a marker that has been detected in fibroblasts and keratinocytes in human skin in vivo in an age-dependent fashion in some reports (Dimri et al., 1995), although not in others (Severino et al., 2000).

The accumulation of non-proliferating cells with altered function, as a consequence of replicative senescence, has been proposed as one possible cause of the observed decline in tissue function and repair during organismal ageing. The contribution of cellular senescence to age-associated dysfunction may be an unselected evolutionary by-product of effective tumour suppression during the organism's reproductive lifespan (Faragher, 2000; Faragher and Kipling, 1998). Although the occurrence of senescent-like cells has been reported in vivo in human (Dimri et al., 1995) and mouse (Paramio et al., 2001) skin, and in rat kidney (Melk et al., 2003), there remains only limited evidence that senescent cells do accumulate in aged tissues, and even less that they do so to a sufficient extent that they make a quantitatively significant impact on the ageing of that tissue. This has largely been the result of a paucity of critical studies, reflecting a lack of reagents to assess or modulate senescent cells, as opposed to a wealth of negative data. There is thus a strong need for the development of new tools, and in particular new animal model systems, in which to probe the link between replicative senescence and ageing.

Testing the link between senescence and ageing: methods to abrogate senescence

One approach to testing the link between senescence and ageing would be to modulate the occurrence of senescence by altering the signalling pathways that trigger it. A reduction in the fraction of senescent cells should then correlate with a reduction in key metrics of tissue ageing postulated to be associated with the accumulation of senescent cells.

Many studies have used viral oncoproteins to dissect the mechanism of senescence. Studies of the actions of small DNA tumour virus genes on cultured human cells have identified several mediators of senescence, and established a model for cellular immortalization. These viruses allow their host cells to escape senescence by expressing gene products that disrupt cell cycle control proteins and inactivate senescence pathways. For example, the human papillomavirus 16 (HPV16) E6 oncoprotein inhibits p53 (amongst other functions), whereas the E7 gene product binds hypophosphorylated pRB and maintains its growth permissive status, as well as disrupting the function of other cell cycle regulators such as $p21^{WAF1/CIP1}$, $p27^{KIP1}$, cyclins A and E and members of the E2F family (Southern and Herrington, 2000). The SV40 large T antigen inhibits both p53 and pRB (Bryan and Reddel, 1994). In a study by Shay et al.

(1991a), transfection of SV40 T antigen into conditionally immortalized human fibroblasts extended their proliferative lifespan beyond that of control untransfected cells, as did treatment with a combination of HPV E6 and E7, demonstrating the necessity of both p53 and pRB in fibroblast senescence. After a further 20–30 PD beyond the onset of senescence in control cells, the transfected cultures entered a state of "crisis" in which mitosis continued, but with an increased rate of cell death, thus resulting in no net increase, and eventually a decline, in cell number. These results illustrate that escape from senescence (also termed mortality stage 1, M1) is not sufficient for immortalization, but a subsequent proliferative lifespan barrier, crisis, or M2, must also be overcome (Shay et al., 1991b). Crisis occurs in fibroblasts as a consequence of continued telomere shortening when normal senescence pathways have been inactivated. Immortalization occurs when crisis is overcome by expression of telomerase or activation of an alternative mechanism of telomere maintenance (Reddel, 1998).

Such studies do not provide a ready route to testing the link between senescence and ageing, though, because of the large number of targets for these viral oncoproteins which thus confounds any interpretation. However, similar studies, together with a wealth of other data on the signalling pathways in senescence (see below), and with recent developments in our ability to modulate gene expression patterns in senescent cells in a very precise, targeted fashion (Beausejour et al., 2003; Narita et al., 2003), will now allow a critical test of this hypothesis.

Telomeres can count cell divisions in human cells

Whereas one approach to senescence is to abrogate it by modulation of the upstream signalling pathways, a second is to define the nature of the cell division counter itself and to alter its function – literally "stopping the clock". In some cell types we now have a sufficient understanding of the nature of the cell division counter involving the progressive erosion of chromosomal telomeres that we are able to experimentally intervene and modulate its function.

Hayflick's observation that cellular lifespan is a function of PD in culture implied that cells have a mechanism by which they "count" the number of mitotic cycles they have completed, and which triggers senescence when a critical point is reached. A candidate for the role of mitotic "clock" was recognized when it was predicted that telomeres would progressively shorten with each round of division because of the nature of the DNA replication machinery (Reddel, 1998). Since DNA is polymerized in the 5′ to 3′ direction, the "leading" strand can be synthesized as a continuous molecule that can replicate the entire length of a linear template. The "lagging" strand, however, must be synthesized as a series of discontinuous Okazaki fragments, each requiring an RNA primer that is later removed, leaving a gap between each Okazaki fragment and the next. Internal gaps are readily filled, but the gap between the most extreme 5′ Okazaki fragment and the end of the chromosome cannot be filled since there is no "upstream" DNA for a priming event. As a result, the newly synthesized lagging strand is shorter than its template and a 3′ overhang is created on the template strand.

This telomere attrition is prevented in some cell types by the action of telomerase (see Xin and Broccoli, this issue), but in cells in which expression of this enzyme is suppressed, telomeres become incrementally shorter at each round of DNA replication and mitosis. According to the telomere hypothesis of senescence, this progressive reduction in telomere length is the factor that defines the replicative lifespan of each cell. Support for this theory was provided by observations that mean telomere length in human fibroblasts decreases with serial passage and is associated with cell division (Harley et al., 1990; Allsopp et al., 1995; Allsopp and Harley, 1995). This hypothesis was confirmed when it was shown that forced expression of hTERT, the catalytic component of telomerase, maintained telomere length in both normal human foreskin fibroblasts and retinal pigmented epithelial cells, and dramatically extended their in vitro lifespan (Bodnar et al., 1998; Counter et al., 1998; Vaziri and Benchimol, 1998). Remarkably, these hTERT-expressing fibroblasts appear to be immortal while remaining diploid and retaining their anchorage-dependent growth characteristics (Jiang et al., 1999; Morales et al., 1999). However, subsequent analysis revealed that hTERT immortalized cells experience a phase of genomic instability (Vaziri et al., 1999) and acquire epigenetic changes such as upregulation of c-Myc (Wang et al., 2000), and cannot be considered to be completely phenotypically identical to their young wild type counterparts. As well as fibroblasts, telomere-driven senescence has been demonstrated in other human cell types, including vascular endothelial cells (Yang et al., 1999), bone marrow stromal cells (Simonsen et al., 2002), oesophageal squamous cells (Morales et al., 2003), and myometrial cells (Condon et al., 2002). In these cell types that can be immortalized by expression of hTERT, telomere attrition appears to be the predominant inducer of replicative senescence, but the telomere hypothesis does not preclude other cell cycle "counting" mechanisms (Reddel, 1998).

The tumour suppressor and checkpoint protein p53 links telomere erosion to cell cycle arrest

Mammalian telomeres consist of several kilobases of hexameric repeats of the sequence TTAGGG which end in a distinctive looped configuration, stabilized by the binding of specific proteins (see Xin and Broccoli, this issue). These telomeric loops are thought to mask the ends of the chromosomes to prevent their detection by the cell cycle checkpoint machinery, and recent studies have suggested that altered structure of shortened telomeres, rather than the DNA loss itself, may initiate senescence (Karlseder et al., 2002; Smogorzewska and de Lange, 2002; Li et al., 2003). Whether it is average telomere length (Martens et al., 2000) or the shortest telomere (Hemann et al., 2001) that triggers senescence currently remains unclear. A simple model is that a critically short telomere is no longer able to provide the key capping function to that chromosome, leading to the natural end of the chromosome being seen as a double-stranded break by the cell and thus triggering a DNA damage like growth arrest response. Indeed, the pattern of post-translational phosphorylation of p53 in senescent HCA2 fibroblasts overlaps, but is not identical to, the p53 phosphorylation pattern induced by UV and γ irradiation. (Webley et al., 2000).

Several lines of evidence demonstrate the importance of p53 in both induction and maintenance of the senescence-associated cell cycle arrest. First, pre-senescent human fibroblasts expressing a retrovirally-delivered dominant negative mutant p53 escape senescence and complete an average of 19 additional PD before reaching a crisis-like state (Bond et al., 1994). These authors later showed, using fibroblasts stably transfected with a p53-driven reporter construct, that the onset of senescence coincides with increased transcriptional activity of p53 (Bond et al., 1996). Furthermore, direct inhibition of p53 by microinjection of anti-p53 antibodies is sufficient to restore DNA synthesis and cell division in senescent human dermal fibroblasts (Gire and Wynford-Thomas, 1998).

At senescence, p53 initiates pRB-mediated growth arrest by transcriptional upregulation of the cyclin-dependent kinase inhibitor (CDKI), p21$^{WAF1/CIP1}$. p53 and pRB are regulators of intersecting pathways that normally act together in senescence, but can each bring about cell cycle arrest individually to compensate for loss of function of the other (Bond et al., 1999).

The role of p16^{INK4A} in senescence

A distinctive feature of replicative ageing in human diploid fibroblasts is the dramatic accumulation of the cyclin-dependent kinase inhibitors p21$^{WAF1/CIP1}$ (p21) and p16^{INK4A} (p16) as they approach the end of their lifespan (Alcorta et al., 1996; Hara et al., 1996; Stein et al., 1999). p21and p16 are the prototype members of two families of CDK inhibitors (CDKIs) that can halt the cell cycle in late G1 by preventing CDKs from phosphorylating pRB. The p21 family includes two related proteins, p27^{KIP1} and p57^{KIP2}, and the INK4 family has four known members: p16^{INK4A}, p15^{INK4B}, p18^{INK4C} and p19^{INK4D}. The CIP/KIP family are multifunctional, and are able to bind cyclin D-, E- and A-dependent kinases. They are potent inhibitors of cyclin E- and A- dependent cdk2, but appear to be positive regulators of cyclin D-dependent kinases. p21 also negatively regulates cell cycle progression by binding to PCNA. In contrast, the INK4 family are very specific inhibitors of the catalytic subunits of cyclin D dependent kinases, CDK4 and CDK6 (Sherr and Roberts, 1999). Both of these CDKIs have been shown to be sufficient to induce features of the senescent phenotype as well as cell cycle arrest when overexpressed in normal fibroblasts (McConnell et al., 1999) and immortalized fibroblast lines (Vogt et al., 1998). However, of the two, the role of p16 is particularly intriguing because loss of p16 function by mutation or promoter methylation is a frequent occurrence in human cancer, whereas loss of p21 is rarely, if ever, seen (Sherr, 1996; Ruas and Peters, 1998).

How does p16 impose cell cycle arrest? The central regulator of the G1 checkpoint is the retinoblastoma susceptibility protein, pRB, the activity of which depends on its phosphorylation state (Sherr and Roberts, 1999; Trimarchi and Lees, 2002). In its hypophosphorylated (active) condition, pRB prevents entry into S phase by binding "activating" members of the E2F family of transcription factors. If released from pRB inhibition, E2Fs initiate expression of essential S phase genes. In the absence of mitogens, pRB is maintained in this hypophosphorylated state while cyclin E-CDK2 complexes are inhibited by p21 or p27^{KIP1}. Mitogens stimulate the synthesis of cyclin D1 and its assembly into active complexes with CDK4 or 6. These D1-CDK4 or 6 complexes are thought to be resistant to CIP/KIP inhibition and titrate p21 and p27^{KIP1}, which in turn releases cyclin E-CDK2 from negative regulation. pRB is phosphorylated by the active CDKs, and E2F is released, allowing transition into S phase. Induction of p16 prevents cell cycle progression by reversal of these events. p16 inactivates cyclin D-CDK4 or 6 by dissociating the complex and sequestering CDK4, with release of the bound p21 or p27^{KIP1}. These CIP/KIP proteins subsequently bind and inhibit cyclin E-CDK2 (Jiang et al., 1998; McConnell et al., 1999). Thus, although p16 is a specific biochemical inhibitor of CDK4 and 6, it effectively inhibits pRB phosphorylation by a pleiotropic mechanism involving CIP/KIP proteins and cyclin E-CDK2.

Investigations of p21 and p16 induction in senescent fibroblasts have suggested differential roles for these two CDKIs. Both p21 and p16 mRNA and protein steadily increase with serial passage. p21 concentration peaks in early senescence and then declines, although it remains higher than in log phase proliferating cells, whereas p16 levels continue to rise after growth arrest has occurred, and remain elevated for at least two months (Alcorta et al., 1996; Stein et al., 1999). At the time the growth rate of a culture starts to diminish (bromodeoxyuridine labelling index (BrdU LI) of approx. 5%), p21 concentrations are typically high but p16 levels are still low. Peak p16 levels occur when the BrdU LI has fallen to less than 1% (Alcorta et al., 1996) and coincides with changes in cell morphology and expression of SA-βgal. For this reason, Stein and colleagues have suggested that p21 is responsible for the initial senescence cell cycle arrest, but that p16 is necessary for late senescence maintenance and phenotype.

p16 has no obvious role in development or normal cell-cycle control, but the extremely long half-life of its mRNA (>24 h; Hara et al., 1996) is consistent with its involvement in maintenance of a long-term phenotype like replicative senescence. Indeed, two recent studies by Lowe and colleagues, in which RNAi technology was used to silence p16 expression in human fibroblasts have revealed that p16 does not appear to be required for initiating cell cycle arrest in senescent human fibroblasts, but is necessary for maintaining it. Replicative senescence was shown to be reversible by forced expression of a dominant negative p53 in BJ fibroblasts but not in WI38 fibroblasts which differ in the degree to which they upregulate p16 at late passage, with WI38 cells expressing significantly more p16 than BJ. To test the role p16 plays in this situation, WI38 were transduced to express short hairpin RNAs (shRNA) that stably suppress p16^{INK4A} expression. This led to a small extension of proliferative lifespan of 2–3 PD after which they senesced with high levels of p21. Subsequent introduction of dominant negative p53 was now able to reverse this p21-dependent p16-independent senescence (Beausejour et al., 2003). In a separate study, p16 or pRB suppression by shRNA in WI38 cells did not prevent Ras-induced growth arrest, but did inhibit senescence-associated heterochromatin formation and the E2F target gene silencing (Narita et al., 2003).

The product of the alternative transcript of the p16 locus, p14[ARF], a positive regulator of p53, which is involved in mouse embryonic fibroblast senescence, is not induced in fibroblast senescence in humans (Wei et al., 2001; Collins and Sedivy, 2003).

Telomere-independent senescence mechanisms also exist

Senescence can also occur in response to non-telomere signals. These can be both acute in nature, with senescence occurring within a few days (e.g. in response to high levels of oxidative stress or aberrant mitogenic stimulation), or after a period of several weeks and many population doublings (e.g. several human cell types under normal growth conditions, and human thyroid epithelium forced to divide following expression of oncogenic RAS).

A senescent-like state can be induced in human fibroblasts before the Hayflick limit is reached by a variety of extra- and intracellular stimuli including DNA damage by γ-irradiation (Di Leonardo et al., 1994), hyperoxia (von Zglinicki et al., 1995), hydrogen peroxide (Chen and Ames, 1994) and overexpression of the H-ras oncogene (Serrano et al., 1997; Ferbeyre et al., 2002; Huot et al., 2002). Some of these appear to be independent of telomere shortening and induce growth arrest and markers of senescence within days of onset. For example, forced expression of H-ras by retroviral transduction causes human fibroblasts to begin to adopt a senescence-like flattened morphology within 2 days and undergo complete G1 growth arrest, without reaching confluence, after 6 days. These growth-arrested cells have elevated levels of p53, p16 and p21 and a reduction in hyperphosphorylated pRB. They also had increased amounts of SA-βgal and PAI-1, and appear identical to fibroblasts that had undergone senescence due to replicative exhaustion by every criterion examined. Ras-induced senescence is dependent on both p53 and p16/pRB, since individual abrogation of either of these pathways by expression of a dominant negative p53, or a p16-binding deficient mutant CDK4 failed to prevent the cell cycle arrest (Serrano et al., 1997). Ras induction of premature senescence is independent of telomere status because hTERT immortalized human fibroblasts are still subject to ras-induced growth inhibition (Morales et al., 1999). The ras family of proto-oncogenes encodes small GTP-binding proteins that transduce mitogenic signals from tyrosine-kinase receptors by activation of MAP kinase pathways, as part of the normal cellular growth regulatory system. As may be expected, the immediate effect of ras in primary cells is mitogenic, as shown by microinjection studies (Lumpkin et al., 1986), but in certain situations continued ras activation leads to growth inhibition mediated by upregulation of p53 and p16.

The observation that "premature" senescence can be induced by oncogenic H-ras and external DNA damaging agents extends the hypothesis of senescence as a tumour suppressive mechanism. It indicates that as well as responding to signals generated by cumulative population doublings, the senescence pathways can be induced by aberrant mitogenic stimulation or by DNA damage as an antiproliferative failsafe mechanism.

Telomere-independent senescence: slow cell division counters

In addition to the acute triggers of senescence described above, there are examples of senescence that are associated with significant cell division prior to arrest. These instances are referred to as telomere-independent senescence (TIS) on the basis of an operational definition, insofar as they do not show extended cellular lifespan following forced expression of telomerase in those cells. Well-studied examples include human mammary epithelial cells (Foster and Galloway, 1996; Brenner et al., 1998; Huschtscha et al., 1998; Farwell et al., 2000), keratinocytes (Kiyono et al., 1998; Ramirez et al., 2001; Rheinwald et al., 2002), astrocytes (Evans et al., 2003), and thyroid epithelium (Jones et al., 2000). Evidence for TIS has been presented in other cell types including corneal endothelial cells (Egan et al., 1998), pancreatic islet β cells (Halvorsen et al., 2000) and chondrocytes (Martin et al., 2002).

Human mammary epithelial cells (HMECs) cultured in chemically-defined medium enter a preliminary senescent phase after 15–20 PD, which has been termed M0 (Foster and Galloway, 1996). M0 is associated with elevated p16[INK4A] transcript and protein levels, expression of SA-βgal and enlarged, flattened cellular morphology (Brenner et al., 1998). The culture remains inactive in M0 for 2–4 weeks, then a process termed self-selection occurs, where actively growing cells with the typical "cobblestone" epithelial cell morphology emerge and soon dominate the culture. This escape from M0 senescence is associated with inactivation of p16 by promoter methylation (Brenner et al., 1998; Huschtscha et al., 1998; Farwell et al., 2000). Proliferation of these "post-selection" epithelial cells continues until a total of 45–100 PD have been completed, when the culture then enters a second stage of growth arrest termed "final arrest" or M1. During the extended lifespan of post-selection HMECs p16 remains inactivated and shows no accumulation at late passage or at M1 (Brenner et al., 1988, Huschtscha et al., 1988), unlike the senescent phenotype of fibroblasts. The observation that M0 can be bypassed by expression of HPV E7 (Wazer et al., 1995), corroborated by the high concentration of p16 at M0, suggests that M0 is regulated by the p16/pRB pathway. On the other hand, E6 alone is sufficient to immortalize post-selection HMECS, implying that final arrest is under the control of p53 (Shay et al., 1993). Studying the effects of hTERT, E6 and E7 on cultured cells, Kiyono and colleagues (Kiyono et al., 1998) reported that HMECs cannot be immortalized by telomerase alone, but require co-expression of E7 (which inactivates p16/pRB). This shows that the p16-associated component of HMEC and keratinocyte senescence occurs independently of telomere length.

Human keratinocytes can be immortalized with telomerase under some (Ramirez et al., 2001) but not all (Kiyono et al., 1998) conditions. The mechanisms of senescence in human keratinocytes have been studied in detail (Rheinwald et al., 2002) using retroviral expression of the p16-binding-deficient mutant CDK4 (CDK4[R24C]) and a dominant negative p53 (p53DD) to inhibit the actions of these regulatory molecules. It was found that CDK4[R24C]-expressing keratinocytes required subsequent transduction with hTERT and p53DD in order to become immortalized, therefore revealing an additional, separate p53-dependent TIS mechanism in these cells. Studies on human astrocytes (Evans et al., 2003) have revealed a similar p53-dependent, telomere-independent senescence.

Finally, human thyroid epithelium provides an example of TIS that may be particularly relevant to tumour progression. Normal thyrocytes will undergo no more than 2–3 PD in culture. However, following forced expression of oncogenic (V12) H-ras they will undergo an additional 15–20 PD. Using retroviral vectors to force telomerase expression in these cells, it has been shown that both the normal growth arrest (after 2–3 PD), and the senescence that occurs in ras-expressing thyrocytes, occurs in a telomere-independent manner (Jones et al., 2000). This is consistent with earlier work (Bond et al., 1999) suggesting there was little if any role for p53 in thyrocyte senescence. H-ras mutations are a common initiating event in thyroid follicular cancer, and may lead to the formation of the earliest lesion (adenoma). In turn, the TIS arrest at the end of 15–20 PD is consistent with it being a barrier to further tumour development, and the observation of so-called "burnt out" adenomas in the intact gland. This raises an important point that we will return to later, namely that TIS may be a particularly powerful model system to dissect the selective pressures that occur during the early stages of tumour development, as opposed to the growth regulation of normal tissue.

How is p16 being regulated in these situations?

A common feature of TIS is that it is inevitably associated with high-level expression of p16 which, where tested, appears to be necessary to trigger, (or at least maintain) growth arrest. A key question, then, is what leads to the up-regulation of p16 in these cells. At the present time, little is known about the pathways that induce p16 expression, but recent studies have revealed some details.

A candidate for a common signalling pathway that might be used in a series of diverse situations (i.e. telomere-dependent replicative senescence, stress-induced premature senescence, and TIS) is that mediated by the MAP kinase p38[MAPK] (Iwasa et al., 2003). p38[MAPK] (a down-stream mediator of Ras signalling) was shown to be activated in all three types of senescence, and its inhibition by SB203580 delayed onset of senescence. Overexpression of p38[MAPK] in normal human fibroblasts resulted in induction of the senescent phenotype in an RB-dependent manner, possibly via TGFβ upregulation of CDKIs (Iwasa et al., 2003).

Downstream of the Ras-raf-Mek pathway, which is activated during Ras-induced senescence, p16 was shown to be under the control of the transcription factors Ets-1 and Ets-2 and the transcriptional suppressor Id-1 that are differentially expressed as human diploid fibroblasts approach senescence (Ohtani et al., 2001). p16 is also subject to transcriptional repression by the polycomb protein Bmi-1, which is downregu-

lated during WI38 fibroblast senescence, and overexpression of which leads to bypass of senescence (Jacobs et al., 1999; Itahana et al., 2003). Much remains to be elucidated regarding the network of pathways involved in p16 regulation.

Murine models of replicative senescence

In contrast with humans, laboratory mice have extremely long telomeres (Kipling and Cooke, 1990) and constitutively express telomerase in many tissues, making it appear unlikely that telomere shortening would present a restriction to the growth of normal mouse cells either in culture or in vivo. Perhaps the clearest demonstration of the number of potential divisions before telomere erosion becomes limiting for murine cell growth comes from studies of telomerase-knockout mouse embryonic stem (ES) cells (Niida et al., 1998), which will divide in excess of 400 PD in culture. However, mouse embryonic fibroblasts (MEF) enter a state of growth arrest after a short time in culture (10–15 PD) which closely resembles human fibroblast senescence, both biochemically and physiologically. Unlike M1 senescence in human cells, MEF senescence does not involve pRb (Zalvide and DeCaprio, 1995) but does require a functional Arf-p53 pathway (Serrano et al., 1996; Kamijo et al., 1999; Sherr, 2001). MEF growth arrest is induced in the absence of telomere shortening (Blasco et al., 1997) and, in this respect, is similar to the telomere-independent proliferative lifespan barriers observed in a variety of human cells, including mammary epithelial cells and epidermal keratinocytes.

It has been argued that wild type mice do not provide a good system in which to model the role of telomere erosion and telomerase expression in humans (Kipling, 1997). Indeed, the combination of short telomeres and stringent telomerase repression in human cells may have evolved as an additional barrier to tumour formation in large, highly cellular and long-lived species (Steinert et al., 2002). However, despite these caveats, the telomerase knockout mouse is starting to provide a useful model (Wynford-Thomas and Kipling, 1997), albeit after "humanizing" the telomere lengths down to those seen in humans, achieved by breeding the mice for several generations in the homozygous null state (see Xin and Broccoli, this issue). In such situations MEFs can now be generated that show telomere-dependent senescence (Espejel and Blasco, 2002). However, it has been shown that the senescence-inducing signalling pathways initiated by telomere dysfunction differ between humans and mice. A model of telomere dysfunction, in which the activity of the telomere binding protein TRF2 was abrogated by retroviral transduction of cells with a dominant negative mutant TRF2, was shown to induce senescence in human fibroblasts and MEFs. Co-expression of HPV E6 or E7, or a dominant negative p53 with the mutant TRF2 in human cells demonstrated that senescence could be imposed by p53 or RB independently, as expected. In contrast, MEFs from genetically engineered p53 null mice failed to senescence upon overexpression of the mutant TRF2, indicating that the p16/pRb pathway is not active in the mouse. (Smogorzewska and de Lange, 2002).

However, the observation that murine cells can be engineered to show telomere-dependent senescence does not address the nature of the growth arrest in wild type MEFs in vitro, nor the trigger to senescence in vivo. Despite mice having long telomeres, there have been reports of SA-β-gal activity being detected in adult mouse skin (Paramio et al., 2001). However, one of the clearest examples of the detection of senescent cells comes from in vivo BrdU labelling performed in combination with formal colony size distribution analyses in vitro to determine replicative potential. In one such study by Wolf and colleagues (Li et al., 1997), an age-dependent increase in senescent cells in the lens epithelium was reported. This was reduced by caloric restriction, a treatment that extends the lifespan and healthspan of mice. It remains unclear whether these senescent cells arose as the consequence of telomere erosion, an issue that can potentially be addressed by use of a constitutively telomerase-positive transgenic mouse (Artandi et al., 2002). There is intriguing recent data from the Campisi laboratory suggesting that oxidative stress is the major factor in regulating MEF lifespan (Parrinello et al., 2003), which is provocative when taken with the observation that mutations in p66[shc] (which results in improved anti-oxidant responses) cause a marked increase in mouse lifespan (Migliaccio et al., 1999). It will be interesting to determine if this lifespan-extending mutation is associated with a decrease in the number of senescent cells in vivo.

Final thoughts – does TIS simply reflect "inadequate cell culture"?

One notable feature of TIS is that is can be modulated, in many cases, by changing the growth conditions; fibroblasts that would otherwise immortalize with telomerase fail to do so when grown using defined medium (Ramirez et al., 2001). This has led some (Ramirez et al., 2001) to explicitly describe TIS as a response to "inadequate" growth conditions. More recently, it has been argued that oxidative stress underlies this response. This has come from studies showing that some human fibroblast strains can also show a TIS-like behaviour. Some fetal strains of fibroblasts, with particularly poor anti-oxidant defences, show TIS when grown under standard (21%) oxygen levels, but have an extended lifespan and are immortalized with telomerase when grown under conditions (around 3% oxygen) that more closely mirror the oxygen tension in the dermis in vivo (Forsyth et al., 2003). How chronic oxidative damage can function as a cell division counter, unless the damage that it causes is directly mutagenic and thus cumulative, is unclear.

A similar issue surrounds data indicating that the presence of feeder cells can modulate TIS. Indeed, the whole relevance of TIS to in vivo cellular physiology was brought into question by a report (Ramirez et al., 2001) that showed that HMECs and keratinocytes grown on fibroblast feeder layers could be immortalized by hTERT expression alone, without inactivation of p16/pRB. Conversely, HMEC grown on feeder layers and immortalized by hTERT undergo p16-associated TIS 10–15 PD after being transferred to plastic and chemically-defined medium (Ramirez et al., 2001). This suggests that TIS can be engaged, despite the presence of telomerase, by modulating the growth conditions.

The observation that certain cell culture conditions can delay or inactivate senescence is not restricted to TIS cells. For example, the cytokine, vascular endothelial growth factor (VEGF) delays telomere-dependent senescence in vascular endothelial cells and can even "rescue" them after the onset of senescence (Watanabe et al., 1997). It is important to remember that cells in their natural environment do not exist alone, but are subject to diverse influences from neighbouring cells, the extracellular matrix, circulating hormones etc.

None of our cell culture conditions could ever be described as truly physiological in nature. All cell culture is attempting to do is provide cells that behave in a way that models a certain situation in vivo, thus giving an experimentally tractable system. It is therefore wrong to ask whether cell culture conditions are "adequate" or not. Instead, we should ask to what extent a specific set of growth conditions provide a good model for the particular situation under study.

These thoughts are particularly relevant when considering modelling the early stages of cancer development using in vitro cell culture systems. It may well be entirely appropriate that some culture "stress" leading to up-regulation of p16 and TIS should occur, as this may model similar stress seen in vivo as a cell is now removed into a different, and potentially more stress-inducing, tissue microenvironment. Indeed, a p16-mediated barrier to proliferation appears to be physiologically important, since there is a high incidence of p16 inactivation in human cancers (Ruas and Peters, 1998). Understanding the physiological signals that trigger this defensive up-regulation of p16 during tumour development, and thus the selective pressure for its inactivation, are key unanswered questions in tumour biology.

Acknowledgements

Many of the ideas in this review are the result of extensive discussions with other workers in the field. In addition to members of our own laboratory, we would like to particularly thank Drs Richard Faragher, Jerry Shay, Judy Campisi, and Ken Parkinson, for their invaluable contributions.

References

Alcorta DA, Xiong Y, Phelps D, Hannon G, Beach D, Barrett JC: Involvement of the cyclin-dependent kinase inhibitor p16 (INK4a) in replicative senescence of normal human fibroblasts. Proc Natl Acad Sci USA 93:13742–13747 (1996).

Allsopp RC, Harley CB: Evidence for a critical telomere length in senescent human fibroblasts. Exp Cell Res 219:130–136 (1995).

Allsopp RC, Chang E, Kashefi-Aazam M, Rogaev EI, Piatyszek MA, Shay JW, Harley CB: Telomere shortening is associated with cell division in vitro and in vivo. Exp Cell Res 220:194–200 (1995).

Artandi SE, Alson S, Tietze MK, Sharpless NE, Ye S, Greenberg RA, Castrillon DH, Horner JW, Weiler SR, Carrasco RD, DePinho RA: Constitutive telomerase expression promotes mammary carcinomas in aging mice. Proc Natl Acad Sci USA 99:8191–8196 (2002).

Beausejour CM, Krtolica A, Galimi F, Narita M, Lowe SW, Yaswen P, Campisi J: Reversal of human cellular senescence: roles of the p53 and p16 pathways. EMBO J 22:4212–4222 (2003).

Blasco MA, Lee HW, Hande MP, Samper E, Lansdorp PM, DePinho RA,Greider C W: Telomere shortening and tumor formation by mouse cells lacking telomerase RNA. Cell 91:25–34 (1997).

Bodnar AG, Ouellette M, Frolkis M, Holt SE, Chiu CP, Morin GB, Harley CB, Shay JW, Lichtsteiner S, Wright WE: Extension of life-span by introduction of telomerase into normal human cells. Science 279:349–352 (1998).

Bond JA, Wyllie FS, Wynford-Thomas D: Escape from senescence in human diploid fibroblasts induced directly by mutant p53. Oncogene 9:1885–1889 (1994).

Bond J, Haughton M, Blaydes J, Gire V, Wynford-Thomas D, Wyllie F: Evidence that transcriptional activation by p53 plays a direct role in the induction of cellular senescence. Oncogene 13:2097–2104 (1996).

Bond JA, Haughton MF, Rowson JM, Smith PJ, Gire V, Wynford-Thomas D, Wyllie FS: Control of replicative life span in human cells: barriers to clonal expansion intermediate between M1 senescence and M2 crisis. Mol Cell Biol 19:3103–3114 (1999).

Brenner AJ, Stampfer MR, Aldaz CM: Increased p16 expression with first senescence arrest in human mammary epithelial cells and extended growth capacity with p16 inactivation. Oncogene 17:199–205 (1998).

Bryan TM, Reddel RR: SV40-induced immortalization of human cells. Crit Rev Oncog 5:331–357 (1994).

Bryan TM, Englezou A, Dalla-Pozza L, Dunham MA, Reddel RR: Evidence for an alternative mechanism for maintaining telomere length in human tumors and tumor-derived cell lines. Nat Med 3:1271–1274 (1997).

Campisi J: The biology of replicative senescence. Eur J Cancer 33:703–709 (1997).

Campisi J: Cellular senescence as a tumor-suppressor mechanism. Trends Cell Biol 11:S27–31 (2001).

Chen Q, Ames BN: Senescence-like growth arrest induced by hydrogen peroxide in human diploid fibroblast F65 cells. Proc Natl Acad Sci USA 91:4130–4134 (1994).

Collins CJ, Sedivy JM: Involvement of the INK4a/Arf gene locus in senescence. Aging Cell 2:145–150 (2003).

Condon J, Yin S, Mayhew B, Word RA, Wright WE, Shay JW, Rainey WE: Telomerase immortalization of human myometrial cells. Biol Reprod 67:506–514 (2002).

Counter CM, Hahn WC, Wei W, Caddle SD, Beijersbergen RL, Lansdorp PM, Sedivy JM, Weinberg RA: Dissociation among in vitro telomerase activity, telomere maintenance and cellular immortalization. Proc Natl Acad Sci USA 95:14723–14728 (1998).

Di Leonardo A, Linke SP, Clarkin K, Wahl GM: DNA damage triggers a prolonged p53-dependent G1 arrest and long-term induction of Cip1 in normal human fibroblasts. Genes Dev 8:2540–2551 (1994).

Dimri GP, Lee X, Basile G, Acosta M, Scott G, Roskelley C, Medrano EE, Linskens M, Rubelj I, Pereira-Smith O, Peacocke M, Campisi J: A biomarker that identifies senescent human cells in culture and in aging skin in vivo. Proc Natl Acad Sci USA 92:9363–9367 (1995).

Egan CA, Savre-Train I, Shay JW, Wilson SE, Bourne WM: Analysis of telomere lengths in human corneal endothelial cells from dozens of different ages. Invest Ophthalmol Vis Sci 39:648–653 (1998).

Espejel S, Blasco MA: Identification of telomere-dependent "senescence-like" arrest in mouse embryonic fibroblasts. Exp Cell Res 276:242–248 (2002).

Evans RJ, Wyllie FS, Wynford-Thomas D, Kipling D, Jones CJ: A P53-dependent telomere-independent proliferative life span barrier in human astrocytes consistent with the molecular genetics of glioma development. Cancer Res 63:4854–4861 (2003).

Faragher RG: Cell senescence and human aging: where's the link? Biochem Soc Trans 28:221–226 (2000).

Faragher RG, Kipling D: How might replicative senescence contribute to human ageing? Bioessays 20:985–991 (1998).

Farwell DG, Shera KA, Koop JI, Bonnet GA, Matthews CP, Reuther GW, Coltrera MD, McDougall JK, Klingelhutz AJ: Genetic and epigenetic changes in human epithelial cells immortalized by telomerase. Am J Pathol 156:1537–1547 (2000).

Ferbeyre G, de Stanchina E, Lin AW, Querido E, McCurrach ME, Hannon GJ, Lowe SW: Oncogenic ras and p53 cooperate to induce cellular senescence. Mol Cell Biol 22:3497–3508 (2002).

Forsyth NR, Evans AP, Shay JW, Wright WE: Developmental differences in the immortalization of lung fibroblasts by telomerase. Aging Cell 2:235–243 (2003).

Foster SA, Galloway DA: Human papillomavirus type 16 E7 alleviates a proliferation block in early passage human mammary epithelial cells. Oncogene 12:1773–1779 (1996).

Gire V, Wynford-Thomas D: Reinitiation of DNA synthesis and cell division in senescent human fibroblasts by microinjection of anti-p53 antibodies. Mol Cell Biol 18:1611–1621 (1998).

Halvorsen TL, Beattie GM, Lopez AD, Hayek A, Levine F: Accelerated telomere shortening and senescence in human pancreatic islet cells stimulated to divide in vitro. J Endocrinol 166:103–109 (2000).

Hara E, Smith R, Parry D, Tahara H, Stone S, Peters G: Regulation of p16CDKN2 expression and its implications for cell immortalization and senescence. Mol Cell Biol 16:859–867 (1996).

Harley CB, Futcher AB, Greider CW: Telomeres shorten during ageing of human fibroblasts. Nature 345:458–460 (1990).

Hayflick L: The limited in vitro lifetime of human diploid cell strains. Exp Cell Res 37:614–636 (1965).

Hayflick L, Moorhead PS: The serial cultivation of human diploid cell strains. Exp Cell Res 25:585–621 (1961).

Hemann MT, Strong MA, Hao LY, Greider CW: The shortest telomere not average telomere length is critical for cell viability and chromosome stability. Cell 107:67–77 (2001).

Huot TJ, Rowe J, Harland M, Drayton S, Brookes S, Gooptu C, Purkis P, Fried M, Bataille V, Hara E, Newton-Bishop J, Peters G: Biallelic mutations in p16(INK4a) confer resistance to Ras- and Ets-induced senescence in human diploid fibroblasts. Mol Cell Biol 22:8135–8143 (2002).

Huschtscha LI, Noble JR, Neumann AA, Moy EL, Barry P, Melki JR, Clark SJ, Reddel RR: Loss of p16INK4 expression by methylation is associated with lifespan extension of human mammary epithelial cells. Cancer Res 58:3508–3512 (1998).

Itahana K, Zou Y, Itahana Y, Martinez JL, Beausejour C, Jacobs JJ, Van Lohuizen M, Band V, Campisi J, Dimri GP: Control of the replicative life span of human fibroblasts by p16 and the polycomb protein Bmi-1. Mol Cell Biol 23:389–401 (2003).

Iwasa H, Han J, Ishikawa F: Mitogen-activated protein kinase p38 defines the common senescence-signalling pathway. Genes Cells 8:131–144 (2003).

Jacobs JJ, Kieboom K, Marino S, DePinho RA, van Lohuizen M: The oncogene and Polycomb-group gene bmi-1 regulates cell proliferation and senescence through the ink4a locus. Nature 397:164–168 (1999).

Jiang H, Chou HS, Zhu L: Requirement of cyclin E-Cdk2 inhibition in p16(INK4a)-mediated growth suppression. Mol Cell Biol 18:5284–5290 (1998).

Jiang XR, Jimenez G, Chang E, Frolkis M, Kusler B, Sage M, Beeche M, Bodnar AG, Wahl GM, Tlsty TD, Chiu CP: Telomerase expression in human somatic cells does not induce changes associated with a transformed phenotype. Nat Genet 21:111–114 (1999).

Jones CJ, Kipling D, Morris M, Hepburn P, Skinner J, Bounacer A, Wyllie FS, Ivan M, Bartek J, Wynford-Thomas D, Bond JA: Evidence for a telomere-independent "clock" limiting RAS oncogene-driven proliferation of human thyroid epithelial cells. Mol Cell Biol 20:5690–5699 (2000).

Kamijo T, van de Kamp E, Chong MJ, Zindy F, Diehl JA, Sherr CJ, McKinnon PJ: Loss of the ARF tumor suppressor reverses premature replicative arrest but not radiation hypersensitivity arising from disabled atm function. Cancer Res 59:2464–2469 (1999).

Karlseder J, Smogorzewska A, de Lange T: Senescence induced by altered telomere state not telomere loss. Science 295:2446–2449 (2002).

Kim NW, Piatyszek MA, Prowse KR, Harley CB, West MD, Ho PL, Coviello GM, Wright WE, Weinrich SL, Shay JW: Specific association of human telomerase activity with immortal cells and cancer. Science 266:2011–2015 (1994).

Kipling D: Mammalian telomerase: catalytic subunit and knockout mice. Hum Mol Genet 6:1999–2004 (1997).

Kipling D: Telomeres replicative senescence and human ageing. Maturitas 38:25–37 (2001).

Kipling D, Cooke HJ: Hypervariable ultra-long telomeres in mice. Nature 347:400–402 (1990).

Kiyono T, Foster SA, Koop JI, McDougall JK, Galloway DA, Klingelhutz AJ: Both Rb/p16INK4a inactivation and telomerase activity are required to immortalize human epithelial cells. Nature 396:84–88 (1998).

Li GZ, Eller MS, Firoozabadi R, Gilchrest BA: Evidence that exposure of the telomere 3′ overhang sequence induces senescence. Proc Natl Acad Sci USA 100:527–531 (2003).

Li Y, Yan Q, Wolf NS: Long-term caloric restriction delays age-related decline in proliferation capacity of murine lens epithelial cells in vitro and in vivo. Invest Ophthalmol Vis Sci 38:100–107 (1997).

Lumpkin CK, Knepper JE, Butel JS, Smith JR, Pereira-Smith OM: Mitogenic effects of the proto-oncogene and oncogene forms of c-H-ras DNA in human diploid fibroblasts. Mol Cell Biol 6:2990–2993 (1986).

Martens UM, Chavez EA, Poon SS, Schmoor C, Lansdorp PM: Accumulation of short telomeres in human fibroblasts prior to replicative senescence. Exp Cell Res 256:291–299 (2000).

Martin JA, Mitchell CJ, Klingelhutz AJ, Buckwalter JA: Effects of telomerase and viral oncogene expression on the in vitro growth of human chondrocytes. J Gerontol A Biol Sci Med Sci 57:B48–53 (2002).

Mathon NF, Lloyd AC: Cell senescence and cancer. Nat Rev Cancer 1:203–213 (2001).

McConnell BB, Gregory FJ, Stott FJ, Hara E, Peters G: Induced expression of p16(INK4a) inhibits both CDK4- and CDK2-associated kinase activity by reassortment of cyclin-CDK-inhibitor complexes. Mol Cell Biol 19:1981–1989 (1999).

Melk A, Kittikowit W, Sandhu I, Halloran KM, Grimm P, Schmidt BM, Halloran PF: Cell senescence in rat kidneys in vivo increases with growth and age despite lack of telomere shortening. Kidney Int 63:2134–2143 (2003).

Migliaccio E, Giorgio M, Mele S, Pelicci G, Reboldi P, Pandolfi PP, Lanfrancone L, Pelicci PG: The p66shc adaptor protein controls oxidative stress response and life span in mammals. Nature 402:309–313 (1999).

Morales CP, Holt SE, Ouellette M, Kaur KJ, Yan Y, Wilson KS, White MA, Wright WE, Shay JW: Absence of cancer-associated changes in human fibroblasts immortalized with telomerase. Nat Genet 21:115–118 (1999).

Morales CP, Gandia KG, Ramirez RD, Wright WE, Shay JW, Spechler SJ: Characterisation of telomerase immortalised normal human oesophageal squamous cells. Gut 52:327–333 (2003).

Narita M, Nunez S, Heard E, Lin AW, Hearn SA, Spector DL, Hannon GJ, Lowe SW: Rb-mediated heterochromatin formation and silencing of E2F target genes during cellular senescence. Cell 113:703–716 (2003).

Niida H, Matsumoto T, Satoh H, Shiwa M, Tokutake Y, Furuichi Y, Shinkai Y: Severe growth defect in mouse cells lacking the telomerase RNA component. Nat Genet 19:203–206 (1998).

Ohtani N, Zebedee Z, Huot TJ, Stinson JA, Sugimoto M, Ohashi Y, Sharrocks AD, Peters G, Hara E: Opposing effects of Ets and Id proteins on p16INK4a expression during cellular senescence. Nature 409:1067–1070 (2001).

Paramio JM, Segrelles C, Ruiz S, Martin-Caballero J, Page A, Martinez J, Serrano M, Jorcano JL: The ink4a/arf tumor suppressors cooperate with p21cip1/waf in the processes of mouse epidermal differentiation senescence and carcinogenesis. J Biol Chem 276:44203–44211 (2001).

Parkinson EK, Newbold RF, Keith WN: The genetic basis of human keratinocyte immortalisation in squamous cell carcinoma development: the role of telomerase reactivation. Eur J Cancer 33:727–734 (1997).

Parrinello S, Samper E, Krtolica A, Goldstein J, Melov S, Campisi J: Oxygen sensitivity severely limits the replicative lifespan of murine fibroblasts. Nat Cell Biol 5:741–747 (2003).

Ramirez RD, Morales CP, Herbert BS, Rohde JM, Passons C, Shay JW, Wright WE: Putative telomere-independent mechanisms of replicative aging reflect inadequate growth conditions. Genes Dev 15:398–403 (2001).

Reddel RR: A reassessment of the telomere hypothesis of senescence. Bioessays 20:977–984 (1998).

Rheinwald JG, Hahn WC, Ramsey MR, Wu JY, Guo Z, Tsao H, De Luca M, Catricala C, O'Toole KM: A two-stage p16(INK4A)- and p53-dependent keratinocyte senescence mechanism that limits replicative potential independent of telomere status. Mol Cell Biol 22:5157–5172 (2002).

Ruas M, Peters G: The p16INK4a/CDKN2A tumor suppressor and its relatives. Biochim Biophys Acta 1378:F115–177 (1998).

Serrano M, Lee H, Chin L, Cordon-Cardo C, Beach D, DePinho RA: Role of the INK4a locus in tumor suppression and cell mortality. Cell 85:27–37 (1996).

Serrano M, Lin AW, McCurrach ME, Beach D, Lowe SW: Oncogenic ras provokes premature cell senescence associated with accumulation of p53 and p16INK4a. Cell 88:593–602 (1997).

Severino J, Allen RG, Balin S, Balin A, Cristofalo VJ: Is beta-galactosidase staining a marker of senescence in vitro and in vivo? Exp Cell Res 257:162–171 (2000).

Shay JW, Pereira-Smith OM, Wright WE: A role for both RB and p53 in the regulation of human cellular senescence. Exp Cell Res 196:33–39 (1991a).

Shay JW, Wright WE, Werbin H: Defining the molecular mechanisms of human cell immortalization. Biochim Biophys Acta 1072:1–7 (1991b).

Shay JW, Wright WE, Brasiskyte D, Van der Haegen BA: E6 of human papillomavirus type 16 can overcome the M1 stage of immortalization in human mammary epithelial cells but not in human fibroblasts. Oncogene 8:1407–1413 (1993).

Shelton DN, Chang E, Whittier PS, Choi D, Funk WD: Microarray analysis of replicative senescence. Curr Biol 9:939–945 (1999).

Sherr CJ: Cancer cell cycles. Science 274:1672–1677 (1996).

Sherr CJ: The Pezcoller lecture: cancer cell cycles revisited. Cancer Res 60:3689–3695 (2000).

Sherr CJ: Parsing Ink4a/Arf: "pure" p16-null mice. Cell 106:531–534 (2001).

Sherr CJ, Roberts JM: CDK inhibitors: positive and negative regulators of G1-phase progression. Genes Dev 13:1501–1512 (1999).

Simonsen JL, Rosada C, Serakinci N, Justesen J, Stenderup K, Rattan SI, Jensen TG, Kassem M: Telomerase expression extends the proliferative lifespan and maintains the osteogenic potential of human bone marrow stromal cells. Nat Biotechnol 20:592–596 (2002).

Smogorzewska A, de Lange T: Different telomere damage signaling pathways in human and mouse cells. EMBO J 21:4338–4348 (2002).

Southern SA, Herrington CS: Disruption of cell cycle control by human papillomaviruses with special reference to cervical carcinoma. Int J Gynecol Cancer 10:263–274 (2000).

Stein GH, Drullinger LF, Robetorye RS, Pereira-Smith OM, Smith JR: Senescent cells fail to express cdc2 cycA and cycB in response to mitogen stimulation. Proc Natl Acad Sci USA 88:11012–11016 (1991).

Stein GH, Drullinger LF, Soulard A, Dulic V: Differential roles for cyclin-dependent kinase inhibitors p21 and p16 in the mechanisms of senescence and differentiation in human fibroblasts. Mol Cell Biol 19:2109–2117 (1999).

Steinert S, White DM, Zou Y, Shay JW, Wright WE: Telomere biology and cellular aging in nonhuman primate cells. Exp Cell Res 272:146–152 (2002).

Trimarchi JM, Lees JA: Sibling rivalry in the E2F family. Nat Rev Mol Cell Biol 3:11–20 (2002).

Vaziri H, Benchimol S: Reconstitution of telomerase activity in normal human cells leads to elongation of telomeres and extended replicative life span. Curr Biol 8:279–282 (1998).

Vaziri H, Squire JA, Pandita TK, Bradley G, Kuba RM, Zhang H, Gulyas S, Hill RP, Nolan GP, Benchimol S: Analysis of genomic integrity and p53-dependent G1 checkpoint in telomerase-induced extended-life-span human fibroblasts. Mol Cell Biol 19:2373–2379 (1999).

Vogt M, Haggblom C, Yeargin J, Christiansen-Weber T, Haas M: Independent induction of senescence by p16INK4a and p21CIP1 in spontaneously immortalized human fibroblasts. Cell Growth Differ 9:139–146 (1998).

von Zglinicki T, Saretzki G, Docke W, Lotze C: Mild hyperoxia shortens telomeres and inhibits proliferation of fibroblasts: a model for senescence? Exp Cell Res 220:186–193 (1995).

Wang E: Senescent human fibroblasts resist programmed cell death and failure to suppress bcl2 is involved. Cancer Res 55:2284–2292 (1995).

Wang J, Hannon GJ, Beach DH: Risky immortalization by telomerase. Nature 405:55–756 (2000).

Watanabe Y, Lee SW, Detmar M, Ajioka I, Dvorak HF: Vascular permeability factor/vascular endothelial growth factor (VPF/VEGF) delays and induces escape from senescence in human dermal microvascular endothelial cells. Oncogene 14:2025–2032 (1997).

Wazer DE, Liu XL, Chu Q, Gao Q, Band V: Immortalization of distinct human mammary epithelial cell types by human papilloma virus 16 E6 or E7. Proc Natl Acad Sci USA 92:3687–3691 (1995).

Webley K, Bond JA, Jones CJ, Blaydes JP, Craig A, Hupp T, Wynford-Thomas D: Posttranslational modifications of p53 in replicative senescence overlapping but distinct from those induced by DNA damage. Mol Cell Biol 20:2803–2808 (2000).

Wei W, Hemmer RM, Sedivy JM: Role of p14(ARF) in replicative and induced senescence of human fibroblasts. Mol Cell Biol 21:6748–6757 (2001).

Wynford-Thomas D: Cellular senescence and cancer. J Pathol 187:100–111 (1999).

Wynford-Thomas D, Kipling D: Telomerase Cancer and the knockout mouse. Nature 389:551–552 (1997).

Xin Z, Broccoli D: Manipulating mouse telomeres: models of tumorigenesis and aging. Cytogenet Genome Res 105:471–478 (2004).

Yang J, Chang E, Cherry AM, Bangs CD, Oei Y, Bodnar A, Bronstein A, Chiu CP, Herron GS: Human endothelial cell life extension by telomerase expression. J Biol Chem 274:26141–26148 (1999).

Zalvide J, DeCaprio JA: Role of pRb-related proteins in simian virus 40 large-T-antigen-mediated transformation. Mol Cell Biol 15:5800–5810 (1995).

Cytogenet Genome Res 105:464–470 (2004)
DOI: 10.1159/000078220

Telomere length measurement in mouse chromosomes by a modified Q-FISH method

H.-P. Wong and P. Slijepcevic

Brunel Institute of Cancer Genetics and Pharmacogenomics, Brunel University, Uxbridge, (UK)

Abstract. Telomeres are physical ends of mammalian chromosomes that dynamically change during the lifetime of a cell or organism. In order to understand mechanisms responsible for telomere dynamics, it is necessary to develop methods for accurate telomere length measurement. The most sensitive method for measuring telomere length in mouse chromosomes is quantitative fluorescence in situ hybridization (Q-FISH). The usual protocol for Q-FISH requires plasmids with variable numbers of telomeric repeats and fluorescence beads as calibration standards. Here, we describe a Q-FISH protocol in which two mouse lymphoma cell lines with well-defined telomere lengths are used as calibration standards. Using this protocol we demonstrate that reproducible results can be obtained in a set of four different mouse cell lines. This method can be adapted so that any pair of mammalian cell lines can serve as an internal calibration standard.

Copyright © 2004 S. Karger AG, Basel

Mammalian chromosomes are complex structures, which require specialized functional elements for their accurate segregation. One such essential functional element is the telomere. The main function of telomeres is to protect chromosome integrity and stability (Blackburn, 2000). During the lifetime of a cell, telomeres undergo significant changes. To fully understand mechanisms of telomere dynamics and the role of individual molecules that regulate this dynamics, it is necessary to be able to measure telomere length with a high degree of accuracy.

Telomere length measurement in mouse chromosomes is not simple. Classical methodology, based on Southern blot analysis, while accurate enough in the case of human chromosomes, is frequently non-informative in the case of mouse chromosomes (Kipling and Cooke, 1990; Blasco et al., 1997). This is most likely because Southern blot analysis of telomeres in mouse chromosomes overestimates telomere length by taking account of telomeric sequences which are present in sub-telomeric regions of chromosomes in relatively large quantities (Zijlmans et al., 1997). Therefore, alternative techniques are required to accurately measure mouse telomeres. Two such techniques have been developed in recent years and include quantitative fluorescence in situ hybridization (Q-FISH) and FISH based on flow cytometry or flow-FISH (Zijlmans et al., 1997; Rufer et al., 1998). Both techniques work on similar principles and this article will focus on Q-FISH.

Q-FISH is probably the most sensitive technique currently available for telomere length measurement in mouse chromosomes. It is based on using fluorescently labeled peptide nucleic acid (PNA) telomeric oligonucleotides, which are more efficient in hybridization experiments than normal DNA oligonucleotides. Signals generated by only ~ 150 nucleotides that hybridize with fluorescent telomeric PNA can be detected under a fluorescence microscope equipped with a CCD camera and these signals can easily be quantified by appropriate software (Zijlmans et al., 1997). Telomere fluorescence values, which are proportional to telomere length, can be generated for each individual telomere. However, Q-FISH is a complex technique and it requires proper calibration protocols to eliminate inherent variations associated with fluorescence microscopy. In this article we describe a method used in our laboratory for telomere length measurement in mouse chromosomes.

Supported in part by a grant from the EURATOM program (contract FIGH-CT-2002-00217) from European Commission.

Received 26 September 2003; accepted 7 November 2003.

Request reprints from: Dr. Predrag Slijepcevic
Brunel Institute of Cancer Genetics and Pharmacogenomics
Brunel University, Kingston Lane, Uxbridge, Middlesex UB8 3PH (UK)
telephone: +44 1895 274 000 ext. 2109; fax: +44 1895 274 000 ext. 2109
e-mail: predrag.slijepcevic@brunel.ac.uk.

Table 1. Unmodified telomere fluorescence in LY-R and LY-S cells observed at five different dates. SEM (standard error of the mean) was calculated by pooling together all telomeres from all cells of a single sample (also applies to Tables 2 and 3).

Date of experiment	LY-R		LY-S		Ratio LY-R/LY-S
	Flu.	SEM	Flu.	SEM	
09/01/01	1446.4	30.8	266.4	8.8	5.4
10/01/01	1167.1	31.9	178.2	11.1	6.5
22/01/01	1848.1	42.5	234.5	5.9	7.9
01/02/01	1930.2	38.4	332.9	10.7	5.8
12/02/01	1598.6	24.3	180.4	7.1	8.9
Average	1598.1	33.6	238.5	8.7	6.9

Materials and methods

Tissue culture

Mouse lymphoma cell lines LY-R and LY-S were grown in Fischer medium (Gibco), 10% foetal calf serum and antibiotics. Mouse fibroblast cell lines CB17, *scid*, *scid* 50D and *scid* 100E were grown in Wymouth medium (Gibco), 10% foetal calf serum and antibiotics. In addition, the growth medium for *scid* 50D and *scid* 100E cells was supplemented with puromycin (Sigma).

Chromosome preparation and in situ hybridization

Chromosome preparation was performed using classical methods. Briefly, cells were treated with colcemid for 1 h when semi-confluent, harvested by trypsinization, treated with a hypotonic solution and fixed using acetic acid/methanol. A few drops of fixative solution containing cells were used for each slide. Slides were allowed to dry and were incubated on a hot plate (55°C) overnight before hybridization.

Slides with chromosome preparations were hybridized with the PNA telomeric oligonucleotide $(CCCTTA)_3$ labeled with Cy3 (PE Biosystems). Digital images were acquired using an Axioskop 2 microscope (Zeiss) equipped with a CCD camera. Telomere fluorescence intensity was analyzed using TFL-Telo software provided by Dr. Peter Lansdorp, Terry Fox Laboratory, Vancouver, B.C., Canada.

Q-FISH calibration procedure

To achieve reproducible results in Q-FISH experiments we developed the following calibration protocol. Mouse lymphoma cell lines LY-R and LY-S that have well defined telomere lengths were analyzed by Q-FISH on five different dates (see Results). The mean values of telomere fluorescence for LY-R and LY-S cells in these five experiments were used as historical values and labelled as Fl_{LY-R} or Fl_{LY-S}. Each time a new sample was analyzed by Q-FISH we also acquired images of at least ten LY-R or LY-S metaphase cells and used these values ($Fl_{LY-R(exp)}$ or $Fl_{LY-S(exp)}$) as internal controls to obtain two correction factors CF1 and CF2 using following formulas:

$$CF1 = Fl_{LY-R}/Fl_{LY-R(exp)}$$

$$CF2 = Fl_{LY-S}/Fl_{LY-S(exp)}.$$

The final correction factor, CF, was the mean of CF1 and CF2. To express telomere fluorescence of a sample we used the following formula:

$$CcFl = CF \times Fl_x.$$

In this formula CcFl represents corrected calibrated fluorescence and Fl_x represents unmodified fluorescence of a sample under investigation.

Results

Telomere fluorescence intensity in mouse lymphoma cell lines

For the experiments described below we used two mouse lymphoma cell lines, LY-R and LY-S, with long and short telomeres, respectively. The LY-S cell line was derived from the parental LY-R cell line and both cell lines have similar karyotypes. The LY-R cell line has telomeres typical of mouse chromosomes estimated to be in the region of ~48 kb, whereas LY-S cells have much shorter telomeres estimated to be in the region of ~7 kb (McIlrath et al., 2001). Telomere length in these cell lines is stable over a period of several months as estimated by the flow-FISH method (Cabuy and Slijepcevic, unpublished observations).

The measurement of fluorescence intensity by fluorescence microscopy is known to produce variable results. This reflects daily variations due to aging of the microscope lamp or problems with the microscope optical alignment (Zijlmans et al., 1997; Poon et al., 1999). In addition, there may be other sources of variation including variable efficiency of hybridization in different parts of the microscope slide or other unknown factors. To examine the extent of variation in telomere fluorescence intensities, we performed Q-FISH on five different dates in the above two cell lines. Results of these experiments are presented in Fig. 1 and Table 1. The mean fluorescence intensity observed in LY-R cells varied between ~1200 and ~1900 units (1.6× difference) and in LY-S cells between ~180 and ~330 units (1.8× difference) (Table 1). Measurements performed on LY-R cells on 22/1/01 and 1/2/01 and measurements performed on LY-S cells on 10/1/01 and 12/2/01 (Tables 2 and 3) did not generate statistically significant differences. However, differences between other measurements were statistically significant. Previous Q-FISH studies revealed considerable differences between measurements of the same metaphase cells at different dates (Poon et al., 1999).

The ratio of the mean fluorescence intensities between the two cell lines varied from 5.4× to 8.9× giving the average ratio of fluorescence intensities LY-R: LY-S = 6.9× (Table 1). This is in line with our previous results, which revealed a seven-fold difference in telomere fluorescence intensities between the two cell lines (McIlrath et al., 2001). Previous results were based on the calibration method developed by the Lansdorp's group (Zijlmans et al., 1997). This suggests that although absolute fluorescence intensity values vary from date to date, relative fluorescence intensity values remain stable.

To verify the accuracy of the above measurements we monitored fluorescence intensity values in sister chromatids in all samples shown in Fig. 1 (i.e. sister chromatids are exact replicas of each other and telomere length should be exactly the same in both chromatids). This analysis revealed that correlation coeffi-

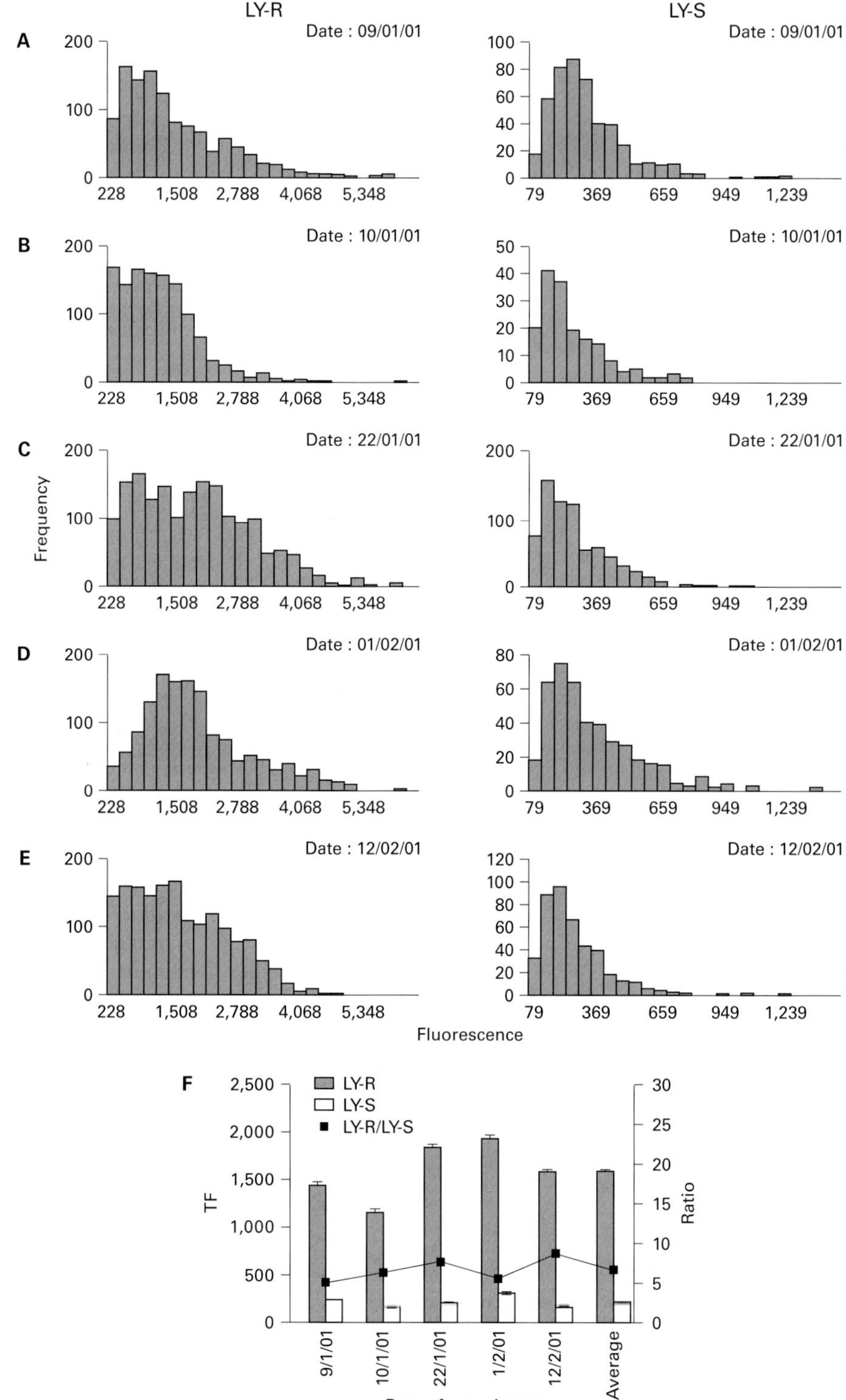

Fig. 1. (**A–E**) Unmodified telomere fluorescence in mouse lymphoma LY-R and LY-S cells observed at five different dates. (**F**) Mean telomere fluorescence values in LY-R and LY-S cells at five different dates. Standard error values were calculated by pooling together all telomeres from all cells of a single sample; as a result of the large number of telomeres analyzed, the standard error is small.

cients (R) are generally close to 1 suggesting that fluorescence intensities of sister chromatids are similar (Fig. 2). However, there was a clear difference in the R factor between two cell lines (lower in LY-S) suggesting that the accuracy of Q-FISH may be reduced when mouse telomeres are shorter (Fig. 2).

Telomere fluorescence variation in individual cells

We monitored distribution of fluorescence intensities in individual cells at each measurement (Tables 2 and 3). This analysis revealed considerable differences between individual cells in the same sample. For example, in the case of LY-R cells mean telomere fluorescence values observed in single cells var-

Table 2. Unmodified telomere fluorescence (mean ± SEM) in single cells (LY-R) measured at different dates

Cell	1	2	3	4	5	6	7	8	9	10	11	12	13
9/1/01	1209.8	2112.2	984.7	829.3	3128.7	958.3	1187.4	2585.2	2588.3	2246.7	2103.8	678.9	3348.6
	±58.3	±99.9	±53.4	±50.7	±175.1	±59.8	±52.4	±107.5	±90.8	±139.2	±87.8	±37.8	±136.1
	14	15	Mean										
	737.0	1132.9	1446.4										
	±47.1	±60.7	±30.8										

Cell	1	2	3	4	5	Mean
10/1/01	1148.0	501.2	1265.6	1745.8	1203.3	1167.1
	±60.3	±33.5	±68.0	±94.1	±42.4	±32.0

Cell	1	2	3	4	5	6	7	8	9	10	11	12	13
22/1/01	1226.2	1637.3	1231.2	719.5	1120.1	748.7	2500.7	2742.1	2416.4	2838.9	580.7	3194.3	2131.2
	±59.7	±112.0	±152.1	±77.8	±97.5	±85.4	±163.6	±178.4	±85.9	±85.9	±47.5	±121.9	±116.6
	14	15	Mean										
	2734.2	1569.3	1848.1										
	±219.2	±114.8	±42.5										

Cell	1	2	3	4	5	6	7	8	9	10	11	12	13
1/2/01	1242.0	3648.2	2434.7	1327.2	1080.6	3576.6	1606.2	2045.7	2915.6	3097.2	2039.9	1641.7	1566.3
	±68.3	±190.6	±106.4	±81.9	±76.6	±200.7	±114.8	±112.3	±87.4	±207.2	±115.7	±90.2	±47.5
	14	15	Mean										
	2840.9	1461.2	1930.2										
	±200.0	±100.2	±38.4										

Cell	1	2	3	4	5	6	7	8	9	10	11	12	13
12/2/01	2482.6	2368.4	889.9	1693.3	490.0	2701.9	643.2	1020.4	2135.4	1747.9	423.7	1800.0	1045.5
	±79.7	±269.4	±53.2	±121.8	±75.8	±92.2	±49.4	±51.3	±127.8	±133.7	±36.3	±198.7	±51.6
	14	15	16	17	18	19	20	21	22	23	24	Mean	
	1638.5	1029.5	705.5	2070.2	2483.5	2247.1	1000.0	1711.2	2242.6	1968.1	2890.8	1598.6	
	±62.3	±35.8	±50.1	±102.8	±125.4	±82.4	±57.1	±166.2	±72.1	±157.7	±66.5	±24.3	

Table 3. Unmodified telomere fluorescence (mean ± SEM) in single cells (LY-S) measured at different dates

Cell	1	2	3	4	5	6	7	8	9	10	Mean
9/1/01	253.2	212.6	210.2	184.9	190.4	353.4	203.7	116.7	370.7	287.6	266.4
	±15.2	±19.1	±37.2	±24.8	±18.0	±22.4	±16.3	±35.0	±25.6	±35.0	±8.8

Cell	1	2	3	4	5	Mean
10/1/01	285.8	147.5	108.2	130.9	334.6	178.2
	±36.0	±16.3	±18.8	±13.7	±37.2	±11.1

Cell	1	2	3	4	5	6	7	8	9	10	11	12	13
22/1/01	319.5	258.7	169.1	199.3	131.7	245.3	176.7	215.8	283.0	169.7	295.8	184.7	222.5
	±24.8	±14.7	±17.9	±23.0	±14.8	±17.6	±21.1	±23.8	±24.1	±40.2	±30.3	±15.7	±14.2
	14	Mean											
	247.3	234.5											
	±18.4	±5.9											

Cell	1	2	3	4	5	6	7	8	9	10	11	12	13
1/2/01	218.0	263.4	296.7	277.3	380.8	374.9	298.4	216.3	316.66	271.8	436.5	580.9	350.8
	±28.8	±25.8	±23.5	±48.1	±41.2	±35.6	±43.4	±20.6	±94.4	±19.6	±89.2	±52.2	±89.4
	14	15	16	Mean									
	214.0	394.9	216.1	332.93									
	±41.4	±27.4	±32.9	±10.7									

Cell	1	2	3	4	5	6	7	8	9	10	11	12	13
12/2/01	217.1	246.5	192.5	412.8	156.7	266.9	164.7	231.8	88.25	128.6	173.8	178.4	236.0
	±46.7	±57.0	±28.2	±63.3	±35.2	±48.3	±36.7	±20.1	±50.6	±34.7	±22.3	±25.8	±21.9
	14	15	16	17	18	19	20	21	Mean				
	161.1	119.6	210.1	135.8	145.8	125.6	153.4	156.0	180.4				
	±25.3	±26.0	±37.1	±28.8	±21.4	±24.1	±24.2	±27.1	±7.1				

ied between ~3.5× (dates of measurement: 10/1/01 and 1/2/01) and ~7× (date of measurement: 12/2/01) (Table 2). In the case of LY-S cells differences were smaller and varied between 2.4× (date of measurement: 22/1/01) and 4.7× (date of measurement: 12/2/01) (Table 3). These results indicate that telomere fluorescence intensity varies considerably between different cells in the same sample.

Telomere fluorescence and colcemid treatment time

Although it is possible that differences in telomere fluorescence values between individual cells measured by Q-FISH reflect genuine differences in telomere length between cells, one cannot rule out the possibility that artefacts associated with the Q-FISH protocol may be the source of these variations. For example, a potential source of telomere fluorescence variation includes variable levels of chromosome condensation. Each

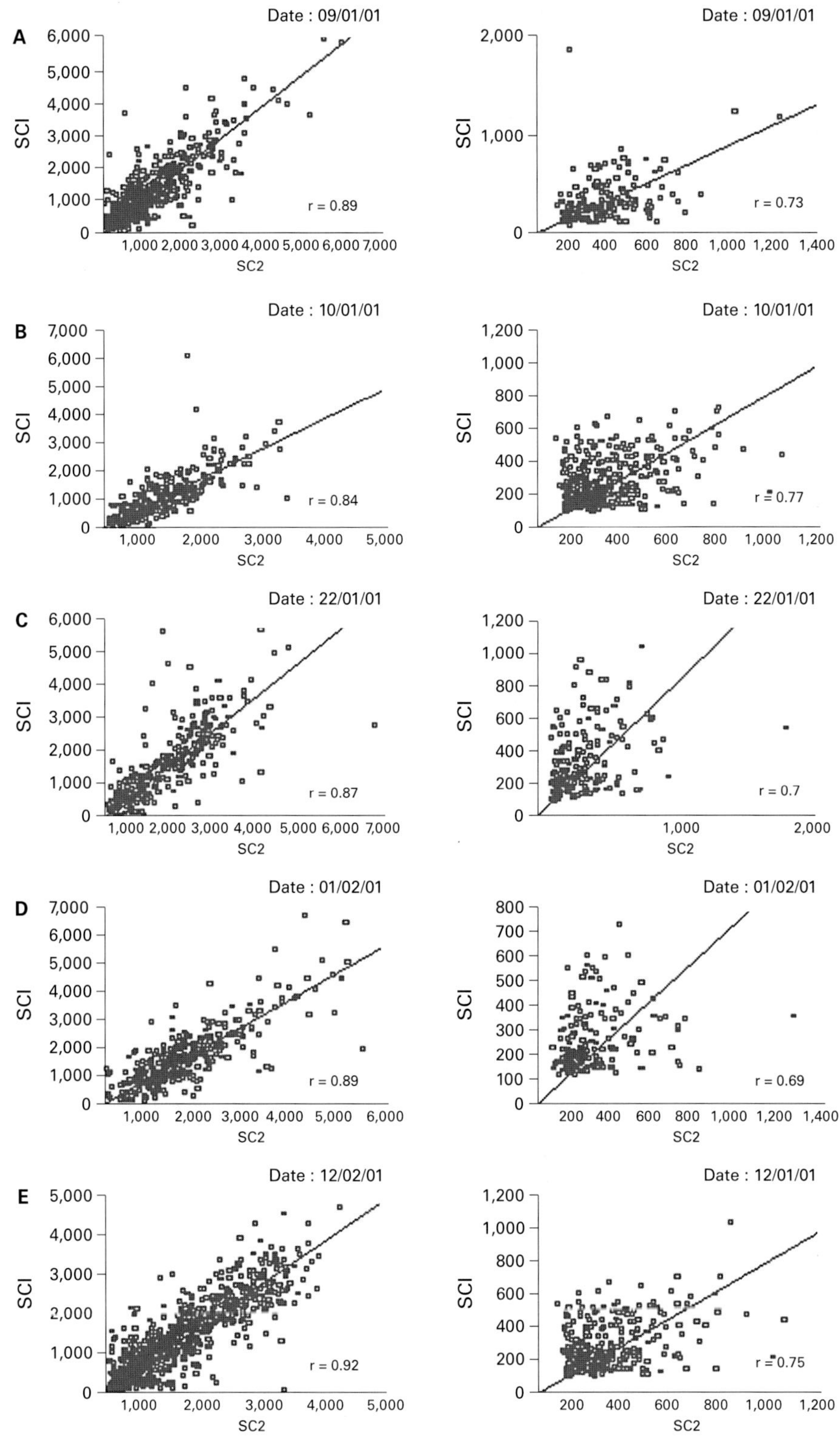

Fig. 2. (A–E) Unmodified telomere fluorescence in sister chromatid 1 (SC1) and sister chromatid 2 (SC2) in LY-R (right side) and LY-S (left side) cells on five different dates. r is the correlation coefficient.

metaphase cell shows different degrees of chromosome condensation and heavily condensed chromosomes may produce more intense fluorescence signal resulting in artificially high telomere fluorescence in comparison with less condensed chromosomes from a different metaphase cell that may have similar telomere lengths. To address this possibility experimentally the following protocol was employed. LY-R cells were treated with colcemid for 1, 3, and 6 h. Apart from arresting cells in metaphase prolonged treatment of cells with colcemid generates hypercondensation of mitotic chromosomes. Following the above treatment, we selected a marker chromosome in LY-R cells which was easy to identify cytogenetically (not shown) and

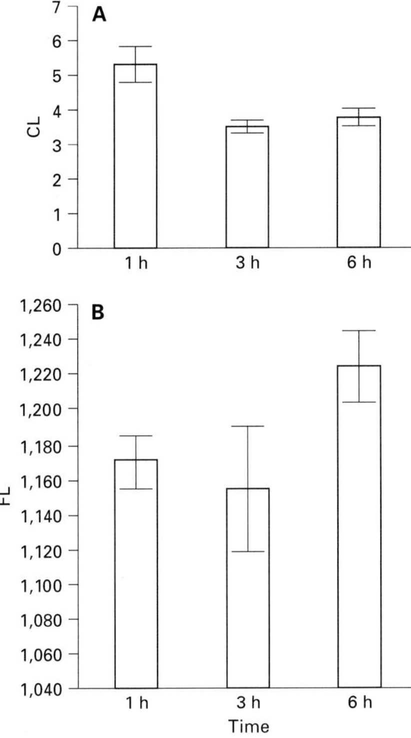

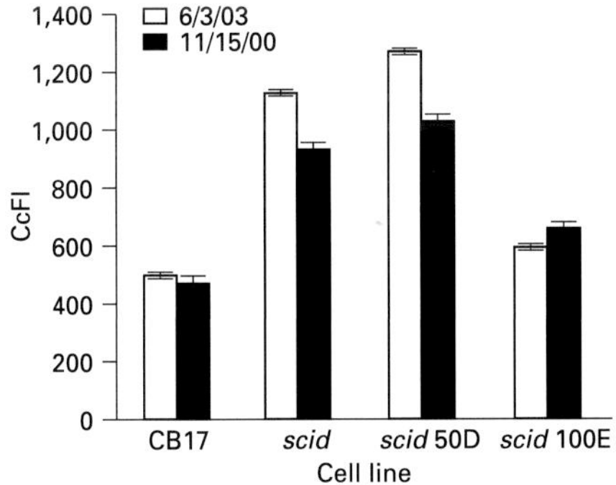

Fig. 3. (**A**) Chromosome length (CL) in LY-R cells expressed in arbitrary units after different incubation times with colcemid. (**B**) Telomere fluorescence (FL) in LY-R cells after different incubation times with colcemid. Note that telomere fluorescence measurements have been performed on the same date, a procedure which generally yields similar fluorescence values when the same sample is analysed twice.

Fig. 4. Corrected calibrated fluorescence (CcFl) in a set of four mouse cell lines observed at two different dates. For details of calibration protocol see Materials and methods.

measured the length of this chromosome digitally in 10 metaphase cells at each time point. Results of this analysis are shown in Fig. 3. The difference in chromosome length between cells treated for 1 h and cells treated for 3 or 6 h was statistically significant (Fig. 3A). There was no difference in chromosome length between cells treated with colcemid for 3 and 6 h, respectively (Fig. 3A). Having established that prolonged treatment of mouse cells with colcemid leads to hypercondensation of mitot-

ic chromosomes, we then measured telomere fluorescence in above cells to examine if chromosome hypercondensation leads to increase in telomere fluorescence intensity. Telomere fluorescence measurements in above three samples were taken on the same date. Our results revealed similar fluorescence intensity values in cells treated with colcemid for 1 and 3 h but higher values in cells treated with colcemid for 6 h (Fig. 3B). However, differences between these values were not statistically significant.

Telomere length in a set of four mouse cell lines

The above results indicate that measurements of mean telomere fluorescence intensities observed in LY-R and LY-S cells are reproducible, i.e. relative values of telomere fluorescence intensities between LY-R and LY-S cells are stable. This suggests that LY-R and LY-S cell lines can be used as internal controls in Q-FISH experiments. For example, fluorescence intensity of any sample can be expressed relative to values of LY-R and LY-S cells. To test validity of this approach we measured telomere length in a set of four different mouse cell lines at two different dates using LY-R and LY-S cells as an internal control (Fig. 4). Details of the protocol used to obtain corrected fluorescence values are described in Materials and methods.

Cell lines included a CB17 cell line, which has a normal response to DNA damage and a *scid* (severe combined immunodeficiency) cell line, which shows sensitivity to ionizing radiation as a result of deficiency in DNA double-strand break repair due to a defective protein DNA-PKcs. In addition, two cell lines were derived from the *scid* cell line: *scid* 50D and *scid* 100E. The *scid* 100E cell line contains a gene for functional DNA-PKcs introduced on a fragment from human chromosome 8. By contrast, the *scid* 50D cell line contains the same fragment but without the gene for DNA-PKcs. The cell lines have been described elsewhere (Kirchgessner et al., 1995). Measurements were performed on 15/11/00 and 03/06/03 (Fig. 4). In spite of almost three years difference between measurements telomere fluorescence values in four different cell lines were similar on both occasions (Fig. 4). For example, on both occasions telomere fluorescence values were significantly lower in *scid* 100E cells than in either parental *scid* cells or *scid* 50D cells suggesting that DNA-PKcs may be involved in control of telomere length. Some of the differences observed between two dates could be attributable to natural variations in telomere length in the same cell lines.

Discussion

The most widely used protocol for Q-FISH is based on the calibration procedure employing plasmids that contain variable numbers of telomeric TTAGGG repeats and fluorescence beads of specific size (Zijlmans et al., 1997; Samper et al., 2000; McIlrath et al., 2001; Goytisolo et al., 2001). In this protocol telomere fluorescence is expressed in telomere fluorescence units (TFUs), each one corresponding to approximately 1 kb of telomeric DNA. Using this protocol we have shown previously that telomere length in LY-R and LY-S cells is ~ 48 and ~ 7 kb, respectively (McIlrath et al., 2001). In the present study we wanted to ask the question if telomere length can reproducibly be measured in mouse cells if the above two cell lines are used as internals controls, instead of fluorescence beads and telomeric plasmids. The reasoning behind this is that we already have two telomere length reference points and if these points are generated reproducibly every time a test sample is measured then the fluorescence intensity of the test sample can be expressed relative to these two points. Since this procedure always compares the cell line with long telomeres (LY-R) against the cell line with short telomeres (LY-S), the procedure is essentially self-correcting and eliminates the need to perform conventional Q-FISH calibration. We measured telomere fluorescence in LY-R and LY-S cells at 5 different dates and obtained an average ratio LY-R:LY-S = 6.9 (Fig. 1). The telomere fluorescence ratio between LY-R and LY-S cells when measured by conventional Q-FISH is 48 kb:7 kb = 6.9 (McIlrath et al., 2001). Therefore, we were satisfied that we can reproducibly generate two telomere length reference points without using the classical Q-FISH calibration procedure which relies on telomeric plasmids and fluorescence beads. To eliminate the chance that day-to-day variation associated with the microscopy set-up may interfere with the results, we always acquired digital images of chromosomes and telomeres from both cell lines at the same time in a single, short microscopy session.

Using LY-R and LY-S as internal controls we then measured telomere fluorescence in a set of four different mouse cell lines on two different dates and in both cases observed similar telomere fluorescence values (Fig. 4). We observed some differences between measurements (see for example fluorescence values for *scid* and *scid* 50D cells in Fig. 4). However, these differences are compatible with published results. For example, there is a natural fluctuation in telomere size after each cell division (Blackburn, 2000) and in our case the difference in replicative histories between the same cell lines at different measurement dates was ~ 5 passages (15 mean population doublings). Similarly, our long-term flow-FISH measurements indicate that telomere fluorescence in LY-R and LY-S cells, although generally stable, varies to some degree after each cell division (Cabuy

and Slijepcevic, unpublished observations). Finally, fluorescence intensity measurement by fluorescence microscopy is known to produce variable results even if the same metaphase cells are analysed on two different occasions (Poon et al., 1999).

The advantage of our protocol is that using mammalian cell lines as calibration standards/internal controls in Q-FISH experiments is biologically more relevant than using plasmids or fluorescence beads. For example, DNA compaction in plasmids is dramatically different in comparison to that in mammalian cells. Also, fluorescence beads have a static size whereas in mammalian cells telomeres of the same size could generate different fluorescence intensities as a result of differential chromatin compaction, i.e. we observed that the increase in mitotic chromosome condensation causes some, though non-significant, increase in telomere fluorescence intensity (Fig. 3).

In order to achieve reproducible results with our protocol we recommend that the following points must be strictly observed. First, any pair of mammalian cell lines which may be suitable as internal controls must be verified by either conventional Q-FISH, flow-FISH (mouse) or Southern blot (human) to obtain telomere length reference points. Second, our results show considerable variations in telomere fluorescence between individual cells of the same sample (Tables 2 and 3). This is in line with previously published results (Poon et al., 1999). Therefore, to control variation in telomere length values between individual cells it is important to analyse a sufficient number of cells. We have been able to obtain reproducible results after analysing at least ten LY-R cells and ten LY-S cells as internal controls when our test samples were measured (Fig. 4). However, we recommend that a minimum of twenty cells for each of the two telomere length reference points should be analysed if there are large differences between individual cells. Third, image acquisition of metaphase cells from reference cell lines and test samples should be performed in a single microscope session and under the same conditions to eliminate day-to-day variation in microscope settings as a source of error.

References

Blackburn EH: Telomere states and cell fates. Nature 408:53–56 (2000).

Blasco MA, Lee H-W, Hande MP, Samper E, Lansdorp PM, DePinho R, Greider CW: Telomere shortening and tumor formation by mouse cells lacking telomerase RNA. Cell 91:25–34 (1997).

Goytisolo FA, Samper E, Edmonson S, Taccioli GE, Blasco MA: The absence of the DNA-dependent protein kinase catalytic subunit in mice results in anaphase bridges and in increased telomeric fusions with normal telomere length and G-strand overhang. Mol Cell Biol 21:3642–3651 (2001).

Kipling D, Cooke HJ: Hypervariable ultra-long telomeres in mice. Nature 347:347–402 (1990).

Kirchgessner CU, Patil CK, Evans JW, Cuomo CA, Fried LM, Carter T, Oettinger MA, Brown JM: DNA-dependent kinase (p350) as a candidate gene for the murine SCID defect. Science 267:1178–1183 (1995).

McIlrath J, Bouffler S, Samper E, Cuthbert A, Wojcik A, Szumiel I, Bryant PE, Riches AC, Thompson A, Blasco MA, Newbold RF, Slijepcevic P: Telomere length abnormalities in mammalian radiosensitive cells. Cancer Res 61:912–915 (2001).

Poon SSS, Martens UM, Ward RK, Lansdorp PM: Telomere length measurements using digital fluorescence microscopy. Cytometry 36:267–278 (1999).

Rufer N, Dragowska W, Thornbury G, Roosnek E, Lansdorp PM: Telomere length dynamics in human lymphocyte subpopulations measured by flow cytometry. Nat Biotechnol 16:743–747 (1998).

Samper E, Goytisolo AF, Slijepcevic P, Van Buul PW, Blasco MA: Mammalian Ku86 protein prevents telomeric fusion independently of the length of TTAGGG repeats and the G-strand overhang. EMBO Rep 1:244–252 (2000).

Zijlmans JMJM, Martens UM, Poon SSS, Raap AK, Tanke HJ, Ward RK, Lansdorp PM: Telomeres in the mouse have large inter-chromosomal variations in the number of T2AG3 repeats. Proc Natl Acad Sci USA 94:7423–7428 (1997).

Cytogenet Genome Res 105:471–478 (2004)
DOI: 10.1159/000078221

Manipulating mouse telomeres: models of tumorigenesis and aging

Z. Xin and D. Broccoli

Department of Medical Oncology, Fox Chase Cancer Center, Philadelphia, PA (USA)

Abstract. Telomeres are capping structures at the ends of chromosomes, composed of a repetitive DNA sequence and associated proteins. Both a minimal length of telomeric repeats and telomere-associated binding proteins are necessary for proper telomere function. Functional telomeres are essential for maintaining the integrity and stability of eukaryotic genomes. The capping structure enables cells to distinguish chromosome ends from double strand breaks (DSBs) in the genome. Uncapped chromosome ends are at great risk for degradation, recombination, or chromosome fusion by cellular DNA repair systems. Dysfunctional telomeres have been proposed to contribute to tumorigenesis and some aging phenotypes. The analysis of mice deficient in telomerase activity and other telomere-associated proteins has allowed the roles of dysfunctional telomeres in tumorigenesis and aging to be directly tested. Here we will focus on the analysis of different mouse models disrupted for proteins that are important for telomere functions and discuss known and proposed consequences of telomere dysfunction in tumorigenesis and aging.

Copyright © 2004 S. Karger AG, Basel

Telomere structure

Telomeres are specialized structures required for chromosome stability that are composed of repetitive DNA and associated proteins (Fig. 1). The DNA element of mammalian telomeres consists of tandem repeats of the simple hexanucleotide sequence, TAAGGG, which is present in hundreds to thousands of copies at each chromosome end. At the very end of the chromosome there is a single-stranded 3′-G-strand overhang. This 3′-overhang can fold back and invade the double-stranded telomeric repeats, displacing one strand and hybridizing to its complementary sequence to form the "T-loop" (Griffith et al., 1999). This structure physically hides and protects the 3′-chromosome end from cellular activities. Human telomeres are 10–15 kb (de Lange et al., 1990; Harley et al., 1990), but the telomeres of the laboratory mouse, *Mus musculus*, are much longer (>30 kb) and more heterogeneous in size (Kipling and Cooke, 1990; Zijlmans et al., 1997). In contrast, the telomeres of a related mouse species, *Mus spretus*, are slightly shorter than those of humans (Starling et al., 1990; Prowse and Greider, 1995).

Telomere-associated proteins are essential for maintaining telomere structure and telomere function. Several telomere-associated proteins have been identified to date. TRF1 and TRF2 bind directly and specifically to double-stranded telomeric DNA (Chong et al., 1995; Bilaud et al., 1996, 1997; Broccoli et al., 1997). Both TRF1 and TRF2 are negative regulators of telomere length. Although TRF1 and TRF2 contain a C-terminal myb type DNA binding domain and a large dimerization domain, the dimerization domains do not heterodimerize, leading to the establishment of two distinct protein complexes at telomeres. TRF1 plays a role in telomere length regulation. Increasing the amount of TRF1 associated with telomeres cause telomere shortening, while removal of TRF1 leads to telomere elongation. The function of TRF1 is regulated by TIN2 (Kim et al., 1999), another double-stranded telomeric DNA binding protein and negative regulator of telomere length. TIN2 binding to TRF1 with its TRF1 binding domain may recruit proteins to telomeres that inhibit telomerase or

Received 2 October 2003; accepted 21 October 2003.

Request reprints from: Dr. Dominique Broccoli, Department of Medical Oncology
  Fox Chase Cancer Center, 333 Cottman Avenue, Philadelphia, PA 19111
  (USA)
  telephone: +1-215-728-7133; fax: +1-215-728-4333;
  e-mail: K_Broccoli@fccc.edu.

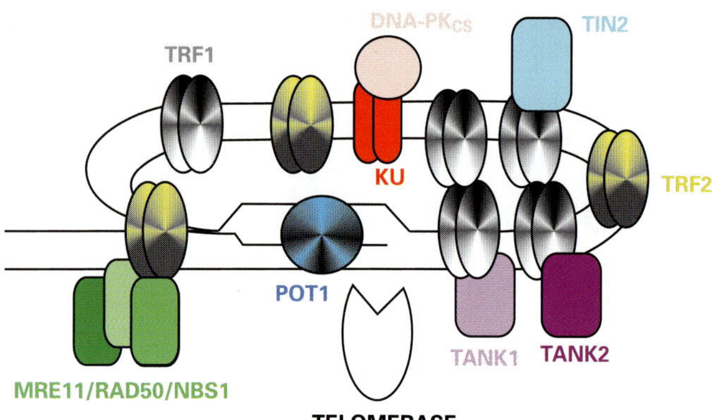

Fig. 1. Telomere-associated proteins and telomerase. Mammalian telomeres end in a T-loop whose structure and stability may depend on an array of proteins. TRF1 and TRF2 recruit a number of proteins such as TIN2, TANK1, TANK2, DNA repair complex MRE11/RAD50/NBS1 and the DNA-dependent protein kinase which is composed of a catalytic subunit (DNA-PKcs) and a DNA end binding heterodimer (Ku) to telomeres. Pot1 binds to the single-strand telomeric DNA overhang and protects it.

change telomeric structure to limit telomerase access to the 3′-telomeric terminus. Targeted deletion of exon1 of the mouse *Trf1* causes early embryonic lethality, suggesting that *Trf1* has an essential function that is independent of telomere length regulation (Karlseder et al., 2003). The early lethality caused by disruption of *Trf1* precludes analysis of the role of alterations in TRF1 in aging and tumorigenesis.

TRF2 is the major protective factor of telomeres. Removal of TRF2 from the telomeres by overexpression of a dominant negative allele causes loss of the G-strand overhang and results in end-to-end chromosome fusion (van Steensel et al., 1998). Loss of telomere function through TRF2 inhibition results in either ATM- and p53-dependent apoptosis or senescence (van Steensel et al., 1998; Karlseder et al., 1999). TRF2 also recruits hRAP1 to human telomeres through the interaction between the dimerization domain of TRF2 and the Rap1p C-terminus domain of hRAP1. Overexpression of the human homologue of the yeast RAP1 protein, hRAP1, causes telomere elongation (Li et al., 2000). Disruptions of TRF2, hRAP1 and TIN2 have not yet been reported.

Two related poly(ADP-ribose) polymerases (PARP), Tankyrase1 (TANK1) and Tankyrase2 (TANK2), have been found in association with TRF1 (Smith et al., 1998; Kaminker et al., 2001; Cook et al., 2002). TANK1 positively regulates telomere length through its interaction with TRF1 (Smith et al., 1998; Smith and Jackson, 1999; Smith and de Lange, 2000). Overexpression of TANK1 in the nucleus promotes telomere elongation in telomerase-positive human cells, suggesting that TANK1 regulates access of telomerase to the telomeric complex probably by inhibiting the binding of TRF1 to telomeres (Smith and de Lange, 2000). TANK1 binds to TRF1 with its Ankyrin-repeat domain and ADP ribosylates TRF1 (Smith et al., 1998; Smith and Jackson, 1999). ADP ribosylation of TRF1 by TANK1 releases TRF1 from telomeres. The telomere-unbound form of TRF1 can be ubiquitinated and degraded by

the proteasome. This subsequent ubiquitination of released TRF1 may prevent its rapid reassociation with telomeres resulting in the telomere elongation phenotype (Chang et al., 2003b). TANK2 also can interact with TRF1. Overexpression of TANK2 in the nucleus released endogenous TRF1 from telomeres, suggesting a possible role for TANK2 at telomeres (Cook et al., 2002).

Pot1 (protection of telomeres 1), a single-stranded telomeric DNA-binding protein (Baumann and Cech, 2001; Baumann et al., 2002; Mitton-Fry et al., 2002), appears to be important for stabilizing telomeric structure and protecting chromosome ends from degradation and fusion. Pot1 can bind to single-stranded telomeric DNA with its OB (oligonucleotide-binding) fold domain, and this binding is regulated by the TRF1 complex (Loayza and de Lange, 2003). A mutant form of Pot1 lacking the DNA binding domain induced rapid telomere elongation regardless of TRF1 levels. Based on these findings it was proposed that an interaction between the TRF1 complex and Pot1 regulates telomere length by affecting access of telomerase to the chromosome terminus.

A number of proteins involved in the repair of double-stranded DNA breaks are also found associated with telomeres. This may be a reflection of the telomere being the end of a linear chromosome. Ku86, Ku70 and DNA-PKcs are components of the DNA-dependent protein kinase complex, which is involved in DNA double-strand break repair by non-homologous end joining (NHEJ) (Smith and Jackson, 1999). These Ku proteins have been demonstrated to interact with TTAGGG repeats (Bianchi and de Lange, 1999; Hsu et al., 1999) and with telomeric proteins TRF1 and TRF2 (Hsu et al., 2000; Song et al., 2000). Both Ku86 and DNA-PKcs deficiency in mice result in telomere fusions, indicating that these proteins have protective roles at the mammalian telomeres (see below). MRE11, a double-strand break repair complex, has been found in association with TRF2 (Zhu et al., 2000). The function of MRE11 at telomeres is not clear yet.

Telomere replication

As normal somatic cells proliferate, telomeric repeats are lost from telomeres due in part to the end-replication problem (Olovnikov, 1973). Thus, telomeres are shorter in older individuals than in younger individuals. These findings led to the proposal that progressive telomere shortening ultimately limits the replicative capacity of cells (Cooke and Smith, 1986; Harley et al., 1990). In contrast to the situation in human cells, telomere shortening has not been seen in mouse cells. Cultured mouse cells senesce probably due to the acquirement of telomere-independent damage.

Telomerase is a reverse transcriptase that elongates telomeres de novo using an associated RNA molecule (Terc, telomerase RNA component) as a template (Collins, 2000). The progressive loss of telomeric DNA accompanying cell divisions can be prevented if cells express sufficiently high levels of telomerase. Most untransformed human somatic cells do not have sufficient telomerase and suffer telomere attrition coupled with cell division (Prowse and Greider, 1995; Holt et al., 1997).

Table 1. Function and knockout phenotype of telomerase and other telomere-associated proteins

Protein	Function	Telomere length of knockout mice	Tumorigenesis of knockout	Aging phenotype of knockout
Telomerase	telomere lengthening	shorter	reduced	enhanced
TRF1	negative regulator of telomere length			
TRF2	protection of telomere			
Ku86	NHEJ	longer	reduced	enhanced
DNA-PKcs	NHEJ	normal	normal	normal
TIN2	negatively regulate telomere length through interaction with TRF1			
RAP1	positive regulator of telomere length			
TANK1	ADP ribosylate TRF1 and diminish TRF1 binding to telomere			
TANK2	inhibit TRF1 binding to telomere			
Pot1	protection of single-strand telomeric DNA			

When telomeres reach a critically short length, cells arrest proliferation and acquire a characteristic enlarged morphology and a variety of altered functions. This response has been termed replicative senescence. It has been suggested that senescence evolved to suppress tumorigenesis because cells with critically shorted telomeres are at high risk for chromosome fusion and, therefore, at high risk for neoplastic transformation (Campisi, 2000).

Compared to normal somatic cells, the majority of human tumors reactivate and express telomerase to allow their immortal growth (Shay and Bacchetti, 1997; Hiyama and Hiyama, 2002). Likewise, inhibition of telomerase in human tumor cell lines leads to telomere shortening and loss of cell viability (Hahn et al., 1999; Zhang et al., 1999). In contrast to humans, laboratory mice have long telomeres and telomere shortening as a result of the end-replication problem has not been seen in mouse cells, either in culture or in vivo (Prowse and Greider, 1995; Hemann and Greider, 2000; Sherr and DePinho, 2000; Wright and Shay, 2000). Although mouse cell lines with short telomeres have been described, the stability of these telomeres over many cell generations points to a rapid deletion mechanism, rather than the gradual telomere shortening that accompanies telomerase insufficiency, as the underlying cause of the length difference (McIlrath et al., 2001). Despite the ultra-long telomeres and apparent independence of the process of senescence from telomere length, telomerase upregulation also appears to play a role in tumorigenesis in mice. First, although many primary tissues of both laboratory mice and *M. spretus* express detectable amounts of telomerase (Chadeneau et al., 1995; Prowse and Greider, 1995), telomerase is upregulated during mouse tumorigenesis (Blasco et al., 1996; Broccoli et al., 1996). In addition, overexpression of mTert results in mice with an increased susceptibility to form skin or breast tumors (Gonzalez-Suarez et al., 2001; Artandi et al., 2002). These observations indicate that increased telomerase expression might enhance survival and promote proliferation independently of telomere length. However, some human and mouse cell lines and tumors that lack telomerase activity are able to

maintain or elongate their telomeres by alternative mechanisms (ALT) (Bryan et al., 1995, 1997; Chang et al., 2003a). Thus, the telomere length-independent role of telomerase is not absolutely required for tumorigenesis.

Due to the long telomeres present in mice, telomerase-deficient mice are relatively asymptomatic for the first few generations. However, after 4–5 generations, the telomeres have eroded to near human telomere length. At this time, the telomerase-deficient mice begin to develop cancer and exhibit certain other pathologies associated with human aging such as hair graying, alopecia, skin lesions, impaired wound healing, cancer, and shortened lifespan (Blasco et al., 1997; Lee et al., 1998; Herrera et al., 1999; Rudolph et al., 1999; Artandi et al., 2000). These findings suggest that short telomere length triggers the same cellular response pathways in mice and humans and, together with the results from knock out mice for other telomere-associated proteins (discussed below), suggest that the study of knockout models for telomeric proteins will enhance our understanding of the role of telomere dynamics in the processes of tumorigenesis and aging (summarized in Table 1).

Mouse models for tumorigenesis

Several studies of mouse models for telomere dysfunction suggest that telomere dysfunction plays a crucial role in tumorigenesis. Dysfunctional telomeres contribute to the initiation and progression of malignant tumors due in part to increased genomic instability. Telomere dysfunction can be caused by replication-mediated shortening, direct damage, or deficiency in telomere-associated proteins. Mice disrupted for telomerase and several telomere-associated proteins allowed modeling of the contribution of telomere dysfunction to tumorigenesis.

Telomerase-deficient mice
Several telomerase-deficient mice have been generated (Blasco et al., 1997; Niida et al., 1998; Yuan et al., 1999; Liu et al., 2000). The mTerc–/– mice in which expression of the telo-

merase RNA, mTerc, has been eliminated are initially viable, but only a few generations can be derived before complete loss of viability. This is due to loss of telomeric DNA and increased end-to-end chromosome fusion. After several generations, the telomeres of the cells in affected tissues become too short to function, and lead to widespread genomic instability. This cellular response to dysfunctional telomeres in Terc–/– mice is initiated by the shortest telomere (Hemann et al., 2001).

Late generation mTerc–/– mice showed reduced tumorigenesis, indicating that critically short telomeres in the absence of telomerase activity can suppress tumor progression in many tissues. On the other hand, there is a high incidence of tumor formation in highly proliferative cell types, such as the bone marrow, in later generation mTerc–/– mice (Rudolph et al., 1999). These observations indicate a link between telomere dysfunction and tumorigenesis. This is also consistent with the finding that the pathologies associated with telomere dysfunction in mTerc–/– mice have been seen mainly in highly proliferative organs, such as the germ line, gut, skin, immune system and bone marrow (Lee et al., 1998). These results suggest that telomerase is essential for the long-term maintenance of diverse cell types with high proliferation profiles. Interestingly, late generation mTerc–/– mice showed significantly fewer skin tumors than wild type controls following chemical carcinogenesis (Gonzalez-Suarez et al., 2000), suggesting that tumor formation in skin requires telomere maintenance above a threshold length. In the tumors that develop in telomerase-deficient mice, telomere maintenance occurs by the telomerase-independent ALT mechanism (Chang et al., 2003a). Thus it is also possible that the reduced tumor incidence reflects an increased difficulty in the activation of the ALT pathway.

The catalytic component of telomerase, mTert, has also been knocked out (Yuan et al., 1999; Liu et al., 2000). Although the first generation of mTert knockout mice are apparently normal, G2 mTert-deficient mice showed significant reduction in relative telomere length compared with wild type and G1 mTert-deficient mice (Liu et al., 2000). Cells derived from mTert-deficient mice also showed progressive loss of telomeric DNA following prolonged growth. mTert-deficient ES cells exhibited genomic instability, aneuploidy, and increased end-to-end chromosome fusion. These results indicate that like mTerc, mTert is not immediately essential for normal development and cell viability but is required for telomere length maintenance. The phenotypes associated with telomere dysfunction occur sooner in mTert–/– mice than in mTerc–/– mice. One possibility is that binding of catalytic component to telomeres attenuates the loss of telomeric DNA resulting from the end-replication problem or the activity of cellular exonucleases. In addition, ES cells deficient in one allele of mTert (mTert+/–) lost telomeric DNA in successive passages, suggesting that mTert is a limiting component for telomere-length maintenance (Liu et al., 2000).

Terc–/–/p53–/– mice and Terc–/–/(p16/p19arf)–/– mice

As mentioned above, senescence is a defensive response to dysfunctional telomeres. This response requires cell cycle and DNA damage checkpoints controlled by the pRB and p53 tumor suppressor pathways. If only the p53 check point is intact, cells with telomere dysfunction generally will undergo p53-mediated cell death (apoptosis). If only the pRB pathway is intact, cells with dysfunctional telomeres can gain a few additional cell divisions before subsequently undergoing senescence. If neither the p53 nor pRB checkpoint is intact, cells may survive with genomic instability, which can lead to cancer.

In mice doubly deficient for mTerc and p53, generated by crossing telomerase-deficient mice with p53-null mice, cancer development and progression are accelerated (Chin et al., 1999). This result indicates that telomere dysfunction acts synergistically with p53 deficiency to promote tumorigenesis, probably by enhancing genomic instability. In this scenario the inability of the cell to respond to unstable chromosome ends allows cells to continue to grow and accumulate growth advantageous rearrangements. Thus, an intact p53 pathway is very important for the cellular defensive response to dysfunctional telomeres.

Double knockout mice for Terc and the INK4A locus (a locus which includes the two tumor suppressors, p16 and p19ARF) showed a 50% reduction in the number of tumors as compared with tumors developed in the (p16/p19arf)–/– mice (Greenberg et al., 1999). This suggests that the presence of telomerase in (p16/p19arf)–/– mice promotes tumor progression, presumably by preventing further telomere shortening and/or elongating critically short telomeres. Thus, telomerase deficiency in combination with deficiency in p16 and/or p19ARF significantly reduced the incidence of fibrosarcomas and B cell lymphomas, suggesting that telomerase inhibitors may be effective in preventing tumor progression. However, the signaling pathways that respond to dysfunctional telomeres in mouse might differ with that in humans. A recent study using a dominant-negative allele of *TRF2* to de-protect telomeres induced growth arrest and senescent morphology in mouse fibroblasts. This growth arrest was completely abrogated by p53 deficiency, indicating that the p16/RB response to telomere dysfunction is not active in mouse fibroblasts (Smogorzewska and de Lange, 2002). Further studies are necessary to address whether there is a fundamental difference in telomere damage signaling pathway between human and mouse. This is also important for evaluating mouse models for the role of telomere dysfunction in tumorigenesis.

The Ku 86 and DNA-PKcs-deficient mice

As mentioned above, components of the cellular response to double-stranded DNA breaks are also found at telomeres. This includes components of the NHEJ complex, the Ku heterodimer and DNA-PKcs. Ku86 has a protective role at mammalian telomeres. Ku86 deficiency in mice results in telomeric fusion between long telomeres (Bailey et al., 1999; Hsu et al., 2000; Samper et al., 2000), indicating that the telomeric fusions are not a result of telomere loss, but rather of a loss of telomere protection. Ku86 deficiency also results in significant telomere lengthening, suggesting that Ku86 impairs the access of elongating activities to the telomeres. This has been strongly supported by the study of mice doubly deficient for telomerase and Ku86 (Espejel et al., 2002). In this study, the investigators compared the telomere length of mouse embryonic fibroblasts (MEFs) derived from Terc+/+/Ku86–/– mice and wild type controls. The Terc+/+/Ku86–/– MEFs showed a significant increase in telomere length compared to wild type controls. In contrast, Terc–/–/Ku86–/– MEFs showed a similar rate of telomere loss as corresponding Terc–/–/Ku86+/+ controls, indicating that K86 deficiency does not result in telomere elongation in a telomerase-deficient background. Taken together, the findings suggest that Ku86 either directly or indirectly impairs the access of telomerase to the telomeres. In addition, the absence of Ku86 prevented end-to-end fusion of critically short telomeres caused by telomerase deficiency in Terc–/–/Ku86–/– mice (Espejel et al., 2002). Similar to late generation telomerase-deficient mice, Ku86 deficiency also can suppress tumor growth in a wild type p53 background (Difilippantonio et al., 2000). DNA-PKcs, another component of the NHEJ complex, is also important for telomere stability. DNA-PKcs deficiency also results in end-to-end chromosomal fusion (Goytisolo and Blasco, 2002), but the frequency of end-to-end chromosomal fusion in DNA-PKcs–/– cells is significantly lower than that in Ku86–/– cells, suggesting that Ku86 is more important for telomere protection than DNA-PKcs (Goytisolo and Blasco, 2002).

Mouse models for aging

Telomeres can contribute to mammalian aging in different ways. Normal human cells cultured in vitro exhibit a finite number of cell divisions before they reach a non-dividing status called replicative senescence. Replicative senescence in human cells may be caused by progressive telomere shortening (Cooke and Smith, 1986; Harley et al., 1990, 1992). This cellular senescence may contribute to organismal aging. Therefore, telomere dysfunction may contribute to aging by virtue of its ability to induce cellular senescence. Critically short or dysfunctional telomeres can also induce apoptosis (Karlseder et al., 1999; Lee et al., 1998), which may contribute to the decline in tissue function and integrity that is a hallmark of aging. Also, senescent cells have an altered behavior and pattern of gene expression, and may alter their local tissue microenvironment to contribute to the shift from young to aged tissue (Campisi, 2000). Because the accumulation of somatic damage is considered a main cause of the aging process, telomeres may contribute to aging by participating in DNA repair with telomeric binding proteins such as Ku86. Although telomere shortening has not been seen in normal mouse cells, recent evidence (discussed below) clearly showed that telomere dysfunction plays an important role in the process of aging. The importance of telomere function in age-related pathology has been demonstrated by several studies of mouse models.

Telomerase-deficient mice
Older mTerc–/– mice demonstrated accelerated aging-related phenotypes (Rudolph et al., 1999). As discussed above, the first few generation mTerc–/– mice are phenotypically normal and have a normal life span. This is probably due to the long telomeres of inbred mice (Kipling and Cooke, 1990). However, the aging population of late generation mTerc–/– mice did have a shortened lifespan. This was accompanied by aging-related

phenotypes compared to age-matched wild type mice, including decreased body weight, decreased stress responses, increased cancer incidence, and an increased incidence of skin lesions. All of these phenotypes correlated with a decline in telomere length compared to normal control mice. In addition, telomerase deficiency has an even stronger impact on life span and severity of phenotypes if the Terc-null mutant is in a genetic background with an initial shorter telomere length such as C57BL6 (Herrera et al., 1999; Rudolph et al., 1999).

As discussed above with respect to tumorigenesis, the effect of mTerc deficiency is greatest on highly proliferative tissues. (Lee et al., 1998). Similarly, mTerc deficiency has the greatest effect on aging of highly proliferative tissues. Late generation Terc–/– mice showed an increase in the incidence of hair graying and alopecia, a decrease of hair follicles in anagen, skin ulcerations, impaired wound healing and depletion of male germ cells (Lee et al., 1998; Rudolph et al., 1999). In addition, the mice exhibited all hallmarks of immunosenescence including splenic atrophy, abnormal hematology, an impaired reaction of B and T cells to mitogen stimulation, and a defective germinal center reaction following antigen immunization (Lee et al., 1998; Herrera et al., 1999, 2000).

Because p53 is critical for both the apoptotic and senescent responses to dysfunctional telomeres, some of the aging-associated phenotypes in Terc–/– mice such as reduction in reproductive competence could be rescued by crossing the null p53 alleles (p53–/–) into the Terc null background (Chin et al., 1999). These results are consistent with the hypothesis that p53 plays an important role in the cellular response to shortened telomeres and may mediate some of the aging-associated phenotypes of Terc–/– mice. The role of p53 in aging has also been supported by a recent study on p53 mutant mice (Tyner et al., 2002). A p53 mutant allele (allele m) produces an N-terminally deleted protein. This truncated p53 interacts with wild type p53 protein and results in increased p53 activity. Mice carrying the mutant allele m and a wild type allele (p53m/+) showed premature development of several aging phenotypes, including osteopenia, organ atrophy, diminished stress tolerance, and shortened lifespan. Since telomere dysfunction can activate p53 and activated p53 can cause premature aging phenotypes, it is clear that the p53-dependent response to telomere dysfunction plays an important role in the process of aging.

Ku86-deficient mice
Ku86 deficiency results in several early age-specific characteristics such as osteoporosis, atrophic skin, hepatocellular degeneration, hepatocellular lesions and shortened lifespan (Vogel et al., 1999). In the Ku86–/– mice chromosomal fusion between long telomeres is observed suggesting that the aging phenotypes in Ku86–/– mice are partly due to telomere dysfunction. In addition, Ku86-deficient mice show similar phenotypes to those of late generation telomerase-deficient mice, which includes small size, infertility and decreased viability (Nussenzweig et al., 1996; Vogel et al., 1999). Mouse embryo fibroblasts derived from Ku86-deficient mice showed premature senescence in culture (Nussenzweig et al., 1996; Espejel and Blasco, 2002). This could be rescued by introduction of p53 nullizygosity (Lim et al., 2000), indicating that the aging

phenotypes were p53 dependent. Thus, telomere dysfunction induced by telomere shortening or deficiency in telomere-associated proteins such as Ku86 can result in some age-related phenotypes through a p53-dependent pathway. Despite a similar telomere dysfunction phenotype, DNA-PKcs knockout mice do not have a premature aging phenotype (Hsu et al., 2000; Samper et al., 2000), suggesting that other factors in addition to telomere dysfunction may contribute to the aging phenotype in Ku86–/– mice.

ATM-deficient mice and Atm–/–/Terc–/– mice

Ataxia-Telangiectasia (A-T) results from the loss of Ataxia-Telangiectasia mutant (ATM) function and is characterized by accelerated telomere loss, genomic instability, progressive neurological degeneration, premature aging and increased neoplasia incidence (Shiloh and Kastan, 2001). Several studies have demonstrated that ATM participates in telomere maintenance (Metcalfe et al., 1996; Pandita, 2002). Loss of Atm function causes increased end-to-end chromosomal fusion in Atm-deficient mice and human cells (Metcalfe et al., 1996; Pandita, 2002). Removal of TRF2 from the telomeres by overexpression of a dominant negative allele causes end-to-end chromosomal fusion, and ATM- and p53-dependent apoptosis in human cells (Karlseder et al., 1999). p53 may sense dysfunctional telomeres through ATM, an upstream kinase that directly phosphorylates p53 (Kastan and Lim, 2000). Several Atm mutant mice showed modest premature aging phenotypes, including slight growth retardation and subtle motor coordination learning defects (Barlow et al., 1996; Elson et al., 1996; Xu et al., 1996; Borghesani et al., 2000). Perhaps the excessively long telomeres of mouse alleviate the aging phenotype in Atm-deficient mice.

Recently, Wong and colleagues (2003) examined the impact of Atm deficiency as a function of progressive telomere attrition in mice doubly null for Atm and Terc. They generated a series of cohorts with varying telomere length and function by breeding a mutant Atm allele through successive generations of mTerc–/–. The late generation Atm–/–/Terc–/– mice showed decreased body weight and significantly shorter life span compared to Terc–/–/Atm+/+ controls. By detailed analysis, late generation Terc–/–/Atm–/– mice showed diminished precursor/stem cell reserve and increased apoptosis in several organ systems. In addition, hair graying and alopecia was increased in the late generation Terc–/–/Atm–/– mice as compared with their Terc–/–/Atm+/+ controls. All of these findings suggest that Atm deficiency and telomere dysfunction act together to impair cellular and whole organism viability, thus contributing to the premature aging phenotypes.

Summary

To fully understand the roles of telomeres in tumorigenesis and aging, a complete map of all telomere-associated proteins needs to be established because both a minimal length of telomeric repeats and the telomere binding proteins are necessary for proper telomere functions. In addition, a complete network of the functional interactions between telomeres and the cellular signaling pathways that respond to dysfunctional telomeres, such as p53, p16/Rb and ATM needs to be constructed. Finally, telomeres and telomere-associated proteins may interact with the nuclear matrix and participate in anchoring chromosomes to their preferred intranuclear positions. Therefore, telomere integrity may be essential for positional stability (Luderus et al., 1996). Telomere dysfunction may affect a variety of cellular events, such as DNA replication, transcription, and cell cycle control. Identification of such specific telomere functional targets is very important to understanding the roles of telomere dysfunction in tumorigenesis and aging. With powerful gene targeting techniques we can use mouse models that are disrupted for the function of telomere-associated proteins, alone or in combination with additional telomere-associated or signaling proteins, to establish the consequence of telomere dysfunction for cancer and aging.

References

Artandi SE, Chang S, Lee SL, Alson S, Gottlieb GJ, Chin L, DePinho RA: Telomere dysfunction promotes non-reciprocal translocations and epithelial cancers in mice. Nature 406:641–645 (2000).

Artandi SE, Alson S, Tietze MK, Sharpless NE, Ye S, Greenberg RA, Castrillon DH, Horner JW, Weiler SR, Carrasco RD, DePinho RA: Constitutive telomerase expression promotes mammary carcinomas in aging mice. Proc Natl Acad Sci USA 99: 8191–8196 (2002).

Bailey SM, Meyne J, Chen DJ, Kurimasa A, Li GC, Lehnert BE, Goodwin EH: DNA double-strand break repair proteins are required to cap the ends of mammalian chromosomes. Proc Natl Acad Sci USA 96:14899–14904 (1999).

Barlow C, Hirotsune S, Paylor R, Liyanage M, Eckhaus M, Collins F, Shiloh Y, Crawley JN, Ried T, Tagle D, Wynshaw-Boris A: Atm-deficient mice: A paradigm of ataxia telangiectasia. Cell 86:159–171 (1996).

Baumann P, Cech TR: Pot1, the putative telomere end-binding protein in fission yeast and humans. Science 292:1171–1175 (2001).

Baumann P, Podell E, Cech TR: Human Pot1 (protection of telomeres) protein: Cytolocalization, gene structure, and alternative splicing. Mol Cell Biol 22:8079–8087 (2002).

Bianchi A, de Lange T: Ku binds telomeric DNA in vitro. J Biol Chem 274:21223–21227 (1999).

Bilaud T, Koering CE, Binet-Brasselet E, Ancelin K, Pollice A, Gasser SM, Gilson E: The telobox, a Myb-related telomeric DNA binding motif found in proteins from yeast, plants and human. Nucleic Acids Res 24:1294–1303 (1996).

Bilaud T, Brun C, Ancelin K, Koering CE, Laroche T, Gilson E: Telomeric localization of TRF2, a novel human telobox protein. Nat Genet 17:236–239 (1997).

Blasco MA, Rizen M, Greider CW, Hanahan D: Differential regulation of telomerase activity and telomerase RNA during multi-stage tumorigenesis. Nat Genet 12:200–204 (1996).

Blasco MA, Lee HW, Hande MP, Samper E, Lansdorp PM, DePinho RA, Greider CW: Telomere shortening and tumor formation by mouse cells lacking telomerase RNA. Cell 91:25–34 (1997).

Borghesani PR, Alt FW, Bottaro A, Davidson L, Aksoy S, Rathbun GA, Roberts TM, Swat W, Segal RA, Gu Y: Abnormal development of Purkinje cells and lymphocytes in Atm mutant mice. Proc Natl Acad Sci USA 97:3336–3341 (2000).

Broccoli D, Godley LA, Donehower LA, Varmus HE, de Lange T: Telomerase activation in mouse mammary tumors: Lack of detectable telomere shortening and evidence for regulation of telomerase RNA with cell proliferation. Mol Cell Biol 16:3765–3772 (1996).

Broccoli D, Smogorzewska A, Chong L, de Lange T: Human telomeres contain two distinct Myb-related proteins, TRF1 and TRF2. Nat Genet 17:231–235 (1997).

Bryan TM, Englezou A, Gupta J, Bacchetti S, Reddel RR: Telomere elongation in immortal human cells without detectable telomerase activity. EMBO J 14:4240–4248 (1995).

Bryan TM, Englezou A, Dalla-Pozza L, Dunham MA, Reddel RR: Evidence for an alternative mechanism for maintaining telomere length in human tumors and tumor-derived cell lines. Nat Med 3:1271–1274 (1997).

Campisi J: Cancer, aging and cellular senescence. In Vivo 14:183–188 (2000).

Chadeneau C, Siegel P, Harley CB, Muller WJ, Bacchetti S: Telomerase activity in normal and malignant murine tissues. Oncogene 11:893–898 (1995).

Chang S, Khoo CM, Naylor ML, Maser RS, DePinho RA: Telomere-based crisis: Functional differences between telomerase activation and ALT in tumor progression. Genes Dev 17:88–100 (2003a).

Chang W, Dynek JN, Smith S: TRF1 is degraded by ubiquitin-mediated proteolysis after release from telomeres. Genes Dev 17:1328–1333 (2003b).

Chin L, Artandi SE, Shen Q, Tam A, Lee SL, Gottlieb GJ, Greider CW, DePinho RA: p53 deficiency rescues the adverse effects of telomere loss and cooperates with telomere dysfunction to accelerate carcinogenesis. Cell 97:527–538 (1999).

Chong L, van Steensel B, Broccoli D, Erdjument-Bromage H, Hanish J, Tempst P, de Lange T: A human telomeric protein. Science 270:1663–1667 (1995).

Collins K: Mammalian telomeres and telomerase. Curr Opin Cell Biol 12:378–383 (2000).

Cook BD, Dynek JN, Chang W, Shostak G, Smith S: Role for the related poly(ADP-Ribose) polymerases tankyrase 1 and 2 at human telomeres. Mol Cell Biol 22:332–342 (2002).

Cooke HJ, Smith BA: Variability at the telomeres of the human X/Y pseudoautosomal region. Cold Spring Harb Symp Quant Biol 51:213–219 (1986).

de Lange T, Shiue L, Myers RM, Cox DR, Naylor SL, Killery AM, Varmus HE: Structure and variability of human chromosome ends. Mol Cell Biol 10:518–527 (1990).

Difilippantonio MJ, Zhu J, Chen HT, Meffre E, Nussenzweig MC, Max EE, Ried T, Nussenzweig A: DNA repair protein Ku80 suppresses chromosomal aberrations and malignant transformation. Nature 404:510–514 (2000).

Elson A, Wang Y, Daugherty CJ, Morton CC, Zhou F, Campos-Torres J, Leder P: Pleiotropic defects in ataxia-telangiectasia protein-deficient mice. Proc Natl Acad Sci USA 93:13084–13089 (1996).

Espejel S, Blasco MA: Identification of telomere-dependent "senescence-like" arrest in mouse embryonic fibroblasts. Exp Cell Res 276:242–248 (2002).

Espejel S, Franco S, Rodriguez-Perales S, Bouffler SD, Cigudosa JC, Blasco MA: Mammalian Ku86 mediates chromosomal fusions and apoptosis caused by critically short telomeres. EMBO J 21:2207–2219 (2002).

Gonzalez-Suarez E, Samper E, Flores JM, Blasco MA: Telomerase-deficient mice with short telomeres are resistant to skin tumorigenesis. Nat Genet 26:114–117 (2000).

Gonzalez-Suarez E, Samper E, Ramirez A, Flores JM, Martin-Caballero J, Jorcano JL, Blasco MA: Increased epidermal tumors and increased skin wound healing in transgenic mice overexpressing the catalytic subunit of telomerase, mTERT, in basal keratinocytes. EMBO J 20:2619–2630 (2001).

Goytisolo FA, Blasco MA: Many ways to telomere dysfunction: In vivo studies using mouse models. Oncogene 21:584–591 (2002).

Greenberg RA, Chin L, Femino A, Lee KH, Gottlieb GJ, Singer RH, Greider CW, DePinho RA: Short dysfunctional telomeres impair tumorigenesis in the INK4a(delta2/3) cancer-prone mouse. Cell 97:515–525 (1999).

Griffith JD, Comeau L, Rosenfield S, Stansel RM, Bianchi A, Moss H, de Lange T: Mammalian telomeres end in a large duplex loop. Cell 97:503–514 (1999).

Hahn W, Stewart SA, Brooks M, York S, Eaton E, Kurachi A, Beijersbergen R, Knoll J, Meyerson M, Weinberg R: Inhibition of telomerase limits the growth of human cancer cells. Nature Med 5:1164–1170 (1999).

Harley CB, Futcher AB, Greider CW: Telomeres shorten during ageing of human fibroblasts. Nature 345:458–460 (1990).

Harley CB, Vaziri H, Counter CM, Allsopp RC: The telomere hypothesis of cellular aging. Exp Gerontol 27:375–382 (1992).

Hemann MT, Greider CW: Wild-derived inbred mouse strains have short telomeres. Nucleic Acids Res 28:4474–4478 (2000).

Hemann MT, Strong MA, Hao LY, Greider CW: The shortest telomere, not average telomere length, is critical for cell viability and chromosome stability. Cell 107:67–77 (2001).

Herrera E, Samper E, Martin-Caballero J, Flores JM, Lee HW, Blasco MA: Disease states associated with telomerase deficiency appear earlier in mice with short telomeres. EMBO J 18:2950–2960 (1999).

Herrera E, Martinez AC, Blasco MA: Impaired germinal center reaction in mice with short telomeres. EMBO J 19:472–481 (2000).

Hiyama E, Hiyama K: Clinical utility of telomerase in cancer. Oncogene 21:643–649 (2002).

Holt SE, Wright WE, Shay JW: Multiple pathways for the regulation of telomerase activity. Eur J Cancer 33:761–766 (1997).

Hsu HL, Gilley D, Blackburn EH, Chen DJ: Ku is associated with the telomere in mammals. Proc Natl Acad Sci USA 96:12454–12458 (1999).

Hsu HL, Gilley D, Galande SA, Hande MP, Allen B, Kim SH, Li GC, Campisi J, Kohwi-Shigematsu T, Chen DJ: Ku acts in a unique way at the mammalian telomere to prevent end joining. Genes Dev 14:2807–2812 (2000).

Kaminker PG, Kim SH, Taylor RD, Zebarjadian Y, Funk WD, Morin GB, Yaswen P, Campisi J: TANK2, a new TRF1-associated poly(ADP-ribose) polymerase, causes rapid induction of cell death upon overexpression. J Biol Chem 276:35891–35899 (2001).

Karlseder J, Broccoli D, Dai Y, Hardy S, de Lange T: p53- and ATM-dependent apoptosis induced by telomeres lacking TRF2. Science 283:1321–1325 (1999).

Karlseder J, Kachatrian L, Takai H, Mercer K, Hingorani S, Jacks T, de Lange T: Targeted deletion reveals an essential function for the telomere length regulator Trf1. Mol Cell Biol 23:6533–6541 (2003).

Kastan MB, Lim DS: The many substrates and functions of ATM. Nat Rev Mol Cell Biol 1:179–186 (2000).

Kim SH, Kaminker P, Campisi J: TIN2, a new regulator of telomere length in human cells. Nat Genet 23:405–412 (1999).

Kipling D, Cooke HJ: Hypervariable ultra-long telomeres in mice. Nature 347:400–402 (1990).

Lee HW, Blasco MA, Gottlieb GJ, Horner JW, 2nd, Greider CW, DePinho RA: Essential role of mouse telomerase in highly proliferative organs. Nature 392:569–574 (1998).

Li B, Oestreich S, de Lange T: Identification of human Rap1: implications for telomere evolution. Cell 101:471–483 (2000).

Lim DS, Vogel H, Willerford DM, Sands AT, Platt KA, Hasty P: Analysis of ku80-mutant mice and cells with deficient levels of p53. Mol Cell Biol 20:3772–3780 (2000).

Liu Y, Snow BE, Hande MP, Yeung D, Erdmann NJ, Wakeham A, Itie A, Siderovski DP, Lansdorp PM, Robinson MO, Harrington L: The telomerase reverse transcriptase is limiting and necessary for telomerase function in vivo. Curr Biol 10:1459–1462 (2000).

Loayza D, de Lange T: POT1 as a terminal transducer of TRF1 telomere length control. Nature 424:1013–1018 (2003).

Luderus ME, van Steensel B, Chong L, Sibon OC, Cremers FF, de Lange T: Structure, subnuclear distribution, and nuclear matrix association of the mammalian telomeric complex. J Cell Biol 135:867–881 (1996).

McIlrath J, Bouffler SD, Samper E, Cuthbert A, Wojcik A, Szumiel I, Bryant PE, Riches AC, Thompson A, Blasco MA, Newbold RF, Slijepcevic P: Telomere length abnormalities in mammalian radiosensitive cells. Cancer Res 61:912–915 (2001).

Metcalfe JA, Parkhill J, Campbell L, Stacey M, Biggs P, Byrd PJ, Taylor AM: Accelerated telomere shortening in ataxia telangiectasia. Nat Genet 13:350–353 (1996).

Mitton-Fry RM, Anderson EM, Hughes TR, Lundblad V, Wuttke DS: Conserved structure for single-stranded telomeric DNA recognition. Science 296:145–147 (2002).

Niida H, Matsumoto T, Satoh H, Shiwa M, Tokutake Y, Furuichi Y, Shinkai Y: Severe growth defect in mouse cells lacking the telomerase RNA component. Nat Genet 19:203–206 (1998).

Nussenzweig A, Chen C, da Costa Soares V, Sanchez M, Sokol K, Nussenzweig MC, Li GC: Requirement for Ku80 in growth and immunoglobulin V(D)J recombination. Nature 382:551–555 (1996).

Olovnikov AM: A theory of marginotomy. The incomplete copying of template margin in enzymic synthesis of polynucleotides and biological significance of the phenomenon. J Theor Biol 41:181–190 (1973).

Pandita TK: ATM function and telomere stability. Oncogene 21:611–618 (2002).

Prowse KR, Greider CW: Developmental and tissue-specific regulation of mouse telomerase and telomere length. Proc Natl Acad Sci USA 92:4818–4822 (1995).

Rudolph KL, Chang S, Lee HW, Blasco M, Gottlieb GJ, Greider C, DePinho RA: Longevity, stress response, and cancer in aging telomerase-deficient mice. Cell 96:701–712 (1999).

Samper E, Goytisolo FA, Slijepcevic P, van Buul PP, Blasco MA: Mammalian Ku86 protein prevents telomeric fusions independently of the length of TTAGGG repeats and the G-strand overhang. EMBO Rep 1:244–252 (2000).

Shay JW, Bacchetti S: A survey of telomerase activity in human cancer. Eur J Cancer 33:787–791 (1997).

Sherr CJ, DePinho RA: Cellular senescence: Mitotic clock or culture shock? Cell 102:407–410 (2000).

Shiloh Y, Kastan MB: ATM: genome stability, neuronal development, and cancer cross paths. Adv Cancer Res 83:209–254 (2001).

Smith GC, Jackson SP: The DNA-dependent protein kinase. Genes Dev 13:916–934 (1999).

Smith S, de Lange T: Tankyrase promotes telomere elongation in human cells. Curr Biol 10:1299–1302 (2000).

Smith S, Giriat I, Schmitt A, de Lange T: Tankyrase, a poly(ADP-ribose) polymerase at human telomeres. Science 282:1484–1487 (1998).

Smogorzewska A, de Lange T: Different telomere damage signaling pathways in human and mouse cells. EMBO J 21:4338–4348 (2002).

Song K, Jung D, Jung Y, Lee SG, Lee I: Interaction of human Ku70 with TRF2. FEBS Lett 481:81–85 (2000).

Starling JA, Maule J, Hastie ND, Allshire RC: Extensive telomere repeat arrays in mouse are hypervariable. Nucleic Acids Res 18:6881–6888 (1990).

Tyner SD, Venkatachalam S, Choi J, Jones S, Ghebranious N, Igelmann H, Lu X, Soron G, Cooper B, Brayton C, Hee Park S, Thompson T, Karsenty G, Bradley A, Donehower LA: p53 mutant mice that display early ageing-associated phenotypes. Nature 415:45–53 (2002).

van Steensel B, Smogorzewska A, de Lange T: TRF2 protects human telomeres from end-to-end fusions. Cell 92:401–413 (1998).

Vogel H, Lim DS, Karsenty G, Finegold M, Hasty P: Deletion of Ku86 causes early onset of senescence in mice. Proc Natl Acad Sci USA 96:10770–10775 (1999).

Wong KK, Maser RS, Bachoo RM, Menon J, Carrasco DR, Gu Y, Alt FW, DePinho RA: Telomere dysfunction and Atm deficiency compromises organ homeostasis and accelerates ageing. Nature 421: 643–648 (2003).

Wright WE, Shay JW: Telomere dynamics in cancer progression and prevention: Fundamental differences in human and mouse telomere biology. Nat Med 6:849–851 (2000).

Xu Y, Ashley T, Brainerd EE, Bronson RT, Meyn MS, Baltimore D: Targeted disruption of ATM leads to growth retardation, chromosomal fragmentation during meiosis, immune defects, and thymic lymphoma. Genes Dev 10:2411–2422 (1996).

Yuan X, Ishibashi S, Hatakeyama S, Saito M, Nakayama J, Nikaido R, Haruyama T, Watanabe Y, Iwata H, Iida M, Sugimura H, Yamada N, Ishikawa F: Presence of telomeric G-strand tails in the telomerase catalytic subunit TERT knockout mice. Genes Cells 4:563–572 (1999).

Zhang X, Mar V, Zhou W, Harrington L, Robinson M: Telomere shortening and apoptosis in telomerase-inhibited human tumor cells. Genes Dev 13:2388–2399 (1999).

Zhu XD, Kuster B, Mann M, Petrini JH, de Lange T: Cell-cycle-regulated association of RAD50/MRE11/NBS1 with TRF2 and human telomeres. Nat Genet 25:347–352 (2000).

Zijlmans JM, Martens UM, Poon SS, Raap AK, Tanke HJ, Ward RK, Lansdorp PM: Telomeres in the mouse have large inter-chromosomal variations in the number of T2AG3 repeats. Proc Natl Acad Sci USA 94:7423–7428 (1997).

Cytogenet Genome Res 105:479–484 (2004)
DOI: 10.1159/000078222

Genetics of rare diseases of the kidney: learning from mouse models

C. Zoja,[a] M. Morigi,[a] A. Benigni,[a] and G. Remuzzi[a,b]

[a] Mario Negri Institute for Pharmacological Research, Bergamo;
[b] Division of Nephrology and Dialysis, Azienda Ospedaliera, Ospedali Riuniti di Bergamo, Bergamo (Italy)

Introduction

Rare diseases are more difficult to study than common disorders for several reasons, which includes the lack of experimental animal models that prevents the understanding of pathogenesis and the testing of potential remedies. These are among the reasons why rare diseases are poorly understood, insufficiently characterized, and most often left untreated. During the last decade, however, the growth of genomic research has offered a new opportunity to identify the molecular abnormalities underlying disease processes. Furthermore, the development of embryonic stem cell technology together with the possibility of inducing targeting mutations has contributed to revolutionize biology by enhancing our ability to manipulate the genome introducing new genetic information into the germ-line of mice and providing new model systems. The current availability of transgenic and gene knock-out mouse models has favored the knowledge of physiologic processes and molecular mechanisms governing common and rare diseases. The present review is aimed at summarizing the most recent findings of the studies based on the exploitation of mouse models in a number of rare kidney diseases.

Polycystic kidney disease

Definition of the human disease
Polycystic kidney disease (PKD) represents a set of hereditary nephropathies characterized by progressive formation of fluid-filled cysts that arise from kidney tubules, leading to renal

enlargement and progressive renal failure. Extrarenal manifestations are common and include cysts in liver and pancreas, cardiovascular abnormalities and cerebral aneurysms.

Autosomal dominant PKD (ADPKD) is the most prevalent of the inherited polycystic kidney diseases occurring in 1:1,000 individuals. It is caused by mutations in at least two different genes, PKD1 present on chromosome 16 and PKD2 on chromosome 4 (Peters and Breuning, 2001; Sutters and Germino, 2003). Mutation of the PKD1 gene accounts for approximately 85% of all ADPKD cases. Disease progression is typically more rapid in patients with PKD1 mutations, but in all other respects ADPKD types 1 and 2 share virtually indistinguishable phenotypes. Polycystin 1 is an integral membrane glycoprotein which contains a subset of structural motifs likely involved in cell-to-cell or cell-to-matrix interaction. Its carboxy terminus interacts with a region in the carboxy terminus of polycystin 2. This second protein shares significant sequence homology with voltage-activated calcium- and sodium-channel proteins. Loss or a reduced level of polycystin 1 and 2 complex results in the formation of tubule-derived cysts. A plausible explanation for the sporadic pattern of cyst formation in ADPKD is provided by the "two-hit model" (Reeders 1992; Qian et al., 1996; Pei et al., 1999). Loss of both polycystin alleles is required for initiation of the processes leading to cyst formation. The germ-line mutation is present in all cells, but cyst generation is not triggered until the second allele is rendered inactive by somatic mutation.

Autosomal recessive PKD (ARPKD) is less frequent (1:20,000) than ADPKD and typically presents in infancy. The clinical phenotype is dominated by renal collecting duct ectasia, biliary dysgenesis and portal tract fibrosis. The principal disease locus, PKHD1, has been mapped to chromosome 6p21.1 → p12 (Guay-Woodford et al., 1995).

Characteristics of the mouse model
Among ADPKD models, knockout mice for Pkd1 (Pkd1[del34] and Pkd1[L]) have been created which developed cysts as a consequence of reduction or loss of polycystin 1. In the Pkd1[del34]

Received 21 October 2003; manuscript accepted 3 November 2003.

Request reprints from: Dr. Marina Morigi
Mario Negri Institute for Pharmacological Research
Via Gavazzeni 11, IT–24125 Bergamo (Italy)
telephone: +39-035 319888; fax: +39-035 319331
e-mail: morigi@marionegri.it

KARGER

Fax + 41 61 306 12 34
E-mail karger@karger.ch
www.karger.com

Accessible online at:
www.karger.com/cgr

model older heterozygotes have occasional cysts, while homozygotes die in the perinatal period with enlarged and cystic kidney (Lu et al., 1997, 1999a). Cystic kidneys are also found in Pkd1^L homozygotes, but they have a more severe vascular phenotype with earlier embryonic lethality (Kim et al., 2000). Germ-line mutations in PKD2 cause autosomal dominant PKD. Studies have shown that the introduction of a mutant exon 1 in tandem with the wild-type exon 1 at the mouse Pkd2 locus generates an unstable allele (WS25) that undergoes somatic inactivation by intragenic homologous recombination to produce a true null allele (Wu et al., 1998). The null homozygotes die at the embryonic stage, the heterozygotes have a small number of cysts. Pkd2^{WS25} homo- and heterozygotes are viable and develop cyst formation in maturing nephrons and pancreatic ducts, suggesting that this clinical manifestation of ADPKD also occurs by a two-hit mechanism.

Mouse models of ARPKD have also been described, in which the mutant phenotypes resemble the human disease. Genetic background affects the disease phenotype in most of these models. Using experimental crosses, the modifying effects can be dissected into discrete genetic factors referred to as quantitative trait loci (QTL). Recently QTL have been mapped to chromosome 1 and 10 for *jck*, 1 and 19 for *kat*^{2j}, 4 and 16 for *pcy*, 4 for *cpk* and 6 and 1 for *bpk* (Iakoubova et al., 1995; Woo et al., 1995, 1997; Upadhya et al., 1999; Guay-Woodford et al., 2000).

Therapeutic approaches: lessons from mouse research
Several dietary and pharmacological intervention strategies including blood pressure control, protein restriction, lipid-lowering agents and antioxidants have been studied in animals and humans with PKD (Qian et al., 2001). The increasing understanding of the abnormalities in cell-matrix and cell-cell communication, endocrine, paracrine and autocrine mechanisms, and signal transduction pathways in PKD has contributed in recent years to provide novel targets for delineation of therapeutic intervention. These findings have established a major role of the epidermal growth factor receptor (EGFR) in promoting epithelia hyperplasia and cyst formation and enlargement. EGFR is overexpressed and mislocated to the apical surface of cystic epithelial cells in the cpk and bpk mouse models, as in humans (Orellana et al., 1995). The ligand of EGFR, transforming growth factor-α (TGF-α), is also abnormally expressed. Interruption of EGFR activity by treatment of bpk mice with a tyrosine kinase inhibitor resulted in a marked reduction of collecting tubule cysts, improved renal function, decreased biliary epithelial abnormalities, and an increased life span (Sweeney et al., 2000). These effects were further maximized when EGFR tyrosine kinase inhibitor was given in combination with an inhibitor of tumor necrosis factor-α converting enzyme that reduced TGF-α bioavailability (Sweeney et al., 2003). Therapies aimed at reducing fluid secretion by cyst-derived epithelial cells have been proposed. Administration of vasopressin V2 receptor antagonist, an inhibitor of adenyl cyclase and cAMP, key regulators of solute secretion and fluid into the cysts, reduced the severity of cystic disease and renal insufficiency in cpk/cpk mice (Qian et al., 2001).

Alport syndrome

Definition of the human disease
Alport syndrome is a form of hereditary nephritis characterized by progressive glomerular basement membranes (GBM) thickening and delamination, proteinuria with subsequent glomerular sclerosis and end-stage renal failure, often accompanied by sensorineural deafness and typical ocular changes (Hudson et al., 2003). The disease is caused by mutations in any one of the genes encoding the alpha 3, alpha 4 and alpha 5 chains of type IV collagen of the basement membranes (Barker et al., 1990; Lemmink et al., 1994; Mochizuchi et al., 1994). A mutation affecting one of these chains forming a putative alpha 3-4-5 network can alter or abolish GBM expression not only of the corresponding chain but also of the other two chains. This confers an increased susceptibility to proteolytic enzymes, accumulation of alpha 1 (IV) and alpha 2 (IV) collagen, which is normally present only in the developing immature GBM, and collagen V and VI as compensatory response. The most common form of the disease is X-linked, and is caused primarily by mutations in collagen alpha 5 (IV) gene, accounting for approximately 80% of the cases. In about 15% of patients, the inheritance is autosomal recessive and the involved genes are located on chromosome 2 respectively encoding alpha 3 and alpha 4 (IV) chains.

Characteristics of the mouse model
Mouse models of Alport disease have been generated through the deletion of the noncollagenous 1 (NC1) domain of alpha 3 (IV) collagen, or an insertional mutation knocking out both the alpha 3 (IV) and alpha 4 (IV) collagen genes (Miner and Sanes, 1996; Cosgrove et al., 1996; Lu et al., 1999b; Heikkila et al., 2000). These mice all exhibited renal phenotypes that closely resemble those seen in autosomal recessive human disease, with development of a delayed onset glomerulonephritis that progresses to end-stage renal failure (ESRF). Notably, the genetic background strongly influences the timing of onset of disease and rate of renal disease progression in these mice, to the extent that 129 *col4a3*–/– mice reached ESRF at 2 months of age while B6 *col4a3*–/– mice lived to greater than 6 months (Andrews et al., 2002). Histopathological analysis showed that GBM lesions were more frequent in homozygotes on the 129X1/SvJ background and could be detected earlier than on the C57BL/6J background, suggesting the existence of modifier genes that influence GBM structure and disease progression. This is consistent with the fact that Alport syndrome patients with an identical mutation can exhibit very different rates of progression to ESRF.

Therapeutic approaches: lessons from mouse research
There is no definitive therapy that may delay the time of dialysis or kidney transplant for patients with Alport syndrome. One preliminary report claims benefit from cyclosporine treatment in decreasing proteinuria and preventing aggravation of histological lesions (Callis et al., 1999). As for other forms of progressive renal disease, angiotensin-converting enzyme inhibitors may offer a specific advantage. Administration of enalapril to few patients with the classic form and two

siblings with the autosomal recessive form transiently reduced proteinuria (Proesmans et al., 2000). A recent study in *col4a3–/–* mice provided the evidence that early administration of the ACE inhibitor ramipril starting at 4 weeks of life markedly delayed the onset of proteinuria, progressive renal damage and uremia, and more importantly, prolonged life span of these mice by several weeks up to 150 days (Gross et al., 2003). In parallel, drug treatment exerted antifibrotic effects attributable to inhibition of excess TGF-β production. Conversely, late therapy starting at 7 weeks when animals were already strongly proteinuric had no effect on renal damage and survival. Thus early diagnosis and a preemptive treatment before onset of proteinuria would appear to be crucial in humans with Alport disease (Abbate and Remuzzi, 2003).

Studies in knockout mice were instrumental to establish distinct roles of transforming growth factor-β and integrin α1β1 in the pathogenesis of glomerular damage of Alport syndrome thus disclosing novel targets for therapy (Cosgrove et al., 2000). Pharmacological blocking of TGF-β by a soluble TGF-β type II receptor inhibited focal GBM thickening without affecting podocyte foot process effacement in Alport mice. On the other hand, deletion of integrin α1 gene produced in Alport mice attenuated changes only in the podocyte morphology. When mice null for both *col4a3* and integrin α1 genes were treated with the TGF-β inhibitor, glomerular foot process and GBM morphology were primarily restored and renal function was markedly improved.

Preliminary experiments of gene therapy in Alport mice have shown rescue of the Alport syndrome phenotype by means of a yeast artificial chromosome containing human COL4A3-COL4A4 genes. *col4a3–/–* mice were crossed with transgenic mice carrying the COL4A3-COL4A4 genes. Whereas *col4a3–/–* mice have hematuria and premature death, littermates *col4a3–/–* mice carrying the transgene displayed normal phenotype, no hematuria and normal renal histology (Heidet et al., 2003). Attempts in transferring a corrected type 4 alpha collagen gene into renal glomerular cells have been accomplished in dog and swine models which mimic the most common X-linked form of the disease (Heikkila et al., 2000, 2001); however, extensive research still needs to be performed to render gene therapy a successful approach.

Fabry disease

Definition of the human disease

Fabry disease is an X-linked metabolic disorder of glycosphingolipid catabolism caused by a wild variety of mutations in the gene encoding the lysosomal hydrolase α-galactosidase A (α-gal A) (Brady et al., 1967; Breunig et al., 2003a). The enzymatic defect leads to the inability of the body to break down globotriaosylceramide (Gb3), also known as ceramide trihexoside in affected males and to a lesser extent in female carriers. The glycolipid progressively accumulates in the endothelial lining of blood vessels within the kidney, heart, liver, spleen. Major clinical manifestations include paresthesias in the extremities, corneal dystrophy, angiokeratoma in skin and mucous membranes, and hypohydrosis in childhood or adoles-

cence (Breunig et al., 2003a). With increasing age, proteinuria, hypostenuria and lymphedema appear. Severe renal impairment leads to hypertension and uremia. Renal failure typically occurs in the third to fifth decades of life with patients needing chronic hemodialysis and/or renal transplantation as life extending procedures (Sessa et al., 2003; Breunig et al., 2003a). Death usually occurs from renal failure or from cardiac or cerebrovascular disease.

Characteristics of the mouse model

The identification of the genomic sequences and cDNAs encoding human and mouse α-gal A genes sharing high homology (Bishop et al., 1986, 1988; Ohshima et al., 1995) led to the generation of mice with targeted disruption of the α-gal A gene (Ohshima et al., 1997). These mice exhibit clinically normal phenotypes at 10–14 weeks of age; however, ultrastructural analysis revealed the presence of lipid inclusions in the liver and kidney. Gb3 accumulation and pathological lesions in the affected organs increase with age (Ohshima et al., 1999).

Therapeutic approaches: lessons from mouse research

Besides symptomatic management, enzyme replacement therapy with recombinant α-galactosidase A represents the only specific treatment currently available for Fabry disease (Breunig et al., 2003b).

α-gal A knockout mice have been used to evaluate enzyme and gene therapy approaches. Infusions of recombinant human α-gal A in knockout mice (Ioannou et al., 2001) resulted in the clearance of accumulated Gb3 in the liver, spleen and heart with concomitant elevation of α-gal activity. These preclinical studies provided the rationale for recent clinical trials that demonstrated the safety and effectiveness of enzyme replacement therapy (Schiffmann et al., 2000, 2001; Breunig et al., 2003b), that, however, may have some limitations due to the long-term need of repeated intravenous injections of large amounts of the enzyme, the possibility of a reduced response to the therapy on account of enzyme antibody development, and high costs.

Gene therapy conceived as gene-mediated enzyme replacement therapy in which the expression vector for α-gal A is inserted into the patient's cells instead of direct infusion of the enzyme, has been widely explored in the Fabry mouse model as an alternative approach. The first attempt to use gene therapy in Fabry disease mice was accomplished by transfecting bone marrow mononuclear cells with a bicistronic retroviral vector engineering the expression of both the α-gal A gene and the human IL-2Rα chain gene, taken as a selectable marker. The latter allowed ex vivo preselection of transduced cells resulting in enhanced intracellular and secreted enzyme activity in vitro as well as in vivo in plasma and organs of Fabry mice for more than 6 months (Qin et al., 2001). Subsequently, a recombinant adeno-associated viral vector encoding human α-gal was constructed and injected in the hepatic portal vein of Fabry mice (Jung et al., 2001). As long as 6 months after treatment the transduced animals showed higher α-gal A levels in liver and other tissues compared to untreated Fabry controls. A significant reduction in Gb3 levels to near normal was seen starting from 2–5 weeks post treatment. More recently, a simple and clinical applicable strategy for gene-based enzyme replacement of Fabry disease has been employed. Injection via the tail vein or into the quadriceps muscle of Fabry knockout mice of adeno-associated viral vector containing the α-gal A gene resulted in a long-term enzymatic and functional correction of the disease (Takahashi et al., 2002; Park et al., 2003).

Systemic lupus erythematosus

Definition of the human disease
Systemic lupus erythematosus (SLE) is a complex, multisystem autoimmune disease characterized by the production of autoantibodies against nuclear and endogenous antigens, among which nucleosomes, DNA complexed to histones, seem to be the most prominent (Berden et al., 1999; Foster and Kelley, 1999). These autoantibodies cause end-organ damage by a variety of mechanisms, notably via immune complex-mediated inflammation, which can result in glomerulonephritis, arthritis and vasculitis. At the molecular level, defects in both immune complex clearance and in T- and B-cell regulation have been implicated in the pathogenesis of the disease.

Characteristics of the mouse model
There are numerous synthetic and spontaneous murine models of lupus disease (Theophilous and Dixon, 1985; Foster, 1999). Knockout and transgenic mice for single genes involved in lymphocyte interactions, apoptosis or antigen clearance have been generated that helped clarify the role of different pathways in autoimmune lupus abnormalities (Mohan, 2001; Wakeland et al., 2001). Classic spontaneous models include the (NZB × NZW)F1 (or NZB/W) mouse and the congenic recombinant NZM2410 strain derived from this cross, the MRL/lpr mouse, and the BXSB/yaa mouse. NZB/W mice develop systemic autoimmunity reminiscent to human lupus disease (Howie and Helyer, 1968; Morel and Wakeland, 1998), with polyclonal B cell activation, excessive synthesis of antibodies to nucleic acid and endogenous antigens, renal immune deposits, and proteinuria. The resulting progressive glomerulonephritis is the primary cause of death in these animals. In the MRL strain, the lpr (lymphoproliferation) mutation of the *Fas* gene causes lymphoproliferation and an accelerated autoimmune glomerulonephritis from which these mice ultimately die (Theophilous and Dixon, 1985). In the case of BXSB mice, the Y-linked autoimmune accelerator gene causes severe disease in males when expressed in the susceptible BXSB genome (Nguyen et al., 2002).

A quite large number of genetic loci associated with lupus predisposition has been identified in murine models as well as in humans (Nguyen et al., 2002; Mageed and Prud'homme, 2003). Evidence is available for multiple susceptibility genes involving major histocompatibility complex (MHC) and non-MHC loci. Among the latter are genes controlling the pattern of cytokine production and genes that control lymphocyte proliferative responses. In addition, deficiencies in the early components of the complement cascade (C1q, C1r, C1s and C2) as well as complement receptors, CR1 and CR2, contributed to cause the phenotype.

Therapeutic approaches: lessons from mouse research
The above murine models have been employed to test new agents of potential interest for the therapy of human lupus, as an alternative to immunosuppressants and steroids that despite excellent therapeutic effects, cause toxicity in the long term. Among modern molecules that can modulate formation and deposition of immune complexes, mycophenolate mofetil (MMF) by virtue of its selective antiproliferative activity on T- and B-lymphocytes and its good tolerability was administered to lupus mice early before the onset of renal disease and found to confer remarkable renoprotection to either NZB/W or MRL/lpr strains (Corna et al., 1997; van Bruggen et al., 1998). In another study, MMF dramatically delayed the onset of proteinuria and improved survival of NZB/W mice even at a phase of established renal disease if it was given in combination with a selective cyclooxygenase-2 inhibitor (Zoja et al., 2001), the rationale for such a treatment being based on the excessive renal production of cyclooxygenase-2-derived thromboxane A2 in lupus nephritis (Tomasoni et al., 1998). Other therapeutic approaches were aimed to intervene on the inflammatory reaction that follows immune complex deposition in the kidney and contribute substantially to renal dysfunction. Bindarit, an indazolic derivative molecule devoid of immunosuppressive effects, retarded the clinical expression of lupus nephritis in NZB/W mice (Zoja et al., 1998), due at least in part to its effect of limiting renal overexpression of monocyte chemoattractant protein-1 (MCP-1), thus reducing the signaling pathway by which inflammatory mononuclear cells are recruited into renal interstitium and glomeruli (Zoja et al., 1997). Targeting cytokines abnormally expressed in lupus mice using blocking antibodies and soluble receptors had beneficial effects in delaying disease progression, as it was the case for IFN-γ, IL-6 and IL-10 (reviewed in Mageed and Prud'homme, 2003). Controversial results are instead available as for therapeutic interventions to limit synthesis or activity of TNFα and IL-1 in lupus mice.

A limited number of studies employing gene therapy in lupus mice are available. Encouraging results have been reported after injections of TGF-β1 plasmid DNA into MRL *lpr/lpr* mice to the extent that animal survival was prolonged, renal function improved and kidney inflammation was reduced (Raz et al., 1995). Treatment with IFN-γR/IgG1 plasmid by intramuscular injections, especially with electroporation, remarkably protected MRL *lpr/lpr* from renal disease progression (Lawson et al., 2000). An alternative approach to using inhibitory cytokines involves blocking co-stimulatory signals necessary for inducing T-cell proliferation after antigen presentation. A single intravenous injection of a recombinant adenovirus vector containing CTLA-4/IgG gene into lupus mice resulted in a marked amelioration of the disease (Takiguchi et al., 2000).

Conclusions

Genomic medicine is producing a profound change in our understanding of common and rare diseases. In this review we have presented evidence that molecular biology has allowed major breakthroughs in the comprehension of the pathophysiological mechanisms responsible for the development of rare genetic diseases of the kidney.

The availability of appropriate animal models is the indispensable step in the development of new therapeutic tools among which gene therapy may represent the ultimate goal. However, gene therapy as a potentially powerful clinical approach has not yet shown unequivocal clinical success. In the early days much was promised, but to date little has been accomplished, due to the hurdles encountered in the gene transfer, duration of expression, and safety.

However, gene therapy for rare (and for that matter, common) kidney diseases is only postponed since extensive basic research and pivotal clinical experience will eventually result in a feasible gene therapy strategy. The integration of the existing pharmaceutical armamentarium with gene-based therapies will help customize treatment for each individual patient rendering the highly anticipated future of pharmacogenomics a reality.

Acknowledgements

We are greatly indebted to Dr. Arrigo Schieppati, Head Laboratory for Coordination of Diagnosis and Information on Rare Diseases, Clinical Research Center for Rare Diseases "Aldo e Cele Daccò", Ranica, Bergamo, for his precious contribution to this manuscript.

References

Abbate M, Remuzzi G: Renoprotection: Clues from knockout models of rare diseases. Kidney Int 63:764–766 (2003).

Andrews KL, Mudd JL, Li C, Miner JH: Quantitative trait loci influence renal disease progression in a mouse model of Alport syndrome. Am J Pathol 160:721–730 (2002).

Barker DE, Hostikka SL, Zhou J, Show LT, Oliphant AR, Gerken SC, Gregory MC, Skolnick MH, Atkin CL, Tryggvason K: Identification of mutations in the COL4A5 collagen gene in Alport syndrome. Science 248:1224–1227 (1990).

Berden JHM, Licht R, van Bruggen MCJ, Tax WJM: Role of nucleosomes for induction and glomerular binding of autoantibodies in lupus nephritis. Curr Opin Nephrol Hypertens 8:299–306 (1999).

Bishop DF, Calhoun DH, Bernstein HS, Hantzopoulos P, Quinn M, Desnick RJ: Human alpha-galactosidase A: nucleotide sequence of a cDNA clone encoding the mature enzyme. Proc Natl Acad Sci USA 83:4859–4863 (1986).

Bishop DF, Kornreich R, Desnick RJ: Structural organization of the human alpha-galactosidase A gene: further evidence for the absence of a 3′ untranslated region. Proc Natl Acad Sci USA 85:3903–3907 (1988).

Brady RO, Gal AE, Bradley RM, Martensson E, Warshaw AL, Laster L: Enzymatic defect in Fabry's disease: Ceramidetrihexosidase deficiency. New Engl J Med 276:1163–1167 (1967).

Breunig F, Weidemann F, Beer M, Eggert A, Krane V, Spindler M, Sandstede J, Strotmann J, Wanner C: Fabry disease: Diagnosis and treatment. Kidney Int 63:S181–S185 (2003a).

Breunig F, Knoll A, Wanner C: Enzyme replacement therapy in Fabry disease: clinical implications. Curr Opin Nephrol Hypertens 12:491–495 (2003b).

Callis l, Vila A, Carrera M: Long-term effects of cyclosporine A in Alport's syndrome. Kidney Int 55:1051–1056 (1999).

Corna D, Morigi M, Facchinetti D, Bertani T, Zoja C, Remuzzi G: Mycophenolate mofetil limits renal damage and prolongs life in murine lupus autoimmune disease. Kidney Int 51:1583–1589 (1997).

Cosgrove D, Meehan DT, Grunkemeyer JA, Kornak JM, Sayers R, Hunter WJ, Samuelson GC: Collagen COL4A3 knockout: A mouse model for autosomal Alport syndrome. Genes Dev 10:2981–2992 (1996).

Cosgrove D, Rodgers K, Meehan D, Miller C, Bovard K, Gilroy A, Gardner H, Kotelianski V, Gotwals P, Amatucci A, Kalluri R: Integrin α1β1 and transforming growth factor-β1 play distinct roles in Alport glomerular pathogenesis and serve as dual targets for metabolic therapy. Am J Pathol 157:1649–1659 (2000).

Eng CM, Banikazemi M, Gordon RE, Goldman M, Phelps R, Kim L, Gass A, Winston J, Dikman S, Fallon JT, Brodie S, Stacy CB, Mehta D, Parsons R, Norton K, O'Callaghan M, Desnick RJ: A phase 1/2 clinical trial of enzyme replacement in Fabry disease: Pharmacokinetic, substrate clearance, and safety studies. Am J Hum Genet 68:711–722 (2001).

Foster MH: Relevance of systemic lupus erythematosus nephritis animal models to human disease. Semin Nephrol 19:12–24 (1999).

Foster MH, Kelley VR: Lupus nephritis: Update on pathogenesis and disease mechanisms. Semin Nephrol 19:173–181 (1999).

Gross O, Beirowski B, Keopke M-L, Kuck J, Reiner M, Addicks K, Smyth N, Schulze-Lohoff E, Weber M: Preemptive ramipril therapy delays renal failure and reduces renal fibrosis in COL4A3-knockout mice with Alport syndrome. Kidney Int 63:438–446 (2003).

Guay-Woodford L, Muecher G, Hopkins S, Avner E, Germino G, Guillot A, Herrin J, Holleman R, Irons D, Primack W, Thomson P, Waldo F, Lunt P, Zerres K: The severe perinatal form of autosomal recessive polycystic kidney disease (ARPKD) maps to chromosome 6p21.1→p12: Implications for genetic counseling. Am J Hum Genet 56:1101–1107 (1995).

Guay-Woodford LM, Wright CJ, Walz G, Churchill GA: Quantitative trait loci modulate renal cystic disease severity in the mouse bpk model. J Am Soc Nephrol 11:1253–1260 (2000).

Heidet L, Borza DB, Jouin M, Sich M, Mattei MG, Sado Y, Hudson BG, Hastie N, Antignac C, Gubler M-C: A human-mouse chimera of the alpha3alpha4alpha5(IV) collagen promoter rescues the renal phenotype in Col4a3-/- Alport mice. Am J Pathol 163:1633–1644 (2003).

Heikkila P, Tryggvason K, Thorner P: Animal models of Alport syndrome: advancing the prospects for effective human gene therapy. Exp Nephrol 8:1–7 (2000).

Heikkila P, Tibell A, Morita T, Chen Y, Wu G, Sado Y, Ninomiya Y, Pettersson E, Tryggvason K: Adenovirus-mediated transfer of type IV collagen alpha 5 chain cDNA into swine kidney in vivo: deposition of the protein into the glomerular basement membrane. Gene Therapy 8:882–890 (2001).

Howie JB, Helyer BJ: The immunology and pathology of NZB mice. Adv Immunol 9:215–266 (1968).

Hudson BG, Tryggvason K, Sundaramoorthy M, Neilson EG: Alport's syndrome, Goodpasture's syndrome, and type IV collagen. New Engl J Med 348:2543–2556 (2003).

Iakoubova O, Dunshkin H, Beier D: Localization of a murine recessive polycystic kidney disease mutation and modifying loci that affect disease severity. Genomics 26:107–114 (1995).

Ioannou YA, Zeidner KM, Gordon RE, Desnick RJ: Fabry disease: Preclinical studies demonstrate the effectiveness of alpha-galactosidase A replacement in enzyme-deficient mice. Am J Hum Genet 68:14–25 (2001).

Jung S-C, Han IH, Limaye A, Xu R, Gelderman MP, Zerfas P, Tirumalai K, Murray GJ, During MJ, Brady RO, Qasba P: Adeno-associated viral vector-mediated gene transfer results in long-term enzymatic and functional correction in multiple organs of Fabry mice. Proc Natl Acad Sci USA 98:2676–2681 (2001).

Kim K, Drummond I, Ibraghimov-Beskrovnaya O, Klinger K, Arnaout MA: Polycystin 1 is required for the structural integrity of blood vessels. Proc Natl Acad Sci USA 97:1731–1736 (2000).

Lawson BR, Prud'homme GJ, Chang Y, Gardner HA, Kuan J, Kono DH, Theofilopoulos AN: Treatment of murine lupus with cDNA encoding IFN-gamma R/Fc. J Clin Invest 106:207–215 (2000).

Lemmink HH, Mochizuki T, van de Heuvel LPWJ, Schroder CH, Barrientos A, Monnens LAH, van Oost BA, Brunner HG, Reeders ST, Smeets HJM: Mutations in the type IV collagen α3 (COL4A3) gene in autosomal recessive Alport syndrome. Hum Mol Genet 3:1269–1273 (1994).

Lu W, Peissel B, Babakhanlou H, Pavlova A, Geng L, Fan X, Larson C, Brent G, Zhou J: Perinatal lethality with kidney and pancreas defects in mice with a targetted Pkd1 mutation. Nat Genet 17:179–181 (1997).

Lu W, Fan X, Basora N, Babakhanlou H, Law T, Rifai N, Harris PC, Perez-Atayde AR, Rennke HG, Zhou J: Late onset of renal and hepatic cysts in Pkd1-targeted heterozygotes. Nat Genet 21:160–161 (1999a).

Lu W, Phillips CL, Killen PD, Hlaing T, Harrison WR, Elder FF, Miner JH, Overbeek PA, Meisler MH: Insertional mutation of the collagen genes Col4a3 and Col4a4 in a mouse model of Alport syndrome. Genomics 61:113–124 (1999b).

Mageed RA, Prud'homme GJ: Immunopathology and the gene therapy of lupus. Gene Therapy 10:861–874 (2003).

Miner JH, Sanes JR: Molecular and functional defects in kidneys of mice lacking alphacollagen3(IV): Implications for Alport syndrome. J Cell Biol 135:1403–1413 (1996).

Mochizuki T, Lemmink HH, Mariyama M, Antignac C, Gubler MC, Pirson Y, Verellen-Dumoulin C, Chan B, Schroder CH, Smeets HJ, Reeders ST: Identification of mutations in the α3(IV) and α4(IV) collagen genes in autosomal recessive Alport syndrome. Nat Genet 8:77–82 (1994).

Mohan C: Murine lupus genetics: Lessons learned. Curr Opin Rheumatol 13:352–360 (2001).

Morel L, Wakeland ED: Susceptibility to lupus nephritis in the NZB/W model system. Curr Opin Immunol 10:718–725 (1998).

Nguyen C, Limaye N, Wakeland EK: Susceptibility genes in the pathogenesis of murine lupus. Arthritis Res 4:S255–S263 (2002).

Ohshima T, Murray GJ, Nagle JW, Quirk JM, Kraus MH, Barton NW, Brady RO, Kulkarni AB: Structural organization and expression of the mouse gene encoding alpha-galactosidase A. Gene 166:277–280 (1995).

Ohshima T, Murray GJ, Swaim WD, Longenecker G, Quirk JM, Cardarelli CO, Sugimoto Y, Pastan I, Gottesman MM, Brady RO, Kulkarni AB: α-Galactosidase A deficient mice: A model of Fabry disease. Proc Natl Acad Sci USA 94:2540–2544 (1997).

Ohshima T, Schiffmann R, Murray GJ, Kopp J, Quirk JM, Stahl S, Chan C-C, Zerfas P, Tao-Cheng J-H, Ward JM, Brady RO, Kulkarni AB: Aging accentuates and bone marrow transplantation ameliorates metabolic defects in Fabry disease mice. Proc Natl Acad Sci USA 96:6423–6427 (1999).

Orellana SA, Sweeney WE, Neff CD, Avner ED: Epidermal growth factor receptor expression is abnormal in murine polycystic kidney. Kidney Int 47:490–499 (1995).

Park J, Murray GJ, Limaye A, Quirk JM, Gelderman MP, Brady RO, Qasba P: Long-term correction of globotriaosylceramide storage in Fabry mice by recombinant adeno-associated virus-mediated gene transfer. Proc Natl Acad Sci USA 100:3450–3454 (2003).

Pei Y, Watnick T, He N, Wang K, Liang Y, Parfrey, P, Germino G, St George-Hyslop P: Somatic PKD2 mutations in individual kidney and liver cysts support a "two-hit" model of cystogenesis in type 2 autosomal dominant polycystic kidney disease. J Am Soc Nephrol 10:1524–1529 (1999).

Peters DJM, Breuning MH: Autosomal dominant polycystic kidney disease: modification of disease progression. Lancet 358:1439–1444 (2001).

Proesmans W, Knockaert H, Trouet D: Enalapril in paediatric patients with Alport syndrome: two years' experience. Eur J Pediatr 159:430–433 (2000).

Qian F, Watnick TJ, Onuchic LF, Germino GG: The molecular basis of focal cyst formation in human autosomal dominant polycystic kidney disease type I. Cell 87:979–987 (1996).

Qian Q, Harris PC, Torres VE: Treatment prospects for autosomal-dominant polycystic kidney disease. Kidney Int 59:2005–2022 (2001).

Qin G, Takenaka T, Telsch K, Kelley L, Howard T, Levade T, Deans R, Howard BH, Malech HL, Brady RO, Medin JA: Preselective gene therapy for Fabry disease. Proc Natl Acad Sci USA 98:3428–3433 (2001).

Raz E, Dudler J, Lotz M, Baird SM, Berry CC, Eisenberg RA, Carson DA: Modulation of disease activity in murine systemic lupus erythematosus by cytokine gene delivery. Lupus 4:286–292 (1995).

Reeders ST: Multilocus polycystic disease. Nat Genet 1:235–237 (1992).

Schiffmann R, Murray GJ, Treco D, Daniel P, Sellos-Moura M, Myers M, Quirk JM, Zirzow GC, Borowski M, Loveday K, Anderson T, Gillespie F, Oliver KL, Jeffries NO, Doo E, Liang TJ, Kreps C, Gunter K, Frei K, Crutchfield K, Selden RF, Brady RO: Infusion of alpha-galactosidase A reduces tissue globotriaosylceramide storage in patients with Fabry disease. Proc Natl Acad Sci USA 97:364–370 (2000).

Schiffmann R, Koop JB, Austin HA, Sabnis S, Moore DF, Weibel T, Balow JE, Brady RO: Enzyme replacement therapy in Fabry disease: A randomized controlled trial. JAMA 285:2743–2776 (2001).

Sessa A, Meroni M, Battini G, Righetti M, Maglio A, Tosoni A, Nebuloni M, Vago G, Giordano F: Renal involvement in Anderson-Fabry disease. J Nephrol 16:310–313 (2003).

Sutters M, Germino GG: Autosomal dominant polycystic kidney disease: Molecular genetics and pathophysiology. J Lab Clin Med 141:91–101 (2003).

Sweeney WE, Chen Y, Nakanishi K, Frost P, Avner ED: Treatment of polycystic kidney disease with a novel tyrosine kinase inhibitor. Kidney Int 57:33–40 (2000).

Sweeney WE, Hamahira K, Sweeney J, Garcia-Gatrell M, Frost P, Avner ED: Combination treatment of PKD utilizing dual inhibition of EGF-receptor activity and ligand bioavailability. Kidney Int 64:1310–1319 (2003).

Takahashi H, Hirai Y, Migita M, Seino Y, Fukuda Y, Sakuraba H, Kase R, Kobayashi T, Hashimoto Y, Shimada T: Long-term systemic therapy of Fabry disease in a knockout mouse by adeno-associated virus-mediated muscle-directed gene transfer. Proc Natl Acad Sci USA 99:13777–13782 (2002).

Takiguchi M, Murakami M, Nakagawa I, Saito I, Hashimoto A, Uede T: CTLA4 IgG gene delivery prevents autoantibody production and lupus nephritis in MRL/lpr mice. Life Sci 66:991–1001 (2000).

Theophiloulos AN, Dixon FJ: Murine models of lupus systemic erythematosus. Adv Immunol 37:269–290 (1985).

Tomasoni S, Noris M, Zappella S, Gotti E, Casiraghi F, Bonazzola S, Benigni A, Remuzzi G: Upregulation of renal and systemic cyclooxygenase-2 in patients with active lupus nephritis. J Am Soc Nephrol 9:1202–1212 (1998).

Upadhya P, Churchill G, Birkenmeier E, Barker J, Frankel W: Genetic modifiers of polycystic kidney disease in intersubspecific KAT2J mutants. Genomics 58:129–137 (1999).

van Bruggen MC, Walgreen B, Rijke TPM, Berden JHM: Attenuation of murine lupus nephritis by mycophenolate mofetil. J Am Soc Nephrol 9:1407–1415 (1998).

Wakeland EK, Liu K, Graham RR, Behrens TW: Delineating the genetic basis of systemic lupus erythematosus. Immunity 15:397–408 (2001).

Woo D, Miao S, Tran T: Progression of polycystic kidney disease in cpk mice is a quantitative trait under polygenic control (Abstract). J Am Soc Nephrol 6:731A (1995).

Woo D, Nguyen D, Khatibi N, Olsen P: Genetic identification of two major modifier loci of polycystic kidney disease progression in pcy mice. J Clin Invest 100:1934–1940 (1997).

Wu G, D'Agati V, Cai Y, Markowitz G, Hoon Park J, Reynolds MD, Maeda Y, Le TC, Hou H, Kucherlapati R, Edelmann W, Somlo S: Somatic inactivation of Pkd2 results in polycystic kidney disease. Cell 93:177–188 (1998).

Zoja C, Liu X-H, Donadelli R, Abbate M, Testa D, Corna D, Taraboletti G, Vecchi A, Dong QG, Rollins BJ, Bertani T, Remuzzi G: Renal expression of monocyte chemoattractant protein-1 in lupus autoimmune mice. J Am Soc Nephrol 8:720–729 (1997).

Zoja C, Corna D, Benedetti G, Morigi M, Donadelli R, Guglielmotti A, Pinza M, Bertani T, Remuzzi G: Bindarit retards renal disease and prolongs survival in murine lupus autoimmune disease. Kidney Int 53:726–734 (1998).

Zoja C, Benigni A, Noris M, Corna D, Casiraghi F, Pagnoncelli M, Rottoli D, Abbate M, Remuzzi G: Mycophenolate mofetil combined with a cyclooxygenase-2 inhibitor ameliorates murine lupus nephritis. Kidney Int 60:653–663 (2001).